Handbook of Zoology
Mammalia

Mammalian Evolution, Diversity and Systematics

Handbook of Zoology

Founded by Willy Kükenthal
continued by M. Beier, M. S. Fischer, J.-G. Helmcke, H. Schliemann,
D. Starck, H. Wermuth
Editor-in-chief Andreas Schmidt-Rhaesa

Mammalia

Edited by Frank E. Zachos

DE GRUYTER

Mammalia

Mammalian Evolution, Diversity and Systematics

Edited by Frank E. Zachos and Robert J. Asher

DE GRUYTER

Scientific Editors
Dr. habil. Frank E. Zachos
Natural History Museum Vienna
Mammal Collection
Burgring 7
1010 Vienna, Austria
and
Department of Integrative Zoology
University of Vienna
Althanstraße 14
1090 Vienna, Austria
frank.zachos@nhm-wien.ac.at

Robert J. Asher, Ph.D.
Museum of Zoology
Downing St. CB2 3EJ or Trinity Hall
CB2 1TJ University of Cambridge
Cambridge, United Kingdom
r.asher@zoo.cam.ac.uk

ISBN 978-3-11-027590-2
e-ISBN (PDF) 978-3-11-034155-3
e-ISBN (EPUB) 978-3-11-038254-9
ISSN 2193-2824

Library of Congress Cataloging-in-Publication Data
Names: Zachos, Frank E., 1974- editor.
Title: Mammalian evolution, diversity and systematics / edited by Frank E. Zachos, Robert J. Asher.
Description: Berlin ; Boston : De Gruyter, [2018] | Includes bibliographical references and index.
Identifiers: LCCN 2018029220 (print) | LCCN 2018031457 (ebook) | ISBN 9783110341553 (electronic Portable Document Format (pdf)) | ISBN 9783110275902 (print : alk. paper) | ISBN 9783110341553 (pdf) | ISBN 9783110382549 (epub)
Subjects: LCSH: Mammals--Evolution.
Classification: LCC QL708.5 (ebook) | LCC QL708.5 .M366 2018 (print) | DDC 599--dc23 LC record available at https://lccn.loc.gov/2018029220

Bibliographic information published by the Deutsche Nationalbibliothek
The Deutsche Nationalbibliothek lists this publication in the Deutsche Nationalbibliografie; detailed bibliographic data are available in the Internet at http://dnb.dnb.de.

Typesetting: Compuscript Ltd. Shannon, Ireland
Printing and Binding: CPI books GmbH, Leck

www.degruyter.com

Preface

This book is the second mammal volume within the *Handbook of Zoology* since its relaunch as an exclusively English series (after the authoritative work on comparative anatomy of the digestive system by Peter Langer published in 2017), but conceptually it is somewhat of an introduction to the new series. The last two decades have seen so much new and exciting research on the diversity and phylogenetic relationships of mammals, both fossil and extant, that we decided to publish a summary volume of these topics. Unfortunately, mammals are also increasingly under human pressure, with up to one in three species threatened with extinction, which is why we have included a chapter on mammal conservation as well (Turvey). We are aware that the selection of topics could have been different, and any such volume will contain some level of arbitrariness with respect to the issues that are covered and those that are not. Apart from the chapter dealing with extinction risk and conservation, we have three chapters on the evolution and systematics of the mammalian lineage – one focusing on the Paleozoic synapsids (Angielczyk and Kammerer), one on Mesozoic mammals (Martin), and one on Cenozoic and particularly extant groups (Asher). There is also a chapter by Robert Asher on the history of mammalian classification, describing the consensus on the interrelationships of the living high-level taxa. Werneburg and Spiekman present a comprehensive treatment of mammalian embryology, an area of research that deals with many of the key morphological, reproductive, and developmental novelties of mammals that explain the group's huge evolutionary and ecological success. The remainder of the book comprises two chapters on species concepts and speciation. Zachos summarizes recent debates in mammalian taxonomy about splitting and lumping and the most commonly accepted species concepts among mammalogists – the Biological Species Concept and the Phylogenetic Species Concept. These debates have triggered a discussion that reaches far beyond the mammalogical community. Lastly, mammals have also been studied in depth with regard to speciation, and chromosomal rearrangements have turned out to be a key element in mammalian diversification. This topic is dealt with in a review chapter by Pavlova and Searle.

Other aspects of mammalian diversity and evolution could have been covered as well, such as, for example, reviews of the diverse ways in which mammals have specialized their anatomy and physiology to exploit the natural world. These and other issues, including authoritative treatments of key high-level mammalian groups, will be the subject of further volumes of the Mammalia series within the *Handbook of Zoology*. Finally, we would like to express our gratitude to the de Gruyter publishing team for giving us the opportunity to continue the mammal series and for their help in producing the present volume.

Vienna and Cambridge,
3 May 2018
Frank E. Zachos and Robert J. Asher

https://doi.org/10.1515/9783110341553-202

Preface

Contents

List of contributing authors

Kenneth Angielczyk
Integrative Research Center
Field Museum of Natural History
1400 South Lake Shore Drive
Chicago, Illinois 60605 U.S.A.
e-mail: kangielczyk@fieldmuseum.org

Robert J. Asher
Museum of Zoology
University of Cambridge
Downing St, Cambridge, CB2 3EJ, United Kingdom
www.robertjasher.com

Christian Kammerer
North Carolina Museum of Natural Sciences
11 West Jones Street, Raleigh, NC 27601, USA
e-mail: christian.kammerer@naturalsciences.org

Thomas Martin
Steinmann-Institut für Geologie, Mineralogie und Paläontologie
University of Bonn
Nussallee 8, 53115 Bonn, Germany
e-mail: tmartin@uni-bonn.de

Svetlana V. Pavlova
A.N. Severtsov Institute of Ecology and Evolution
Russian Academy of Sciences
33 Leninsky pr., Moscow, 119071, Russia
e-mail: svetpavlova@yandex.ru

Jeremy B. Searle
Department of Ecology and Evolutionary Biology
Cornell University
E139 Corson Hall, Ithaca, NY 14853, USA
e-mail: jeremy.searle@cornell.edu

Stephan Spiekman
Paläontologisches Institut und Museum Zürich
University of Zurich – UHZ
Karl-Schmid-Strasse 4, 8006 Zurich, Switzerland
e-mail: stephan.spiekman@pim.uzh.ch

Samuel Turvey
Institute of Zoology
Zoological Society of London
Regent's Park, London, NW1 4RY, United Kingdom
e-mail: samuel.turvey@ioz.ac.uk

Ingmar Werneburg
Palaeontological Collection, Senckenberg Center for Human Evolution and Palaeoenvironment (HEP)
Eberhard Karls Universität
Sigwartstr. 10, 72076 Tübingen, Germany
and
Fachbereich für Geowissenschaften, Eberhard Karls Universität,
Hölderlinstraße 12, 72074 Tübingen, Germany
e-mail: ingmar.werneburg@senckenberg.de

Frank E. Zachos
Natural History Museum Vienna
Mammal Collection
Burgring 7, 1010 Vienna, Austria
and
Department of Integrative Zoology
University of Vienna
Althanstraße 14
1090 Vienna, Austria
e-mail: frank.zachos@nhm-wien.ac.at

Frank E. Zachos

1 Species concepts and species delimitation in mammals

1.1 Introduction

Few topics in biology are as contentious and vexed as the 'species problem'—a group of issues relating to what biologists over the past centuries have been calling species, such as Do species exist? What is a species? Is there a one-size-fits-all species concept applicable in a satisfactory manner to all groups of living beings? How can species be delimited from one another, and is this possible in an objective way? Debates on these questions can be intense, reflecting the importance of the species as a conceptual unit in biology and because every biologist has (or feels he or she should have) some idea of what makes a species. Each single question given above to circumscribe the species problem is in itself worth a whole treatise, and a detailed, let alone complete, discussion is far beyond the scope of this chapter. Yet, mammalian taxonomy is not only deeply affected by the answers to these questions but has also triggered some of the recent debates concerning species concepts. The present chapter highlights and summarizes these developments with a particular focus on mammal species, points out strengths and weaknesses of different taxonomic approaches, and also briefly addresses the practical consequences of our taxonomic decisions.

1.2 Species concepts — a short outline

In this section, I will only highlight a few aspects and basic ideas of the general debate inasmuch as they are relevant to the more particular issues of this chapter. This will necessarily be short and at times superficial; for a more detailed discussion of the historical, philosophical, and theoretical aspects of species concepts, there are a number of books and review articles to which I refer the interested reader and from which I have drawn in the following discussion (e.g., Wilson 1999, Wheeler and Meier 2000, Ghiselin 2001, Mallet 2001, Stamos 2003, Rieppel 2007, Wilkins 2009a,b, 2018, Richards 2010, Kunz 2012, Zachos 2016, 2018, and references therein).

1.2.1 Terminology

There are a few terminological issues in the species debate that are worth clarifying up front:

Species taxa vs. the species category. There is a fundamental difference between a species taxon, e.g., *Homo sapiens*, and the species as a category (and rank), i.e., a general unit with certain properties that we use to define what we call species. It is the latter that is aimed at in the definitions of the various species concepts. Species taxa are historical entities; they come into being through speciation and they cease to exist when they become extinct (or, perhaps, when they give rise to daughter species). Philosophically speaking, species taxa are *individuals* (Ghiselin 1974), and the various organisms that make up a species (all human beings in the case of *H. sapiens*) are *parts* of that individual, just like all the kings and queens and their immediate relatives are parts of a royal dynasty. The species category, on the contrary, is what philosophers call a *class*, a group whose members (which are usually called *elements*) show the defining characters of that class. These characters are essential, i.e., they are both necessary and sufficient: the class of red triangles is defined by color (red) and shape (triangular)—all its elements are red and triangular, and all red triangular objects are members of that class. Although individuals are bound in space and time (they have a beginning and an end), classes do not have any spatiotemporal limits—they exist regardless of space and time. *H. sapiens* is an ephemeral historical entity; the class of red triangles is not. When the last human dies, *H. sapiens* will no longer exist, and it will never again exist; the class of red triangles can be empty (no red triangles in the universe), but it can never cease to exist. When all life on earth dies out, none of the species *taxa* (humans, tigers, dandelions) will ever exist again, but if life evolves again (on earth or elsewhere), entities (i.e., groups of organisms) might come into existence that fall under one or more of our definitions of the species *category*.

For example, there may be least inclusive monophyletic groups of organisms (one version of the phylogenetic species concept [PSC]) or groups of organisms that produce fertile offspring among each other but infertile or no offspring with other such groups (biological species concept [BSC]). An analogy for the difference between

https://doi.org/10.1515/9783110341553-001

the species taxa and the species category is that of a particular neighbor and neighbors in general: my neighbor Mr. Miller is an individual and, in the analogy, equivalent to the species taxon, whereas neighbors in general, defined as people living next door, are equivalent to the species category. From this, another difference becomes clear: the species category can be defined (as in the various species concepts), but species taxa can only be pointed out ostensively, and their names are therefore proper names ('Mr. Miller'). The same is true for higher taxa: one can define the class of higher taxa as monophyla (a group containing a stem species and all and only its descendants), but each actual monophylum in the Tree of Life, Mammalia for example, is an individual and can only be discovered and delimited through phylogenetic analysis, as an actual instance of the class of monophyla by pointing out (or rather inferring) a stem species and all its evolutionary descendants. The fact that species taxa cannot be defined means that there is no character that is necessary and sufficient to make an organism part of a certain species. Morphological or genetic definitions of species taxa are, strictly speaking, invalid. A tiger that is born without stripes is still a tiger, and that holds for any and every character that one might choose to try and define a species taxon. An organism is not part of a species taxon due to its characters but only due to it being part of the historical entity that came into being in a speciation event at some particular point in time. This directly leads to the next important distinction, that of ontology and operationalism.

Ontology/definition vs. operationalism/identification. The difference between these two with respect to species concepts is a fundamental one, and the solution (in theory) of the species problem as presented below is a direct consequence of recognizing this difference. Ontology is about what things are and, consequently, how they are defined; operationalism (in this context) means being applicable and useful in practice. In the species debate, operational species 'concepts' (or rather criteria, see below) are those that help us identify species. Choosing human siblings as an analogy, the difference immediately becomes clear. Siblings are humans who share the same parents. This is a definition of what siblings are, but it is often useless in identifying siblings because 'having the same parents' is usually not directly observable. There are many ways of identifying siblings, though, morphologically and genetically. However, to say siblings are humans who share (on average) 50% of their genome is, strictly speaking, wrong. Apart from the fact that the same is true of parents and their children, it is not what *makes* two people siblings but simply a consequence of their being siblings that we use to identify them as such.

1.2.2 A hierarchy of species concepts

There are more than 30 published species concepts (Mayden 1997, http://scienceblogs.com/evolvingthoughts/2006/10/01/a-list-of-26-species-concepts/ by J. S. Wilkins, Wilkins 2009b, 2018, Appendix B, Zachos 2016, chapter 4.). Identifying the best of those 30 concepts naturally depends on the purpose to which a species concept is put. It was probably Mayden (1997) who first explicitly distinguished two conceptual levels in species concepts. According to this view – the division-of-conceptual-labor solution to the species problem according to Richards (2010) – there is only one primary species concept that gives a true *definition* of what species ontologically are, whereas all the others function as secondary operational concepts to *identify* species, i.e., as criteria that must be fulfilled to satisfy the definition of the primary concept. The single primary species concept, according to Mayden, is the evolutionary species concept (ESC) that defines species as lineages of ancestral-descendant populations that evolve independently of other such lineages (Simpson 1951, 1961, Wiley 1978, Wiley and Mayden 2000a,b, and c; see Tab. 1.1). The ESC has often been criticized for not being operational, but ironically, this is exactly what makes it (ontologically) superior to all the other concepts. The same line of reasoning can be found in a number of publications by Kevin de Queiroz under the names of General Lineage Concept (de Queiroz 1998, 1999) and Unified Species Concept (de Queiroz 2005, 2007)[1]. Again, separately evolving population (or meta-population) lineages are viewed as common to all species concepts, and the secondary concepts or criteria only highlight certain thresholds that will be crossed sooner or later by diverging lineages, e.g., occupying different niches (Ecological Species Concept), producing sterile hybrids (BSC), being reciprocally monophyletic (monophyly version of the PSC), etc. (see Fig. 5.4 in de Queiroz 1998 and Fig. 5.2 in Zachos 2016). Viewed in this light, the existence of so many (secondary) species concepts that function as operational criteria to identify independent population lineages (= species as defined by

1 The General Lineage Species Concept holds that being a separately evolving (meta-)population lineage is the common denominator of all species concepts. The Unified Species Concept posits that this is the only necessary condition for species status – in other words that all such lineages are species.

Tab. 1.1: Definitions of selected species concepts dealt with in this chapter. For all of the listed concepts, more definitions than the ones given here exist, although differences are subtle and not relevant for the present discussion. Dobzhansky (1935) quoted from Wilkins (2009b). * The reference to populations in the definition of Wiley and Mayden (2000a) compared to that of Wiley (1978) was dropped to include asexual taxa. For a longer list of the 30+ published species concepts and their definitions, see chapter 4 of Zachos (2016) and Appendix B in Wilkins (2018).

Species concept	Species definition	Reference
Evolutionary species concept	"a lineage of ancestral descendant populations which maintains its identity from other such lineages and which has its own evolutionary tendencies and historical fate"	Wiley (1978)
	"an entity composed of organisms that maintains its identity from other such entities through time and over space and that has its own independent evolutionary fate and historical tendencies"*	Wiley and Mayden (2000a)
Biological species concept	"a group of individuals fully fertile inter se, but barred from interbreeding with other similar groups by its physiological properties (producing either incompatibility of parents, or sterility of the hybrids, or both)"	Dobhzansky (1935)
	"groups of interbreeding natural populations that are reproductively isolated from other such groups. Alternatively, one can say that a biological species is a reproductively cohesive assemblage of populations"	Mayr (2000)
Genetic species concept	"a group of genetically compatible interbreeding natural populations that is genetically isolated from other such groups"	Baker and Bradley (2006)
Phylogenetic species concept (diagnosability version)	"the smallest diagnosable cluster of individual organisms within which there is a parental pattern of ancestry and descent"	Cracraft (1983)
	"the smallest aggregation of (sexual) populations or (asexual) lineages diagnosable by a unique combination of character states"	Wheeler and Platnick (2000a)
Phylogenetic species concept (monophyly version)	"the smallest monophyletic groups deemed worthy of formal recognition, because of the amount of support for their monophyly and/or because of their importance in biological processes operating on the lineage in question"	Mishler and Theriot (2000a)

the primary ESC) is actually an advantage rather than a nuisance. From this, it becomes clear that *all* species concepts are based on biological realities. Although they may be inconsistent as practiced by some of their adherents or not applicable to all taxa, they cannot simply be wrong. The decisive question is whether the biological reality a certain species concept highlights (sterile hybrids, diagnosable differences, separate niches, etc.) is what we think deserves the label 'species'. Below, I argue that this is ultimately a question of convention.

Mayden (1997) presents an analogy from cladistics to illustrate the conceptual difference between the primary and the secondary species concepts: the primary concept, the ESC, is equivalent to the concept of a monophylum that is not directly observable; the secondary concepts function as identification criteria and correspond to the search for synapomorphies that are used to identify or infer monophyletic groups. Unfortunately, this division-of-conceptual-labor solution (Richards 2010) to the species problem pertains only to the theoretical aspects of the species problem. The challenge lies in the realm of taxonomic practice where one has to make concrete decisions whether two populations or groups of populations are one or two species, i.e., it is not so much the definition of the species category, but rather the delimitation of species taxa that is most contentious (and most difficult).

1.2.3 Taxonomy's basic problems: continuous evolution and the fractal nature of the Tree of Life

The species problem for practicing biologists is mainly a problem of delimitation. This holds regardless of the underlying species concept, as boundaries will always be fuzzy (for a detailed discussion, see Hey 2001a,b), and it is concrete species *taxa* that are delimited, not theoretical categories. A first and very basic reason for this is that evolution proceeds continuously, whereas taxonomy is discrete. A certain group of organisms is either assigned a taxonomic name or not. Sublevels like subspecies notwithstanding, there is no such thing as 'half a species', and a group of organisms can either be one or two, three, etc. but not, say, 1.378 species. There is only one big Tree of Life, and any two organisms in it are connected by a single line of parent-offspring relations. This is true for two elephants as well as for an elephant and a *Caenorhabditis elegans* nematode. In the latter pair, the parent-offspring line is only much longer. The total amount of morphological (and other) divergence between elephant and *C. elegans* is nothing but the sum of all the minimal differences between parents and offspring in each of the parent-offspring pairs along the branches connecting the elephant and the nematode. In fact, it is only because almost all organisms that have ever lived are now dead that we find morphological, genetic, etc. gaps between living organisms in the first place. However, even when only considering the data at hand – extant organisms and fossils – the continuousness of the evolutionary process causes boundaries between closely related taxa to be fuzzy, and thus delimitation is expected to be contentious and ambiguous around what we conceive as the 'species level'.

This grey area is one of the best pieces of evidence in favor of evolution; if all taxa were unequivocally separate from one another, independent creation would be a more parsimonious explanation than common descent with modification. However, how then can we split up the Tree of Life into objective units ('at its joints' as it is called since Plato) that we want to refer to as species? The hierarchical view of species concepts argues that we have to find independent population lineages, but this makes obvious the second major difficulty: the whole Tree of Life is nothing but a huge structure made up of lineages at all levels. Like a matryoshka doll, nodes on the Tree of Life from its tips to its root are lineages made up of lineages again. The nested hierarchy of all life shows something akin to a fractal pattern (Mishler and Theriot 2000a, Hey 2001a). In such a system, is there any one level that is different from all the others and can objectively be called species?

It is important that even a negative answer to this question does not imply strict species nominalism, i.e., the notion that species taxa only exist in our minds. All these levels are very real, but they might all be *equally* real, and then it would be impossible to objectively claim that one of them stands out and deserves the name species (for a discussion of this, see Mishler 1999, 2010, Mishler and Wilkins 2018 and Zachos 2016, chapter 3.6, and references therein). In that case, the species as a category would be no different from the higher Linnaean categories that are known to be arbitrary and artificial: although Carnivora, Proboscidea, Pholidota, etc. are real as taxa, their categorial rank 'order' is not, i.e., what is called 'order' is not the same in these cases but lineages of different and incommensurable levels in the hierarchy of the Tree of Life (see Laurin 2010, Zachos 2011, Lambertz and Perry 2015, and references therein).

The solution for the species category, at least for sexually reproducing taxa, may be the tokogeny/phylogeny divide sensu Hennig (1966, particularly fig. 6; see also, among others, Wheeler and Platnick 2000b). This is the level where interindividual relationships (reproduction) within populations dissolve into interpopulational or phylogenetic relationships, in other words, the level where, looking up from 'below', reticulate relationships dissolve into dichotomous (or polytomous) relationships or, looking down from 'above', vice versa. The boundaries may be fuzzy, but ontologically that is not very relevant. For a discussion of the tokogeny/phylogeny divide and the species rank, see Zachos (2016, chapter 6.2).

What follows from this can briefly be summarized as follows: (1) the very nature of evolution and the structure of the Tree of Life make species delimitation difficult, and (2) it is not clear and perhaps even unlikely whether the species as a category is an objective level in the hierarchical Tree of Life that is comparable across different groups of organisms.

1.3 Species and mammalian taxonomy

The debate about species delimitation is of course a general one and not specific to mammalian taxonomy. However, the discussion within the mammalogical community has been centre-stage in the wider debate recently, and it goes back a long way: in the very first issue of the *Journal of Mammalogy*, there is a paper and a rejoinder to it on how to delimit species, subspecies, and genera (Merriam 1919, Taverner 1920). Mammals are regarded as well studied, and they certainly are compared to insects. New mammal species are described much less frequently than new insects, but there are nevertheless many new mammal species published each decade (Reeder *et al.* 2007, Burgin *et al.* 2018), and although the most authoritative taxonomic reference work comprises 5,416 species (Wilson and Reeder 2005), total numbers of about 7,300–9,000 species that will eventually be recognized have been suggested (Reeder *et al.* 2007, Burgin *et al.* 2018).

Although there is no doubt that there are many new species of mammals 'out there' (or in museum collections), mammalian taxonomy has seen a recent debate on what has been named 'taxonomic inflation'—the increase in species numbers due to the allegedly unwarranted splitting of formerly single species into two or more. The term was coined by Isaac *et al.* (2004) in reaction to the increase in recognized primate species which, after a period of stability up to the mid-1980s, doubled until the early 2000s from 150 to 200 species to more than 350 in Groves (2001). Yet, only about 30 new discoveries of primates were made during that time; the rest is due to splitting. The recent primate volume within the *Handbook of the Mammals of the World* even lists more than 480 species (Mittermeier *et al.* 2013), and Burgin *et al.* (2018) give a number of 518. The same trend holds for the Bovidae: an increase of almost 100% from the Wilson and Reeder reference (Grubb 2005, 143 species) to Groves and Grubb (2011, 279 species), the *Handbook of the Mammals of the World* (Groves and Leslie 2011, 279 species) and Burgin *et al.* (2018, 297 species). There have always been 'splitters' and 'lumpers' in taxonomy, but this recent trend is in large parts due to a shift from the BSC to the PSC, particularly a version of the PSC based on diagnosability (dPSC), and this has obviously hit a nerve with many as can be seen from the multitude of recent publications, commentaries, and rejoinders on the topic (Meiri and Mace 2007, Frankham *et al.* 2012, Gippoliti and Groves 2012, Groves 2012, 2013, Gippoliti *et al.* 2013, 2018, Gutiérrez and Helgen 2013, Heller *et al.* 2013, 2014, Zachos and Lovari 2013, Zachos *et al.* 2013a,b, Cotterill *et al.* 2014, Zachos 2015, Groves *et al.* 2017). Such active debate should be viewed as a healthy discussion on the theoretical foundations of taxonomy. The term taxonomic inflation has been countered with 'taxonomic inertia' by the adherents of the dPSC (e.g., Gippoliti *et al.* 2018), but elsewhere I have argued that maybe both terms should be avoided due to their negative connotations, particularly since splitting and lumping are positions along a continuum that are equally right or wrong as a consequence of a grey area after lineage sundering where taxonomic decisions necessarily contain an element of arbitrariness (Zachos 2018).

The number of newly described mammal species considered valid since Linnaeus's 10th edition of his *Systema Naturae* in 1758 averages almost 250 per decade (Burgin *et al.* 2018). In the 14 years between July 1992 and June 2006, it was 341 (Reeder *et al.* 2007), and from 2000 to 2009, it was 359 (IISE 2011). As expected, these new species are not evenly or randomly distributed with respect to taxonomy, size, and geography; there are biases toward smaller species in diverse taxa (e.g., rodents and bats), species with restricted ranges, and toward islands and the tropics. One important factor, however, that also contributes to which species are newly described is the propensity of experts on the respective groups to 'lump' or to 'split', and that propensity results from the author's preferred species concept. In the following, I will give a short overview of the three species concepts most relevant to mammalian taxonomy as practiced today: the BSC, the genetic species concept (GSC), and the PSC (Tab. 1.1). None of the three is a primary or ontological concept in the sense described above. All three of them function as operational delimitation criteria.

1.3.1 The BSC and polytypic species

The BSC is arguably the best-known species concept. It dominates undergraduate and school textbooks, and it is probably the one that comes to mind first among biologists when asked what a species is. This is not primarily due to its scientific merits but because it is (a) intuitive and (b) promoted most successfully. By intuitive, I mean that the BSC is appealing because the highest level of interfertility often coincides with our intuitive classification of similar groups of organisms ('folk taxonomy'), and that the notion of species has always centered around reproduction (sheep produce sheep, humans produce humans, etc.). Wilkins (2009a, p. 10) calls this the "marriage of generation or reproduction, with form" or the "generative conception" of the species. By successful promotion, I refer to the fact that the BSC is the species concept of the Modern Synthesis and as such was supported by highly influential evolutionary biologists, most importantly Ernst Mayr. Mayr was neither the first nor the only one among the architects of the Modern Synthesis to promote the BSC (see Dobzhansky 1935, 1937), but his definition of species as "groups of interbreeding natural populations that are reproductively isolated from other such groups" (Mayr 2000, p. 17; see also Mayr 1940, 1942) became the BSC which, alongside the Synthetic Theory of Evolution, rose to fame in the second half of the 20th century. In a modified version adjusted to cladistic theory, it was incorporated into what has been called the Hennigian species concept (e.g., Meier and Willmann 2000), and it is also related to the GSC discussed below.

One major criticism of the BSC is that it only applies to taxa with sexual reproduction. There are, however, other serious limitations of the BSC, for example, that it is only applicable in sympatry and synchrony. Corbet (1997) gives an overview of how to deal with cases of sympatry, parapatry, and allopatry, and it becomes clear that, once more, it is difficult where exactly to draw the line between two species unless they occur sympatrically, synchronically, and without evidence of interbreeding. Even in sympatry or parapatry, however, interbreeding is a continuum, and between closely related forms, the extremes of full interfertility on the one side and complete lack of successful interbreeding on the other are probably the exception rather than the rule. For parapatry, Corbet (1997) distinguishes (i) diagnosable parapatric forms with minimal hybridization (two species), (ii) diagnosable parapatric forms with substantial hybridization (two subspecies), and (iii) parapatric forms with minimal differences and some hybridization (two subspecies). For (i), he gives the European hedgehogs *Erinaceus europaeus* and *Erinaceus concolor* (now *Erinaceus roumanicus*) as examples; for (ii), the house mice *Mus musculus musculus* and *Mus musculus domesticus*; and for (iii), the mole rat *Nannospalax ehrenbergi* that is comprised of groups that differ in karyotype. Clearly, there are bound to be many borderline cases, and often a decision could be made either way.

It gets even worse when it comes to allopatric populations. The usual way of treating these within the framework of the BSC is to compare the differences among allopatric forms with those found in interbreeding sympatric forms. If they are of a kind regularly found in interbreeding populations, the allopatric forms are classified as conspecific; if not, they are treated as two different species. This may, at first glance, sound plausible, but it is biologically inadequate. Leaving aside the possibility that reproductive barriers can be due to single or few 'speciation genes' in the absence of other detectable differences, the main error in this line of thought is the implicit assumption that evolution proceeds in the same way in sympatry and allopatry. However, it is only in sympatry that there is selection pressure against hybrids of reduced viability, and consequently, isolation mechanisms can evolve quickly and differences pertaining to mating behavior, reproductive anatomy, etc. will be reinforced. In allopatry, there is no need (or opportunity) for this. The two populations will diverge with respect to all sorts of characters, but those involved in potential interbreeding will often only be governed by drift, not by selection: "Hybrid sterility or inviability might therefore be a simple byproduct of the divergence of genomes that are geographically isolated" (Coyne and Orr 2004, p. 269; see also Kunz 2012, p. 150). Therefore, it is not uncommon that species can successfully interbreed even after millions of years of separation (Mallet 2005).

Hybrids of different species of big cats are regularly produced in captivity, and sometimes females (in line with Haldane's rule)[2] will even be fertile (e.g., in crosses of lions and tigers), but it seems unlikely that this would be possible or at least as easy if lions and tigers had not been occurring allopatrically and without selection against hybrids for so long. In sympatry, regular interbreeding with fertile hybrids is good evidence of there being only a single lineage and species, and using genetic standard methods, one can easily test this. In allopatry, however, the potential capacity for interbreeding is just another trait, and one that is not directly selected for or against. The guideline for the classification of allopatric populations (comparison with sympatric forms) is therefore, at best, a taxonomic convention for the sake of convenience; from an evolutionary viewpoint, its basis is doubtful. In addition, the question remains of how much hybridization must occur to view the two groups as evolutionarily independent. In most parts of their natural distribution ranges (and in many of the regions where both species co-occur through introductions), red deer (*Cervus elaphus*) and sika deer (*Cervus nippon*) do not hybridize, but in some they do, sometimes extensively so in Europe (where sika have been introduced) (Geist 1998, McDevitt *et al.* 2009, Pérez-Espona *et al.* 2009, Senn *et al.* 2010). Should they therefore be treated as a single species? The taxonomic consensus is that they are two species, so even hybridization is not as clear-cut a criterion (unless it is absent) as one might think.

Another contentious point about the BSC is the subspecies issue. Most species concepts acknowledge polytypic species, but the BSC probably makes more use of the subspecies category than any other concept. Hence, there are many described subspecies of birds and mammals, many of which are considered invalid and "virtually meaningless in that they cannot be related to discrete diagnosable taxa" (Corbet 1997, p. 352). A standard definition of the subspecies is that of "an aggregate of phenotypically similar populations of a species, inhabiting a geographic subdivision of the range of a species, and differing taxonomically from other populations of the species" (Mayr 1969, p. 41). The part "differing taxonomically" implies that because all local populations differ from each other in some way, only those that differ "by sufficient diagnostic morphological characters" (p. 42) should be given subspecific trinomials. The problem about the subspecies therefore is not so much that they do not reflect some biological reality, but that this reality is often a trivial one.

More importantly, subspecific designations are often inconsistent. This inconsistency is due to the fact that subspecies are often described based on external characters that are readily perceived by taxonomists (plumage or coat patterns, body size, etc.). Whether these are biologically important is a different question. Populations that show, for example, distinctly different adaptive physiological traits but are externally alike will not be acknowledged taxonomically, whereas others will be even if their differences are adaptively irrelevant. All these differences are real, but as it now stands, subspecies taxonomy is biased toward a limited number of characters, and that makes it highly inconsistent, which is why alternative concepts have been developed to capture intraspecific variability more objectively, particularly in a conservation biological framework (e.g., the Evolutionarily

2 Haldane's rule states that if only one sex is sterile in hybrids, it will be the heterogametic one (i.e., males in the case of mammals).

Significant Unit, see Ryder 1986, Moritz 1994, Crandall *et al.* 2000; also see Zachos 2018 for 'meaningful taxonomic units' in conservation).

This is not just an academic issue because subspecies are listed as official taxa in many international inventories and treaties such as the IUCN Red List of Threatened Species or CITES (the Convention on International Trade in Endangered Species of Wild Fauna and Flora). A case in point is again the red deer (*C. elaphus*). Many subspecies have been described even for the European part of the distribution range, most of them rightfully forgotten now, but ecological, genetic, and phylogeographic research has shown some populations to be evolutionarily distinct and unique, e.g., the North African Barbary stag (*Cervus elaphus barbarus*) and the Tyrrhenian red deer (*Cervus elaphus corsicanus*), which is confined to the islands of Sardinia and Corsica (Zachos and Hartl 2011 and references therein). These populations are classified as subspecies and are included in the Red List (http://www.iucnredlist.org/) and/or CITES. However, it has long been known that the relict population of red deer in Mesola in the Po delta region in Italy is equally distinct and unique, yet for a long time it was lumped with the majority of European mainland red deer into a single subspecies (usually called *Cervus elaphus hippelaphus*) until it was finally described as a subspecies of its own (*Cervus elaphus italicus*, Zachos *et al.* 2014). In Zachos *et al.* (2014), we emphasized the problems of the subspecies category but argued that if any of the intraspecific groups of red deer deserve subspecies status, then the Mesola red deer are certainly one of them.

Although many subspecies are of doubtful validity (validity defined as comprising a distinct and unique part of a species' evolutionary legacy), there are examples where subspecies coincide with well-differentiated units and where the 'traditional' subspecies classification captured all or nearly all such units. One such case is the tiger (*Panthera tigris*) whose living subspecies (three more became extinct in the 20th century) are clearly recognizable also based on molecular phylogenetic data, which also uncovered an additional distinct group now suggested to be a new subspecies (the Malayan tiger, *Panthera tigris jacksoni*) (Luo *et al.* 2004)[3]. Nevertheless, delimitation of subspecies remains a fuzzy and often arbitrary business, and many taxonomists, in particular adherents of versions of the PSC, ask whether the line between subspecies of polytypic species of the BSC and 'true' species is not largely artificial. In a way, the subspecies issue repeats the species taxon vs. species category distinction at a lower level. Although all subspecies taxa that are based on hereditary differences capture some biological reality, it is very doubtful that they do so in a way that makes different subspecies comparable. Different subspecies taxa, just like different species taxa, may therefore simply be incommensurable.

Fully aware of the BSC's limited applicability, Edward O. Wilson still thought that it "works well enough in enough studies on most kinds of organisms, most of the time" (Wilson 1992, p. 45). Mishler and Theriot (2000b, p. 123f.) come to a very different conclusion when they consider the BSC both 'unapologetically nonuniversal' and 'unapologetically nondimensional'. In spite of the BSC's popularity it has to be said that, most of the time, they are right.

1.3.2 The GSC sensu Baker and Bradley

Baker and Bradley (2001, 2006) favor the GSC based on earlier views by Bateson (1909), Dobzhansky (1937), and Muller (1942). The GSC is reminiscent of the BSC, and it is similar in that genetic isolation and reproductive isolation are of course correlated. Baker and Bradley (2006, p. 645) define a species as "a group of genetically compatible interbreeding natural populations that is genetically isolated from other such groups". In the process of speciation, genetic changes accumulate in two diverging lineages, creating "genetic isolation and protection of the integrity of the 2 respective gene pools that have independent evolutionary fates" (ibid.). From this, it becomes clear that the ESC theoretically underpins the GSC. The emphasis on integrated gene pools that would suffer a fitness decrease through genic incompatibilities when disrupted (by hybridization) goes back to Bateson, Dobzhansky, and Muller. Baker and Bradly call the GSC the Bateson-Dobzhansky-Muller model, but it is most often just called the Dobzhansky-Muller model (Coyne and Orr 2004), and it is also at the heart of the differential fitness species concept (Hausdorf 2011). What separates the GSC from the BSC is the fact that, according to the GSC, as long as the integrity of two gene pools is maintained, each is considered to be a different species regardless of reproductive isolation. That also holds if hybridization does occur and fertile hybrids can or could be produced.

3 There is subspecies lumping with regard to tigers as well: Wilting *et al.* (2015) hold that there are only two subspecies, one continental and one on the Sunda islands. A similar trend of lumping can be found for other cat species, among them lions, in the new felid taxonomy adopted by the IUCN Cat Specialist group (Cat News Special Issue 11, 2017).

Baker and Bradley (2006) give a number of examples where genetic but not reproductive isolation occurs between two groups (i.e., species according to the GSC), among them African savanna and forest elephants (*Loxodonta africana* and *Loxodonta cyclotis*), mule deer and white-tailed deer (*Odocoileus hemionus* and *Odocoileus virginianus*), two pocket gopher species pairs (*Thomomys bottae* and *Thomomys townsendii* and *Geomys bursarius* and *Geomys knoxjonesi*), and tent-making bats (*Uroderma bilobatum* and *Uroderma davisi*). Another particularly well-known case is found among cervids: red deer, sika, and wapiti (*C. elaphus/nippon/canadensis*) whose phylogenetic and taxonomic status is not yet finally settled. The difference between genetic and reproductive isolation is that the former refers to two separated gene pools with distinct evolutionary fates and adaptive peaks regardless of occasional hybridization and gene flow, but in practice the boundaries are once more often fuzzy. The important aspect is that simple interfertility or the lack of it is no longer regarded as the silver bullet in deciding whether two populations form one or two species. Many of the species recognized under the GSC will be morphologically very similar or nearly identical (cryptic species). Baker and Bradley 2006 mention examples among bats (*Myotis*), rodents (*Peromyscus*), and lipotyphlans (*Sorex* and *Crocidura*) and estimate over 2,000 unrecognized species that exist in addition to the 4,629 species in Wilson and Reeder (1993).

Baker and Bradley (2006) also suggest genetic distance values as a quantitative proxy of species status. However, they are fully aware that this is bound to be error prone, and they regard this only as a first step, particularly if only a single gene or marker is used as is the case with genetic barcoding. Using cytochrome *b* sequence data in a number of bat and rodent species, Bradley and Baker (2001) concluded that pairwise distance values <2% were indicative of intraspecific variation, whereas values >11% were indicative of species status; anything in between could be either. It is important not to misunderstand the authors on this point. These values are mere indications, not a definition of the species category: "if the goal is to efficiently recognize cryptic or currently unrecognized genetic species of mammals, examples of phylogroups within a single species of mammal where the genetic divergence in the cytochrome *b* gene is >10% [in their 2001 paper it is 11%] will be the best group for study, and additional data (nuclear genes or morphological variation) should be collected to determine if an unrecognized species exist [sic]" (Baker and Bradley 2006, p. 653). The same reasoning was followed by Osmers *et al.* (2012) when they concluded that the average genetic distance of 39.9% at mitochondrial control region sequences between the southern African gemsbok (*Oryx gazella*) and the East African beisa oryx (*Oryx beisa*), compared to intraspecific divergence values of 2.7% and 10.8% in *O. gazella* and *O. beisa*, respectively, supported (not established) the classification as two distinct species.

Operationally, Baker and Bradley (2006) distinguish between the application of the GSC in sympatric and allopatric groups. For the latter, they suggest to use the magnitude of genetic distance typically found in known sister species, ideally derived from different molecular marker systems, but "[m]inimally, there should be ≥1 marker each for mtDNA and nuclear DNA" (ibid., p. 655). For sympatric groups, matters are more complicated. If there are no hybrids (or only sterile ones), then there are obviously two species, but even the presence of fertile hybrids does not preclude the existence of two separate species under the GSC, particularly in the case of parapatric hybrid zones that are limited geographically and beyond which hybridization is insignificant. The justification for this is again the integrity or independence of the two gene pools, and this is equivalent to the species as lineages notion of the ESC: "Unless the hybrid zone is of recent origin, a narrow geographically restricted hybrid zone is evidence of genetic isolation and consequently both phylogroups have a high probability of an independent evolutionary fate" (ibid., p. 654, but for an example in crows where the hybrid zone is permeable for large parts of the genome, see Poelstra *et al.* 2014). Obviously, there is a continuum here and what is considered "insignificant hybridization" has to be determined either on a case-by-case basis or through convention by a threshold value. As outlined above, however, this is not a weakness of species concepts themselves but a direct consequence of the continuousness of the evolutionary process.

1.3.3 PSCs and what makes a lineage

There are two species concepts that are most often subsumed under the PSC, one based on diagnosability (hereafter dPSC) and another based on monophyly (hereafter mPSC) (see Tab. 1.1). Sometimes, however, what is otherwise known as the Hennigian or cladistic species concept is also called a PSC (e.g., de Queiroz 1998; for details on the use of the term phylogenetic in species concepts, see Zachos 2016, p. 78f and chapter 5.6). The adjective 'phylogenetic' is therefore somewhat imprecise and also misleading (as if other species concepts were 'unphylogenetic'), but it is probably fair to say that the same could be said about any of the other terms used to specify species concepts ('biological', 'genetic', 'evolutionary', 'ecological', etc.). As far as mammalian taxonomy and species splitting in mammals are concerned, it is mainly the dPSC that is relevant, particularly in the recent debate triggered by taxonomic changes within primates, artiodactyls, and perissodactyls (see above). I will therefore focus on the dPSC. Whether monophyly is applicable to species is contentious because at this level there are still reticulated tokogenetic (horizontal) relationships among individuals whereas monophyly requires strictly vertical relationships (Wheeler and Meier 2000, Funk and Omland 2003, Rieppel 2010, Zachos 2016, chapter 5.6.1). Hennig (1966) similarly applied monophyly only to the supraspecific level.

There are a number of definitions of the dPSC. The best known is perhaps that of Cracraft (1983, p. 170) who defines species as "the smallest diagnosable cluster of individual organisms within which there is a parental pattern of ancestry and descent". Adherents of the dPSC usually aim at the level of the population so that criticisms pointing out that under this concept single family groups would have to be granted species status (e.g., Avise 2000) are mostly unfounded. Wheeler and Platnick (2000a, p. 58) are more precise in that they define species as "the smallest aggregation of (sexual) populations or (asexual) lineages diagnosable by a unique combination of character states". Whether asexual taxa do indeed form populations and/or species ('agamospecies') is a controversial point in the species debate (see Zachos 2016, chapter 5.1) but irrelevant for mammalian taxonomy. It should be mentioned that even among adherents of the dPSC, there is considerable disagreement as to the theoretical underpinning of this species concept. Although some proponents seem to regard the dPSC as a basic concept (a primary species concept sensu Mayden, see above), others explicitly apply the dPSC in the framework introduced by Mayden (1997): "The ESC frames operations employing the [d]PSC to compare diagnostic characters to test whether or not candidate populations represent distinct lineages" (Cotterill *et al.* 2014, p. 820f).

Regardless of whether the dPSC is viewed as a primary species concept or as a secondary species identification criterion, what it boils down to in taxonomic practice is this: as soon as a population is diagnosably distinct, it is to be classified as a species in its own right. Diagnosability can be tested for in at least two different ways: either by searching for fixed unique characters (such as, but not restricted to, autapomorphies) or through statistical tests of overlap based on molecular or morphometric data such as a discriminant analysis or a principal component analysis. If the populations under study form non-overlapping clouds in character space, they are viewed as distinct species. These two approaches are quite different, and elsewhere (Zachos 2016) I have called them qualitative and quantitative diagnosability, respectively. In the first, there has to be a qualitative character that is found in all members of a phylogenetic species but is absent in all other organisms; the second, 'statistical' approach does not need fixed diagnostic characters. If, for example, two populations show different allele frequencies at a number of loci, they may well be non-overlapping, although all alleles may be present in both groups. The dPSC has been criticized heavily (see Heller *et al.* 2013, 2014, Zachos and Lovari 2013, Zachos *et al.* 2013a,b, Zachos 2015, and references therein; for authors in favor of the dPSC, see particularly Gippoliti *et al.* 2013, Groves 2013, Cotterill *et al.* 2014). Most recently, Groves *et al.* (2017) and Gippoliti *et al.* (2018), adherents of the dPSC, claimed that what they consider conservative and imprecise taxonomy (i.e., more inclusive species delimitation such as that based on the BSC) had hampered conservation in African ungulate species, but in a reply I have argued that the theoretical underpinning of their claims is flawed and that the dPSC is no more objective than most other species concepts (Zachos 2018; see also Zachos 2015, Gippoliti in press, Zachos in press a and the following paragraphs).

Although the presence of a diagnostic character or non-overlapping clouds in genetic space or morphospace is of course a biological reality (if statistical artifacts can be ruled out), it is a different issue if this is the kind of reality that we want to designate as species. Mishler and Theriot (2000b), among others, have rightfully stated that the dPSC has just shifted the question 'What is a species?' to 'What is a population?', because given high enough resolution of the characters or markers analyzed, almost every population will be diagnosable and species will be equated with single populations. Zachos and Lovari (2013) have pointed out that anthropogenic habitat fragmentation will render many relict populations diagnosable simply through genetic drift and thus make them phylogenetic species. For example, Indian tigers will undergo many such 'speciation' events before finally going extinct. Of course, once a de novo mutation produces an allele that was previously confined to another population and served as the diagnostic character for their species status, that kind of 'speciation' would immediately be reversed. The tiger (*P. tigris*) example was chosen for a reason—Cracraft *et al.* (1998) and Mazák and Groves (2006) had introduced one and two new tiger species, respectively, based on diagnostic mtDNA alleles (Sumatran tiger *Panthera sumatrae*, Cracraft *et al.* 1998) and non-overlapping craniometric data (Sumatran tiger and Javan tiger *Panthera sondaica*, Mazák and Groves 2006). Virtually, the same argument as the one brought forward by Zachos and Lovari (2013) on tigers had previously been made by Avise (2000) based on threatened tiger beetles (*Cicindela dorsalis*). Divergence at the population level is an interesting and relevant phenomenon – indeed, a large part of population genetics is devoted to it – but turning it into the yardstick by which to assign species status might also trivialize the species category in both evolutionary and biodiversity research and conservation biology. This holds even when not taking into account the statistical shortcomings of some of the most recent species splitting. For example, splitting of the single klipspringer species (*Oreotragus oreotragus*) into 11 distinct species by Groves and Grubb (2011) was based on morphometric data derived from sample sizes often lower than n = 5 (see independent criticism in Heller *et al.* 2013 and Zachos *et al.* 2013a). To emphasize the excessive splitting that would result from a rigorous application of the dPSC, several authors independently pointed out that our own species, *H. sapiens*, would almost certainly have to be split into several species as well (Ghiselin 2001, Zachos *et al.* 2013a). To many, this splitting trend is reminiscent of Merriam (1918) who described 82 species of brown bear in North America alone. On the other hand, the underestimation of species numbers in many taxa (i.e., unwarranted lumping) has also been explicitly acknowledged by critics of the dPSC. Zachos *et al.* (2013a), for example, give a number of examples where they think that splitting is justified (African elephants, clouded leopards, and Eurasian badgers) or potentially justified pending further data (giraffe) (Buckley-Beason *et al.* 2006, Kitchener *et al.* 2006, Brown *et al.* 2007, Wilting *et al.* 2007, Christiansen 2008, Del Cerro *et al.* 2010, Rohland *et al.* 2010; for a recent study presenting further evidence in favor of four different giraffe species, see Fennessy *et al.* 2016).

It is true that diagnosability is more easily testable than, for instance, the sterility of hybrids, but this operational advantage is deceptive because of the inherent trivialization of the species category (see above). All species are lineages, but not all lineages are species. Another shortcoming of the diagnosability approach as carried out and defended by, for example, Groves and Grubb (2011), Cotterill *et al.* (2014), Groves *et al.* (2017) and Gippoliti *et al.* (2018), is its theoretical inconsistency (Zachos 2015). According to the above-quoted passage from Cotterill *et al.* (2014), diagnosability is explicitly used as a proxy to detect lineages, and these lineages are considered species sensu ESC. It is the lineage that makes a species, not diagnosability. Consequently, even if the characters analyzed do not confirm diagnosable distinctness of two or more populations, any evidence of these populations being separate lineages besides diagnosability should be enough to grant species status. This, however, is the case with all truly allopatric populations. No isolated island population, according to this logic, can be the same species as the mainland or another island population (except in very mobile species such as birds) because by definition they are different lineages. Why, then, are some island populations granted species status (e.g., the Tyrrhenian red deer, *C.* (*elaphus*) *corsicanus*), whereas others which are equally isolated (such as the red deer in the British Isles) are not (Groves and Grubb 2011)? The answer is of course because of the existence of diagnosability (or the lack thereof). Instead of being a proxy for lineages in the framework of the ESC, diagnosability thus becomes the yardstick for which level of divergence a lineage must show to be assigned species status. This, however, is exactly the same that all other secondary species concepts (as species identification criteria, see above) also do. The claim of the theoretical superiority of the dPSC is therefore untenable. Rather, the only real difference in practice between the dPSC and the other secondary species concepts is that it delimits species less inclusively, or, in other words, is more prone to splitting.

1.4 Discussion and conclusions

It might be said that the good news is that the species problem is largely solved as far as the theoretical dimension is concerned. Adopting Mayden's (1997) hierarchy of species concepts (the division-of-conceptual-labor solution, Richards 2010), species can be truly defined as evolutionarily independent population lineages. The bad news is that in so doing, little is gained in biological and taxonomic practice, and at least some philosophers of science are not satisfied with it either. The continuousness of the evolutionary process and the fractal nature of the Tree of Life, which consists of lineages nested within lineages, seem to preclude an unequivocal and objective cutoff criterion to carve the Tree of Life at its joints. This problem persists despite ever more sophisticated tests for species status or methods of species delimitation (e.g., Sites and Marshall 2003, Flot *et al.* 2010, Yang and Rannala 2010, Ence and Carstens 2011, Zhang *et al.* 2013, Sukumaran and Knowles 2017; for a recent application to mammals, see Giarla *et al.* 2014). In fact, unambiguous species designations for closely related populations are only possible under very limited conditions, e.g., when there is no hybridization among sexually reproducing sympatric and synchronic organisms. These are the ideal conditions for the application of the BSC, but even here it is only the lack of hybridization that is easy to deal with; the occurrence of occasional hybrids is more difficult to evaluate, and a species concept would ultimately require an arbitrary cutoff criterion. According to the species-as-lineages consensus, speciation is the ultimately irreversible divergence of population lineages, but where exactly along that process the threshold to irreversibility is crossed is a grey area, which is why there will always be some level of subjectivity and convention in species delimitation. It is of utmost importance to realize that this has serious ramifications for the common practice of using species as the currency in ecology, evolutionary biology and conservation, e.g., when diversification rates through time are calculated or biodiversity is quantified as a basis for conservation decisions (ecological triage). The fact that species taxa cannot objectively be delimited results in more and less inclusive entities called species, depending on whether the taxonomist is more prone to lumping or splitting, and this introduces an 'apples-and-oranges' problem to large parts of comparative biology (for more, see Riddle and Hafner 1999, Faurby *et al.* 2016, Zachos 2016, chapter 7, and references therein).

It is worth noting that this grey area is only perceived as a problem with closely related and similar groups. There may be heated debates on whether red deer, tigers, or klipspringers should be considered single or several species, but it is hardly contentious that elephants and mosquitoes, hippos and rhinos or even tigers and lions, and red and fallow deer are separate species. The continuous nature of the divergence process is why Dobzhansky (1937, p. 312) famously said that "species is a stage in a process, not a static unit", whereas Mayr (1942, p. 119) insisted that species were the result of a process (not a stage). De Queiroz (1998, p. 71) sides with Dobzhansky when he says that the secondary species concepts or identification criteria "would no longer be species criteria – at least not in the sense of standards for granting lineages taxonomic status as species. Instead, they would be criteria for different stages in the existence of species – the diagnosable stage, the monophyletic stage, the reproductively isolated stage, and so on." As a consequence, "there would still be problems related to determining the limits of species in practice, but there would no longer be any greater controversy about the concept of species than currently exists for the concept of organism" (p. 72). The species problem, therefore, is more of a practical problem of taxonomy, comparative biology and conservation than a theoretical problem of evolutionary biology.

It is also important to acknowledge that species have both a synchronic dimension in a single time horizon as cohesive communities and a diachronic dimension as lineages through time (see Zachos 2016, chapter 1.4 for a short discussion). The first is explicitly addressed in the view of species as gene flow communities (see Kunz 2012). The synthesis of both dimensions – species as cohesive synchronic gene flow communities in or along diachronic meta-population lineages – offers a sound theoretical backbone, but its transformation into taxonomic practice is again hampered by the danger of trivializing the species category. The only objective break in the pattern of lineages in the Tree of Life is the level where horizontal (reproductive) relationships in sexual taxa dissolve into dichotomous (or polytomous) phylogeny. However, this would mean that any two isolated populations would have to be considered different species because of the lack of tokogenetic relationships between them. Kunz (2012, pp. 165–167) argues exactly this when he says that populations have to be considered different species when there is no gene flow between them; as soon as the populations exchange genes again, the two species would merge into one. Population lineages are demographic units, and isolation is defined by the lack of tokogenetic relationships among them whether or not they are diagnosably distinct. Logically, this is sound and consistent, but it would mean that every breeding group in a zoo that is not part of a larger breeding program, and every group of introduced tortoises on Mediterranean islands would, strictly speaking, have to be given species status. It goes without saying that this kind of logical consistency is neither feasible nor desirable—one would simply be throwing out the taxonomic baby with the logical bathwater. It is precisely due to this predicament that practically every taxonomist and evolutionary biologist wants to avoid naming ephemeral or temporary groups although that means that some qualitative judgment of significance will be inevitable (Mishler and Theriot 2000a, Tab. 1.1, Mishler and Theriot 2000b, Sites and Marshall 2004, Zachos 2015, 2016, chapter 6, 2018, in press a, b). In his chapter on species concepts in mammalian taxonomy, Corbet (1997, p. 350) agrees with this when he says that "where two forms coexist with little or no hybridization they should be treated as species, however subtle the differences. However, formal naming is best delayed until it is reasonably certain that the differences are not ephemeral, and care needs to be taken to avoid extrapolation beyond the area from which samples have been examined". This level of significance is usually measured by some kind of similarity that creeps back in again through the backdoor. This does not mean that typological or pheneticist arguments have to be revived, but nonetheless it is ultimately similarity (or the lack of it) that makes us classify African lions and the last relict population of Asiatic lions in India as conspecifics but Indian lions and Indian tigers as two distinct species. The second part of the ESC definition where species are required to "maintain their identity" from other lineages and to have their "own evolutionary tendencies and historical fate" (see Tab. 1.1) aims at exactly this—the non-ephemeral nature of lineages that are deemed worthy of species status. It is important to acknowledge the inherent limitations of taxonomy due to nature's fuzzy boundaries (Zachos in press b). In *practice*, and contrary to de Queiroz's *theoretically* sound argumentation, being a (meta)population lineage is only a necessary, but not at the same time a sufficient condition for species status (see also Freudenstein *et al.* 2017). As a consequence, the decision that two populations deserve species status can only be made in hindsight, well after the actual sundering of the lineages in question when (and if) it has become clear that they have "maintained their identity" and have their "own evolutionary fate" (Zachos 2015). Without this necessary judgment, a small propagule that gets isolated from its population of origin but becomes extinct soon after that would, ironically, attain species status exactly because it did not last longer. Similar thoughts were already expressed more than 30 years ago by Elliott Sober (1984, p. 339): "species individuation is retrospective [...] The founders were founders of a new species precisely because of what happened later, and not in virtue of anything special about them."

Where does all this leave the taxonomic practitioner then? The description of a new species or the splitting of a single species into two or more (but also the lumping of two or more into one) should in each case be justified (i) on the basis of factual evidence as to whether there is only one or two/several independent lineages and (ii) on an explicit justification why these lineages are or are not deemed significantly different. Although the first and necessary condition is more or less objectively testable, the second, sufficient condition is a matter of consensus and convention. It could be argued that diagnosability could be used as a guideline for this, but this would (and already does) inevitably lead to an explosive increase in the number of species that many evolutionary biologists will consider trivial, thus devaluing the arguably most important unit and currency in biodiversity and evolutionary research (but see Sangster 2014 who states that for avian taxonomy the consequences are much less pronounced in practice than often thought). There have been attempts at reconciling the fuzziness of species boundaries and the principal discrepancy between a continuous process (evolution) and a discrete taxonomic nomenclature. Amadon (1966), drawing on work by Rensch, Mayr, and others, proposed a formal nomenclature for "superspecies", "semispecies", and "allospecies", which increased the arsenal available to make more subtle distinctions than just species and subspecies. However, his suggestion did not solve the basic underlying problem and has never been systematically adopted by taxonomists (although, in a modified version, it was implemented in the guidelines of the British Ornithologists' Union BOU, Helbig *et al.* 2002). A different approach is to accept that species delimitation cannot be completely non-arbitrary and introduce criteria that, if followed, at least make it consistent (as has long been done in microbiology based on genetic distance).

Perhaps the best-known such attempt comes from ornithology, the so-called Tobias criteria (Tobias *et al.* 2010). Species are delimited based on a point system where scores are allocated for five classes of taxonomic characters (morphology/biometrics, acoustics, plumage and bare parts, ecology and behavior, and geographic relationships). The less similar the two populations, the higher the score, and species status is assigned if the threshold of a total score of seven points is reached or exceeded. Genetic data were not considered as they are not present for many if not most bird species. This system has already been adopted in a new authoritative checklist (del Hoyo and Collar 2014), but it has also been criticized for oversplitting (Cheke 2015). The general idea of Tobias *et al.* (2010) is certainly inspiring, but it is obvious that their criteria are tailored for birds, and different criteria would have to be developed for other taxa (Brooks and Helgen 2010). This means that species numbers in different branches of the Tree of Life will perhaps never really be comparable.

There are also very pragmatic arguments against oversplitting. Splitting leads to smaller population numbers of the single species many of which are threatened (or will become threatened through splitting if only then they meet one of the criteria to be classified as such by the IUCN). This can create an unnecessary burden on management and conservation (Frankham *et al.* 2012, Zachos *et al.* 2013a,b, Zachos 2015). Agapow *et al.* (2004) estimated that over a number of diverse taxa, including vertebrates, plants, fungi, and insects, a PSC-based approach in species delimitation (not limited to the dPSC) would result in an increase of species numbers by at least 48.7%. Consequently, the average number of mature individuals (one of the IUCN criteria to be classified as threatened) decreased by 32.8%, which was equivalent to a reclassification from 'Vulnerable' to 'Endangered' in ca. 11% of all cases. The costs associated with the protection of threatened species would also increase considerably. All these points in themselves of course would not justify the rejection of an otherwise scientifically sound and superior species concept, but such a concept is elusive. All species concepts suffer from nature being messy (Mishler and Theriot 2000b, Hey 2001a,b, Zachos in press b); there is no silver bullet to the species problem. Once we recognize that a certain level of convention is inevitable as a direct consequence of the way evolution proceeds, we can focus more on biological phenomena and less on names and nomenclature. The grey area between what we call one or two species is where many exciting evolutionary phenomena and processes can and should be studied. It is, as it were, the evolutionary biologist's dream and the taxonomist's nightmare. But the species debate, as interesting and important as it is, should not detract us from empirical research into the processes governing the divergence between lineages in the Tree of Life.

Acknowledgments

I am grateful to Nicole Grunstra and Rob Asher for their helpful comments on earlier versions of this chapter.

Literature

Agapow, P.-M., Bininda-Emonds, O.R.P., Crandall K.A., Gittleman, J.L., Mace, G.M., Marshall, J.C. & Purvis A. (2004): The impact of species concept on biodiversity studies. *Quarterly Review of Biology* 79: 161–179.

Amadon, D. (1966): The superspecies concept. *Systematic Zoology* 15: 245–249.

Avise, J.C. (2000): Cladists in wonderland. *Evolution* 54: 1828–1832.

Baker, R.J. & Bradley, R.D. (2006): Speciation in mammals and the genetic species concept. *Journal of Mammalogy* 87: 643–662.

Bateson, W. (1909): Heredity and variation in modern lights. In: Seward, A.C. (ed.): *Darwin and Modern Science*. Cambridge University Press, Cambridge, UK.: 85–101.

Bradley, R.D. & Baker, R.J. (2001): A test of the genetic species concept: cytochrome-b sequences and mammals. *Journal of Mammalogy* 82: 960–973.

Brooks, T.M. & Helgen, K.M. (2010): A standard for species. *Nature* 467: 540–541.

Brown, D.M., Brenneman, R.A., Koepfli, K.-P., Pollinger, J.P., Milá, B., Georgiadis, N.J., Louis, E.E.L., Jr., Grether, G.F., Jacobs, D.K. & Wayne R.K. (2007): Extensive population genetic structure in the giraffe. *BMC Biology* 5: 57.

Buckley-Beason, V.A., Johnson, W.E., Nash, W.G., Stanyon, R., Menninger, J.C., Driscoll, C.A., Howard, J., Bush, M., Page, J.E., Roelke, M.E., Stone, G., Martelli, P.P., Wen, C., Ling, L., Duraisingam, R.K., Lam, P.V. & O'Brien, S.J. (2006): Molecular evidence for species-level distinctions in clouded leopards. *Current Biology* 16: 2371–2376.

Burgin, C.J., Colella, J.P., Kahn, P.L. & Upham, N.S. (2018): How many species of mammals are there? Journal of Mammalogy 99: 1–14.

Cat News Special Issue 11 (2017), http://www.catsg.org/fileadmin/filesharing/5.Cat_News/5.3._Special_Issues/5.3.10._SI_11/CN_Special_Issue_11_Revised_taxonomy_of_the_Felidae.pdf

Cheke, A. (2015): Book review of del Hoyo J. & Collar N.J. *Illustrated Checklist of the Birds of the World. Volume 1: Non-passerines*. Ibis 157: 205–207.

Christiansen, P. (2008): Species distinction and evolutionary differences in the clouded leopard (*Neofelis nebulosa*) and Diard's clouded leopard (*Neofelis diardi*). *Journal of Mammalogy* 89: 1435–1446.

Corbet, G.B. (1997): The species in mammals. In: Claridge, M.F., Dawah, H.A. & Wilson, M.R. (eds.): *Species: The Units of Biodiversity*. Chapman & Hall, London: 341–356.

Cotterill, F.P.D., Taylor, P.J., Gippoliti, S., Bishop, J.M. & Groves, C.P. (2014): Why one century of phenetics is enough: response to 'Are there really twice as many bovid species as we thought?' *Systematic Biology* 63: 819–832.

Coyne, J.A. & Orr, H.A. (2004): *Speciation*. Sinauer Associates, Inc., Sunderland, Massachusetts.

Cracraft, J. (1983): Species concepts and speciation analysis. In: Johnston, R.F. (ed.): *Current Ornithology*. Vol. 1. Plenum Press, New York: 159–187.

Cracraft, J., Feinstein, J., Vaughn, J. & Helm-Bychowski, K. (1998): Sorting out tigers (*Panthera tigris*): mitochondrial sequences, nuclear inserts, systematics, and conservation genetics. *Animal Conservation* 1: 139–150.

Crandall, K.A., Bininda-Emonds, O.R.P., Mace, G.M. & Wayne, R.K. (2000): Considering evolutionary processes in conservation biology. *Trends in Ecology and Evolution* 15: 290–295.

Del Cerro, I., Marmi, J., Ferrando, A., Chashchin, P., Taberlet, P. & Bosch, M. (2010): Nuclear and mitochondrial phylogenies provide evidence for four species of Eurasian badgers (Carnivora). *Zoologica Scripta* 39: 415–425.

del Hoyo, J. & Collar, N.J. (2014): *Illustrated Checklist of the Birds of the World. Vol. 1: Non-passerines*. Lynx Edicions, Barcelona.

de Queiroz, K. (1998): The general lineage concept of species, species criteria, and the process of speciation. In: Howard, D. & Berlocher, S.H. (eds.): *Endless Forms: Species and speciation*. Oxford University Press, Oxford: 57–75.

de Queiroz, K. (1999): The general lineage concept of species and the defining properties of the species category. In: Wilson, R.A. (ed.): *Species. New Interdisciplinary Essays*. The MIT Press, Cambridge, 49–89.

de Queiroz, K. (2005): A unified concept of species and its consequences for the future of taxonomy. Proceedings of the California Academy of Sciences 56: 196–215.

de Queiroz, K. (2007): Species concepts and species delimitation. *Systematic Biology* 56: 879–886.

Dobzhansky, T. (1935): A critique of the species concept in biology. *Philosophy of Science* 2: 344–355.

Dobzhansky, T. (1937): *Genetics and the Origin of Species*. Columbia University Press, New York.

Ence, D.D. & Carstens, B.C. (2011): SpedeSTEM: a rapid and accurate method for species delimitation. *Molecular Ecology Resources* 11: 473–480.

Faurby, S., Eiserhardt, W.L. & Svenning, J.-C. (2016): Strong effect of variation in taxonomic opinion on diversification analyses. Methods in Ecology and Evolution 7: 4–13.

Fennessy, J., Bidon, T., Reuss, F., Kumar, V., Elkan, P., Nilsson, M.A., Vamberger, M., Fritz, U. & Janke, A. (2016): Multi-locus analyses reveal four giraffe species instead of one. *Current Biology* 26: 1–7.

Flot, J.-F., Couloux, A. & Tillier, S. (2010): Haplowebs as a graphical tool for delimiting species: a revival of Doyle's "field for recombination" approach and its application to the coral genus *Pocillopora* in Clipperton. *BMC Evolutionary Biology* 10: 372.

Frankham, R., Ballou, J.D., Dudash, M.R., Eldridge, M.D.B., Fenster, C.B., Lacy, R.C., Mendelson, J.R. III, Porton, I.J., Ralls, K. & Ryder, O.A. (2012): Implications of different species concepts for conserving biodiversity. *Biological Conservation* 153: 25–31.

Freudenstein, J.V., Broe, M.B., Folk, R.A. & Sinn, B.T. (2017): Biodiversity and the species concept – lineages are not enough. Systematic Biology 66: 644–656.

Funk, D.J. & Omland, K.E. (2003): Species-level paraphyly and polyphyly: frequency, causes, and consequences, with insights from animal mitochondrial DNA. *Annual Review of Ecology, Evolution and Systematics* 34: 397–423.

Geist, V. (1998): *Deer of the World. Their Evolution, Behavior, and Ecology*. Stackpole Books, Mechanicsburg, PA.

Ghiselin, M.T. (1974): A radical solution to the species problem. *Systematic Zoology* 23: 536–544.

Ghiselin, M.T. (2001): Species concepts. In: *Encyclopedia of Life Sciences*. John Wiley & Sons, Hoboken, NJ: 1–5.

Giarla, T.C., Voss, R.S. & Jansa, S.A. (2014): Hidden diversity in the Andes: comparison of species delimitation methods in montane marsupials. *Molecular Phylogenetics and Evolution* 70: 137–151.

Gippoliti, S. (in press): Species delimitation in mammals: A comment on Zachos (2018). *Mammalian Biology*.

Gippoliti, S. & Groves, C.P. (2012): "Taxonomic inflation" in the historical context of mammalogy and conservation. *Hystrix, Italian Journal of Mammalogy* 23: 6–9.

Gippoliti, S., Cotterill, F.P.D. & Groves, C. (2013): Mammal taxonomy without taxonomists: a reply to Zachos and Lovari. *Hystrix, Italian Journal of Mammalogy* 24: 3–5.

Gippoliti, S., Cotterill, F.P.D., Zinner, D. & Groves, C.P. (2018): Impacts of taxonomic inertia for the conservation of African ungulate diversity: an overview. Biological Reviews 93: 115–130.

Groves, C. (2012): Species concepts in primates. *American Journal of Primatology* 74: 687–691.

Groves, C.P. (2013): The nature of species: a rejoinder to Zachos et al. *Mammalian Biology* 78: 7–9.

Groves, C. (2001): *Primate Taxonomy*. Smithsonian Institution Press, Washington and London.

Groves, C.P., Cotterill, F.P.D., Gippoliti, S., Robovský, J., Roos, C., Taylor, P.J. & Zinner, D. (2017): Species definitions and conservation: a review and case studies from African mammals. Conservation Genetics 18: 1247–1256.

Groves, C. & Grubb, P. (2011): *Ungulate Taxonomy*. The Johns Hopkins University Press, Baltimore.

Groves, C. & Leslie, Jr. D.M. (2011): Family Bovidae (Hollow-horned Ruminants). In: Wilson, D.E. & Mittermeier, R.A. (eds.): *Handbook of the Mammals of the World. Vol. 2. Hoofed Mammals*. Lynx Edicions, Barcelona: 444–779.

Grubb, P. (2005): Order Artiodactyla. In: Wilson, D.E. & Reeder, D.M. (eds.): *Mammal Species of the World, a Taxonomic and Geographic Reference*. Third edition. The Johns Hopkins University Press, Baltimore: 637–722.

Gutiérrez, E.E. & Helgen, K.M. (2013): Outdated taxonomy blocks conservation. *Nature* 495: 314.

Hausdorf, B. (2011): Progress toward a general species concept. *Evolution* 65: 923–931.

Helbig, A.J., Knox, A.G., Parkin, D.T., Sangster, G. & Collinson, M. (2002): Guidelines for assigning species rank. *Ibis* 144: 518–525.

Heller, R., Frandsen, P., Lorenzen, E.D. & Siegismund, H.R. (2013): Are there really twice as many bovid species as we thought? *Systematic Biology* 62: 490–493.

Heller, R., Frandsen, P., Lorenzen, E.D. & Siegismund, H.R. (2014): Is diagnosability an indicator of speciation? Response to "Why one century of phenetics is enough". *Systematic Biology* 63: 833–837.

Hennig, W. (1966): *Phylogenetic Systematics*. University of Illinois Press, Champaign, IL.

Hey, J. (2001a): *Genes, Categories, and Species. The Evolutionary and Cognitive Causes of the Species Problem*. Oxford University Press, New York.

Hey, J. (2001b): The mind of the species problem. *Trends in Ecology and Evolution* 16: 326–329.

IISE (2011): *Retro SOS 2000–2009: A Decade of Species Discovery in Review*. International Institute for Species Exploration, Tempe, AZ. http://www.esf.edu/species/documents/sosretro.pdf

Isaac, J.B., Mallet, J. & Mace, G.M. (2004): Taxonomic inflation: its influence on macroecology and conservation. *Trends in Ecology and Evolution* 19: 464–469.

Kitchener, A.C., Beaumont, M.A. & Richardson, D. (2006): Geographical variation in the clouded leopard, *Neofelis nebulosa*, reveals two species. *Current Biology* 16: 2377–2383.

Kunz, W. (2012): *Do Species Exist? Principles of Taxonomic Classification*. Wiley-Blackwell, Weinheim.

Lambertz, M. & Perry, S.F. (2015): Chordate phylogeny and the meaning of categorial ranks in modern evolutionary biology. *Proceedings of the Royal Society B* 282: 20142327.

Laurin, M. (2010): The subjective nature of Linnaean categories and its impact in evolutionary biology and biodiversity studies. *Contributions to Zoology* 79: 131–146.

Luo, S.-J., Kim, J.-H., Johnson, W.E., van der Walt, J., Martenson, J., Yuhki, N., Miquelle, D.G., Uphyrkina, O., Goodrich, J.M., Quigley, H.B., Tilson, R., Brady, G., Martelli, P., Subramaniam, V., McDougal, C., Hean, S., Huang, S.-Q., Pan, W., Karanth, U.K., Sunquist, M., Smith, J.L.D. & O'Brien, S.J. (2004): Phylogeography and genetic ancestry of tigers (*Panthera tigris*). *PLoS Biology* 2(12): e442.

Mallet, J. (2001): Species, concepts of. In: *Encyclopedia of Biodiversity*, volume 5. Academic Press, Waltham, MA: 427–440.

Mallet, J. (2005): Hybridization as an invasion of the genome. *Trends in Ecology and Evolution* 20: 229–237.

Mayden, R.L. (1997): A hierarchy of species concepts: the denouement in the saga of the species problem. In: Claridge, M.F., Dawah, H.A. & Wilson, M.R. (eds.): *Species: The Units of Biodiversity*. Chapman & Hall, London: 381–424.

Mayr, E. (1940): Speciation phenomena in birds. *American Naturalist* 74: 249–278.

Mayr, E. (1942): *Systematics and the Origin of Species*. Columbia University Press, New York.

Mayr, E. (1969): *Principles of Systematic Zoology*. McGraw-Hill, New York.

Mayr, E. (2000): The biological species concept. In: Wheeler, Q.D. & Meier, R. (eds.): *Species Concepts and Phylogenetic Theory—A Debate*. Columbia University Press, New York: 17–29.

Mazák, J.H. & Groves, C.P. (2006): A taxonomic revision of tigers (*Panthera tigris*) of Southeast Asia. *Mammalian Biology* 71: 268–287.

McDevitt, A.D., Edwards, C.J., O'Toole, P., O'Sullivan, P., O'Reilly, C. & Carden, R.F. (2009): Genetic structure of, and hybridisation between, red (*Cervus elaphus*) and sika (*Cervus nippon*) deer in Ireland. *Mammalian Biology* 74: 263–273.

Meier, R. & Willmann, R. (2000): The Hennigian species concept. In: Wheeler, Q.D. & Meier, R. (eds.): *Species Concepts and Phylogenetic Theory—A Debate*. Columbia University Press, New York: 30–43.

Meiri, S. & Mace, G.M. (2007): New taxonomy and the origin of species. *PLoS Biology* 5: e194.

Merriam, C.H. (1918): Review of the grizzly and big brown bears of North America (genus *Ursus*) with the description of a new genus, *Vetularctos*. *North American Fauna* 41: 1–136.

Merriam, C.H. (1919): Criteria for the recognition of species and genera. *Journal of Mammalogy* 1: 6–9.

Mishler, B.D. (1999): Getting rid of species? In: Wilson, R.A. (ed.): *Species. New Interdisciplinary Essays*. The MIT Press, Cambridge, MA: 307–315.

Mishler, B.D. (2010): Species are not uniquely real biological entities. In: Ayala, F.J., Arp, R. (eds.): Contemporary Debates in Philosophy of Biology. Blackwell Publishing Ltd.: 110–122.

Mishler, B.D. & Theriot, E.C. (2000a): The phylogenetic species concept (sensu Mishler and Theriot): monophyly, apomorphy, and phylogenetic species concepts. In: Wheeler, Q.D. & Meier, R. (eds.): *Species Concepts and Phylogenetic Theory—A Debate*. Columbia University Press, New York: 44–55.

Mishler, B.D. & Theriot, E.C. (2000b): A critique from the Mishler and Theriot phylogenetic species concept perspective: monophyly, apomorphy, and phylogenetic species concepts. In: Wheeler, Q.D. & Meier, R. (eds.): *Species Concepts and Phylogenetic Theory—A Debate*. Columbia University Press, New York: 119–132.

Mishler, B.D., Wilkins, J.S. (2018): The hunting of the SNaRC: A snarky solution to the species problem. *Philosophy, Theory, and Practice in Biology* 10: 1.

Mittermeier, R.A., Rylands, A.B. & Wilson, D.E. (eds.) (2013): *Handbook of the Mammals of the World. Vol. 3 Primates*. Lynx Edicions, Barcelona.

Moritz, C. (1994): Defining 'evolutionary significant units' for conservation. *Trends in Ecology and Evolution* 9: 373–375.

Muller, H.J. (1942): Isolating mechanisms, evolution, and temperature. *Biology Symposium* 6: 71–125.

Osmers, B., Petersen, B.-S., Hartl, G.B., Grobler, J.P., Kotze, A., Van Aswegen, E. & Zachos, F.E. (2012): Genetic analysis of southern African gemsbok (*Oryx gazella*) reveals high variability, distinct lineages and strong divergence from the East African *Oryx beisa*. *Mammalian Biology* 77: 60–66.

Pérez-Espona, S., Pemberton, J.M. & Putman, R. (2009): Red and sika deer in the British Isles, current management issues and management policy. *Mammalian Biology* 74: 247–262.

Poelstra, J.W., Vijay, N., Bossu, C.M., Lantz, H., Ryll, B., Müller, I., Baglione, V., Unneberg, P., Wikelski, M., Grabherr, M.G. & Wolf, J.B.W. (2014): The genomic landscape underlying phenotypic integrity in the face of gene flow in crows. *Science* 344: 1410–1414.

Reeder, D.M., Helgen, K.M. & Wilson, D.E. (2007): Global trends and biases in new mammal species discoveries. *Occasional Papers, Museum of Texas Tech University* 269: 1–35.

Richards, R.A. (2010): *The Species Problem—A Philosophical Analysis*. Cambridge University Press, Cambridge.

Riddle, B.R. & Hafner, D.J. (1999): Species as units of analysis in ecology and biogeography: time to take the blinders off. Global Ecology and Biogeography 8: 433–441.

Rieppel, O. (2007): Species: kinds of individuals or individuals of a kind. *Cladistics* 23: 373–384.

Rieppel, O. (2010): Species monophyly. Journal of Zoological Systematics and Evolutionary Research 48: 1–8.

Rohland, N., Reich, D., Mallick, S., Meyer, M., Green, R.E., Georgiadis, N.J., Roca, A.L. & Hofreiter, M. (2010): Genomic DNA sequences from Mastodon and Woolly Mammoth reveal deep speciation of Forest and Savanna elephants. *PLoS Biology* 8(12): e1000564.

Ryder, O.A. (1986) Species conservation and systematics: the dilemma of subspecies. *Trends in Ecology and Evolution* 1: 9–10.

Sangster, G. (2014): The application of species criteria in avian taxonomy and its implications for the debate over species concepts. *Biological Reviews* 89: 199–214.

Senn, H.V., Barton, N. H., Goodman, S.J., Swanson, G.M., Abernethy, K.A. & Pemberton, J.M. (2010): Investigating temporal changes in hybridization and introgression in a predominantly bimodal hybridizing population of invasive sika (*Cervus nippon*) and native red deer (*C. elaphus*) on the Kintyre Peninsula, Scotland. *Molecular Ecology* 19: 910–924.

Simpson, G.G. (1951): The species concept. *Evolution* 5: 285–298.

Simpson, G.G. (1961): *Principles of Animal Taxonomy*. Columbia University Press, New York.

Sites, J.W. Jr. & Marshall, J.C. (2003): Delimiting species: a Renaissance issue in systematic biology. *Trends in Ecology and Evolution* 18: 462–470.

Sites, J.W. Jr. & Marshall, J.C. (2004): Operational criteria for delimiting species. *Annual Review of Ecology, Evolution and Systematics* 35: 199–227.

Sober, E. (1984): Sets, species, and evolution: comments on Philip Kitcher's "species". *Philosophy of Science* 51: 334–341.

Stamos, D.N. (2003): *The Species Problem. Biological species, Ontology, and the Metaphysics of Biology*. Lexington Books, Lanham.

Sukumaran, J. & Knowles, L.L. (2017): Multispecies coalescent delimits structure, not species. Proceedings of the National Academy of Sciences USA 114: 1607–1612.

Taverner, P.A. (1920): The test of the subspecies. *Journal of Mammalogy* 1: 124–127.

Tobias, J.A., Seddon, N., Spottiswoode, C.N., Pilgrim, J.D., Fishpool, L.D. & Collar, N.J. (2010): Quantitative criteria for species delimitation. *Ibis* 152: 724–746.

Wheeler, Q.D. & Meier, R. (ed.) (2000): *Species Concepts and Phylogenetic Theory—A Debate*. Columbia University Press, New York.

Wheeler, Q.D. & Platnick, N.I. (2000a): The phylogenetic species concept (sensu Wheeler and Platnick). In: Wheeler, Q.D. & Meier, R. (eds.): *Species Concepts and Phylogenetic Theory—A Debate*. Columbia University Press, New York: 55–69.

Wheeler, Q.D. & Platnick, N.I. (2000b): A critique from the Wheeler and Platnick phylogenetic species concept perspective: problems with alternative concepts of species. In: Wheeler, Q.D. & Meier, R. (eds.): *Species Concepts and Phylogenetic Theory—A Debate*. Columbia University Press, New York: 133–145.

Wiley, E.O. (1978): The evolutionary species concept reconsidered. *Systematic Zoology* 27: 17–26.

Wiley, E.O. & Mayden, R.L. (2000a): The evolutionary species concept. In: Wheeler, Q.D. & Meier, R. (eds.): *Species Concepts and Phylogenetic Theory—A Debate*. Columbia University Press, New York: 70–89.

Wiley, E.O. & Mayden, R.L. (2000b): A critique from the evolutionary species concept perspective. In: Wheeler, Q.D. & Meier, R. (eds.): *Species Concepts and Phylogenetic Theory—A Debate*. Columbia University Press, New York: 146–158.

Wiley, E.O. & Mayden, R.L. (2000c): A defence of the evolutionary species concept. In: Wheeler, Q.D. & Meier, R. (eds.): *Species Concepts and Phylogenetic Theory—A Debate*. Columbia University Press, New York: 198–208.

Wilkins, J.S. (2009a): *Species. A History of the Idea*. University of California Press, Berkeley, Los Angeles, London.

Wilkins, J.S. (2009b): *Defining Species. A Sourcebook from Antiquity to Today*. Peter Lang Publishing, New York.

Wilkins, J.S. (2018): Species. The Evolution of the Idea. 2nd edition. CRC Press, Boca Raton, London, New York.

Wilson, E.O. (1992): The diversity of life. Penguin Books, London.

Wilson, D.E. & Reeder, D.M. (eds.) (1993): *Mammal Species of the World, a Taxonomic and Geographic Reference*. Second edition. Smithsonian Institution Press, Washington, D.C.

Wilson, D.E. & Reeder, D.M. (eds.) (2005): *Mammal Species of the World, a Taxonomic and Geographic Reference*. Third edition. The Johns Hopkins University Press, Baltimore.

Wilson, R.A. (ed.) (1999): *Species. New Interdisciplinary Essays*. The MIT Press, Cambridge, Mass., London.

Wilting, A., Buckley-Beason, V.A., Feldhaar, H., Gadau, J., O'Brien, S.J. & Linsenmair, K.E. (2007): Clouded leopard phylogeny revisited: support for species recognition and population division between Borneo and Sumatra. *Frontiers in Zoology* 4: 15.

Wilting, A., Courtiol, A., Christiansen, P., Niedballa, J., Scharf, A.K., Orlando, L., Balkenhol, N., Hofer, H., Kramer-Schadt, S., Fickel, J. & Kitchener, A.C. (2015): Planning tiger recovery: Understanding intraspecific variation for effective conservation. Science Advances 1: e1400175.

Yang, Z. & Rannala, B. (2010): Bayesian species delimitation using multilocus sequence data. *Proceedings of the National Academy of Sciences USA* 107: 9264–9269.

Zachos, F.E. (2011): Linnean ranks, temporal banding, and time-clipping: why not slaughter the sacred cow? *Biological Journal of the Linnean Society* 103:732–734.

Zachos, F.E. (2015): Taxonomic inflation, the phylogenetic species concept and lineages in the Tree of Life—a cautionary comment on species splitting. *Journal of Zoological Systematics and Evolutionary Research* 53: 180–184.

Zachos, F.E. (2016): *Species Concepts in Biology. Historical Development, Theoretical Foundations and Practical Relevance*. Springer International Publishing Switzerland.

Zachos, F.E. (2018): Mammals and meaningful taxonomic units: the debate about species concepts and conservation. *Mammal Review* 48: 153–159.

Zachos, F.E. (in press a): Species delimitation in mammals: A response to Gippoliti. *Mammalian Biology*.

Zachos, F.E. (in press b): (New) Species concepts, species delimitation and the inherent limitations of taxonomy. *Journal of Genetics*.

Zachos, F.E., Apollonio, M., Bärmann, E.V., Festa-Bianchet, M., Göhlich, U., Habel, J.C., Haring, E., Kruckenhauser, L., Lovari, S., McDevitt, A.D., Pertoldi, C., Rössner, G.E., Sánchez-Villagra, M.R., Scandura, M. & Suchentrunk, F. (2013a): Species inflation and taxonomic artefacts—a critical comment on recent trends in mammalian classification. *Mammalian Biology* 78: 1–6.

Zachos, F.E., Clutton-Brock, T.H., Festa-Bianchet, M., Lovari, S., Macdonald, D.W., & Schaller, G.B. (2013b): Species splitting puts conservation at risk. *Nature* 494: 35.

Zachos, F.E. & Hartl, G.B. (2011): Phylogeography, population genetics and conservation of the European red deer *Cervus elaphus*. Mammal Review 41: 138–150.

Zachos, F.E. & Lovari, S. (2013): Taxonomic inflation and the poverty of the phylogenetic species concept—a reply to Gippoliti and Groves. *Hystrix, Italian Journal of Mammalogy* 24: 142–144.

Zachos, F.E., Mattioli, S., Ferretti, F. & Lorenzini, R. (2014): The unique Mesola red deer of Italy: taxonomic recognition (*Cervus elaphus italicus nova ssp.*, Cervidae) would endorse conservation. *Italian Journal of Zoology* 81: 136–143.

Zhang, J., Kapli, P., Pavlidis, P. & Stamatakis, A. (2013): A general species delimitation method with applications to phylogenetic placements. *Bioinformatics* 29: 2869–2876.

Svetlana V. Pavlova and Jeremy B. Searle

2 Chromosomes and speciation in mammals

2.1 Introduction

Chromosomes are the hereditary material: DNA together with a histone scaffold (Annunziato 2008). Genes and their regulatory regions being linked together on chromosomes make regular segregation at somatic cell division (mitosis) possible. Coupled with mechanisms of recombination in the generation of gametes (at meiosis), chromosomes represent an exquisite vehicle for transmission of genes and for the shuffling of variants to a sufficient extent to provide a key advantage to sexual reproduction (Crow 1994). In terms of DNA, chromosomes are more than just a string of genes and regulatory regions; they have structures that are important for their function (centromeres and telomeres; Blackburn and Szostak 1984), and they are susceptible to invasion by "selfish DNA", which has no advantage to the organism but which spreads according the principles of natural selection (Orgel and Crick 1980).

Chromosomes can be visualized at mitosis and meiosis because they need to be maximally contracted at these stages of segregation (Annunziato 2008). They can be examined under the light microscope and are astonishingly beautiful, as is evident even from the publication that can be considered the foundation of the chromosomal theory of inheritance (see Fig. 7 in Sutton 1902, Crow and Crow 2002), and in modern images, both in monochrome after conventional staining and in multicolor after using fluorescently labeled DNA probes (Fig. 2.1). It is using chromosome morphology, meiotic pairing, and various ways to visualize differentiation along chromosomes (all illustrated in Fig. 2.1) that allows the complement of chromosomes (the "karyotype"; see Glossary section) to be characterized, including the identification of individual chromosomes. When individuals, species, or major geographic forms within species ("races") differ in their karyotype, this can be due to difference in ploidy, addition/deletion of specific chromosomes or major amounts of chromosomal material ("chromatin") within chromosomes, or differences due to chromosomal rearrangements (CRs) of which Robertsonian (Rb) fusions, tandem fusions, and inversions are illustrated in Fig. 2.1. CRs do not lead to any but the most minor changes in the amount of chromatin; they just repackage the existing chromatin in different ways, although always changing linkage relationships. As shown in Fig. 2.2, Rb fusions and fissions and tandem fusions change the number of chromosomes and generate chromosomes of a notably different size compared with previous, inversions and reciprocal translocations interfere with the sequence of genes within chromosome arms, centric shifts and most pericentric inversions change the positioning of centromeres within chromosomes, and whole-arm reciprocal translocations (WARTs) change the association of chromosome arms without changing the number of chromosomes.

Of the changes to karyotype in mammals, it is CRs that are the most important to think about regarding the role of chromosomes in speciation. Therefore, the focus of this article will be on CRs (although other forms of chromosomal variability will also be mentioned). Why might it be suggested that CRs could promote speciation in mammals? It is (a) because species of mammal, including closely related ones, often differ in CRs; (b) because there are clearly understood ways in which it can be suggested that CRs may lead to reproductive isolation; and (c) because there are races that differ by CRs within some species of mammal. These facts indicate a possible process of chromosomal speciation in mammals: the progression from occurrence of chromosomal mutations within species, through the generation of races distinguished by those mutations, and then resulting in separate chromosomally differentiated species. In this chapter, we will consider all these aspects of the possible role of CRs in the speciation of mammals.

Glossary

biological species concept: two forms are considered two species if they are "reproductively isolated", i.e., they would be unable to interbreed effectively if brought into contact, not being able to produce fertile hybrids. In this way, there is no gene flow between species defined on the biological species concept, and they are on separate evolutionary trajectories.

chromosome painting: cross-species chromosome painting is a method whereby the chromosomes of one species are sorted by flow cytometry, and chromosome-specific probes are designed which can then be hybridized onto the chromosomes of other species, using multicolor fluorescence to identify where the hybridizations of the different chromosomal probes lie, i.e., regions of chromosomal homology (Scherthan *et al.* 1994, Speicher and Carter 2005).

effective population size: considering what a population size of a species (or other unit) would be under idealized conditions (from

https://doi.org/10.1515/9783110341553-002

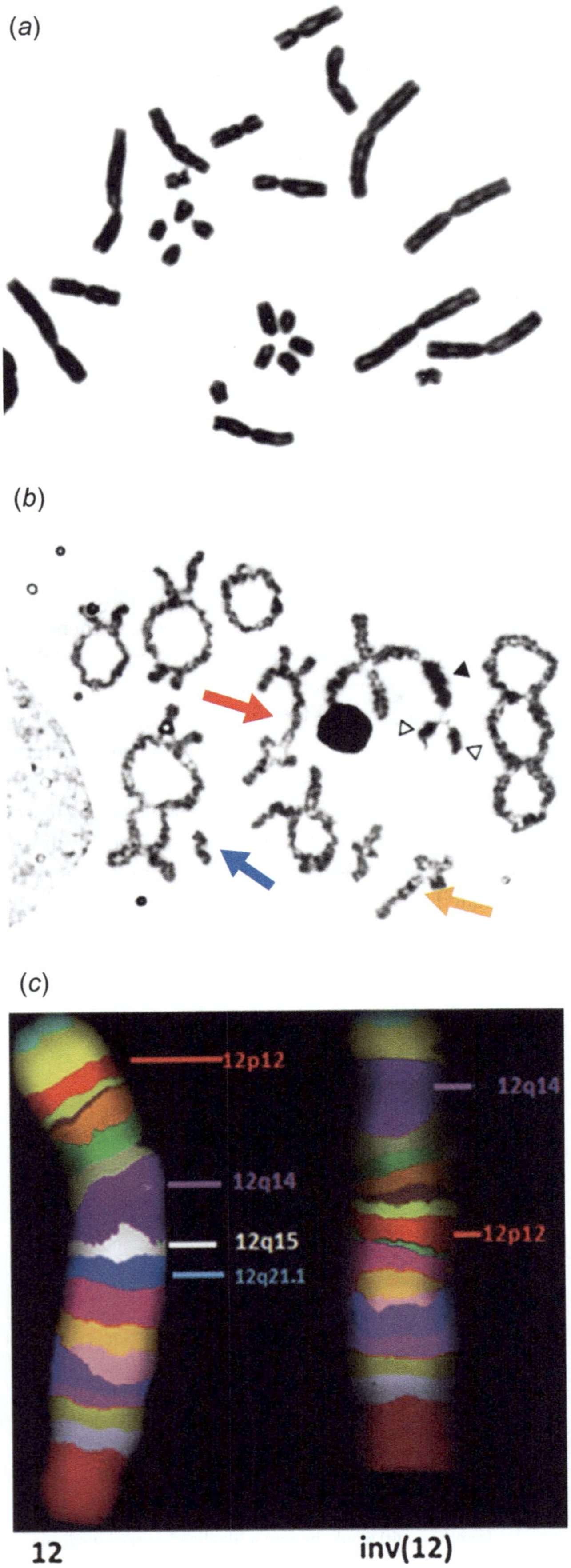

Fig. 2.1: Beautiful chromosomes, beautiful in appearance and for the complex mechanistic and evolutionary stories they tell (see Fig. 2.2 for help with interpretation). (*a*) Conventionally stained mitotic preparation of chromosomes of the common shrew *Sorex araneus* showing both acrocentric and metacentric chromosomes (centromeres at the end and within the chromosome, respectively), the metacentrics generally having been generated by Robertsonian (Rb) fusion mutations, the fusion of acrocentrics at their centromeres (Searle and Wójcik 1998). (Reproduced with permission from Searle 1983.) (*b*) Conventionally stained meiotic preparation of common shrew chromosomes at diakinesis, illustrating an autosomal trivalent (red arrow) due to heterozygosity for an Rb fusion, autosomal univalency (blue arrow) deriving from breakdown of another autosomal trivalent, whose vestiges remain (orange arrow), and a sex multivalent due to a tandem fusion between the original X and an autosome and due to Y chromosome aneuploidy (an extra Y chromosome). Within the sex multivalent, the heteropycnotic true X segment (black arrowhead) pairs with the two heteropycnotic Y chromosomes (white arrowheads). (Reproduced with permission from Searle and Wilkinson 1986.) (*c*) High-resolution multicolor banding on human chromosomes from one mitotic cell, showing a normal chromosome 12 on the left and a chromosome 12 with an inversion and a deletion on the right. (Reproduced with permission from Genesio *et al.* 2013.)

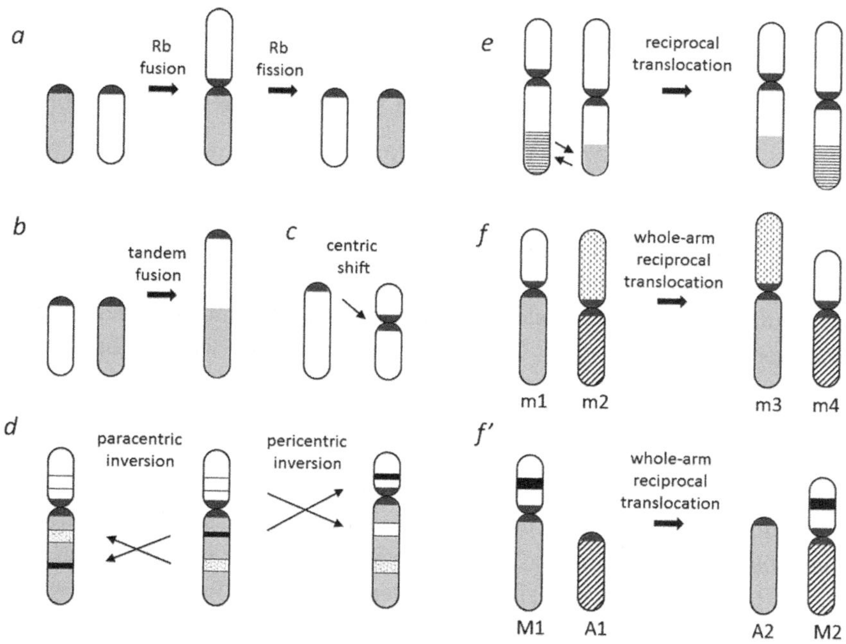

Fig. 2.2: Chromosomal rearrangements. Illustrations use different shading to indicate different chromosome arms or regions within chromosome arms. Dark grey shading indicates the position of the centromere in each chromosome, with an acrocentric chromosome being a chromosome consisting of a single chromosome arm and a centromere at the end of that arm, and a metacentric being a chromosome with an internal centromere. (*a*) Robertsonian (Rb) fusion: the joining of two acrocentric chromosomes at the centromere; Rb fission: the reverse process of separation of a metacentric into two acrocentrics (these rearrangements are also known as centric fusions and fissions, respectively). (*b*) Tandem fusion: the joining of two chromosomes such that at least one of the chromosomes does not join at its centromere (illustrated: two acrocentric chromosomes join at the centromere of one chromosome and telomere of the other). (*c*) Centric shift (or centromeric shift): involving relocation of the centromere from one position within a chromosome to another, without detectable rearrangement of other parts of the chromosome. (*d*) Inversion: involving parts of the chromosome becoming inverted, either excluding or including the centromere (paracentric and pericentric, respectively). (*e*) Reciprocal translocation: the swapping of chromosomal regions between two chromosomes; (*f, f′*) whole-arm reciprocal translocation (WART): the swapping of whole chromosome arms either between two metacentrics (m1, m2) or a metacentric and an acrocentric (M1, A1) to generate new products (m3 and m4 and A2 and M2, respectively).

a population genetic perspective), thereby creating a universal unit of population size allowing comparison of species with very different life histories and demographies (see Charlesworth 2009).

G-banding: a protocol typically applied to mitotic chromosome spreads, which results (in mammals) in chromosome-specific sequences of light and dark bands along the chromosomes, which allow different chromosomes in the complement, including rearranged chromosomes, to be identified.

homosequential: when chromosomes of different taxa (e.g., different species) are compared and do not differ overtly, e.g., in G-banding pattern.

karyotype: the diploid complement of chromosomes, referring to individuals, populations, or species; often described in terms of morphological characteristics of the chromosomes (including, e.g., G-banding pattern), and the photomicrograph of a chromosome spread is often known as a "karyotype".

karyotypic orthoselection: the phenomenon of a particular category of chromosomal rearrangement being fixed in the same phylogenetic lineage (e.g., repeated fixation of Robertsonian fusions in gerbils; Aniskin *et al.* 2006).

meiotic drive: biased segregation of chromosomes at meiosis in female heterozygotes for chromosomal rearrangements, such that either the rearranged or the ancestral chromosome(s) are more predominant in the haploid oocytes.

recombination suppression: recombination occurring at considerably lower frequency than normal in heterozygotes for chromosomal rearrangements, in the vicinity of the chromosomal breakpoint (e.g., close to the centromere in heterozygotes for Robertsonian fusions), reflecting an absence of homologous pairing in the regions concerned—homologous pairing being needed for recombination.

subspecies: a traditional taxonomical subdivision of species, based on morphology (although more recently also supported by genetics). Given as a Linnean trinomial (e.g., *Mus musculus domesticus*).

underdominance: in the context of chromosomal variation, it is the unfitness of hybrids of chromosomal races, relating to the meiotic aberrations associated with heterozygosity for chromosomal rearrangements.

2.2 Species of mammal differ in karyotype (but not always)

From the outset, it should be said that we do not believe that CRs are essential for speciation in mammals. There are situations where an ancestral species will be subdivided into two or more geographical isolates, and the populations in the different isolates will accumulate genic differences that will be sufficient to cause reproductive isolation from the populations in the other isolates. There may also be accumulation of CRs between the populations, but even so, it could be that the genic differences are the primary source of reproductive isolation, e.g., promoting differences in behavior such that individuals from the different populations do not perceive members of the other populations as potential mates.

An acid test that demonstrates a lack of importance of CRs in a speciation event is if the descendent species ("sister species") do not differ by CRs. Carson *et al.* (1967) demonstrated this in Hawaiian picture-wing *Drosophila*, showing not only for sister pairs but also larger lineages that are homosequential (see Glossary section) in terms of banding sequence on salivary chromosomes, i.e., no indication of CRs. There are also closely related mammals where karyotypes fail to reveal CRs distinguishing them. Karyotypes in mammals are usually studied by G-banding (see Glossary section), a method that typically results in around 200–300 chromosome bands (both dark and light) along the chromosomes of the haploid mitotic complement (e.g., Levan 1974). Examples where closely related species are homosequential in G-banding pattern include cats (Felidae; Dobigny *et al.* 2017) and bats (*Myotis*; Bickham 1979).

However, we have said that closely related species of mammal do often differ in karyotype and have given this as a reason to think that CRs may be important in speciation in mammals. To counter the examples given above, there are cases of extreme difference in karyotype between closely related species of mammal. The best known example is the deer genus *Muntiacus*, which consists of species with different chromosome numbers between 6 and 46 due to tandem and Rb fusions (Figs. 2.2 and 2.3; Yang *et al.* 1995). Another example is the rodent genus *Gerbillus*, where it has been suggested that as many as 70 Rb fusions, 2 pericentric inversions, 1 tandem fusion, and 13 other CRs may differentiate four species (Aniskin *et al.* 2006). A further case is the shrew genus *Sorex*, for which representatives of 12 species were investigated by cross-species "chromosome painting" (see Glossary section) and for which 54 fusions, 11 fissions, and 20 centric shifts were described (Biltueva *et al.* 2011). Other very dramatic examples of extreme and rapid evolution of mammalian karyotypes are given by Dobigny *et al.* (2005).

Bringing these two observations together – situations of large difference in karyotype between related forms and situations of little or no difference in karyotype – there is actually extraordinary heterogeneity in mammals in the extent of chromosomal difference between related species. This was shown nicely with cross-species chromosome painting among multiple members of the Artiodactyla, mapping the results onto a morphological/ molecular supertree (Kulemzina *et al.* 2009). This procedure allowed the rate of CR to be estimated in different parts of the phylogeny, demonstrating striking disparities, with 1.76 CRs per million years in the Suina (the pigs and peccaries) at one extreme and 0.07 CRs per million years in the cetaceans (whales, dolphins, and porpoises) at the other extreme. Clearly, this is a massive difference in rates of observed fixation of CRs, which tallies with the long-held knowledge that cetaceans are remarkably homogeneous in karyotype (Árnason 1972).

How can we explain instances of a high rate of CR in a mammalian lineage, such that closely related species tend to differ in karyotype, sometimes substantially? Mutation is, of course, an essential prerequisite of this. A high rate of occurrence of fixed CRs in a lineage would seem to imply at least a moderate chromosomal mutation rate, sometimes a very high one. Often the same type of mutation occurs recurrently. Thus, in the muntjac case (Fig. 2.3), tandem fusions (and occasional Rb fusions) have to occur repeatedly in order for the Indian muntjac karyotype to evolve from a karyotype like that of the Chinese muntjac (Yang *et al.* 1995). There must be a strong tendency for simultaneous breaks near the centromeres and distal telomeres of the ancestral chromosomes (Fig. 2.2). Tandem fusions are not a particularly commonly detected CR in mammalian evolution, which suggests that there must be some greater tendency than normal for chromosome breakages that promote this sort of CR in the muntjac lineage. Low copy number DNA repeats are preferentially distributed at centromeres and telomeres, and so one way of explaining repeat tandem fusions could be non-allelic homologous recombination (NAHR) involving these repeats (see Stankiewicz and Lupski 2002). Because low copy number repeats can derive from endogenous retroviruses, this could be the source of massive heterogeneity between species and between lineages (which ultimately derive from single species) because of large differences in insertion sites of endogenous retroviruses and insertion number between

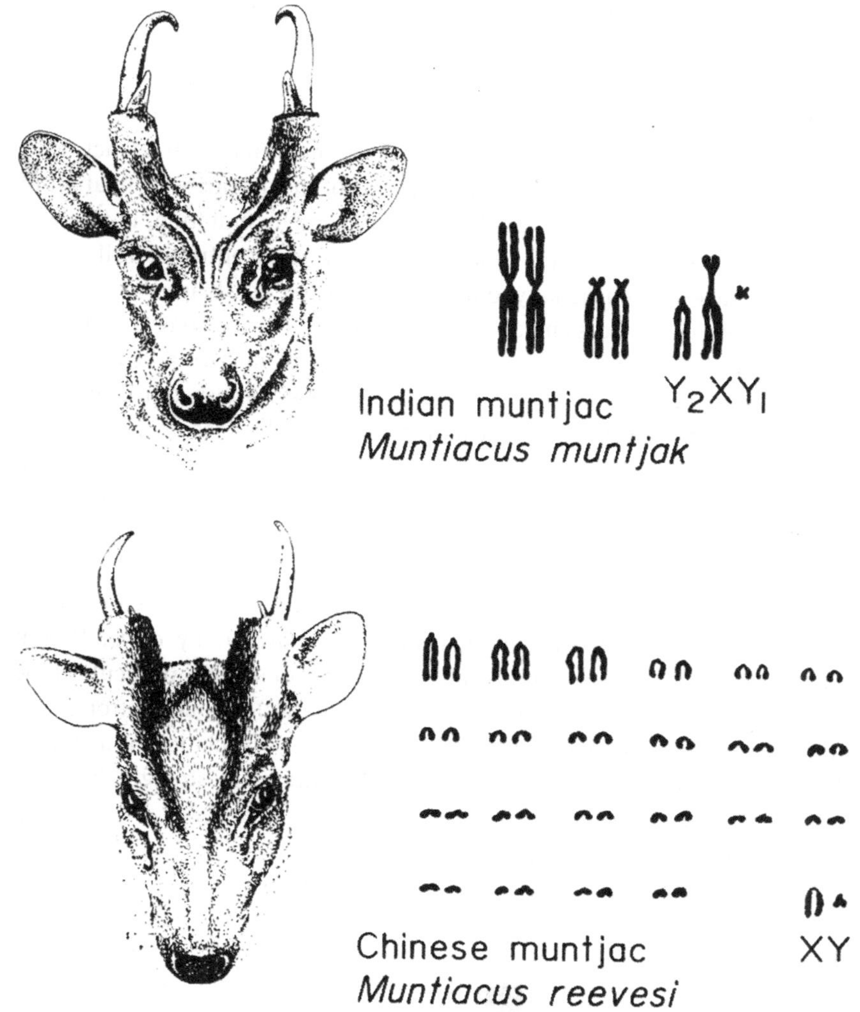

Fig. 2.3: Muntjac karyotypes. (Reproduced with permission from Short 1976.)

species (Gifford and Tristem 2003). Assuming NAHR as a general process of CR (not just tandem fusions) can help explain why CRs occur more frequently in some mammalian lineages than others and why particular types of CR may occur more frequently in one lineage than another. CR by NAHR may also occur, and indeed be more easily detected, in cases where blocks of high copy number repeats (forming "heterochromatin", see below; also known as "satellite DNA") are the template (Garagna *et al.* 2014, Dobigny *et al.* 2017, Martinez *et al.* 2017). The dynamics of spreading of "selfish DNA", i.e., whether few or many locations for the repeated sequences, and whether there are large or small blocks of the repeated sequences, and the effect that these variables may have on NAHR and CRs, is not fully understood. Indeed, as well as DNA repeats being locations for CRs, the CRs may also move repeats around and so may be part of the equation of selfish DNA dynamics. Another factor is that nuclear architecture may help determine which chromosome regions come into proximity to allow the movement of DNA repeats and to promote NAHR and CRs; nuclear architecture is another factor that can be species specific (Cremer and Cremer 2001) and, therefore, lineage specific.

This tendency of the same CR reoccurring within a lineage has been termed "karyotypic orthoselection" (White 1973; see Glossary section). In the case of mammals, we have described how tandem fusions are prevalent in the muntjac lineage, but different types of CR dominate other lineages. For instance, in the primates, the lemur diversification is characterized by Rb fusions, the cercopithecid (old world monkey) lineage displays many Rb fissions, and pericentric inversions are common in the hominids (Dutrillaux 1979) (see Fig. 2.2). It has been estimated that 38 different pericentric inversions have occurred in deer mice *Peromyscus* (Greenbaum *et al.* 1986).

Mutation is necessary for chromosomal evolution and may "steer" the direction of chromosomal evolution (i.e., predisposition for certain types of CR). However, it does not appear that mutation is the sole determining factor for tempo of chromosomal evolution. Thus, it would be expected that mutation rate (per unit time) would relate to generation time, with small mammals tending to have a shorter generation time than large mammals (Martin and Palumbi 1993, Martinez *et al.* 2017). Rodents do have a higher rate of CR than carnivorans and whales (larger) or bats (similar size, but longer generation time); however, primates buck that trend (they are large and show rapid chromosomal evolution) (Bush *et al.* 1977). Conditions that promote fixation are likely to be very important. Thus, smaller species of mammals are more likely to have a subdivided population structure than larger species (geographically subdivided, easily affected by barriers), but there are also some large mammals that are subdivided (socially—into clans or harems) and primates and horses are examples of those and have high rates of CR (Bush *et al.* 1977).

If population structure is important, then this points to the involvement of genetic drift in fixation of CRs (Charlesworth 2009). If there is a small effective population size (as expected in a subdivided population; see Glossary section), then genetic drift may even overcome the underdominance that may be associated with CRs (see below; Lande 1979). However, small populations may also be important for other evolutionary processes. Thus, local adaptation is something associated with small populations, and CRs through their effect on linkage of genes, and levels of recombination, can lead to alleles at two loci being inherited as essentially a single coadapted unit, and selectively favored on those grounds, leading to fixation (Navarro and Barton 2003, Kirkpatrick and Barton 2006, Guerrero and Kirkpatrick 2014). Also, meiotic drive (see Glossary section and below), which can be very effective to promote fixation of CRs (Lande 1979, Walsh 1982), may only promote the spread of that fixed rearrangement at a local level because of presence or evolution of drive suppressors, halting wider spread (Ardlie 1998).

Thus, making comparisons among species of mammal, we have described two interesting tendencies: first, variation in rate of CR among lineages, with a higher rate in lineages characterized by small populations, and, second, the tendency for particular types of CR in particular lineages. A third tendency can be seen on comparison of species level karyotypes in mammals. Here, there is a clear trend that species have either fully (or nearly fully) acrocentric karyotypes or fully (or nearly fully) metacentric karyotypes (Fig. 2.4). Pardo-Manuel de Villena and Sapienza (2001), who discovered this trend, argue that this relates to fixation of Rb fusion or fission mutations (Fig. 2.2). Karyotypes of particular species (and lineages deriving from them) have a tendency, they believe, for either fixation of metacentrics or fixation of acrocentrics when they arise by Rb mutation, in this way promoting evolution of fully metacentric or fully acrocentric karyotypes. Meiotic drive, i.e., biased segregation of dissimilar but homologous chromosomes in female meiosis (a metacentric vs. the homologous acrocentrics in this case; see Fig. 2.2 and Glossary section), could be weighted in favor of metacentric chromosomes or in favor of acrocentric chromosomes, and this could lead to the extraordinary bimodal distribution in Fig. 2.4. Chmátal *et al.* (2014) provide a mechanism that could explain that meiotic drive. They found that house mouse chromosomes could have either "strong" or "weak" centromeres, with the strong centromeres favored by drive. Within mouse strains, all the chromosomes of a particular type (either metacentric or acrocentric) can have strong centromeres. If such a tendency for association of metacentrics or acrocentrics with centromere strength happens

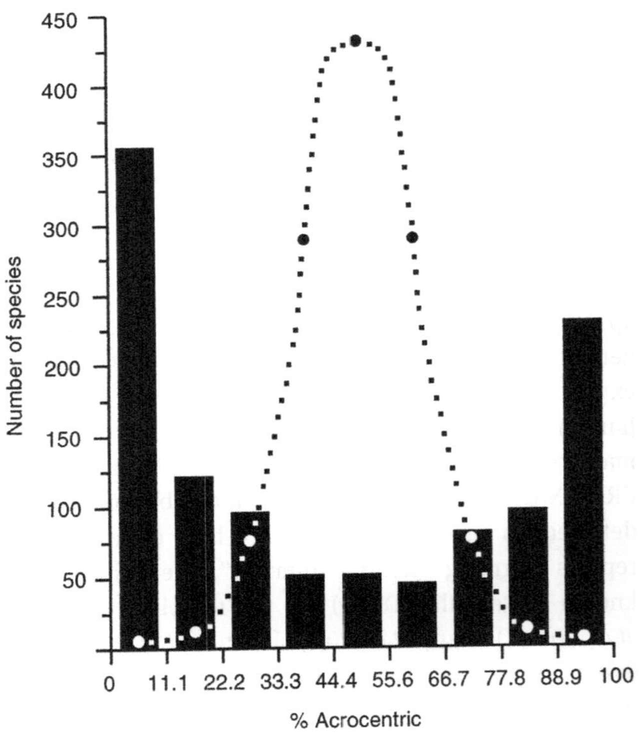

Fig. 2.4: Chromosome morphology in mammals. The bars tally the numbers of species with a particular percentage of acrocentrics in their karyotypes. The expected distribution under a binomial distribution centered on the mean is shown with a dotted line. (Reproduced with permission from Pardo-Manuel de Villena and Sapienza 2001.)

more generally, with possibilities of switches from metacentric-strength to acrocentric-strength, then the bimodal distribution in Fig. 2.4 can be explained.

2.3 Chromosomes as agents of reproductive isolation

There are numerous species concepts (Zachos 2016). Here, we follow the biological species concept (Mayr 1963, Coyne and Orr 2004; see Glossary section) and with regard to making decisions about whether mammalian chromosomal forms are conspecific or distinct species it works reasonably well. For instance, when a geographic form of house mouse (*Mus musculus*) with an unusual chromosomal complement was described, it was first considered a separate species (*Mus poschiavinus*); it had 2n = 26 and lived in a very specific part of the Alps and differed by seven Rb fusions (Fig. 2.2) from the 40-chromosome, all-acrocentric standard house mouse (Gropp *et al.* 1969). However, even that first description showed that the two forms could interbreed. Subsequently, 101 chromosomal forms of the house mouse have been found (Piálek *et al.* 2005, Hauffe *et al.* 2012). It would be a complete nomenclature nightmare if each of those chromosomal forms were described as a separate "species". Although some of the chromosomal forms are reproductively isolated from each other, to our knowledge, all are able to interbreed with the standard race, and there is an expectation that this should be so. Also, those 101 chromosomal forms are all within the *Mus musculus domesticus* "subspecies" of the house mouse (see Glossary section), defined morphologically and genetically (Boursot *et al.* 1993, Piálek *et al.* 2005). Again, to define the chromosomal forms as "species" within "subspecies" would be unwieldy. Therefore, on practical grounds, in this instance, and others, the biological species concept is appropriate. Also, heuristically, in terms of the relevance of CRs to the speciation process, it makes sense to follow the biological species concept. There is an interest in the interruption to gene flow among hybridizing chromosomal forms, as caused by CRs, and this interruption can be considered "partial reproductive isolation". Partial reproductive isolation among hybridizing chromosomal forms may progress to total reproductive isolation.

Therefore, the nomenclature that we will use is this: Forms that are reproductively isolated from each other (which may or may not differ chromosomally) will be considered separate species. Chromosomally different geographic forms within species will be called "chromosomal races", following the detailed definition provided by Hausser *et al.* (1994).

We have discussed the fixation of CRs and, therefore, the formation of chromosomal races. However, how might it be that these chromosomal races may become separate species?

In other words, we need to address in more detail in what ways CRs could contribute to reproductive isolation. This has long been an area of active discussion (for recent considerations, see Faria and Navarro 2010). At face value, as different packaging for the same genome, CRs could be viewed as rather trivial. They need not, in themselves, make any changes to DNA sequences at genes or regulatory regions within that genome (although see below). This is, however, looking at the somatic effect of CRs, which is not, in general, of greatest importance in speciation. Instead, the most important effect of CRs is at meiosis. In mammals, and other sexually reproducing diploids, the paternal and maternal copy of each chromosome has to pair and segregate at meiosis. In a heterozygote for a CR, the paternal and maternal copies of the heterozygous chromosomes differ. They are homologous so must still pair at meiosis, but that pairing and segregation is complicated because of the chromosomal difference. Consider each of the CRs shown in Fig. 2.2, the heterozygote will have one copy of the ancestral chromosome and one copy of the rearranged chromosome, and they have to pair and segregate at meiosis. Not surprisingly there is a tendency for abnormalities associated with that pairing and segregation because the norm is pairing and segregation of identical-looking homologues. These meiotic abnormalities have enormous significance with regard to the possible role of CRs in speciation.

The first type of meiotic abnormality is that the heterozygous chromosomes fail to pair completely. This unpairing, that can occur around the position of chromosome breakpoints in heterozygous configurations, leads to germ cell death because complete pairing is needed for normal gene expression (Searle 1993, Garagna *et al.* 2014). Thus, a contribution to infertility of males and females is a possible outcome of meiotic unpairing on heterozygous chromosomes, and a clear example of how CRs may cause partial reproductive isolation between hybridizing chromosomal races of mammals. At an extreme, the chromosome unpairing may lead to sterility, particularly in males (an example of Haldane"s Rule: Orr 1997). Malsegregation (also known as "non-disjunction") of the heterozygous chromosomes is another possible source of hybrid unfitness (underdominance; see Glossary section), leading to chromosomally unbalanced gametes and then embryos, which usually die during gestation (Searle 1993). For some rearrangements (e.g., within inversions), crossovers in heterozygous configurations can lead to chromosomal unbalance as well (Giménez *et al.* 2017). Thus, CRs in mammals may promote underdominance and therefore partial reproductive isolation between hybridizing chromosomal races. It has been a conundrum that CRs may become fixed to form chromosomal races, and yet those races may show a degree of reproductive isolation from one another. One of the reasons is that the underdominance is not high for single rearrangements but can be very high for multiple rearrangements (Searle 1993). This model has been particularly developed for Rb fusions and WARTs (Fig. 2.2) where multiple metacentrics may accumulate in two races with monobrachial homology, such that the hybrids have very long chain or ring configurations (Fig. 2.5). Such hybrids may be expected to suffer much reduced fertility; hence this process being proposed as a route to speciation (Baker and Bickham 1986).

Therefore, chromosomal heterozygotes can have reduced fertility due to germ cell death from inappropriate gene expression associated with unpairing, and death of their embryos from inappropriate gene expression due to chromosomal unbalance. Thus, there can be underdominance attributable to CRs in mammals. Accumulation of CRs (e.g., in allopatry) could, at an extreme, result in speciation purely through chromosomally induced underdominance. A lower degree of underdominance could also promote speciation more indirectly (see below).

Another attribute associated with chromosomal heterozygosity is disruption of normal crossing-over. If recombination does not occur in a particular region, that could help retain any genetic differentiation that already exists for that region between the hybridizing forms and allow new genetic differentiation to build up. This could ultimately lead to speciation (Rieseberg 2001). For short inversions in mammals, it has been shown that instead of an inversion loop, there is straight, non-homologous pairing (Hale 1986, Searle 1993, Dobigny *et al.* 2017). Clearly, without homologous pairing, there cannot be recombination. This has led to the recombination suppression model of speciation (see Glossary section) first promulgated by Rieseberg (2001), Noor *et al.* (2001), and Navarro and Barton (2003) for CRs, in particular inversions, but which, as a concept, has been developed more broadly in response to accumulating population genomic data from many systems (see below).

Regarding CRs promoting speciation through either underdominance and/or recombination suppression, it is only underdominance that can lead directly to speciation by reinforcement (Servedio and Noor 2003). In hybrid zones between chromosomal races, it can be the hybrid unfitness associated with reduced fertility resulting from heterozygosity of CRs that can promote assortative mating, which may ultimately lead to complete behavioral isolation of the chromosomal races. There is a proposed example of this reinforcement process in the house mouse in response to underdominance of CRs (Giménez *et al.* 2017). This was documented in an isolated village in the Italian Alps. Only mice within that one village stopped interbreeding; the two chromosomal races continued to hybridize in other villages (Hauffe and Searle 1992). It appears that assortative mating evolved in a very localized context, and a combination of hybrid unfitness and very small population sizes of the two races, may have allowed fixation of behavioral differences between them.

In some cases, it would seem almost inconceivable that underdominance is not part of the speciation process. The extraordinary difference in karyotypes of some closely related species of mammal (Dobigny *et al.* 2005) argues for them having gained reproductive isolation through chromosomal underdominance, and although there may be genic differences unrelated to the chromosomal differences, it is clearly unwarranted to ignore potential effects of the CRs. Under the biological species concept, it should not be forgotten that the attainment of complete reproductive isolation could occur while the speciating forms are in allopatry. Thus, the accumulation of CRs in two populations in isolation from each other can lead to speciation. Surely, the massive chromosomal divergence of closely related species of muntjac (Fig. 2.3) is an example of chromosomal speciation through underdominance, given the low fertility expected with heterozygosity of tandem fusions (Searle 1993, Dobigny *et al.* 2017). Therefore, we believe that there are cases where speciation can occur without CRs being important, even when CRs are present, and other cases where the CRs are preeminent, particularly if there is a very dramatic change in karyotype in the speciating forms that would lead to substantial underdominance.

In cases where speciating forms differ by CRs and are in contact and hybridizing with each other, either underdominance or recombination suppression associated with CRs or both could be important in the maintenance and buildup of genic differences. In both cases, it is to be expected that the greatest effect of underdominance or recombination suppression will be close to the breakpoints of the CRs (Panithanarak *et al.* 2004, Giménez *et al.* 2017). Thus, in studies of chromosomal hybrid zones, it can be difficult to decide whether underdominance or recombination suppression is having the greatest effect on gene flow around the chromosomal breakpoints. This was precisely the conclusion in a series of studies

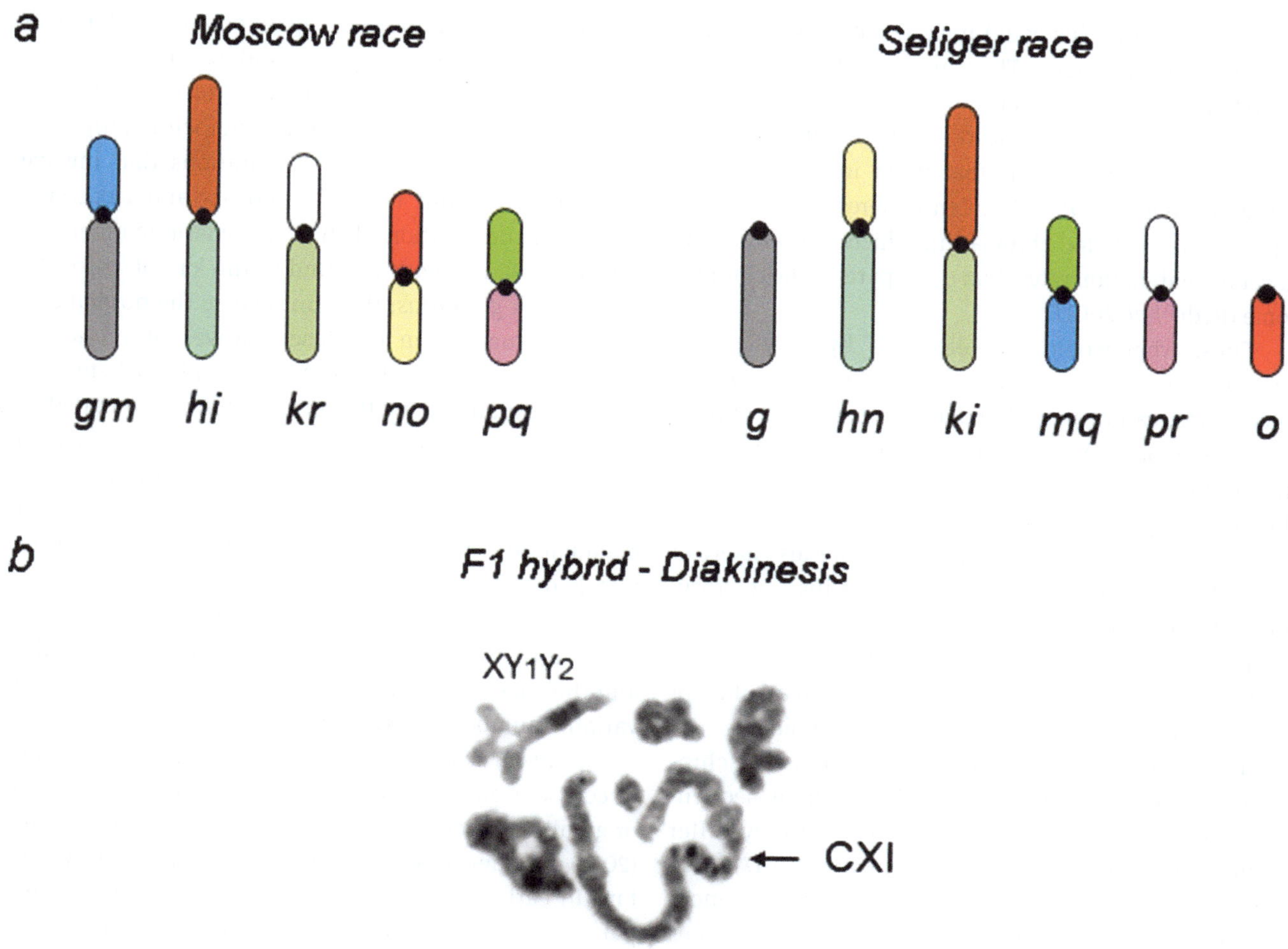

Fig. 2.5: An extraordinary chromosomal heterozygote. An example of a naturally occurring hybrid between two chromosomal races of the common shrew (*Sorex araneus*): an F_1 hybrid between the Moscow and Seliger races. (*a*) These races differ such that 10 chromosome arms are joined into metacentrics in different ways in the two races. Thus, the Moscow race has five metacentrics that have single-arm (monobrachial) homology with four metacentrics and two acrocentrics of the Seliger race. This chromosomal difference has arisen through Rb fusions and/or WARTs (Fig. 2.2). Chromosome arms are labeled by letters of the alphabet in the common shrew (metacentrics *xy* and acrocentrics *x*, *y*). This diagram uses different colors for each chromosome arm. (*b*) The karyotypic differences mean that an F_1 hybrid will form a meiotic chain configuration consisting of the five race-specific chromosomes of the Moscow race and the six race-specific chromosomes of the Seliger race, resulting in a chain-of-eleven (CXI) chromosomes visible at the diakinesis stage of meiosis. Overall, the diakinesis spread has the CXI configuration, four bivalents, and a sex trivalent (see Fig. 2.1 b for another diakinesis in the common shrew, with a more normal situation of many more configurations). (Reproduced with permission from Pavlova *et al.* 2008.)

examining gene flow across a chromosomal hybrid zone of the house mouse in northern Italy, using mapped markers around the chromosomal breakpoints and elsewhere in the genome (Panithanarak *et al.* 2004, Giménez *et al.* 2013, 2017, Merico *et al.* 2013, Förster *et al.* 2016). In this case, the CRs involved were Rb fusions and possibly WARTs. Potentially, if there is a buildup of genetic divergence between hybridizing chromosomal forms and an increase in genetic incompatibility, then this leads to underdominance that has a genic rather than a chromosomal basis. Such underdominance could be another route that leads to selection for assortative mating, i.e., reinforcement.

Where does this leave our understanding of the role of CRs in speciation? First of all, we need to step aside from variation at the gross karyotypic level and consider much finer level variation that would not be detectable with conventional cytology. Following Nachman and Payseur (2012) and Ortiz-Barrientos *et al.* (2016), recent genomic data on a number of systems, mammalian and non-mammalian, have suggested that regions of low recombination tend to be areas of differentiation at the level of DNA sequence, when considering interbreeding of genetically differentiated major geographic forms within species ("races"). These areas of low recombination include regions close to the centromere, but also regions around the breakpoints of inversions, and this could include small cytologically undetectable inversions. Thus, regions of low recombination can maintain and accumulate genetic differences that have the potential to lead to reproductive isolation between hybridizing forms that differ in these regions (Ortiz-Barrientos *et al.* 2016). Centromeres are regions of repetitive DNA (heterochromatin), which typically show lower recombination rates than other parts of the chromosome (Redi *et al.* 2001).

Thus, what would be called CRs by a cytogeneticist, because they are detectable using a compound microscope, can be considered at the coarse end of rearrangements of DNA sequence that actually happen. It is important to consider whether a large rearrangement has different properties from a small one. Submicroscopic rearrangements in a heterozygous state are less likely to affect the segregation process at meiosis than heterozygosity for CRs as defined by cytogeneticists. However, both large and small rearrangements are likely to affect gene flow and genetic differentiation through a failure of recombination near the chromosomal breakpoints.

Thus, we have to be careful that when the chromosomes of two related mammals are homosequential according to G-banding, they may actually differ by small inversions (see also Dobigny *et al.* 2017). Also G-banding is not a good method to show differences in repetitive DNA (a method known as C-banding does that, as does fluorescent in situ hybridization [FISH] using probes for repeated sequences). Large blocks of repetitive DNA are known at heterochromatin (Redi *et al.* 2001) because they have specific properties within the karyotype (e.g., a greater tendency for condensation). Differences in quantities of heterochromatin can be substantial between closely related species of mammal, but the role of this differentiation in speciation is not clear. Some closely related species have large differences in amount of heterochromatin on the sex chromosomes (Redi *et al.* 2001); and given the disproportionate effect of the sex chromosomes on reproductive isolation (Coyne 1992), it is conceivable that these alterations to the sex chromosomes have been drivers of speciation. The *Microtus* voles are extraordinary examples of sex chromosome evolution. The genus is young (about 2 million years old; Jaarola *et al.* 2004) and is a mix of species with normal mammalian sex chromosomes, some with minor heterochromatin on the sex chromosomes and others with giant sex chromosomes (Fig. 2.6) due to massive expansions of different repetitive elements (Marchal *et al.* 2004, Acosta *et al.* 2011). Heterochromatin variation in mammals is not limited to the sex chromosomes. Expansion of repetitive DNA can also occur in the autosomes, again creating differences between closely related species and again of uncertain importance during the speciation process (although see above for a discussion how heterochromatin may promote CRs). An example of mutations that involve expansions of both telomeric repeats and ribosomal repeats at the position of the centromeric telomere on acrocentric autosomes is found in the karyotype of the shrew *Sorex granarius*; this contrasts to the normal size of such telomeres in its close relative *Sorex araneus* (Zhdanova *et al.* 2007). Another example of heterochromatin expansion is the occurrence of so-called B chromosomes, which are additional chromosomes to the normal complement and usually highly heterochromatic and have been observed in over 50 species of mammal (Vujošević and Blagojević 2004). They are ultimately derived from the standard chromosome complement but have become independent entities that behave as large blocks of selfish DNA. They are dispensable, and their meiotic transmission is disorderly, leading to variable numbers between individuals.

In this article, the focus is on recent speciation, but of course even the earliest splits in the tree of life were originally associated with speciation events. Graves (2016) considers that chromosomal mutation was important in early mammalian evolution. She highlights the particular evolutionary potency of the sex chromosomes and argues that CRs and other changes of those chromosomes may have been important in the origins of the three major lineages of mammals: monotremes, marsupials, and placentals.

Here we have focused on the meiotic effects of chromosomal heterozygosity and how that may lead to reproductive isolation between hybridizing forms. The effect of CRs on the speciation process may not be limited to underdominance and recombination suppression, however. There are two other long-standing ideas on the effects of CRs that should be borne in mind as possible factors in speciation. First, because CRs can change numbers of chromosomes, and therefore numbers of segregating units, this can have implications on genetic variability in populations (Qumsiyeh 1994), and genetic variability is the foundation of any genetic change, including speciation. Second, CRs, through breakages in chromosomes may, in some instances, affect gene function. Such breakage causing damage to structural genes is likely to deleterious, but more likely is a change in the vicinity of genes, which may alter gene regulation (Wilson *et al.* 1974, 1975). Thus, CRs could, in this way, promote important genic differences that could contribute to reproductive isolation.

Fig. 2.6: Giant sex chromosomes. Karyotypes of the field vole, *Microtus agrestis*, female above, male below, illustrating sex chromosomes whose large size is the result of expansion of heterochromatin. (Reproduced with permission from Giménez *et al.* 2012.)

2.4 The landscape of future chromosomal species

Species of mammal can differ dramatically in karyotype from their closely relatives. Given the continuity of the evolutionary process, there is an expectation of instances of species being subdivided into chromosomally different forms that could, in time or with the right circumstances, evolve into separate chromosomally differentiated species. This subdivision of species is, of course, seen in some mammals, with differentiation into so-called chromosomal races. These chromosomal races are really critical to our understanding of the role of CRs in speciation. From their characteristics and the way they interact, it should be possible to infer whether the CRs that define the races are crucial in promoting reproductive isolation.

Table 2.1 presents a list of species of mammal that are each subdivided into at least three chromosomal races. These, therefore, are instances of species that show a substantial tendency for CRs to become fixed in parts of their range. These cases of multiple division into chromosomal races should provide particular insights. Cases where new chromosomal races keep on arising within species are surely going to be the most informative about the circumstances that lead to fixation of CRs and therefore origin of chromosomal races (and ultimately origin of chromosomally distinct species). The list of species that are subdivided into multiple chromosomal races is extensive, but we do not wish to hide the biases inherent in that list. We do not claim to have been comprehensive in our coverage of the literature. In addition, there are undoubtedly biases in the types of mammal that people tend to study cytogenetically. Nevertheless, we believe that this list does provide an indication of the types of mammal that are the most prone to be subdivided chromosomally.

With regard to the effect of CRs on reproductive isolation, the instances where hybridization has been studied in multiple comparisons between chromosomal races are particularly informative. There is a clear tendency for underdominance to be greater when the chromosomal difference between the hybridizing forms is greatest. In the case of hybrid zones between races, this is manifested in reduced width of hybrid zones the greater the chromosomal difference (see Barton and Hewitt 1985).

Taxonomically, the list consists almost entirely of rodents. Other than that, there is only one shrew, two bats, and one small artiodactyl on the list. This fits well with chromosomal variation seen at the species level. Rodents and other small mammals show a high rate of chromosomal evolution and with closely related species often differing in karyotype. We have already seen muntjac as an example of a small artiodactyl with exceptional chromosomal differences between species (also involving tandem fusions as for the intraspecific variation in the Kirk's dik-dik in Tab. 2.1). Whales and carnivorans are not represented in Tab. 2.1 and show low rates of chromosomal evolution with a tendency for closely related species to have identical or similar karyotypes. The same is true for bats, which are only represented by two species in Tab. 2.1 despite, within mammals, Chiroptera being the second-most speciose Linnean order after Rodentia.

Thus, small body size, high generation time, and a tendency to be subdivided into small geographic areas and small populations are all features that are associated with high rates of chromosomal evolution, a high tendency for chromosomal difference between closely related species, and a high tendency for subdivision of species into multiple chromosomal races. The subdivision into small populations is interesting because rodents have a high reproductive capacity and therefore are capable of having extremely large populations if conditions are right.

Tab. 2.1: Examples of 24 species of eutherian mammals subdivided into three or more chromosomal races, giving information about the diploid number (2n) and number of chromosome arms (nombre fondamental, NF) of the defining karyotypes, and chromosomal rearrangements that led to those (see Fig. 2.2), the areas occupied by the races (large = several hundreds of square kilometers), and the information available on hybridization between the races. The species are listed in a taxonomic sequence.

Common name	Scientific name	Distribution	No. races	Area per race	Chromosomal rearrangements	2n variation	NF variation	Hybrids and hybrid zones	Ref.
Common shrew	*Sorex araneus* (Eulipotyphla, Soricidae)	Eurasia	76	Small to Large	Rb fusions, WARTs and Rb fissions also proposed (XX/XY_1Y_2 sex chromosome system)	20–33	40	Thirty-six interracial hybrid zones described. Hybrid zone width shows inverse relationship with degree of chromosomal difference; hybrids low to high infertility although not complete sterility and behavioural isolation not observed	1–3
Lesser horseshoe bat	*Rhinolophus hipposideros* (Chiroptera, Rhinolophidae)	Western Europe to southwest Asia	3	Large	Rb fusions, variation in Y chromosome	54, 56, 58	66	No clear evidence of hybridization	4–5
Peters' tent-making bat	*Uroderma bilobatum* (Chiroptera, Phyllostomidae)	Central and South America	3	Large	Rb fusions, tandem fusions, pericentric inversion	38, 42, 44	44, 48, 50	2n = 38 and 2n = 44 races hybridize, forming a wide hybrid zone	6–8
Kirk's dik-dik	*Madoqua kirkii* (Artiodactyla, Bovidae)	Central Africa (Somalia/Tanzania and Angola/Namibia)	4	Large	Tandem fusions, pericentric inversions (XX/XY_1Y_2 sex chromosome system)	46–48	48, 50, 52	No clear evidence of hybridisation in nature, laboratory hybridisation led to sterile males between races which differ by normal X chromosome and a compound X	9, 10
Long-tailed spiny rat	*Proechimys longicaudatus* (Rodentia, Echimyidae)	Central South America	4	Large	Unclear, probably includes variation in heterochromatin	28, 30	46, 48, 50, 52	No clear evidence of hybridization	11, 12
Lami tuco-tuco	*Ctenomys lami* (Rodentia, Ctenomyidae)	Rio Grande do Sul (southern Brazil)	4	Large	Rb fusions, pericentric inversions	54–58	78–88	Hybridization inferred in two of the three interracial contacts studied (i.e., presence of 2n = 55 and 2n = 57) but not in the third	13
Tiny tuco-tuco	*Ctenomys minutus* (Rodentia, Ctenomyidae)	Coast of southern Brazil	8	Small	Rb fusions, tandem fusions, pericentric inversion	42–50	72–84	Six hybrid zones between races differing by single Rb fusions	14, 15
Northern pocket gopher	*Thomomys talpoides* (Rodentia, Geomyidae)	Midwest of United States	8+	Small	Rb fusions, other rearrangements	40–60	70–82	At contacts that have been studied in nature varies between broad intergradation and no interbreeding. Degree of hybridization shows inverse relationship with degree of chromosomal difference. Behavioral isolation in one zone (2n = 48 and 2n = 56)?	16–18

Tab. 2.1 (continued)

Common name	Scientific name	Distribution	No. races	Area per race	Chromosomal rearrangements	2n variation	NF variation	Hybrids and hybrid zones	Ref.
Palestine mole rat	*Nannospalax ehrenbergi* (Rodentia, Spalacidae)	Southern Anatolia to Israel	20	Many small	Rb fusions, centromeric shifts, other rearrangements	48–62	62–86	Hybrid zone width in three zones in Israel shows inverse relationship with degree of chromosomal difference; narrow hybrid zone in Upper Galilee with high variety of hybrid karyotypes (2n = 50–60)	19–22
Lesser mole rat	*Nannospalax leucodon* (Rodentia, Spalacidae)	The Balkans and nearby areas	25	Many small	Not clear, probably Rb fusions, other rearrangements	46–58	76–98	No clear evidence of hybridization	21, 22
Anatolian mole rat	*Nannospalax xanthodon* (Rodentia, Spalacidae)	Turkish Anatolia	28	Many small	Not clear, probably Rb fusions, other rearrangements	36–62	68–90	Scant evidence of hybrids, 2n = 49 individuals found in one locality	21, 22
Goodwin's brush-tailed mouse	*Calomyscus elburzensis* (Rodentia, Calomyscidae)	Afghanistan, Iran, Turkmenistan	6+	Small	Unclear, probably including heterochromatin variation	44	62–76	Hybrids detected in nature and can hybridize in captivity	23–26
Hotson's mouse-like hamster	*Calomyscus hotsoni* (Rodentia, Calomyscidae)	Iran, Pakistan	3	Small	Unclear, including an Rb fusion or fission	48–50	50–52	No clear evidence of hybridization	25–26
Eastern mole vole	*Ellobius tancrei* (Rodentia, Cricetidae)	China, Mongolia, southern Siberia, Tajikistan, Uzbekistan, Kyrgyzstan, southeastern Kazakhstan	25	Many small; 2n = 54 widespread	Rb fusions (unusual sex chromosome system: XX in both sexes)	30–54	56	Small area of substantial chromosomal variation and hybridization in Pamiro-Alay and Tien Shan	27–29
Thomas's pine vole	*Microtus thomasi* (Rodentia, Cricetidae)	Southeastern Balkans	9	Generally small	Numerous sex chromosomal variants due to heterochromatin additions, also Rb fusions	38–44	40–46	Hybrids detected in nature and can hybridize in captivity	30–33
Michoacán deer mouse	*Osgoodomys banderanus* (Rodentia, Cricetidae)	Mexico	4	Two large, especially the NF = 52, two small	Probably most importantly heterochromatin additions	48	52, 62, 78	No clear evidence of hybridization	34
Deer mouse	*Peromyscus maniculatus* (northeastern) (Rodentia, Cricetidae)	North America (here only north-eastern part of range considered)	5	Large	Pericentric inversions, heterochomatin additions	48	80–86	No clear evidence of hybridization	35

Tab. 2.1 (continued)

Common name	Scientific name	Distribution	No. races	Area per race	Chromosomal rearrangements	2n variation	NF variation	Hybrids and hybrid zones	Ref.
Western house mouse	*Mus musculus domesticus* (Rodentia, Muridae)	Global, chromosomal variation largely limited to western Europe	101	Many small; 2n = 40 very widespread	Rb fusions, WARTs	22–40	40	Hybrid zone width shows inverse relationship with degree of chromosomal difference; hybrids low to high infertility; evidence of behavioral isolation in some zones	36, 37
African pygmy mouse	*Mus (Nannomys) minutoides* (Rodentia, Muridae)	Southern half of Africa	3	Large	Rb fusions, WARTs, variation in sex chromosomes (presence/absence of sex chromosome-autosome rearrangements involving both X and Y)	18–36	35–36	No clear evidence of hybridization	38–41
Black rat	*Rattus rattus* (Rodentia, Muridae)	Global, chromosomal variation largely limited to Indian and Pacific Ocean regions	5	Large	Rb fusions, other rearrangements	38–42	58–62	Laboratory studies showed high fertility for crosses between 2n = 38 and 2n = 40, and between 2n = 38 and one (Mauritius) type of 2n = 42, but low fertility for cross between 2n = 38 and a different (common) type of 2n = 42. Hybrids between 2n = 38 and the common 2n = 42 found in nature, and introgression detected	42, 43
Pygmy wood mouse	*Sylvaemus uralensis* (Rodentia, Muridae)	Central Europe to central Asia	3	Large	Variation in centromeric heterochromatin	48	48	No clear evidence of hybridization	44–46
Cretan spiny mouse	*Acomys minous* (Rodentia, Muridae)	Crete (Greece)	3	Small	Rb fusions	38, 40, 42	68	Hybrids found between 2n = 38 and 2n = 40, 2n = 38 and 2n = 42, 2n = 40 and 2n = 42	47
Nigerian gerbil	*Gerbillus nigeriae* (Rodentia, Muridae)	West Africa	14+	Many small	Rb fusions, heterochromatin variation	60–74	116–144	Regional differences in range of chromosome number but polymorphism in each area; high levels of heterozygosity	48–50
Garden dormouse	*Eliomys quercinus* (Rodentia, Gliridae)	Western Europe (Russia to Iberia), Mediterranean islands	4	Large, especially 2n = 48	Rb fissions (ancestrally 2n = 48), other rearrangements	48, 50, 52, 54	86, 90	Scant data, 2n = 49 individuals found	51–55

References:
1. Searle and Wójcik 1998.
2. Wójcik *et al.* 2002.
3. Searle *et al.* in press.
4. Zima 2004.
5. Volleth *et al.* 2013.
6. Baker 1981.
7. Hoffmann *et al.* 2003.
8. Da Silva *et al.* 2005.
9. Ryder *et al.* 1989.
10. Cernohorska *et al.* 2010.
11. Machado *et al.* 2005.
12. Amaral *et al.* 2013.
13. De Freitas 2007.
14. De Freitas 1997.
15. Lopes *et al.* 2013.
16. Thaeler 1968.
17. Thaeler 1974.
18. Thaeler 1980.
19. Nevo and Bar-El 1976.
20. Ivanitskaya *et al.* 2010.
21. Kryštufek *et al.* 2012.
22. Arslan *et al.* 2016.
23. Meyer and Malikov 2000.
24. Graphodatsky *et al.* 2000.
25. Shahabi *et al.* 2010.
26. Akbarirad *et al.* 2016.
27. Lyapunova *et al.* 1980.
28. Lyapunova *et al.* 2010.
29. Bakloushinskaya *et al.* 2013.
30. Giagia-Athanasopoulou and Stamatopoulos 1997.
31. Mitsainas *et al.* 2009.
32. Rovatsos *et al.* 2011.
33. Rovatsos and Giagia-Athanasopoulou 2012.
34. Núñez-Garduño *et al.* 1999.
35. Unice *et al.* 1998.
36. Piálek *et al.* 2005.
37. Hauffe *et al.* 2012.
38. Veyrunes *et al.* 2010.
39. Britton-Davidian *et al.* 2012.
40. Chevret *et al.* 2014.
41. Veyrunes *et al.* 2014.
42. Baverstock *et al.* 1983.
43. Aplin *et al.* 2011.
44. Bogdanov 2001.
45. Bogdanov and Rozanov 2005.
46. Karamysheva *et al.* 2010.
47. Giagia-Athanasopoulou *et al.* 2011.
48. Dobigny *et al.* 2002.
49. Gauthier *et al.* 2010.
50. Hima *et al.* 2011.
51. Filippucci *et al.* 1988.
52. Filippucci *et al.* 1990.
53. Perez *et al.* 2013.
54. Gornung *et al.* 2010.
55. Libois *et al.* 2012.

However, often habitats are not continuous, and rodents can persist in small areas with temporary small populations because with their high reproductive capacity they can expand sufficiently for population survival. Rodents may often have a metapopulation structure (e.g., Telfer *et al.* 2001, Pita *et al.* 2007, Schooley and Branch 2009) providing a greater resilience to local extinction than may be the case in other mammals.

One of the striking aspects of Tab. 2.1 is the number of burrowing, subterranean rodents in the list. This includes three blind mole rats, one pine vole, one mole vole, one pocket gopher, and two tuco-tucos. It is also striking how burrowing rodents are subdivided into chromosomally differentiated species, suggesting a continuity in the generation of chromosomal races and new species. Thus, tuco-tucos of the genus *Ctenomys* are a fine example of a large genus consisting of chromosomally distinct species (Reig *et al.* 1990). Pocket gophers of the genera *Thomomys* and *Geomys* are examples of burrowing rodents at the cusp of being subdivided into chromosomal races or into chromosomal species, with hybrid zones between forms revealing examples of full interbreeding and other examples of behavioral isolation (Tab. 2.1; Patton 1993). Another beautifully documented case of chromosomal variation involving a fossorial species, not in Tab. 2.1, is the aptly named *Microtus subterraneus* (the European pine vole), which is subdivided over its wide European range into two chromosomal races differing by an Rb fusion, with the ancestral 2n = 54 race having a peripheral distribution, and the derived 2n = 52 race being central (Zima 2004). Burrowing rodents are dependent on particular soil types, which can be distributed in a very patchy way. Thus, populations are likely to be isolated and small, but in typical rodent fashion able somehow to persist. It is not only their subdivision into chromosomal races that makes subterranean rodents of interest in terms of their karyotype. Another extraordinary chromosomal characteristic of mole voles is their sex chromosome constitution with different species of *Ellobius* having normal XX/XY sex chromosomes, XO/XO and XX/XX (Graves 2002). As a footnote, returning to the theme of heterogeneity in chromosomal evolution in mammals, it is notable that golden moles, afrotherian burrowers, do not show the high rates of chromosomal evolution evident in some of the fossorial rodents (Gilbert *et al.* 2006).

Also evident from Tab. 2.1 is that among the species that are subdivided into multiple chromosomal races, it is commonly Rb fusions that distinguish the races. This is particularly the case for species that are subdivided into very large numbers of chromosomal races exemplified by western house mice *M. m. domesticus* (101 races) and common shrew *S. araneus* (76 races). These two species have been studied particularly well, and it has been demonstrated that there is a low fertility cost associated with heterozygosity for a single Rb fusion (Searle 1993). Thus, if there is a high mutation rate for Rb fusions (and WARTs) in these species (as appears likely), then it is relatively easy for the mutations to be fixed. This fixation process is likely enhanced by meiotic drive, with a particular mechanism proposed for Rb fusions (Pardo-Manuel de Villena and Sapienza 2001, Chmátal *et al.* 2014). Other processes such as the generation of new races in the hybrid zones between existing races can further enhance diversity without the need of new mutation (White *et al.* 2010).

The comparison of *M. m. domesticus* and *S. araneus* with related taxa is also very interesting. *M. m. domesticus* is one of three major subspecies of *Mus musculus* (the others being *M. m. musculus* and *M. m. castaneus*), and yet it is only *domesticus* that shows fixation of CRs and formation of chromosomal races. All these subspecies are commensal and have spread around the world with humans (Bonhomme and Searle 2012). In all cases, there is the typical rodent tendency for subdivision into small populations. It appears unlikely that any differences in population structure between the subspecies can fully explain enhanced fixation of CRs in *domesticus* compared with the other subspecies. This reiterates the likelihood that the mutation frequency and/or tendency for meiotic drive are higher in *domesticus* than in the other subspecies.

For *M. m. domesticus*, the occurrence of chromosomal races is not mirrored by closely related chromosomal distinct species. In *S. araneus*, there are several sibling species that were first identified from their chromosomal difference, constituting the *S. araneus* group (Basset *et al.* 2008). Thus, it can clearly be seen that what were chromosomal races in this system ultimately became new species. In studies of these sibling species based on microsatellite loci mapped to chromosomes, the microsatellite loci on those chromosomes that have been subject to CR ("rearranged chromosomes") are more likely to show strong differentiation than the loci on chromosomes that have not been subject to rearrangement ("homosequential chromosomes") (Basset *et al.* 2008). By contrast, at the contacts of chromosomal races in *S. araneus*, there is no difference in levels of differentiation between loci on rearranged chromosomes and homosequential chromosomes (Horn *et al.* 2012). In this study, the microsatellite loci were mapped to chromosomes, but it is not clear how close they are to the breakpoint. Based on studies in the house mouse, it is only loci close to the centromere (mutation breakpoint for Rb fusion and WART) that reliably show high differentiation between chromosomal races due to interruption to gene flow. Thus, the results for the chromosomal races in *S. araneus* are not surprising. The results with closely related species are interesting. It suggests that at late stages of the speciation process, there have been occasional hybridizations between the chromosomally distinctive sibling species that have successfully backcrossed to the parental forms. Under these circumstances, gene flow between the hybridizing forms may be disrupted along a substantial length of the rearranged chromosomes, but not on the homosequential chromosomes. It is very unlikely that the disruption to gene flow along a substantial length along the chromosome will be a consequence of recombination suppression. Instead, this is likely to be due to underdominance.

2.5 Conclusions: the steps in chromosomal speciation

Are CRs important in mammalian speciation? There are indications that they can be involved. This is supported on two main fronts:

First, there is a general perception at the current time that inversions (Fig. 2.2) are important in speciation of many types of organism (Ortiz-Barrientos *et al.* 2016). Fixation of inversions may be relatively easy in many instances in mammals. The underdominance associated with inversions can be minimal if there is non-homologous meiotic pairing through the inversion in heterozygotes (Hale 1986, Searle 1993, Rieseberg 2001). Thus, chromosomal races may rather easily become characterized by inversions. The inverted regions thus become "genomic islands of speciation" (Feder *et al.* 2012), allowing the chromosomal races to differentiate for that region of the genome as if the races were in allopatry. Thus, chromosomal races distinguished by inversions that are in contact with each other are able to become differentiated, and ultimately reproductively isolated, in the face of gene flow. This means that it is not essential to have forms to become reproductively isolated in allopatry. During the speciation process, the races may become differentiated in a way to stop interbreeding through genic changes in the inverted region.

Second, there are instances in mammals of an extreme rate of chromosomal evolution, either observable among closely related species or among races within species. Those multiple rearrangements are likely to cause an interruption to gene flow either through underdominance or recombination suppression or both. It is reasonable to believe that the effect of the CRs in these instances will be sufficient to promote reproductive isolation. In mammals, the best documented examples of these fast rates of chromosomal evolution are Rb fusions in rodents (and shrews), but less easily detected inversions may also be occurring commonly in rodents. Also notable are instances of multiple tandem fusions in artiodactyls (detectable both between closely related species and among chromosomal races within species). Tandem fusions are expected to be associated with severe underdominance (Searle 1993, Dobigny *et al.* 2015).

We can suggest a sequence of events in chromosomal speciation in mammals.

First there needs to be mutation, generating CRs, and this is not a homogeneous process. Karyotypic orthoselection almost certainly at least partially reflects a tendency for different types of mutation in different lineages. Mutation rates may be very high in some species, e.g., as estimated for Rb fusions in the western house mouse (Nachman and Searle 1995), a mammal with an exceptional number of chromosomal races defined by Rb fusions and WARTs (Tab. 2.1).

Second, there needs to be fixation and there may be multiple "special processes" that enhance the fixation of CRs, generating chromosomal races. As already suggested by Bush *et al.* (1977), subdivision into small populations appears to predispose to a high rate of chromosomal evolution. This subdivision can be microgeographic (as exemplified by burrowing rodents) or behavioral (Bush *et al.* give the example of rapid chromosomal evolution in horses). Small populations may enhance fixation whether through genetic drift, local adaptation (the CR bringing together genes into close linkage, including favorable combinations of alleles at those genes), or meiotic drive. Different processes or a combination of processes may occur in different situations. In the case of Rb fusions, a mechanism has been suggested for meiotic drive (Pardo-Manuel de Villena and Sapienza 2001, Chmátal *et al.* 2014). The extraordinary examples of rapid multiple fixation of Rb fusion, particularly in the house mouse and common shrew, fit well with a high rate of mutation and meiotic drive.

Third, there needs to be the attainment of reproductive isolation between chromosomal races. Again a combination of multiple factors can contribute to this. There can be chromosomal underdominance, resulting from meiotic aberrations associated with chromosomal heterozygosity. Alternatively, there can be recombination suppression associated with heterozygosity of CRs. Underdominance and recombination suppression are both expected to lead to an interruption to gene flow around the breakpoints of the CRs. They may act in combination. The geographic context also needs to be considered. In allopatry, there may be an accumulation of CRs differentiating isolated populations, but those populations may also accumulate genic differences. Thus, the CRs may be contributing to speciation but may not be the only process. Not only may the genic differences lead to underdominance in themselves, but they may influence the level of underdominance shown by the CRs, e.g., enhancing the level of aneuploidy associated with heterozygosity of CRs (Everett *et al.* 1996). Some combination of effect of CRs and accumulated genic difference appears to explain the hybrid unfitness in one of the hybrid zones between *S. araneus* and its closest relative, *Sorex antinorii*. The properties of the Les Houches hybrid zone, involving the Cordon chromosomal race of *S. araneus* (Brünner and Hausser 1996), suggest that the hybrids are much less fit than would be expected on chromosomal grounds alone (comparing the characteristics of the zone with chromosomal hybrid zones within *S. araneus*).

Probably because of the massive influence of studies on hybrid zones for our understanding of the process of speciation (Barton and Hewitt 1985, Harrison 1990), there tends to be a focus on neighboring races and how they interact. However, speciation may be a much less localized process in time or space. In the house mouse, "a disappearing speciation event" has been described (Hauffe and Searle 1992). This was when two chromosomal races occurred in the same village without interbreeding—a case of local reproductive isolation, believed to be the result of a reinforcement in response to chromosomal underdominance (see above; Giménez *et al.* 2017). However, one of those races went extinct and so the tableau of a recent speciation event was lost. However, that was just one event in a certain place at a certain time. In a species that is subdivided into multiple chromosomal races, there may be all manner of local extinction events until there is a situation of two surviving races, which have a large enough difference in complement of CRs that the CRs in themselves or in combination with genic differences promote reproductive isolation. Those forms will be species that have (at least partly) resulted from fixation of CRs.

Whatever way speciation occurs, once a single species is subdivided into two, the process can "stabilize". According to the biological species concept, those forms are no longer able to exchange genes. It is this severing of genetic continuity that allows ecological differentiation to go unimpeded (Cain 1993). Initially, small differences between the chromosomally different sister species can magnify through time.

Clearly, there is still great uncertainty about the importance of CRs in speciation in mammals. However, it appears that rodents and some other mammals have a tendency to be subdivided into small populations and that this is associated with high rates of chromosomal evolution. Here, multiple CRs may differentiate populations within a single species, and genic differences may be involved as well. The process can be considered to a degree allopatric. In these cases, we suspect that it is underdominance associated with the CRs that is most important in reproductive isolation. However, there are other mammals, like whales and carnivorans, which have low rates of chromosomal evolution, at least in terms of easily discernible CRs, not much tendency for species subdivision, and not a clearly allopatric process. In these cases, if CRs are involved in speciation, we suspect it is cytologically undetectable inversions that are involved, inversions that spread through part but not all of the species distribution because of advantage that they provide. Thus, chromosomal speciation can be viewed as a spectrum between these extremes. This viewpoint of chromosomal speciation is speculative but provides testable scenarios for suitably targeted genomic studies in the future.

Literature

Acosta, M.J., Romero-Fernández, I., Sánchez, A. & Marchal, J.A. (2011): Comparative analysis by chromosome painting of the sex chromosomes in arvicolid rodents. *Cytogenetic and Genome Research* 132: 47–54.

Akbarirad, S., Darvish, J. & Aliabadian, M. (2016): Molecular, chromosomal and morphometric variation in *Calomyscus hotsoni* and *C. elburzensis* (Calomyscidae, Rodentia) in the east of Iran. *Folia Zoologica* 65: 27–37.

Amaral, P.J.S., Nagamachi, C.Y., Noronha, R.C.R., Costa, M.J.R., Pereira, A.L., Rossi, R.V., Mendes-Oliveira, A.C. & Pieczarka, J.C. (2013): *Proechimys* (Rodentia, Echimyidae): characterization and taxonomic considerations of a form with a very low diploid number and a multiple sex chromosome system. *BMC Genetics* 14: 21.

Aniskin, V.M., Benazzou, T., Biltueva, L., Dobigny, G., Granjon, L. & Volobouev, V. (2006): Unusually extensive karyotype reorganization in four congeneric *Gerbillus* species (Muridae: Gerbillinae). *Cytogenetic and Genome Research* 112: 131–140.

Annunziato, A. (2008): DNA packaging: nucleosomes and chromatin. *Nature Education* 1: 26.

Aplin, K.P., Suzuki, H., Chinen, A.A., *et al.* (2011): Multiple geographic origins of commensalism and complex dispersal history of black rats. *PLoS One* 6: e26357.

Ardlie, K.G. (1998): Putting the brake on drive: meiotic drive of t haplotypes in natural populations of mice. *Trends in Genetics* 14: 189–193.

Árnason, Ú. (1972): The role of chromosomal rearrangement in mammalian speciation with special reference to Cetacea and Pinnipedia. *Hereditas* 70: 113–118.

Arslan, A., Kryštufek, B., Matur, F. & Zima, J. (2016): Review of chromosome races in blind mole rats (*Spalax* and *Nannospalax*). *Folia Zoologica* 65: 249–301.

Baker, R.J. (1981): Chromosome flow between chromosomally characterized taxa of a volant mammal, *Uroderma bilobatum* (Chiroptera: Phyllostomatidae). *Evolution* 35: 296–305.

Baker, R.J. & Bickham, J.W. (1986): Speciation by monobrachial centric fusions. *Proceedings of the National Academy of Sciences USA* 83: 8245–8248.

Bakloushinskaya, I., Romanenko, S.A., Serdukova, N.A., Graphodatsky, A.S. & Lyapunova, E.A. (2013): A new form of the mole vole *Ellobius tancrei* Blasius, 1884 (Mammalia, Rodentia) with the lowest chromosome number. *Comparative Cytogenetics* 7: 163–169.

Barton, N.H. & Hewitt, G.M. (1985): Analysis of hybrid zones. *Annual Review of Ecology and Systematics* 16: 113–148.

Basset, P., Yannic, G. & Hausser, J. (2008): Chromosomal rearrangements and genetic structure at different evolutionary levels of the *Sorex araneus* group. *Journal of Evolutionary Biology* 21: 842–852.

Baverstock, P.R., Adams, M., Maxson, L.R. & Yosida, T.H. (1983): Genetic differentiation among karyotypic forms of the black rat, *Rattus rattus*. *Genetics* 105: 969–983.

Bickham, J.W. (1979): Banded karyotypes of 11 species of American bats (genus *Myotis*). *Cytologia* 44: 789–797.

Biltueva, L., Vorobieva, N., Perelman, P., *et al.* (2011): Karyotype evolution of Eulipotyphla (Insectivora): the genome homology of seven *Sorex* species revealed by comparative chromosome painting and banding data. *Cytogenetic and Genome Research* 135: 51–64.

Blackburn, E.H. & Szostak, J.W. (1984): The molecular structure of centromeres and telomeres. *Annual Review of Biochemistry* 53: 163–194.

Bogdanov, A.S. (2001): Chromosome differentiation of the pygmy wood mouse *Sylvaemus uralensis* populations in the eastern part of its range. *Entomological Review* 81(Suppl. 2): S303–S314.

Bogdanov, A.S. & Rozanov, Y.M. (2005): Variability in size of the nuclear genome in pygmy wood mouse *Sylvaemus uralensis* (Rodentia, Muridae). *Russian Journal of Genetics* 41: 1123–1129.

Bonhomme, F. & Searle, J.B. (2012): House mouse phylogeography. In: Macholán, M., Baird, S.J.E., Munclinger, P. & Piálek, J. (eds.): *Evolution of the House Mouse*. Cambridge University Press, Cambridge: 278–296.

Boursot, P., Auffray, J.C., Britton-Davidian, J. & Bonhomme, F. (1993): The evolution of house mice. *Annual Review of Ecology and Systematics* 24: 119–152.

Britton-Davidian, J., Robinson, T.J. & Veyrunes, F. (2012): Systematics and evolution of the African pygmy mice, subgenus *Nannomys*: a review. *Acta Oecologica* 42: 41–49.

Brünner, H. & Hausser, J. (1996): Genetic and karyotypic structure of a hybrid zone between chromosomal races Cordon and Valais in the common shrew, *Sorex araneus*. *Hereditas* 125: 147–158.

Bush, G.L., Case, S.M., Wilson, A.C. & Patton, J.L. (1977): Rapid speciation and chromosomal evolution in mammals. *Proceedings of the National Academy of Sciences USA* 74: 3942–3946.

Cain, A.J. (1993): *Animal Species and their Evolution*. Princeton University Press, Princeton, NJ.

Carson, H.L., Clayton, F.E. & Stalker, H.D. (1967): Karyotypic stability and speciation in Hawaiian *Drosophila*. *Proceedings of the National Academy of Sciences USA* 57: 1280–1285.

Cernohorska, H., Kubickova, S., Vahala, J., Robinson, T.J. & Rubes, J. (2010): Cytotypes of Kirk's dik-dik (*Madoqua kirkii*, Bovidae) show multiple tandem fusions. *Cytogenetic and Genome Research* 132: 255–263.

Charlesworth, B. (2009): Effective population size and patterns of molecular evolution and variation. *Nature Reviews Genetics* 10: 195–205.

Chevret, P., Robinson, T.J., Perez, J., Veyrunes, F. & Britton-Davidian, J. (2014): A phylogeographic survey of the pygmy mouse *Mus minutoides* in South Africa: taxonomic and karyotypic inference from cytochrome b sequences of museum specimens. *PLoS One* 9: e98499.

Chmátal, L., Gabriel, S.I., Mitsainas, G.P., Martínez-Vargas, J., Ventura, J., Searle, J.B., Schultz, R.M. & Lampson, M.A. (2014): Centromere strength provides the cell biological basis for meiotic drive and karyotype evolution in mice. *Current Biology* 24: 2295–2300.

Coyne, J.A. (1992): Genetics and speciation. *Nature* 355: 511–515.

Coyne, J.A. & Orr, H.A. (2004): *Speciation*. Sinauer, Sunderland, MA.

Cremer, T. & Cremer, C. (2001): Chromosome territories, nuclear architecture and gene regulation in mammalian cells. *Nature Reviews Genetics* 2: 292–301.

Crow, E.W. & Crow, J.F. (2002): 100 years ago: Walter Sutton and the chromosome theory of heredity. *Genetics* 61: 1–4.

Crow, J.F. (1994): Advantages of sexual reproduction. *Developmental Genetics* 15: 205–213.

Da Silva, A.M., Marques-Aguiar, S.A., Barros, R.M.D.S., Nagamachi, C.Y. & Pieczarka, J.C. (2005): Comparative cytogenetic analysis in the species *Uroderma magnirostrum* and *U. bilobatum* (cytotype 2n = 42) (Phyllostomidae, Stenodermatinae) in the Brazilian Amazon. *Genetics and Molecular Biology* 28: 248–253.

De Freitas, T.R.O. (1997): Chromosome polymorphism in *Ctenomys minutus* (Rodentia-Octodontidae). *Brazilian Journal of Genetics* 20: 1–7.

De Freitas, T.R.O. (2007): *Ctenomys lami*: the highest chromosome variability in *Ctenomys* (Rodentia, Ctenomyidae) due to a centric fusion/fission and pericentric inversion system. *Acta Theriologica* 52: 171–180.

Dobigny, G., Aniskin, V., Granjon, L., Cornette, R. & Volobouev, V. (2005): Recent radiation in West African *Taterillus* (Rodentia, Gerbillinae): the concerted role of chromosome and climatic changes. *Heredity* 95: 358–368.

Dobigny, G., Britton-Davidian, J. & Robinson, T.J. (2017): Chromosomal polymorphism in mammals: an evolutionary perspective. *Biological Reviews* 92: 1–21.

Dobigny, G., Nomao, A. & Gautun, J.-C. (2002): A cytotaxonomic survey of rodents from Niger: implications for systematics, biodiversity and biogeography. *Mammalia* 66: 495–524.

Dutrillaux, B. (1979): Chromosomal evolution in primates: tentative phylogeny from *Microcebus murinus* (Prosimian) to man. *Human Genetics* 48: 251–314.

Everett, C.A., Searle, J.B. & Wallace, B.M.N. (1996): A study of meiotic pairing, nondisjunction and germ cell death in laboratory mice carrying Robertsonian translocations. *Genetical Research* 67: 239–247.

Faria, R. & Navarro, A. (2010): Chromosomal speciation revisited: rearranging theory with pieces of evidence. *Trends in Ecology & Evolution* 25: 660–669.

Feder, J.L., Egan, S.P. & Nosil, P. (2012): The genomics of speciation-with-gene-flow. *Trends in Genetics* 28: 342–350.

Filippucci, M.G., Catzeflis, F. & Capanna, E. (1990): Evolutionary genetics and systematics of the garden dormouse, *Eliomys* Wagner, 1840 (Gliridae, Mammalia): 3. Further karyological data. *Bolletino di Zoologia* 57: 149–152.

Filippucci, M.G., Civitelli, M.V. & Capanna, E. (1988): Evolutionary genetics and systematics of the garden dormouse, *Eliomys* Wagner 1840: 1 Karyotype divergence. *Bolletino di Zoologia* 55: 35–45.

Förster, D.W., Jones, E.P., Jóhannesdóttir, F., Gabriel, S.I., Giménez, M.D., Panithanarak, T., Hauffe, H.C. & Searle, J.B. (2016): Genetic differentiation within and away from the chromosomal rearrangements characterising hybridising chromosomal races of the western house mouse (*Mus musculus domesticus*). *Chromosome Research* 24: 271–280.

Garagna, S., Page, J., Fernandez-Donoso, R., Zuccotti, M. & Searle, J.B. (2014): The Robertsonian phenomenon in the house mouse: mutation, meiosis and speciation. *Chromosoma* 123: 529–544.

Gauthier, P., Hima, K. & Dobigny, G. (2010): Robertsonian fusions, pericentromeric repeat organization and evolution: a case study within a highly polymorphic rodent species, *Gerbillus nigeriae*. *Chromosome Research* 18: 473–486.

Genesio, R., Ronga, V., Castelluccio, P., Fioretti, G., Mormile, A., Leone, G., Conti, A., Cavaliere, M.L. & Nitsch, L. (2013): Pure 16q21q22.1 deletion in a complex rearrangement possibly caused by a chromothripsis event. *Molecular Cytogenetics* 6: 29.

Giagia-Athanasopoulou, E.B., Rovatsos, M.T.H., Mitsainas, G.P., Martimianakis, S., Lymberakis, P., Angelou, L.-X.D., Marchal, J.A. & Sánchez, A. (2011): New data on the evolution of the Cretan spiny mouse, *Acomys minous* (Rodentia: Murinae), shed light on the phylogenetic relationships in the *cahirinus* group. *Biological Journal of the Linnean Society* 102: 498–509.

Giagia-Athanasopoulou, E.B. & Stamatopoulos, C. (1997): Geographical distribution and interpopulation variation in the karyotypes of *Microtus* (*Terricola*) *thomasi* (Rodentia, Arvicolidae) in Greece. *Caryologia* 50: 303–315.

Gifford, R. & Tristem, M. (2003): The evolution, distribution and diversity of endogenous retroviruses. *Virus Genes* 26: 291–315.

Gilbert, C., O'Brien, P.C., Bronner, G., Yang, F., Hassanin, A., Ferguson-Smith, M.A. & Robinson, T.J. (2006): Chromosome painting and molecular dating indicate a low rate of chromosomal evolution in golden moles (Mammalia, Chrysochloridae). *Chromosome Research* 14: 793–803.

Giménez, M.D., Paupério, J., Alves, P.C. & Searle, J.B. (2012): Giant sex chromosomes retained within the Portuguese lineage of the field vole (*Microtus agrestis*). *Acta Theriologica* 57: 377–382.

Giménez, M.D., White, T.A., Hauffe, H.C., Panithanarak, T. & Searle, J.B. (2013): Understanding the basis of diminished gene flow between hybridizing chromosome races of the house mouse. *Evolution* 67: 1446–1462.

Giménez, M.D., Förster, D.W., Jones, E.P., Jóhannesdóttir, F., Gabriel, S.I., Panithanarak, T., Scascitelli, M., Merico, V., Garagna, S., Searle, J.B. & Hauffe, H.C. (2017): A half-century of studies on

a chromosomal hybrid zone of the house mouse. *Journal of Heredity* 108: 25–35.
Gornung, E., Bizzoco, D., Colangelo, P. & Castiglia, R. (2010): Comparative cytogenetic and genetic study of two Italian populations of the garden dormouse *Eliomys quercinus* L. (Sciuromorpha: Gliridae). *Bolletino di Zoologia* 77: 137–143.
Graphodatsky, A.S., Sablina, O.V., Meyer, M.N., *et al.* (2000): Comparative cytogenetics of hamsters of the genus *Calomyscus*. *Cytogenetics and Cell Genetics* 88: 296–306.
Graves, J.A.M. (2002): Sex chromosomes and sex determination in weird mammals. *Cytogenetic and Genome Research* 96: 161–168.
Graves, J.A.M. (2016): Did sex chromosome turnover promote divergence of the major mammal groups? *BioEssays* 38: 734–743.
Greenbaum, I.F., Hale, D.W. & Fuxa, K.P. (1986): Synaptic adaptation in deer mice: a cellular mechanism for karyotypic orthoselection. *Evolution* 40: 208–213.
Gropp, A., Tettenborn, U. & Von Lehmann, E. (1969): Chromosomenuntersuchungen bei der Tabakmaus (*M. poschiavinus*) und bei Tabakmaus-Hybriden. *Experientia* 25: 875–876.
Guerrero, R.F. & Kirkpatrick, M. (2014): Local adaptation and the evolution of chromosome fusions. *Evolution* 68: 2747–2756.
Hale, D.W. (1986): Heterosynapsis and suppression of chiasmata within heterozygous pericentric inversions of the Sitka deer mouse. *Chromosoma* 94: 425–432.
Harrison, R.G. (1990): Hybrid zones: windows on evolutionary process. *Oxford Surveys in Evolutionary Biology* 7: 69–128.
Hauffe, H.C., Giménez, M.D. & Searle, J.B. (2012): Chromosomal hybrid zones in the house mouse. In: Macholán, M., Baird, S.J.E., Munclinger P. & Piálek J. (eds.): *Evolution of the House Mouse*. Cambridge University Press, Cambridge: 407–430.
Hauffe, H.C. & Searle, J.B. (1992): A disappearing speciation event? *Nature* 357: 26.
Hausser, J., Fedyk, S., Fredga, K., Searle, J.B., Volobouev, V., Wójcik, J.M. & Zima, J. (1994): Definition and nomenclature of chromosome races of *Sorex araneus*. *Folia Zoologica* 43 (Suppl. 1): 1–9.
Hima, K., Thiam, M., Catalan, J., Gauthier, P., Duplantier, J.M., Piry, S., Sembène, M., Britton-Davidian, J., Granjon, L. & Dobigny, G. (2011): Extensive Robertsonian polymorphism in the African rodent *Gerbillus nigeriae*: geographic aspects and meiotic data. *Journal of Zoology* 284: 276–285.
Hoffmann, F.G., Owen, J.G. & Baker, R.J. (2003): mtDNA perspective of chromosomal diversification and hybridization in Peters' tent-making bat (*Uroderma bilobatum*: Phyllostomidae). *Molecular Ecology* 12: 2981–2993.
Horn, A., Basset, P., Yannic, G., *et al.* (2012): Chromosomal rearrangements do not seem to affect the gene flow in hybrid zones between karyotypic races of the common shrew (*Sorex araneus*). *Evolution* 66: 882–889.
Ivanitskaya, E., Rashkovetsky, L. & Nevo, E. (2010): Chromosomes in a hybrid zone of Israeli mole rats (*Spalax*, Radentia). *Russian Journal of Genetics* 46: 1149–1151.
Jaarola, M., Martínková, N., Gündüz, I., *et al.* (2004): Molecular phylogeny of the speciose vole genus *Microtus* (Arvicolinae, Rodentia) inferred from mitochondrial DNA sequences. *Molecular Phylogenetics and Evolution* 33: 647–663.
Karamysheva, T.V., Bogdanov, A.S., Kartavtseva, I.V., Likhoshvay, T.V., Bochkarev, M.N., Kolcheva, N.E., Marochkina, V.V. & Rubtsov, N.B. (2010): Comparative FISH analysis of C-positive blocks of centromeric chromosomal regions of pygmy wood mice *Sylvaemus uralensis* (Rodentia, Muridae). *Russian Journal of Genetics* 46: 712–724.
Kirkpatrick, M. & Barton, N. (2006): Chromosome inversions, local adaptation and speciation. *Genetics* 173: 419–434.
Kryštufek, B., Ivanitskaya, E., Arslan, A., Arslan, E. & Bužan, E.V. (2012): Evolutionary history of mole rats (genus *Nannospalax*) inferred from mitochondrial cytochrome b sequence. *Biological Journal of the Linnean Society* 105: 446–455.
Kulemzina, A.I., Trifonov, V.A., Perelman, P.L., Rubtsova, N.V., Volobuev, V., Ferguson-Smith, M.A., Stanyon, R., Yang, F. & Graphodatsky, A.S. (2009): Cross-species chromosome painting in Cetartiodactyla: reconstructing the karyotype evolution in key phylogenetic lineages. *Chromosome Research* 17: 419–436.
Lande, R. (1979): Effective deme sizes during long-term evolution estimated from rates of chromosomal rearrangement. *Evolution* 33: 234–251.
Levan, G. (1974): Nomenclature for G-bands in rat chromosomes. *Hereditas* 77: 37–52.
Libois, R., Ramalhinho, M.G. & Rosoux, R. (2012): Evidence for a differentiated chromosomal race north of classical south European refuge areas in the garden dormouse *Eliomys quercinus*. *Acta Theriologica* 57: 313–320.
Lopes, C.M., Ximenes, S.S.F., Gava, A. & De Freitas, T.R.O. (2013): The role of chromosomal rearrangements and geographical barriers in the divergence of lineages in a South American subterranean rodent (Rodentia: Ctenomyidae: *Ctenomys minutus*). *Heredity* 111: 293–305.
Lyapunova, E.A., Bakloushinskaya, I.Y., Saidov, A.S. & Saidov, K.K. (2010): Dynamics of chromosome variation in mole voles *Ellobius tancrei* (Mammalia, Rodentia) in Pamiro-Alay in the period from 1982 to 2008. *Russian Journal of Genetics* 46: 566–571.
Lyapunova, E.A., Vorontsov, N.N., Korobitsyna, K.V., Ivanitskaya, E.Y., Borisov, Y.M., Yakimenko, L.V. & Dovgal, V.Y. (1980): A Robertsonian fan in *Ellobius talpinus*. *Genetica* 52/53: 239–247.
Machado, T., Silva, M.J.D.J., Leal-Mesquita, E.R., Carmignotto, A.P. & Yonenaga-Yassuda, Y. (2005): Nine karyomorphs for spiny rats of the genus *Proechimys* (Echimyidae, Rodentia) from north and central Brazil. *Genetics and Molecular Biology* 28: 682–692.
Marchal, J.A., Acosta, M.J., Nietzel, H., Sperling, K., Bullejos, M., Díaz de la Guardia, R. & Sánchez, A. (2004): X chromosome painting in *Microtus*: origin and evolution of the giant sex chromosomes. *Chromosome Research* 12: 767–776.
Martin, A.P. & Palumbi, S.R. (1993): Body size, metabolic rate, generation time, and the molecular clock. *Proceedings of the National Academy of Sciences USA* 90: 4087–4091.
Martinez, P.A., Jacobina, U.P., Fernandes, R.V., Brito, C., Penone, C., Amado, T.F., Fonseca, C.R. & Bidau, C.J. (2017): A comparative study on karyotypic diversification rate in mammals. *Heredity* 118: 366–373.
Mayr, E. (1963): *Animal Species and Evolution*. Harvard University Press, Cambridge, MA.
Merico, V., Giménez, M.D., Vasco, C., Zuccotti, M., Searle, J.B., Hauffe, H.C. & Garagna, S. (2013): Chromosomal speciation in mice: a cytogenetic analysis of recombination. *Chromosome Research* 21: 523–533.

Meyer, M.N. & Malikov, V.G. (2000): New species and subspecies of mouse-like hamsters of the genus *Calomyscus* (Rodentia, Cricetidae) from Southern Turkmenistan. *Zoologicheskii Zhurnal* 79: 219–223. [in Russian with English summary]

Mitsainas, G.P., Rovatsos, M.T., Rizou, E.I. & Giagia-Athanasopoulou, E.B. (2009): Sex chromosome variability outlines the pathway to the chromosomal evolution in *Microtus thomasi* (Rodentia, Arvicolinae). *Biological Journal of the Linnean Society* 96: 685–695.

Nachman, M.W. & Payseur, B.A. (2012): Recombination rate variation and speciation: theoretical predictions and empirical results from rabbits and mice. *Philosophical Transactions of the Royal Society B* 367: 409–421.

Nachman, M.W. & Searle, J.B. (1995): Why is the house mouse karyotype so variable? *Trends in Ecology & Evolution* 10: 397–402.

Navarro, A. & Barton, N.H. (2003): Accumulating postzygotic isolation genes in parapatry: a new twist on chromosomal speciation. *Evolution* 57: 447–459.

Nevo, E. & Bar-El, H. (1976): Hybridization and speciation in fossorial mole rats. *Evolution* 30: 831–840.

Noor, M.A., Grams, K.L., Bertucci, L.A. & Reiland, J. (2001): Chromosomal inversions and the reproductive isolation of species. *Proceedings of the National Academy of Sciences USA* 98: 12084–12088.

Núñez-Garduño, A., López-Wilchis, R. & Cervantes, F.A. (1999): Chromosomal variation of *Osgoodomys banderanus* (Rodentia: Muridae). *Cytologia* 64: 319–326.

Orgel, L.E. & Crick, F.H.C. (1980): Selfish DNA: the ultimate parasite. *Nature* 284: 604–607.

Orr, H.A. (1997): Haldane's rule. *Annual Review of Ecology and Systematics* 28: 195–218.

Ortiz-Barrientos, D., Engelstädter, J. & Rieseberg, L.H. (2016): Recombination rate evolution and the origin of species. *Trends in Ecology & Evolution* 31: 226–236.

Panithanarak, T., Hauffe, H.C., Dallas, J.F., Glover, A., Ward, R.G. & Searle, J.B. (2004): Linkage-dependent gene flow in a house mouse chromosomal hybrid zone. *Evolution* 58: 184–192.

Pardo-Manuel de Villena, F. & Sapienza, C. (2001): Female meiosis drives karyotypic evolution in mammals. *Genetics* 159: 1179–1189.

Patton, J.L. (1993): Hybridization and hybrid zones in pocket gophers (Rodentia, Geomyidae). In: Harrison, R.G. (ed.): *Hybrid Zones and the Evolutionary Process.* Oxford University Press, New York: 290–308.

Pavlova, S.V., Kolomiets, O.L., Bulatova, N. & Searle J.B. (2008): Demonstration of a WART in a hybrid zone of the common shrew (*Sorex araneus* Linnaeus, 1758). *Comparative Cytogenetics* 2: 115–120.

Perez, G.C.L., Libois, R. & Nieberding, C.M. (2013): Phylogeography of the garden dormouse *Eliomys quercinus* in the western Palearctic region. *Journal of Mammalogy* 94: 202–217.

Piálek, J., Hauffe, H.C. & Searle, J.B. (2005): Chromosomal variation in the house mouse: a review. *Biological Journal of the Linnean Society* 84: 535–563.

Pita, R., Beja, P. & Mira, A. (2007): Spatial population structure of the Cabrera vole in Mediterranean farmland: the relative role of patch and matrix effects. *Biological Conservation* 134: 383–392.

Qumsiyeh, M.B. (1994): Evolution of number and morphology of mammalian chromosomes. *Journal of Heredity* 85: 455–465.

Redi, C., Garagna, S., Zacharias, H., Zuccotti, M. & Capanna, E. (2001): The other chromatin. *Chromosoma* 110: 136–147.

Reig, O.A., Busch, C., Ortells, M.O. & Contreras, J.R. (1990): An overview of evolution, systematics, population biology, cytogenetics, molecular biology and speciation in *Ctenomys*. In: Nevo, E. & Reig, O.A. (eds): *Evolution of Subterranean Mammals at the Organismal and Molecular Levels.* Alan R. Liss, New York: 71–96.

Rieseberg, L.H. (2001): Chromosomal rearrangements and speciation. *Trends in Ecology & Evolution* 16: 351–358.

Rovatsos, M.T. & Giagia-Athanasopoulou, E.B. (2012): Taxonomical status and phylogenetic relations between the "thomasi" and "atticus" chromosomal races of the underground vole *Microtus thomasi* (Rodentia, Arvicolinae). *Mammalian Biology* 77: 6–12.

Rovatsos, M.T., Mitsainas, G.P., Paspali, G., Oruci, S. & Giagia-Athanasopoulou, E.B. (2011): Geographical distribution and chromosomal study of the underground vole *Microtus thomasi* in Albania and Montenegro. *Mammalian Biology* 76: 22–27.

Ryder, O.A., Kumamoto, A.T., Durrant, B.S. & Benirschke, K. (1989): Chromosomal divergence and reproductive isolation in dik-diks. In: Otte, D. & Endler, J.A. (eds): *Speciation and its Consequences.* Sinauer, Sunderland, MA: 208–225.

Scherthan, H., Cremer, T., Arnason, U., Weier, H.U., Lima-de-Faria, A. & Fönicke, L. (1994): Comparative chromosome painting discloses homologous segments in distantly related mammals. *Nature Genetics* 6: 342–347.

Schooley, R.L. & Branch, L.C. (2009): Enhancing the area–isolation paradigm: habitat heterogeneity and metapopulation dynamics of a rare wetland mammal. *Ecological Applications* 19: 1708–1722.

Searle, J.B. (1983): *Robertsonian Chromosomal Variation in the Common Shrew Sorex araneus L.* PhD Thesis, University of Aberdeen.

Searle, J.B. (1993): Chromosomal hybrid zones in eutherian mammals. In: Harrison, R.G. (ed.): *Hybrid Zones and the Evolutionary Process.* Oxford University Press, New York: 309–353.

Searle, J.B. & Wilkinson, P.J. (1986): The XYY condition in a wild mammal: an XY/XYY mosaic common shrew (*Sorex araneus*). *Cytogenetics and Cell Genetics* 41: 225–233.

Searle, J.B. & Wójcik, J.M. (1998): Chromosomal evolution: the case of *Sorex araneus*. In: J.M. Wójcik & M. Wolsan (eds.): *Evolution of Shrews.* Mammal Research Institute, Polish Academy of Sciences, Białowieża, Poland: 219–268.

Searle, J.B., Zima, J. & Polly, P.D., eds (in press): *Shrews, Chromosomes and Speciation.* Cambridge University Press, Cambridge.

Servedio, M.R. & Noor, M.A. (2003): The role of reinforcement in speciation: theory and data. *Annual Review of Ecology, Evolution, and Systematics* 34: 339–364.

Shahabi, S. (2013): Karyologic study of three species of *Calomyscus* (Rodentia: Calomyscidae) from Iran. *Iranian Journal of Animal Biosystematics* 6: 55–60.

Short, R.V. (1976): The origin of species. In: Austin, C.R. & Short, R.V. (eds.): *Reproduction in Mammals. 6, The Evolution of Reproduction.* Cambridge University Press, Cambridge: 110–148.

Speicher, M.R. & Carter, N.P. (2005): The new cytogenetics: blurring the boundaries with molecular biology. *Nature Reviews Genetics* 6: 782–792.

Stankiewicz, P. & Lupski, J.R. (2002): Genome architecture, rearrangements and genomic disorders. *Trends in Genetics* 18: 74–82.

Sutton, W.S. (1902): On the morphology of the chromosome group in *Brachystola magna*. *Biological Bulletin* 4: 24–39.

Telfer, S., Holt, A., Donaldson, R. & Lambin, X. (2001): Metapopulation processes and persistence in remnant water vole populations. *Oikos* 95: 31–42.

Thaeler Jr, C.S. (1968): Karyotypes of sixteen populations of the *Thomomys talpoides* complex of pocket gophers (Rodentia-Geomyidae). *Chromosoma* 25: 172–183.

Thaeler Jr, C.S. (1974): Four contacts between ranges of different chromosome forms of the *Thomomys talpoides* complex (Rodentia: Geomyidae). *Systematic Zoology* 23: 343–354.

Thaeler Jr, C.S. (1980): Chromosome numbers and systematic relations in the genus *Thomomys* (Rodentia: Geomyidae). *Journal of Mammalogy* 61: 414–422.

Unice, S.M.M., Hale, D.W. & Greenbaum, I.F. (1998): Karyotypic variation in populations of deer mice (*Peromyscus maniculatus*) from eastern Canada and the northeastern United States. *Canadian Journal of Zoology* 76: 584–588.

Veyrunes, F., Catalan, J., Tatard, C., Cellier-Holzem, E., Watson, J., Chevret, P., Robinson, T.J. & Britton-Davidian, J. (2010): Mitochondrial and chromosomal insights into karyotypic evolution of the pygmy mouse, *Mus minutoides*, in South Africa. *Chromosome Research* 18: 563–574.

Veyrunes, F., Perez, J., Borremans, B., Gryseels, S., Richards, L.R., Duran, A., Chevret, P., Robinson, T.J. & Britton-Davidian, J. (2014): A new cytotype of the African pygmy mouse *Mus minutoides* in Eastern Africa. Implications for the evolution of sex-autosome translocations. *Chromosome Research* 22: 533–543.

Volleth, M., Biedermann, M., Schorcht, W. & Heller K.-G. (2013): Evidence for two karyotypic variants of the lesser horseshoe bat (*Rhinolophus hipposideros*, Chiroptera, Mammalia) in Central Europe. *Cytogenetic and Genome Research* 140: 55–61.

Vujošević, M. & Blagojević, J. (2004): B chromosomes in populations of mammals. *Cytogenetic and Genome Research* 106: 247–256.

Walsh, J.B. (1982): Rate of accumulation of reproductive isolation by chromosome rearrangements. *American Naturalist* 120: 510–532.

White, M.J.D. (1973): *Animal Cytology and Evolution*, 3rd edn. Cambridge University Press, Cambridge.

White, T.A., Bordewich, M. & Searle, J.B. (2010): A network approach to study karyotypic evolution: the chromosomal races of the common shrew (*Sorex araneus*) and house mouse (*Mus musculus*) as model systems. *Systematic Biology* 59: 262–276.

Wilson, A.C., Bush, G.L., Case, S.M. & King, M.-C. (1975): Social structuring of mammalian populations and rate of chromosomal evolution. *Proceedings of the National Academy of Sciences USA* 72: 5061–5065.

Wilson, A.C., Sarich, V.M. & Maxson, L.R. (1974): The importance of gene rearrangement in evolution: evidence from studies on rates of chromosomal, protein, and anatomical evolution. *Proceedings of the National Academy of Sciences USA* 71: 3028–3030.

Wójcik, J.M., Ratkiewicz, M. & Searle, J.B. (2002): Evolution of the common shrew, *Sorex araneus*: chromosomal and molecular aspects. *Acta Theriologica* 47(Suppl. 1): 139–167.

Yang, F., Carter, N.P., Shi, L. & Ferguson-Smith, M.A. (1995): A comparative study of karyotypes of muntjacs by chromosome painting. *Chromosoma* 103: 642–652.

Zachos, F.E. (2016): *Species Concepts in Biology: Historical Development, Theoretical Foundations and Practical Relevance*. Springer, Cham, Switzerland.

Zhdanova, N.S., Minina, J.M., Karamisheva, T.V., Draskovic, I., Rubtsov, N.B. & Londoño-Vallejo, J.A. (2007): The very long telomeres in *Sorex granarius* (Soricidae, Eulipotyphla) contain ribosomal DNA. *Chromosome Research* 15: 881–890.

Zima, J. (2004): Karyotypic variation in mammals of the Balkan Peninsula. In: Griffiths H.L., Kryštufek B. & Reed J.M. (eds.): *Balkan Biodiversity: Pattern and Process in the European Hotspot*. Kluwer Academic Publishers, Dordrecht, Netherlands: 109–133.

Robert J. Asher

3 Taxonomy, trees, and truth in historical mammalogy

3.1 Introduction

Categorization is integral to rationality. A rational being observes its environment and tries to make sense of it: safe vs. dangerous, edible vs. indigestible, interesting vs. boring. These intuitive dichotomies are among the most basic forms of taxonomy, or systems for categorizing the patterns we observe. Taxonomies pervade human cultures and are not only a result of real patterns but also influenced by our social and psychological biases. Taxonomies in some fields are widely held and of great importance (e.g., law and economics) but are heavily influenced by arbitrary social convention, often to a greater extent than by any intrinsic reality to the things being categorized. In principle, biological patterns are amenable to categorization based on intrinsic reality. Anyone can perceive nature, and perceptions of biological patterns are often consistent across observers. Therefore, taxonomies that represent biological patterns can also be consistent because they represent a reality that is accessible to independent observers, despite cultural and linguistic differences (Hunn 1975). When they do not represent reality, it may be because of limitations to symbolic language, economic or political motivations that are not biological, or misunderstandings of the process by which the phenomena to be categorized have become diverse. There may also be distinct aspects of reality that a given classification may strive to represent, for example grades of adaptation vs. historical patterns of interrelatedness.

We expect that when the process, or mechanism, behind a given phenomenon is well understood, taxonomies of that phenomenon will be more accurate than taxonomies based on a poorly understood process. This is not to imply that any particular theory of process needs to be assumed, endorsed, or even understood by those engaged in taxonomy, but it recognizes a logical independence between efforts to categorize patterns, such as taxonomy, and the processes that may underlie those patterns, such as evolution (Patterson 1988). In those cases when a causal relationship really does exist between the two, understanding the process should make categorizations of its resulting patterns less susceptible to bias and more rooted in reality. For example, medical treatments based on the zodiac are less effective that those based on the germ theory of disease because the latter entails not just claims of correlation or agency but also well-corroborated theories of cause and effect. Both astrology and medical study of infectious disease entail descriptions of pattern (or "what"), but the latter goes further to explain the process (or "how"), and thereby leads to categories of treatment firmly rooted in empirical reality. An understanding of the process behind infection underlies a medical taxonomy including hygiene and antibiotics, whereas no demonstrable process is behind a medical taxonomy including Jupiter and the pituitary gland.

Similarly, the discovery of a process by which livings things have become diverse over time (i.e., evolution) should lead to an improvement in taxonomists' ability to categorize organisms. Animal taxonomies are ubiquitous throughout recorded human history (Gregory 1910), whereas a widely accepted theory of biological evolution is fairly recent. This makes it possible to ask two interrelated questions: (1) are taxonomies based on an evolutionary understanding of nature more accurate than non-evolutionary taxonomies, and (2) has better access to biological patterns, and the methods to quantify them, led to improved taxonomic accuracy?

3.2 Goals of this study

In this chapter, I address these questions with a focus on mammals and rely heavily on Gregory's (1910) history of mammalian classification, supplemented by a number of additional classifications and phylogenetic studies. If it is true that taxonomies of mammals have become better at representing reality since the widespread acceptance of evolution, and as access to pattern documenting that reality has improved, then we would expect two things to occur over time: (1) taxonomies should converge toward the currently most well-corroborated hypothesis of mammalian interrelationships, and (2) taxonomies published by authors using methods that provide a clearer grasp of reality should resemble each other more than those using inferior methods.

I argue that both expectations hold although past taxonomies were generally not intended to represent evolutionary history. Al-Jahiz in the 9th century was an

https://doi.org/10.1515/9783110341553-003

advocate of monotheistic creation (Eisenstein 1991; Stott 2012; Montgomery 2013). (Incidentally, contemporary religious creationists who offer explanations in the form of an agent, e.g., "god created species", often have views that are incompatible with a process-oriented explanation such as descent with modification [Numbers 2006]. However, without invoking any processes or mechanism by which their agent did the creating, and religious creationists do not provide much detail on this point, simple assertion of agency does not necessarily conflict with a theory of process such as evolution; see Sober 2008; Asher 2012.) Georges Cuvier was an empiricist and skeptical of Lamarckian evolution, but he did not clearly advocate any specific theory of process either (Rudwick 1997). William King Gregory, in contrast, did accept evidence for evolutionary descent with modification as the process behind biodiversity (Gregory 1910). Despite this variation, Al-Jahiz, Cuvier, and Gregory all used, at least in part, biological patterns intrinsic to the organisms they observed to construct their taxonomies. Whatever processes they thought were driving the patterns they observed, each author was constrained by the reality of those patterns. A bovid's split hoof was no less obvious to Al-Jahiz than it was to Cuvier or Gregory. A process like evolution that proposes to explain differing levels of similarity across living things is, therefore, at least partly manifest in any taxonomy based on similarity. We expect more recent taxonomies to be better than older ones for two reasons: First, recent investigators benefit from more technology with which they can appreciate pattern; and second, recent investigators have a better understanding of the now well-corroborated process behind biological diversity (evolution).

Perhaps more interestingly, we do not have to assume that Gregory's understanding of an evolutionary process is accurate, but based on the patterns he interpreted to make claims about relationships, we can test whether it is or not (Penny *et al.* 1982; Patterson 1988). If he was right, then we would expect patterns of biological data unknown to Gregory (e.g., DNA) to yield taxonomies closer to his than to others who did not subscribe to an evolutionary process, such as Cuvier (1817) or Owen (1868). We would also expect that, if they are indeed better, novel methodologies such as cladistic taxonomy (Hennig 1966) would lead to improved accuracy and more agreement on classification among authors who use such methods.

In order to use the patterns evident in classifications to test a given process behind them, one cannot logically presuppose that process to have caused the pattern in the first place (Rieppel 1986; Patterson 1988). Hence, I use the terms like "cladogram", "branching diagram", "classification", and "taxonomy" to indicate past attempts at summarizing biological patterns, regardless of a given author's views on a theory of process. A "tree" or "phylogeny", on the other hand, denotes a pattern generated by an evolutionary mechanism (Patterson 1988). Therefore, comparisons of cladograms can test evolution; comparisons among trees cannot, in the sense that "trees" are cladograms already assumed or shown to be the result of an evolutionary process. This may seem like an esoteric point, and to some extent it is. Evolutionary theory has already been reasonably demonstrated; just as civil engineers may assume that gravity is relevant to their hypotheses on optimal bridge construction, contemporary biologists may assume that descent with modification is relevant to their hypotheses on how animals should be classified. On the other hand, terminological precision is important, and 21st century biologists should not forget the reasons why evolutionary descent with modification is a compelling theory of process. This theory entails postulates such as common descent that can be demonstrated by making predictions about cladograms (Penny *et al.* 1982), which do not have to presuppose anything about evolution (Patterson 1988). This makes the distinction between "cladogram" as a statement about pattern, and "tree" as a statement about process, worthwhile.

3.3 Methods

I extract cladograms from a number of historical classifications (Tab. 3.1; Appendix 3.1) and quantify the extent to which they resemble each other and the current, well-corroborated tree of mammals (Fig. 3.1). By "well corroborated", I mean a hypothesis of interrelationships among the ca. 5500 species of extant monotremes, marsupial, and placental mammals that has received support from a wide variety of methods and data (Asher *et al.* 2009). This tree is nearly but not completely resolved, and it is based on consistency across topologies generated by alignments of 35,603 base pairs (bp) of nuclear DNA for 169 mammals (Meredith *et al.* 2011), up to 15,535 bp of nuclear and mitochondrial DNA for 1,265 rodents (Fabre *et al.* 2012), 43,616 bp of nuclear and mitochondrial DNA for 193 marsupials and 10 other mammals (Mitchell *et al.* 2014), and 32,116,455 bp of nuclear DNA for 36 mammals and 16,050 bp of micro-RNA for 39 taxa (Tarver *et al.* 2016). The latter study also incorporated large, amino acid sequence data sets from Hallström and Janke (2010), Romiguer *et al.* (2013), and O'Leary *et al.* (2013). Thus, Fig. 3.1 shows areas of agreement among the optimal topologies published by Meredith

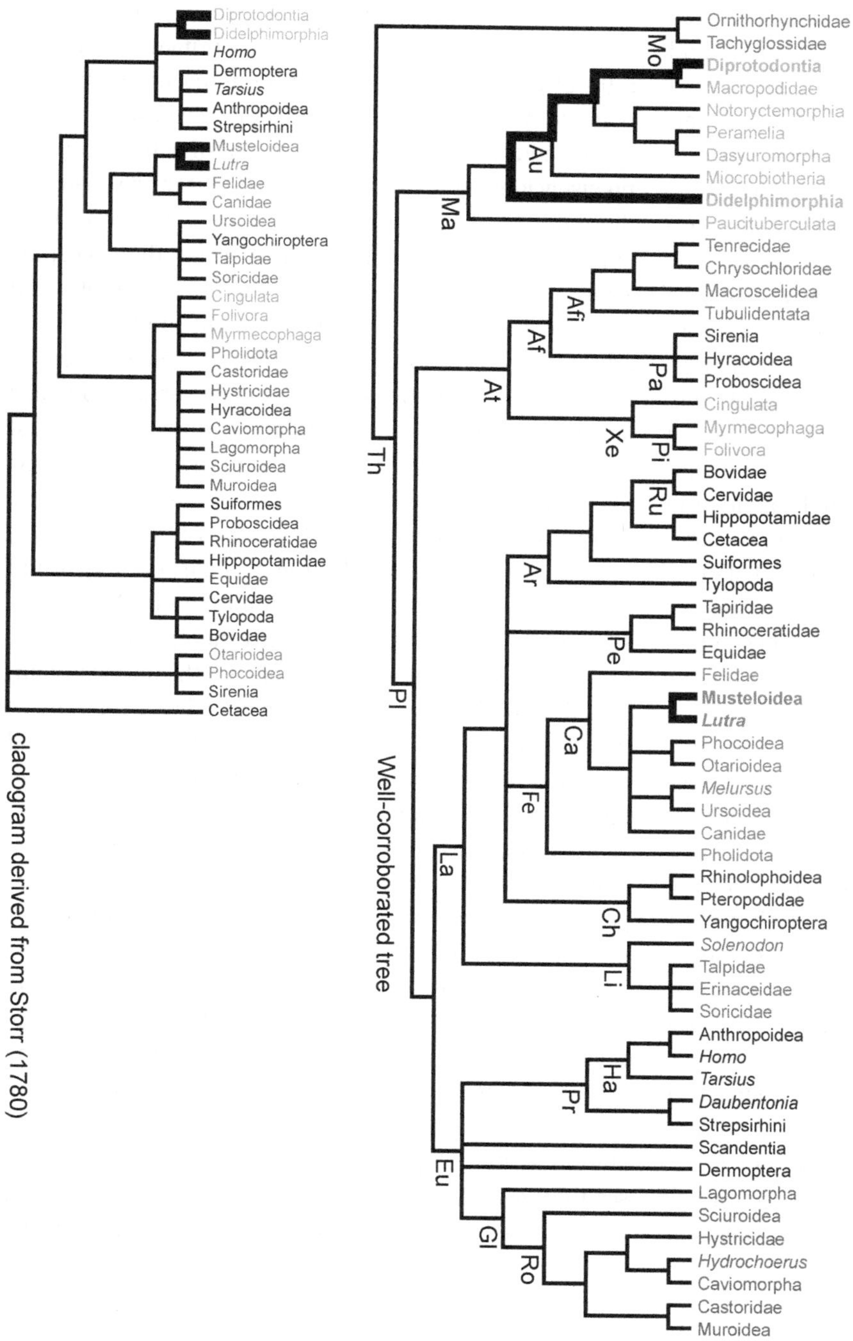

Fig. 3.1: Right: Well-corroborated tree of mammalian relationships based on congruence between optimal topologies from Meredith *et al.* (2011), Fabre *et al.* (2012), Mitchell *et al.* (2014), and Tarver *et al.* (2016). Left: branching diagram based on classification of Storr (1780); groups in agreement with the well-corroborated tree have thick branches. Abbreviations (below) represent high-level taxa to the right of the respective node (nomenclature following Asher and Helgen 2010): Af = Afrotheria, Afi = Afroinsectiphilia, Ar = Artiodactyla, At = Atlantogenata, Au = Australidelphia, Ca = Carnivora, Ch = Chiroptera, Eu = Euarchontoglires, Fe = Ferae, Gl = Glires, Ha = Haplorhini, La = Laurasiatheria, Li = Lipotyphla, Ma = Marsupialia, Mo = Monotremata, Pa = Paenungulata, Pe = Perrisodactyla, Pi = Pilosa, Pl = Placentalia, Pr = Primates, Ro = Rodentia, Ru = Ruminantia, Th = Theria, Xe = Xenarthra. Selected groups are color coded (red = Euarchontoglires, blue = Laurasiatheria, green = Atlantogenata, yellow = monotremes and marsupials) for ease of reference to subsequent figures. Dark red indicates "Archonta" (i.e., Primates, Dermoptera, and Scandentia), the monophyly of which is uncertain (as indicated in the tree).

et al. (2011), Fabre *et al.* (2012), Mitchell *et al.* (2014), and Tarver *et al.* (2016). Polytomies are shown where these studies disagree (e.g., on sister taxa of Scandentia, Perissodactyla, and Chiroptera).

3.3.1 Quantifying similarity

I measured similarity by counting the number of "shared partitions" among rooted cladograms in Mesquite (Maddison and Maddison 2015), excluding the two basal-most nodes. This enables comparisons of cladograms that share similarly grouped taxa but have different taxon samples. For example, as shown in Fig. 3.1, Storr (1780) places the otter (*Lutra*) close to other carnivorans, such as *Mustela*. This contrasts with authors such as Scopoli (1777), who placed *Lutra* with other aquatic species such as seals and hippos. A mustelid-*Lutra* group is therefore one area of agreement (i.e., a "shared partition") of Storr's classification with the well-corroborated phylogeny shown in Fig. 3.1. Similarly, John Ray (1693) accurately recognized humans with "simians" and bovids with cervids (Fig. 3.2). Slightly more complicated is Storr's group "Plantares" (in turn a part of his "Manuati" and "Primates"), consisting of *Didelphis* and *Phalanger*. I understand both terms to mean modern genera of marsupials, and Storr classifies no other marsupial taxon in his 1780 classification (Gregory 1910, pp. 49–50). Thus, although our modern understanding of marsupial interrelations does not have a *Didelphis-Phalanger* clade (Fig. 3.1), it does have a group containing the parent taxa of those two species (Didelphimorphia and Diprotodontia) to the exclusion of all other species classified by Storr. The "shared partitions" tool in Mesquite accurately represents this similarity, as depicted in Fig. 3.1, and I use it to compare each of the cladograms derived from historical classifications to the well-corroborated topology in Fig. 3.1, and also all pairwise comparisons across cladograms extracted from the 55 classifications listed in Tab. 3.1. All cladograms derived from these studies are available in nexus format in Appendix 1. For data comparisons and summaries, I used R version 3.1.0 (R Core Team 2014) and LibreOffice/Microsoft Excel.

Comparisons among classifications need to account for differing taxon samples and the extent to which a given study made testable claims about affinity. For example, Prothero (2007) effectively sampled all of the 59 mammalian taxa represented here (as did McKenna 1975; see below), but only made precise statements of affinity for 37, leaving the rest in polytomies. By contrast, Song *et al.* (2012) sampled 27 of these taxa, resolved all of them, among which 20 groups are held in common with the well-corroborated topology (Fig. 3.1). In absolute terms, Prothero (2007) has more (24) partitions in common with Fig. 3.1, but as a proportion of the number he could have correctly identified given his sample (equivalent to the number of sampled taxa minus three), his accuracy is lower: 24 of a maximum possible of 56, or 0.43. Song *et al.* (2012) recovered 20 of a possible 24 in their sample, or 0.83. Hence, I evaluate historical topologies primarily using this ratio of correctly identified partitions by the number of potentially identified partitions for any given study.

In addition, many historical classifications did not precisely identify all of their terminal taxa, and I exercised discretion to give credit (or not) to authors who would have agreed with certain relationships given then-accepted knowledge of mammalian systematics, for example, that *Lutra* is a mustelid, *Melursus* an ursid, and *Hydrochoerus* a caviid. McKenna (1975) did not explicitly classify *Daubentonia* as a strepsirhine primate, or macropodids as diprotodont marsupials, but as a leading mammalogist of the latter half of the 20th century, he certainly knew about these taxa and would not have questioned these relationships. His 1975 classification is thus here represented as containing *Daubentonia* as part of Strepsirhini and Macropodidae as part of Diprotodontia. McKenna's 1975 classification did not explicitly mention *Dromiciops*, but he was of course aware of this South American marsupial as well. However, agreement did not yet exist in 1975 about the affinity of *Dromiciops* with Australian and not other South American marsupials; therefore, I credit his 1975 classification only with recognizing that *Dromiciops* is a marsupial, with uncertain affinities (in the form of a polytomy) to other marsupial groups. In other cases, past assumptions of monophyly among terminal taxa in older classifications were wrong, such as terrestrial Artiodactyla to the exclusion of cetaceans (Gatesy *et al.* 1996), "Insectivora" including tenrecids and chrysochlorids (Stanhope *et al.* 1998), or Microchiroptera including rhinolophoids (Teeling *et al.* 2005). Therefore, I represent the relevant terminals (e.g., Cetacea, Tenrecidae, Chrysochloridae, Rhinolophoidea) so as to identify those analyses that, even into the 21st century (e.g., Prothero 2007), misidentified several well-corroborated branches on the

Tab. 3. 1: Sources of branching diagrams used to represent historical classifications.

Author	Year	Publication	Source	Method
Leviticus 11	~-500	*The Bible*	Gregory 1910, pp. 7–8	Ancient
Al-Jahiz	~850	*Kitab al Hayawan*	Eisenstein 1991	Ancient
Ray	1693	*Synopsis Methodica Animalium Quadrupedum et Serpentini Generis*	Gregory 1910, pp. 18–19	Preevolutionary
Linnaeus	1735	*Systema Naturae*, I ed.	Gregory 1910, p. 102	Preevolutionary
Klein	1751	*Quadrupedum Dispositio Brevisque Historia Naturalis*	Gregory 1910, pp. 26–27	Preevolutionary
Brisson	1762	*Regnum Animale...*, II ed.	Gregory 1910, pp. 42–43	Preevolutionary
Linnaeus	1766	*Systema Naturae*, XII ed.	Gregory 1910, pp. 29–30	Preevolutionary
Scopoli	1777	*Introductio ad Historium*	Gregory 1910, p. 37	Preevolutionary
Blumenbach	1779	*Handbuch der Naturgeschichte*	Gregory 1910, pp. 45–46	Preevolutionary
Storr	1780	*Prodromus Methodi Mammalium*	Gregory 1910, pp. 49–50	Preevolutionary

Tab. 3.1 (continued)

Author	Year	Publication	Source	Method
Pennant	1781	*History of Quadrupeds*	Gregory 1910, pp. 51–52	Preevolutionary
Lacépède	1799	*Tableau des Divisions, Sousdivisions, Ordres et Genres des Mammifères, des Cétacés et des Oiseaux*	Gregory 1910, pp. 62–63	Preevolutionary
Cuvier	1800	*Leçons d'Anatomie Comparée*	Gregory 1910, pp. 65–66	Preevolutionary
Illiger	1811	*Prodomus Systematis Mammalium et Avium Additis Terminis Zoographicii Utriusque Classis Eorumque Versione Germanica*	Gregory 1910, pp. 70–71	Preevolutionary
de Blainville	1816	*Prodrome d'une nouvelle distribution...*	Gregory 1910, pp. 77–78	Preevolutionary
Cuvier	1817	*Le Règne Animal*	Gregory 1910, p. 80	Preevolutionary
Blumenbach	1830	*Handbuch der Naturgeschichte*, 12th ed.	Gregory 1910, p. 81	Preevolutionary
de Blainville	1834	Gervais, *Mammalogie ou Mastologie...*	Gregory 1910, pp. 82–83; Guerin 1836, p. 619	Preevolutionary
Bonaparte	1837	*Trans Linn Soc Lond* 18: 247–304	Gregory 1910, p. 84; Bonaparte 1837	Preevolutionary
Wagner	1855	*Schreber's Säugetiere*	Gregory 1910, pp. 86–87	Preevolutionary
Haeckel	1866	*Anthropogenie...*	Pietsch 2012, p. 119	Evolutionary
Owen	1868	*On the Anatomy of Vertebrates*	Gregory 1910, p. 90	Preevolutionary
Gill	1870	*On the Relations of the Orders of Mammals*	Gregory 1910, pp. 92–93	Evolutionary
Huxley	1872	*Anatomy of Vertebrated Animals*	Gregory 1910, p. 93	Evolutionary
Flower	1883	*Proc Zool Soc Lond* 1883: 178–186	Gregory 1910, pp. 96–97	Evolutionary
Cope	1898	*Syllabus of Lectures on Geology and Paleontology...*	Gregory 1910, pp. 98–99	Evolutionary
Gadow	1898	*A Classification of Vertebrata, Recent and Extinct*	Gadow 1898, pp. 39–54	Evolutionary
Weber	1904	*Die Säugetiere*	Gregory 1910, pp. 100–101	Evolutionary
Haeckel	1905	*Der Kampf um den Entwicklungsgedanken*	Pietsch 2012, p. 121	Evolutionary
Gregory	1910	*The Orders of Mammals*	Gregory 1910, pp. 466–467	Evolutionary
Osborn	1917	*Origin and Evolution of Life*	Pietsch 2012, p. 169	Evolutionary
Winge	1921*	*The Interrelationships of the Mammalian Genera*	Winge 1941	Evolutionary
Cabrera	1922	*Manual de Mastozoologia*	Cabrera 1922	Evolutionary
Simpson	1931	*A New Classification of Mammals*	Simpson 1931	Evolutionary
Romer	1945	*Vertebrate Paleontology*, 1st (1945) and 3rd (1966) ed.	Pietsch 2012, pp. 190–194	Evolutionary
Simpson	1945	*Principles of Classification...*	Simpson 1945	Evolutionary
Grassé	1955	*Traité de Zoologie*, vol. 17	Piveteau 1955, p. 8	Evolutionary
Romer	1959	*The Vertebrate Story*	Romer 1959	Evolutionary
McKenna	1975	*Phylogeny of the Primates*, chapter 2	McKenna 1975	Cladistic
Miyamoto	1986	*Syst Zool* 35(2):230–240	Miyamoto and Goodman 1986: fig. 3	Molecular
Novacek	1986	*Bull Am Mus Nat Hist* 183: 1–112	Novacek 1986: fig. 35	Cladistic
Novacek	1992	*Nature* 356:121–125	Novacek 1992: fig. 1	Cladistic
McKenna	1997	*Classification of Mammals*	McKenna and Bell 1997	Cladistic
Shoshani	1998	*Mol Phy Evol* 9(3):572–584	Shoshani and McKenna 1998: fig. 1	Cladistic
Stanhope	1998	*Proc Nat Acad Sci USA* 95:9967–9972	Stanhope *et al.* 1998: fig. 1	Molecular
Murphy	2001	*Science* 294:2348–2351	Murphy *et al.* 2001: fig. 1	Molecular
Arnason	2002	*Proc Nat Acad Sci USA* 99:8151–8156	Arnason *et al.* 2002: fig. 1	Molecular
Asher	2003	*J Mamm Evol* 10:131–194	Asher *et al.* 2003: fig. 5 (left)	Combined
Kjer	2007	*BMC Evol Biol* 7:8	Kjer and Honeycutt 2007: fig. 1	Molecular
Prothero	2007	*Evolution: What the Fossils Say and Why It Matters*	Prothero 2007, pp. 285,291, 298, 309, 338, 312, 324	Cladistic
Arnason	2008	*Gene* 412:37–51	Arnason *et al.* 2008: fig. 2	Molecular
Meredith	2011	*Science* 334:521–524	Meredith *et al.* 2011: fig. 1	Defines correct tree
McCormack	2012	*Genome Res* 22:746–754	McCormack *et al.* 2012: fig. 2	Molecular
Song	2012	*Proc Nat Acad Sci USA* 109:14942–47	Song *et al.* 2012: fig. 1	Molecular
Tarver	2016	*Genome Biol Evol* 8:330–344	Tarver *et al.* 2016: fig. 2	Defines correct tree

* As noted in the text, Winge's taxonomic views were formulated at latest during the early 1920s, not the (posthumous) 1941 date of publication.

mammalian tree. In every case, my goal was to identify the biological groups stated or implied by a given author and make them comparable across cladograms inferred from historical classifications (Tab. 3.1).

3.3.2 Source classifications

Table 3.1 outlines the 55 publications that contain the taxonomies considered here, dating to Leviticus in (or near) the 6th century BC to Tarver *et al.* (2016). Classifications require a consistent and reasonably diverse number of animal groups in order to be compared across authors. For most pre-20th century classifications, I rely on Gregory (1910). For example, the summary and tables given for Ray (1693), both in his original publication (Fig. 3.2, Tab. 3.2) and in Gregory (1910, pp. 18–19), are sufficiently detailed so as to compare Ray's ideas with his successors in later centuries. By contrast, Gregory's summaries for the Assyrians, Aristotle, Gessner, Caesalpinus, Wotton, Erxleben, Perrault, Buffon and Daubenton, Boddaert, Vic D'azir, Geoffroy St. Hilaire, and Duméril do not provide enough information to represent their ideas in a form that is sufficiently comparable with other studies; these authors are therefore excluded. In sum, I examine 55 branching diagrams that represent the patterns observed by 45 different authors (Appendix 1).

Most of the pre-1910 citations are derived from summaries thereof in Gregory (1910); the exceptions (e.g., Al-Jahiz [Eisenstein 1991]; Haeckel 1866, 1905; Gadow 1898) are given in Tab. 3.1. Several premodern and 20th century cladograms were also drawn from Pietsch (2012). Transforming older classifications into explicit branching diagrams is not a straightforward task; as noted above, they did not intend to formulate evolutionary trees. Ancient classifications such as those in Leviticus (ca. 6th century BC) or *The Book of Animals* by Al-Jahiz (9th century AD) were based on function and/or perceived utility for humans. Al-Jahiz used categories such as swimming, flying, and crawling to group, for example, bats with birds (Eisenstein 1991). The authors of Leviticus sought to distinguish clean vs. unclean animals for human consumption. Nor were later natural philosophers primarily concerned with representing patterns of common ancestry, which were either unknown (e.g., John Ray in the 17th century) maligned as unempirical (Cuvier reacting to Lamarck in the early 19th century), or otherwise disregarded until 1859. Even explicitly evolutionary taxonomies (e.g., Gregory 1910 or Simpson 1945) did not adhere to the modern convention of naming monophyletic taxa. Nonetheless, all of these classifications were based on at least some then-current ideas of animal similarity, whether in an evolutionary sense (e.g., Huxley 1872) or as empirical (Cuvier 1817) or philosophical (Owen 1868) attempts to document "natural" groups. Whatever the motivations behind any historical author may have been, and whatever their understandings of natural or supernatural processes were, there appears to have been (at least from the Renaissance onward) a genuine desire to uncover "true" similarities across animal groups and to use classification to reflect this similarity. This makes premodern classifications fair game to investigate how the methods and theories of their day compared with those of the present.

3.3.3 Interpreting historical classifications

Early taxonomists sometimes used terms that imply different animals relative to today's standard. For example, Lacépède (1799) used *Cynocephalus* to refer to baboons (Primates) and *Galeopithecus* for colugos (Dermoptera)."Manati" or "Manatus" were probably used by Ray (1693), Klein (1751), and Scopoli (1777) in reference to pinnipeds, not sirenians, as these authors classified "Manati" using features of the pes (completely lacking in Recent sirenians). For Ray, "Manati" was part of his "Quadrupedia" (Fig. 3.2, Tab. 3.2), and Klein referred to its hindfeet (along with those of the otter, beaver, seal, and walrus) as duck-like: "pedibus quibuscumque anserinis", presumably in reference to the hindfoot webbing of aquatic carnivorans and rodents that would have been obvious in the descriptions and/or cadavers available to 17th and 18th century observers. However, Pennant (1781) and Bonaparte (1837) probably used "Manati" for Sirenia, as they describe them as "herbivorous" and "phytophagous", respectively. To make matters yet more interesting, authors such as Lacépède (1799) used "Manatus" in reference to Atlantic sirenians (i.e., manatees) but understood their current genus name (*Trichechus*) to be a walrus (*Trichechus rosmarus* instead of *Odobenus rosmarus*). Cuvier (1800) and Illiger (1811) also identified "Manatus" among manatees ("lamatins") and *Trichechus* among walruses ("morses").

More generally, we take for granted the affinity of the otter with mustelid carnivorans, of kangaroos with diprotodont marsupials, of the capybara with caviomorph rodents, and of the aye-aye and human with other primates. By contrast, early taxonomists could not. In the 17th century, John Ray would never have heard of, much less seen or dissected, an elephant shrew, golden mole, colugo, numbat, or platypus. The novelty of these animals is sometimes reflected in early classifications, such as Lacépède's (1799) placement of the hyrax among rodents and the aye-aye with kangaroos, or Illiger's (1811) placement of "thumbed" marsupials with primates in his "Pollicata".

General Table of Animals

- *Animals are either*
  - *with blood*
    - *and breathing with lungs and possessing in the heart*
      - *two ventricles*
        - *and live birth*
          - *and are aquatic (Cetacea)*
          - *or terrestrial and quadrupedal, and also possess hair, as in ["Manati"]. Animals whose genera we count as terrestrial amphibians.*
        - *or are egg-laying (Aves)*
      - *one ventricle of the heart: quadrupeds that lay eggs* and serpents*
    - *gills for breathing: also all marine fish with blood*
  - *without blood*

Table of quadrupedal animals with live birth

- *Hairy animals with live birth or four legs are either*
- *Ungulates, those with either*
  - *one-toed: horse, ass, zebra*
  - *double-toed: bisulca, or split hoof, or those which*
    - *Ruminantia with horns*
      - *perpetual: cows, sheep, goats*
      - *periodic: deer*
    - *non-ruminant: swine*
  - *four-toed: rhino, hippo, etc.*
- *Unguiculata, whose feet are either*
  - *only two toes: camel*
  - *many toes: which are either*
    - *undivided digits attached to a common roof of skin with the margins of the foot bearing weight, strong and blunt digits: elephant*
    - *when divided one from the other are either*
      - *flat-nailed and anthropomorphous: simians*
      - *narrow-nailed, teeth or incisors in each maxilla, either*
        - *several or all carnivorous and predatory, at least insectivora or common diet of insects and plants*
          - *greater rostrum*
            - *short, round head: cats*
            - *long: dogs*
          - *lesser, long slender body, short legs: weasels*
        - *two important kinds, whose diet is herbivorous: hares*

Among the live-bearing, multi-toed, anomalous quadrupeds are the hedgehog, tatou or armadillo, mole, mouse, anteater, bat, and sloth. The former five have a lengthened rostrum like the dog, but the verminous species have teeth of different form and disposition. Indeed the anteater completely lacks teeth. In the latter two the rostrum is short.

*Gregory (1910:18) notes that Ray's original text states "*viviparous*" (see Figure 2) but points out that this is a lapsus and that Ray's intent is to identify egg-laying animals.

Tab. 3. 2: Translation of Ray 1693, pp. 53, 60–61 (original in Fig 3.2).

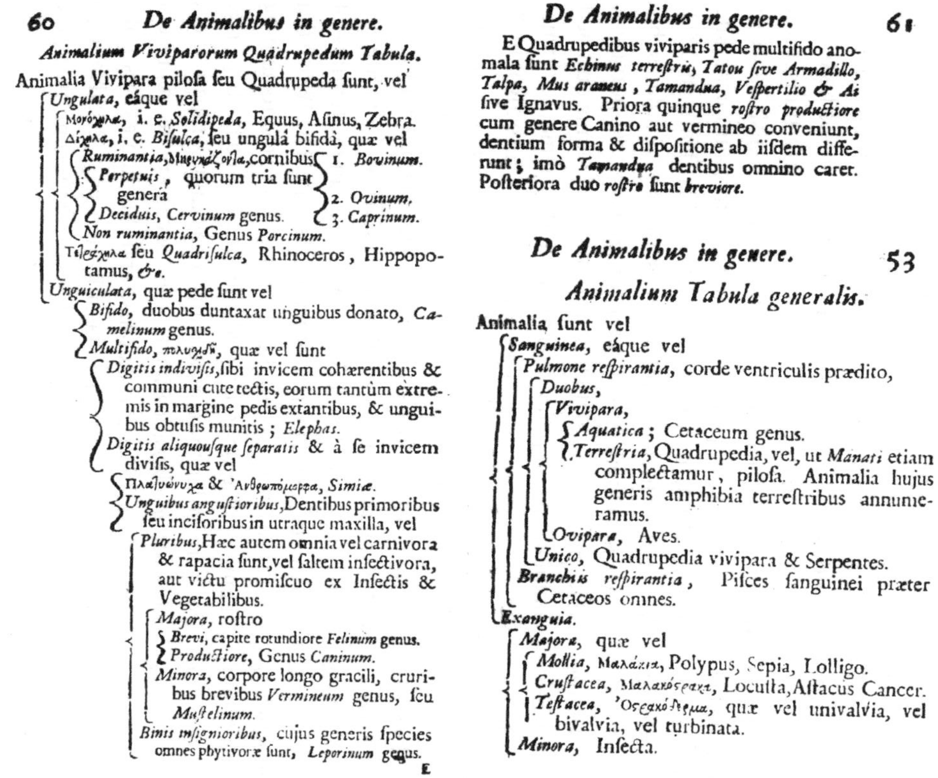

60 *De Animalibus in genere.*

Animalium Viviparorum Quadrupedum Tabula.

Animalia Vivipara pilosa seu Quadrupeda sunt, vel
Ungulata, eáque vel
Μονώνυχα, i. e. *Solidipeda*, Equus, Asinus, Zebra.
Δίχηλα, i. e. *Bisulca*, seu ungulâ bifidâ, quæ vel
Ruminantia, Μηρυκάζοντα, cornibus
Perpetuis, quorum tria sunt genera 1. *Bovinum*. 2. *Ovinum*. 3. *Caprinum*.
Deciduis, *Cervinum* genus.
Non ruminantia, Genus *Porcinum*.
Τετράχηλα seu *Quadrisulca*, Rhinoceros, Hippopotamus, *&c.*
Unguiculata, quæ pede sunt vel
Bifido, duobus duntaxat unguibus donato, *Camelinum* genus.
Multifido, πολυσχιδῆ, quæ vel sunt
Digitis indivisis, sibi invicem cohærentibus & communi cute tectis, eorum tantùm extremis in margine pedis extantibus, & unguibus obtusis munitis; *Elephas*.
Digitis aliquousque separatis & à se invicem divisis, quæ vel
Πλατυώνυχα & Ἀνθρωπόμορφα, *Simiæ*.
Unguibus angustioribus, Dentibus primoribus seu incisoribus in utraque maxilla, vel
Pluribus, Hæc autem omnia vel carnivora & rapacia sunt, vel saltem insectivora, aut victu promiscuo ex Insectis & Vegetabilibus.
Majora, rostro
Brevi, capite rotundiore *Felinum* genus.
Productiore, Genus *Caninum*.
Minora, corpore longo gracili, cruribus brevibus *Vermineum* genus, seu *Mustelinum*.
Binis insignioribus, cujus generis species omnes phytivoræ sunt, *Leporinum* genus.

De Animalibus in genere. 61

E Quadrupedibus viviparis pede multifido anomala sunt *Echinus terrestris*, *Tatou sive Armadillo*, *Talpa*, *Mus araneus*, *Tamandua*, *Vespertilio & Ai* sive Ignavus. Priora quinque *rostro productiore* cum genere Canino aut vermineo conveniunt, dentium forma & dispositione ab iisdem differunt; imò *Tamandua* dentibus omnino caret. Posteriora duo *rostro* sunt *breviore*.

De Animalibus in genere. 53

Animalium Tabula generalis.

Animalia sunt vel
Sanguinea, eáque vel
Pulmone respirantia, corde ventriculis prædito,
Duobus,
Vivipara,
Aquatica; Cetaceum genus.
Terrestria, Quadrupedia, vel, ut *Manati* etiam complectamur, pilosa. Animalia hujus generis amphibia terrestribus annumeramus.
Ovipara, Aves.
Unico, Quadrupedia vivipara & Serpentes.
Branchiis respirantia, Pisces sanguinei præter Cetaceos omnes.
Exanguia.
Majora, quæ vel
Mollia, Μαλάκια, Polypus, Sepia, Lolligo.
Crustacea, Μαλακόστρακα, Locusta, Astacus Cancer.
Testacea, Ὀστρακόδερμα, quæ vel univalvia, vel bivalvia, vel turbinata.
Minora, Insecta.

Fig. 3.2: Reproduction of Ray (1693), pp. 60–61 ("Animalium Viviparorum Quadrupedum Tabula") and p. 53 ("Animalium Tabula Generalis"), reprinted in Gregory (1910, pp. 18–19) and translated in Table 3.2.

Yet the anatomical patterns that form the basis of most historical classification were sufficient to enable some investigators to recognize startlingly accurate groups, even in the absence of a theory of process as to how those patterns arose. Ray (1693) classified humans and at least some apes as many-toed unguiculate animals with flat-nails together in his category "simians" (Fig. 3.2, Tab. 3.2). Subsequent authors, even into the 20th century, tended to place humanity in its own category apart from other animals (e.g., the Kingdom "Psychozoa" of Huxley 1957, p. 91). Cuvier (1800) was among the first to recognize the affinities of the hyrax with elephants (albeit with a variety of other "ungulates" mixed in), and Illiger (1811) and de Blainville (1834) placed the aye-aye with other primates. To give due credit to these prescient categorizations, I have represented the taxa most frequently misplaced by early authors (e.g., *Daubentonia*, *Lutra*, Macropodidae) as entities apart from the high-level taxa to which they belong (Strepsirhini, Musteloidea, Diprotodontia). This enables me to graphically acknowledge taxonomies in which these species are misplaced (e.g., Illiger's "Pollicata") and give due credit to those which are correct. I do not thereby intend to imply that the Strepsirhini of, say, Meredith *et al.* (2011) is paraphyletic by showing it apart from *Daubentonia* (Fig. 3.1).

As graphic depictions of evolutionary lineages became more common in the 19th and early 20th centuries (Pietsch 2012), authors did not necessarily make their branching diagrams consistent with their classifications. For example, Gregory (1910) illustrated his preferred branching diagram in Figs. 31 and 32 of his 1910 monograph. By contemporary standards, his depiction of "diprotodont" and "polyprotodont" marsupials in his Fig. 32 implies marsupial paraphyly, as the latter branch closer to placental mammals than the former. Yet his Fig. 31 shows the two immediately adjacent, reflected also in his classification (1910, p. 464). Based on his classification and Fig. 31, I have represented Gregory's view as supporting marsupial monophyly. Similarly, Gregory uses the separate taxa Pinnipedia and Fissipedia to delineate aquatic vs. terrestrial carnivorans, but his figure of their "genetic relations" (his Fig. 31) shows a more modern placement of seals and walruses (i.e., pinnipeds) among ursoids and other caniforms to the exclusion of feliforms. Hence, the cladogram representing his views (Appendix 1) reflects this evolutionary understanding more than his taxonomy, which of course predated the widespread acceptance of cladistic taxonomy and the use of monophyletic taxa. In addition, Gregory's classification (1910, pp. 464–466) delineates groups such as "Therictoidea", containing his "Insectivora" plus Carnivora. Yet in his depiction of the "genetic relations of the orders" (1910: fig. 31), he places lipotyphlan families on a branch adjacent to Archonta. Again, I rely on the latter to represent his views, and where possible, I use authors' own figures (as indicated in Tab. 3.1) to represent their views here.

Winge (1941) is the posthumous, English translation of several papers originally published in Danish between 1887 and 1919, apparently edited by the author as late as 1921 (1941, p. vii), shortly before his death in 1923. The 1941 volume therefore substantially postdates his actual taxonomic views, and for comparative purposes, I date his work at 1921 (Tab. 3.1). Winge's volume outlined mammalian comparative anatomy and systematics, focusing on monotremes, marsupials, bats, "insectivorans", and "edentates". His ideas on the interrelations among "edentates" are figured on p. 329, bats on p. 261, "insectivorans" (including scandentians and dermopterans) on p. 145, marsupials on p. 70, and therians vs. monotremes on pp. 16–17, and he briefly shows his understanding of "Ungulata" in relation to other orders on p. 136. Winge did not detail his views on the classification of the remaining mammalian groups beyond the single sketch on p. 136. Thus, in order to represent his ideas and compare them with other classifications, it is necessary to represent Glires, Primates, Carnivorans, and "ungulates" as polytomies. This places his classification at a disadvantage compared with others from the early 20th century, as the number of potentially recognized clades in common with the well-corroborated tree is high, but he does not give details on what he thought the relations were within these four groups, thereby missing potentially correct groups recognized by others in the late 19th and early 20th century (e.g., pinnipeds, ruminants, perissodactyls). Nonetheless, his classification improves the sample of those from the early 20th century, so I include his work in Tab. 3.1.

Unlike Gregory (1910), Simpson (1945) did not present a figure of his views on mammalian phylogeny. Despite the many qualifications in his text, and the fact that as an "evolutionary taxonomist" (Van Valen 1978) Simpson did not follow the standard of monophyletic taxa generally practiced today, his classification (pp. 39–162) is a definitive and unambiguous statement based largely on his understanding of mammalian phylogenetic history. The cladogram representing his views is therefore based on his nested classification, not on the many more nuanced statements about relationships in his text. In his classification, for example, he places tupaiids and tarsiiforms within "Prosimii", and aquatic carnivorans in "Pinnipedia" to the exclusion of terrestrial carnivorans. Both concepts were widely held in the middle 20th century, although Simpson was often more pragmatic than dogmatic about the biological reality of any given taxon.

In *Vertebrate Paleontology* (1st edition 1945, 3rd edition 1966), Alfred Sherwood Romer figured his ideas on mammalian systematics. These were republished in Pietsch (2012, pp. 190–194) for Placentalia, Carnivora, and Artiodactyla (1945) and Rodentia (1966). The classification from Romer (1959) is derived from *The Vertebrate Story* (fourth edition), using his figures for synapsids (p. 231), placentals (p. 237), carnivorans (p. 240), perissodactyls (p. 264), artiodactyls (p. 265), and primates (p. 310).

The final volume (no. 17) of the comprehensive *Traité de Zoologie* (Grassé 1955) consisted of two subvolumes (or "fascicules") dedicated to systematics and comparative anatomy of mammals. Pierre Paul Grassé (1895–1985) was editor of this massive series; its 17 volumes contained from one to seven fascicules and were written by a number of additional authors, published between 1952 and 2007. Grassé (1973) was known for his sympathy for Lamarckism, as described in his book *L'évolution du vivant* (published in English as *Evolution of Living Organisms*). The first figure in the opening chapter of *Traité de Zoologie* vol. 17 fasc. 1 is an outline of what was then understood about the relationships among the extant eutherian orders, but it was composed by the evolutionary paleontologist Jean Piveteau, not Grassé. This figure forms the basis of the cladogram extracted here. Further details on the interrelationships within specific orders follow the classifications given in the subsections therein, such as "ungulates", carnivorans, and whales in part I and "edentates", "insectivorans", primates, glires, and bats in part II.

Novacek (1986) published a classification and figured four different cladograms based on varying combinations of his cranial data set, and he used parsimony as his optimality criterion to figure those cladograms. His Fig. 35 is based on the most characters and is used here to represent his views. Prothero (2007) is a popular exposition of paleontological evidence that demonstrates evolution. In his book, he figures a cladogram of placental orders (p. 285) and others for carnivorans (p. 291), primates (p. 338), artiodactyls (p. 312), tethytheres (p. 324), perissodactyls (p. 309), and "ungulates" (p. 298). Where these differ (e.g., sirenians and hyracoids), I follow his Fig. 13.9 (p. 285).

3.3.4 Defining methodologies

I categorize the studies listed in Tab. 3.1 into one of seven types: ancient, preevolutionary, evolutionary, cladistic, molecular, combined, and correct. Two molecular studies used to define the well-corroborated tree (Meredith *et al.* 2011 and Tarver *et al.* 2016) are defined as "correct", apart from "molecular", in order to measure the latter category without circularity. The two oldest classifications (Leviticus and Al-Jahiz) are defined as "ancient", not "preevolutionary", as they were based more on perceptions of function and human utility than other factors (e.g., anatomy). All of these categories are practical groupings, and I acknowledge that they are not mutually exclusive and that authors in each were not singular practitioners of their respective methodology. For example, McKenna (1975) applied "cladistic" principles to nomenclature, but he built his classifications in a similar manner as "evolutionary" authors such as Simpson (1945).

Authors in this sample from the Renaissance to the mid-19th century used "preevolutionary" criteria to build their classifications. Although such criteria are diverse, and in later cases (e.g., de Blainville 1834; Owen 1868) were probably influenced by proto-evolutionary ideas, even the oldest preevolutionary study (Ray 1693) consisted of more than just dietary prescriptions and superficial functional categories. Ray (1693) recognized that mammals, including humans and whales, breathed air, possessed two ventricles in the heart, and had live birth ("*pulmone respirantia, corde ventriculis praedito duobus, vivipara*"; Fig. 3.2), demonstrating that Ray valued comparative anatomy as a basis for classification. Authors (except Owen) from 1859 through Romer (1959) used "evolutionary" criteria, i.e., they understood their taxonomic groups to reflect snapshots of evolving mammalian lineages. Although some variation exists among these authors (e.g., Grassé as noted above), those behind the classifications used here (Tab. 3.1) largely or entirely endorsed biological evolution in the form of the modern synthesis (Simpson 1944). Importantly, further elaborations and improvements upon it have never disproven its basic mechanism of descent with modification (Wray *et al.* 2014). Moreover, authors in the cladistic, combined, and molecular categories are also "evolutionary" in the broad sense, but they incorporated additional methodological improvements, the utility of which is worth testing here.

McKenna (1975) is widely credited (e.g., Prothero 2007) as the first author to apply cladistic methods to a classification of mammals. In fact, he built his classification in a similar fashion as previous authors, using experience and intuition without applying an optimality criterion to discrete characters. However, unlike Romer or Simpson, McKenna used the now widely accepted standard of naming monophyletic taxa; his classification therefore qualifies as "cladistic". McKenna and Bell (1997) and Prothero (2007) followed suit, whereas other cladistic studies (Novacek 1986, 1992; Shoshani and McKenna 1998) favored cladograms that minimized character change, defined by a matrix of discrete morphological characters. My 2003 study (Asher *et al.* 2003) is also cladistic in both its taxonomy and phylogeny-reconstruction method, but it combines a 22-kb molecular data set with 196 morphological characters and, thus, is categorized as a "combined" analysis. Studies in the "molecular" category also name only monophyletic taxa but differ in that their favored hypotheses are based on DNA characters alone, evaluated primarily with probabilistic optimality criteria (e.g., maximum likelihood, Bayes' theorem) using models of sequence evolution (Felsenstein 2004).

3.4 Results

The main question posed in the introduction was whether or not taxonomies of mammals have become biologically more accurate with an evolutionary understanding of biological process, and with improvements in technologies and methods to recognize patterns. If they have, we would expect that (1) taxonomies should converge toward the currently most well-corroborated hypothesis of mammalian interrelationships, and (2) taxonomies published by authors using improved methods should become more similar to each other over time, as we would expect improved methods to enable taxonomies to more clearly reflect biological reality. By and large, the historical classifications quantified here bear out both expectations. The two periods of time that show the greatest increases in both similarity of historical classifications to the currently best corroborated tree (Fig. 3.3), and in the agreement of authors with each other (Fig. 3.4), are the mid-1800s and the late 1990s to early 2000s. This corresponds to the increased acceptance of evolution post-1859 and the widespread development of systematic methods for molecular data in the late 1990s to early 2000s.

3.4.1 Similarity to the known tree over time

A best fit line to data points falling into each of the methodological categories with adequate samples (preevolutionary, evolutionary, cladistic, and molecular) shows that there is increased accuracy over time in each (Figs. 3.3 and 3.5), although not necessarily with statistical significance (Fig. 3.3). Nonetheless, the preevolutionary trend line is below the accuracy recovered in the four earliest evolutionary studies (Haeckel 1866, Gill 1870, Huxley 1872, and Flower 1883), suggesting that early evolutionary studies performed better than expected based simply on improvement over time. The four least accurate evolutionary studies (Cope 1898, Haeckel 1905; Osborn 1917; Winge 1921) are close to the preevolutionary trend line (Fig. 3.3) extended into the early 20th century. Trend lines for evolutionary and cladistic studies intersect the trend lines of their successors (cladistic and molecular, respectively). This suggests two interpretations: first, that evolutionary studies were fundamentally better at recovering a biologically real mammalian taxonomy than preevolutionary studies (as expected if evolution actually happened), and second, that at least some of the improvements from

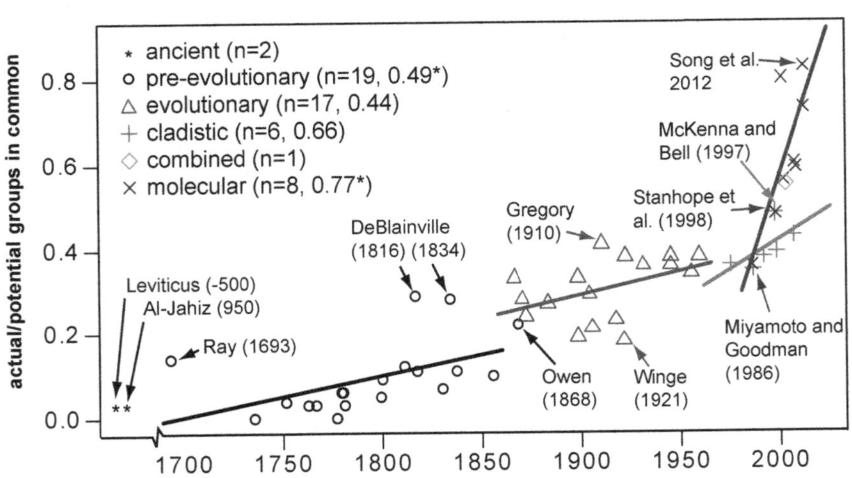

Fig. 3.3: Ratio of actual by potential number of groups (*y* axis) held in common with the well-corroborated tree (Fig. 3.1) by year (*x* axis), distinguishing data points and trend lines for different methods (asterisk = ancient, black circle = preevolutionary, red triangle = evolutionary, green plus = cladistic, aqua diamond = combined, blue *x* = molecular). Selected authors are indicated. Sample sizes (n) and Pearson's correlation coefficients are given for each category in parentheses at top left; parametric significance of the correlation below 0.05 is indicated with an asterisk. Note the discontinuity in the *x* axis prior to 1693.

cladistic and molecular methods might have been accessible to those using just evolutionary and cladistic methods, respectively, at least initially. However, cladistic and molecular methods exhibit steeper trend lines (Fig. 3.3), showing that the rate at which improvements happened using these latter methods, particularly molecular, was faster than rates in preevolutionary and evolutionary methods.

Of the 20 classifications in this study that predate 1859, only de Blainville (1816, 1834) approaches evolutionary classifications in the number of accurately/potentially reconstructed groups. He was the first to recognize that monotremes, marsupials, and placentals are distinct; he also recognized the integrity of glires and primates, and that suiforms and ruminants are closer to each other than to, say, perissodactyls. de Blainville's 1834 classification (Figs. 3.6 B and 3.7) was among the first to recognize seals and walruses (pinnipeds) in a group near felids and canids. His ratios of correct/potential groups (1816, 0.29; 1834, 0.28) are better than several derived from classifications published decades later, including those by Huxley (1872, 0.24), Flower (1883, 0.27), Cope (1898, 0.19), Haeckel (1905, 0.21), Osborn (1917, 0.23), and Winge (1921, 0.18). Otherwise, the most accurate preevolutionary classification is that of Owen (1868, 0.22), followed by John Ray (1693, 0.14).

The most accurate evolutionary classification is Gregory (1910), who identified 23 groups in common out of a maximum possible of 56 (0.41, Fig. 3.6 C). This included strepsirhine and haplorhine primates together to the exclusion of scandentians, pinnipeds closer to caniform than feliform carnivorans, hyracoids near proboscideans and sirenians, and rodents and lagomorphs together in Glires. The most accurate cladistic classification is McKenna and Bell (1997), who recovered 27 out of 56 possible groups

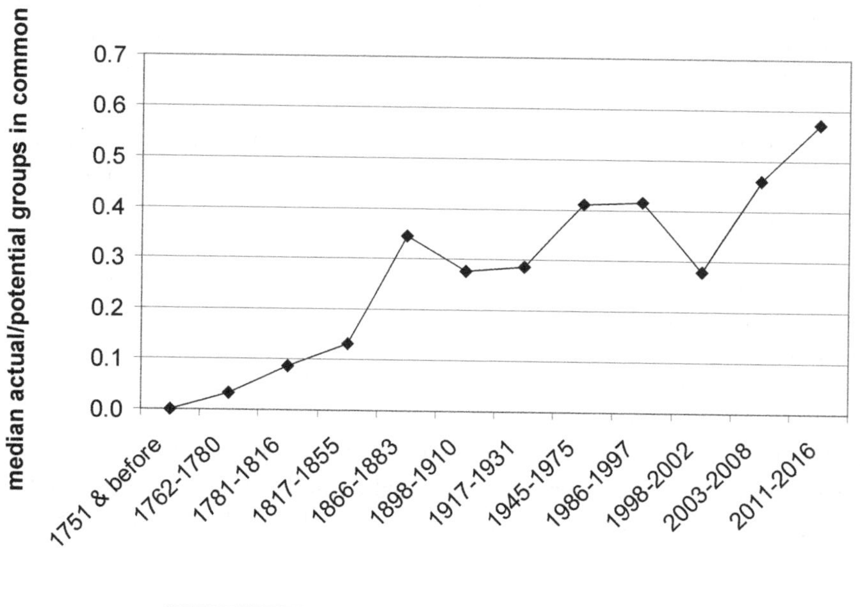

Fig. 3.4: Agreement among authors within 12 time bins, calculated as the median value of each pairwise comparison of actual/potential number of groups in common with each other (Tab. 3.1). Higher values on the *y* axis represent greater levels of inter-author agreement.

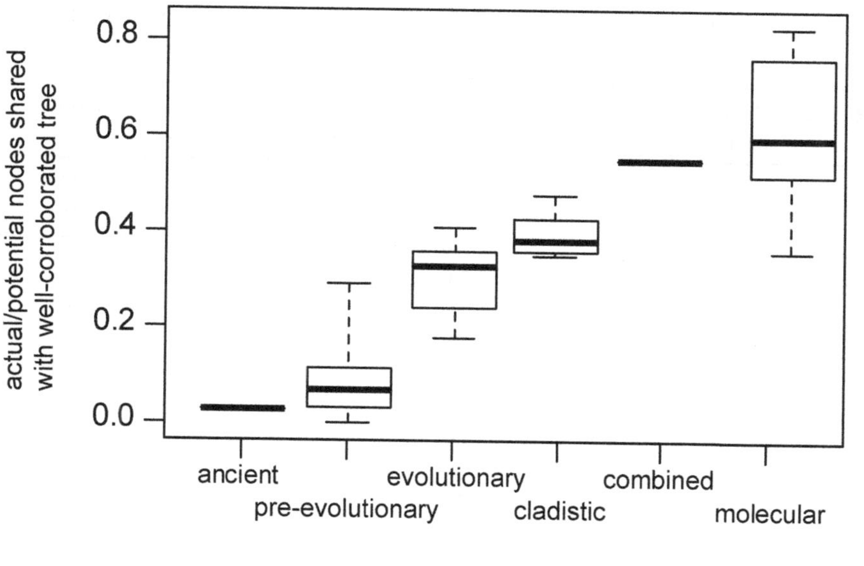

Fig. 3.5: Box plots showing average (thick horizontal line), middle quartiles (box), and range (dashed vertical lines) of ratio of actual/potential number of groups held in common with the well-corroborated tree (*y* axis) across methods. Higher values on the *y* axis represent greater levels of agreement with the well-corroborated tree (Fig. 3.1).

(Fig. 3.6 D). They improved on Gregory (1910) by placing tarsiers among haplorhine primates, pholidotans near carnivorans, dermopterans closer to primates than bats, the common ancestor of xenarthrans close to the placental root, and *Dromiciops* closer to Australian marsupials than to other South American taxa.

3.4.2 Interauthor agreement and methodological improvements

In order to measure the extent to which authors publishing in a given period of time agreed with one another (as would be expected if they were using effective methods and/or an improved understanding of process), I created 12 time bins with four to five classifications each, from 500 BC to 2016 (Tab. 3.3, Fig. 3.4). Figure 3.4 shows the extent to which the authors within each of these time bins agreed with one another by showing the median value of all pairwise combinations of groups held in common within each time bin. For example, from 1866 to 1883, I sampled five classifications (Haeckel, Owen, Gill, Huxley, and Flower). There are 10 pairwise combinations of these studies, ranging from 12 (e.g., Huxley-Flower) to 16 (e.g., Gill-Owen) groups in common with each other. The median of all 10 comparisons is 13 groups in common. When standardized by the ratio of observed to potential groups in common (see Methods), the median is 0.343 groups in common. This is substantially higher than the previous bin (0.129 in 1817–1855) and slightly higher than the following (0.276 in 1898–1910).

The two greatest increases in the median number of groups held in common among authors within each time bin took place in the mid-19th and near the beginning of the 21st century (Fig. 3.4). Two decreases in interauthor agreement also occur: from 1883 into the 20th century and from 1997 into the 21st. These decreases might be explained as follows: Toward the end of the 19th century and into the 20th, authors recognized the need to categorize then-novel and obscure mammalian taxa (e.g., *Solenodon*, *Dromiciops*, *Notoryctes*, macroscelidids, and tupaiids), but they were not sure how to do so. Hence, the number of potentially recognized groups in common with the well-corroborated tree rose in the form of polytomies, reducing the number of resolved groups in common with their peers. Stated differently, resolution generally dropped and polytomies increased from the 1860s into the 1900s. An explanation of the drop in interauthor agreement during the 1990s to early 2000s is that several molecular studies from the 1990s favored a root position for Placentalia within Glires (e.g., Stanhope *et al.* 1998) or erinaceids (Arnason *et al.* 2002). This led to a marked reduction in agreement with other branching diagrams (e.g., Murphy *et al.* 2001) that favored a root position closer to the common ancestors of Afrotheria (elephant, sea cow, hyrax, tenrec, golden mole, sengi, and aardvark) and Xenarthra (sloth, anteater, and armadillo), which resembles the correct root position near the clade uniting these two groups in Atlantogenata (Tarver *et al.* 2016; see Fig. 3.1).

During the time bin (1986–1997) when cladistic methods in nomenclature and morphology dominated, the authors sampled here did not show a greatly increased level of agreement compared with the previous bin (1945–1975; see Fig. 3.4), implying that cladistic methods were not a great improvement over evolutionary ones in terms of enabling investigators to create accurate taxonomies. However, although slight, similarity of cladistic authors' taxonomies to the well-corroborated tree does increase

Tab. 3.3: Median similarity (using "partitions in common" in Mesquite [Maddison and Maddison 2015]) among cladograms to each other, derived from classifications within 12 discrete time bins.

Time	Dominant method	No. studies	Median no. of partitions in common with each other	Median actual/possible partitions in common with each other
1751 and before	Preevolutionary	5	0.0	0.000
1762–1780	Preevolutionary	5	1.0	0.033
1781–1816	Preevolutionary	5	3.0	0.085
1817–1855	Preevolutionary	5	5.0	0.129
1866–1883	Evolutionary	5	13.0	0.343
1898–1910	Evolutionary	5	12.5	0.276
1917–1931	Evolutionary	4	15.5	0.286
1945–1975	Evolutionary	5	23.0	0.411
1986–1997	Cladistic	4	17.5	0.417
1998–2002	Molecular	4	12.5	0.279
2003–2008	Molecular	4	19.0	0.461
2011–2016	Molecular	4	16.0	0.572

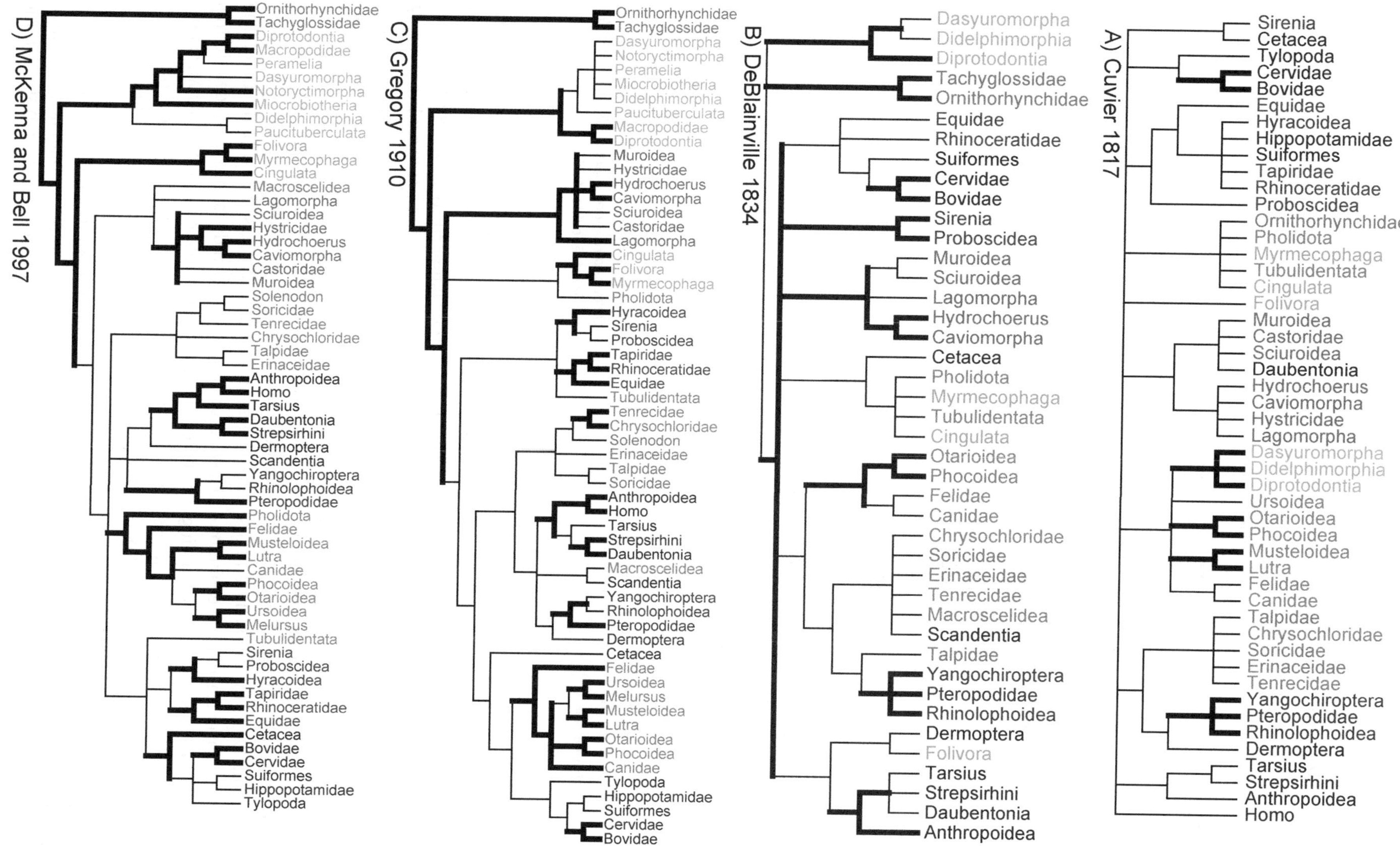

Fig. 3.6: Cladograms extracted from classifications by (A) Cuvier 1817, (B) de Blainville 1834 (see also Fig. 3.7), (C) Gregory 1910, and (D) McKenna and Bell 1997. Selected groups (as defined in Fig. 3.1) are color coded (red = Euarchontoglires, blue = Laurasiatheria, green = Atlantogenata, yellow = monotremes and marsupials). Groups present in the well-corroborated tree (Fig. 3.1) are shown with thick branches.

Sous-classe I. MONADELPHES.

I. *Quadrumanes.*

Normaux, narines rapprochées, *Pitheci* (Chimpanzé, Gibbon et Orang, Cercopithèque, Macaque, Cynocéphale).
Normaux, narines éloignées. . . *Neopitheci* (Atèle, Sajou, Nycticèbe, Ouistiti).
Normaux, ... *Pseudopitheci* (Maki, Indri, Lori, Galago, Tarsier, Aye-Aye).
Anomaux, pour voler. . . . . *Galeopitheci.*
Anomaux, pour grimper . . . . *Tardigradi*, Paresseux.

II. *Carnassiers.*

Claviculés. . . anomaux, pour le vol. *Cheiropteri* (1° Méganyctères, 2° Nyctères (Noctilio, Vespertilio, etc.), 3° Rhinonyctères).
Claviculés. . . anomaux, pour nager. *Orycteri* (Taupes, etc.).
Claviculés. . . normaux. . . . . *Insectivori* (Tenrec, Hérisson, Musaraigne, Macroscélide, Gymnure, Tupaia).
Non claviculés... normaux. . . . . *Plantigradi*, 1° ou omnivores.
Non claviculés... normaux. . . . . *Digitigradi*, 2° Digitigrades.
Non claviculés... anomaux pour nager . . . *Punigradi*, 3° ou Phoques, auxquels on est conduit par les Chiens et les Protèles.

III. *Édentés.*

Claviculés et terrestres, *Brutes*, Tatou, Oryctérope, Fourmilier, Pangolin.
Non claviculés nageurs, *Cétacés*, Dauphin, Narwal, Physéter, Baleine.

IV. *Rongeurs.*

Claviculés. *Grimpeurs*, *Sciurei*, Écureuil, Marmotte, Spermophile, etc.
Claviculés. *Fouisseurs* ou *Murini* : 1° dents molaires, $\frac{2}{2}$ ou $\frac{3}{3}$.
* tuberculeuses.
** non tuberculeuses.
2° dents molaires, $\frac{4}{4}$.
* à replis non compliqués.
** à replis compliqués.
*** lamelleuses.
Subclaviculés, *Coureurs* ou *Lepores*, les Lièvres.
Non claviculés, *Marcheurs* ou *Cavia*, Chloromys, Cavia, Hydrochœrus.

V. *Gravigrades.*

Normaux terrestres, *Proboscidei*, Eléphans.
Anomaux pour nager, *Sirenei*, Lamantins.

VI. *Ongulogrades.*

A doigts. impairs . . . . 3 doitgs, Pachydermes. 1 doigt, Solipèdes : Cheval.
A doigts. pairs, antérieurement au moins 4 doigts, Brutæ : Cochon, etc. 2 doigts, Ruminantia.

Sous-classe II. DIDELPHES.

Ier degré.

A doigts postérieurs libres. *Pédimanes.* *Phascogales.*

IIe degré.

Deux des doigts postérieurs réunis.
Animaux disposés pour grimper, *Phalangers* (Phalanger, Phascolarctos).
Animaux disposés pour sauter, *Sauteurs* (Potoron, Kanguroo).
Animaux disposés pour fouir, *Fouisseurs*, Phascolome.

Sous-classe III. ORNITHODELPHES.

Disposés pour fouir, Échidné.
Disposés pour nager, Ornithorhynque.

Fig. 3.7: Facsimile of de Blainville's 1834 classification, from p. 619 of the *Dictionnaire Pittoresque d'Histoire Naturelle* (Guérin-Méneville 1836).

(Fig. 3.5, Tab. 3.4). Moreover, some time bins contain multiple methods. For example, Miyamoto and Goodman (1986) is a molecular study in a "cladistic" age; Prothero (2007) is a cladistic study in a "molecular" age. The (admittedly small) sample of authors that applied cladistic nomenclature and/or phylogeny reconstruction methods to morphological data (e.g., Shoshani and McKenna 1998) exhibited, on average, higher levels of similarity to the known tree compared with evolutionary authors (e.g., Haeckel 1866; Gadow 1898; Simpson 1931, 1945). Evolutionary authors were in turn more accurate, on average, than preevolutionary studies (e.g., Owen 1868; see Figs. 3.3 and 3.5). Ignoring for a moment the unavoidable non-independence across authors in Tab. 3.1 (see Discussion), and using a non-parametric Wilcoxon rank sum test (excluding the studies in Tab. 3.1 used to define the well-corroborated tree; Meredith *et al.* 2011 and Tarver *et al.* 2016), the better performance of evolutionary, cladistic, and molecular methods over preevolutionary studies is highly significant ($p << 0.01$); the higher accuracy of cladistic over evolutionary studies is suggestive but not significant ($p = 0.068$); the higher accuracy of molecular over evolutionary studies is highly significant ($p = 0.002$); and the higher accuracy of molecular over cladistic studies is suggestive but not significant ($p = 0.100$; see Fig. 3.5, Tab. 3.4).

Molecular studies show considerable overlap with the well-corroborated tree, which is of course based on the largest molecular data sets (see Introduction). The least accurate molecular study is also the oldest in this sample: the analysis of protein sequences by Miyamoto and Goodman (1986), which recovered 13 out of 39 possible (for their sample) well-corroborated groups, yielding a ratio of 0.36. They are followed by the three-gene (12S, 16S, and valine) analysis of mitochondrial RNA of Stanhope *et al.* (1998, 0.48), a combined DNA-morphology study of mine (Asher *et al.* 2003, 0.55), and the

Tab. 3.4: Significance values of a pairwise Wilcoxon rank sum test across mean similarities of cladograms derived from classifications based on different methodologies (with sample size over two) to the well-corroborated tree. Significant differences are shown in bold.

	Molecular	Cladistic	Evolutionary
Preevolutionary	**0.00036**	**0.00191**	**0.00003**
Evolutionary	**0.00159**	0.06846	–
Cladistic	0.10010	–	

mitogenomic analyses of Arnason *et al.* (2002, 0.56; 2008, 0.59) and Kjer and Honeycutt (2007, 0.6). The branching diagram of McCormack *et al.* (2012) is among the most accurate, recovering 11 out of a possible total of 15 partitions (or 0.73). Notably, all of the preceding nucleotide analyses (Miyamoto and Goodman 1986 concern protein sequences) exceed the accuracy of those based on cladistic morphology methods. (Stanhope *et al.* 1998 has a ratio of shared/potentially shared nodes of 0.48387; McKenna and Bell 1997 has 0.48214.) Studies by Stanhope *et al.* (1998), Arnason (2002, 2008), and Kjer and Honeycutt (2007) are based entirely on mitochondrial data; the data set of McCormack *et al.* (2012) is composed of ultraconserved elements (UCEs). These data did not contribute to the two studies of greatest relevance in defining the well-corroborated tree (Meredith *et al.* 2011, Tarver *et al.* 2016; see Fig. 3.1). Song *et al.* (2012) used genomic data that do overlap with the Tarver *et al.* (2016) data set, but they used different phylogenetic methods (coalescence) to identify their optimal branching diagram, which was very similar to the well-corroborated tree (0.83). Because mitochondrial data were used to help define the correct tree for marsupials and rodents by Mitchell *et al.* (2014) and Fabre *et al.* (2012), respectively, and because some investigators (e.g., Song *et al.* 2012) have used genomic data sets that overlap those used to identify the topology in Fig. 3.1 (Meredith *et al.* 2011; Tarver *et al.* 2016), I am compelled to avoid the circular conclusion that accuracy among molecular phylogenies improves over time, when that accuracy is in turn defined by some of those same molecular data. Yet the independence of mitochondrial data (e.g., Kjer and Honeycutt 2007) and UCEs (McCormack *et al.* 2012) from the largely nuclear coding regions behind the well-corroborated tree (Fig. 3.1) enables an only slightly less-strong version of this conclusion, as these independent, molecular studies are clearly more successful in recovering the well-corroborated tree compared with all of the branching diagrams implied or stated by authors prior to 2000, regardless of data or method (Fig. 3.3).

3.5 Discussion

Assuming that the current, well-corroborated hypothesis of mammalian interrelationships (Fig. 3.1) is in fact "the true tree", or reasonably close to it, then there is no question that human taxonomies of mammalian diversity have improved greatly since the authors of Leviticus. Long before the advent of PCR and methods to reconstruct evolutionary history, and even before the widespread acceptance of evolution itself, human classifications of mammalian diversity show improvement (Fig. 3.3). Furthermore, there are correlations in the timing of this improvement with the development of specific advances in biology, particularly the first, clear articulation of the evolutionary mechanism in the mid-19th century, and the widespread application of molecular data to systematics in the late 20th century.

The improvement over time evident among evolutionary studies yields a trend line that fits squarely among cladistic studies in the 1980s, suggesting that other aspects of improved technology and access to pattern may have led to the better accuracy among studies in the 1980s, rather than any intrinsic advantages of cladistic methods. Relatedly, all authors considered here enjoyed insights and methods not considered in this study, and their appreciation of such here-unquantified factors undoubtedly played some role in the observed improvement over time (Fig. 3.3). Notably, at least some of the early 20th century authors behind the classifications shown in Tab. 3.1 were sympathetic to Lamarckian mechanisms (e.g., Grassé 1955 and possibly Winge 1941). Osborn (1917) was notoriously unclear about his preferred evolutionary mechanisms. (However, according to Pietsch [2012], the figure in Osborn's 1917 volume was by Gregory.) The 19th century witnessed greatly improved anatomical knowledge, such as the dissemination of microscopic techniques that facilitated discoveries about previously inaccessible areas such as histology and embryology. All of this would potentially have helped taxonomists reach better conclusions about "natural" groups during the course of the 19th century. Nonetheless, as summarized above, correlations of improvement over time for a given method (Figs. 3.3 and 3.7) show that the advent of evolutionary studies post-1859 was above where accuracy was predicted to have been given the trend evident among preevolutionary studies. Notably, the preevolutionary taxonomist Richard Owen (1868) is also above the trend line derived using other preevolutionary studies, but still below that derived from evolutionary studies (Fig. 3.3).

3.5.1 H.M.D. de Blainville

The accuracy of Henri Marie Ducrotay de Blainville (1816, 1834; Figs. 3.6 B and 3.7) was even more impressive; his 1834 classification recognized nearly twice as many accurate clades compared with Illiger (1811), and nearly three times as many as Cuvier (1817; see Fig. 3.6 A), despite the fact that both Illiger and Cuvier classified more taxa than de Blainville. Gregory (1910, pp. 81, 82) described de Blainville's 1834 work as "brilliant, prophetic" and "far superior to any hitherto met with". As noted above, de Blainville was more accurate than several evolutionary authors, and his ratio of actual to potentially recognized groups is even above an evolutionary trend line extended backward into the 1830s (Fig. 3.3).

de Blainville (1777–1850) was about eight years younger than Georges Cuvier (1769–1832). His personality and relationship to Cuvier may shed light on why he stands out from others of his time. First, although he was initially a student of Cuvier, and spent his career in Paris, de Blainville was too independent-minded to serve as a technician and eventually reached his intellectual independence with some animosity toward his former mentor (Appel 1980). de Blainville was also sympathetic to the "philosophical zoology" of Lamarck and Geoffroy St. Hilaire, who endorsed idealistic theories which Cuvier famously held in low regard (Appel 1987). Thus, de Blainville was keen to set himself apart from Cuvier, who was notoriously anti-evolutionary, or at least anti-Lamarckian (Rudwick 1997).

Moreover, Georges Cuvier was arguably the most politically influential biologist of all time and would have likely exercised substantial influence (deliberately or not) toward contemporaries who lacked de Blainville's apparent independence. Other late 18th and early 19th century taxonomists such as Lacépède (1799), Illiger (1811), and Blumenbach (1830) agreed with Cuvier (1800, 1817) in not recognizing marsupials, monotremes, or placentals as distinctive groups, and in emphasizing hand and foot structure in classification, thereby placing such taxa as didelphid marsupials near primates and/or carnivorans (Fig. 3.6 A). Thus, Blumenbach's "Multungulata" (repeated by Illiger [1811]) resembled Cuvier's "pachydermes ordinaires" in joining tapirs, suiforms, and rhinoceratids with proboscideans and/or hyraxes. These authors also placed one or more aquatic mammalian groups together, variably including sirenians, cetaceans, pinnipeds, ornithorhynchids, and/or castorids. Bonaparte (1837) recognized placental mammals as distinct from monotremes and marsupials, but followed Cuvier in using a "pachyderm"-like group and in placing sirenians with cetaceans and the aye-aye (*Daubentonia*) with rodents (Appendix 1). de Blainville (1816, 1834) made none of these mistakes. It remains for future studies to decipher if this was because of his antipathy toward his senior rival, his affinity for Lamarckian ideas that Cuvier rejected, and/or because of his other insights into comparative anatomy.

3.5.2 Non-independence across studies

In addition to the observation above that some of the molecular studies share data with those used to define the well-corroborated tree, there is further interdependence among the studies considered here. The authors listed in Tab. 3.1 all sought to categorize nature, and any later author would have undoubtedly been influenced by those who came before. de Blainville was a successor of Cuvier, and would also have read Brisson, Linnaeus, Ray, Leviticus, and possibly even Al-Jahiz. All but the authors of Leviticus (ca. 6th century BC) would have potentially been influenced by Aristotle (4th century BC), and later authors will always have had more predecessors by whom to be influenced. In a similar vein, later authors in the modern era would have been more likely to use overlapping data sets, i.e., the same or similar character complexes or genes, to generate their branching diagrams. Indeed, this data set (Tab. 3.1) contains 10 pairs of branching diagrams derived from the same first author (Arnason, Blumenbach, Cuvier, de Blainville, Haeckel, Linnaeus, McKenna, Novacek, Romer, and Simpson). Furthermore, as previously noted, the figure in Osborn (1917) was in fact drawn by Gregory (Pietsch 2012), a long-time protégée of Osborn at the American Museum of Natural History.

Overall, the median number of well-corroborated trees agreed upon by different authors increases over time. However, as noted above, there are dips from 1883 into the 20th century, and from 1997 into the 21st (Fig. 3.4), showing that authors are not simply repeating each other over time. Furthermore, regardless of how authors are influenced by their predecessors in deciding upon their favored classifications, similarity to the now well-corroborated mammalian tree (Fig. 3.1), derived from data that played no role in creating any of the branching diagrams in Tab. 3.1 prior to Miyamoto and Goodman (1986), is a discrete, independent metric by which to judge accuracy. From a purely chronological view, there are spans of time (e.g., 1920–1992) when similarity to the well-corroborated tree does not trend upward (Fig. 3.3). Nonetheless, the average success of classifications based on preevolutionary criteria was lower than those based on evolution, which was in turn lower than those based on cladistic morphology, which was in turn lower than those based on molecular data (Fig. 3.5, Tab. 3.4). This is consistent with the hypothesis that widespread acceptance of the evolutionary process following 1859 led to a genuine improvement in our ability to categorize mammals, as did the increased availability of molecular data starting in the 1990s. A smaller improvement in the ability of classifications to recover the known tree took place with the adoption of cladistic methods in nomenclature, although (as noted above) it is not clear if this improvement is due to cladistics itself or to other advances that took place during the mid-20th century, not directly accounted for in this study.

3.6 Future directions and conclusions

The sample considered here could undoubtedly be improved, for example, by including more than six cladistic morphology studies and more than one pre-1990 molecular study. In addition, I cannot rule out the possibility that authors such as McKenna and Bell (1997) were influenced by the then extensive molecular literature, for example, on comparisons of protein sequences (e.g., Miyamoto and Goodman 1986) and/or DNA hybridization (Kirsch *et al.* 1991). It is difficult to quantify the extent to which McKenna's relatively accurate classification (by 1997 standards) was the result of molecular data sets "leaking" into it. McKenna was himself a student of molecular phylogenetics (McKenna 1991), but his classification was more accurate than (and far from identical to) the protein-sequence branching diagrams published by, for example, Miyamoto and Goodman (1986). McKenna had no reservations about naming groups unsupported by molecular studies (cf. Glires with macroscelidids or hyracoids close to perissodactyls), and he rejected claims made by some early molecular analyses (e.g., paraphyletic Rodentia and Odontoceti). Hence, neither he nor other cladistic authors were simply repeating results from molecular studies in their classifications.

In summary, the success of all evolutionary methods in recovering much of the well-corroborated mammalian tree is significant relative to preevolutionary methods (Fig. 3.5, Tab. 3.4). This is consistent with the hypothesis that the widespread acceptance of evolution in the mid-19th century, and application of further evolutionary methods and data in the 20th, helped mammalian taxonomies to accurately reflect reality, as would be expected if evolution actually happened.

Acknowledgments

I am grateful to Roberto Inchingolo and Adrian Friday for access to literature. I thank Paul Braterman, Joe Felsenstein, Adrian Friday, Nick Matzke, Marcelo Sánchez-Villagra, Caroline Willich, and Frank Zachos for helpful comments on the text, James Montgomery for insightful discussion of Al-Jahiz, and Frank Zachos for his encouragement and patience with this chapter.

Appendix

Appendix 3.1: Treefile in nexus format with topologies derived from all 55 studies listed in Tab. 3.1. This appendix is available online via the following links:
http://people.ds.cam.ac.uk/rja58/trees-appen1.nex
or
https://www.degruyter.com/view/supplement/9783110341553_Appendix_treefile.txt

Literature

Appel, T.A. (1980): Henri de Blainville and the animal series: a nineteenth-century chain of being. *Journal of the History of Biology* 13: 291–319.

Appel, T.A. (1987). *The Cuvier-Geoffroy Debate: French Biology in the Decades before Darwin*. Oxford University Press, USA.

Arnason, U., Adegoke, J.A., Bodin, K., Born, E.W., Esa, Y. B., Gullberg, A., Nilsson, M., Short, R.V., Xu, X. & Janke, A. (2002). Mammalian mitogenomic relationships and the root of the eutherian tree. *Proceedings of the National Academy of Sciences of the United States of America* 99: 8151–8156.

Arnason, U., Adegoke, J.A., Gullberg, A., Harley, E.H., Janke, A. & Kullberg, M. (2008). Mitogenomic relationships of placental mammals and molecular estimates of their divergences. *Gene* 421: 37–51.

Asher, R.J., Bennett, N. & Lehmann, T. (2009). The new framework for understanding placental mammal evolution. *Bioessays* 31: 853–864.

Asher, R.J. & Helgen, K.M. (2010). Nomenclature and placental mammal phylogeny. *BMC Evolutionary Biology* 10: 1.

Asher, R.J., Novacek, M.J. & Geisler, J.H. (2003). Relationships of endemic African mammals and their fossil relatives based on morphological and molecular evidence. *Journal of Mammalian Evolution* 10: 131–194.

Asher, R.J. (2012). *Evolution and Belief: Confessions of a Religious Paleontologist*. Cambridge University Press, Cambridge.

Blumenbach, J.F. (1779). *Handbuch der Naturgeschichte: mit kupfern*. Johann Christian Dieterich, Göttingen.

Bonaparte, C.L. (1837). XVI. A new systematic arrangement of vertebrated animals. *Transactions of the Linnean Society of London* 18: 247–304.

Brisson, M.J. (1762). *Regnum animale in classes IX distributum, sive synopsis methodica sistens classium, quadripedum scilicet & cetaceorum, particularum divisionem in ordines, sectiones, genera & species*. Theodorum Haak, Lugduni Batavorum, Leiden: 296.

Cabrera, A. (1922). *Manual de mastozoología*. Calpe, Madrid.

Cope, E.D. (1898). *Syllabus of Lectures on the Vertebrata*. Published for the University of Pennsylvania, Philadelphia.

Cuvier, G. (1817): *Le règne animale*, 4 vols. Paris: Déterville.

Cuvier, G. (1800): *Leçons d'anatomie comparée*, 5 vols. Baudouin, Paris.

Eisenstein, H. (1991). *Einführung in die arabische Zoographie. Das tierkundliche Wissen in der arabisch-islamischen Literatur*. Dietrich Reimer, Berlin.

Fabre, P.H., Hautier, L., Dimitrov, D. & Douzery, E.J. (2012). A glimpse on the pattern of rodent diversification: a phylogenetic approach. *BMC Evolutionary Biology* 12: 1.

Felsenstein, J. (2004): *Inferring Phylogenies*. Sinauer Associates, Sunderland, MA.

Flower, W.H. (1883): On the arrangement of the orders and families of existing Mammalia. *Proceedings of the Zoological Society of London* 1883: 178–186.

Gadow, H. (1898): *A Classification of Vertebrata, Recent and Extinct*. Black, London.

Gatesy, J., Hayashi, C., Cronin, M.A. & Arctander, P. (1996). Evidence from milk casein genes that cetaceans are close relatives of hippopotamid artiodactyls. *Molecular Biology and Evolution* 13: 954–963.

Gill, T. (1870): On the relations of the orders of mammals. *Proceedings of the American Association for the Advancement of Science* 19: 267–270.

Grassé, P.-P. (1973): *L'évolution du vivant*. Albin Michel, Paris.

Grassé, P.-P. (1955): *Mammifères: les ordres: anatomie, éthologie, systématique*. Tome XVII, Fascicule 1&2. Masson, Paris.

Gregory, W.K. (1910): The orders of mammals. *Bulletin of the American Museum of Natural History* 27: 1–524.

Guérin-Méneville, F.-E. (1836): *Dictionnaire pittoresque d'histoire naturelle et des phénomènes de la nature*. Au Bureau de souscription, Paris.

Haeckel, E. (1866): *General Morphologie der Organismen*. G. Reimer, Berlin.

Haeckel, E. (1905): *Der Kampf um den Entwickelungs-Gedanken*. G. Reimer, Berlin.

Hennig, W. (1966): *Phylogenetic Systematics*. Translated by D. Dwight Davis and Rainer Zangerl. University of Illinois Press, Urbana.

Hunn, E. (1975): A measure of the degree of correspondence of folk to scientific biological classification. *American Ethnologist* 2: 309–327.

Huxley, J. (1957): *New Bottles for New Wine: Essays*. Chatto & Windus, London.

Huxley, T.H. (1872): *Anatomy of Vertebrated Animals*. Appleton, New York.

Illiger, J.K.W. (1811): *Prodomus Systematis Mammalium et Avium Additis Terminis Zoographicii Utriusque Classis Eorumque Versione Germanica*. C. Salfeld, Berlin.

Kirsch, J.A., Dickerman, A.W., Reig, O.A. & Springer, M.S. (1991). DNA hybridization evidence for the Australasian affinity of the American marsupial *Dromiciops australis*. *Proceedings of the National Academy of Sciences of the United States of America* 88: 10465–10469.

Kjer, K.M., Honeycutt, R.L. (2007): Site specific rates of mitochondrial genomes and the phylogeny of eutheria. *BMC Evolutionary Biology* 7: 8.

Klein, J.T. (1751): *Quadrupedum Dispositio Brevisque Historia Naturalis*. BC Bretikopf, Leiden.

Lacépède, B. (1799): *Tableau des Divisions, Sousdivisions, Ordres et Genres des Mammifères, des Cétacés et des Oiseaux*. Plassan, Paris.

Linnaeus, C. von. (1766): *Systema Naturae. Editio Duodecima, Reformata*. Laurentii Salvii, Stockholm.

Maddison, W.P., Maddison, D.R. (2015): *Mesquite: A Modular System for Evolutionary Analysis*. http://mesquiteproject.org

McCormack, J.E., Faircloth, B.C., Crawford, N.G., Gowaty, P.A., Brumfield, R.T. & Glenn, T.C. (2012): Ultraconserved elements are novel phylogenomic markers that resolve placental mammal phylogeny when combined with species-tree analysis. *Genome Research* 22: 746–754.

McKenna, M.C. (1975): Toward a phylogenetic classification of the Mammalia. In W.P. Luckett & F.S. Szalay (eds.): *Phylogeny of the Primates*. Springer, US: 21–46.

McKenna, M.C. (1991): The alpha crystallin A chain of the eye lens and mammalian phylogeny. *Annales Zoologici Fennici* 28: 349–360.

McKenna, M.C. & Bell, S.K. (1997): *Classification of Mammals Above the Species Level*. Columbia University Press, New York.

Meredith, R.W., Janečka, J.E., Gatesy, J., Ryder, O.A., Fisher, C.A., Teeling, E.C., Goodbla, A., Eizirik, E., Simão, T.L.L., Stadler, T. & Rabosky, D.L. (2011): Impacts of the Cretaceous Terrestrial Revolution and KPg extinction on mammal diversification. *Science* 334: 521–524.

Mitchell, K.J., Pratt, R.C., Watson, L.N., Gibb, G.C., Llamas, B., Kasper, M., Edson, J., Hopwood, B., Male, D., Armstrong, K.N. & Meyer, M. (2014): Molecular phylogeny, biogeography, and habitat preference evolution of marsupials. *Molecular Biology and Evolution* 31: 2322–2330.

Miyamoto, M.M. & Goodman, M. (1986): Biomolecular systematics of eutherian mammals: phylogenetic patterns and classification. *Systematic Biology* 35: 230–240.

Montgomery, J.E. (2013): *Al-Jahiz: In Praise of Books*. Edinburgh University Press, Edinburgh.

Murphy, W.J., Eizirik, E., O'Brien, S.J., Madsen, O., Scally, M., Douady, C.J., Teeling, E., Ryder, O.A., Stanhope, M.J., de Jong, W.W. & Springer, M.S. (2001): Resolution of the early placental mammal radiation using Bayesian phylogenetics. *Science* 294: 2348–2351.

Novacek, M.J. (1992): Mammalian phylogeny: shaking the tree. *Nature* 356: 121–125.

Novacek, M.J. (1986): The skull of leptictid insectivorans and the higher-level classification of eutherian mammals. *Bulletin of the American Museum of Natural History* 183: 1–112.

Numbers, R.L. (2006): *The Creationists, Expanded Edition*. Harvard University Press, Cambridge, MA

Osborn, H.F. (1917): *The Origin and Evolution of Life*. Scribners, New York.

Owen, R. (1868): *On the Anatomy of Vertebrates, vol. 3 Mammals*. Longman Green, London.

Patterson, C. (1988): The impact of evolutionary theories on systematics. In D.L. Hawksworth (ed.): *Prospects in Systematics*. Clarendon Press, Oxford: 59–91.

Pennant, T. (1781): *History of Quadrupeds*. B. White, London.

Penny, D., Foulds, L.R., Hendy, M.D. (1982): Testing the theory of evolution by comparing phylogenetic trees constructed from five different protein sequences. *Nature* 297: 197–200.
Pietsch, T.W. (2012): *Trees of Life: A Visual History of Evolution.* Johns Hopkins University Press, Baltimore.
Prothero, D.R. (2007): *Evolution: What the Fossils Say and Why It Matters.* Columbia University Press, New York.
R Core Team (2014): *R: A Language and Environment for Statistical Computing.* R Foundation for Statistical Computing, Vienna, Austria.
Ray, J. (1693): *Synopsis Methodica Animalium Quadrupedum et Serpentini Generis.* Smith & Walford, London.
Rieppel, O. (1986): Species are individuals. *Evolutionary Biology* 20: 283–317.
Romer, A.S. (1959): *The Vertebrate Story.* University of Chicago Press, Chicago, IL.
Romer, A.S. (1945): *Vertebrate Paleontology.* University of Chicago Press, Chicago, IL.
Rudwick, M.J.S. (1997): *Fossil Bones & Geological Catastrophes.* University of Chicago Press, Chicago, IL.
Scopoli, G.A. (1777): *Introductio ad historiam naturalem: sistens genera lapidum, plantarum, et animalium hactenus detecta, caracteribus essentialibus donata in tribus divisa, subinde ad leges naturae.* Wolfgang Gerle, Prague.
Shoshani, J. & McKenna, M.C. (1998): Higher taxonomic relationships among extant mammals based on morphology, with selected comparisons of results from molecular data. *Molecular Phylogenetics and Evolution* 9: 572–584.
Simpson, G.G. (1931): A new classification of mammals. *Bulletin of the American Museum of Natural History* 59: 259–293.
Simpson, G.G. (1944): *Tempo and Mode in Evolution.* Columbia University Press, New York.
Simpson, G.G. (1945): The principles of classification and a classification of mammals. *Bulletin of the American Museum of Natural History* 85: 1–350.
Sober, E. (2008): *Evidence and Evolution: the Logic behind the Science.* Cambridge University Press, Cambridge.
Song, S., Liu, L., Edwards, S.V. & Wu, S. (2012): Resolving conflict in eutherian mammal phylogeny using phylogenomics and the multispecies coalescent model. *Proceedings of the National Academy of Sciences of the United States of America* 109: 14942–14947.
Stanhope, M.J., Waddell, V.G., Madsen, O., de Jong, W., Hedges, S.B., Cleven, G.C., Kao, D. & Springer, M.S. (1998): Molecular evidence for multiple origins of Insectivora and for a new order of endemic African insectivore mammals. *Proceedings of the National Academy of Sciences of the United States of America* 95: 9967–9972.
Storr, G.C.C. (1780): *Prodromus methodi mammalium ... inauguralem disputationem.* F. Wolffer, Tübingen.
Stott, R. (2012): *Darwin's Ghosts: the Secret History of Evolution.* Bloomsbury, London.
Tarver, J.E., dos Reis, M., Mirarab, S., Moran, R.J., Parker, S., O'Reilly, J.E., King, B.L., O'Connell, M.J., Asher, R.J., Warnow, T., Peterson, K.J., Donoghue, P.C.J. & Pisani, D. (2016): The interrelationships of placental mammals and the limits of phylogenetic inference. *Genome Biology and Evolution* 8(2):330–344.
Teeling, E.C., Springer, M.S., Madsen, O., Bates, P., O'Brien, S.J. & Murphy, W.J. (2005): A molecular phylogeny for bats illuminates biogeography and the fossil record. *Science* 307: 580–584.
Van Valen, L. (1978): Why not to be a cladist. *Evolutionary Theory* 3: 285–299.
Wagner, J.A. (1855): *Schreber's Die Säugethiere*, suppl. Vol. 5. Weigel, Leipzig.
Weber, M.W.C. (1904): *Die Säugetiere: Einführung in die Anatomie und Systematik der Recenten und Fossilen Mammalia.* Gustav Fischer, Jena.
Winge, H. (1941): *The Interrelationships of the Mammalian Genera.* CA Reitzels, Copenhagen.
Wray, G.A., Hoekstra, H.E., Futuyma, D.J., Lenski, R.E., Mackay, T.F.C., Schluter, D. & Strassmann, J.E. (2014): Does evolutionary theory need a rethink? *Nature* 514: 161–164.

Ingmar Werneburg, Stephan N. F. Spiekman

4 Mammalian embryology and organogenesis

4.1 Historical introduction

The origin of the human being is one of the major philosophical questions. One way of answering it is through an ontogenetic and evolutionary approach. Theories about the developmental origin in the mother's womb were first documented by Greek scholars in the 5th century BC (Breidbach 2015). Being dependent on the success of reproduction, fertility has triggered human thinking for a much longer period of human evolution as is documented in the art of the Paleolithic, the Mesolithic, the Neolithic, and later phases of human cultural evolution (Fig. 4.1). Fertility was manifested by the representation of stout mother goddesses with a demonstrative focus on their hips and breasts (Fig. 4.1 C) and by the presentation of erect penises in men (Fig. 4.1 G). In some cases, the process of birth was depicted as a key event of life (Fig. 4.1 D). Birth, pregnancy, and lactation of other vertebrate species were also of interest to ancient cultures and were illustrated, particularly, among species that played an important role in everyday life, such as prey and domesticated, mostly mammalian, animals (Fig. 4.1 A and I). Fertility rites, including hierogamy (Fig. 4.1 H) and sodomy (Fig. 4.1 J), are thought to have played a major role in prehistoric cults and were meant to increase the number of offspring and to guarantee their health (Frazer 1922, Campbell 1960).

These attempts to understand and to influence our own development culminated in the artificial representation of homunculi in the Renaissance age, in which a human was "created" by alchemists in a jar. The *mental* construct of a test-tube baby (Fig. 4.1 K) was a symbol for the development of the human mind itself, which alchemists tried to raise to a higher level of cognition (Gebelein 1996, Wiesing 2004). The historical representation of homunculi also corresponds to the ideas of preformationists in the 17th century that the human being is fully present as a miniature in the male's sperm and only has to unfold to progress through embryological development to a fully formed human (Fig. 4.1 F; Hartsoeker 1694).

With the rise of evolutionary thinking (Bell *et al.* 2010), a broad scientific discourse about the unique anatomy and development of mammalian species emerged. Mammalian embryology received particular attention in comparative anatomy and physiology because of its relevance in understanding our own biological history (Haeckel 1874, 1877, Keibel 1906).

Recent scientific approaches try to manipulate the early embryology of mammals, including man, to prevent diseases (Gilbert 2006) or to select for particular forms through domestication (Clutton-Brock 1999). Experimental biological studies mainly focus on model organisms that are easy to breed, have a large number of offspring, and have short reproductive cycles. For placental embryology, the house mouse *Mus musculus* is most often used (Cook 1965, Green 1968, Theiler 1989, Hedrich 2012). Among marsupials, the opossum *Monodelphis domestica* and the wallaby *Macropus eugenii* are two of the most important model organisms (e.g., Smith 2002, Hickford *et al.* 2009, Keyte and Smith 2009, Rousmaniere *et al.* 2010). However, recent studies have highlighted the value of non-model organisms for testing the taxonomic applicability of experimental studies, and more evolutionary questions are being asked that require extensive knowledge of non-model organisms all over the tree of life (Butler and Juurlink 1987, Jenner and Wills 2007, Milinkovitch and Tzika 2007, Tzika and Milinkovitch 2008), including mammals (Werneburg *et al.* 2013).

The aim of the present chapter is to provide a general overview of the main developmental differences among the three major mammalian groups, to characterize mammals among amniotes, and to present some crucial aspects of the anatomy and physiology of mammalian embryogenesis. We cover different aspects of embryology in a chronological order. We first discuss particular attributes of monotremes, followed by marsupials and lastly placental mammals. We also focus on the developmental peculiarities in humans. Finally, we summarize some discussions on the potential relationships between life history parameters and evolutionary patterns. The selection of topics is mainly anatomically oriented and reflects our personal morphological interest. It does not aim to represent an in-depth review of all aspects of mammalian embryology, nor does it aim to be complete. As new data, we present a reconstruction of the ancestral developmental sequence of external organ characters for Placentalia, which is based on a phylogenetic comparison of about 80 tetrapod species.

https://doi.org/10.1515/9783110341553-004

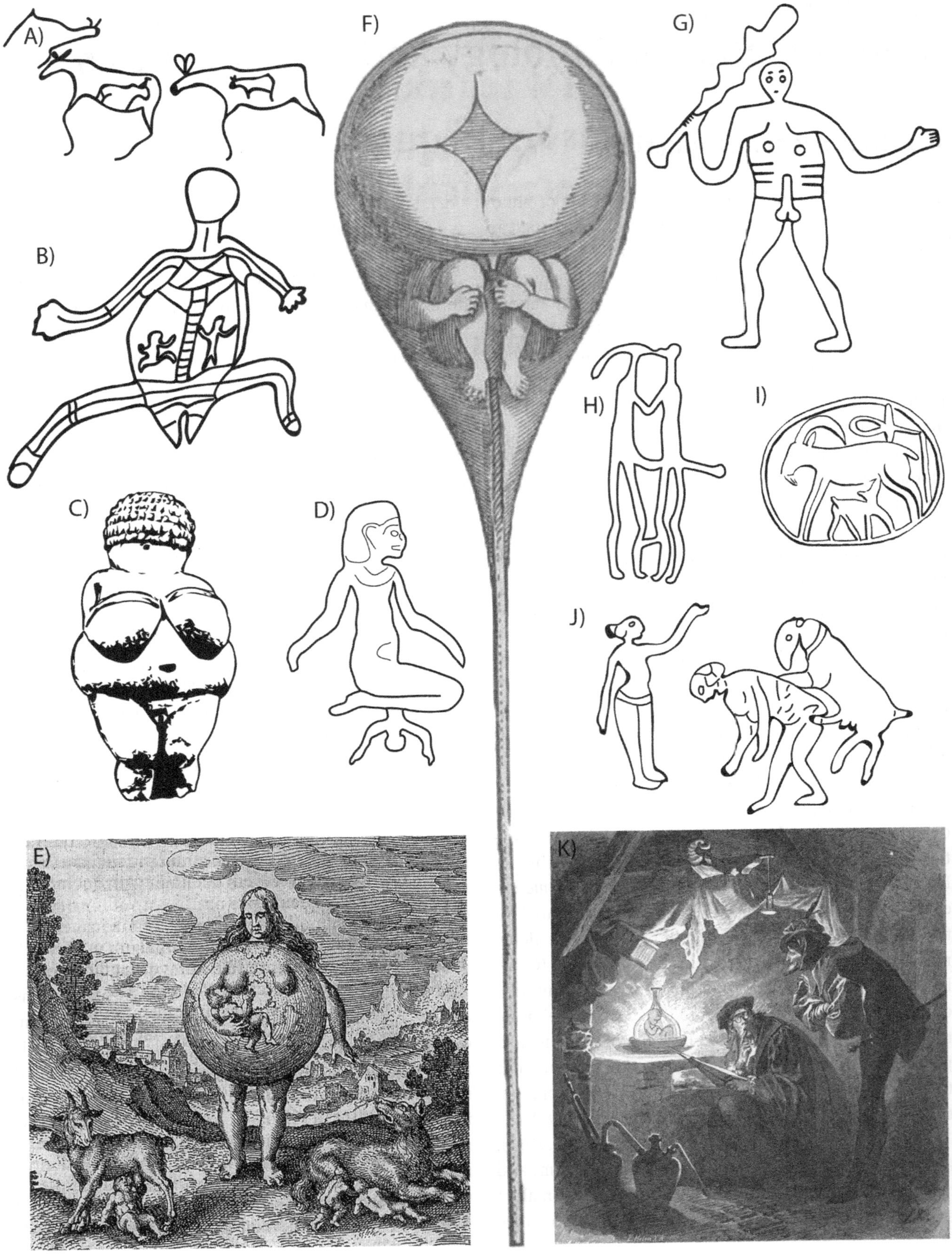
A)
B)
C)
D)
E)
F)
G)
H)
I)
J)
K)

4.2 An overview of mammalian embryology

Mammals are distinguished from sauropsids by a number of autapomorphic characters, which are documented in stepwise evolutionary transformations in the synapsid stem lineage leading to extant mammals (Luo 2007, Chinsamy-Turan 2011). Many of these distinguishing characters relate to life history and mode of reproduction (Sánchez-Villagra 2010).

Extant mammals are characterized by the presence of complex glands in the skin including milk glands to nourish the young. Early suckling behavior of therian mammals resulted in a complex developmental modification of the skull, mainly in the areas surrounding the mouth cavity (Maier 1999). Lips and cheeks (exclusively mammalian characters) are fully formed at birth. In addition, the hyoid is also fully developed and is formed by the hyoid bone and the first branchial (first posthyoidal) arch. This structure functions as a strong cartilaginous bar at which the muscles of the tongue (hyoglossus), as well as other important muscles of the mouth and throat (e.g., sternohyoideus, thyrohyoideus, and geniohyoideus), insert. The tongue is the most important structure for suckling and is well developed in all mammalian neonates (newborns). The discovery of mammalian-like hyoids in synapsid ancestors (Kemp 2006) indicates that its general morphology is ancestral to all mammals. Furthermore, the complex mammalian larynx has important respiratory functions in altricial neonates, the development of which is enabled by the presence of three posthyoideal branchial arches, compared with only two in sauropsids.

The formation of a secondary palate can also be observed in Permian synapsids (Maier *et al.* 1996), and it most likely evolved to enable continued breathing during suckling in neonates. Although most of the secondary palate is formed by bone (hard palate), the posterior part is formed by soft tissue (velum or soft palate), which can be moved by muscles during suckling. In extant mammals, the importance of the secondary palate is highlighted by the problems that occur in neonates with misformed secondary palates.

Together, cheeks, tongue, and secondary palate restricted the movement of food inside the mouth, which may have favored the development of advanced chewing in mammals. That resulted in highly specialized postcanine teeth such as the tribosphenic molar. This specialized dentition, which is now believed to have played a very important role in the evolutionary success of mammals, is therefore probably also indirectly the result of adaptations to suckling.

Another important morphological feature of nearly all mammals is their non-kinetic skull, which allows for relatively stronger bite forces. This akinetism (such as solid fusion of the upper jaw articulation to the braincase) is related to the formation of the alisphenoid bone, which braces the surrounding elements of the skull and forms a secondary lateral skull wall (Maier 1987).

Finally, mammalian crania are also distinguished from those of other amniotes by the presence of an extensive network of facial muscles, which are innervated by the facial nerve (n. VII). In sauropsids, facialis-innervated muscles are restricted to the throat and neck area only (Maier 1999).

It has been hypothesized that specialized maternal care and an increased activity of the animal were linked to an increased anaerobic threshold and an increased aerobic activity in mammalian ancestors. Endothermy was a key feature to enable maximal aerobic metabolism. As a result, the basal metabolism of these animals would have increased, which was an essential prerequisite for the evolution of almost all derived features known for mammals. Correlated to endothermy, important transformations developed in all organ systems, which are crucial for optimized breathing (e.g., diaphragm), exploration of the environment (e.g., ear, nose, and brain anatomy), and food uptake (e.g., skull stiffening, jaw, and teeth anatomy) (Kemp 2006, Ruf 2014).

Many studies have illustrated the different anatomical transformations through ontogeny that have led to the peculiar morphotype of mammals. One of the classic examples is the evolution and development of the middle ear, which is highly derived in mammals. One auditory ossicle is derived from the ancestral tetrapod ear bone, the

◂ **Fig. 4.1: Prehistorical and historical illustrations.** (A) Two pregnant elks; rock carvings from Alta/Norway (Mesolithic age). (B) Aboriginal X-ray style figure of the pregnant Mother of Creation, Northern Territory, Australia (redrawn from a photograph). (C) Venus from Willendorf/Austria (Upper Palaeolithic, about 29,500 years old). (D) Egyptian hieroglyph of a woman giving birth; relief from the temple of Kom Ombo (redrawn from a photograph). (E) "Nutrix ejus terra est" (The Earth is his nurse) from Maier (1617/1618). (F) Homunculus in a sperm by Hartsoeker (1694). (G) Cerne Abbas Giant with erect penis, a turf-cut outline filled with chalk; Dorset/United Kindom, uncertain age. (H) Hierogamy in the rock carvings of Bohuslän/Sweden. (I) Suckling caprid; seal from Sidon/Lebanon (8th century BC), redrawn after Keel (2013). (J) Zoophilia in a rock carving at Kedareshwara Temple, Belligavi/India, redrawn from a photograph. (K) Homunculus develops in a jar; a scene from Goethe's Faust with Wagner (sitting) and Mephistopheles (von Goethe 1899).

columella (= stapes in mammals), and several other middle ear components are derived from bones of the ancestral lower jaw of early synapsids. Associated with that, a secondary jaw joint has formed between the remaining dermatocranial bones in the posterior jaw region, the dentary and squamosal (Luo 2007, Ji *et al.* 2009, Martin and Ruf 2009, Maier and Ruf 2016). Another famous case study concentrates on the ontogenetic emergence of the secondary lateral skull wall formed by the alisphenoid, which resulted from the modification of the ancestral tetrapod epipterygoid (Maier 1987).

More recent ontogenetic studies concentrate on taxonomic differences in the developmental timing of embryonic characters, and ancestral sequences of character appearance are reconstructed (Sánchez-Villagra *et al.* 2008, Weisbecker *et al.* 2008, Hautier *et al.* 2010, 2011, Werneburg *et al.* 2016, Spiekman and Werneburg 2017). One of those studies discovered an early appearance of the supraoccipital bone in the ontogeny of all mammals, which can be correlated to the increased growth and expansion of the mammalian brain when compared with non-mammalian vertebrates (Koyabu *et al.* 2014).

All those studies concentrate on the correlation between ontogenetic and phylogenetic transformations. As such, comparative studies on organismic aspects of embryology help in understanding morphological aspects of mammalian evolution.

Although embryonic development usually is defined to start with fertilization, its finalization is rather difficult to define. The term "embryology" is often restricted to intra-egg or intrauterine development of an animal, whereas birth or hatching are usually defined as the end point of embryonic life. A further distinction between embryonic and fetal developmental phases can be made prior to birth or hatching.

The start of fetal development is usually assigned to the point of development when most organs are already established and recognizable and extensive allometric growth of the animal occurs. Given the historical lack of a clear definition, we propose the disappearance of the mesencephalic vaulting of the brain in the embryo as a homologous landmark and define it as the end of embryonic development and as the start point of fetal growth (see also Werneburg *et al.* 2016). At this point in development, the brain and the bones of the cranial vault are well developed; i.e., the frontal and parietal bones of both body sides have started closure.

However, both during fetal development and after hatching or birth, organs such as claws and hairs in mammals continue to emerge, making the end point of development hard to define.

At birth or hatching, most mammals are extremely immature (Tabs. 4.1–4.3). Compared with other vertebrates, one can state that, in most cases, development continues after birth through the processes of lactation and weaning, which can take many months. After that, the animal might still rely on the help of the mother to survive, and important ontogenetic changes still occur, including the replacement of milk teeth. The maturation of gametes can be largely delayed, and some other permanent organs first develop at or after sexual maturation (e.g., antlers). To a certain extent, the animals still grow after reaching sexual maturation. In contrast to sauropsids, however, growth usually stops at a particular age in mammals, a fact that can be recognized via the fossil record (O'Meara and Asher 2016).

We follow a holistic organism concept, in which ontogeny is a process that starts at fertilization and ends with the death of the organism (Maier 1999, Maier and Werneburg 2014). An organism has to be understood as an integration of different organs and processes that change their characteristics over time. Factors that influence those changes include anatomical, functional, and ecological demands, which need to be considered in any integrative morphological and evolutionary study. As such, in the present overview on mammalian "embryology", we consider different aspects that influence embryonic development such as reproduction-related morphology of the parents and early life of the organisms after hatching or birth. Body size, litter size, and gestation length are dependent on different anatomical and physiological conditions of the mother, which also need to be considered for a comprehensive discussion on mammalian embryology.

Although embryonic or fetal development is usually concealed to general human awareness, postnatal life is more obvious. Three major extant mammalian groups can be distinguished. The Monotremata (i.e., as crown group of "Prototheria") form the sister group to Theria, which are composed of Marsupialia (i.e., as crown group of Metatheria) and Placentalia (i.e., as crown group of Eutheria). The three groups drastically differ in their mode of reproduction and life history (Tabs. 4.1–4.3). The first and most obvious difference is that monotremes have an oviparous (egg-laying) lifestyle, whereas marsupials and placentals are viviparous (live bearing). Second, marsupials have extremely immature so-called altricial neonates, whereas the other groups show a slightly (monotremes) or largely (placentals) more developed anatomy at birth with precocial groups as extremes among Placentalia. The terms altriciality (e.g., naked and blind and dependent on parents for food) and precociality (e.g., early independent, nidifugous) were coined by Haeckel (1866) and are

Tab. 4.1: Comparison of important life history features in sauropsids and the three major mammalian groups.

Taxon	Sauropsida (in general)	Monotremata	Marsupialia	Placentalia
Extant diversity	>20.000 described species	1 platypus, 4 echidna species	>330 described species	>4.500 described species
Endothermic	Only Aves (possibly some non-avian dinosaurs)	Yes	Yes	Yes
Cloaca	Yes	Yes	No	Rare
Perineum	No	No	Yes	Yes
Reproduction	Primarily oviparous (egg laying)	Oviparous (egg laying)	Viviparous (live birth)	Viviparous (live birth)
Scrotum	No	No	Yes	Yes
Eggs	Polylecithal	Polylecithal	Alecithal	Alecithal
Cleavage	Meroblastic-discoidal	Meroblastic-discoidal	Holoblastic-rotational	Holoblastic-rotational
Placenta	Some squamates	No	Yes	Yes
Placenta type	If existing: choriovitelline	–	Mainly choriovitelline	Choriovitelline and/or allantochorial
Embryonic nutrition	Yolk, placenta in some squamates	Histotrophs, yolk	Histotrophs, placenta	Placenta
Gestation length	Medium	Short	Short	Long
Maturity level at delivery	Generally precocial	Altricial	Altricial, neonates have to crawl to mother's teat	Wide range from altricial to precocial neonates
Mammary glands	No	Yes	Yes	Yes
Teats	No	No	Yes	Yes
Uniform milk composition during lactation	–	Yes	No	Yes
Pouch	No	Only in echidna	Yes	No
Pouch bones	No	Yes	Yes	No
Continued growth throughout life	Yes	No	No	No

derived from the Latin words altrix (i.e., foster mother or dependent on care) and praecox (i.e., precipitate, in relation to the mechanisms of the nervous system and the locomotory apparatus).

After birth or hatching, mammals show a great investment in parental care. This is particularly visible in the form of guarding and lactation. After weaning, food is often supplied to the young.

Early development of monotremes is difficult to study because the animals only have a small litter size of one or two young. Nowadays, they are protected by national conservation laws in Australia and New Guinea, and it is difficult to access new embryos. One famous researcher of monotreme development was Richard Semon (1859–1918) from Jena, Germany (Hoßfeld and Olsson 2003a), who collected a large egg sample of the short-beaked echidna *Tachyglossus aculeatus* and described the early development of this species (Semon 1894a–d, 1897a, b, 1904; expanded by Werneburg and Sánchez-Villagra 2011). This material was used as a source for further developmental studies of particular organs (e.g., Klaatsch 1895). Detailed studies on the earliest development were performed by James Peter Hill (1873–1954), whose monotreme and marsupial collections are currently stored at Museum für Naturkunde Berlin, Germany (Richardson and Narraway 1999, Giere and Zeller 2006).

In their reproductive biology, monotremes represent a mosaic of ancestral amniote, derived mammalian, and unique monotreme characters (Griffiths 1989, Werneburg and Sánchez-Villagra 2011), as mirrored in discoveries of whole-genome studies (Warren *et al.* 2008). Like all amniotes, therian mammals have an internal fertilization, but they evolved a specialized reproductive system. Therian cleavage is unique because it develops a specialized cell population in the embryo that allows integration in the uterine tissue of the mother. With this integration, a new organ, the placenta, is formed that permits elongated embryonic development in the mother's womb. Intrauterine development enables the viviparity of therian mammals. No extrauterine egg is needed.

Like placental mammals, marsupials are viviparous. However, they give birth to highly altricial and immature young and much of organogenesis occurs during lactation. The gestation, during which maternal nutrition is supplied by a placenta, is very short (1–2 weeks). To overcome the extreme conditions to which the neonate is exposed at birth, marsupials evolved highly specialized anatomical adaptations.

In contrast to marsupials, placental mammals have, on average, a relatively longer gestation length, and the young are born at a more developed stage. There is a wide range between altricial and precocial conditions at birth. However, all placentals are born at a more developed stage than marsupials, so there is no overlap between the altricial state found in marsupials and that in placentals. Placentals are characterized by a great extant diversity and include arboreal, fossorial, terrestrial, gliding, flying, diving, and permanently aquatic forms. All morphological adaptations are mirrored by the relative timing of related prenatal character development (Müller 1973). These changes in timing result in different relative sizes and degrees of differentiation in the adults (Werneburg *et al.* 2015). For example, the fossorial mole or the bats develop their specialized and large forelimbs relatively early when compared with other placental mammals (Sears 2006, 2008, Richardson *et al.* 2009, Mitgutsch *et al.* 2012, Bickelmann *et al.* 2012, Koyabu *et al.* 2014).

Relatively few organic changes appear in later phases of life (e.g., aging processes and temporary changes) (Portmann 1944), although changes in bone density and shape can be traced quantitatively (e.g., Wilson 2011, Urban *et al.* 2016). These aspects of postnatal development are not discussed in the present overview.

4.3 Reproduction and related organs

Unlike marsupials and placentals, monotremes are oviparous. Like sauropsids (and a few placental mammals), they have a cloaca (Romer 1976). The cloaca is a single opening ("Monotremata" from Greek μόνος and τρῆμα, "single hole") in which feces, urine, and eggs are deposited before leaving the body. During embryonic development of therian mammals, the opening of the urogenital system gets separated from the anus (Fig. 4.2 A–D) (Liem *et al.* 2001). Monotremes show an intermediate anatomy in this regard. Although they possess a cloaca (Fig. 4.3 D), like sauropsids, the proximal part is separated into a rectal region (coprodaeum) and a more ventral urodaeum, which contains the urinal and genital products (Fig. 4.2 E–F and Q). Therian mammals have a perineum, which separates the anus and the urogenital tract (Fig. 4.2 G, L, and R–S).

In marsupials, a small diverticulum represents the last trace of the cloaca. It is fully reduced in most placental mammals (Riedelsheimer *et al.* 2007), and the coprodaeum (cranial-most part of the ancestral cloaca) becomes the terminal part of the gut and opens into the anus (Fig. 4.2 S).

In monotremes and marsupials, the whole urogenital system is paired. The ovaries of the female monotremes resemble those of the sauropsids. In the monotreme species *Ornithorhynchus anatinus*, the platypus, only the left ovary is functional (Fig. 4.2 F; in birds the right one is functional). In the right oviduct, which is about the same size as the left one, the gamete only reaches the ovocyst stage (Zeller 2004a).

In the monotreme species *T. aculeatus*, the short-beaked echidna, both ovaries are evenly developed (Fig. 4.2 E). The egg is transported to one of the paired uteri, which open separately into the sinus urogenitalis that enters the cloaca (Greven 2004). The egg then enters the single urogenital sinus and then the cloaca (Fig. 4.2 E). A vagina is absent.

In marsupials, the paired uteri open into the vaginal sinus (Fig. 4.2 L–P). The Müllerian ducts do not fuse during development, which results in two vaginas (didelphy) (Starck 1995). Their posterior parts are fused in certain groups (Macropodidae). The anterior parts of the vaginae are sometimes fused to an unpaired sinus vaginalis and enter a united urogenital canal (Fig. 4.2 N; Greven 2004). In peramelids, the sinus vaginalis forms an extended anterior diverticulum ("caecum vaginalia"), which can serve as a spermatozoa reservoir (Fig. 4.2 P). In many forms,

▸ **Fig. 4.2: Reproductive organs.** (A–D) Lateral diagrams of the division of the embryonic cloaca in placental mammals (sexually homologous urethral segments are identified by the same number): (A–B) early and later sexually indifferent stages, (C) differentiation of the female, (D) differentiation of the male. (E) Genital apparatus of the female short-beaked echidna, *Tachyglossus aculeatus*. (F) Female organs of the monotreme platypus, *Ornithorhynchus anatinus*. (G–K) Genital apparatus of female placental mammals: (G) general structure of the female apparatus in a placental mammal, (H) uterus bicornis, (I) uterus bipartitus, (J) uterus simplex, (K) uterus duplex. (L–P) Genital apparatus of female marsupial mammals: (L) general structure of the female apparatus in a marsupial mammal; vaginal modification in (M) the large American opossum *Didelphis*, (N) the kangaroo *Macropus*, (O) the rat-kangaroo *Hypsiprymnodon*, and (P) bandicoots (Peramelidae). (Q–S) Genital apparatus of males: (Q) monotreme, (R) marsupial, (S) placental mammal condition. A–D after Liem *et al.* (2001); E, G, L after Starck (1995) and Mickoleit (2004); F after Grant (2013); H–K after Romer (1976); M–P after Starck (1995), Pough *et al.* (2012), and Mickoleit (2004); Q–S after Mickoleit (2004) and Liem *et al.* (2001).

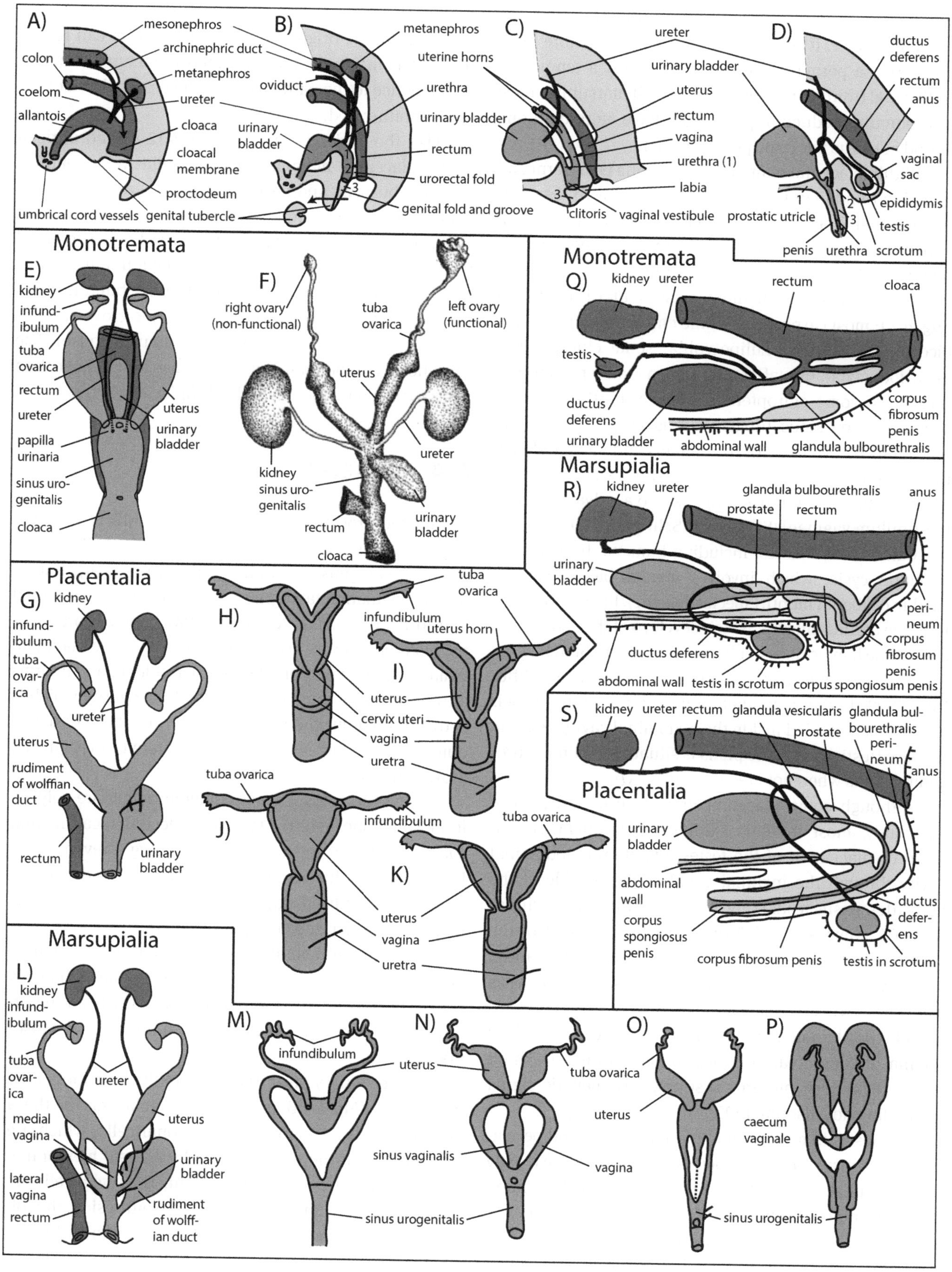
A)
mesonephros
archinephric duct
metanephros
colon
coelom
allantois
ureter
cloaca
cloacal membrane
proctodeum
umbrical cord vessels
genital tubercle
B)
oviduct
urinary bladder
metanephros
uterine horns
urethra
urinary bladder
rectum
urorectal fold
genital fold and groove
C)
ureter
urinary bladder
uterus
rectum
vagina
urethra (1)
labia
clitoris
vaginal vestibule
D)
ductus deferens
rectum
anus
vaginal sac
epididymis
testis
prostatic utricle
penis
urethra
scrotum
Monotremata
E)
kidney
infund-ibulum
tuba ovarica
rectum
ureter
papilla urinaria
sinus uro-genitalis
cloaca
uterus
urinary bladder
F)
right ovary (non-functional)
tuba ovarica
left ovary (functional)
uterus
kidney
sinus uro-genitalis
ureter
rectum
urinary bladder
cloaca
Monotremata
Q)
kidney
ureter
rectum
cloaca
testis
ductus deferens
urinary bladder
abdominal wall
glandula bulbourethralis
corpus fibrosum penis
Marsupialia
R)
kidney
ureter
glandula bulbourethralis
prostate
rectum
anus
urinary bladder
peri-neum
corpus fibrosum penis
ductus deferens
abdominal wall
testis in scrotum
corpus spongiosum penis
Placentalia
G)
kidney
infund-ibulum
tuba ovar-ica
ureter
uterus
rudiment of wolffian duct
rectum
urinary bladder
H)
tuba ovarica
infundibulum
uterus horn
I)
uterus
cervix uteri
vagina
uretra
tuba ovarica
J)
infundibulum
tuba ovarica
K)
uterus
vagina
uretra
S)
kidney
ureter
rectum
glandula vesicularis
glandula bul-bourethralis
prostate
peri-neum
anus
Placentalia
urinary bladder
abdominal wall
corpus spongiosus penis
corpus fibrosum penis
ductus deferens
testis in scrotum
Marsupialia
L)
kidney
infund-ibulum
tuba ovar-ica
ureter
medial vagina
uterus
urinary bladder
lateral vagina
rectum
rudiment of wolff-ian duct
M)
infundibulum
sinus urogenitalis
N)
uterus
tuba ovarica
sinus vaginalis
vagina
sinus urogenitalis
O)
uterus
sinus urogenitalis
P)
caecum vaginale

the sinus vaginalis continues caudally until it reaches the urogenital duct. It breaks through that duct before birth and forms a permanent or, in most cases, a temporary birth canal (third or pseudovagina). Most marsupials have a seasonal reproduction period and have more than one period of estrus per year (polyoestrous).

Within placental mammals, the vagina is uniform and there is a trend to unite both uteri (Fig. 4.2 G–K). As such, the uteri can open separately (Fig. 4.2 K; uterus duplex: Rodentia, some Chiroptera) or with only one opening (Fig. 4.2 G–J) into the vagina. Depending on the degree of fusion, some placental mammals have a uterus bipartitus (Fig. 4.2 I; most carnivorans, some chiropterans), a uterus bicornis (Fig. 4.2 H; Eulipotyphla, some chiropterans, Cetartiodactyla), or – after total fusion – a uterus simplex (Fig. 4.2 J; some chiropterans, some xenarthrans, most primates). The sinus urogenitalis is largely reduced in placentals (Greven 2004). In primates and some rodents, urinary and genital tracts open separately. The vagina is unpaired (monodelphy) and opens into a vulval vestibule (vestibulum vaginae) (Storch and Schröpfer 2004).

Catarrhine primates, including humans, have a menstruation cycle, which involves regular bleedings in a rhythm of about 1 month (species specific). Bleeding derives from the uterine mucosa, which develops at the end of each cycle if the female does not become pregnant (Starck 1995).

The reproductive tracts of monotreme males differ from most therian mammals in having no scrotum; their testes are embedded inside the abdominal cavity near the kidneys (testicondy). The penis is situated within the urogenital sinus when not erected (Fig. 4.2 Q).

Although some placental groups secondarily show testicondy, most therians exhibit a permanent or seasonal displacement of the testes outside the abdominal cavity through the inguinal canal to the outside of the body (descensus testis) (Fig. 4.2 R–S) (Storch and Schröpfer 2004, Kleisner *et al.* 2010). The displacement of the testes is most likely caused by the high body temperature, which would have a mutagenic effect on spermatogenesis (Greven 2004). Most placental mammals have their testes outside the abdomen; only Xenarthra, Cetacea, as well as Sirenia, Proboscidea, and most other Afrotheria exhibit testicondy, which presents a strong phylogenetic signal (Werdelin and Nilsonne 1999).

The testes in marsupials experience a full and permanent "descensus" and lie subcutaneously. They are mostly situated in a scrotum, which can be pediculate. Compared with placentals, scrotum and testes are situated anterior to the penis in marsupials (except for the marsupial mole *Notoryctes typhlops*). The penis lies inside a penis pocket and has a kinked shape in its dormant phase (Fig. 4.2 R). A musculus levator penis can compensate the kink and a retractor muscle originating from the sacrum can retract the muscle inside the pocket. In most marsupials, in correspondence to the paired vagina, the terminal end of the penis is bifurcated.

Like the altricial platypus (Fig. 4.12 A) and altricial birds, placental mammals partly build very complex breeding burrows and nest constructions. Among others, subterranean nests are known for moles, common hamsters, and marmots. Epigaeic nests are known for field voles and boars. Nests in the soil can be found in harvest mice and hazel dormouse. Edible dormice build nests in tree holes and squirrels have nests in tree crowns. Nests can be complex (squirrel) or simple (house mouse). The litter nests can be more comprehensive and denser than nests for sleeping (Storch and Schröpfer 2004).

4.4 Gametes

During the proestrus, in all mammals, follicle cells are formed in the ovary of the female (Fig. 4.4 B), which produce estrogens and induce the estrus (heat). Subsequently, ovulation takes place. The empty follicle is folded and forms the corpus luteum (yellow body). It produces progesterone and is necessary to maintain pregnancy in females. After ovulation, the follicle is folded, and it forms a true corpus luteum (yellow body of the ovary), which is active throughout the whole intrauterine phase of development. Size increase is only facilitated through cell growth and not through cell division.

As in sauropsids, the eggs of monotremes are polylecithal, sometimes referred to as macrolecital, meaning that a large amount of yolk is present in the egg (Greven 2004). Inside the oviduct, tertiary egg layers are developed around the fertilized egg (zygote). These are homologous to those of the sauropsid egg (Fig. 4.5 G).

The primary membrane of the ovum is the vitelline layer, which is called zona pellucida in mammals (Fig. 4.4 Db-1). One of the glycoproteins of this layer is an important binding site for the spermatozoa during the acrosomal reaction.

The secondary layer, the theca folliculi, is formed by follicle cells (Fig. 4.4 B) in one or two sublayers. The theca externa consists of large and polygonal cells, which are separated from the theca interna by a basal membrane; both are formed by the stroma ovarii. The lutein cells are solely formed by follicle cells. Like in sauropsids, the follicle does not form a cavity in monotremes (Starck 1995).

The tertiary layer, the actual eggshell and albumen, is formed by secretions of the oviduct and uterine glands

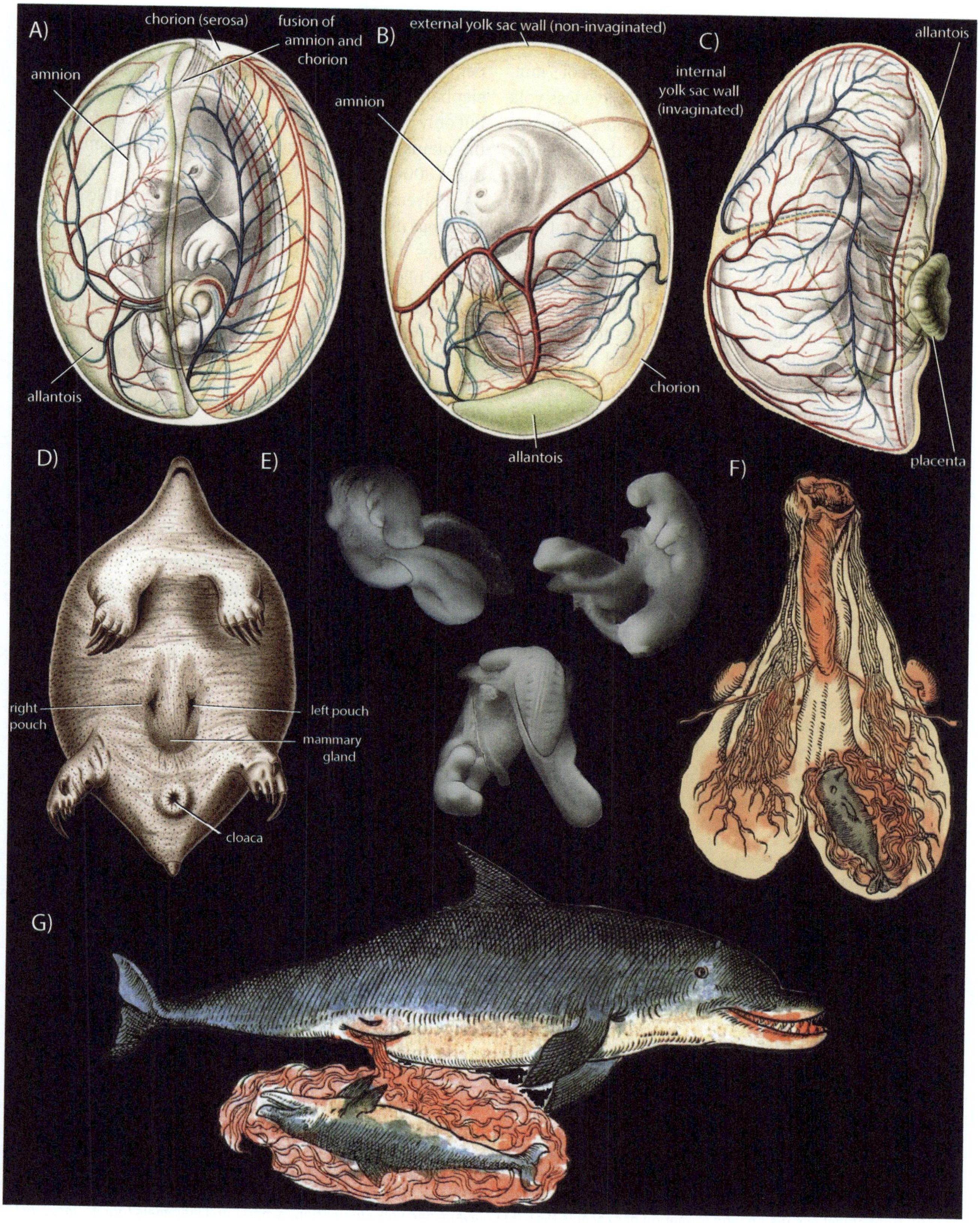

Fig. 4.3: Historical images on mammalian development. (A) Embryo of the monotreme short-beaked echidna *Tachyglossus aculeatus* within its extraembryonic membranes. (B) Embryo of the koala *Phascolarctos cinereus*, slightly shifted position within the yolk sac. (C) Late rabbit embryo inside its extraembryonic membranes; chorion not shown. (D) External sexual organs of a female *T. aculeatus*. (E) Embryo of *T. aculeatus* with a coiling trunk in three different views. Common bottlenose dolphin *Tursiops truncatus* in utero (F) and at birth (G). A–C, after Semon (1894b), D after Klaatsch (1895), E, from Embryological collection Berlin (M153); see also Werneburg and Sánchez-Villagra (2011). F and G, after Belon (1551).

(Zeller 2004b). The external layers form the shell and are parchmentlike in monotremes (containing keratin).

At a total diameter of 15 mm and an ovum size of 5 mm, monotreme eggs are relatively smaller and have less yolk when compared with the eggs of sauropsids. However, compared with those of therian mammals, they are enormous in size. During ovulation, the egg grows to up to four times its initial size. In contrast to sauropsids, the egg is still growing during the passage through the genital tract by receiving nutrition in the form of embryotroph and uterine milk (histotrophy) (Storch and Schröpfer 2004).

In monotremes, the albumen layer of the egg is shed shortly before fertilization and then serves to trap spermatozoa. The eggshells of monotremes and marsupials are very dense and spermatozoa cannot penetrate them. Therefore, insemination has to occur before shell formation (Greven 2004). After fertilization, the egg is quickly transmitted through the oviduct and fertilization stages can be found in the uterus (Starck 1995).

In contrast to those of other amniotes, therian follicle cells form a cavity (Fig. 4.4 B-1), which appears to be related to size reduction and the transition from poly- to alecithal eggs (Starck 1995). The tenrecid placentals are an exception in this regard (Strauss 1938, Enders *et al.* 2005).

Marsupial eggs are covered by a hull of albumen and shell for over two thirds of their intrauterine development but are nourished by uterine secretions during the final stages of intrauterine development (Fig. 4.4 Da). The eggshell may serve as an immunological barrier to the uterus. The external albumin layer of the marsupial egg is homologous to that of monotremes (Zeller 2004b) and consists of mucopolysaccharide. An external, keratinized shell membrane forms in the uterus. The corpus luteum is fully developed three days after ovulation and persists during the whole lactation phase.

The eggs of therian mammals are relatively small with a diameter of about 0.1 to 0.2 mm (human: 0.15 mm).

In general, the number of egg cells during one ovulation corresponds with the maximum number of offspring per litter in mammals. However, in few species (e.g., the elephant shrew *Elephantulus*), up to 120 eggs are released but only one or two are implanted in the uterus. Moreover, the plains viscacha *Lagostomus maximus* (Rodentia) produces 300–800 eggs (Weir 1971).

Polyembryony is known in two armadillo species (*Dasypus hybridus* et *novemcinctus*), in which only one egg ovulates (with one corpus luteum), and through division of one blastocyst, four or nine embryos of the same sex develop (Fernandez 1909, Newman and Patterson 1910).

Male gametes can be produced throughout the entire lifecycle, whereas all female gametes are developed around birth in mammals. As in sauropsids, sperm of monotremes have a long, filamentous head (Storch and Schröpfer 2004). In contrast to sauropsids and some marsupials, however, no physiological polyspermy (egg fertilization with more than one sperm) is found in monotremes (Starck 1995).

The spermatozoans have disklike "heads" in marsupials (partly barlike in the quoll *Dasyurus*) and have two processes at their terminal ends, which are deeply invaginated at the insertion site of the middle part. The terminal part of the sperm "tail" is very short and thin. Mature spermatozoa of South American marsupials (Didelphidae and Caenolestidae) attach to each other and form units while swimming. They separate again in the oviduct, where fertilization takes place.

4.5 Blastogenesis

Monotremes show a meroblastic (incomplete) type of cleavage (Caldwell 1884), which is an ancestral amniotic feature. A large amount of yolk is concentrated at one

▸ **Fig. 4.4: Early development in mammals.** (A) Yolk elimination in the form of small yolk particles in the opossum *Didelphis*. (B) Early embryonic development in the house mouse *Mus musculus* (Placentalia) until implantation. Graafian follicle with oocyte (1), ovulation (2), fertilization (3), cleavage: 2-cell stage (4), 4-cell stage (5), morula (6), compaction of morula (7), blastocyst (8), hatching (9), and implantation (10). (C) General steps in successive differentiation of the mesoderm and the neural tube in vertebrates. Mesoderm initially comes to lie between ectoderm and entoderm, and neurulation begins with a dorsal thickening of the ectoderm into a neural plate (1). Mesoderm differentiates into three major regions: epimere, mesomere, and hypomere, and the neural plate folds (2). Each mesoderm layer gives rise to specific layers and populations of mesodermally derived cell populations. The neural folds fuse and form a hollow neural tube, and neural crest cells emerge from the edges of the original neural plate (3). (D) Comparison of early ontogenetic processes of Marsupialia and Placentalia: yolk elimination, blastocyst formation (1–3), yolk sac formation (4a), entypy of the embryoblast (4b), mesoderm formation (5a, 6b), trophoblast expansion, and formation of exocoele (5b). (E) Cleavage and yolk elimination in marsupial mammals (quoll, *Dasyurus*). Non-fertilized egg cell with yolk vacuoles (1); 2-cell stage, the yolk vacuole is eliminated (2); 8-cell stage, formation of trophoblast cells (3); formation of blastula, the yolk body is resorbed (4); further development of the blastocyst (5–6). (F) Differentiation of the blastocyst in therian mammals. Blastocyst (1); early integration of the embryoblast (i.e., inner cell mass) into the trophoblast in Marsupialia (2); entypy: trophoblast grows over embryoblast in Placentalia (3). A, after Starck (1975); B, after Wehner and Gehring (2007); C, after Kardong (2008); D and F, after Starck (1995); E, after Starck (1975).

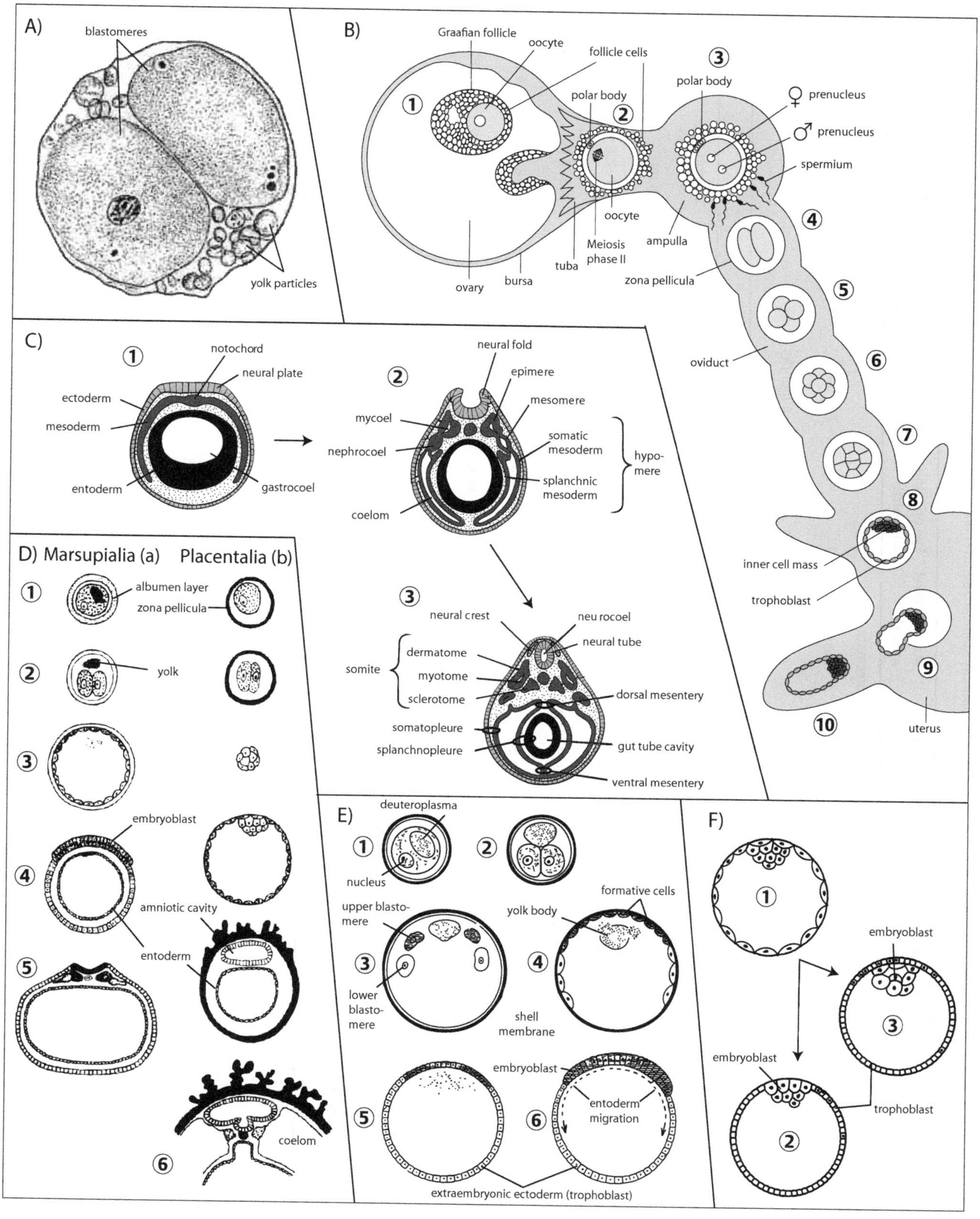
A)
blastomeres
yolk particles
B)
Graafian follicle
oocyte
follicle cells
polar body
polar body
♀ prenucleus
♂ prenucleus
spermium
oocyte
Meiosis phase II
ampulla
tuba
bursa
ovary
zona pellicula
oviduct
inner cell mass
trophoblast
uterus
C)
notochord
neural plate
ectoderm
mesoderm
entoderm
gastrocoel
neural fold
epimere
mesomere
mycoel
nephrocoel
somatic mesoderm
splanchnic mesoderm
hypo-mere
coelom
neural crest
neu rocoel
neural tube
somite
dermatome
myotome
sclerotome
dorsal mesentery
somatopleure
splanchnopleure
gut tube cavity
ventral mesentery
D) Marsupialia (a) Placentalia (b)
albumen layer
zona pellicula
yolk
embryoblast
amniotic cavity
entoderm
coelom
E)
deuteroplasma
nucleus
upper blasto-mere
lower blasto-mere
yolk body
formative cells
shell membrane
embryoblast
entoderm migration
extraembryonic ectoderm (trophoblast)
F)
embryoblast
embryoblast
trophoblast

Tab. 4.2: Comparison of selected life history traits in monotreme and marsupial mammals. Mainly summarized after Hayssen *et al.* (1993).

Major taxon	Species	Common name	Gestation length	Litter size	Maturity level at birth	Eye opening after birth	Weaning	Neonatal mass	Neonatal size	Sexual maturity
					Monotremata					
Tachyglossidae	*Tachyglossus aculeatus*	Echidna	27 days	1 or 2	Altricial	?	~200 days	378–380 mg	12–17 mm	~1 year
Ornithorhynchidae	*Ornithorhynchus anatinus*	Platypus	12–30 days	2	Altricial	>11 weeks	3–4 months	?	~25 mm	1–2 years
					Marsupialia					
Didelphidae	*Monodelphis domestica*	Grey short-tailed opossum	14–15 days	1 to 11	Altricial	2–5 weeks	49–56 days	100 mg	10 mm	3–7 months
Didelphidae	*Didelphis virginiana*	Virginia opossum	11–13 days	6 to 14	Altricial	50–72 days	80–104 days	130–160 mg	14 mm	6–8 months
Dasyuridae	*Dasyurus viverrinus*	Eastern quoll	~20 days	1 to 6	Altricial	72–91 days	5.5–6 months	20 mg	4.7–7.6 mm	11 months
Dasyuridae	*Sminthopsis crassicaudata*	Fat-tailed dunnart	13–18 days	6 to 9 (down to 1 rare)	Altricial	48–50 days	~70 days	~10 mg	3.5–4 mm	100–160 days
Peramelidae	*Perameles nasuta*	Long-nosed bandicoot	11–13 days	1 to 6	Altricial	44–49 days	~67–75 days	237 mg	12.8 mm	~4 months
Peramelidae	*Isoodon macrourus*	Northern brown bandicoot	12 days	1 to 7	Altricial	44–49 days	59 days	188.3 mg	13.83 mm	~4 months
Tarsipedidae	*Tarsipes rostratus*	Honey possum	21–28 days	1 to 4	Altricial	56 days	90 days	4.3 mg	?	5–6 months
Macropodidae	*Macropus rufus*	Red kangaroo	31–33 days	1	Altricial	115–150 days	~1 year	817 mg	?	1.5–3 years
Phascolarctidae	*Phascolarctos cinereus*	Koala	25–36 days	1 (2 occasional)	Altricial	?	10–12 months	360 mg	?	2–3 years

pole (i.e., telolecithal, from Greek τέλος, telos, meaning "end"), and only the animal pole shows cleavage (Storch and Schröpfer 2004). This is similar to the cleavage mode of fishes, sauropsids, and cephalopods (Fioroni 1987). Cleavage already takes place inside the uterus. The early embryo shows a so-called discoidal cleavage, which results in a flat blastodisk. The blastomeres (i.e., dividing cells) are attached to the yolk and separate themselves by strangulation and not by horizontal cleavage division. From the margin of the blastodisk, some cells (vitellocytes) migrate to the yolk. Later on, they fuse and form a syncytium, which surrounds the blastodisk.

The cleavage in therian mammals is total and equatorial (i.e., rotational cleavage). Therian mammals show holoblastic development, i.e., all cells show cleavage. In this regard, it mirrors more the non-amniote vertebrate than the sauropsid development. In contrast to other vertebrates with holoblastic development (in which both yolk and embryo cells show cleavage), early therian development occurs inside the mother. Therefore, the ancestral food supply via yolk cells is reduced, because much of the nutrition comes from the placenta. The development is hence secondarily holoblastic, as only embryonic cells show cleavage. In fact, this is rudimentary meroblastic cleavage because it is derived from the ancestral amniote mode of oviparous, meroblastic reproduction (Starck 1995).

Inside the blastocyst of marsupials, yolk particles are present, which are eliminated from the egg during the first cleavage stage (Fig. 4.4 A and D–E). They persist for a while as a yolk body between the blastomeres in most species until they are disintegrated and digested (phagocytosed) by blastocyst cells (Blüm 1986, Greven 2004).

Cleavage itself is relatively consistent in marsupials. First, a meridian cleavage takes place (radial orientation of cells). The fourth cleavage step is horizontal and results in the formation of two cell rings, which are superimposed upon each other. In contrast to placental mammals, no morula is formed.

In contrast to placentals, gestation and lactation do not result in the suppression of the estrous cycle in marsupials, and it is important to note that the estrous cycle is longer than gestation (Tyndale-Biscoe 2005). Directly after giving birth, kangaroo females can conceive again. If one kangaroo young is still in the pouch, the cleavage of a new embryo stops at the 100-cell stage (diapause) (Renfree 1993, Renfree and Shaw 2002, Hickford *et al.* 2009), and it continues to develop when the older sibling leaves the pouch.

Usually, placental mammals have oligolecithal eggs with no or only very little yolk inside (Starck 1995). The albumen layer is missing in placental mammals (except for rabbits, *Oryctolagus cuniculus*).

Cleavage of placental mammals is extremely slow and is not radial (Fig. 4.4 B and Db). The first cleavage is meridional (Fig. 4.4 B-4). In the second cleavage, one of the blastomeres shows a meridional cleavage, the other one an equatorial cleavage (Fig. 4.4 B-5), and so on. The cells do not separate completely from each other (Blüm 1986). Beginning with the 16-cell stage, the blastomeres form a morula (mulberry) inside the uterus (Fig. 4.4 B-6), which has the size of a pinhead and is comparable with the blastula of monotremes. Subsequently, however, the cells undergo a compaction with the help of adhesion molecules (Fig. 4.4 B-7). During the subsequent multiplication of cells, a blastocyst develops with a fluid filled cavity (blastocoel) inside.

4.6 Gastrulation

During gastrulation, three germ layers emerge from the blastocyst: ectoderm, mesoderm, and entoderm (Figs. 4.4 C and 4.5 A, C, I). The ectoderm differentiates into organs and tissues such as the epidermis, the brain (through neurulation), and the facial skeleton (Hall 2009). The mesoderm develops into the heart, blood vessels, and most musculature, among other tissues. The entoderm develops into the gut, gut-related organs such as the liver, and other structures. All three germ layers can participate in the formation of extraembryonic membranes, which variably contribute to the placenta.

During cleavage, a layer of eight cells forms in monotremes. After closure of the syncytial ring (i.e., an outer layer of multinucleate cells), and through active cell migration, early growth of the yolk and a flattening of the blastocyst occur. The so-called blastoderm is able to ingest the maternal nutrients inside the uterus (Zeller 2004b). Now, a subgerminal cavity is present between embryo and yolk. This blastoderm of monotremes is a bit larger than in sauropsids and already envelops the yolk at the beginning of gastrulation (Blüm 1986). Around the yolk-navel, the syncytial ring now shrinks into a plasma mass. This process is finished before the primitive streak is formed.

Early in monotreme development, two cell layers are distinguishable in the blastoderm. The external one consists of large cells, the prospective ectoderm cells. The layer of smaller cells will become entoderm cells. Later on, both layers integrate in the unilaminar blastoderm.

Afterward, the entoderm cells amoeboidally migrate down and condensate into the entoderm (hypoblast); the ectoderm cells stay to form the ectoderm (epiblast) (Starck 1995).

In this bilaminar stage, the embryonic body begins to form with the appearance of a primitive streak at the posterior end of the embryo. Always separated from it, a primitive knot develops in the center of the embryo. Here the invagination of the primitive gut (archenteron) takes place. Around the primitive knot, which is homologous to the primitive plate of sauropsids, the chordal plate (dorsomedial), the protochordal plate (rostal), and the gastral mesoderm (lateral) form. Formation of gastral mesoderm occurs earlier in monotremes than in sauropsids. The thickened ectodermal area anterior to the primitive knot/plate (Keimschild), however, appears earlier in reptiles.

The primitive streak of monotremes serves as an independent area of mesoderm formation and is homologous to the primitive streak found in therian mammals (Starck 1995).

The further tissue formation of monotremes (neurulation, somite formation, and schizamnion) is similar to the processes of all amniotes (Starck 1995), but differences occur in the relative timing of organ appearance and growth rates.

In marsupials, one cell ring of the forth cleavage stage forms the embryonic part (embryoblast) and the other ring forms the extraembryonic ectoderm (trophoblast). The embryoblast (i.e., formed by blastomeres) is parietal and forms a one-layered protoderm. This protoderm is often called blastula because it is embedded in its epithelium and no primitive knot is formed. However, because a major part of its parietal wall persists as extraembryonic ectoderm, the term "blastocyst" is more adequate (Starck 1995). It takes about 1 to 2 weeks from fertilization to blastocyst formation. Afterward, the embryo continues intrauterine development for a further 1 or 2 weeks until birth (Zeller 2004a).

In contrast to Placentalia, no compaction (morula) of the embryoblast takes place, and the trophoblast does not grow over the embryoblast (i.e., no entypy) in marsupials. Consequently, the embryoblast is never covered by the trophoblast in marsupials (Fig. 4.4 E-4-6 and F2-3) (Starck 1995, Lillegraven 2003, Greven 2004, Zeller 2004a). The unilaminar embryoblast (blastocyst layer) separates entoderm cells. The now bilaminar blastocyst forms the neural plate with a primitive streak.

Through entypy of embryoblast cells, the morula of placental mammals differentiates into two distinct parts, the external trophoblast and the more internal embryoblast (Fig. 4.4 B-8 and F-3). The trophoblast corresponds to the extraembryonic ectoderm, which is formed in the case of meroblastic development in monotremes and sauropsids (Blüm 1986, Smith 2015). The embryoblast (the actual embryo) consists of totipotent, embryonic stem cells and can form the source for creating chimaeras and transgenic animals in experimental biology (Houdebine 1997). A separation of the embryoblast can result in polyembryony (Newman and Patterson 1910).

4.7 Extraembryonic membranes

The vitelline (yolk) sac represents the ancestral extraembryonic membrane, which is plesiomorphically present in all vertebrates (Greven 2004). In mammals, the vitelline sac is initially formed by the enclosure of yolk or the hollow blastocyst by mesodermal cells from the embryonic disk (i.e., the hypoblast) (Fig. 4.5 B/D). A migration of entoderm cells forms a bilaminar yolk sac (Fig. 4.4 E-6 and 5G). The vitelline sac is slightly (Marsupialia) or largely (Placentalia) reduced in therian mammals and is filled with fluid (Figs. 4.5 F, I and 4.6 E–J). The anlage of the vitelline sac is strong evidence for the telolecithal origin of therian eggs (i.e., a large amount of yolk is concentrated at one pole) and that the yolk was secondarily lost. It is likely that the vitelline sac initially had a nutritional (Luckett 1975, Smith 2015) and a respiratory function (Houillon 1972). At the beginning of vitelline sac formation, a comprehensive circulatory system develops. As in most vertebrates, the first blood islands and blood vessels are formed on the surface of the vitelline sac or near the food resources. The vitelline sac is connected to the middle gut of the embryo via a yolk stalk (Greven 2004). Blood from the mesodermal vitelline veins (blood vessels of the umbilical cord) transport the blood to the heart of the embryo. Later on, the vitelline sac loses this initial function. However, it continues to have a functional circulation during yolk sac placentation. The vitelline sac serves for the

▸ **Fig. 4.5: Extraembryonic membranes.** Formation of the plect-/pleuramnion (A, B), schizamnion (C, D), and allantois (E, F). (G) Egg layers and formation of extraembryonic membranes in a chicken (*Gallus gallus*). (H) Extraembryonic membranes of the platypus, *Ornithorhynchus anatinus* (Monotremata). (I) Extraembryonic membranes of the golden mole (Chrysochloridae, Placentalia). A–F, after Houillon (1972); H, after Luckett (1975); I, after Kardong (2008).

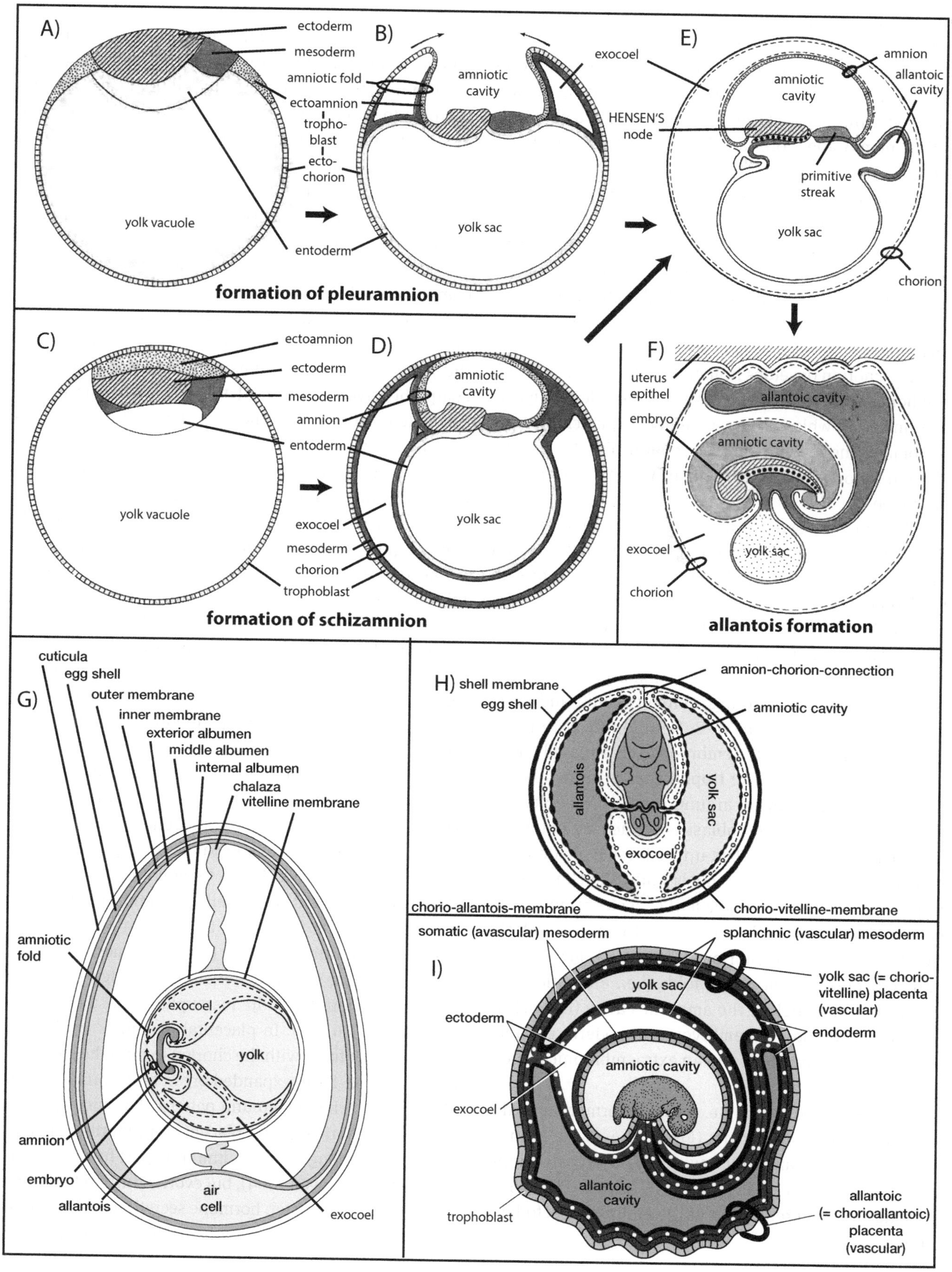
A)
ectoderm
mesoderm
amniotic fold
ectoamnion
tropho-
blast
ecto-
chorion
yolk vacuole
entoderm
B)
amniotic
cavity
yolk sac
formation of pleuramnion
exocoel
E)
amnion
allantoic
cavity
amniotic
cavity
HENSEN'S
node
primitive
streak
yolk sac
chorion
C)
ectoamnion
ectoderm
mesoderm
amnion
entoderm
yolk vacuole
D)
amniotic
cavity
yolk sac
exocoel
mesoderm
chorion
trophoblast
formation of schizamnion
F)
uterus
epithel
embryo
allantoic cavity
amniotic cavity
exocoel
yolk sac
chorion
allantois formation
G)
cuticula
egg shell
outer membrane
inner membrane
exterior albumen
middle albumen
internal albumen
chalaza
vitelline membrane
amniotic
fold
exocoel
yolk
amnion
embryo
allantois
air
cell
exocoel
H) shell membrane
egg shell
amnion-chorion-connection
amniotic cavity
allantois
yolk sac
exocoel
chorio-allantois-membrane
chorio-vitelline-membrane
somatic (avascular) mesoderm
splanchnic (vascular) mesoderm
I)
yolk sac
yolk sac (= chorio-
vitelline) placenta
(vascular)
ectoderm
endoderm
amniotic cavity
exocoel
allantoic
cavity
trophoblast
allantoic
(= chorioallantoic)
placenta
(vascular)

formation of the primary blood elements in haplorhine primates, in which it does not form a placenta (Carter 2015). As soon as it is almost fully reduced, hematopoiesis is transferred to the liver (Houillon 1972). The fusion of the vitelline sac and the chorion results in a choriovitelline membrane (yolk sac placenta) (Figs. 4.5 H–I and 4.6 E, G–J).

Characteristic for all amniotes, including monotremes and therian mammals, is the presence of three additional extraembryonic (fetal) membranes, which develop from the ventral side of the embryo's trunk: chorion (or serosa in sauropsids), amnion, and allantois. This condition is called cleidoic egg (from Greek Κλειστός, kleidos, meaning "enclosed").

The chorion develops from trophectoderm (synonym: trophoblast) and from extraembryonic mesoderm (Fig. 4.5 A–F and H–I), which is derived from the splanchno- and/or somatogenetic embryonic mesoderm (Fig. 4.4 C). The chorion is non-vascular and primarily responsible for gas exchange.

The amnion serves as a shock absorber and protects the embryo against desiccation (Gilbert 2006). It can form in two different ways: by folding (Fig. 4.5 A–B) or through the formation of a cavity (Fig. 4.5 C–D).

The folding process (plect- or pleuramnion) is most likely ancestral, and it is visible in sauropsids (Fig. 4.5G), monotremes, marsupials, lagomorphs, artio- and perissodactyls, all carnivorans, eulipotyphlans, and strepsirrhine primates (Fig. 4.5 A–B; Houillon 1972, Tyndale-Biscoe 2005). Initially, the embryoblast is situated on the top of the blastocyst. The trophoblast is formed below the embryoblast, namely around the yolk vacuole. The lower layer of the embryoblast is formed by entoderm (here called entophyll). The upper layer is made up of ectoderm and mesoderm and is embedded inside the upper part of the trophoblast (i.e., ectoamnion). The ectoamnion is ventrally connected to the rest of the trophoblast (i.e., ectochorion) (Fig. 4.5 A). Ectoamnion and ectochorion form dorsal folds (i.e., amniotic fold), which fuse and enclose the amniotic cavity (Fig. 4.5 B). Subsequently, the extraembryonic mesoderm expands inside the amniotic folds and an extraembryonic coelom (exocoel) appears.

In the pig, for example, amnion formation is finished when ten somite pairs (Fig. 4.10-4–5) are formed. At the base of the amniotic cavity, the embryo develops, and the formation of the primitive streak (Fig. 4.5 E) indicates the differentiation of the embryonic ectoderm (cf. Fig. 4.4 C-2).

Compared with marsupials, most erinaceids, soricids, tenrecids, chiropterans, and primates (including humans) show a full coverage of the embryoblast by the trophoblast (Fig. 4.5 C). The embryoblast remains compact for some time. Later on, vacuoles develop and fuse in the upper part of the embryoblast (ectoamnion) and other cell elements. Hence, a cleft is formed that expands to the amniotic cavity (Fig. 4.5 C–D; schizamnion). The roof of this cavity represents the ectoamnion, whereas the bottom is formed by the embryo proper. The mesoderm is more involved in amnion formation when compared with the pleuramnion and aligns early to the trophoblast (Fig. 4.5 D). The early formation of the amnion, which appears even before the formation of the primitive streak in humans, is associated with the comprehensive reduction of the yolk sac. This is also correlated to the size of the mesoderm. The yolk sac is surrounded by exocoel from its first appearance on the entoderm. The yolk sac does not take part in the formation of the inner wall of the trophoblast.

In most rodents, the embryoblast sinks into the yolk vacuole (like the clapper of a bell) but keeps its upper connection to the trophoblast. The trophoblast persists, thickens, and later fuses to the maternal epithelium. The yolk sac is well developed and aligns closely to the rest of the trophoblast, which is finally degenerated in this area. The formation of the amnion is a complex mechanism (ectochorial cyst formation), which can be derived from the schizamnion, namely from delamination of the embryoblast.

The allantois develops from extraembryonic entoderm and mesoderm (Fig. 4.5 E–F). It is vascularized and is responsible for the removal of nitrogenous waste. During development, the allantois expands and suppresses the exocoelomic cavity (Fig. 4.5 F). In monotremes, the allantois begins to form at the 27-cell stage and rapidly grows into the exocoel. It fuses with the chorion and forms an expanded respiratory surface. In monotremes, the allantois has a similar size as the vitelline sac (Figs. 4.3 A and 4.5 H) (Houillon 1972). In marsupials, the allantois is smaller (Fig. 4.6 G–J). In placental mammals (and some marsupials), it fuses with the chorion to form the chorioallantois (Fig. 4.5 I), an expanded organ for respiration (Fig. 4.5 H–I) (following the ground pattern of amniotes; Zeller 2004b). The chorioallantoic placenta is primarily responsible for gas exchange and calcium transport from the shell to the embryo (Greven 2004), but evolved further functions such as nutrient transfer, hormone secretion, and immune responses (Benirschke and Kaufmann 1995, Carter 2012).

Tab. 4.3: Comparison of selected life history features of different groups of placental mammals. Mainly summarized after Hayssen *et al.* (1993) and Wildman *et al.* (2006). Legend and other references within table.

Major taxon	Species	Common name	Placental interface	Placental shape	Maternofetal interdigitation	Gestation length	Litter size	Maturity level at birth	Eye opening after birth (0 = at birth)	Weaning	Neonatal mass	Neonatal size	Sexual maturity
Afrotheria													
Sirenia	*Dugong dugon*	Dugong	?	Zonary	Villous	~1 year	1 (2 rare)	Precocial	0	~1.5 year	20–35 kg	100–150 cm	9–10 years
Sirenia	*Trichechus manatus*	West Indian manatee	Endotheliochorial (1)	Zonary	Labyrinthine	~1 year	1 (2 rare)	Precocial	0	1–2 years	11–27 kg	100–120 cm	3–5 years
Hyracoidea	*Procavia capensis*	Rock hyrax	Hemochorial	Zonary	Labyrinthine	6–8 months	1 to 3	Precocial	0	3–7 months	140–405 g	190–204 mm	~15 months
Proboscidea	*Loxodonta africana*	African elephant	Endotheliochorial	Zonary	Labyrinthine	20–24 months	1 (2 rare)	Precocial	0	6–? Months	73–120 kg	?	7–15 years
Tubulidentata	*Orycteropus afer*	Aardvark	Endotheliochorial	Zonary	Labyrinthine	~7 months	1 (2 rare)	Naked with open eyes	0	?	1–2 kg	~550 mm	3 years
Afroinsectiphilia	*Elephantulus rufescens*	Elephant shrew	Hemochorial	Discoid	Labyrinthine	50–65 days	1 or 2	Precocial (not explicitly mobile)	0	15 days–1 month	9.3–13.0 g	?	~2 months
Chrysochloridae	*Amblysomus hottentotus*	Hottentot golden mole	Hemochorial	Discoid	Labyrinthine	?	1 or 2	Naked and blind, but adults also blind	n/a	?	~4.5 g	47 mm	?
Afrosoricida	*Potamogale velox*	Giant otter shrew	Endotheliochorial (2)	Discoid	Labyrinthine	?	1 or 2	Altricial	?	?	?	?	?
Afrosoricida	*Tenrec ecaudatus*	Tailless tenrec	Hemochorial	Discoid	Labyrinthine	53–64 days	1 to 4	Altricial	9–15 days	25–30 days	10–27.4 g	84–92 mm	~6 months
Xenarthra													
Cingulata	*Dasypus novemcinctus*	Nine-banded armadillo	Hemochorial	Discoid	Villous (6)	3–5 months delay, 4–4.5 months development	4	Altricial	?	4–5 months	28.6–133 g	25 cm	~1 year
Pilosa	*Myrmecophaga tridactyla*	Giant anteater	Hemochorial	Discoid	Villous (7–8)	~6 months	1	Precocial (not explicitly mobile)	0	3 months	1480–1720 g	?	?
Pilosa	*Choloepus didactylus*	Linnaeus's two-toed sloth	Hemochorial	Discoid	Trabecular	5.5–7 months	1	Precocial (not explicitly mobile)	0	5 months	364.3 g	16–18 cm	3–4.5 years

Tab. 4.3 (continued)

Major taxon	Species	Common name	Placental interface	Placental shape	Maternofetal interdigitation	Gestation length	Litter size	Maturity level at birth	Eye opening after birth (0 = at birth)	Weaning	Neonatal mass	Neonatal size	Sexual maturity
Euarchontoglires													
Primates	*Daubentonia madagascariensis*	Aye-aye	Epitheliochorial	Diffuse	Villous	?	1	?	?	6–7 months	?	?	?
Primates	*Homo sapiens*	Human	Hemochorial	Discoid	Villous	38 weeks	1 (2 rare)	Altricial	0	0.5–3 years	2500–4000 g	ca. 50 cm	12–14 years
Dermoptera	*Cynocephalus volans*	Philippine flying lemur	Hemochorial	Discoid	Labyrinthine	~150 days	1	?	?	~200 days	35.8 g	254 mm	?
Scandentia	*Tupaia glis*	Common tree shrew	Endotheliochorial	Bidiscoid	Labyrinthine	43–50 days	2 to 5	Altricial	13–25 days	~30 days	10–15 g	79–140 mm	3–6 months
Rodentia	*Marmota monax*	Groundhog	Hemochorial	Discoid	Labyrinthine	30–40 days	4 to 9	Altricial	20–28 days	4–5 weeks	23.7–32.3 g	85–89 mm	~1 year
Rodentia	*Rattus norvegicus*	Brown rat	Hemochorial	Discoid	Labyrinthine	20–24 days	2 to 12	Altricial	5–22 days	4 weeks	5.27–7.35 g	52 mm	27–47 days
Rodentia	*Mus musculus*	House mouse	Hemochorial	Discoid	Labyrinthine	18–21 days	2 to 8	Altricial	12–15 days	3 weeks	0.8–1.14 g	22 mm	2 months
Rodentia	*Myocastor coypus*	Nutria	Hemochorial	Discoid	Labyrinthine	120–150 days	1 to 9 (up to 14 rare)	Precocial (not explicitly mobile)	0	5–10 weeks	100–250 g	?	2–9 months
Rodentia	*Hystrix africaeaustralis*	Cape porcupine	Hemochorial	Discoid	Labyrinthine	93–112 days	1 to 4	Precocial (not explicitly mobile)	0	~7 weeks	~300 g	?	8–18 months
Lagomorpha	*Oryctolagus cuniculus*	European rabbit	Hemochorial	Discoid	Labyrinthine	28–37 days	4 to 7	Altricial	10 days	22–25 days	30–45 g	117,70–119,07 mm	5–8 months
Lagomorpha	*Ochotona alpina*	Alpine pika	Hemochorial	Discoid	Labyrinthine	30 days	3 to 8	Altricial	8–10 days	20–22 days	6–10 g	?	25–30 days
Laurasiatheria													
Eulipotyphla	*Talpa europaea*	European Mole	Endotheliochorial	Discoid	Labyrinthine	30–40 days	2 to 6	Altricial	20–22 days	3–5 weeks	3–3.5 g	30–44 mm	~1 year
Eulipotyphla	*Erinaceus europaeus*	European hedgehog	Hemochorial	Discoid	Labyrinthine	5–6 weeks	2 to 8	Altricial	8–22 days	38–45 days	10.7–17.17 g	25–91.5 mm	10–12 months

Tab. 4.3 (continued)

Major taxon	Species	Common name	Placental interface	Placental shape	Maternofetal interdigitation	Gestation length	Litter size	Maturity level at birth	Eye opening after birth (0 = at birth)	Weaning	Neonatal mass	Neonatal size	Sexual maturity
Eulipotyphla	*Crocidura leucodon*	Bicolored white-toothed shrew	Hemochorial	Discoid	Labyrinthine	20–35 days	2 to 4	Altricial	9–13 days	16–26 days	0.8–1.0 g	?	3–4 months
Chiroptera	*Pteropus poliocephalus*	Grey-headed flying fox	Hemochorial (3)	Discoid	Labyrinthine	6 months	1	Altricial	1–10 days	?	70–90 g	60 mm	18 months
Chiroptera	*Megaderma lyra*	Greater false vampire bat	Hemochorial (4)	Discoid	Labyrinthine	150–160 days	1 (2 rare)	?	0	2 months	7–8 g	?	15–18 months
Chiroptera	*Molossus molossus*	Velvety free-tailed bat	Hemochorial	Discoid	Labyrinthine	~3.5 months	1	?	0	1.5–4 months	3.3–3.9 g	31.67–57.5 mm	3 months ?
Chiroptera	*Macrotus californicus*	California leaf-nosed bat	Hemochorial (5)	Discoid	Labyrinthine	8–9 months with about 4.5 months diapause	1	?	0	1 month	?	?	3–16 months
Pholidota	*Manis tricuspis*	Tree pangolin	Epitheliochorial	Diffuse	Villous	>5 months	1	Precocial (not explicitly mobile)	0–9 days	3–7.5 months	90–150 g	300–350 mm	2 years
Carnivora	*Vulpes vulpes*	Red fox	Endotheliochorial	Zonary	Lamellar	51–60 days	1 to 7	Altricial	8–14 days	5–10 weeks	50–150 g	131–150 mm	9–12 months
Carnivora	*Felis silvestris*	Wild cat	Endotheliochorial	Zonary	Lamellar	56–69 days	3–5 (up to 10 rare)	Altricial	1–15 days	3–4 months	82–149 g	126.7–155 mm	8–14 months
Perissodactyla	*Rhinoceros unicornis*	Indian rhinoceros	Epitheliochorial	Diffuse	Villous	463–488 days	1 (2 rare)	Precocial	0	12–18 months	33.75–71.3 kg	96.5–122 cm	4–7 years
Perissodactyla	*Tapirus* spp.	Tapir	Epitheliochorial	Diffuse	Villous	~13 months	1	Precocial	0	6–8 months	6–10.20 kg	72 cm	3–4.5 years
Perissodactyla	*Equus caballus*	Horse	Epitheliochorial	Diffuse	Villous	320–340 days	1	Precocial	0	6–12 months	?	?	2 years
Cetartiodactyla	*Lama glama*	Llama	Epitheliochorial	Diffuse	Villous	11–13 months	1	Precocial	0	4–6 months	8–16 kg	?	?
Cetartiodactyla	*Sus scrofa*	Domestic pig/ wild boar	Epitheliochorial	Diffuse	Folded	~4 months	4 to 12	Precocial	0	2.5–4 months	500–1534 g	24 cm	0.75–1.75 years
Cetartiodactyla	*Balaenoptera musculus*	Blue whale	Epitheliochorial	Diffuse	Villous	10–12 months	1	Precocial	0	6–7 months	2500 kg	700–800 cm	8–10 years

Tab. 4.3 (continued)

Major taxon	Species	Common name	Placental interface	Placental shape	Maternofetal interdigitation	Gestation length	Litter size	Maturity level at birth	Eye opening after birth (0 = at birth)	Weaning	Neonatal mass	Neonatal size	Sexual maturity
Cetartiodactyla	*Delphinus delphis*	Short-beaked common dolphin	Epitheliochorial	Diffuse	Villous	~11 months	1	Precocial	0	10 months	6.8–7.35 kg	75–90 cm	3–7 years
Cetartiodactyla	*Hippopotamus amphibius*	Hippopotamus	Epitheliochorial	Diffuse	Villous	~8 months	1 (2 very rare)	Precocial	0	9–14 months	25–50 kg	748–1270 mm	3–9 years
Cetartiodactyla	*Bos frontalis*	Gayal	Epitheliochorial	Cotyledonary	Villous	8–9 months	1 or 2	Precocial	0	4.5 months	13–33 kg ?		1–2 years

Placenta types, cited after Wildman *et al.* 2006, supplement.
Placental interface: The placental interface describes the degree of invasiveness of fetal (i.e., placental) tissue into maternal tissue, with epitheliochorial being least invasive and hemochorial being most invasive. There are six types of placental interface:
a. Epitheliochorial: the fetal chorion is in contact with the endometrium epithelium, and the blastocyst does not invade the maternal endometrium.
b. Syndesmochorial: the fetal chorion destroys the uterine epithelium; this character state was thought to exist in sheep and goats, but its existence has been refuted more recently (2).
c. Endotheliochorial: the fetal chorion erodes the endometrial epithelium and connective tissue, resulting in apposition to the uterine endothelium.
d. Hemochorial: the most deeply invasive form of placental interface. The trophoblast erodes all of the layers until the maternal vessels and the fetal trophoblast are directly in contact. This type of interface has been subdivided into three subtypes according to the number of layers of the cytotrophoblast: (i) hemomonochorial, the outer trophoblast layer is constructed of syncytiotrophoblast only; (ii) hemodichorial, only one side of the trophoblast is covered by cytotrophoblast; (iii) hemotrichorial, both sides of the trophoblast are covered by the cytotrophoblast.
Placental shape: The shape and area of the maternal side of the placenta where it interfaces with maternal tissue. There are five types of placental shape: (a) *Diffuse:* maternal-fetal interdigitation extends over the entire surface of the chorionic sac. (b) *Cotyledonary:* many spotlike areas of maternal-fetal interdigitation. (c) *Zonary:* ringlike area of maternal-fetal interdigitation. (d) *Bidiscoid:* two disklike areas of maternal-fetal interdigitation. (e) *Discoid:* a single disklike area of maternal-fetal interdigitation.
Maternofetal interdigitation: The form of contact between maternal and fetal tissues and/or blood. There are five types of maternofetal interdigitation: (a) *Folded:* ridgelike folds of the chorion that fit into grooves of the uterine mucosa. (b) *Lamellar:* ridgelike folds multiply in branch to form complicated columnlike folds. (c) *Trabecular:* branching folds in which leaflike and fingerlike villi branch off. (d) *Villous:* treelike branching of the chorion. (e) *Labyrinthine:* tissue block of trophoblast penetrated by weblike channels that are filled with either maternal blood or fetal capillaries.
Additional references: 1, Carter *et al.* (2008); 2, Carter *et al.* (2006); 3, Karim and Bhatnaga (1996); 4, Bhiwgade (1990); 5, Bodley (1974); 6, Enders (1960); 7, Becher (1931); 8, Mess *et al.* (2012).

4.8 Placentation

In monotremes, the vitelline sac is fused with the chorion (Figs. 4.3 A and 4.5 H; synonyms: choriovitelline, omphalopleura) and is able to resorb histotrophs, which are nutricial secretions from the mother's placenta, via its parietal (chorion) wall. One could hypothesize this form of nutrition to be an ancestral kind of placentation in amniotes because sauropsids also resorb nutrients via their vitelline sac (Mossman 1987, Zeller 2004b).

In some vertebrate taxa, viviparity is accomplished through the presence of a placenta, a specialized interface between maternal and fetal tissues. A true vitelline placenta is known from some viviparous squamates, in which the choriovitelline membrane is largely associated to the uterine epithelium (Greven 2004, Kardong 2008, Blackburn 2015). A close association of the chorion and the vitelline membrane seems to be the ancestral amniotic condition with the potential to evolve a true placenta as seen in viviparous squamates and possibly characterizing the ancestral condition for therian mammals. In other groups, viviparity is the result of retaining the egg in the body and only little or no maternal nutrition is provided, and the embryo is nourished by the yolk supplied in the egg. In some vertebrate species, embryonic consumption of maternal tissue, litter mates, or specialized eggs takes place. Viviparity evolved more than 100 times within squamates (Blackburn 2015). Most of those species rely on yolk nutrition and have some specialized fetal membranes for gas, water, and mineral exchange. All mammals and many squamates provide maternal nutrition in the form of uterine secretions, which is absorbed by extraembryonic or placental tissue (histotrophy), or by exchange of nutrients, waste products, and gas by maternal and fetal blood supplies (hemotrophy) (Smith 2015).

In therian mammals, matrotrophic viviparity involves maternal nutrition beyond the ancestral yolk supply (Mess and Carter 2006). This involves the formation of a placenta but results in a reduction of yolk content and litter size (Greven 2004). In amniotes, the chorion may fuse with the yolk sac (= vitelline membrane) to form the choriovitelline placenta (synonyms: yolk sac placenta, omphaloplacenta) and/or with the allantois to form the chorioallantoic placenta (Fig. 4.5 I). In monotremes and many squamates, choriovitelline and chorioallantoic membranes are well developed (Fig. 4.5 H). The first one serves for nutrition, the second one is responsible for gas exchange (respiration).

With the help of proteases, the trophoblast (chorioectoderm) of many therian mammals invades the endometrium, which is formed by the gland-rich uterine epithelium and the vascular connective tissue. As such, the trophoblast enables the implantation and embedding (nidation) of the embryoblast into the uterine wall and becomes a part of the embryonic aspect of the placenta (Greven 2004).

In addition to early nutrient exchange, this basic placenta type is important for the production of growth factors, binding proteins, and receptors, as well as for blood formation (i.e., hematopoiesis) and cholesterol production (Carter 2012).

Correlated with the small yolk sac, an association of the chorioallantoic membrane and the maternal epithelium (chorioallantoic placenta) is mainly found in placental mammals (Fig. 4.6 F). However, a wide range of both placenta types occurs among all clades of therian mammals (Smith 2015) (Figs. 4.5 I and 4.6 E–J, M). The chorioallantoic placenta is highly effective in that embryonic waste is extensively discharged.

Independent of the specific fetal membrane contribution to the placenta, different types, subtypes, variations, and homoplastic developments of invading the uterine lining exist (Smith 2015) (Tab. 4.3).

The least invasive type is the epitheliochorial placenta (in some cases it is not even invasive). Here the epithelia of the chorion and the uterus align to each other without invasion or erosion of uterine tissue. Nutrition is exchanged through the epithelia (Fig. 4.6 A). An invasive placenta type was certainly present in the ground pattern of placental mammals (Mess and Carter 2006, Elliot and Crespi 2009). The uterine epithelium can be partly destroyed, and the chorionic villi can be covered by epithelium and have a large contact surface to the connective tissue of the uterus (formerly known as syndesmochorial placenta). This type is restricted to small areas (cotyledonary or multiplex placenta) (Fig. 4.6 B).

The endotheliochorial placenta involves an invasion of extraembryonic tissue through the uterine lining and a direct contact of the maternal blood vessels with the endothelium (Fig. 4.6 C).

The hemochorial placenta involves the erosion of maternal blood vessels. This results in the accumulation of maternal blood around the chorioallantoic tissue (Fig. 4.6 D).

The ways in which extraembryonic and maternal tissue interdigitate and increase the surface of contact (villous, folded or lamellar trabecular, or labyrinthine) differ among taxa and homoplastic developments exist.

In some cases, multinucleate cells (syncytia) from mother and fetus can develop at their interface to increase the efficiency of nutritional exchange. Syncytia can

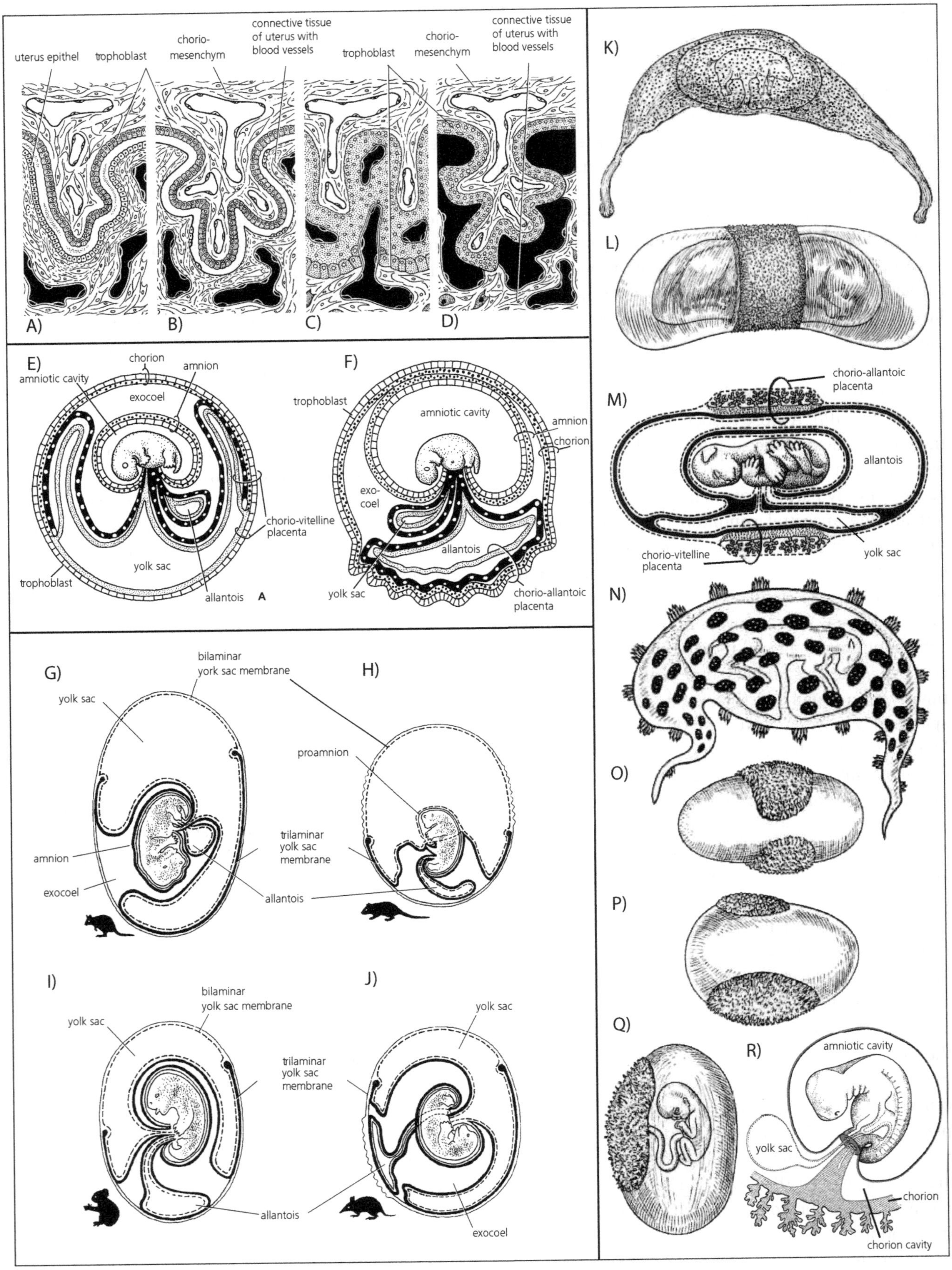
uterus epithel
trophoblast
chorio-mesenchym
connective tissue of uterus with blood vessels
trophoblast
chorio-mesenchym
connective tissue of uterus with blood vessels
A)
B)
C)
D)
E)
chorion
amnion
amniotic cavity
exocoel
chorio-vitelline placenta
yolk sac
trophoblast
allantois
A
F)
trophoblast
amniotic cavity
amnion
chorion
exo-coel
allantois
yolk sac
chorio-allantoic placenta
G)
bilaminar york sac membrane
H)
yolk sac
proamnion
trilaminar yolk sac membrane
amnion
exocoel
allantois
I)
J)
bilaminar yolk sac membrane
yolk sac
yolk sac
trilaminar yolk sac membrane
allantois
exocoel
K)
L)
M)
chorio-allantoic placenta
allantois
chorio-vitelline placenta
yolk sac
N)
O)
P)
Q)
R)
amniotic cavity
yolk sac
chorion
chorion cavity

develop on the maternal and/or the fetal side; however, in some groups (placental ruminants and marsupial peramelids), a syncytium of both tissues occurs (e.g., Wooding and Burton 2008). Different types of external anatomy of the placenta exist among placental mammals (Tab. 4.3, Fig. 4.6 K–Q).

The embryonic part of the epitheliochorial placenta (afterbirth, decidua) is delivered without bleeding (placenta indeciduata). Because of the close association to the maternal tissue of the uterus, the multiplex and the hemochorial placentas are delivered with bleeding (placenta deciduata) (Greven 2004).

Marsupial eggs are small and only yolk vacuoles (i.e., yolk particles in Fig. 4.4 A) are present (Fig. 4.4 A, Da, E). They serve as nutrition to the cleavage cells (Starck 1995). It appears that initially there is significant histotrophic nutrition in all marsupial groups. Highly efficient maternal uterine glands can be present to nourish the embryo (tammar wallaby, *Macropus eugenii*). Hemotrophic nutrition in marsupials as well as placentals is independent of the degree of invasiveness.

After resorption of the shell in marsupials (Fig. 4.4 Da-4), the vitelline sac expands and reaches the chorion to form the choriovitelline placenta (Greven 2004). The vitelline sac is only partly covered by mesoderm, so that the marsupial yolk sac wall is bilaminar in its lower half and trilaminar in its upper pole (Fig. 4.6 G–J; Tyndale-Biscoe and Renfree 1987, Tyndale-Biscoe 2005). In some cases, the allantois reaches the chorion and a second chorioallantoic placenta can also be present (bandicoot, *Perameles*) (Greven 2004, Smith 2015) (Fig. 4.6 J). The choriovitelline placenta forms the final functional placenta in most marsupials (Fig. 4.6 G–J). A chorioallantoic placenta can also appear transitionally in several marsupials and contributes to the final functional placenta in peramelid marsupials (Tyndale-Biscoe and Renfree 1987: fig. 7-17; Smith 2015). Among marsupials, the most common type is the large yolk sac placenta, in which the allantois does not reach the chorion (quokka, *Setonix brachyurus*) (Fig. 4.6 G). In another type, the allantois reaches the chorion but is later reduced and a yolk sac placenta is formed (eastern quoll, *Dasyurus viverrinus*) (Fig. 4.6 H). The allantois can also strongly attach to the chorion to form a large yolk sac placenta (koala, *Phascolarctos cinereus*) (Fig. 4.6 I). Finally, both an enlarged allantois and a yolk sac placenta can be present at the same time (long-nosed bandicoot, *Perameles nasuta*) (Fig. 4.6 J). Most marsupials have an epitheliochorial contact between fetal and maternal parts of the placenta, and no erosion of maternal tissue occurs (many macropodids) (Fig. 4.6 A). However, endotheliochorial invasion of extraembryonic tissue to the maternal blood cells and maternal-fetal syncytia are also known (Roberts and Breed 1994, Zeller and Freyer 2002) (Fig. 4.6 C).

In contrast to marsupials, the placenta develops very early in placental mammals. Before it is implanted in the uterine wall, the blastocyst hatches from the zona pellucida (Greven 2004). The trophoblast, which enables attachment of the embryoblast to the uterine wall, was hypothesized to be a unique feature in placental mammals (Lillegraven *et al.* 1987, Lillegraven 2004). However, most authors (e.g., Johnson and Selwood 1996, Selwood and Hickford 1999, Selwood and Johnson 2006, Smith 2015) highlighted that this external layer of the chorion membrane is homologous to the external layer of the chorion (trophectoderm) of non-placental amniotes. In many placentals (and also some marsupials), however, it appears to have a generally high potential to enable strong implantation of the embryo in the uterus by forming very complex chorionic villi.

As shown above, a great diversity of placenta types, forms of implantation, and fetal-maternal integration evolved among placental mammal clades (Tab. 4.3, Fig. 4.6 A–D and K–Q) (Wildman *et al.* 2006). In general, two major types in the constellation of characters of placenta morphology and life history traits can be identified (Lewitus and Soligo 2011, Smith 2015).

◂ **Fig. 4.6: Placentation.** (A–D) Placenta types among placental mammals. (A, B) epitheliochorial, (C) endotheliochorial, and (D) hemochorial placenta. (E) Yolk sac placenta (opossum, Marsupialia). (F) Chorioallantoic placenta (bush baby, *Galago*; Placentalia). (G, H) Placentation types among marsupials; most common type is the large yolk sac placenta, allantois does not reach the chorion (quokka, *Setonix brachyurus*) (G); allantois reaches the chorion but is later reduced, yolk sac placenta (eastern quoll, *Dasyurus viverrinus*) (H); allantois closely attaches to the chorion, large yolk sac placenta (koala, *Phascolarctos cinereus*) (I); enlarged allantois and yolk sac placenta (*Perameles nasuta*) (J). (K–Q) External shape of the placenta in different placental mammals; placenta diffusa (domestic pig, *Sus scrofa domestica*) (K); placenta zonaria (carnivoran) (L, M), two basic types of placental structures as seen in a transitional stage of an implanted domestic cat embryo (*Felis catus*). Yolk sac and chorioallantoic placenta are present. The latter grows outward and takes over the function of the earlier, primary yolk sac placenta (M); placenta cotyledonaria (cattle, *Bos primigenius taurus*) (N); incomplete zonary placenta (raccoon, *Procyon lotor*) (O); placenta bidiscoidalis (many primates) (P); placenta discoidalis (human, *Homo sapiens*) (Q). (R) Schematic diagram of a human embryo of about 10 mm in length and its relation to the extraembryonic membranes. A–D, after Westheide and Rieger (2010); E–F, after Kardong (2008); G–J, after Tyndale-Biscoe and Renfree (1987); K, L, N–R, after Starck (1965); M, after Pough *et al.* (2012).

One type is represented by an epitheliochorial placenta interface (Fig. 4.6 A), a diffuse placenta shape, and a villous to trabecular maternofetal interdigitation. The yolk sac is free. Placental mammals with such a placental anatomy are generally characterized by a long life span, precocial and large neonates, small litter size, long gestation, late weaning and maturity, small social group size, and high interbirth intervals (e.g., the brown greater galago *Otolemur crassicaudatus* and the horse *Equus caballus*).

Another type is represented by a hemochorial placenta interface (Fig. 4.6 D), a discoid placenta shape (Fig. 4.6 Q), and a labyrinthine maternofetal interdigitation. The yolk sac is inverted. These placental mammals are generally characterized by a short life span, altricial and small neonates, large litter size, short gestation, early weaning and maturity, large social group size, and low interbirth interval (e.g., the rabbit *Oryctolagus cuniculus*; Lewitus and Soligo 2011). There are some obvious exceptions, such as humans who have the latter placenta shape (Martin 2007).

The chorioectoderm (trophoblast) was thought to be a unique feature in placental mammals, and to enable a particularly efficient immunological barrier in placental mammals. For that reason, placental mammalian fetuses would be longer retained in the womb and would reach a more advanced anatomical state at birth than marsupial mammals do (e.g., Lillegraven *et al.* 1987, Lillegraven 2004). However, more recently, several authors highlighted that the chorioectoderm is homologous among amniotes (see Smith 2015). Apparently, in most placental mammals, it has a particularly strong ability to invade the maternal tissue and to form complex integrations of the embryo/fetus and the mother. However, complex placentas are also found in some marsupials, which possibly evolved the ability of trophoblast invasion secondarily.

It is worth noting that the placental mammal embryo already invades the maternal tissue at a very early stage of development, whereas in marsupials implantation first appears at the beginning of the last trimester of intrauterine development. This means that marsupials, in general, have less time to build a complex placenta. No data are available on whether marsupial species with complex placentas show a relatively earlier implantation than species with less complex placentas. If so, not the immunological barrier but more strikingly the degree of embryonic nourishment enabled by the deep integration of maternal and fetal tissue could be a reason for the longer intrauterine development of placental mammals.

The chorioallantoic placenta, which is particularly, but not exclusively, present in placental mammals was thought to be highly effective for embryonic/fetal development. The integration of the allantois to the fetal part of the placenta suggests a highly effective removal of embryonic nitrogenic waste via the maternal circulatory system (Storch and Schröpfer 2004). It has been suggested that the choriovitelline placenta in marsupial mammals and other viviparous amniotes is less effective for waste removal because the allantois is largely or completely separated from the fetal/embryonic part of the placenta.

4.9 Evolution of organogenesis

After gastrulation, organ systems develop and differentiate (Fig. 4.4 C), including the skeletal, muscular, circulatory, digestive, reproductive, and neural systems (Keibel 1897, Starck 1965). The major mammalian groups only differ

▸ **Fig. 4.7: Embryonic series of *Ornithorhynchus anatinus* (Monotremata, Ornithorhynchidae) (Part 1/2).** Previously unpublished drawings originating from the Embryological collection of the Museum für Naturkunde Berlin, Germany. The drawings are part of the extensive collection of James Peter Hill (1873–1954). The origins of the drawings remain unknown, although Hill is known to have collected and studied *O. anatinus* specimens with colleagues at the end of the 19th century and the beginning of the 20th century (Watson 1955). A number of these drawings were previously published. Previously published images are cited below. We scanned the original drawings from the J.P. Hill Collection. (1a, b) M15 (unpublished), egg after removal of shell, 4.3 × 4.2 mm; (2) M16 (unpublished), egg after removal of shell; (3a, b, c) M17 (unpublished), egg after removal of shell, 5 × 4.5 mm; (4a) M18 (published as fig. 1 in Hughes 1993), 8-celled egg, 4-mm diameter; (4b, c, d, e) M18 (unpublished), 8-celled egg, 4-mm diameter; (5) M19 (unpublished), egg; (6a, b, c, d, e, f) M22 (unpublished), egg after removal of shell, 4.2 mm, blastodisk, 0.36 × 0.31 mm; (7a, b) M31 (unpublished), early embryo?; (8a) M32 (unpublished), cephalic region; (8b) M32 (unpublished), early embryo?; (9a) M43 (unpublished), embryo, TL about 6.5 mm; (9b) M43 (published as fig. 3 in Hughes and Hall 1998), embryo, TL about 6.5 mm, drawn by A. Cronin; (9c) M43 (unpublished), embryo, dorsal view, egg 17 × 14.5 mm, drawn by A. Cronin; (10a) M37 (unpublished), embryo; (10b) M37 (unpublished), allantois, 16.5 × 14 mm; (10c) M37 (unpublished), two eggs, egg size, 16.5 × 14 mm and 18 × 14 mm, embryo size 8 mm and 8.5 mm, respectively; (11a, b, c) M38 (unpublished), embryo; (12a, b, c, d, e, f, g, h, i) M39 (unpublished), embryo, egg size 16.5 × 15 mm; (13) M41 (unpublished), embryo, documented by G. Wilson; (14a) M40 (unpublished), embryo, TL: 9 mm, TL of egg: 15 mm, documented by G. Wilson, 1898; (14b) M40 (published as fig. 6 in Hughes and Hall 1998 and as fig. 12 in Hughes 1993), embryo, TL 9 mm, TL of egg 15 mm, documented by G. Wilson, 1898; (14c) M40 (published as fig. 16 in Hughes 1984), embryo, TL 9 mm, TL of egg 15 mm, documented by G. Wilson, 1898. This figure is continued in Fig. 4.8.

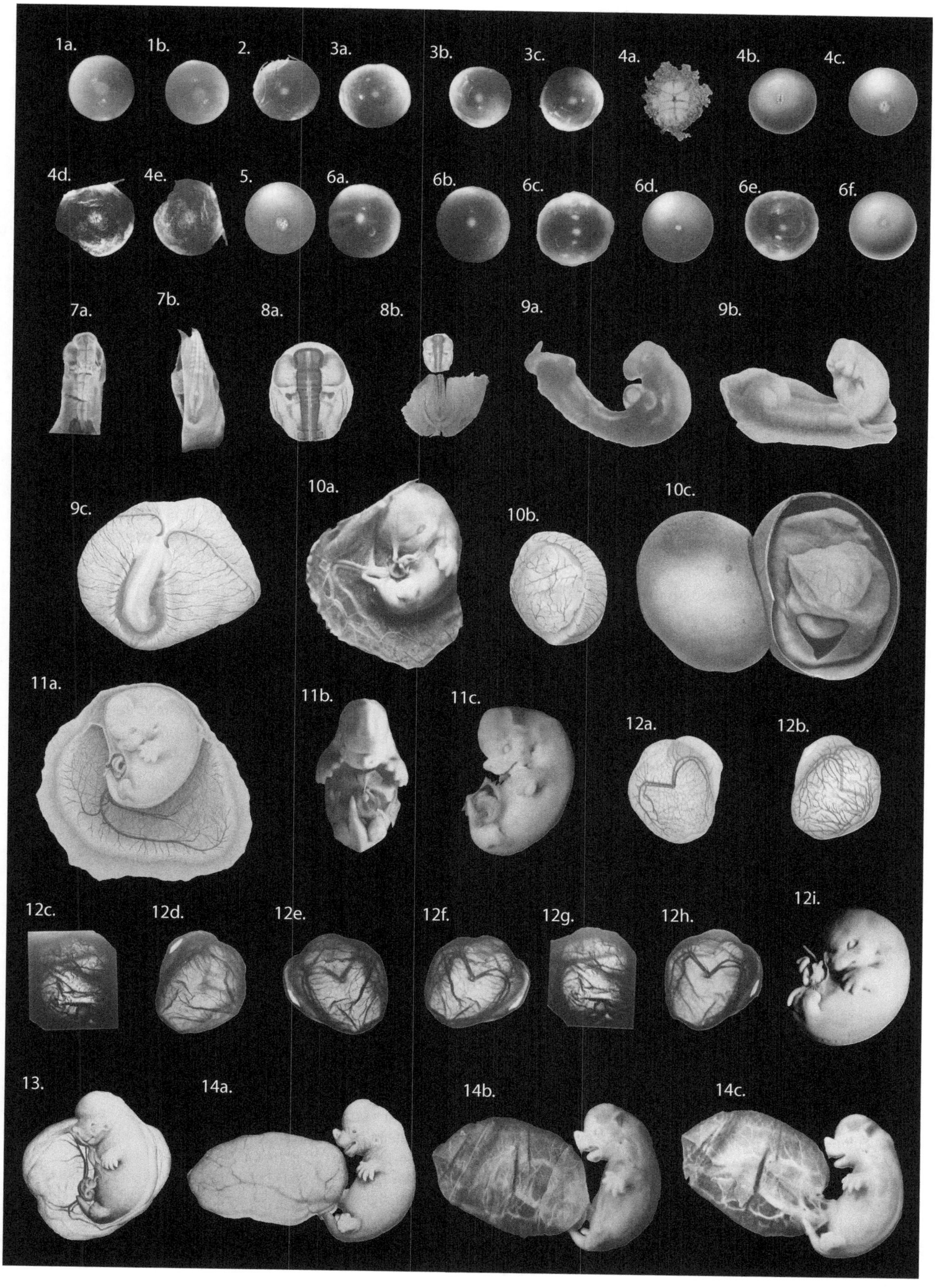
1a.
1b.
2.
3a.
3b.
3c.
4a.
4b.
4c.
4d.
4e.
5.
6a.
6b.
6c.
6d.
6e.
6f.
7a.
7b.
8a.
8b.
9a.
9b.
9c.
10a.
10b.
10c.
11a.
11b.
11c.
12a.
12b.
12c.
12d.
12e.
12f.
12g.
12h.
12i.
13.
14a.
14b.
14c.

slightly in their general anatomical construction, and mainly proportional developmental changes distinguish the different clades. In the following, we only concentrate on external organ development due to comprehensive data availability.

4.9.1 Methodological framework

External characteristics are initially used to stage embryos in the progress of their development before investigating internal structures. External anatomy gives an initial impression of how far organs could be developed underneath the covering skin. For example, a paddlelike limb will not yet have fully developed phalanges and the eye anlage will not yet be able to process visual information. In recent years, developmental characters were the subject of several evolutionary studies in amniotes (e.g., Bininda-Emonds *et al.* 2002, 2003), mainly using two different methods. On the one hand, specific larval characters of lissamphibians were compared in a traditional cladistic sense (e.g., Haas 2003). On the other hand, new evolutionary approaches were developed on how to quantify the relative timing of ontogenetic characters through ontogeny in different species (e.g., Smith 1997, Jeffery *et al.* 2005, Harrison and Larsson 2008, Germain and Laurin 2009). Both approaches help to reconstruct ancestral ontogenies throughout evolutionary history. In the following, we use the second approach and provide new data in reconstructing the ancestral timing of organogenesis in the last common ancestor of placental mammals.

4.9.2 Historical background

In order to perform comparative studies on embryonic development, it is necessary to work with a comparable and consistent anatomical nomenclature. The first systematic scientific documentation of external embryonic characteristics dates back to von Soemmerring (1799), who documented the external anatomical development of 18 aborted human embryos ordered after the last menstrual cycles of their mothers (Hopwood 2007). The figures represented an idealistic illustration rather than actual specimens. During the 19th century, more comprehensive embryonic studies were provided. Franz Keibel edited a 16-volume series from 1897 to 1938, called "Normentafeln zur Entwicklungsgeschichte der Wirbeltiere" (normal plates of vertebrate embryonic development), on a variety of vertebrate embryos, including several mammalian species. He introduced his series with the development of the domestic pig (Fig. 4.11) (Keibel 1897). In this series, the author developed the approach of normal plates ("Normentafeln"), which uses a combination of illustrations and tables to describe the anatomy of individual embryos separately. In these, non-aborted embryos, without malformations (hence "normal"), were described and depicted. Although an enormous data set was provided by the authors of this book series and a conceptual framework was available in the form of evolutionary theory, the methodology was not yet developed to analyze different patterns of embryology quantitatively. Today, this can be done using new comparative methods that take into account phylogeny and shape changes (e.g., Richardson and Keuck 2002, Schmidt and Starck 2004, Maxwell and Harrison 2009). To this end, the relative or absolute timing of developmental characters is compared among species and ancestral developmental sequences are reconstructed (see below).

During the 20th century, embryological research developed more toward an experimental science, and important contributions to our understanding of embryonic mechanisms emerged (Hoßfeld and Olsson 2003b). A particular interest for experimental model organisms developed. For these species (e.g., chicken, mouse, fruit fly, and round worm), comprehensive staging tables were

▸ **Fig. 4.8: Embryonic series of *Ornithorhynchus anatinus* (Monotremata, Ornithorhynchidae) (Part 2/2).** Continuation of Fig. 4.7. (15a, b) NMA (National Museum of Australia, Canberra, Australia) 684 (modified from Manger *et al.* 1998: fig. 1), hatchling, CR length 13.5 mm; (16a, b) M44 (unpublished), hatchling, TL (total length) 16 mm; (16c) M44 (published as fig. 7 in Hughes and Hall 1998), hatchling, TL 16 mm; (16d, e, f, g, h) M44 (unpublished), hatchling, TL 16 mm; (17a, b) NMA 685 (modified from Manger *et al.* 1998: fig. 1), hatchling, CRL (crown-rump-length) 16.2 mm; (18a, b) NMA 686 (modified from Manger *et al.* 1998: fig. 1), hatchling, CRL 21.1 mm; (19a, b) NMA 687 (modified from Manger *et al.* 1998: fig. 1), hatchling, CRL 26.1 mm; (20a, b) NMA 689 (modified from Manger *et al.* 1998: fig. 2), hatchling, CRL 31.6 mm; (21) M42 (unpublished), young, documented by G. Wilson, 1898; (22) M45 (unpublished), young, TL 33 mm; (23a, b) NMA 690 (modified from Manger *et al.* 1998: fig. 2), young, CRL 37.6 mm; (24a, b) NMA 691 (modified from Manger *et al.* 1998: fig. 2), young, CRL 43.2 mm; (25a, b) NMA 692 (modified from Manger *et al.* 1998: fig. 3), young, CRL 56.6 mm; (26a, b) NMA 694 (modified from Manger *et al.* 1998: fig. 3), young, CRL 68.0 mm; (27a, b) NMA 695 (modified from Manger *et al.* 1998: fig. 3), young, CRL 68.3 mm; (28a, b) NMA 696 (modified from Manger *et al.* 1998: fig. 4), young, CRL 70.9 mm; (29) M- (unpublished), young, TL 80 mm; (30a, b) NMA 697 (modified from Manger *et al.* 1998: fig. 4), young, CRL 92.2 mm; (31a, b) MVM (Museum of Victoria, Melbourne, Australia) C27576 (modified from Manger *et al.* 1998: fig. 4), young, CRL 101.3 mm; (32a, b) NMA 698 (modified from Manger *et al.* 1998: fig. 5), young, CRL 150.0 mm.

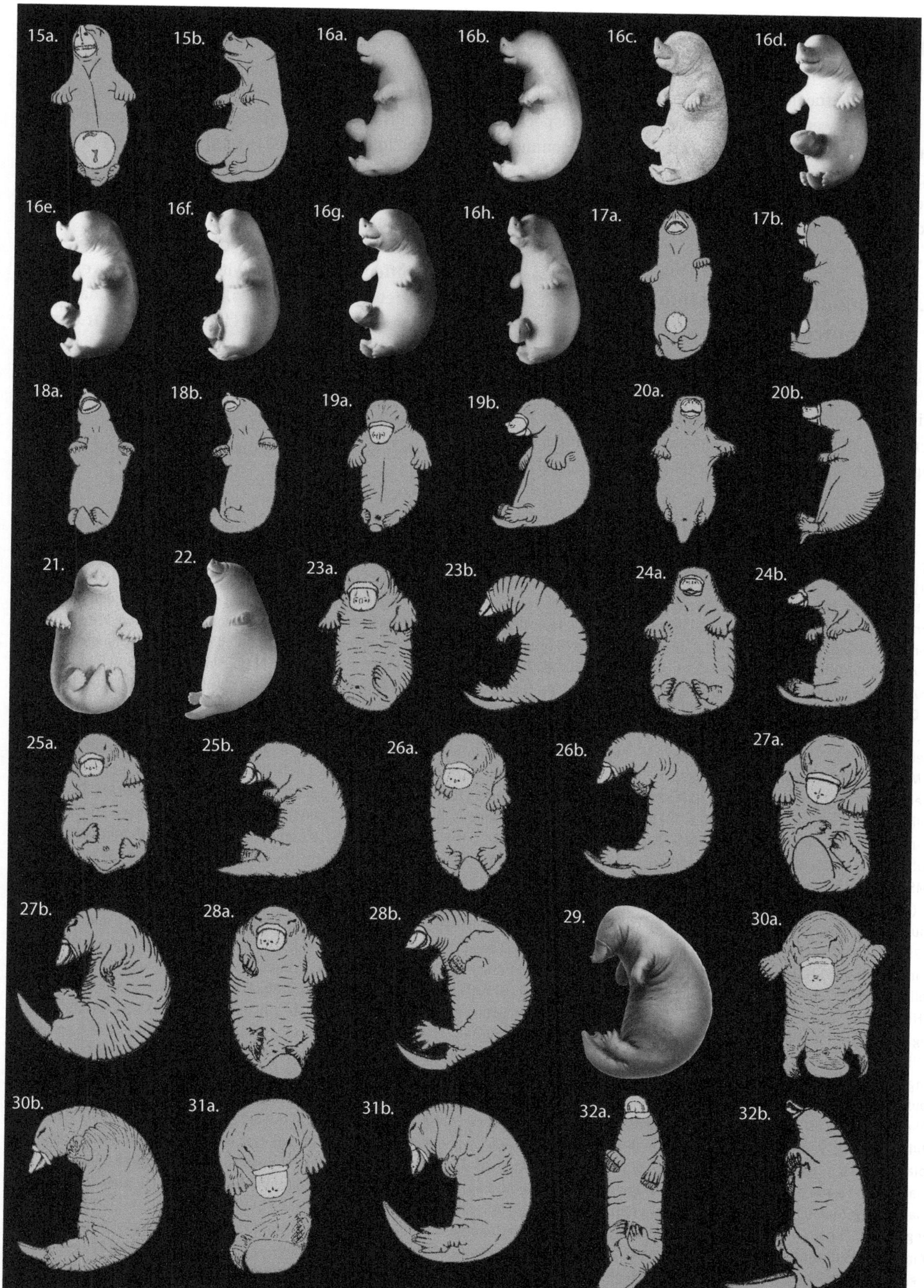
15a.
15b.
16a.
16b.
16c.
16d.
16e.
16f.
16g.
16h.
17a.
17b.
18a.
18b.
19a.
19b.
20a.
20b.
21.
22.
23a.
23b.
24a.
24b.
25a.
25b.
26a.
26b.
27a.
27b.
28a.
28b.
29.
30a.
30b.
31a.
31b.
32a.
32b.

developed (Hamburger and Hamilton 1951, Harrison 1969, Theiler 1989). Based on a large number of embryonic specimens, artificially delimited stages of development, which illustrate the most common embryonic features at a defined period, were described and depicted. Intraspecific variability was neglected by presenting the most common morphotype. As such, embryonic stages do not represent real embryos but idealistic schemes for initial orientation in experimental biology.

4.9.3 Standard event system and heterochrony

The above-mentioned staging tables were created for particular species only and are difficult to compare with other, even closely related, species, making comparisons complicated. To resolve this problem, Werneburg (2009) developed the "Standard-Event-System" (SES) to study vertebrate embryos comparatively. An initial set of 104 discrete embryonic characters was defined that one can easily recognize in any vertebrate species independent of the age of the individual specimen or its taxonomic identity. In the following years, several species were described using the SES system and the count of characters has increased to 166 (updated data set: https://en.wikipedia.org/wiki/Standard_Event_System).

The relative appearance of particular organs and tissues differs among and within all major vertebrate groups. Heterochronic effects, evolutionary changes in the timing of developmental characters, appear that are characteristic for particular clades. In general, the earlier an organ appears in development, the longer it can grow, suppress, and relocate the expansion of other organs (Werneburg *et al.* 2015). This has been called the heterotopic effect (Zelditch and Fink 1996, McNamara 2002).

▸ **Fig. 4.9: Embryonic series of the eastern quoll *Dasyurus viverrinus*** (Marsupialia, Dasyuromorpha) based on drawings from the Embryological collection of the Museum für Naturkunde Berlin, Germany. The drawings are part of the extensive collection of James Peter Hill (1873–1954). To our knowledge, the drawings represent specimens from the captive breeding population of the eastern quoll *Dasyurus viverrinus* that Hill kept between 1899–1901 and 1904 (Watson 1955). A number of these drawings were previously published. We scanned the original drawings from the J.P. Hill Collection. (1a) Ms78 (published as fig. 53 on plate 6 in Hill 1910), 6-celled egg, 0.34-mm diameter, side view; (1b) Ms78 (published as fig. 54 on plate 6 in Hill 1910), 6-celled egg, 0.34-mm diameter, lower pole view; (2a) MS79 (published as fig. 57 on plate 6 in Hill 1910), 12-celled egg, 0.38-mm diameter, side view; (2b) MS79 (published as fig. 58 on plate 6 in Hill 1910), 12-celled egg, 0.38-mm diameter, lower pole view; (2c) MS79 (published as fig. 55 on plate 6 in Hill 1910), 16-celled egg, 0.37-mm diameter, side view; (2d) MS79 (published as fig. 56 on plate 6 in Hill 1910), 16-celled egg, 0.37-mm diameter, lower pole view; (2e) MS 79 (unpublished), 16-celled egg, 0.36-mm diameter; (2f) MS79 (published as fig. 60 on plate 6 in Hill 1910), 31-celled egg, 0.375-mm diameter, side view; (2g) MS79 (published as fig. 59 on plate 6 in Hill 1910), 31-celled egg, 0.375-mm diameter, lower pole view; (2h) MS 79 (unpublished), 32-celled egg, 0.375-mm diameter; (2i) MS79 (published as fig. 61 on plate 6 in Hill 1910) blastocyst, 0.39-mm diameter; (2j) MS79 (published as fig. 62 on plate 6 in Hill 1910) blastocyst, 0.4-mm diameter; (3) MS109 (unpublished), blastula 5.5-mm diameter; (4) MS115 (unpublished), 7.5-mm diameter; (5a) MS118 (unpublished), 15 somites, TL 6.37 mm; (5b) MS118 (unpublished); (5c) MS118 (unpublished), about 11 somites; (6) MS119 (unpublished); (7a) MS120 (unpublished); (7b) MS120 (unpublished); (7c) MS120 (unpublished), TL? 4.7 mm; (8) MS123 (unpublished, TL 4 mm; (9a) MS124 (unpublished); (9b) MS124 (unpublished), TL 4.8 mm; (10a) MS125 (unpublished), TL 3.7 mm; (10b) MS125 (unpublished), TL 3.6 mm; (11) MS126 (unpublished), TL 3.5 mm; (12a) MS127 (unpublished), TL 5.16 mm; (12b) MS127 (unpublished); (12c) MS127 (unpublished), TL 4.2 mm; (13) MS128 (unpublished), TL 5 mm; (14a) MS129 (unpublished), TL 5.4 mm; (14b) MS129 (unpublished), TL 5.5 mm; (14c) MS129 (unpublished), TL 5.4 mm; (15) MS194 (unpublished), TL 5 mm; (16a) MS131 (unpublished), TL 6 mm; (16b) MS131 (unpublished), TL 5.8 mm; (16c) MS131 (unpublished), TL 6 mm; (17) MS132 (unpublished), TL 6.5 mm; (18a) MS193 (published as fig. 1 on plate 1 in Hill and Osman-Hill 1955) embryo shortly before birth, TL 5.5 mm; (18b) MS193 (published as fig. 3 on plate 1 in Hill and Osman-Hill 1955 and as fig. 32B in Klima 1987) pouch young several hours old/neonate?, TL 6/5.5 mm?; (19) MS135 (published as fig. 2 on plate 1 in Hill and Osman-Hill 1955 and as fig. 32A in Klima 1987) neonate, TL 5.6 mm; (20) MS138 (published as fig. 4 on plate 1 in Hill and Osman-Hill 1955) neonate, 3 hours old, TL 6 mm; (21) MS 142 (published as fig. 5 on plate 1 in Hill and Osman-Hill 1955) pouch young attached to teat, 26 hours old, TL 6 mm; (22) MS148 (published as fig. 6 on plate 1 in Hill and Osman-Hill 1955) pouch young, about 3 days old, TL 7 mm; (23) MS149 (published as fig. 7 on plate 1 in Hill and Osman-Hill 1955) pouch young, 5–6 days old, TL 8 mm; (24) MS154 (published as fig. 8 on plate 1 in Hill and Osman-Hill 1955) pouch young attached to teat, about 7 days old, TL 8.5–9 mm; (25) MS158 (published as fig. 9 on plate 1 in Hill and Osman-Hill 1955) pouch young, about 10 days old, TL 10 mm; (26) MS164 (published as fig. 10 on plate 2 in Hill and Osman-Hill 1955) pouch young, about 14 days old, TL 13.5 mm; (27) MS173 (published as fig. 11 on plate 2 in Hill and Osman-Hill 1955 and as fig. 32C in Klima 1987) pouch young, 19 days old, TL 16.5 mm; (28) MS176 (published as fig. 12 on plate 2 in Hill and Osman-Hill 1955) pouch young, about 25 days old, TL 20 mm; (29) MS- (published as fig. 13 on plate 2 in Hill and Osman-Hill 1955) pouch young, about 35 days old, TL 24 mm; (30) MS179 (published as fig. 14 on plate 2 in Hill and Osman-Hill 1955) pouch young, about 41 old, TL 29 mm; (31) MS181 (published as fig. 15 on plate 2 in Hill and Osman-Hill 1955) pouch young, 46 days old, TL 42 mm; (32) MS182 (published as fig. 16 on plate 2 in Hill and Osman-Hill 1955) pouch young, just over 2 months old, TL 59 mm; (33) MS183 (published as fig. 17 on plate 2 in Hill and Osman-Hill 1955) pouch young, about 2.5 months old, TL 65 mm.

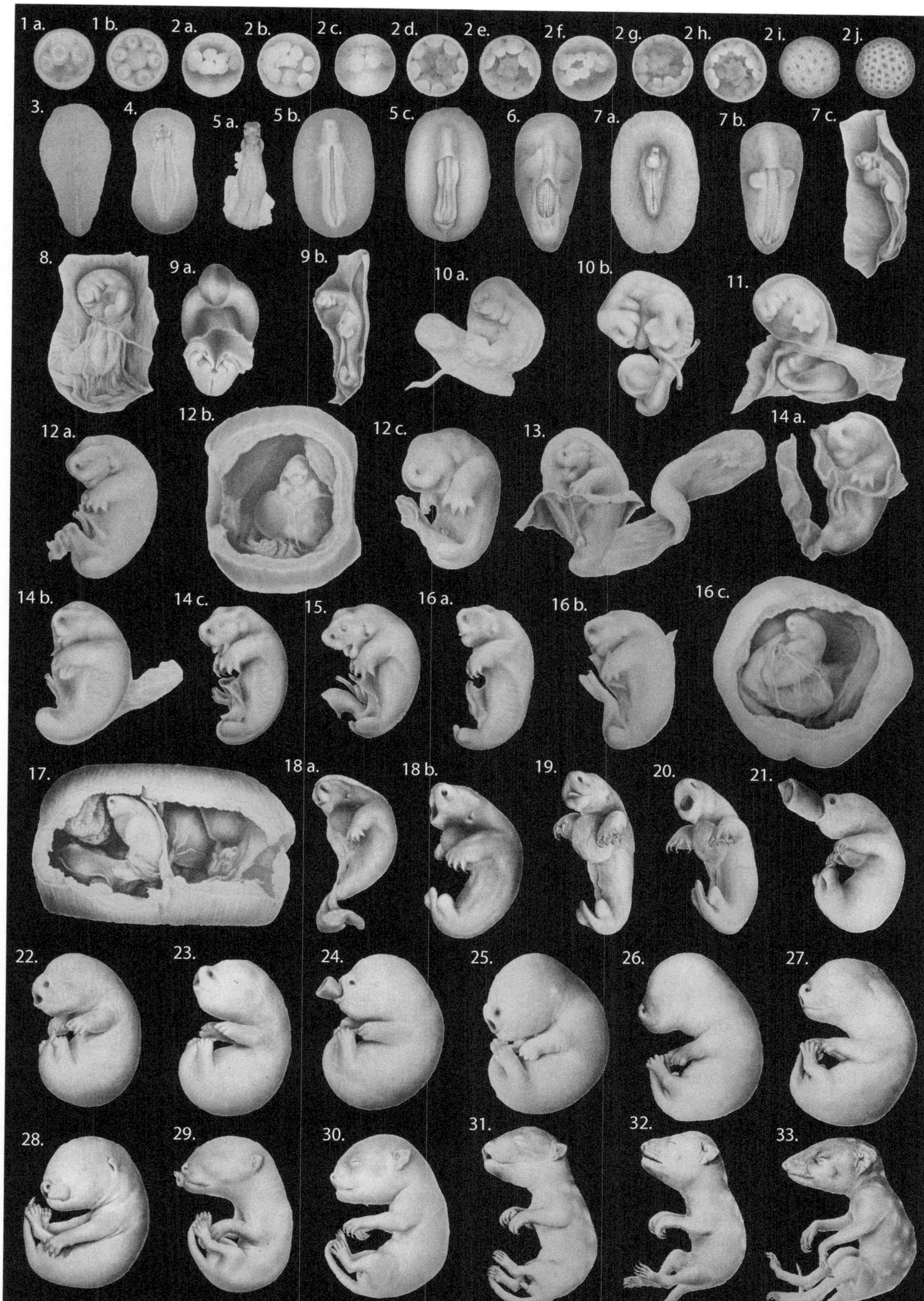
1 a.
1 b.
2 a.
2 b.
2 c.
2 d.
2 e.
2 f.
2 g.
2 h.
2 i.
2 j.
3.
4.
5 a.
5 b.
5 c.
6.
7 a.
7 b.
7 c.
8.
9 a.
9 b.
10 a.
10 b.
11.
12 a.
12 b.
12 c.
13.
14 a.
14 b.
14 c.
15.
16 a.
16 b.
16 c.
17.
18 a.
18 b.
19.
20.
21.
22.
23.
24.
25.
26.
27.
28.
29.
30.
31.
32.
33.

4.9.4 Data collection and evaluation

In a recent study (Werneburg *et al.* 2016), embryonic series of about 80 mammalian and non-mammalian tetrapod species were compared, and the relative timing of 166 external developmental characters (SES characters) was documented. All species in the study exhibit a different length of embryonic development with a gestation ranging from a few days (marsupials) to about 2 years (elephants). To make these developmental series comparable, development was scaled between 0 and 1 (Germain and Laurin 2009). Fertilization was defined as "0" and birth/hatching as "1". This means that each SES character was assigned a score between 0 and 1. These scores were mapped onto a molecular and evolutionary time scaled phylogeny. Using squared-change parsimony methodology (Felsenstein 1985), ancestral values were reconstructed for these continuous characters. The length of each branch in the phylogeny was considered for the calculation of the ancestral value, which finally represents a weighted mean of two values. Using this approach, the ancestral values were reconstructed for all characters of all nodes within the phylogeny.

4.9.5 Evolutionary patterns

The major mammalian clades, Monotremata, Marsupialia, and Placentalia, differ in the relative reconstructed timing of various characters. These heterochronies can be interpreted by the different body proportions that the adults show – the earlier a character appears in ontogeny, the larger or more differentiated it is in the adult (Werneburg *et al.* 2015). Major differences occur around the time of birth/hatching, which can be interpreted by the specialized adaptations that the different life histories of the major mammalian groups require (see below and Werneburg and Geiger 2017).

The ancestral patterns of monotreme and marsupial organogenesis do not differ drastically from the developmental sequences known for the extant species (Werneburg *et al.* 2016). Therefore, we present here one representative embryonic series for each of these two clades, which include images never published before. Although the monotreme *Tachyglossus aculeatus*, the short-beaked echidna, is well documented in its external embryonic anatomy (e.g., Fig. 4.3 A, E; Semon 1894a–d; Werneburg and Sánchez-Villagra 2011), only little is known about the prehatching organogenesis of the platypus, *Ornithorhynchus anatinus* (Hughes and Hall 1988), because the eggs are incubated in breeding burrows, which are up to 30 m in length and difficult to access (Grant 1989). To illustrate marsupial development, we selected the eastern quoll, *Dasyurus viverrinus*. Only part of its embryonic anatomy was documented so far (e.g., Hill 1910, Hill and Osman-Hill 1955). We collected all available images for the platypus (Figs. 4.7–8) and eastern quoll (Fig. 4.9) housed in the collection of James P. Hill, which is now part of the Embryological Collection at the Museum für Naturkunde in Berlin, Germany. We ordered the specimens by the progress of development (following Werneburg 2009).

4.9.6 Embryology of the last common ancestor of Placentalia

In addition to monotremes and marsupials, we were particularly interested in the ancestral organogenesis of the last common ancestor of Placentalia because a variety of body shapes evolved within this clade, including diverse terrestrial, flying, aquatic, fossorial, and arboreal forms. This pattern is reflected in the diversity of embryonic shapes, different life history modes, and anatomical conditions at birth. We present the reconstructed timing of developmental characters for the placental mammal ancestor and created a staging table for this ancestor (Tab. 4.4). When a new state of a character complex such as limb development appeared in the developmental sequence, a new SES stage was assigned. We were able to create 18 SES-stages. Finally, the character sequence was the basis for a graphical reconstruction. For illustration, we mostly relied on the staging system available for the tree shrew, *Tupaia javanica* (de Lange and Nierstrasz 1932), because the adults of this species look similar to the last common ancestor of placentals reconstructed by O'Leary *et al.* (2013). We modified the external anatomy of these embryos based on our phylogenetic reconstruction for

▸ **Fig. 4.10: Schematic illustration of organ development in the last common ancestor of Placentalia.** For general orientation of body proportion, the embryogenesis of the tree shrew *Tupaia javanica* was used (de Lange and Nierstrasz 1932), with the exception of SES stage 4 (modified after a comparable stage of the house mouse *Mus musculus;* Theiler 1989) and SES stage 17 (after the reconstruction of the adult ancestral placental mammal; O'Leary *et al.* 2013). The original drawings were modified to illustrate the sequence of embryonic character appearance in the last common ancestor of placental mammals. Numbers in each picture refer to the fine grade sequence of character appearance in each stage. SES stage 18, weaning, is not illustrated. Embryos not to scale (see Tab. 4.4).

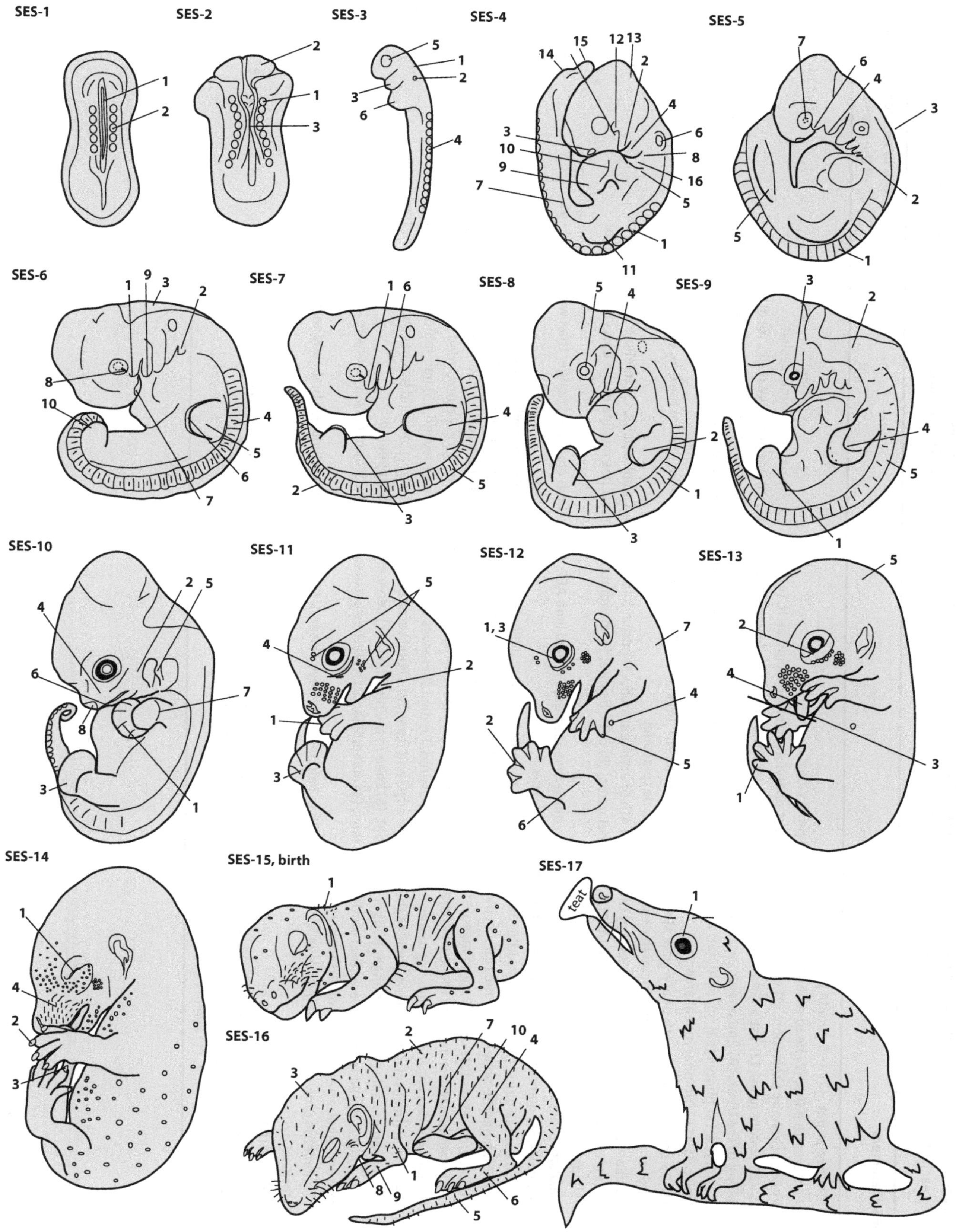
SES-1
SES-2
SES-3
SES-4
SES-5
SES-6
SES-7
SES-8
SES-9
SES-10
SES-11
SES-12
SES-13
SES-14
SES-15, birth
SES-16
SES-17
teat

Tab. 4.4: Embryonic development of the last common ancestor of Placentalia as reconstructed for this chapter. Fine grade sequence of the appearance of SES-characters is indicated by numbers within each stage. For illustration, see Fig. 4.9. For character descriptions, see Werneburg (2009) and https://en.wikipedia.org/wiki/Standard_Event_System.

SES stage	SES characters	% in relation to opening of the eyelid
1.	(1) Primitive streak, (2) 1–5 somite pairs	11.4–14.7
2.	(1) 6–10 somite pairs, (2) head bulbus, (3) neural fold begin to close	14.8–15.7
3.	(1) Anterior cephalic projection, (2) otic pit, (3) bud of the 1st pharyngeal (mandibular) arch, (4) 11–15 somite pairs, (5) optic vesicle	17.2–18.5
4.	(1) 21–25 somite pairs, (2) 1st pharyngeal slit, (3) olfactory pit (placode), (4) 2nd pharyngeal (hyoid) arch, (5) 3rd pharyngeal slit, (6) otic vesicle, (7) coiling of the trunk, (8) 2nd pharyngeal slit, (9) ventricle bulbus, (10) ventricle is S-shaped, (11) forelimb ridge (wider than long), (12) maxillary process of the 1st pharyngeal (mandibular) arch reaches posterior level of the eye, (13) posterior neuropore closed, (14) anterior neuropore closed, (15) tail bud, (16) 3rd pharyngeal arch	20.3–23.4
5.	(1) 31–35 somite pairs, (2) 4th pharyngeal slit, (3) cervical flexure of 90°, (4) mandibular arch reaches posterior level of the eye, (5) hind limb ridge, (6) maxillary process reaches midline of the eye, (7) lens vesicle (placode)	23.4–25.0
6.	(1) Maxillary process reaches anterior level of the lens, (2) 4th pharyngeal arch, (3) nuchal fold, (4) 36–40 somite pairs, (5) forelimb bud (as wide as long), (6) forelimb apical epidermal ridge (AER), (7) external nares, (8) optic fissure, (9) mandibular arch reaches midline of the eye, (10) hind limb bud-shaped	25.0–27.5
7.	(1) Maxillary process at anterior level of the eye, (2) trunk coiling disappeared, (3) hind limb AER, (4) forelimb elongated (longer than wide), (5) 41–45 somite pairs, (6) mandibular arch reaches anterior level of the lens	27.8–28.6
8.	(1) 46–50 somite pairs, (2) forelimb paddle-shaped, (3) hind limb elongated, (4) mandibular arch reaches anterior level of the eye, (5) contour lens/iris	28.7–30.1
9.	(1) Hind limb paddle-shaped, (2) otic capsule inconspicuous, (3) pupil forms, (4) digital plate at forelimb, (5) somites become hard to count	30.2–32.2
10.	(1) Digital grooves on forelimb, (2) pharyngeal slits closed, (3) digital plate at hind limb, (4) maxillary process fuses with frontonasal process, (5) pinna fold, (6) mandibular arch reaches the level of the frontonasal process, (7) elbow in the forelimb, (8) nose openings are surrounded as nostrils	33.1–34.5
11.	(1) Digital serrations at the forelimb, (2) thoracal bulbus disappeared, (3) digital grooves on hind limb, (4) lower eyelid, (5) hair follicles on the head	34.5–36.3
12.	(1) Eyelid is at the level of the scleral papillae (between eye edge and lens), (2) digital serration at the hind limb, (3) eyelid has begun to overgrow the eye, (4) mammary anlage, (5) first finger (longer than wide), (6) knee in the hind limb, (7) cervical flexure disappeared	36.6–39.5
13.	(1) First toe, (2) eyelid reaches the ventral border of the lens, (3) mandibular arch reaches occlusion point with upper jaw, (4) tongue is protruding, (5) mesencephalic head projection disappeared	39.9–42.7
14.	*Fetal allometric growth increasing.* (1) eyelid covers more than half of the eye (eye closure), (2) first claw on forelimb, (3) first claw on hind limb, (4) vibrissal hairs on the snout	43.3–65.7
15.	(1) Hairs on the neck, (2) birth	85.4–86.4
16.	*First postnatal stage.* (1) hairs appear on forelimbs (proximal), (2) back, (3) top of the head, (4) hind limbs (proximal), (5) tail, (6) hind limbs (distal), (7) belly (dorsal part), (8) throat, (9) forelimbs (distal), (10) belly (ventral part).	87.4–97.8
17.	*Second postnatal stage.* (1) eyelids open	100

the placental mammal ancestor. We found, for example, that the neonates of the earliest placental mammals were altricial, almost naked and the eyes were still closed. In contrast to all marsupials, the limbs were evenly developed at birth (Fig. 4.10, SES-stage 15; see also Werneburg *et al.* 2016).

4.10 Gestation

At oviposition, monotreme egg size is less than two centimeters in diameter. Total gestation in the short-beaked echidna, *Tachyglossus aculeatus*, lasts for 17–27 days. At oviposition, the embryo has already developed an embryonic stage comparable with that of a chicken embryo at about 50 hours of incubation, taking a structured neural anlage, formed eye vesicles, 19–20 somite pairs, and a formed head shield of the proamnion as references (Semon 1894a–d, Hamburger and Hamilton 1951, Starck 1995, Werneburg and Sánchez-Villagra 2011). After oviposition, one or two eggs (Tab. 4.2) are incubated in the pouch(es) (Figs. 4.3 D and 4.13 B) for about 10 days; a similar amount of time can be assumed for the western long-beaked echidna, *Zaglossus bruijni* (Hayssen *et al.* 1993).

Intrauterine gestation usually lasts for 10–14 days in the platypus, *Ornithorhynchus anatinus*. After egg lay, normally two eggs are incubated in hollows for a further 7–14 days (Hayssen *et al.* 1993, Hughes 1993, Manger *et al.* 1998). Those hollows end in a chamber, which can have some grass inside (Fig. 13A). The platypus has an incubatorium, in which the egg is laid (Zeller 2004b). The fact that monotremes lay eggs was first recognized 93 years after their discovery by Europeans in 1884 (Caldwell 1884; see also Griffiths 1978). Australian Aborigines knew but were disbelieved by early European settlers (Moyal 2004). In contrast to sauropsids, a mineralized outer egg layer is assumed never to have been present in mammalian ancestors (Stewart 1997, Laurin *et al.* 2000), although monotreme oviparity is certainly inherited from amniote ancestors (Starck 1995). The ancestral amniote mode of reproduction is hard to characterize. Furthermore, it is not clear to what degree reproduction and development can be homologized between monotremes and sauropsids (Storch and Schröpfer 2004).

Incubation takes about 10 days in all monotreme species. The exact length of gestation of the platypus is unknown, partly because it is thought that the female can store sperm after copulation.

Marsupials have a very short gestation period, varying between 11 and 45 days (de Magalhães and Costa 2009, Clauss *et al.* 2014), with a mean of about 28 days (Tab. 4.2). Consequently, the marsupial neonates are born at a very early developmental stage, and a part of morphogenesis does not occur in utero but while the young is attached to the teat of the mother, often within a pouch (marsupium) (Tyndale-Biscoe 2005). In most cases, gestation is relatively shorter in marsupials when compared with placental mammals of similar size (Starck 1995).

In placental mammals, the gestation length varies strongly among species (Clauss *et al.* 2014). The shortest gestation occurs in the golden hamster, *Mesocricetus auratus*, and the common shrew, *Sorex araneus*, at about 2 weeks and the longest is documented for the African elephant, *Loxodonta africana*, at about 2 years (Tab. 4.3).

4.11 Delivery

Monotreme hatchlings have a size of 10–15 mm, are naked, and have closed eyes. The forelimbs are further developed than the hind limbs; however, monotreme hatchlings are not as altricial as marsupials are. Like marsupials, monotremes have large heads and forelimbs at birth, and urine is produced in their mesonephros (Zeller 2004b). They breathe via the primary branching of the bronchia, which is covered by respiratory epithelium before the alveoli and the bronchial branching are fully formed, which occurs postnatally (Starck 1995). Eyelids open about 12 weeks after hatching in the short-beaked echidna *T. aculeatus* (Vaughan *et al.* 2011) and after 11 weeks in the platypus *O. anatinus* (Manger *et al.* 1998).

Around hatching, monotremes have a true egg tooth and a keratinous egg tooth (caruncle) with an ossified part (os carunculae). The egg tooth is formed by the enamel organ and tooth papillae, and it is made up of dentine covered by a thin enamel layer. The caruncle develops as an independent new organ and later fuses with the processus ascendens of the premaxillary. Monotremes are the only vertebrates in which both a true egg tooth and caruncle are present. Squamates have just a true egg tooth; the other sauropsids have a horny egg tooth and their caruncle is only an epidermal structure (Starck 1995). Egg tooth and/or caruncle are used by the hatchling to tear up the eggshell from inside. In marsupials, rudiments of both structures can be found in embryos of some species.

Contrary to mammals, viviparity evolved several times independently in fishes and squamates, permitting inferences on the evolutionary adaptations and constraints involved using comparative methods (Blackburn 2015).

The sudden transition from fetal to postnatal life in therian mammals is correlated with crucial physiological changes in, for instance, the circulatory system

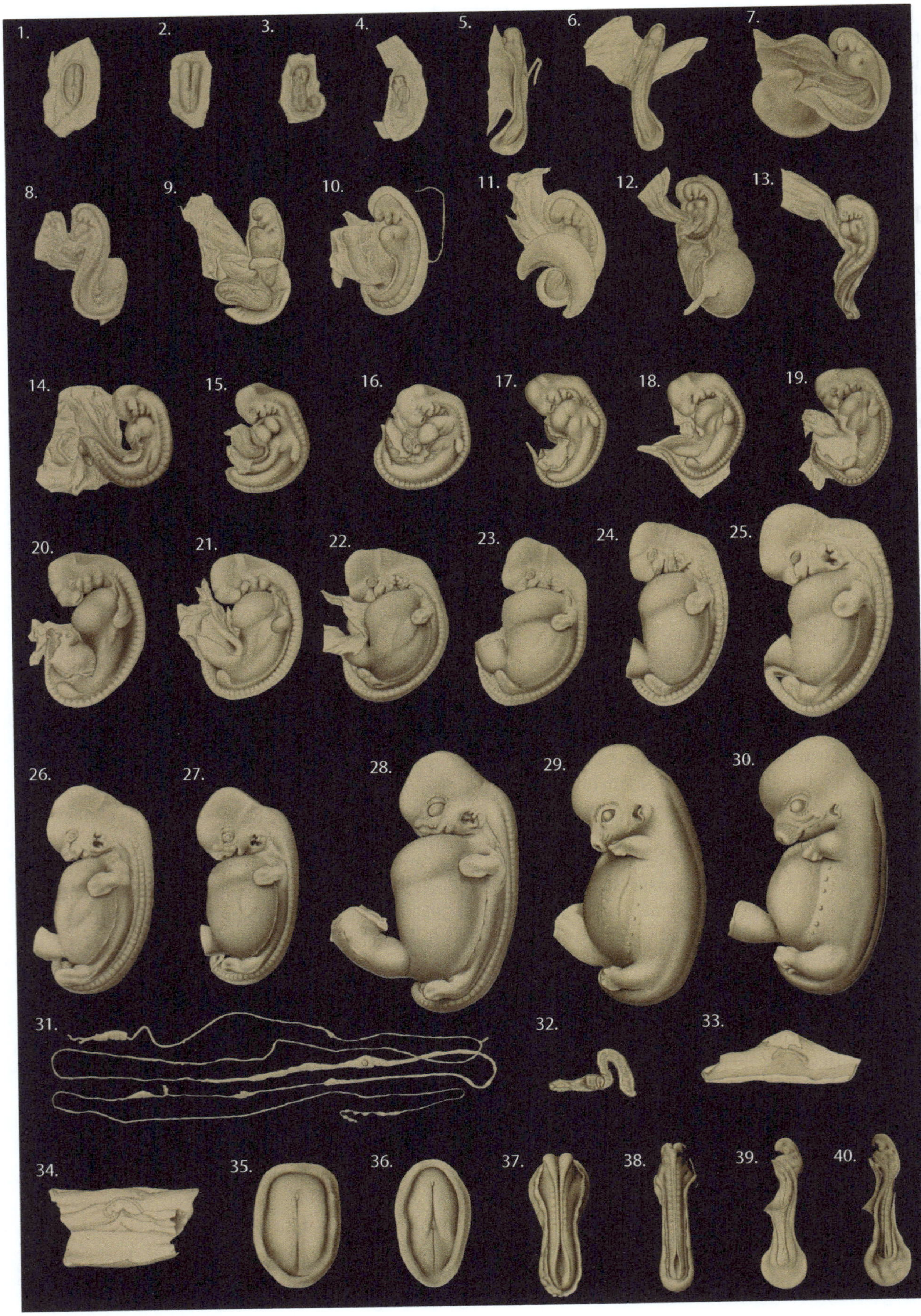
1.
2.
3.
4.
5.
6.
7.
8.
9.
10.
11.
12.
13.
14.
15.
16.
17.
18.
19.
20.
21.
22.
23.
24.
25.
26.
27.
28.
29.
30.
31.
32.
33.
34.
35.
36.
37.
38.
39.
40.

(Fig. 4.12 A–B) and basic metabolism. The neonate has to be prepared prenatally to perform all required functions at birth.

In the embryo, absorption of oxygen and nutrition is enabled through maternal blood circulation in the placenta (Fig. 4.12 A). Conversely, carbon dioxide and waste products are excreted through the placenta. The lungs are not yet unfolded in the fetus, and the blood circulation is not fully developed. Oxygenated blood in the fetus comes from the placenta through the vena umbilicalis, the ductus venosus, and the vena cava caudalis and enters the right atrium of the heart. Afterward, the blood passes through the foramen ovale, an opening in the septum of the right atrium. It floods into the left atrium, then into the left ventricle, and finally into the aorta. Because of the high resistance of the pulmonary vessels, only a minor fraction passes through the right ventricle and the lung circulatory system. Arteries originate from the aorta that direct toward the anterior part of the body, which facilitates a comprehensive blood support of the brain. On its way back from the head region, the right auricle receives deoxygenated blood from the vena cava cranialis. This blood does not enter the foramen ovale and mixes only slightly with the blood of the umbilical vein. Instead, it enters the aorta pulmonalis through the left ventricle. Throughout fetal life, the pulmonary artery has an open connection with the aorta through the ductus arteriosis Botalli. The lungs are still unfolded, and their capillaries provide a large resistance to flow. As a result, most of the blood reaching the first part of the aorta pulmonalis enters the aorta descendens through the ductus arteriosus. From there, it reaches the placenta through the umbilical arteries.

At birth, the pressure of the umbilical vein and the right atrium decreases due to the tearing of the umbilical cord and the fall in pulmonary resistance following expansion of the lungs. The pressure on the left side of the heart increases beyond that on the right side. This results in the closure of the foramen ovale by a preformed valve (Fig. 4.12 B). This valvula foraminis ovalis is pushed from the left side onto the thickened edge of the foramen. During the first breathing cycles of the neonate, triggered by the increased oxygen content of the blood, the ductus arteriosus is closed by the contraction of its smooth arterial musculature. The blood of the right ventricle now completely enters the capillary net of the lungs through the aorta pulmonalis. Here it becomes oxygenated and flows back to the left atrium through the pulmonary veins.

The therian neonate has to be able to breathe directly after birth. Therefore, the musculature of the thoracic wall and the diaphragm need to be fully developed at birth. The bronchi and the alveolar tree of the fetus are filled with liquid during gestation. This liquid is excreted by the alveolar epithelium. The amniotic fluid is separated from the lungs by the ventilator mechanism of the larynx. At birth, the cylindrical alveolar epithelium is flattened by the pressure of the incoming air. The remaining liquid is quickly resorbed.

The urinal waste of the fetus passes the placenta and is excreted along with the maternal urine. During fetal life, kidney function is not required. At the end of gestation, excretion of urine into the amniotic fluid can occur. In therians with a large allantois, urine is excreted through the urachus, which is a specific part of the allantois, particularly in early phases. Later on, urine excretion may also occur through the urethra into the amniotic fluid.

Amniotic fluid is ingested by the fetus (500 ml/day in humans). The fluid quickly passes the esophagus, stomach, and small intestine, and most of the water is resorbed. Other contents, shed epithelial cells and bile pigments, are excreted as dark meconium at the first excretion after birth.

In marsupials, birth takes place through the unpaired birth diverticulum (not through the paired vaginae), which opens into the urogenital duct.

The closure of the eyes before birth or hatching in amniotes is an evolutionarily ancestral feature (Fig. 4.14 B–G). Compared with non-amniote vertebrates, mammals and sauropsids experience a drastic environmental change from an aquatic milieu in the egg or the uterus to atmospheric conditions after delivery. Therefore, the eyelids are usually closed during fetal development to protect the incompletely developed eye at birth (for exceptions, see Müller 1972–1973; Fig. 4.14 B–G). Tear glands still need to be developed, and the visual system has to be able to process visual information. The transitory closure of eyelids, similar to the closure of the ear and mouth openings, is enabled by cells of the intermedial layer of the epidermis (Fig. 4.14 B). The secondary opening later in development is facilitated by the cornification of those cells. The transitory closure of the nose is based on peridermal adherence (similar to the epitrichium, a peridermal layer covering the sense organs in marsupial pouch young).

◂ **Fig. 4.11: Historical drawings of the embryogenesis of the domestic pig, *Sus scrofa*.** Compiled from Keibel (1897). Numbers of embryos correspond to the original publication.

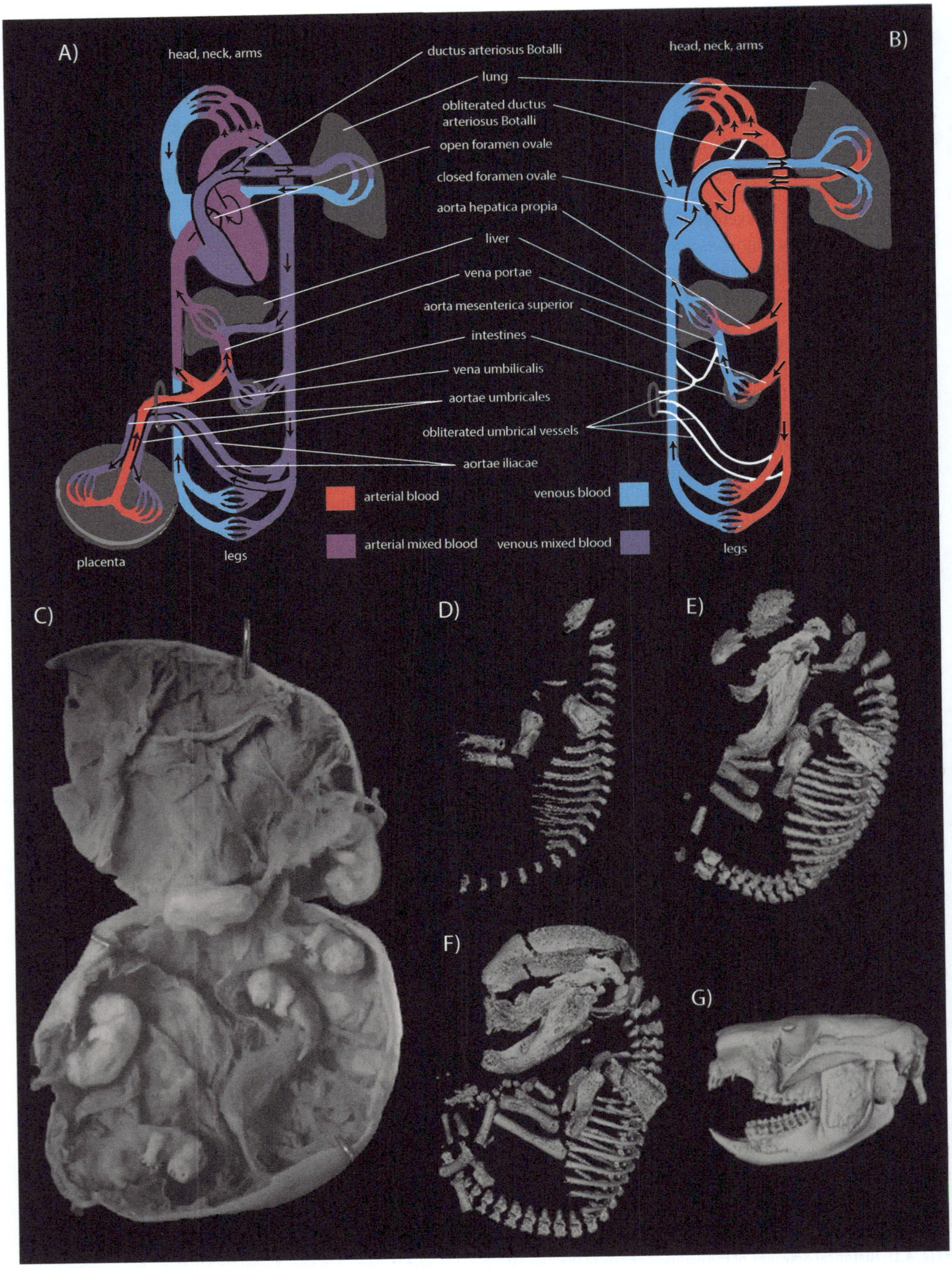
A)
head, neck, arms
ductus arteriosus Botalli
lung
obliterated ductus arteriosus Botalli
open foramen ovale
closed foramen ovale
aorta hepatica propia
liver
vena portae
aorta mesenterica superior
intestines
vena umbilicalis
aortae umbricales
obliterated umbrical vessels
aortae iliacae
arterial blood
venous blood
arterial mixed blood
venous mixed blood
placenta
legs
B)
head, neck, arms
legs
C)
D)
E)
F)
G)

4.12 Early marsupial postnatal life

Although all marsupials are altricial at birth, different degrees of altriciality can be distinguished (Fig. 4.14 E–H) (Hughes and Hall 1988, Smith 2015). In dasyurids and the honey possum *Tarsipes rostratus* (Diprotodontia), neonates have a neonatal weight of 3–20 mg (Tab. 4.2). Neither eye primordia, eyelids, nor retinal pigmentation are visible at birth. The tongue consists of multinucleate tubes and immature striations. The lungs show few partitions with superficial capillaries, and the metanephridic kidney is only represented as a ureteric bud. The metacarpals of the forelimbs are just differentiated and the primordial hind limb is an undifferentiated paddle (grade 1) (Fig. 4.14 E, Fig. 4.9–18 b).

In peramelids, didelphids, and brushtail possums *Trichosurus* (Diprotodontia), neonates have a neonatal weight of 100–300 mg (Tab. 4.2). Eye primordia and pigmentation are visible. The tongue muscles show striations, lung development has progressed, and the metanephric kidney is represented by a primitive ureter and a secondary branching (grade 2) (Fig. 4.13 F).

In kangaroo species (Macropodidae, *Macropus*), neonates have a birth weight of 300–900 mg (Tab. 4.2). Their eyes are prominently primordial, and a pigmented ring has developed. The tongue muscles show mature striations, and the lungs are large and highly subdivided and vascularized. The metanephric kidney has developed terminal branches and collecting ducts (grade 3).

The different grades of altriciality that can be found among marsupials do not represent a phylogenetic gradation (Smith 2015). The intermediate grade (grade 2) appears to be ancestral and is the most widespread marsupial condition. The ultra-altricial condition (grade 1) evolved twice, whereas the most advanced condition is highly derived (grade 3).

It is important to note that neonate marsupials are not only characterized by immature development. As outlined below, several anatomical features (temporary closure of eyes and ears, deciduous claws on the forelimbs, and mesonephric kidneys) represent characters that are absent in the adult. Instead, these are functional adaptations to early postnatal life conditions, and they belong to the ancestral condition inferred for marsupial ontogeny (Maier 1999, Zeller 2004a).

Marsupial birth is characterized by the transfer of the neonate from the mother's urogenital sinus to the teat, which in most species is located inside a pouch (Hughes and Hall 1988, Gemmell *et al.* 2002). The mother has only little influence on the journey of the neonate (Zeller 2004a). In macropodids and the brushtail possums *Trichosurus*, the mother moves into a sitting position during labor. The tail turns ventrally, and the genital opening is directed toward the pouch (respectively toward the teats). In other marsupials, such as bandicoots and dasyurids, the mothers are also known to position the genital opening in such a way as to facilitate the neonate's transfer to the pouch. The neonate frees itself from the egg membranes with the help of the keratinous, deciduous claws of its forelimbs, which are replaced by permanent claws later in development. Some marsupials do not climb at birth (Ashwell and Shulruf 2014).

Altricial marsupial neonates are highly vulnerable after birth and, therefore, need to reach the teat as soon as possible to survive. This is accomplished through sinuous contractions of the para-axial musculature of the torso during which the head swings in a trial and error mode to find the teat. These movements are already performed in utero (Drews *et al.* 2013). The neonates exclusively maneuver themselves using their forelimbs, in which functioning chondrified elements are present and which are at a much further stage of development than the paddlelike hind limbs. Digitopalmar prehension and the above-mentioned deciduous claws in the forelimbs are further adaptations for this critical journey directly after birth. In certain marsupials (some didelphids), the requirement for the neonates to reach a teat as quickly as possible is further increased by the litter size being larger than the number of teats, meaning that neonates that are not able to attach to a teat before they are all occupied by siblings die soon thereafter (Hughes and Hall 1988, Tyndale-Biscoe 2005, Bininda-Emonds *et al.* 2007). In macropodids, the neonate crawls up the fur of the mother and reaches a teat within about five minutes (Starck 1995). The neonate most likely

◂ **Fig. 4.12: Perinatal anatomy.** Blood circulatory system in fetal (A) and postnatal mammals (B). Arrows indicate the direction of blood flow. Note the integration of the placenta into the circulatory system, the open foramen ovale, and the admixture of blood in the fetal condition (A). After birth, the foramen ovale is closed, the umbilical vessels and the ductus arteriosis Botalli are obliterated, and a clear distinction between arterial and venous blood becomes apparent (B). Modified after Deetjen and Speckmann (1999). (C) Opened uterus of the brown four-eyed opossum, *Metachirus nudicaudatus*, with six fetuses connected to their placentas. (D–G) Four developmental stages of postnatal skeletal development in the koala, *Phascolarctos cinereus*, visualized using microcomputed tomography; D–F = pouch young, G = adult skull. Note that in marsupials, the forelimbs ossify much earlier than the hind limb. Specimens are from the Embryological Collection Berlin (C: MS38, D: 487A, E: 485A, F: 484) and the Zoological Collection Berlin (ZMB: 36036).

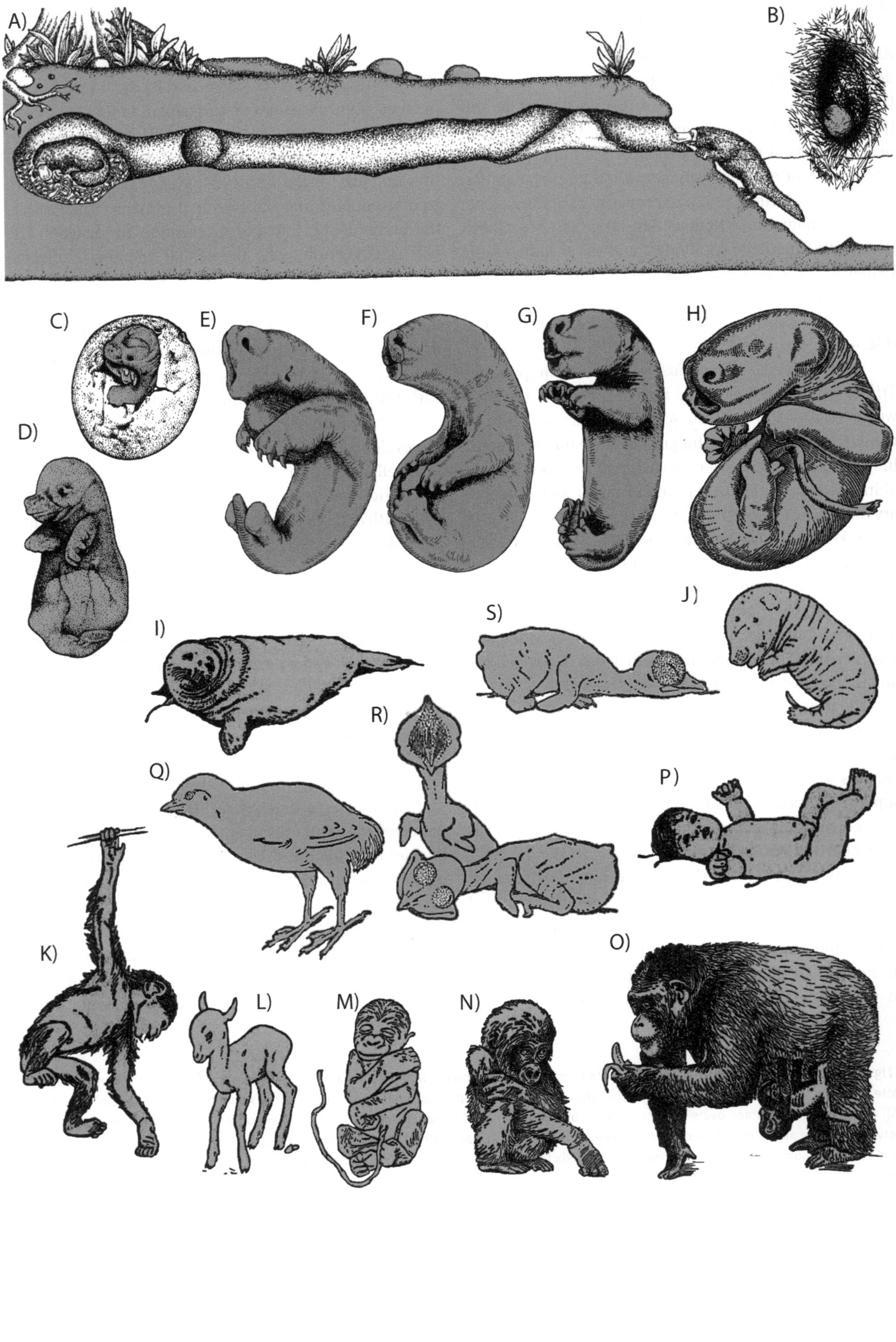
A)
B)
C)
D)
E)
F)
G)
H)
I)
S)
J)
R)
Q)
P)
K)
O)
L)
M)
N)

finds its way to the pouch through olfaction, smelling compounds produced by apocrine sweat glands at the mother's teats. Only the teats to which young are attached will keep up milk production, and the other teats return to a resting mode (Fig. 4.14 Jd). The secreting glands grow up to six times their resting size during lactation, and the teats also grow more than double in size (Fig. 4.14 I, Ja, c).

Embryos of placental mammals develop within the regulated environment of the uterus, and vital functions such as breathing, blood circulation, digestion, and excretion are enabled by the placenta. In marsupials on the other hand, the very short gestation period, the high altriciality at birth, the journey of the neonate to the mother's teat, and the life in the pouch result in several specific adaptations in the neonate in order to manage these functions and survive outside the uterus. These specific adaptations that enable the extreme marsupial life history will receive particular attention in the following pages and should not be dismissed as simple results of an abbreviated intrauterine development and acceleration of some organ structures (Starck 1995, Weisbecker 2015, Weisbecker and Beck 2015).

In marsupials, eyes and ears are at an embryonic stage at birth (Fig. 4.13 E–H, Fig. 4.9–18 b). However, the olfactory system is well developed with a functional epithelium and innervation to detect the mother's teat. Taste buds have only been found in certain species (Hughes and Hall 1988). Eyelids and eye glands are not yet developed at birth. The jaws are relatively well developed and can open widely before reaching the teat. After that, the eyes, ears, and mouth angles are covered by a peridermal epitrichium for protection. This also enables a lateral closure of the mouth for effective suckling and prevents the pouch young from detaching from the teat, which is further prevented by a bulbous swelling of the teat inside the mouth, effectively anchoring the neonate to the teat (Fig. 4.14 H–I, Jb). The epitrichium starts disappearing in the third part of pouch life (Hill and Osman-Hill 1955, Starck 1995) (around Fig. 4.9–29).

The cortex cerebri of the brain is not yet fully developed, and the pyramid pathways are not yet differentiated. At birth, marsupials (studied in the mouse opossum, *Marmosa*, and the quoll, *Dasyurus*) lack cranial nerves II–IV and VI, cerebral commissures, eye pigments, and eyelids (Müller 1972–1973; Smith 1997).

In the lungs of marsupial neonates, vascularized sacs (pseudo-alveoli) are present at the perinatal stage, and only initial branching of the bronchi is present. The lungs can vary in development between a small respiratory cavity divided by simple partitions with superficial vascularization and much larger lungs, with many subdivisions and richly vascularized and cartilaginous trachea. In groups with large lungs, which correlate with neonatal mass, such as in the kangaroos (*Macropus*), the chest is barrel shaped (Hughes and Hall 1988). Coupled with large vascularization of the rump wall, breathing can also occur via the skin, which is very thin and moist. This has been well studied for some dasyurids (Mortola *et al.* 1999).

The heart ventricles are incompletely separated in marsupial neonates but are able to maintain their circulatory function. Temperature regulation starts relatively late during postnatal development and is first fully developed at weaning. Therefore, a close association to the mother is needed for the neonates to maintain their body temperature.

At birth, the metanephros of the kidney is not yet developed in marsupials. Urine excretion happens via the mesonephric (Wolffian) duct.

The pharyngeal area is well developed in marsupial neonates, including cartilaginous and musculatory structures and an epiglottis. The tongue and the mouth floor musculature are highly developed at birth. Together with the peridermal epitrichium, these are adaptations for attachment to the teat and controlled suckling. As the esophagus is not functionally developed, it is believed that swallowing at this stage is also performed by these structures. The stomach is the primary absorptive organ at birth. Although the small intestines seem functional, the large intestines are not as they are lined by an unspecialized epithelium (Hughes and Hall 1988). The nasal skeleton, larynx, and bronchial cartilages are very stable in the neonate and prevent the closure of the airway. As suckling

◂ **Fig. 4.13: Hatching and birth in different mammalian species.** Perinatal life of monotreme mammals: (A) Breeding burrow of the platypus, *Ornithorhynchus anatinus*, with the parents; (B–D) egg in the pouch (B), hatching process (C), and neonate (D) of the short-beaked echidna *Tachyglossus aculeatus*. Neonate marsupial mammals: (E) "ultra-altricial" marsupial *Dasyurus viverrinus* (eastern quoll) (see also Fig. 4.9; (F) "intermediate" marsupial *Monodelphis domestica* (grey short-tailed opossum); (G, H) "advanced altricial" marsupials *Macropus eugenii* (tammar wallaby; G) and *Macropus rufus* (red kangaroo; H). Altricial placental mammals: the neonate harbor seal *Phoca vitulina* (I) and the house mouse *Mus musculus* (J). Precocial placental mammals: a very young chimpanzee *Pan troglodytes* already climbing (K), a neonate domestic goat *Capra hircus* (L), a late fetus of *Gorilla gorilla* with well-developed limbs (M), gorilla neonate with adultlike limb proportions (N), gorilla young clinging to the mother's fur (O). Secondarily altricial placental mammal: the human, *Homo sapiens* (P). Precocial (Q) and altricial (R, S) birds. A, modified after Zeller (2004a, b); B–D, after Maier (1999), E–G, after Smith (2015), H, after Portmann (1959); I–J, L–Q, after Portmann (1944).

and breathing must work in parallel, the opening of the larynx extends above the muscular palate (velum) and reaches the nasopharyngeal space. In that way, the milk can flow laterally to the larynx into the esophagus (Zeller 2004a). The tongue is grooved to allow suckling through a so-called pump-sucking motion, which basically works like a piston. The secondary palate, diaphragm, and epiglottis are fully formed at birth (Lillegraven 1975, Maier 1999).

The short gestation period means that at birth marsupials have no ossification centers, and in most postcranial skeletal elements (Clark and Smith 1993), ossification only starts in the first postnatal days of development (Gemmell *et al.* 1988, Weisbecker *et al.* 2008, Spiekman and Werneburg 2017) (Fig. 4.12 D–G).

The forelimbs, the facial region, and the rest of the anterior part of the postcranium ossify earlier in marsupials compared with placentals. Conversely, the ossification of the hind limbs and posterior part of the postcranium occurs later in development. This is in part a form of heterochrony, in which the development of the entire hind limb is delayed (Weisbecker *et al.* 2008, Sears 2009). The molecular basis of such differences is currently being studied (Keyte and Smith 2010, Sears *et al.* 2012). A comparison between monotreme, placental, and marsupial postcranial skeletogenesis showed that monotremes are more similar in this respect to placentals than to marsupials, indicating that marsupial skeletogenesis (and its hypothesized constraints) does not form the ancestral mammalian or therian condition (Weisbecker 2010, Werneburg *et al.* 2016). Because it is known that muscular activity can influence bone formation (Rot-Nikcevic *et al.* 2006), it has also been hypothesized that the high muscular effort in the anterior part of the body during the climb to the teat has a role in the earlier ossification of this part of the postcranium (Weisbecker *et al.* 2008).

In the marsupial neonate, the shoulder and thoracic apparatus, including shoulder and neck musculature, are already well differentiated. The ventral (coracoid) plate of the shoulder girdle forms a rigid brace for the free forelimbs and it starts to be reduced some time after birth.

A certain similarity of young monotremes and marsupials concerns the anatomy of the limbs (cf. Figs. 4.7–4.8). The forelimbs are stronger than the hind limbs; however, the monotreme hind limbs are only a little (about 1 day) behind the forelimbs in their development, in contrast to the highly immature stage of marsupial hind limbs (Manger *et al.* 1998). It is worth noting that also in placental mammals, the hind limbs are usually somewhat delayed in their development when compared with the forelimbs, which appears to be a general pattern of amniote development and may be derived from the early tetrapod mode of limb development, in which the forelimb ancestrally develops faster than the hind limb (Harrison 1969, Werneburg *et al.* 2016).

The orientation of monotreme limbs is very similar to that of marsupials at birth and also permits a digito-palmar prehension. After reaching the teat, the forelimbs are not used for holding on in marsupials. Monotremes, in contrast, lack teats and slurp the milk from milk fields of the mother and use their strong forelimbs to hold themselves on to the mother's fur (Manger *et al.* 1998). Therefore, the similarities of limb anatomy and proportion in monotremes and marsupials might be caused by different developmental constraints.

It has been suggested that the climbing behavior in neonate marsupials represents the major constraint for marsupial anatomy (Weisbecker 2015). The forelimbs are highly adapted for this unique mode of locomotion, which therefore prevents a great radiation of forelimb anatomy in marsupials. Placental mammals, in contrast, have little early developmental constraints on the forelimbs and a much higher degree of shape variance has evolved, including fins, wings, and elongated limbs with hooves. Consistently, in marsupial species with little climbing activity as newborns (e.g., bandicoots), more variation in forelimb (Weisbecker *et al.* 2008, Weisbecker and Nilsson 2008) and shoulder (Sears 2004) anatomy is recorded. The hind limbs of marsupials do not experience a comparable constraint in early life and, as such, a much greater variation of hind limbs can be found, even compared with the highly diverse placentals (e.g., jumping legs in macropodids).

Initial cranial ossification in marsupials seems to be influenced by another main functional constraint of the neonate, namely the attachment to the mother's teat (Goswami *et al.* 2016, but see Spiekman and Werneburg 2017). The premaxilla, maxilla, dentary, palatine, and pterygoid are generally the first cranial bones to ossify and subsequently show a faster growth rate than any other cranial bones. The endochondral elements of the skull ossify later, starting on average posteriorly and ending anteriorly in the following order: exoccipital, basioccipital, supraoccipital, basisphenoid, and orbitosphenoid. Skeletal elements of the visceral arches on the other hand first ossify anteriorly and then in a caudal direction, with the alisphenoid ossifying first, followed by the malleus, incus, and stapes. Compared with placental mammals, the bones surrounding the oral cavity (premaxillary, maxilla, dentary, palatine, and jugal) are accelerated, and the bones surrounding the braincase are decelerated (Clark and Smith 1993, Sánchez-Villagra and

Sultan 2002, Koyabu *et al.* 2014, Spiekman and Werneburg 2017). Consequently, neurogenesis is also delayed compared with placentals, both in its onset and in its progress of development (Smith 1994). Around birth, the premaxillae of both sides temporarily fuse above the palatal elements. In a few cases, an ascending process of connective tissue forms, which can also ossify in the wooly opossum *Caluromys* (Zeller 2004a). This pattern represents a recapitulation of an os carunculae.

The relatively late ossification of the middle ear bones and the periotic (Sánchez-Villagra *et al.* 2008) are likely a result of a striking developmental pattern in marsupials. At birth, marsupials resemble the jaw articulation of adult sauropsids between the quadrate and articular. However, it does not serve as a functional joint (Sánchez-Villagra and Sultan 2002). After attachment to the mother's teat, this jaw joint is eventually replaced by the mammalian, secondary jaw joint between the squamosal and the dentary. The quadrate and articular are redirected to form the incus and malleus, respectively (Maier and Ruf 2016). The strong attachment to the teat and consequent immobility of the jaw joint is probably essential for this development to take place (Lillegraven 1975).

The function of the primary jaw joint in marsupial neonates is not fully understood. The genuine anatomical similarities between marsupial neonates and adult sauropsids are certainly rooted in the evolutionary mechanics of heterochrony, without dredging up the historical baggage of "recapitulation" (Robert J. Asher, personal communication). It does not have a synchondrotic or syndesmotic connection with any intra-articular space. A movement of the primary jaw joint would be inexpedient as the neonate has to attach tightly to the teat and the lips are partly fused. When weaning occurs, the secondary jaw joint between dentary and squamosal is already well developed.

There are numerous anatomical autapomorphies of Marsupialia (e.g., fenestration of the palate, which therefore incompletely separates the mouth and nasal cavities; a large jugal bone that extends posteriorly all the way to the glenoid fossa of the skull), some of which may relate to the developmental adaptations and/or constraints of these animals (e.g., Goswami *et al.* 2016). Noteworthy is the median inflection of the angular process of the dentary bone, unique to marsupials, which has been discussed in the context of potential hearing function around the time of birth (Sánchez-Villagra and Smith 1997).

To buffer the forces experienced during suckling in marsupial and placental mammals, a nearly complete closure of the secondary palate at birth is developed. An early and strong chondrification of the nasal capsule and the formation of a relatively complete cartilaginous cranial side wall (commisura orbitoparietalis) can also be seen (Hüppi 2018). The latter supports the developing processus ascendens of the ala temporalis (secondary cranial side wall and precursor of the alisphenoid bone), which serves as an attachment site for jaw musculature and to stabilize the skull (Maier 1987). In that way, the large trigeminus ganglion is also protected (cavum epipterygum).

Cranial muscle development occurs quickly compared with cranial bone development in marsupials. Development starts shortly before birth, but most myogenesis occurs postnatally as the muscles start to be functional in the neonate. The onset of myogenesis precedes the onset of both skeletal development and the development of the central nervous system. On the other hand, the development of the peripheral nervous system and certain sensory nerves are also, like the craniofacial musculature, well developed shortly after birth (Smith 1994). The actual innervation of the face by terminal branches of the trigeminal nerve varies among marsupials, and therefore the amount of tactile sense in the face and its role in finding the pouch remains to a large degree unknown (Hughes and Hall 1988). Despite the short period of myogenesis, cranial muscle development can be divided in three groups. Before birth, the tongue, the mylohyoideus and pharyngeal muscles, and some muscles of the neck and shoulder develop first. The muscles of the first pharyngeal arch and most other craniofacial muscles follow about 12 hours later (in the case of the short-tailed opossum *Monodelphis*). The facial and ocular muscles form about 24 hours after this. Taken together, it seems that there is little heterochrony in craniofacial myogenesis among marsupials, and there are also few differences with the development seen in placental mammals (Smith 1994). In the neonate jaw, muscles develop, stabilize the region of the developing jaw joint, and support the tongue for suction pumping.

As a result of its developmental constraint, the facial diversity of marsupials is relatively low, as major snout-related bones are already differentiated to enable teat fixation and suckling at early stages, although prominent exceptions exist (Goswami *et al.* 2011). A faster growth of the snout (Sánchez-Villagra *et al.* 2008) and a delayed brain development (Smith 2001) appear to have no impact on adult brain size, which is very similar in marsupials and placentals, although unique exceptions exist among placentals (primates, cetaceans) (Weisbecker 2015).

4.13 Early life history of placental mammals

Like monotremes and marsupials, many placental mammals are primarily altricial. However, compared with the former, placental mammals have a more or less fully developed body at birth and the forelimbs and hind limbs are more evenly developed (Werneburg *et al.* 2016). In contrast to monotremes and marsupials, a greater variety of conditions are known at birth for placental mammals (e.g., Starck 1995, Gilbert 2006, Westheide and Rieger 2010), which might be correlated to the great diversity and ecological adaptations of that group. In extremely altricial placental species, the neonates have no fur. No species-specific coordination of locomotion has developed, closed eyes are present until several days after birth, incomplete thermoregulation as the fur is not fully developed, and fast postnatal growth takes place. In addition to altricial forms, precocial conditions exist, in which the neonate has operable eyelids, fully developed fur, and locomotion, and sense organs are already functional (Portmann 1944, 1959). There is no clear distinction between altricial and precocial species, as a broad continuum exists (Martin and MacLarnon 1985, Sánchez-Villagra and Sultan 2002, Werneburg *et al.* 2016).

Among placental mammals, an altricial condition at birth can be found in tupaiids, terrestrial carnivorans, lipothyphlans, most rodents, and lagomorphs (Fig. 4.13 I–J). The young are kept inside burrows and are periodically fed. Some precocial animals have no nest, but in the first days, the neonates do not follow their mother and rest in relatively open places (the hare *Lepus*, macroscelids, some rodents and pigs). This condition appears to represent a transitional form to full precocial behavior. True precocial young experience a long gestation period and follow their mother directly after birth. This behavior can be found in the fully aquatic taxa: whales (Cetacea; Fig. 4.3 G) and manatees (Sirenia), whose life in water requires complete adaptation to the aquatic milieu directly after birth. Large herbivores (Bovidae, Proboscidea, and Perissodactyla; Fig. 4.13 L), all hystricognath rodents (e.g., guinea pigs, *Cavia*), and some other rodents (spiny mouse, *Acomys*) exhibit precocity. In even- and odd-toed mammals (Cetartiodactyla and Perissodactyla, respectively), the necessity for long-distance travel in savannas favored neonates that can keep up with the step of the adults directly after birth. Hystricognaths are among the most precocial rodent species and in general rank among the most precocial placental mammals (Sekulić *et al.* 2006).

Clinging young are known for some arboreal and aerial placentals without a nesting site (Dermoptera, Chiroptera, and Primates). The young are always carried by their mother (Fig. 4.13 O). The same is true for the terrestrial pangolins and the giant anteater (*Myrmecophaga*). Clinging is a special case with both altricial aspects, namely relative dependency on the mother, and precocial aspects such as refined limb adaptation to hold on to the mother's fur.

Bats (Chiroptera) are intermediate between altricial and precocial conditions at birth (Kurta and Kunz 1987, Rasweiler and Badwaik 2000, Mess and Carter 2008, Koyabu and Son 2014). The young are carried by their mother in the first phase of life. Afterward, they are deposited in nursery roosts. The young usually attach to the teats or hold on to the fur of the mother. To clasp to the fur of the adult, hind limb ossification is accelerated in bat development (Koyabu and Son 2014). In some taxa, specialized teats are present in the inguinal region specifically developed for the young to cling on to. They do not secrete milk and the young have particularly formed milk teeth that enables them to attach to these specialized teats. Some male bats lactate (Francis *et al.* 1994).

Human children are born as secondarily altricial (Fig. 4.13 P) (Portmann 1944, 1959). Other primates give birth to relatively mature young, with fully developed fur, open eyes, and refined motor function (Fig. 4.13 K, M–O) and are more precocial than altricial in their birth condition. That humans derived from precocial ancestors is recognizable when comparing their neonate anatomy to typical, primarily altricial mammals. Most organs in humans are fully developed, including head hair, and the eyes are open at birth. Moreover, infants show the grasp reflex, a precocial feature, which is inherited from primate ancestors, as outgroup comparison in a cladistics sense prove (Fig. 4.13 P).

Truly precocial mammals, such as the cow, close their eyelids very early during embryogenesis and open their eyelids within the extended period of gestation some time before birth (Fig. 4.14 F). Humans close their eyelids at the end of the first trimester (Fig. 4.14 G). Within the fifth month, four months before birth, the eyelids open again. This is different to other primates. As such, the human has to be declared secondarily altricial at birth. It has been argued by Portmann (1944, 1959) that the first year of human development has to be understood as the extrauterine phase of fetal development, in which species-specific modes of locomotion and communication develop. This is related to the progressive process of brain folding and the establishment of an elongated learning phase in the infant after birth.

There seems to be a tradeoff between the size of the pelvic opening of the mother and the head size of the fetus, which resulted in a physiologically preterm delivery. It has been argued that the pelvic opening could have coevolved with the fetus head size (Van Schaik 2016, Trevathan 2015). To some degree, this seems to be the case as documented by the fossil record (Zollikofer and Ponce de León 2010, Fischer and Mitteroecker 2015). Nonetheless, the effective bipedal locomotion of females was maintained (Gruss and Schmitt 2015), for which a particular pelvic anatomy was required and the pelvic opening therefore could not have widened extensively.

When compared to chimpanzees (*Pan*) as our closest living relatives, different accelerations (neotenic features) and extensions (hypermophoses) of postnatal ontogenetic traits can be observed in the human fossil record. Whereas one of the earliest hominin representatives, *Sahelanthropus tchadensis*, had an adult skull shape comparable to *Pan*, a progressing reduction of the facial skeleton evolved on the stem line to *Homo sapiens* (Fig. 4.15: boxes, Zollikofer and Ponce de León 2010). An absolute reduction is likely related to the progressing use of fire to process food, which then became easier to chew and less jaw muscle insertion area on the skull was needed. A relative reduction of the facial skeleton is also related to increased brain size for better cognition (tool processing, social interaction, language, metaphysical cognition) with *Pan troglydytes* having 365 cm^3 and *Homo sapiens* 1350 cm^3 brain volume (Fig. 4.15: boxes). Whereas skull shape of hominin newborns are relatively similar to *Pan*, the facial skeleton received its final shape relatively faster and brain growth was also accelerated to reach adult size (Zollikofer and Ponce de León 2010). Compared to *Pan*, the human brain, however, develops much longer and also body growth is extended. As a result, humans reach sexual maturity relatively late, which results in less evolutionary fitness. To circumvent this issue, humans wean their young relatively early compared to other great apes and intervals between pregnancies are shorter. This however, results in many non-independent young at the same time. To resolve this fitness disadvantage, hominins evolved a higher degree of allopatric behavior and particularly the grandmothers help raising the premature young (extended family). Largely increased social interactions and communications are among the key innovations for human evolutionary success (Van Schaik 2016). Life history traits are nicely mirrored in the sequence of dental eruption with the appearance of the first molar indicating the principal end of brain growth (followed by brain maturation), the appearance of the second molar indicating the end of the major somatic growth, and the appearance of the third molar indicating the begin of sexual reproduction (Fig. 4.15; Zollikofer and Ponce de León 2010).

4.14 Life of the infant

As mentioned before, milk glands are present in all mammals (even in some males as noted above). Milk glands are lobular, alveolar, apocrine glands, which are derived from sebaceous follicles. In monotremes, they are permanently associated with the hair follicles and they do not form teats but lactation fields instead, in which the excreted milk is collected (Fig. 4.3 D). The gland ducts develop into wide milk ducts in which milk is accumulated between lactation phases (Starck 1995).

The milk of monotremes has similar contents as the milk of placental mammals. However, like marsupials, there is a high concentration of iron in the milk of monotremes, which is much lower in placentals. Compared with placental mammals, both groups have small hatchlings or neonates with a correspondingly small liver. There is no change in the composition of milk constituents during the lactation period in monotremes (Jackson 2003).

The young of echidnas leave the furry pouch of the mother at an age of two months at the earliest. After that they are regularly fed for four to six days (Zeller 2004b). The platypus does not have a pouch (Hayssen *et al.* 1993). Its postnatal development is well documented (Manger *et al.* 1988, Fig. 4.8).

Lactation occurs for up to 200 days in the echidna and 105 days in the platypus (Zeller 2004b). However, weaning is a gradual process, which already starts a few months after hatching. During several weeks of lactation, the caruncle persists in the platypus, and it has been proposed that a rubbing of the caruncle on the mother's belly stimulates milk production (Griffiths 1978, Manger *et al.* 1998).

In therian mammals, the milk glands are separated from the hair follicles, although a connection is still visible in the ontogeny of some species. The number of teats generally corresponds to litter size.

After birth, the young searches for the teat and attaches to it via trial-and-error. The hairline of the mother helps during searching. Suckling on the teats (papillae mammae) stimulates the secretion of milk in the milk glands. Suckling is an active process of the young, which is enabled by a suction pumping mechanism (Maier 1999, German *et al.* 2000, Zeller 2004a, Krockenberger 2006).

As shown by fossil metatherians, the ancestral therian mode of tooth replacement after lactation was similar to that of extant placental mammals. As such, the unique mode of tooth replacement in extant marsupials is highly derived, and it has been correlated to the unique fixation to the mother's teat (Cifelli *et al.* 1996, Martin 1997; but see van Nievelt and Smith 2005).

The pouch of marsupials (Fig. 4.14 K) develops from the paired anlagen of the cranial ends of the genital buds (separated from the scrotum anlagen) and forms a permanent invagination of the skin through the dermal muscular tunic (panniculus carnosus) (Fig. 4.2 B). It is not homologous to the incubatorium (often also referred to as a pouch or marsupium) of female echidnas (Monotremata), as the latter is only a temporary formation, which develops from an unpaired folding of the abdominal wall. In marsupials, the pouch opening can be oriented in a ventral (Didelphidae and Dasyuridae), caudal (toward the cloaca; Peramelidae, Dasyuridae, and Phascolarctidae), or cranial (Macropodidae and Phalangeridae) direction. It is surrounded by the musculus sphincter marsupii, which fully closes during diving in the only semiaquatic marsupial (the water opossum, *Chironectes minimus*). Several small marsupials have no pouch or only a small crest around the teat field. In those species, the young are protected by the mother, and they are also tightly attached to the teats. It is not clear if a pouch belonged to the common ancestor of crown marsupials. It appears that it may have evolved several times independently (Zeller 2004a).

Males and females of all marsupial species have epipubic bones (ossa epipubica). These bones are also present in monotremes. As integrated elements of the abdominal wall, they play a role in locomotion and respiration (Reilly *et al.* 2009) and are fully independent of pouch formation. Epipubic bones and pouches are absent in placentals, but the former have been reported among stem eutherians (Novacek *et al.* 1997) and among mammaliaforms generally (Kielan-Jaworowska *et al.* 2004).

The lactation period of marsupials is an extreme multiplication of the gestation length (Fig. 4.14 A, Tab. 4.2). In marsupials, the milk glands are always situated on the ventral side of the body and consist of balled-up monoptychic gland ducts inside connective tissue with striated musculature (musculus ilio-marsupialis). One or more gland ducts exit into each teat. Plesiomorphically, a large number of teats is present (e.g., 25 in Didelphidae), but the count can be as few as two (southern marsupial mole, *Notoryctes typhlops*) (Zeller 2004a). In contrast to placentals, the teats in marsupials do not develop along a teat crest, but by a thickening of the epidermis, which contains up to six hair follicles (quoll, *Dasyurus*) with sebaceous gland and milk gland anlagen. In contrast to monotremes, the milk hairs fall out when sexual maturity is reached. In some marsupial species teat anlagen can also be found in males.

Marsupial lactation provides less nutritional energy per time unit (Krockenberger 2006) and correspondingly reduces maternal energy investment, which may be suited for harsh environments (Tyndale-Biscoe 2001). Much of organogenesis occurs while the young are attached to teats. At every growth stage, a different composition of the milk is necessary. Three phases of milk production are distinguished (Tyndale-Biscoe 2005, Krockenberger 2006). The older young (Fig. 4.14 K) get more protein- and fat-rich milk than the younger ones (Fig. 4.14 H–I and Jb). Marsupials can nourish several young of different ages in their pouch, each of which is associated to a different teat and receives specific milk contents (Fig. 4.14 J).

Lactation behavior is different among placental mammals. For example, the domestic pig *Sus scrofa* tends to lie down during lactation. Many precocial placental mammals, on the other hand, stand during lactation, whereas primate mothers sit and hold their young near the body (Storch and Schröpfer 2004). The contents of the milk vary among placental mammals; however, in contrast

▸ **Fig. 4.14 Postnatal life.** (A) Absolute time of the preweaning developmental periods as reconstructed for the major tetrapod and mammalian clades by Werneburg *et al.* (2016); legend in the right upper corner. The human is shown as a secondarily altricial species among placental mammals (* = timing of eye opening in the human occurs before birth). (B) Embryonic eye closure in altricial mammals with microscopic close-up of the lid adhesion. (C, D) Late embryonic heads in altricial mammals (C, kitten; D, shrew). The eyelids are grown together. The tip of the ear pinna, which points backward later on, is folded forward and fully fuses with the skin of the head. In panel D, this process is more advanced than in panel C. (E–G) Comparison of head development in mammals with different life history strategies. Primarily altricial mammals show closure of the eye and ear at birth (E). In precocial mammals, the eyelids open shortly before or around birth (F). Humans (G) have an extended prenatal development (see A*), but eye opening ancestrally occurs at an earlier point of development. Although humans have opened eyes at birth, they are not able to perform species-specific locomotion and communication and are altricial for the first year of postnatal life see Fig. 4.13P. Because humans evolved from precocial primates, they are called secondarily altricial. (H) Neonate of the red kangaroo *Macropus rufus* fixed to the teat. (I) Young of the opposum (*Didelphis*) suckling on teats. (J) Inside the pouch of an adult red kangaroo (*Macropus rufus*). (a) milk gland used by a young that already left the pouch, (b) milk gland with a neonate, (c) regression of the milk gland after weaning of a young at over 400 days, (d) reduction of an unused teat after progesterone stimulation during gestation. (K) Suckling young of the yellow-footed rock wallaby *Petrogale xanthopus* after leaving the pouch. B–D, after Portmann (1959); H, after Maier (1999); I, after a drawing of W. Maier; J, after Tyndale-Biscoe and Renfree (1987); K, after Brehm (1927).

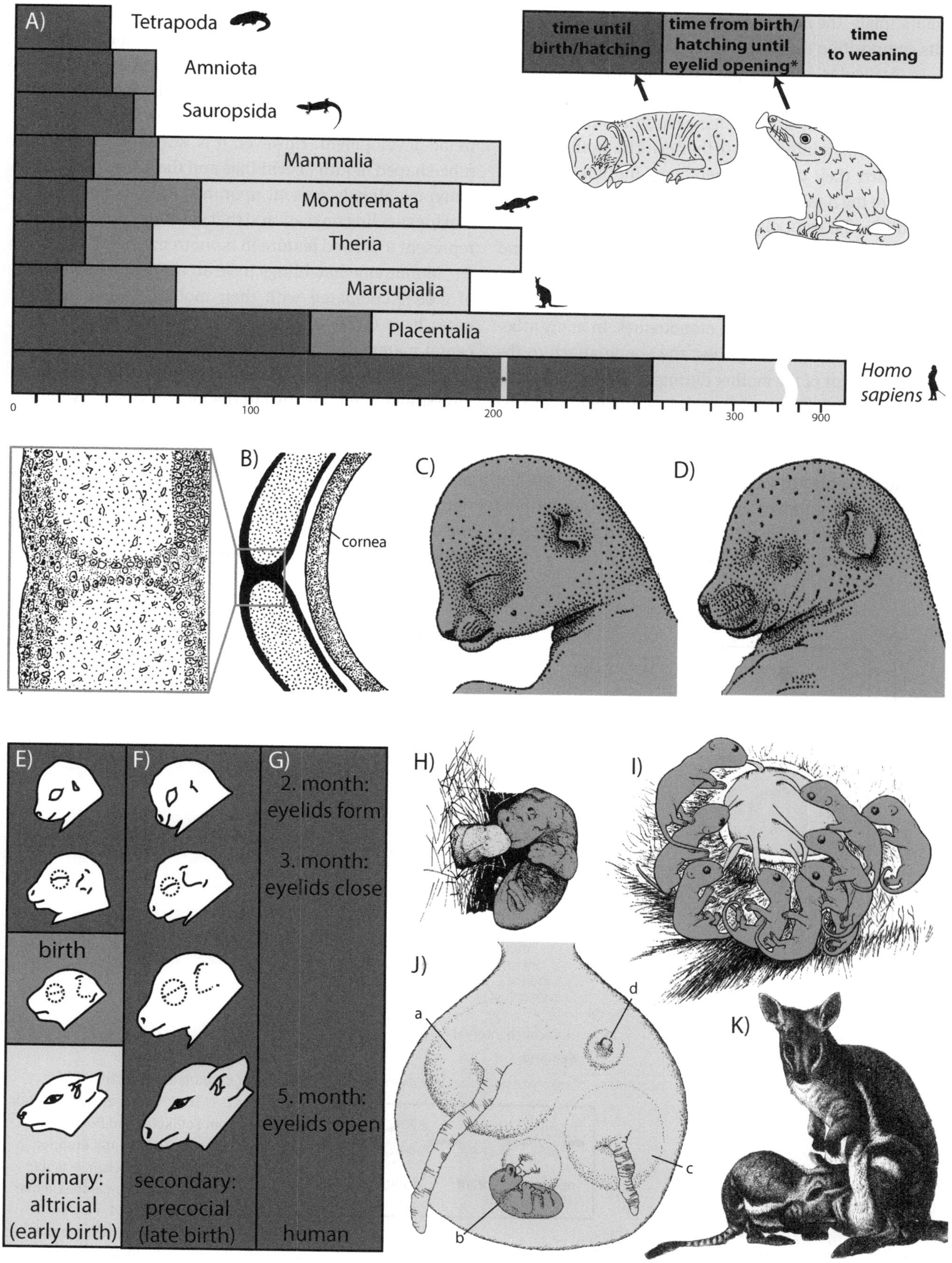
A)
Tetrapoda
Amniota
Sauropsida
Mammalia
Monotremata
Theria
Marsupialia
Placentalia
Homo sapiens
time until birth/hatching
time from birth/ hatching until eyelid opening*
time to weaning
0
100
200
300
900
B)
cornea
C)
D)
E)
F)
G)
2. month: eyelids form
3. month: eyelids close
5. month: eyelids open
birth
primary: altricial (early birth)
secondary: precocial (late birth)
human
H)
I)
J)
a
b
c
d
K)

to marsupials, the milk of a species does not change much in its composition during lactation (Zeller 2004a).

In particular, altricial placentals (many murids, terrestrial carnivores, and strepsirrhines) show a complex retrieval behavior in order to change nesting sites in case of danger or parasite infestation of the nest. The mother or both parents carry their young by their head, fur of the nape, or ventrally at the flanks. In order to enable an effective weight distribution, the young fall into a passive carrying posture (at least in particular phases of development) (Storch and Schröpfer 2004). Whether a comparable retrieval behavior is present in marsupials is not clear (Hunsaker 1977), but it is most likely absent as it is in monotremes. In many arboreal and some terrestrial species, the young attach to the teats or to the fur of the mother during transport.

As in monotremes and marsupials, the period during which lactation ceases (weaning) is a continuous process, which can last for several weeks or months in placental mammals. Even after weaning, many placentals are not completely independent and may be supported by food supply, for example, in the form of shared resources by the parents (e.g., carnivorans and humans).

It is generally considered that teats first evolved in therian mammals and that the monotreme mode of lactation, slurping milk from the milk fields, represents a plesiomorphic condition. This is supported by the fact that no rudimentary trace of a teat has ever been found in monotreme development. However, it is worth noting that the tube-shaped mouth of echidnas and the horny beak of the platypus, already present upon hatching, is not convenient for suckling on teats and that slurping could therefore represent a derived feature in monotremes.

Monotreme hatchlings have an average relative mass of 0.08% compared with their mother. Marsupial neonates have 0.12% of the adult body mass, whereas placental mammal neonates already make up about 15% of the maternal mass. At weaning, the total average mass of monotreme litters relative to the mother is 50%, whereas it is 55% in marsupials and 59% in placental mammals. This shows that maternal investment of monotremes and marsupials is greater during lactation, whereas placental mammals invest most energy in placentation. However, the total investment is very similar between all three clades (Hayssen *et al.* 1985, Renfree *et al.* 2009).

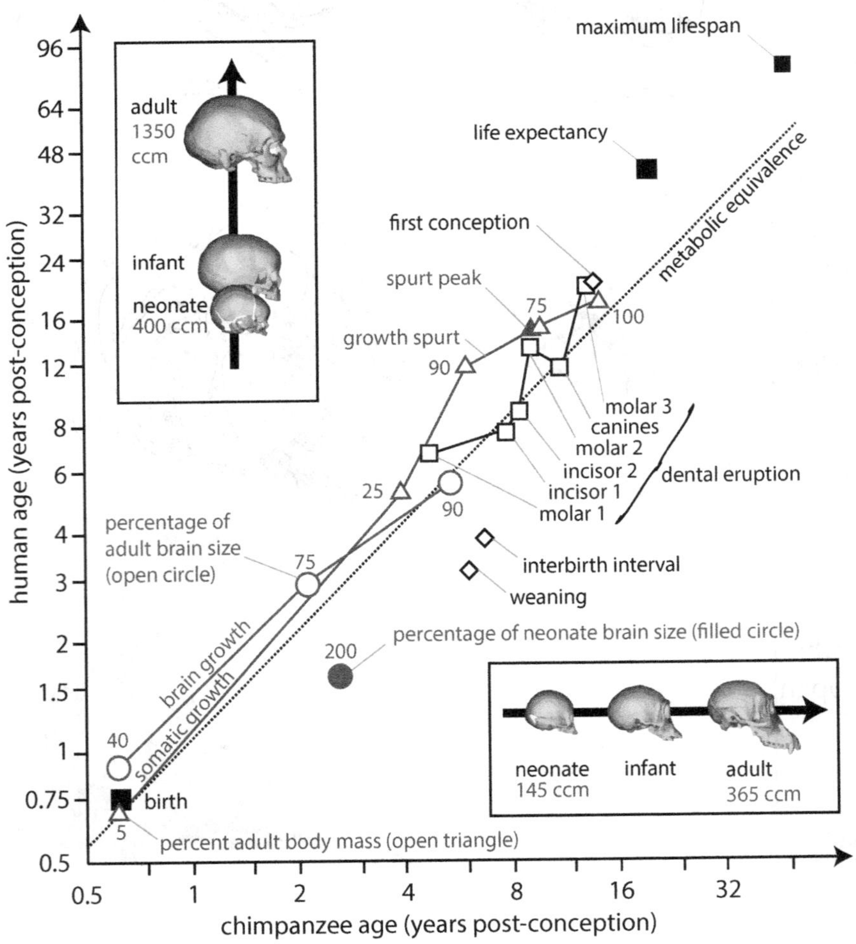

Fig. 4.15. Comparison of postnatal life history traits in the chimpanzee and human (modified from Zollikofer and Ponce de León 2010). Sequence of dental eruption in apes is: molar 1-incisor 1-incisor 2- molar 2 – (premolar 3, premolar 4)-canines-molar 3. In the human, it is: molar 1 – incisor 1 – incisor 2 – (premolar 3, canines, premolar 4) – molar 2 – molar 3. The sequence is dependent on shape of the face and food. ccm = cubic centimeter.

4.15 Life history evolution

Altriciality appears to be ancestral for all mammals because monotremes, marsupials, and most placental mammals have immature hatchlings or neonates (Portmann 1944, 1959, Starck 1995, Werneburg *et al.* 2016). Delivery at a premature stage and nourishing of the young with maternal histotrophs/placental nutrition and milk is tightly correlated with the inceptive reduction (Monotremata) and the elimination (Theria) of yolk in the eggs. Originally, the altricial lifestyle of mammals was apparently linked to a moderate litter size and breeding in a nest or burrow. Two modes of reproduction evolved, resulting from two types of selection, namely *k*-selection and *r*-selection (Pianka 1970), representing ends of a continuum. *R*-selection favors a large litter size and a trend toward fewer pregnancies in a life cycle (with an extreme of only one pregnancy in life, semelparity, as found within the broad-footed marsupial mouse *Antechinus*, for example). *K*-selection, on the other hand, favors a small litter size, which is compensated by intensive parental care and a long childhood, resulting in a relatively high proportion of young reaching sexual maturity. Litter size is correlated with individual longevity, body size, ecological factors, climate, food availability, and social behavior (Starck 1995).

The precocial lifestyle of several placental mammals has to be considered as derived given the well-nested position of these animals in the mammalian tree of life (Mess and Carter 2006, Meredith *et al.* 2011, O'Leary *et al.* 2013).

As mentioned above, a measure for the starting point of independency is the opening of the eyelids. In all precocial sauropsids, the eyelids open at hatching. In contrast, eyelids open long after birth or hatching in most mammals and in some birds, which is associated to their altricial life. Given that information, Werneburg *et al.* (2016) reconstructed the timing of eye opening in the last common ancestor of all amniotes in their phylogenetic reconstruction. They hypothesized that eye opening occurred several weeks after hatching in the amniote ancestor. A late eye opening does not necessarily mean that the young is underdeveloped at hatching (Sánchez-Villagra and Werneburg 2016). On the contrary, Werneburg *et al.* (2016) reconstructed well-developed limb, body, and head proportions. The closed eyes at hatching simply mean that the tear glands have not been developed enough to protect the eyes from drying out, and/or it means that the brain was not yet developed enough to process visual information.

Blind hatchlings require parental care and external food supply. In fact, there is some indication in the fossil record that parental care could have been present in early synapsids (Botha-Brink and Modesto 2007, 2009) and reptilian taxa (Piñeiro *et al.* 2012). Under this hypothesized reconstruction of the amniote ancestor, altriciality and parental care are plesiomorphic features for mammals.

It is most likely that polylecithal eggs, as in sauropsids and monotremes, were present in the amniote ancestor. Furthermore, the coincidental presence of polylecithal eggs and an altricial lifestyle in early amniotes is also not unlikely given that altricial (like precocial) birds also have polylecithal eggs.

Because of the morphological similarities between marsupial neonates and monotreme hatchlings, it has been suggested that oviparity was still present early in marsupial evolution and that the incubation of the eggs was later incorporated into the intrauterine period (Griffiths 1978, Starck 1995).

Viviparity never evolved in birds, probably because their highly vascularized lungs need some time to unfold before delivery (Duncker 1978). For that reason, they breathe in the air cell of the eggs (Fig. 4.5 G) shortly before hatching to permit a smooth "pulmonary transition" to the atmosphere outside the egg. Furthermore, the fact that birds and other non-marine archosaurs (and turtles, which may be archosaurs or at least very close to them) did not develop viviparity is likely related to their calcium metabolism (Gilbert 2006, Liu *et al.* 2017). Mammals, in contrast, do not have such limitations. Oviparity involves a large energetic investment of the mother at the moment when the eggs are produced. In viviparous forms, on the other hand, energetic investment is evenly distributed over a longer term (although huge variation exists, e.g., between marsupials and placental mammals). Monotremes represent an intermediate form in this regard, as they nourish their embryos with placental secretes (histotrophs) and the eggs are retained in the uterus for a longer embryonic time when compared with sauropsids. Viviparity is also beneficial to the endothermic physiology of therian mammals and protects the embryos from the danger of temperature fluctuations. In ectothermic reptiles, the retention of oviparity might have been the only possibility to survive in most cases because the body temperature of the mother fluctuates too much.

The strikingly different reproduction strategies of marsupial and placental mammals have been the source of debate on the origin of both groups and on their consequences for diversity and disparity patterns in evolution (e.g., Krockenberger 2006, Bennett and Goswami 2013, Sánchez-Villagra 2013). Placental mammals produce relatively mature young after a long gestation, whereas marsupial neonates are highly immature after a short gestation, although the anatomical and physiological precondition is very similar in both groups, including the presence of an invasive trophoblastic tissue to initiate a highly effective placenta with the mother.

It has been hypothesized that the short gestation of marsupials should be interpreted as a reaction of the mother's immune system, which cannot cope with the fetal tissue that is genetically different from that of the mother (Lillegraven 1975, Amoroso and Perry 1975). However, it was later shown that maternal immune systems do not reject fetuses in macropodids (Walker and Tyndale-Biscoe 1978, Rodger *et al.* 1985). Instead, it seems that birth is induced by physiological processes of the fetus, and the short gestation and consequently longer lactational period are now considered to be a reproductive strategy to decrease the energetic investment of the mother in her young (Morton *et al.* 1982, Hsu *et al.* 1999).

The specialized trophoblastic tissue (Renfree 1993) of the last common ancestor of Placentalia likely possessed a similar kind of trophoblast as seen in living placental mammals, which may have allowed an enormous extension of intrauterine morphogenesis in eutherian mammals (Lillegraven *et al.* 1987). The elongated prenatal life (Hayssen *et al.* 1993, Vaughan *et al.* 2011) presumably resulted in the relative reduction of lactation time, faster reproduction, and ultimately increased population growth. Simultaneously with the gain of highly effective trophoblastic tissue, mammotrophic hormones may have enabled the mammary glands to nourish larger neonates. In a second phase of placental mammal evolution, higher metabolic rates may have evolved, which resulted in increased pre- and postnatal growth rates, further shortening reproduction time from conception to weaning and increasing population growth. An extended growth period facilitated more cerebralization among various clades of placental mammals, leading to larger relative brain sizes and more neurons and association pathways, which resulted in increased intelligence, enhanced longevity, and extensive play activities (e.g., Lillegraven *et al.* 1987, Barrickman *et al.* 2008, Isler and Van Schaik 2009).

An elongated gestation length in placental mammals, when compared with the mammalian ancestors, has the advantage of protecting the young for a long time in a fully separated and safe environment. The embryo/fetus can be nourished by the placenta for a long time and can reach, on average, a large prenatal body size. An obvious disadvantage is that in fluctuating environmental conditions, the pregnant mother is handicapped for a long time, and the intensive investment in uterine reproduction could result in the death of the mother and her unborn young (Tyndale Biscoe 2001).

Marsupials, at least in Australia, are exposed to a very fluctuating environment, and they can react more easily to changing conditions because of their reproductive strategy (i.e., short gestation and longer lactation). If the young die, a fast reproductive cycle enables the quick recovery of the population (Kirsch 1977). This is particularly obvious in kangaroos, which are able to "store" a dormant embryo in the uterus (diapause) and activate its development to replace a lost pouch young (Renfree 1993, Starck 1995). However, it has been pointed out that the ability to discard pouch young is irrelevant in many small marsupial species because they only raise one litter in their lifetime (Morton *et al.* 1982, Russell 1982).

The efficiency of reproduction is comparable among all mammalian groups, including monotremes. That is to say that per unit body mass, an equal or very similar number of offspring is produced (Hamilton *et al.* 2011). The greater *current* diversity of placental mammals over marsupials (Tab. 4.1) is not a representative time slot for the whole mammalian diversity produced throughout their evolutionary history. Evidence from the fossil record illustrates a great diversity over millions of years in both placentals and marsupials (Sánchez-Villagra 2013). Different life history strategies evolved to cope with different environmental conditions, which are not fully understood yet. Nevertheless, the evolution of different ontogenetic traits resulted in a fascinating variety of body forms found among monotreme, marsupial, and placental mammals.

4.16 Summarizing remarks

Mammalian biology is strongly characterized by features related to life history and reproduction. After cleavage, which is equatorial in therians but only occurs at the animal pole in monotremes, a blastocyst is formed. The trophoblast forms the outer layer of the blastocyst in all mammals (Fig. 4.4 F) and allows for the implantation of the embryo to the uterine wall in therians (Fig. 4.4 B–9). This establishes an intimate physiological connection between the embryo and the mother through the formation of a placenta and therefore supports therian viviparity.

The evolution of viviparity has had strong implications for early development in therians. The tertiary layer of the embryo, the non-mineralized eggshell, and the albumen that are present in monotremes are only present in part of marsupial intrauterine development and are completely absent in placental development with the exception of the rabbit. In addition, the amount of yolk in the egg cells of both marsupials and placentals is greatly reduced compared with monotremes as nutrition is supplied directly in the womb.

In monotremes, the chorion fuses both with the vitelline sac and the allantois to form a choriovitelline and chorioallantoic membrane, respectively, which exchange gas and nutrition with the mother early in development. These membranes could be considered ancestral. The placenta of therian mammals is also formed by extraembryonic membranes (Fig. 4.5). Although a wide range of placenta types can be found in therians, a chorioallantoic placenta, which is especially effective in discharging nitrogenic waste from the embryo, is the most common among placentals (Fig. 4.6 E–J). Therians also show much diversity in the invasiveness of their placentas, ranging from the epitheliochorial placenta, in which almost no invasion of the uterine wall occurs, to endotheliochorial and hemochorial placenta types, in which the placenta is in direct contact with the maternal blood vessels (Fig. 4.6 A–D). Although the placental types overlap between marsupials and placentals, the placenta of placentals develops much earlier and generally shows more placental invasion.

Mammals also show a large diversity in their reproductive organs. Therians have a perineum, which separates the anus and the vagina (Fig. 4.2). Monotremes lack this structure and instead have a cloaca that is divided into a coprodaeum and a urodaeum. A scrotum is also only found in extant therians. Many marsupials have developed a pouch in which the young are kept in relative safety for a large part of their postnatal development. Monotremes and many placentals use a different approach to protecting their offspring by building nests.

Placentals show most diversity, both in life history strategy and morphology; gestation length ranges from short to very long, whereas it is always short in both monotremes and marsupials. Consequently, all monotremes and marsupials are altricial, but placental neonates range from highly altricial to precocial. As a result, placental developmental patterns have likely changed most from its ancestral origin. Reconstructing ancestral values by using squared-change parsimony analysis suggests that ancestral placental neonates were altricial, almost naked and blind, with evenly developed front and hind limbs (Tab. 4.4, Fig. 4.10).

Whereas most other primates are quite precocial, humans are secondarily altricial. This might be a result of the above-mentioned constraint in the size of the pelvic opening through which the head of the fetus must fit at birth (Fig. 4.13 P).

Marsupials are characterized by being the most altricial of all mammals. Because they are born at such an early stage of development, marsupials show specialized adaptations in many structures, including the limbs, the olfactory system, and the skull anatomy. In monotremes, the forelimbs are also at a much further stage of development, but their function is to cling to the fur of the mother, whereas marsupial neonates use the forelimbs to climb toward the mother's teat directly after birth. This adaptation is thought to pose some constraints on morphological diversity in marsupials. Marsupial forelimb morphology is generally conserved, and in contrast to placentals, no flippers or wings have developed in marsupials. Suckling at a very early stage of development also led to a conserved cranial ossification sequence and a highly derived tooth replacement pattern. However, although cerebral development is delayed in marsupials, this does not affect adult brain size, which is generally comparable with placentals.

Monotreme milk composition is generally similar to that of placentals, although it shares a high concentration of iron with marsupial milk, which is not the case in placentals. Marsupial lactation is highly specialized and the milk changes its composition through different phases of the development of the pouch young. If young of different ages suckle at the same time, both teats have a different milk composition. Furthermore, the nutritional value of marsupial milk is relatively low. Both are probably adaptations that reduce maternal investment. Weaning is a continuous process in all mammals and may take up to several months in certain placentals. Although the focus of maternal investment differs, with most investment occurring in utero in placentals and postnatally in marsupials and monotremes, the total amount of energetic investment is very similar across the three mammalian groups.

Overall, two approaches to reproduction can be distinguished in mammals. The first approach favors a large litter size with a high mortality rate and a relatively small parental investment per young. This altricial approach is exhibited most strongly by marsupials. The second approach, precociality, is present in certain placentals and favors a litter size which is quite small and a long childhood with intensive parental care, which therefore gives each individual young a much better chance of reaching sexual maturity. The striking differences between these two approaches have received much attention in the light of their respective diversity and evolutionary success. It has often been hypothesized that marsupial gestation is short due to the maternal immune system rejecting the fetus as it develops. However, this was tested and not confirmed for macropodids. Most likely, short gestation and a longer period of lactation represent a highly specific reproductive strategy that decreases maternal investment.

Therefore, it represents an extreme example of the diverse life histories adopted by mammals, facilitated by their unique developmental characteristics.

Acknowledgments

We wish to thank Marcelo R. Sánchez-Villagra for his generous and continuous support and for comments on the manuscript. In addition, Anthony M. Carter, Robert J. Asher, and Frank E. Zachos provided comprehensive suggestions to improve the manuscript. I.W. thanks Wolfgang Maier for general discussions on mammalian embryology. Peter Giere enabled access to the Embryological Collection in Berlin, and Christiane Funk enabled access to the Zoological collection in Berlin. We thank Frank E. Zachos and Robert J. Asher for the invitation to write this chapter. I.W. was financed by a Swiss National Science Foundation Advanced Postdoc Mobility Grant (P300P3_158526). S.N.F.S. got funding from the Institute of Biology Leiden Travel Grant for Master of Science research project abroad, Curatorenfonds Leiden University, Quintusfonds by student society Algemene Leidse Studentenvereniging Quintus, and the Erasmus Programme.

Literature

Amoroso, E.C. & Perry, J.S. (1975): The existence during gestation of an immunological buffer zone at the interface between maternal and foetal tissues. *Philosophical Transactions of the Royal Society of London. Series B, Biological Sciences* 271: 343–361.

Ashwell, K.W.S. & Shulruf, B. (2014): Vestibular development in marsupials and monotremes. *Journal of Anatomy* 224: 447–458.

Barrickman, N.L., Bastian, M.L., Isler, K. & Van Schaik, C.P. (2008): Life history costs and benefits of encephalization: a comparative test using data from long-term studies of primates in the wild. *Journal of Human Evolution* 54(5): 468–590.

Becher, H. (1931): Placenta und Uterusschleimhaut von *Tamandua tetradactyla* (Myrmecophaga). *Gegenbauer Morphologisches Jahrbuch* 67: 381–458.

Bell, M.A., Futuyma, D.J., Eanes, W.F. & Levinton, J.S. (2010). *Evolution Since Darwin: The First 150 Years*. Sinauer Associations, Sunderland.

Belon, P. (1551): *L'histoire naturelle des estranges poissons marins, avec la vraie peincture et description du dauphin, et de plusieurs autres de son espèce*. Regnaud Chaudiere, Paris.

Bennett, C.V. & Goswami, A. (2013): Statistical support for the hypothesis of developmental constraint in marsupial skull evolution. *BMC Biology* 11: 12 (14 pp).

Benirschke, K. & Kaufmann, P. (1995): *Pathology of the Human Placenta*. Springer Science+Business Media, LCC, New York.

Bhiwgade, D.A. (1990): Comparative electron microscopy of chorio-allantoic placental barrier in some Indian Chiroptera. *Acta Anatomica* 138: 302–17.

Bickelmann, C., Mitgutsch, C., Richardson, M.K., Jimenez, R., de Bakker, M.A.G. & Sanchez-Villagra, M.R. (2012): Transcriptional heterochrony in talpid mole autopods. *EvoDevo* 3.

Bininda-Emonds, O.R.P., Jeffery, J.E., Coates, M.I. & Richardson, M.K. (2002): From Haeckel to event-pairing: the evolution of developmental sequences. *Theory Bioscience* 121: 297–320.

Bininda-Emonds, O.R.P., Jeffery, J.E. & Richardson, M.K. (2003): Is sequence heterochrony an important evolutionary mechanism in mammals? *Journal of Mammalian Evolution* 10(4): 335–361.

Bininda-Emonds, O.R.P., Jeffery, J.E., Sánchez-Villagra, M.R., Hanken, J., Colbert, M., Pieau, C., Selwood, L., Ten Cate, C., Raynaud, A., Osabutey, C.K. & Richardson, M.K. (2007): Forelimb-hind limb developmental timing across tetrapods. *BMC Evolutionary Biology* 7(1).

Blackburn, D.C. (2015): Evolution of vertebrate viviparity and specializations for fetal nutrition: a quantitative and qualitative analysis. *Journal of Morphology* 276(8): 961–990.

Blüm, V. (1986): *Vertebrate Reproduction. A Textbook.* Springer-Verlag, Berlin, Heidelberg, New York, Tokyo.

Bodley, H.D. (1974): Ultrastructural development of the chorioallantoic placental barrier in the bat Macrotus waterhousii. *Anatomical Record* 180(2): 351–367.

Botha-Brink, J. & Modesto, S.P. (2007): A mixed-age classed 'Pelycosaur' aggregation from South Africa: earliest evidence of parental care in amniotes? *Proceedings: Biological Sciences* 274(1627): 2829–2834.

Botha-Brink, J. & Modesto, S.P. (2009): Anatomy and relationships of the Middle Permian varanopid *Heleosaurus scholtzi* based on a social aggregation from the Karoo Basin of South Africa. *Journal of Vertebrate Paleontology* 29(2): 389–400.

Brehm, A. (1927): *Brehms Tierleben, Small Edition.* Gutenberg Verlag, Hamburg.

Breidbach, O. (2015): *Geschichte der Naturwissenschaften. I: Die Antike.* Springer, Berlin, Heidelberg.

Butler, H. & Juurlink, B.H.J. (1987): *An Atlas for Staging Mammalian and Chick Embryos.* CRC Press, Inc., Boca Raton.

Caldwell, W.H. (1884): Telegram, "Monotremes oviparous, ovum meroblastic". Read at meeting of the British Association at Montreal, Sept. 2, 1884. Brit. Assoc. Rep. Montreal Meeting.

Campbell, J. (1960): *The Masks of God: Primitive Mythology.* Secker and Warburg.

Carter, A.M. (2012). Evolution of placental function in mammals: the molecular basis of gas and nutrient transfer, hormone secretion, and immune responses. *Physiological Reviews* 92: 1543–1576.

Carter, A.M. (2008): Evolution of the placenta and associated reproductive characters in bats. *Journal of Experimental Zoology Part B: Molecular and Developmental Evolution* 310B(5): 428–449.

Carter, A.M., Blankenship, T.N., Enders, A.C. & Vogel, P. (2006): The fetal membranes of the otter shrews and a synapomorphy for Afrotheria. *Placenta* 27(2–3): 258–68.

Carter, A.M., Miglino, M.A., Ambrosio, C.E., Santos, T.C., Rosas, F.C., Neto, J.A., Lazzarini, S.M., Carvalho, A.F. & da Silva, V.M. (2008): Placentation in the Amazonian manatee (*Trichechus inunguis*). *Reproduction, Fertility and Development* 20(4): 537–545.

Carter, A.M. (2015): IFPA Senior Award Lecture: mammalian fetal membranes. *Placenta*.

Chinsamy-Turan, A. (2011): Forerunners of mammals. radiation, histology, biology. In: Farlow, J.O. (eds.). *Life of the Past*. University Press, Indiana: 352.

Cifelli, R.L., Rowe, T.B., Luckett, W.P., Banta, J., Reyes, R. & Howes, R.I. (1996): Fossil evidence for the origin of the marsupial pattern of tooth replacement. *Nature* 379(6567): 715–718.

Clark, C.T. & Smith, K.K. (1993): Cranial osteogenesis in *Monodelphis domestica* (Didelphidae) and *Macropus eugenii* (Macropodidae). *Journal of Morphology* 215: 119–149.

Clauss, M., Dittmann, M.T., Mueller, D.W.H., Zerbe, P. & Codron, D. (2014): Low scaling of a life history variable: analysing eutherian gestation periods with and without phylogeny-informed statistics. *Mammalian Biology* 79: 9–16.

Clutton-Brock, J. (1999). *A Natural History of Domesticated Mammals*. Academic Press, Cambridge.

Cook, M.J. (1965): *The Anatomy of the Laboratory Mouse*. Academic Press, Surrey.

de Lange, D. & Nierstrasz, H.F. (1932): *Tabellarische Uebersicht der Entwicklung von Tupaia javanica HORSF.* A. Oosthoek Verlag, Utrecht.

de Magalhães, J.P. & Costa, J. (2009): A database of vertebrate longevity records and their relation to other life-history traits. *Journal of Evolutionary Biology* 22(8): 1770–1774.

Deetjen, P. & Speckmann, E.-J. (1999): *Physiologie*. In: (eds.). Urban and Fischer, München.

Drews, B., Roellig, K., Menzies, B.R., Shaw, G., Buentjen, I., Herbert, C.A., Hildebrandt, T.B. & Renfree, M.B. (2013): Ultrasonography of wallaby prenatal development shows that the climb to the pouch begins in utero. *Scientific Reports* 3(1458).

Duncker, H.-R. (1978): Funktionsmorphologie des Atemapparates und Coelomgliederung bei Reptilien, Vöglen und Säugern. *Verhandlungen der Deutschen Zoologischen Gesellschaft* 99–132.

Elliot, M.G. & Crespi, B.J. (2009): Phylogenetic evidence for early hemochorial placentation in eutheria. *Placenta* 30: 949–967.

Enders, A.C., Carter, A.M., Künzle, H. & Vogel, P. (2005). Structure of the ovaries of the Nimba otter shrew, *Micropotamogale lamottei*, and the Madagascar hedgehog tenrec, *Echinops telfairi*. *Cells Tissue Organs* 179: 179–191.

Enders, A.C. (1960): Development and structure of the villous haemochorial placenta of the nine-banded armadillo (*Dasypus novemcinctus*). *Journal of Anatomy* 94: 34–45.

Felsenstein, J. (1985): Phylogenies and the comparative method. *American Naturalist* 125(1):1–15.

Fernandez, M. (1909). *Beiträge zur Embryologie der Gürteltiere. Morphologisches Jahrbuch* 39: 302–333.

Fioroni, P. (1987): *Allgemeine und vergleichende Embryologie der Tiere*. Springer-Verlag, Berlin.

Fischer, B. & Mitteroecker, P. (2015): Covariation between human pelvis shape, stature, and head size alleviates the obstetric dilemma. *Proceedings of the National Academy of Sciences of the United States of America* 112(18): 5655–5660.

Francis, M.C., Anthony, E.L.P., Brunton, J.A. & Kunz, T.H. (1994): Lactation in male fruit bats. *Nature* 367: 691–692.

Frazer, J.G. (1922): *The Golden Bough: A Study in Magic and Religion*. Abridged ed. The Macmillan Co., New York.

Gebelein, H. (1996): *Alchemie*. Dieterichs, München.

Gemmell, R.T., Johnston, G. & Bryden, M.M. (1988): Osteogenesis in two marsupial species, the bandicoot *Isoodon macrourus* and the possum *Trichosurus vulpecula*. *Journal of Anatomy* 159: 155–164.

Gemmell, R.T., Veitch, C. & Nelson, J. (2002): Birth in marsupials. *Comparative Biochemistry and Physiology Part B* 131: 621–630.

Germain, D. & Laurin, M. (2009). Evolution of ossification sequences in salamanders and urodele origins assessed through event-pairing and new methods. *Evolution & Development* 11(2): 170–190.

German, R.Z., Crompton, A.W. & Thexton, A.J. (2000). The ontogeny of feeding in mammals. In: Schwenk, K. *Feeding: Form, Function and Evolution in Tetrapod Vertebrates*. New York: Academic Press; 359–408.

Giere, P. & Zeller, U. (2006): *Die Embryologische Sammlung [The embryological collection]*. Museum für Naturkunde der Humboldt-Universität zu Berlin. Annual Report Jahresbericht 2006(15).

Gilbert, S.F. (2006): *Developmental Biology*. Sinauer Associates, Sunderland, Mass.

Goswami, A., Milne, N. & Wroe, S. (2011): Biting through constraints: cranial morphology, disparity and convergence across living and fossil carnivorous mammals. *Proceedings of the Royal Society of London B: Biological Sciences* 278: 1831–1839.

Goswami, A., Randau, M., Polly, P.D., Weisbecker, V., Bennett, C.V., Hautier, L. & Sánchez-Villagra, M.R. (2016). Do developmental constraints and high integration limit the evolution of the marsupial oral apparatus? *Integrative and Comparative Biology* 56(3):404–415.

Grant, T. (2013): *Platypus*. CSIRO Publishing, Collingwood.

Grant, T.R. (1989): Ornithorhynchidae. In: (eds.). *Fauna of Australia*. AGPS, Canberra: 436–450.

Green, E.L. (1968): *Biology of the Laboratory Mouse*. Dover Publications, New York.

Greven, H. (2004): Fortpflanzung und Entwicklung. In: Westheide, W. & Rieger, R. (eds.). *Spezielle Zoologie. Teil 2: Wirbel- oder Schädeltiere*. Spektrum Akademischer Verlag, Heidelberg, Berlin: 153–167.

Griffiths, M. (1978): Reproduction and Embryology. In: Griffiths, M. (eds.). *The Biology of the Monotremes*. Academic Press, New York, San Francisco, London: 209–254.

Griffiths, M. (1989): Tachyglossidae. In: (eds.). *Fauna of Australia*. AGPS, Canberra: 407–435.

Gruss, L.T. & Schmitt, D. (2015): The evolution of the human pelvis: changing adaptations to bipedalism, obstetrics and thermoregulation. *Philosophical Transactions of the Royal Society B*, 13 pp.

Haas, A. (2003): Phylogeny of frogs as inferred from primarily larval characters (Amphibia: Anura). *Cladistics* 19: 23–89.

Haeckel, E. (1866): *Generelle Morphologie der Organismen*. Georg Reimer, Berlin.

Haeckel, E. (1874): *Natürliche Schöpfungsgeschichte*. Georg Reimer, Berlin.

Haeckel, E. (1877): *Anthropologie oder Entwickelungsgeschichte des Menschen*. Verlag von Wilhelm Engelmann, Leipzig.

Hall, B.K. (2009): *The Neural Crest and Neural Crest Cells in Vertebrate Development and Evolution*. Springer, New York.

Hamburger, V. & Hamilton, H.L. (1951): A series of normal stages in the development of the chick embryo. *Journal of Morphology* 88: 49–92.

Hamilton, M.J., Davidson, A.D., Sibly, R.M. & Brown, J.H. (2011): Universal scaling of production rates across mammalian lineages. *Proceedings of the Royal Society of London B: Biological Sciences* 278(1705): 560–566.

Harrison, L. & Larsson, H. (2008): Estimating evolution of temporal sequence changes: a practical approach to inferring ancestral developmental sequences and sequence heterochrony. *Systematic Biology* 57(3): 378–387.

Harrison, R.G. (1969): Harrison stages and description of the normal development of the spotted salamander, *Amblystoma puctatum* (Linn.). In: Wilens, S. (eds.). *Organisation and Development of the Embryo.* Yale University Press, New Haven, London: 44–66.

Hartsoeker, N. (1694): *Essay de dioptrique.* J. Anisson, Paris.

Hautier, L., Weisbecker, V., Goswami, A., Knight, F., Kardjilov, N. & Asher, R.J. (2011): Skeletal ossification and sequence heterochrony in xenarthran evolution. *Evolution and Development* 13(5): 460–476.

Hautier, L., Weisbecker, V., Sanchez-Villagra, M.R., Goswami, A. & Asher, R.J. (2010): Skeletal development in sloths and the evolution of mammalian vertebral patterning. *Proceedings of the National Academy of Sciences of the United States of America* 107(44): 18903–18908.

Hayssen, V., Lacy, R.C. & Parker, P.J. (1985): Metatherian reproduction: transitional or transcending? *American Naturalist* 126(5): 617–632.

Hayssen, V., Van Tienhoven, A. & Van Tienhoven, A. (1993): *Asdell's Patterns of Mammalian Reproduction: A Compendium of Species-Specific Data.* Cornell University Press, Ithaca and London.

Hedrich, H.J. (2012): The laboratory mouse. In: (eds.). *The Laboratory Animal Pocket Reference Series.* Academic Press, Amsterdam.

Hickford, D., Frankenberg, S. & Renfree, M.B. (2009): The tammar wallaby, *Macropus eugenii*: a model kangaroo for the study of developmental and reproductive biology. *Cold Spring Harbour Protocols,* pdb-emo137.

Hill, J.P. (1910): The early development of the Marsupialia. IV. The early development of the Marsupialia with special reference to the native cat (*Dasyurus viverrinus*). *Quarterly Journal of Microscopical Science* 56: 1–134.

Hill, J.P. & Osman-Hill, W.C. (1955): The growth-stages of the pouch-young of the native cat (*Dasyurus viverrinus*) together with observations on the anatomy of the new-born young. *Transactions of the Zoological Society of London* 28(5): 349–352.

Hopwood, N. (2007): A history of normal plates, tables and stages in vertebrate embryology. *International Journal of Developmental Biology* 51: 1–26.

Hoßfeld, U. & Olsson, L. (2003a): The history of comparative anatomy in Jena-an overview. *Theory in Bioscience* 122: 109–126.

Hoßfeld, U. & Olsson, L. (2003b): The road from Haeckel: the Jena tradition in evolutionary morphology and the origins of "Evo-Devo". *Biology and Philosophy* 18: 285–307.

Houdebine, L.-M. (1997): *Transgenic Animals.* CRC Press, Boca Bacon.

Houillon, C. (1972): *Embryologie.* Springer, Wiesbaden.

Hsu, M.J., Garton, D.W. & Harder, J.D. (1999): Energetics of offspring production: a comparison of a marsupial (*Monodelphis domestica*) and a eutherian (*Mesocricetus auratus*). *Journal of Comparative Physiology B: Biochemical Systemic and Environmental Physiology* 169(1): 67–76.

Hughes, R.L. (1993): Monotreme development with particular reference to the extraembryonic membranes. *Journal of Experimental Zoology* 266(6): 480–494.

Hughes, R.L. & Hall, L.S. (1988): Structural Adaptations of the Neonate Marsupial. In: Tyndal-Biscoe, C.H. & Janssens, P.A. (eds.). *The Developing Marsupial: Models for Biomedical Research,* Springer-Verlag, Berlin.

Hunsaker, II, D. (1977, ed.): *The Biology of Marsupials.* Academic Press, New York.

Hüppi, E. (2018): External morphology and chondrocranial development in therian mammals. Master's thesis. University of Zürich.

Isler, K. & Van Schaik, C.P. (2009): The expensive brain: a framework for explaining evolutionary changes in brain size. *Journal of Human Evolution* 57(4): 392–400.

Jackson, S.M. (2003): *Australian Mammals: Biology and Captive Management.* CSIRO Publishing, Collingwood.

Jeffery, J.E., Bininda-Emonds, O.R.P., Coates, M.I. & Richardson, M.K. (2005): A new technique for identifying sequence heterochrony. *Systematic Biology* 54(2): 230–240.

Jenner, R.A. & Wills, M.A. (2007): The choice of model organisms in evo–devo. *Nature Reviews Genetics* 8: 311–319.

Ji, Q., Luo, Z.-X., Zhang, X., Yuan, C.-X. & Xu, L. (2009): Evolutionary development of the middle ear in Mesozoic therian mammals. *Science* 326: 278–281.

Johnson, M.H. & Selwood, L. (1996): Nomenclature of early development in mammals. *Reproduction Fertility and Development* 8(4): 759–764.

Kardong, K.V. (2008): *Vertebrates: Comparative Anatomy, Function, Evolution.* McGraw–Hill Primis.

Karim, K.B. & Bhatnagar, K.P. (1996): Early embryology, fetal membranes and placentation. *Annals of Anatomy* 178: 523–530.

Keel, O. (2013): *Das Böcklein in der Milch seiner Mutter und Verwandtes im Lichte eines altorientalischen Bildmotivs.* Veröffentlichungen der Ideagora für Religionsgeschichte, Altertumswissenschaften and Theologie http://hdl.handle.net/10900/43651(V.IRAT III-KEE 1980.1 (Böcklein): 1–163.

Keibel, F. (1897): *Normentafel zur Entwicklungsgeschichte des Schweines* (Sus scrofa domesticus). Verlag von Gustav Fischer, Jena.

Keibel, F. (1906): Die Entwickelung der äußeren Körperform der Wirbeltierembryonen, insbesondere der menschlichen Embryonen aus den ersten 2 Monaten. In: Hertwig, O. (eds.). *Handbuch der vergleichenden und experimentellen Entwicklungslehre der Wirbeltiere.* Verlag von Gustav Fischer, Jena: 1–174.

Kemp, T.S. (2006): The origin of mammalian endothermy: a paradigm for the evolution of complex biological structure. *Zoological Journal of the Linnean Society* 147: 473–488.

Keyte, A.L. & Smith, K.K. (2009): Opossum (*Monodelphis domestica*). A Marsupial Developmental Model. In: Gann, A. & Crotty, D. (eds.). *Emerging Model Organisms. A Laboratory Manual.* Cold Spring Harbour Laboratory Press, New York.

Keyte, A.L. & Smith, K.K. (2010): Developmental origins of precocial forelimbs in marsupial neonates. *Development* 137(24): 4283–4294.

Kielan-Jaworowska, Z., Cifelli, R.L. & Luo, Z.X. (2004). *Mammals from the Age of Dinosaurs*. Columbia University Press, New York.

Kirsch, J.A. (1977): Biological aspects of the marsupial-placental dichotomy: a reply to Lillegraven. *Evolution* 31(4): 898–900.

Klaatsch, H. (1895): Studien zur Geschichte der Mammarorgane. I. Theil. Die Taschen- und Beutelbildungen am Drüsenfeld der Monotremen. In: Semon, R. (eds.). *Monotremen und Marsupialier* 2(2). Zoologische Forschungsreisen in Australien und dem Malayischen Archipel mit Unterstützung des Herrn Dr. Paul von Ritter ausgeführt in den Jahren 1891–1893 von Dr Richard Semon, Professor in Jena. Gustav Fischer, Jena.

Kleisner, K., Ivell, R. & Flegr, J. (2010): The evolutionary history of testicular externalization and the origin of the scrotum. *Journal of Bioscience* 35: 27–37.

Klima, M. (1987): *Early Development of the Shoulder Girdle and Sternum in Marsupials (Mammalia: Metatheria)*. Springer-Verlag, Berlin, Heidelberg, New York, London, Paris, Tokyo.

Koyabu, D. & Son, N.T. (2014). Patterns of postcranial ossification and sequence heterochrony in bats: life histories and developmental trade-offs. *Journal of Experimental Zoology (Molecular and Developmental Evolution)* 322B: 607–618.

Koyabu, D., Werneburg, I., Morimoto, N., Zollikofer, C.P.E., Forasiepi, A.M., Endo, H., Kimura, J., Ohdachi, S.D., Truong, S.N. & Sánchez-Villagra, M.R. (2014): Mammalian skull heterochrony reveals modular evolution and a link between cranial development and brain size. *Nature Communications* 5(3625): 1–9.

Krockenberger, A. (2006): Lactation. In: Armati, P.J., Dickman, C. & Hume, I.D. (eds.). *Marsupials*. University Press, Cambridge: 108–136.

Kurta, A. & Kunz, T.H. (1987). Size of bats at birth and maternal investment during pregnancy. *Symposium of the Zoological Society London* 57: 79–106.

Laurin, I., Girondot, I. & de Ricqlès, I. (2000): Early tetrapod evolution. *Trends in Ecology and Evolution (Personal edition)* 15: 118–123.

Lewitus, E. & Soligo, C. (2011): Life-history correlates of placental structure in Eutherian evolution. *Evolutionary Biology* 38(3): 287–305.

Liem, K.F., Bemis, W.E., Walker, W.F. & Grande, L. (2001): *Functional Anatomy of the Vertebrates*. Harcourt, Philadelphia.

Lillegraven, J.A. (1975): Biological consideration of the marsupial-placental dichotomy. *Evolution* 29: 707–722.

Lillegraven, J.A. (2003). Polarities in mammalian evolution seen through homology of the inner cell mass. *Journal of Mammalian Evolution* 10: 277–333.

Lillegraven, J.A. (2004): Polarities in mammalian evolution seen through homologs of the inner cell mass. *Journal of Mammalian Evolution* 11(2): 143–202.

Lillegraven, J.A., Thompson, S.D., McNab, B.K. & Patton, J.L. (1987): The origin of eutherian mammals. *Biological Journal of the Linnean Society* 32: 281–336.

Liu, J., Organ, L.C., Benton, M.J., Brandley, M.C. & Aitchison J.C. (2017). Live birth in an archosauromorph reptile. *Nature Communications* 8: 14445.

Luckett, W.P. (1975): Ontogeny of amniotic fetal membranes and their application to phylogeny. In: Hecht, M.K., Goody, P.C. & Hecht, B.M. (eds.). *Major Patterns in Vertebrate Evolution*. Plenum Press, New York, London.

Luo, Z.X. (2007): Transformation and diversification in early mammal evolution. *Nature* 450: 1011–1019.

Maier, M. (1617/1618): *Atalanta Fugiens, hoc est emblemata nova de secretis naturae chymica*. Johann Theodor de Bry, Oppenheim.

Maier, W. (1987): The ontogenetic development of the orbitotemporal region in the skull of *Monodelphis domestica* (Didelphidae, Marsupialia), and the problem of the mammalian alisphenoid. In: Kuhn, H.-J. & Zeller, U. (eds.). *Morphogenesis of the Mammalian Skull*. Verlag Paul Parey, Kiel, Hannover.

Maier, W. (1999): On the evolutionary biology of early mammals-with methodological remarks on the interaction between ontogenetic adaptation and phylogenetic transformation. *Zoologischer Anzeiger* 238(1–2): 55–74.

Maier, W. & Ruf, I. (2016): Evolution of the mammalian middle ear: a historical review. *Journal of Anatomy* 228(2): 270–283.

Maier, W. & Werneburg, I. (2014): *Schlüsselereignisse der organismischen Makroevolution*. Scidinge Hall Verlag, Zürich.

Maier, W., Van den Heever, J. & Durand, F. (1996): New therapsid specimens and the origin of the secondary hard and soft palate. *Journal of Zoological Systematics and Evolutionary Research* 34: 9–19.

Manger, P.R., Hall, L.S. & Pettigrew, J.D. (1998): The development of the external features of the platypus (*Ornithorhynchus anatinus*). *Philosophical Transactions of the Royal Society of London. Series B, Biological Sciences* 353(1372): 1115–1125.

Martin, R.D. (2007): The evolution of human reproduction: a primatological perspective. *Yearbook of Physical Anthropology* 50: 59–84.

Martin, R.D. & MacLarnon, A. (1985): Gestation period, neonatal size and maternal investment in placental mammals. *Nature* 313: 220–223.

Martin, T. (1997): Tooth replacement in Late Jurassic Dryolestidae (Eupantotheria, Mammalia). *Journal of Mammalian Evolution* 4(1): 1–18.

Martin, T. & Ruf, I. (2009): Perspectives: on the mammalian ear. *Science* 326: 243–244.

Maxwell, E.E. & Harrison, L.B. (2009): Methods for the analysis of developmental sequence data. *Evolution and Development* 11(1): 109–119.

McNamara, K.J. (2002): Changing times, changing places: heterochrony and heterotopy. *Paleobiology* 28(4): 551–558.

Meredith, R.W., Janečka, J.E., Gatesy, J., Ryder, O.A., Fisher, C.A., Teeling, E.C., Goodbla, A., Eizirik, E., Simão, T.L.L., Stadler, T., Rabosky, D.L., Honeycutt, R.L., Flynn, J.J., Ingram, C.M., Steiner, C., Williams, T.L., Robinson, T.J., Burk-Herrick, A., Westerman, M., Ayoub, N.A., Springer, M.S. & Murphy, W.J. (2011): Impacts of the Cretaceous terrestrial revolution and KPg extinction on mammal diversification. *Science* 334: 521–524.

Mess, A. & Carter, A.M. (2006): Evolutionary transformations of fetal membrane characters in Eutheria with special reference to Afrotheria. *Journal of Experimental Zoology (Molecular and Developmental Evolution)* 306B: 140–163.

Mess, A.M., Favaron, P.O., Pfarrer, C., Osmann, C., Melo, A.P., Rodrigues, R.F., Ambrósio, C.E., Bevilacqua, E. & Miglino, M.A. (2012): Placentation in the anteaters *Myrmecophaga tridactyla* and *Tamandua tetradactyla* (Eutheria, Xenarthra). *Reproductive Biology and Endocrinology* 10(102): 1–7.

Mickoleit, G. (2004). Phylogenetische Systematik der Wirbeltiere. München, Verlag Dr. Friedrich Pfeil.

Milinkovitch, M.C. & Tzika, A.C. (2007): Escaping the mouse trap; the selection of new evo-devo model species. *Journal of Experimental Zoology B: Molecular and Developmental Evolution* 308B: 337–346.

Mitgutsch, C., Richardson, M.K., Jiménez, R., Martín, J.E., Kondrashov, P., de Bakker, M.A.G. & Sánchez-Villagra, M.R. (2012): Circumventing the pentadactyly 'constraint'—autopodial recruitment of pre-axial structures in true moles. *Biology Letters* 8: 74–77.

Morton, S.R., Recher, H.F., Thompson, S.D. & Braithwaite, R.W. (1982): Comments on the relative advantages of marsupial and eutherian reproduction. *American Naturalist* 120(1): 128–134.

Mossman, H.W. (1987): *Vertebrate Fetal Membranes*. MacMillan Press, London.

Moyal, A. (2004): *Platypus: The Extraordinary Story of How a Curious Creature Baffled the World*. Johns Hopkins University Press, Baltimore.

Müller, F. (1972–1973): Zur stammesgeschichtlichen Veränderung der Eutheria-Ontogenesen. Versuch einer Übersicht aufgrund vergleichend morphologischer Studien an Marsupialia und Eutheria. 1. Teil. Zur Evolution der Geburtsgestalt: Gestaltstadien der Eutheria. / 2. Teil. Ontogenesetypus und Cerebralisation. / 3. Teil. Zeitliche Aspekte in der Evolution der Ontogenesetypen. / 4. Spezieller Teil. *Revue Suisse de Zoologie* 79(1,2,4): 1–97, 501–566, 567–612, 1599–1685.

Müller, F. (1973): Zur stammesgeschichtlichen Veränderung der Eutheria-Ontogenesen. Versuch einer Übersicht aufgrund vergleichend morphologischer Studien an Marsupialia und Eutheria. 4. Spezieller Teil. *Revue Suisse de Zoologie* 79(4): 1599–1685.

Newman, H.H. & Patterson, J.T. (1910): The development of the nine-banded armadillo from the primitive streak stage to birth; with especial reference to the question of specific polyembryology. *Journal of Morphology* 21: 359–423.

Mortola, J.P., Frappell, P.B. & Woolley, P.A. (1999): Breathing through skin in a newborn mammal. *Nature* 397: 660–660.

Novacek, M.J., Rougier, G.W., Wible ,J.R., McKenna, M.C., Dashzeveg, D. & Horovitz, I. (1997): Epipubic bones in eutherian mammals from the late Cretaceous of Mongolia. *Nature* 389(6650): 483–486.

O'Leary, M.A., Bloch, J.I., Flynn, J.J., Gaudin, T.J., Giallombardo, A., Giannini, N.P., Goldberg, S.L., Kraatz, B.P., Luo, Z.X., Meng, J., Ni, X.J., Novacek, M.J., Perini, F.A., Randall, Z.S., Rougier, G.W., Sargis, E.J., Silcox, M.T., Simmons, N.B., Spaulding, M., Velazco, P.M., Weksler, M., Wible, J.R. & Cirranello, A.L. (2013): The placental mammal ancestor and the post-K-Pg radiation of placentals. *Science* 339(6120): 662–667.

O'Meara, R.N. & Asher, R.J. (2016, accepted): The evolution of growth patterns in mammalian versus non-mammalian cynodonts. *Paleobiology*.

Pianka, E.R. (1970): On r-and k-selection. *American Naturalist* 140: 592–597.

Piñeiro, G., Ferigolob, J., Meneghel, M. & Laurin, M. (2012): The oldest known amniotic embryos suggest viviparity in mesosaurs. *Historical Biology* 24(6): 620–630.

Portmann, A. (1944): *Biologische Fragmente zu einer Lehre vom Menschen*. Schwebe-Verlag, Basel.

Portmann, A. (1959): *Zoologie und das neue Bild des Menschen*. Rowohlt, Hamburg.

Pough, F.H., Janis, C.M. & Heiser, J.B. (2012): *Vertebrate Life*. Benjamin Cummings, San Francisco.

Rasweiler, VI, J.J. & Badwaik, N.K. (2000). Anatomy and physiology of the female reproductive tract. In: Crichton, E.G. & Krutzsch, P.H. (ed.): *Reproductive Biology of Bats*. Academic Press, San Diego, p. 157–219.

Reilly, S.M., McElroy, E.J. & White, T.D. (2009): Abdominal muscle function in ventilation and locomotion in new world opossums and basal eutherians: breathing and running with and without epipubic bones. *Journal of Morphology* 270: 1014–1028.

Renfree, M. (1993): Diapause, pregnancy, and parturition in Australian marsupials. *Journal of Experimental Zoology* 266: 450–462.

Renfree, M. and Shaw, G. (2001) Reproduction in Monotremes and Marsupials. In *Encyclopedia of Life Sciences*. pp 1–5. Ed. Knobil, E. and Neill, J. John Wiley & Sons, Ltd.

Renfree, M.B., Hore, T.A., Shaw, G., Graves, J.A.M. & Pask, A.J. (2009): Evolution of genomic imprinting: insights from marsupials and monotremes. *Annual Review of Genomics and Human Genetics* 10: 241–262.

Richardson, M.K. & Keuck, G. (2002): Haeckel's ABC of evolution and development. *Biological Reviews* 77: 495–528.

Richardson, M.K. & Narraway, J. (1999): A treasure house of comparative embryology. *International Journal of Developmental Biology* 43: 591–602.

Richardson, M.K., Gobes, S.M.H., Van Leeuwen, A.C., Polman, J.A.E., Pieau, C. & Sánchez-Villagra, M.R. (2009): Heterochrony in limb evolution: developmental mechanisms and natural selection. *Journal of Experimental Zoology B: Molecular and Developmental Evolution* 312B: 639–664.

Riedelsheimer, B., Unterberger, P., Künzle, H. & Welsch, U. (2007). Histological study of the cloacal region and associated structures in the hedgehog tenrec *Echinops telfairi*. *Mammalian Biology* 72(6): 330–341.

Roberts, C.T. & Breed, W.G. (1994) Placentation in the dasyurid marsupial, Sminthopsis crassicaudata, the fat-tailed dunnart, and notes on placentation of the didelphid, *Monodelphis domestica*. *Journal of Reproduction and Fertility* 100: 105–113.

Romer, A.S. (1976): *Vergleichende Anatomie der Wirbeltiere*. Verlag Paul Parey, Hamburg, Berlin.

Rot-Nikcevic, I., Reddy, T., Downing, K.J., Belliveau, A.C., Hallgrímsson, B., Hall, B.K. & Kablar, B. (2006): Myf5–/–:MyoD–/–amyogenic fetuses reveal the importance of early contraction and static loading by striated muscle in mouse skeletogenesis. *Development Genes and Evolution* 216(1): 1–9.

Rousmaniere, H., Silverman, R., White, R.A., Sasaki, M.M., Wilson, S.D., Morrison, J.T. & Cruz, Y.P. (2010): Husbandry of *Monodelphis domestica* in the study of mammalian embryogenesis. *Lab Animal* 37(7): 219–226.

Ruf, I. (2014): Im Schatten der Dinosaurier?-Die Evolution der Säugetiere im Mesozoikum. In: Maier, W. & Werneburg, I. (eds.). *Schlüsselereignisse der organismischen Makroevolution*. Scidinge Hall Verlag, Zürich: 333–360.

Russell, E.M. (1982): Parental investment and desertion of young in marsupials. *American Naturalist* 119(5): 744–748.

Sánchez-Villagra, M.R. (2010): Developmental palaeontology in synapsids: the fossil record of ontogeny in mammals and their closest relatives. *Proceedings of the Royal Society of London B: Biological Sciences* 277(1685): 1139–1147.

Sánchez-Villagra, M.R. (2013): Why are there fewer marsupials than placentals? On the relevance of geography and physiology to evolutionary patterns of mammalian diversity and disparity. *Journal of Mammalian Evolution* 20: 279–290.

Sánchez-Villagra, M.R. & Smith, K.K. (1997). Diversity and evolution of the marsupial mandibular angular process. *Journal of Mammalian Evolution* 4: 119–144.

Sánchez-Villagra, M.R., Goswami, A., Weisbecker, V., Mock, O. & Kuratani, S. (2008): Conserved relative timing of cranial ossification patterns in early mammalian evolution. *Evolution and Development* 10: 519–530.

Sánchez-Villagra, M.R. & Sultan, F. (2002): The cerebellum at birth in therian mammals, with special reference to rodents. *Brain, Behavior and Evolution* 59: 101–113.

Sánchez-Villagra, M.R. & Werneburg, I. (2016). Mammalian organogenesis in deep time: tools for teaching and outreach. *Evolution: Education and Outreach* 9:11, p. 1–15.

Schmidt, K. & Starck, J.M. (2004): Developmental variability during early embryonic development of zebra fish, *Danio rerio*. *Journal of Experimental Zoology B* 302: 446–457.

Schmidt, K. & Starck, J.M. (2011): Testing evolutionary hypotheses about the phylotypic period of zebrafish. *Journal of Experimental Zoology B: Molecular and Developmental Evolution* 316B(5): 319–329.

Sears, K.E. (2004): Constraints on the morphological evolution of marsupial shoulder girdles. *Evolution* 58(10): 2353–2370.

Sears, K.E. (2006): Development of bat flight: morphologic and molecular evolution of bat wing digits. *Proceedings of the National Academy of Sciences* 103(17): 6581–6586.

Sears, K.E. (2008): Molecular determinants of bat wing development. *Cells Tissues Organs* 187(1): 6–12.

Sears, K.E. (2009): Differences in the timing of prechrondrogenic limb in mammals: the Marsupial-Placental dichotomy resolved. *Evolution* 63(8): 2193–2200.

Sears, K.E., Doroba, C., Cao, X., Xie, D. and Zhong, S. (2012): Molecular determinants of marsupial integration and constraint. In: Asher, R.J. & Müller, J. (eds.). *From Clone to Bone: The Synergy of Morphological and Molecular Tools in Palaeobiology*. University Press, Cambridge: 257–278.

Sekulić, S., Božić, K., Bozić, A., Borota, J. & Ćulić, M. (2006): Precocial rodents as new experimental model to study the effects of altered gravitational conditions on fetal development. *Microgravity-Science and Technology* 18(3): 223–225.

Selwood, L. & Hickford, D. (1999): Early cell lineages in marsupial embryos. In: Moody, S.A. (eds.). *Cell Linage and Fate Determination*. Academic Press, New York.

Selwood, L. & Johnson, M.H. (2006): Trophoblast and hypoblast in the monotreme, marsupial and eutherian mammal: evolution and origins. *BioEssays* 28(2): 128–145.

Semon, R. (1894a): Beobachtungen über die Lebensweise und Fortpflanzung der Monotremen nebst Notizen über ihre Körpertemperatur. In: Semon, R. (eds.). *Monotremen und Marsupialier* 2(1) Zoologische Forschungsreisen in Australien und dem Malayischen Archipel mit Unterstützung des Herrn Dr Paul von Ritter ausgeführt in den Jahren 1891–1893 von Dr Richard Semon, Professor in Jena. Gustav Fischer, Jena.

Semon, R. (1894b): Die Embryonalhüllen der Monotremen und Marsupialier. In: Semon, R. (eds.). *Monotremen und Marsupialier* 2(1). Zoologische Forschungsreisen in Australien und dem Malayischen Archipel mit Unterstützung des Herrn Dr Paul von Ritter ausgeführt in den Jahren 1891–1893 von Dr Richard Semon, Professor in Jena. Gustav Fischer, Jena.

Semon, R. (1894c): *Monotremen und Marsupialier I*. Zoologische Forschungsreisen in Australien und dem Malayischen Archipel mit Unterstützung des Herrn Dr Paul von Ritter ausgeführt in den Jahren 1891–1893 von Dr Richard Semon, Professor in Jena. Gustav Fischer, Jena.

Semon, R. (1894d): Zur Entwickelungsgeschichte der Monotremen. In: Semon, R. (eds.). *Monotremen und Marsupialier* 2(1) Zoologische Forschungsreisen in Australien und dem Malayischen Archipel mit Unterstützung des Herrn Dr Paul von Ritter ausgeführt in den Jahren 1891–1893 von Dr Richard Semon, Professor in Jena. Gustav Fischer, Jena.

Semon, R. (1897a): Monotremen und Marsupialier II, 1. Teil. Zoologische Forschungsreisen in Australien und dem Malayischen Archipel mit Unterstützung des Herrn Dr Paul von Ritter ausgeführt in den Jahren 1891–1893 von Dr Richard Semon, Professor in Jena. Gustav Fischer, Jena *(II.1 / Atlas)*.

Semon, R. (1897b): Monotremen und Marsupialier II, 1. Teil. Zoologische Forschungsreisen in Australien und dem Malayischen Archipel mit Unterstützung des Herrn Dr Paul von Ritter ausgeführt in den Jahren 1891–1893 von Dr Richard Semon, Professor in Jena. Gustav Fischer, Jena *(II.1 / Text)*.

Semon, R. (1904): Monotremen und Marsupialier II, 2. Teil. Zoologische Forschungsreisen in Australien und dem Malayischen Archipel mit Unterstützung des Herrn Dr Paul von Ritter ausgeführt in den Jahren 1891–1893 von Dr Richard Semon, Professor in Jena. Gustav Fischer, Jena *(II.2 / Atlas)*.

Smith, K.K. (1994): Development of craniofacial musculature in *Monodelphis domestica* (Marsupialia, Didelphidae). *Journal of Morphology* 222: 149–173.

Smith, K.K. (1997): Comparative patterns of craniofacial development in eutherian and metatherian mammals. *Evolution* 51(5): 1663–1678.

Smith, K.K. (2001): Early development of the neural plate, neural crest and facial region of marsupials. *Journal of Anatomy* 199: 121–131.

Smith, K.K. (2002): *Monodelphis Staging Reference Series*. http://www.biology.duke.edu/kksmithlab (accessed: 5.11.2011).

Smith, K.K. (2015): Placental evolution in therian mammals. In: Dial, K.P., Shubin, N. & Brainerd, E.L. (eds.). *Great Transformations in Vertebrate Evolution*. University of Chicago Press, Chicago.

Spiekman, S.N.F., Werneburg I. (2017). Patterns in the bony skull development of marsupials: high variation in onset of ossification and conserved regions of bone contact. *Scientific Reports* 7: 43197, pp. 1–11

Starck, D. (1965): *Embryologie. Ein Lehrbuch auf allgemein biologischer Grundlage*. Thieme, Stuttgart.

Starck, D. (1995): *5. Teil: Säugetiere. 5/1: Allgemeines, Ordo 1–9*. Gustav Fischer Verlag, Jena, Stuttgart, New York.

Stewart, J.R. (1997): Morphology and evolution of the egg of oviparous amniotes. In: Sumida, S. & Martin, K. (eds.). *Amniote Origins. Completing the Transition to Land*. Academic Press, London: 291–326.

Storch, G. & Schröpfer, R. (2004): Mammalia, Säugetiere. In: Westheide, W. & Rieger, R. (eds.). *Spezielle Zoologie. Teil 2: Wirbel- oder Schädeltiere*. Spektrum Akademischer Verlag, Heidelberg, Berlin: 445–471.

Strauss, F. (1938). Die Befruchtung und der Vorgang der Ovulation bei Ericulus aus der Familie der Centetiden. *Biomorphosis* 1: 281–312.
Theiler, K. (1989): *The House Mouse: Atlas of Embryonic Development.* Springer Verlag, New York.
Trevathan, W. (2015): Primate pelvic anatomy and implications for birth. *Philosophical Transactions of the Royal Society of London. Series B, Biological Sciences.*
Tyndale-Biscoe, C.H. (2001): Australasian marsupials-to cherish and to hold. *Reproduction Fertility and Development* 13(7–8): 477–485.
Tyndale-Biscoe, C.H. (2005): *Life of Marsupials.* CSIRO Publishing, Collingwood.
Tyndale-Biscoe, C.H. & Renfree, M. (1987): *Reproductive Physiology of Marsupials.* Cambridge University Press, Cambridge.
Tzika, A. & Milinkovitch, M.C. (2008): A pragmatic approach for selecting evo-devo model species in amniotes. In: Minelli, A. & Fusco, G. (eds.). *Evolving Pathways. Key Themes in Evolutionary Developmental Biology.* Cambridge University Press, Cambridge: 123–143.
Urban, J.E., Weaver, A.A., Lillie, E.M., Maldjian, J.A., Whitlow, C.T. & Stitzel, J.D. (2016): Evaluation of morphological changes in the adult skull with age and sex. *Journal of Anatomy.* 229(6):838–846
Van Nievelt, A.F.H. & Smith, K.K. (2005). Tooth eruption in *Monodelphis domestica* and its significance for phylogeny and natural history, *Journal of Mammalogy* 86(2): 333–341.
Van Schaik, C. (2016). The Primate Origins of Human Nature. Hoboken, New Jersey, John Wiley & Sons, Inc.
Vaughan, T.A., Ryan, J.M. & Czaplewski, N.J. (2011): *Mammalogy.* Jones and Bartlett Publ., USA.
von Goethe, J.W., (1899): *Goethes Faust. Mit Bildern von F. Simm.* Deutsche Verlagsanstalt, Stuttgart.
von Soemmerring, S.T. (1799): *Icones Embryonum Humanorum.* Varrentrapp und Wenner, Frankfurt am Main.
Warren, W.C., Hillier, L.W., Graves, J.A.M., Birney, E., Ponting, C.P., Grutzner, F., Belov, K., Miller, W., Clarke, L., Chinwalla, A.T., Yang, S.P., Heger, A., Locke, D.P., Miethke, P., Waters, P.D., Veyrunes, F., Fulton, L., Fulton, B., Graves, T., Wallis, J., Puente, X.S., Lopez-Otin, C., Ordonez, G.R., Eichler, E.E., Chen, L., Cheng, Z., Deakin, J.E., Alsop, A., Thompson, K., Kirby, P., Papenfuss, A.T., Wakefield, M.J., Olender, T., Lancet, D., Huttley, G.A., Smit, A.F.A., Pask, A., Temple-Smith, P., Batzer, M.A., Walker, J.A., Konkel, M.K., Harris, R.S., Whittington, C.M., Wong, E.S.W., Gemmell, N.J., Buschiazzo, E., Jentzsch, I.M.V., Merkel, A., Schmitz, J., Zemann, A., Churakov, G., Kriegs, J.O., Brosius, J., Murchison, E.P., Sachidanandam, R., Smith, C., Hannon, G.J., Tsend-Ayush, E., McMillan, D., Attenborough, R., Rens, W., Ferguson-Smith, M., Lefevre, C.M., Sharp, J.A., Nicholas, K.R., Ray, D.A., Kube, M., Reinhardt, R., Pringle, T.H., Taylor, J., Jones, R.C., Nixon, B., Dacheux, J.L., Niwa, H., Sekita, Y., Huang, X.Q., Stark, A., Kheradpour, P., Kellis, M., Flicek, P., Chen, Y., Webber, C., Hardison, R., Nelson, J., Hallsworth-Pepin, K., Delehaunty, K., Markovic, C., Minx, P., Feng, Y.C., Kremitzki, C., Mitreva, M., Glasscock, J., Wylie, T., Wohldmann, P., Thiru, P., Nhan, M.N., Pohl, C.S., Smith, S.M., Hou, S.F., Renfree, M.B. & *et al.* (2008): Genome analysis of the platypus reveals unique signatures of evolution. *Nature* 453(7192): 175–183.
Watson, D.M.S. (1955): James Peter Hill. 1873–1954. *Biographical Memoirs of Fellows of the Royal Society* 1:100–117.
Wehner, R. & Gehring, W. (2007). Zoologie. Stuttgart, New York, Georg Thieme Verlag.
Weir, B.J. (1971). The reproductive physiology of the plains viscacha, *Lagostomus maximus. Journal of Reproduction and Fertility* 25: 355–363.
Weisbecker, V. (2010): Monotreme ossification sequences and the riddle of mammalian skeletal development. *Evolution* 65(5): 1323–1335.
Weisbecker, V. (2015): Are monotremes primitive and marsupials inferior? In: Klieve, A., Hogan, L., Johnston, S, Murray, P. (eds.). *Marsupials and Monotremes—Nature's Enigmatic Mammals.* Nova Science Publishers, New York: pp. 397–411.
Weisbecker, V., Goswami, A., Wroe, S. & Sánchez-Villagra, M.R. (2008): Ossification heterochrony in the therian postcranial skeleton and the marsupial-placental dichotomy. *Evolution* 62(8): 2027–2041.
Weisbecker, V. & Beck, R.M.D. (2015). Marsupial and monotreme evolution and biogeography. In: Klieve, A., Hogan, L., Johnston, S, Murray, P. (eds.). *Marsupials and Monotremes—Nature's Enigmatic Mammals.* Nova Science Publishers, New York: pp. 1–32.
Weisbecker, V. & Nilsson, M. (2008): Integration, heterochrony, and adaptation in pedal digits of syndactylous marsupials. *BMC Evolutionary Biology* 8(1): 160.
Werdelin, L. (1987): Jaw geometry and molar morphology in marsupial carnivores: analysis of a constraint and its macroevolutionary consequences. *Paleobiology* 13(3): 342–350.
Werdelin, L. & Nilsonne, A. (1999): The evolution of the scrotum and testicular descent in mammals: a phylogenetic view. *Journal of Theoretical Biology* 196(1): 61–72.
Werneburg, I. (2009): A standard system to study vertebrate embryos. *PLoS One* 4(6).
Werneburg, I. & Geiger, M. (2017). Ontogeny of domestic dogs and the developmental foundations of carnivoran domestication. *Journal of Mammalian Evolution.* 24(3): 323–343
Werneburg, I., Polachowski, K.M. & Hutchinson, M.N. (2015): Bony skull development in the *Argus monitor* (Squamata, Varanidae, *Varanus panoptes*) with comments on developmental timing and adult anatomy. *Zoology* 118: 255–280.
Werneburg, I., Laurin, M., Koyabu, D., & Sánchez-Villagra, M.R. (2016): Evolution of organogenesis and the origin of altriciality in mammals. *Evolution & Development* 18(4): 229–244.
Werneburg, I. & Sánchez-Villagra, M.R. (2011): The early development of the echidna, *Tachyglossus aculeatus* (Mammalia: Monotremata), and patterns of mammalian development. *Acta Zoologica* 82(1): 75–88.
Werneburg, I., Tzika, A.C., Hautier, L., Asher, R.J., Milinkovitch, M.C. & Sánchez-Villagra, M.R. (2013): Development and embryonic staging in non-model organisms: the case of an afrotherian mammal. *Journal of Anatomy* 222: 2–18.
Westheide, W. & Rieger, R. (2010): *Wirbel- oder Schädeltiere.* Spektrum Akademischer Verlag, Heidelberg, Berlin.
Wiesing, L. (2004): *Philosophie der Wahrnehmung.* Suhrkamp Verlag, Frankfurt am Main.
Wildman, D.E., Chen, C.Y., Erez, O., Grossman, L.I., Goodman, M. & Romero, R. (2006): Evolution of the mammalian placenta

revealed by phylogenetic analysis. *Proceedings of the National Academy of Sciences of the United States of America* 103(9): 3203–3208.

Wilson, L.A.B. (2011): Comparison of prenatal and postnatal ontogeny: cranial allometry in the African striped mouse (*Rhabdomys pumilio*). *Journal of Mammology* 92(2): 407–420.

Wooding, P. & Burton, G. (2008): *Comparative Placentation: Structures, Functions and Evolution.* Springer-Verlag, Berlin, Heidelberg.

Zelditch, M.L. & Fink, W.L. (1996): Heterochrony and heterotopy: stability and innovation in the evolution of form. *Paleobiology* 22(2): 241–254.

Zeller, U. (2004a): Monotremata (Prototheria), Kloakentiere. In: Westheide, W. & Rieger, R. (eds.). *Spezielle Zoologie. Teil 2: Wirbel- oder Schädeltiere.* Spektrum Akademischer Verlag, Heidelberg, Berlin: 471–484.

Zeller, U. (2004b): Marsupialia (Metatheria, Didelphia), Beuteltiere. In: Westheide, W. & Rieger, R. (eds.). *Spezielle Zoologie. Teil 2: Wirbel- oder Schädeltiere.* Spektrum Akademischer Verlag, Heidelberg, Berlin: 484–499.

Zeller, U. & Freyer, C. (2002). Early ontogeny and placentation of the grey short-tailed opossum, *Monodelphis domestica* (Didelphidae: Marsupialia): contribution to the reconstruction of the marsupial morphotype. *Journal of Zoological Systematics and Evolutionary Research* 39: 137–158.

Zollikofer, C.P.E. & Ponce de León, M.S. (2010): The evolution of hominin ontogenies. *Seminars in Cell and Developmental Biology* 21(4): 441–452.

Kenneth D. Angielczyk and Christian F. Kammerer

5 Non-Mammalian synapsids: the deep roots of the mammalian family tree

5.1 Introduction

Mammals are arguably the most conspicuous tetrapods in the modern biota. Although there are fewer extant mammal species (~5,500) than birds (~10,000) or squamates (~10,000), the morphological disparity and ecological diversity attained by mammals is striking. Living mammals include what is probably the largest tetrapod of all time (the blue whale, *Balaenoptera musculus*) and the largest extant terrestrial tetrapod (the African bush elephant, *Loxodonta africana*), as well as highly miniaturized species such as Kitti's hog-nosed bat (*Craseonycteris thonglongyai*) and the Etruscan shrew (*Suncus etruscus*). From an ecomorphological perspective, mammals display an impressive array of specializations, ranging from distinctive limb morphologies for cursoriality, powered flight, climbing, brachiating, pelagic swimming, and digging, to extensive modifications of the skull and dentition associated with high-fiber herbivory, insectivory, carnivory, durophagy, and gnawing, to unique sensory adaptations such as echolocation and high levels of encephalization. When combined with the extensive Cenozoic mammalian fossil record, it is not surprising that the last 66 million years are generally considered the "Age of Mammals".

Mammalian history does not begin at the start of the "Age of Mammals", though. The oldest known members of the mammalian crown-group date from at least the Middle to Late Jurassic, and possibly the Late Triassic, with the record of very mammal-like mammaliaforms extending back into the Late Triassic (Kielan-Jaworowska *et al.* 2004, Luo 2007, Luo *et al.* 2011, Zheng *et al.* 2013, Bi *et al.* 2014, Krause *et al.* 2014, Luo *et al.* 2015). In turn, mammals and mammaliaforms are part of a larger clade, Synapsida, that diverged from its sister clade Sauropsida during the Carboniferous Period, between about 318 and 332 million years ago (Benton *et al.* 2015). Synapsida includes a number of extinct lineages that were diverse and abundant components of terrestrial ecosystems during the Pennsylvanian (late Carboniferous), Permian, and Triassic periods, and members of many of these lineages are highly morphologically disparate from living mammals.

Non-mammalian synapsids (i.e., those synapsids that fall outside of the subclade Mammaliaformes, equivalent to Mammalia sensu Luo *et al.* 2002; see Fig. 5.3) represent the deep roots of the mammal phylogenetic tree, and their fossil record is extensive. For example, over 18,000 non-mammalian synapsid specimens have been collected from the middle Permian (Guadalupian) to the Middle Triassic rocks of the Beaufort Group in the Karoo Basin of South Africa (Nicolas and Rubidge 2009), and Brocklehurst (2015) compiled a data set of over 1,300 primarily North American and European specimens of basal ("pelycosaur"-grade) synapsids for an analysis of the completeness of their fossil record. This density of fossils enables the study of many synapsid character systems and outlines synapsid phylogeny, despite the many differences between the first synapsids and their mammalian descendants. It also has caused synapsids to become a textbook example of how the fossil record documents major evolutionary transitions (Hopson 1969, 1987, 1991, 1994, 2001, Kemp 1982, 1985, 2005, 2012, Cain 1988, Hotton 1991, Rubidge and Sidor 2001, Prothero 2007, Angielczyk 2009, Asher 2012, Benton 2015a). Beyond their importance to understanding mammalian evolution, non-mammalian synapsids are simply fascinating animals. Among their numbers are the largest animals of their times, some of the first tetrapod herbivores, survivors of the largest mass extinction in Earth history, and animals that do not have obvious analogs in the modern biota (e.g., sail-backed carnivores and tusked, beaked herbivores).

Our primary goal in this chapter is to provide an introduction to the diversity of non-mammalian synapsids and to the long history of work that surrounds the group. The discussion of each major synapsid clade is relatively brief, but we have tried to include a mixture of recent and historical references to give interested readers an entry into the literature, so that they can track how perceptions of certain synapsid clades have changed over time and how various ideas about synapsid evolution have developed. Following the review of synapsid diversity, we discuss two areas that we think should be high priorities for future synapsid research. First is the continued need to investigate the full diversity of

https://doi.org/10.1515/9783110341553-005

non-mammalian synapsids as interesting and important animals in their own right, independent of mammal origins. Second is a reassessment of our understanding of the evolution of mammalian characters during synapsid history, using the full methodological toolkit of modern paleontology and evolutionary biology combined with more detailed sampling of non-mammalian synapsid taxa. Over the last three decades, a great deal of effort has been expended revising synapsid taxonomy, describing new species, and clarifying phylogenetic relationships within the group. Yet much of what we "know" about synapsid paleobiology and paleoecology is based on work that was done over 50 years ago, and which often considered limited amounts of species diversity and morphological disparity. Revisiting both of these topics with new specimens, new data, and new analytical techniques will doubtlessly lead to breakthroughs in our understanding of the earliest parts of our history.

Institutional abbreviations—AMNH, American Museum of Natural History, New York City, USA; **BP**, Evolutionary Studies Institute, University of the Witwatersrand, Johannesburg, South Africa; **BSPG**, Bayerische Staatssammlung für Paläontologie und Historische Geologie, Munich, Germany; **CGP**, **CGS**, Council for Geoscience, Pretoria, South Africa; **CM**, Carnegie Museum of Natural History, Pittsburgh, USA; **FMNH**, Field Museum of Natural History, Chicago, USA; **GPIT**, Institut und Museum für Geologie und Paläontologie der Universität Tübingen, Tübingen, Germany; **MB**, Museum für Naturkunde, Berlin, Germany; **MCP**, Museu de Ciências e Tecnologia, Pontificia Universidade Católica do Rio Grande do Sul, Porto Alegre, Brazil; **MCZ**, Museum of Comparative Zoology, Harvard University, Cambridge, USA; **MNHN**, Muséum national d'Histoire naturelle, Paris, France; **NHMUK**, Natural History Museum, London, United Kingdom; **NMMNHS**, New Mexico Museum of Natural History and Science, Albuquerque, USA; **NMQR**, National Museum, Bloemfontein, South Africa; **NTM**, Navajo Tribal Museum, Window Rock, USA; **OMNH**, Sam Noble Museum, University of Oklahoma, Norman, USA, **PIN**, Paleontological Institute of the Russian Academy of Sciences, Moscow, Russia; **ROM**, Royal Ontario Museum, Toronto, Canada; **SAM**, Iziko Museums of South Africa, Cape Town, South Africa; **TM**, Ditsong, the National Museum of Natural History, Pretoria, South Africa; **TMM**, Vertebrate Paleontology Laboratory of the University of Texas, Austin, USA; **UCMP**, University of California Museum of Paleontology, Berkeley, USA; **US**, University of Stellenbosch, Stellenbosch, South Africa; **USNM**, National Museum of Natural History (Smithsonian Institution), Washington D.C., USA.

5.2 Introduction to Synapsida

Synapsida is one of the two great clades of amniote tetrapods (the other being its sister group Sauropsida) (e.g., Kemp 1980a, 1982, Gauthier *et al.* 1988, Hopson 1991, Laurin and Reisz 1995, 1997, Müller and Reisz 2006), and it can be defined as all taxa more closely related to *Dicynodon lacerticeps* Owen (1845) than to *Lacerta agilis* Linnæus (1758). Non-mammalian synapsids are frequently referred to as "mammal-like reptiles" in both popular and scientific literature on account of their superficial similarity to some reptiles and the precladistic tendency to lump any amniote species that lacked certain "key" mammalian characters, such as the presence of three middle ear ossicles, within reptiles (for discussion of the frustrations such practices caused, see Simpson 1959 and van Valen 1960). However, all synapsids share a more recent common ancestor with each other than they do with any member of Sauropsida. Therefore, the term "mammal-like reptile" is misleading because it incorrectly implies that some synapsids are more closely related to modern reptiles than to other synapsids, and that reptiles such as lizards are ancestors of mammals. Such imprecision is especially confusing when discussing synapsid evolution with the general public (Angielczyk 2009), so phylogenetically accurate terms such as "non-mammalian synapsid", "synapsid", or "ancient mammal relative" should always be used instead of "mammal-like reptile".

Considerable uncertainty surrounded the phylogenetic affinities of the first non-mammalian synapsid fossils to be discovered, and the nature of their relationship to mammals was not grasped initially. (A highly accessible account of the discovery of the first synapsid fossils, disagreements over their interpretation, and the surrounding historical and scientific context can be found in Aulie 1974a, b, 1975.) Although the very first synapsid fossils (from Russia) were initially assigned to extant mammalian groups (Kutorga 1838), and Owen noted similarities between characters found in mammals and fossil synapsids collected in South Africa (e.g., Owen 1876), Cope (1878a, b, 1880, 1884a, b, c, 1892) was the first to explicitly suggest a phylogenetic relationship between early ("pelycosaur"-grade) synapsids known from North America, more advanced taxa (therapsids) from South Africa, and mammals. Debate over this issue continued for the next three decades (e.g., Huxley 1878, 1880, Baur 1887, Baur and Case 1897, 1899, Osborn 1898, 1903, Case 1907), even as the very mammal-like character of many fossil synapsids was clearly recognized (e.g., Seeley

1888, 1889, 1894a, b, 1895a, b). Broom (1905a, b, 1907, 1910, 1914a, b, 1932; also see Watson 1914a) made more extensive comparisons between North American and South African synapsid fossils, and he played a key role in solidifying the argument that they were related to each other and to mammals. Following Cope, Broom classified the older and more "primitive" synapsids as members of Pelycosauria and placed the younger and more mammal-like species into a group he named Therapsida (Broom 1905a).

Broom's distinction between "pelycosaurs" and therapsids was maintained by most workers through the mid-20th century (e.g., Romer and Price 1940, Watson and Romer 1956, Olson 1962). Subsequent phylogenetic analyses have recognized "pelycosaurs" as a paraphyletic grade at the base of Synapsida, whereas Therapsida is a monophyletic clade that includes mammals as its living representatives (Reisz 1980, 1986, Hopson and Barghusen 1986, Gauthier *et al.* 1988, Hopson 1994, Sidor and Hopson 1998). Although some aspects of synapsid phylogeny are still debated (e.g., compare Benson 2012 with Reisz and Fröbisch 2014, and Sidor and Hopson 1998 with Kemp 2009 and Kammerer *et al.* 2013a; also see below), this broad pattern is widely accepted.

The great deal of morphological disparity that exists among modern mammals, and between mammals and many non-mammalian synapsids, makes it difficult to compile a list of synapomorphies that can be found in all synapsids. Perhaps the most conspicuous potential synapomorphy of Synapsida is a single lateral temporal fenestra in the skull, primitively bounded on the external surface of the skull by the postorbital, jugal, and squamosal bones (Fig. 5.1) (e.g., Broom 1910, Romer and Price 1940, Kemp 1980a, 1988a, Reisz 1986, Benson 2012). All synapsids retain some manifestation of this temporal opening (Fig. 5.2), although factors such as the loss of the postorbital bone as part of a general trend for simplification of the synapsid skull (Sidor 2001), increased ossification of the braincase, emargination of the fenestra, and reduction of the zygomatic arch in various mammals mean that it often bears little obvious similarity to the morphology present in the earliest synapsids. A long-standing hypothesis for the origin of the temporal fenestra suggests that it reflects optimization of the skull for resisting strains imposed by the actions of the jaw adductor muscles while also minimizing the metabolic expenses associated with deposition and maintenance of bone (Gregory and Adams 1915, Case 1924, Olson 1961, Fox 1964, Frazetta 1968, Preuschoft and Witzel 2005; also see Curtis *et al.* 2011). Kemp (1980a, b, 1982, 2005; although see Reisz and Heaton 1980) focused more on the fenestra providing an expanded area of muscle attachment, hypothesizing that it evolved initially as a muscular insertion on connective tissues associated with the postorbital-squamosal suture that later expanded into an aponeurotic sheet. Other frequently cited synapomorphies that diagnose the basal node of Synapsida, such as a single median postparietal (or interparietal) and aspects of the morphology of the septomaxilla and posttemporal fenestra, have complex evolutionary histories (e.g., Novacek and Wyss 1986, Wible 1987, Wible

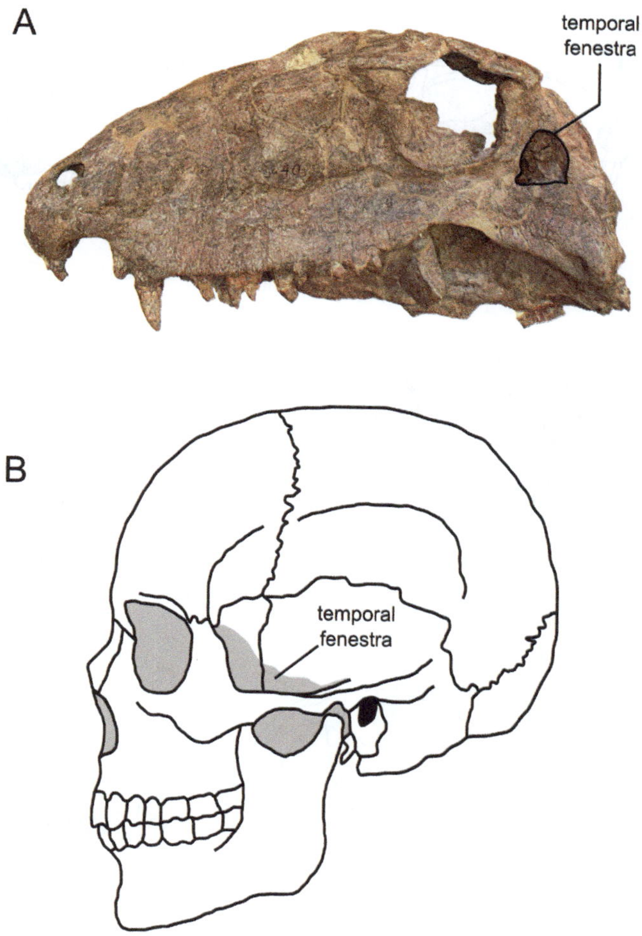

Fig. 5.1: A single lateral temporal fenestra, primitively bounded on the external surface of the skull by the postorbital, jugal, and squamosal bones, can be found (in a sometimes modified form) in all synapsids. (A) Left lateral view of the skull of *Dimetrodon loomisi* (FMNH UC 40) showing the location and relatively small size of the lateral temporal fenestra in a "pelycosaur"-grade synapsid. (B) Line drawing of a human skull showing the location of the lateral temporal fenestra in a modern mammal. The mammalian temporal fenestra is highly emarginated compared with that of an early synapsid like *Dimetrodon*, and the mammalian braincase is greatly expanded and more extensively ossified.

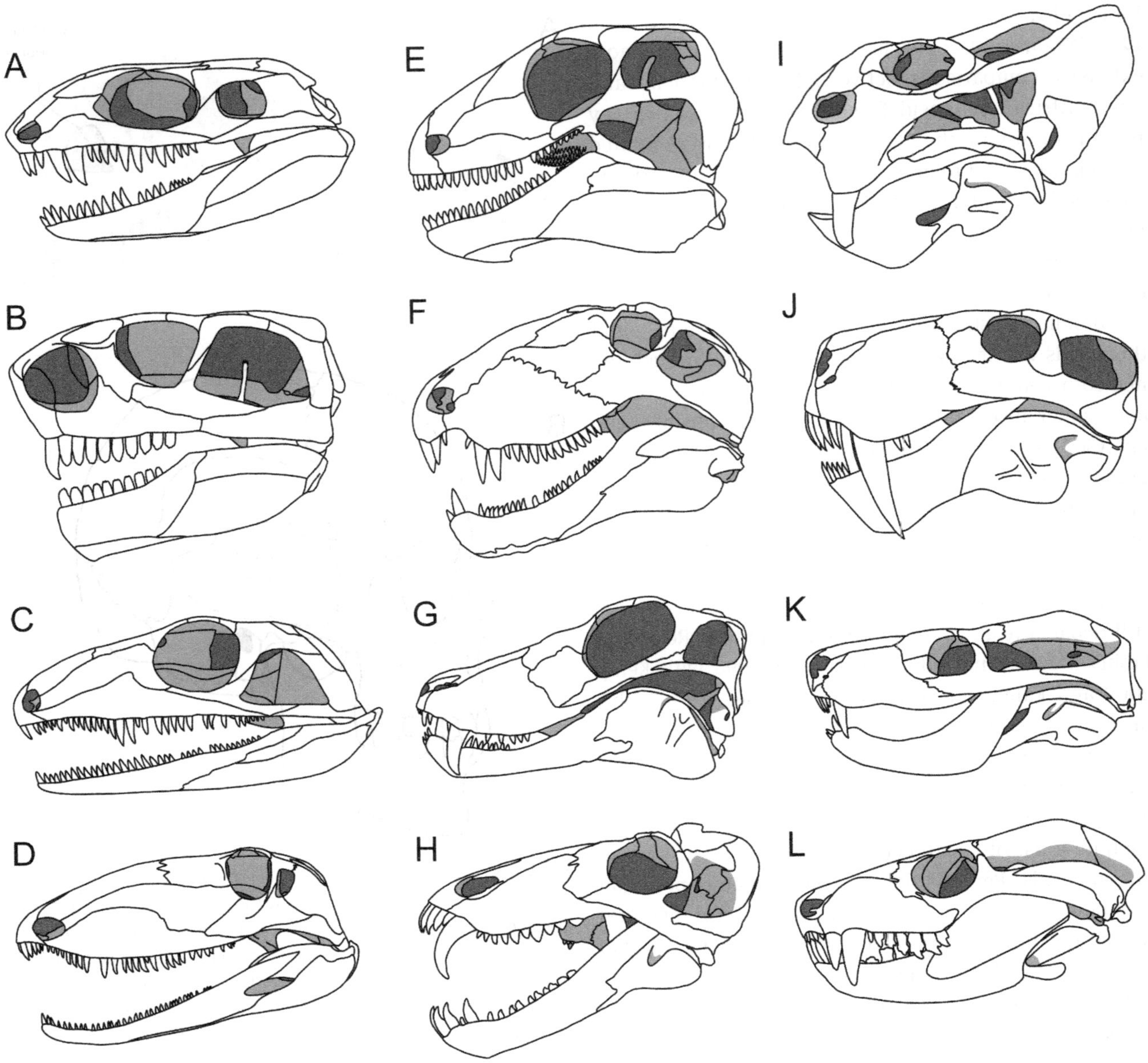

Fig. 5.2: Simplified line drawings of selected synapsid skulls in left lateral view showing the disparity that exists within the clade. (A) The eothyridid *Eothyris parkeyi* (redrawn and modified from Reisz *et al.* 2009). (B) The caseid *Ennatosaurus tecton* (modified from Maddin *et al.* 2008). (C) The varanopid *Varanops brevirostris* (redrawn and modified from Romer and Price 1940). (D) The ophiacodontid *Ophiacodon uniformis* (redrawn and modified from Romer and Price 1940). (E) The edaphosaurid *Edaphosaurus boanerges* (redrawn and modified from Modesto 1995). (F) The sphenacodontid *Dimetrodon limbatus* (redrawn and modified from Romer and Price 1940). (G) The biarmosuchian *Herpetoskylax hopsoni* (redrawn and modified from Sidor and Rubidge 2006). (H) The dinocephalian *Syodon biarmicum* (redrawn and modified from Orlov 1958). (I) The dicynodont *Dicynodon lacerticeps* (redrawn and modified from Kammerer *et al.* 2011 and Cluver and Hotton 1981). (J) The gorgonopsian *Smilesaurus ferox* (redrawn and modified from Kammerer 2016c). (K) The therocephalian *Theriognathus microps* (redrawn and modified from Huttenlocker and Abdala 2015). (L) The cynodont *Lumkuia fuzzi* (redrawn and modified from Hopson and Kitching 2001). The skulls are shown at the same snout-occiput length and are not to scale.

et al. 1990, Koyabu *et al.* 2014) such that they are not readily evident in all synapsid species. Fortunately, the fossil record of non-mammalian synapsids documents these evolutionary transitions, providing transformational homologies (Kemp 1988a, b) that link early synapsids possessing primitive character states and their later relatives with different morphologies.

The divergence between synapsids and sauropsids occurred by the Bashkirian Stage of the Early Pennsylvanian (late Carboniferous Period). This divergence is calibrated by the presence of the sauropsid *Hylonomus lyelli* (Dawson 1860, Carroll 1964) and the putative synapsid *Protoclepsydrops haplous* (Carroll 1964), which occur in the same stratigraphic horizon in the Joggins Formation of Nova Scotia. *Protoclepsydrops* is known from limited, comparatively fragmentary material, and its identity as a synapsid has been questioned (Reisz 1972, 1986). The sauropsid identity of *Hylonomus* is better supported (Müller and Reisz 2006), so its presence in the fossil record implies that synapsids and sauropsids must have diverged by this time, even if *Protoclepsydrops* is not a synapsid. Age estimates for the Joggins Formation range from about 319 to 310 million years ago (Reisz and Müller 2004, van Tuinen and Hadly 2004), with Benton *et al.* (2015) favoring a value of 318 million years ago. Trackway evidence from the Grande Anse Formation in New Brunswick could push this minimum divergence back by about one million years (Falcon-Lang *et al.* 2007), but the precise identities of the trackmakers are more difficult to determine than is the case for body fossils that preserve more morphological information. More certain synapsid tracks from the Bochum Formation of Germany (Voigt and Ganzelewski 2010) and controversial tracks from the Pottsville Formation of Alabama (Hunt *et al.* 2004, 2005, Haubold *et al.* 2005) provide additional evidence for the presence of the clade by about 315 million years ago. Phylogenetic relationships and implied ghost lineages suggest that the actual synapsid-sauropsid divergence is even older than these estimates (Laurin and Reisz 1995, DeBraga and Rieppel 1997, Reisz and Müller 2004, Müller and Reisz 2006), perhaps closer to the suggested soft maximum constraint of 332.9 million years ago of Benton *et al.* (2015) (based on the absence of either clade at the East Kirkton locality in Scotland). The ophiacodontid *Archaeothyris florensis* and *Echinerpeton intermedium*, a synapsid of less certain affinities (Reisz 1986, Benson 2012), from the younger (about 307 million years ago) Morien Group of Nova Scotia are the oldest unequivocal synapsid body fossils (Reisz 1972, 1986, van Tuinen and Hadly 2004, Benson 2012; although see Lee 1999). The presence of several Late Pennsylvanian localities that preserve members of multiple synapsid lineages (e.g., Reisz *et al.* 1982, Sumida and Berman 1993, Harris *et al.* 2004, Kissel and Reisz 2004, Reisz and Fröbisch 2014) demonstrates that the clade's diversification was well underway shortly after its first appearance in the fossil record.

The evolutionary history of non-mammalian synapsids can be thought of as a series of three radiations of increasingly mammal-like animals: the early "pelycosaur"-grade synapsids, the non-cynodont therapsids, and the non-mammaliaform cynodonts (Fig. 5.3) (Kemp 1982, Hopson 1994; also see Sidor and Hopson 1998, although note that the latter authors define the radiations somewhat differently). The transition between the radiations of non-cynodont therapsids and non-mammaliaform cynodonts is marked by the Permo-Triassic mass extinction, the largest mass extinction in Earth history (e.g., Chen and Benton 2012, Benton and Newell 2014, Benton 2015b), which occurred about 252 million years ago. The oldest known cynodonts occur in Wuchiapingian-age (early late Permian) rocks in South Africa (Botha *et al.* 2007, Botha-Brink and Abdala 2008, Kammerer 2016a, date from Rubidge *et al.* 2013), and the presence of the sister group of cynodonts (Therocephalia) in middle Permian-age (Wordian) rocks (Abdala *et al.* 2008) implies that cynodonts must have originated by that time. The meager Permian cynodont fossil record suggests that they were minor components of tetrapod communities at the time, but unlike other therapsid clades, which were strongly affected by the extinction, cynodonts were barely affected and underwent a major radiation during the Triassic recovery (Abdala and Ribeiro 2010, Irmis and Whiteside 2012, Botha-Brink *et al.* 2012, Fröbisch 2013, 2014, Irmis *et al.* 2013, Ruta *et al.* 2013a).

The transition between "pelycosaur" and therapsid radiations is more enigmatic. There is a pronounced geographic, and possibly temporal, dichotomy between the locations where most "pelycosaur" fossils are found and the primary therapsid-bearing localities. Nearly all of the "pelycosaur" records occur in Late Pennsylvanian to early or early middle Permian rocks deposited in a paleoequatorial band extending from New Mexico and Texas, through maritime Canada, and into western Europe (Fig. 5.4) (Anderson and Cruickshank 1978, Parrish *et al.* 1986, Reisz 1986, Berman *et al.* 1997, Brocklehurst *et al.* 2013). By contrast, middle and late Permian therapsid fossils are found predominantly in high-latitude regions, especially southern and eastern Africa and the fore-Ural region of Russia (Anderson and Cruickshank 1978, Parrish *et al.* 1986, Rubidge 2005), and there is little taxonomic overlap between the high-latitude synapsid faunas and those of the equator.

The nature of this apparent gap has long been a cause of uncertainty (e.g., Romer and Price 1940). In a series of papers, Everett Olson described fossils from the middle Permian of Oklahoma and Texas that he thought bridged the evolutionary gap between the radiations of "pelycosaurs" and therapsids, and he proposed biostratigraphic correlations between the Permian of Russia and North America that implied that his North American "therapsids" were contemporaries of early therapsids in Russia (Olson and Beerbower 1953, Olson 1955, 1957, 1962, 1974, 1975, 1986, Olson and Chudinov 1992; also see Efremov 1956, Chudinov 1965, 1983). However, Olson's putative therapsid specimens have since been reinterpreted as the remains of "pelycosaurs" (Sigogneau-Russell 1989, Sidor and Hopson 1995, Battail 2000, Reisz and Laurin

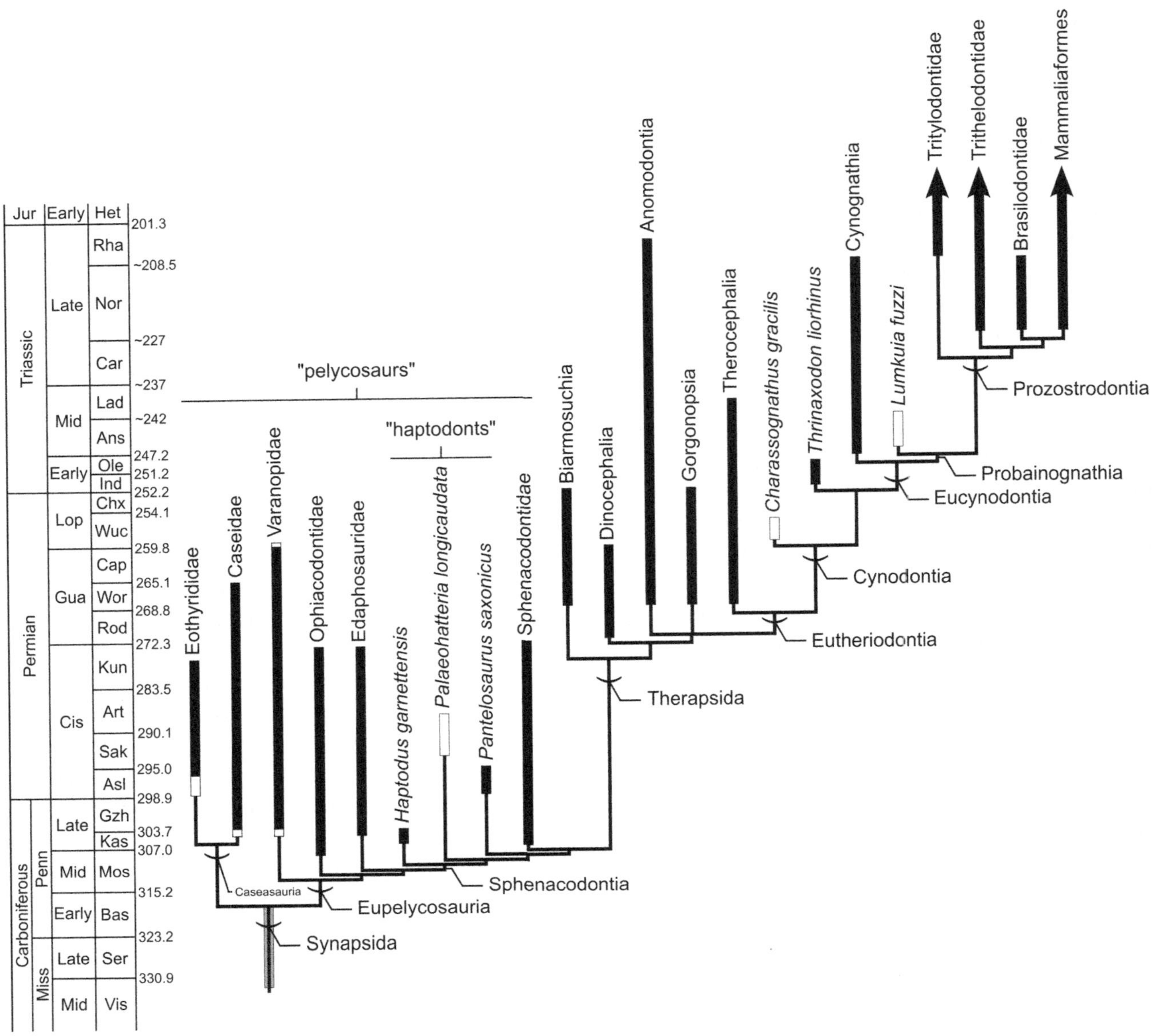

Fig. 5.3: Simplified time-calibrated phylogeny of Synapsida. The grey bar at root of the tree spans the minimum divergence date (318 Mya) and the soft maximum divergence date (332.9 Mya) of Benton *et al.* (2015). Topology for "pelycosaur"-grade synapsids primarily based on Reisz (1986), Laurin (1993), and Sidor and Hopson (1998) (but for a recent alternative, see Benson 2012). Topology for Therapsida primarily based on Hopson and Barghusen (1986) and Sidor and Hopson (1998). Topology for Cynodontia reflects the basal position of *Charassognathus* posited by Botha *et al.* (2007), the placement of *Lumkuia* within Probainognathia suggested by Hopson and Kitching (2001), and the close relationship of Tritylodontidae and Mammaliaformes suggested by e.g., Liu and Olsen (2010) (an alternative hypothesis posits that tritylodontids are derived members of Cynognathia; see e.g., Hopson and Kitching 2001, Sues and Jenkins 2006). Stratigraphic ranges based on Brocklehurst (2015). Black bars represent well-constrained stratigraphic ranges; white bars represent more poorly constrained stratigraphic ranges. Time scale based on the International Commission on Stratigraphy Chronostratigraphic Chart version 2016-04 (Cohen *et al.*, 2013). Abbreviations: Ans, Anisian; Art, Artinskian; Asl, Asselian; Bas, Bashkirian; Cap, Capitanian; Car, Carnian; Chx, Changhsingian; Cis, Cisuralian; Gua, Guadalupian; Gzh, Gzhelian; Het, Hettangian; Ind, Induan; Kas, Kasimovian; Kun, Kungurian; Lad, Ladinian; Lop, Lopingian; Mid, Middle; Miss, Mississippian; Mos, Moscovian; Nor, Norian; Ole, Olenekian; Penn, Pennsylvanian; Rha, Rhaetian; Rod, Roadian; Sak, Sakmarian; Ser, Serpukhovian; Vis, Visean; Wor, Wordian; Wuc, Wuchiapingian.

2004, Kammerer 2011), and biostratigraphic correlations between the tetrapod faunas of the middle Permian of North America and Russia have been debated (Reisz and Laurin 2001, 2002, Lucas 2001, 2002, 2004, 2005, 2013, 2018, Lozovsky 2005, Benton 2012, 2013). The early Permian *Tetraceratops insignis* from Texas is another potential therapsid that could represent a bridge between the two geographic regions and radiations, but the only

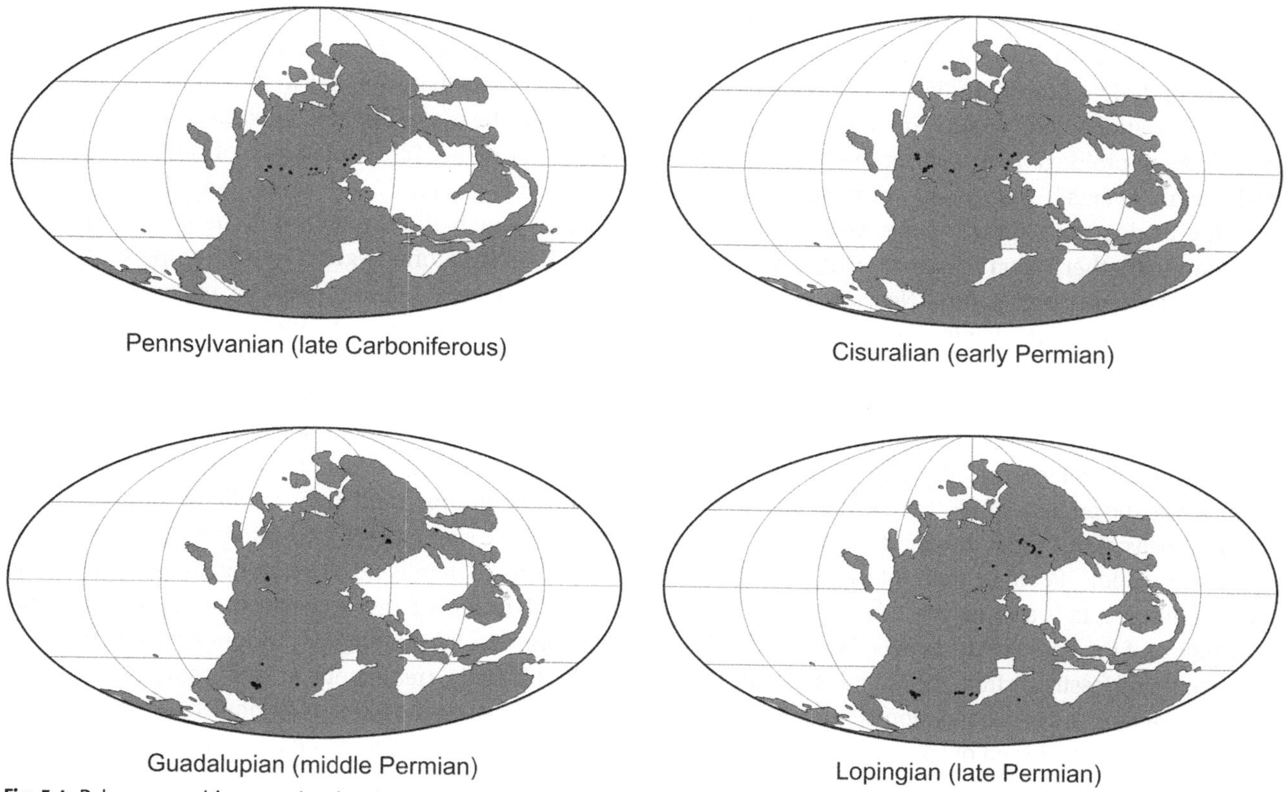

Fig. 5.4: Paleogeographic maps showing the approximate distribution of synapsid-bearing localities in the Pennsylvanian (late Carboniferous), Cisuralian (early Permian), Guadalupian (middle Permian), and Lopingian (late Permian). Note that Permo-Carboniferous localities cluster near the paleoequator, whereas middle and late Permian localities are mostly in higher latitude regions. The clusters of points are intended to provide a sense of the location and density of localities in different areas, but individual localities in a specific area generally are too small to be resolved as individual points. The maps were drafted using the Fossilworks Online Paleogeographic Map Generator (Alroy, J. 2013. Online paleogeographic map generator. http://paleodb.org/?a=mapForm).

known specimen is badly preserved (Fig. 5.10 F) and has been variously interpreted as an eothyridid "pelycosaur", a sphenacodont "pelycosaur", or an early therapsid (Matthew 1908, Romer and Price 1940, Laurin and Reisz 1990, 1996, Reisz *et al.* 1998, Conrad and Sidor 2001, Amson and Laurin 2011). *Raranimus dashankouensis* from the Roadian (early middle Permian) helps to span the evolutionary gap (it was described as the most stemward therapsid, and retains "pelycosaur"-style dentition), but it is part of a relatively high-latitude faunal assemblage from China (Liu *et al.* 2009).

Lucas (2001, 2002, 2004, 2005, 2013, 2017, 2018, Lucas and Heckert 2001) has been a forceful proponent of the hypothesis that the gap between "pelycosaur" and therapsid radiations represents a real hiatus in the fossil record of two to three million years when no known synapsid fossils were preserved, which he termed Olson's Gap. This view has been challenged (Reisz and Laurin 2001, 2002, Lozovsky 2005, Liu *et al.* 2009, Benton 2012, 2013), with much of the debate focusing on the interpretation and relevance of particular marine biostratigraphic data for inferring the ages of terrestrial strata in Texas and Oklahoma.

Sahney and Benton (2008) recently suggested that the transition between "pelycosaur"-dominated and therapsid-dominated communities was accompanied by a previously unrecognized mass extinction across the Cisuralian-Guadalupian (early-middle Permian) boundary in the terrestrial realm, termed Olson's Extinction. A similar pattern has since been confirmed individually for synapsids and other tetrapod clades even when sampling is considered (Ruta and Benton 2008, Ruta *et al.* 2011, Brocklehurst *et al.* 2013), although the shift in the main geographic areas where tetrapods are found surrounding the event complicates fully accounting for sampling artifacts (Benson and Upchurch 2013, but see Brocklehurst *et al.* 2017). The effects of the hypothesized extinction also are not simply a replacement of "pelycosaurs" by therapsids. Along with therapsids, caseid "pelycosaurs" were little affected by the event (Brocklehurst *et al.* 2013), and herbivores in general do not show a substantial drop in richness at this time (Sahney and Benton 2008, Pearson *et al.* 2013). Likewise, the turnover also appears to have occurred more rapidly at high latitudes than in regions closer to the equator (Brocklehurst *et al.* 2017).

Recently, Lucas (2017) has criticized the biostratigraphic and geochronological framework used in the data compilations underlying most studies of Olson's Extinction and has questioned the reality of the event. He suggested a more prolonged turnover in the Kungurian (his "Redtankian Events") associated with climatic drying occurring at this time. Clearly more work is needed to understand whether Olson's Extinction is a real event, and if so, its causes and nature. Nevertheless, it would be quite interesting if the beginnings of two of the main radiations of non-mammalian synapsids were precipitated by mass extinctions, presaging the radiation of therian mammals following the extinction of non-avian dinosaurs in the Cretaceous-Paleogene extinction.

However, the problem of Olson's Gap and the transition between "pelycosaur" and therapsid radiations extends beyond the middle Permian time interval in which it is usually framed, and it also involves deeper issues related to the timing of the origin of therapsids. The oldest definite sphenacodontid fossils are from the Kasimovian (Late Pennsylvanian) Sangre de Cristo Formation of Colorado (Sumida and Berman 1993), and even older, but less certain, potential sphenacodontid material has been found in Nova Scotia (Reisz 1972). Because Sphenacodontidae is the sister group of Therapsida, the presence of sphenacodontids in the Late Pennsylvanian implies a ghost lineage for therapsids extending to that point in time (Abdala *et al.* 2008, Spindler *et al.* 2015). If *Tetraceratops* is a therapsid, it would shorten this ghost lineage, but not completely eliminate it.

There are four potential explanations for the large amount of missing therapsid history: (1) Therapsida diverged from Sphenacodontidae in the Pennsylvanian but did not undergo significant morphological differentiation until the middle Permian, causing early therapsid fossils to be unrecognized; (2) early therapsids preferred habitats that were not conducive to fossil preservation, resulting in little of their initial history being recorded; (3) early therapsids lived in geographic areas that might preserve a fossil record, but that have not been the subject of much paleontological exploration; or (4) Sphenacodontidae is paraphyletic with respect to Therapsida, such that therapsids are descended from an early or middle Permian sphenacodontid ancestor. Variations of all of these hypotheses have received attention in the literature. For example, hypothesis 1 is similar to Kemp's (2009) scenario in which therapsids underwent an extremely rapid morphological diversification, although he was agnostic as to whether this diversification had a long fuse extending back into the Pennsylvanian. Hypotheses 2 and 3 are not mutually exclusive because they both rely on early therapsid evolution occurring "offstage". Olson (e.g., Olson 1974, 1986) was long a proponent of the initial therapsid diversification having occurred in more upland habitats that were not well sampled by the fossil record, and he discussed this hypothesis extensively in the context of his "chronofauna" concept (see below). More recently, Kemp (2006) presented a biogeographic model in which the ancestors of therapsids first dispersed from equatorial regions into the tropics, and then underwent a major radiation that accompanied a further biogeographic shift into temperate regions at higher latitudes. Precladistic treatments of synapsid phylogeny did not discuss the apparent long therapsid ghost lineage because they were framed in a context allowing ancestor-descendant relationships between members of different groups, and they implicitly assumed that hypothesis 4 was correct (e.g., Romer and Price 1940 p. 194 state: "There is no chronologic or geographic bar to consideration of the sphenacodontids as therapsid ancestors; the morphological evidence is strongly positive".) Cladistic treatments of synapsid relationships generally have recovered a monophyletic Sphenacodontidae that excludes Therapsida (see below), but Kemp (2006, 2009) discussed the possibility that Sphenacodontidae is a grade at the base of Therapsida.

A central problem with the first three hypotheses is that they rely heavily on negative evidence, but discoveries of synapsid fossils in previously unsampled areas, or the reidentification of known specimens (e.g., Spindler 2015), offer potential opportunities for testing them. Recent phylogenetic studies that support sphenacodontid monophyly make the fourth hypothesis seem unlikely (Reisz *et al.* 1992a, Fröbisch *et al.* 2011, Benson 2012, Reisz and Fröbisch 2014, Brink *et al.* 2015, Spindler *et al.* 2015, Brocklehurst *et al.* 2016a, b). Regardless, the long Permo-Carboniferous therapsid ghost lineage is of key importance because it represents a much longer period of time than "Olson's Gap" in the middle Permian. Indeed, if we want to fully document the early history of therapsids, then further exploration of Permo-Carboniferous rocks and fossils should be of at least equal, if not higher, priority than work in the middle Permian.

5.3 Diversity of Non-Mammalian Synapsids

"Pelycosaur"-grade synapsids can be grouped into six major clades (Fig. 5.3): Eothyrididae, Caseidae, Varanopidae, Ophiacodontidae, Edaphosauridae, and Sphenacodontidae, as well as the "haptodonts", a paraphyletic assemblage of taxa closely related to Sphenacodontidae (e.g., Reisz 1980, 1986, Kemp 1982, 2005, Reisz *et al.* 1992a, Hopson 1991, 1994, Laurin 1993, 1994, Sidor and Hopson 1998, Benson 2012). Over the past three decades, most phylogenetic work on "pelycosaurs" has recognized a basal split into two main clades: the Caseasauria (eothyridids and caseids) and the Eupelycosauria (the remaining "pelycosaurs" plus Therapsida) (e.g., Reisz 1980, 1986, Gauthier *et al.* 1988, Hopson 1991, Berman *et al.* 1995, Sidor and Hopson 1998, Reisz and Fröbisch 2014; the trees including *Eocasea* in Romano and Nicosia 2015 also are consistent with this hypothesis). The analysis of Benson (2012) challenged the prevailing hypothesis by suggesting a basal dichotomy between a clade including ophiacodontids and varanopids and a second clade including all other "pelycosaurs" and therapsids, echoing precladistic treatments of pelycosaur phylogeny that considered ophiacodontids to be the most basal synapsids (e.g., Case 1907, Romer and Price 1940, Olson 1962, Kemp 1982; also see Brocklehurst *et al.* 2016a). The difference in results seems to be driven by conflicting phylogenetic signals preserved in the cranial and postcranial skeletons (Benson 2012). Reisz and Fröbisch (2014; also see Brocklehurst *et al.* 2016b) disagreed with many of the codings presented in Benson's (2012) data set, however, and found that recoding the characters produced a tree similar to the prevailing hypothesis of a basal split between caseasaurs and eupelycosaurs. We use the latter hypothesis to structure the following systematic review.

5.3.1 Caseasauria

Recent phylogenetic analyses have identified several cranial characters that diagnose Caseasauria (Maddin *et al.* 2008, Reisz *et al.* 2009, Reisz and Fröbisch 2014, Spindler *et al.* 2018; also see Langston 1965 and Reisz 1980, 1986). Some of the most conspicuous of these features are characters of the snout, such as an anteriorly pointed rostrum that overhangs the tooth row, and an anteroposteriorly elongate nostril. The extremely limited amount of postcranial material available for eothyridid caseasaurs and the derived postcrania of most caseids makes it difficult to assess support for Caseasauria in that portion of the skeleton. However, Benson (2012), Sumida *et al.* (2014), and Romano and Nicosia (2015) found evidence suggesting that cranial and postcranial characters provide conflicting information about the relationships of caseasaurs both to other synapsids and amongst themselves.

5.3.1.1 Eothyrididae

Eothyrididae is a clade of small, faunivorous caseasaurs known from the Cisuralian (early Permian) of Texas, New Mexico, and Colorado (Figs. 5.2 A and 5.5) (Reisz *et al.* 2009, Brocklehurst *et al.* 2016b). Members of the clade appear to have been extremely rare components of the communities in which they lived. Of the three recognized species, two (*Eothyris parkeyi* and *Vaughnictis smithae*) are represented by only one specimen (Romer 1937, Romer and Price 1940, Reisz *et al.* 2009, Brocklehurst *et al.* 2016a, b) (Fig. 5.5 B, C), whereas the third (*Oedaleops campi*) is represented by a handful of specimens from a single locality (Langston 1965, Reisz *et al.* 2009, Sumida *et al.* 2014) (Fig. 5.5 A). Spindler *et al.* (2016) suggested that the poorly known caseasaurs *Callibrachion gaudryi* (early Permian of France) and *Datheosaurus macrourus* (Late Pennsylvanian of Poland) may represent eothyridids, but Brocklehurst *et al.* (2016b) found them to group consistently with caseids in their analyses. The authors who have worked on eothyridids consistently comment on their highly plesiomorphic skull morphology compared with other synapsids (Romer 1937, 1946, Romer and Price 1940, Watson 1954, Langston 1965, Reisz 1986, Reisz *et al.* 2009, Benson 2012). Reisz *et al.* (2009) were able to identify seven synapomorphies of the dentition, maxilla, jugal, and squamosal that serve to unite *Eothyris* and *Oedaleops* within Eothyrididae. One of these characters, the presence of two pairs of enlarged caniniform teeth, one near the anterior end of the maxilla and the other just below the orbit, is particularly interesting because it indicates that synapsids were capable of some differentiation of their dentition even at a very early stage in their history. However, Sumida *et al.* (2014), Brocklehurst *et al.* (2016b), and Spindler *et al.* (2018) had difficulty recovering a monophyletic Eothyrididae. Sumida *et al.* (2014) and Brocklehurst *et al.* (2016b) concluded that more complete material will be needed to robustly reconstruct eothyridid relationships, but Spindler *et al.* (2018) found that consideration of postcranial data seemed to play little role in their recovery of a paraphyletic Eothyrididae. The traditionally recognized first appearance of eothyridids in the early Permian implies a long ghost lineage stretching back to the Early Pennsylvanian origin of synapsids, which is almost certainly attributable to the apparent rarity of eothyridids. If *Datheosaurus* is an eothyridid, however, it would fill in some of this ghost lineage. Regardless of these uncertainties, the eothyridids likely provide good models for the morphology and ecology of the first synapsids.

Fig. 5.5: Eothyrididae. (A) Block bearing crushed skull and partial skeleton of *Oedaleops campi* (UCMP 40093), skull exposed dorsally. (B, C) Skull of *Eothyris parkeyi* (MCZ 1161) in dorsal and lateral views. Scale bars equal 5 cm.

5.3.1.2 Caseidae

Caseidae is the second subclade of Caseasauria (Figs. 5.2 B and 5.6), and it is predominantly composed of distinctively specialized herbivores. The first caseid to appear in the fossil record is the recently described *Eocasea martini* from the Late Pennsylvanian of Kansas (Reisz and Fröbisch 2014; although see Spindler *et al.* 2018 for an alternative placement of *Eocasea*), but the clade is best known from the Cisuralian and Guadalupian (early to early middle Permian) of North America and Europe (e.g., Olson 1968, Reisz 1986, Sigogneau-Russell and Russell 1974, Reisz *et al.* 2010a, 2011, Romano and Nicosia 2014, Brocklehurst and Fröbisch 2017, Romano *et al.* 2017). Footprints and body fossils reported from the Lodève Basin in France and footprints from the Paraná Basin of Brazil have been taken as evidence that caseids survived into the late Permian (Schneider *et al.* 2006, da Silva *et al.* 2012). Although the identity of the Brazilian tracks is somewhat uncertain and the French body fossils have not been described in detail, if correctly identified they indicate that the caseids were the longest-surviving clade of "pelycosaurs". Even if the Lodève and Paraná material is ignored, however, *Ennatosaurus tecton* and *Phreatophasma aenigmaticum* from Russia show that caseids were one of only two clades of "pelycosaurs" (the other being varanopids; see below) to persist during the radiation of therapsids (Maddin *et al.* 2008, Brocklehurst and Fröbisch 2017).

Most caseids present a suite of features associated with high-fiber herbivory, including a comparatively short preorbital region of the skull; relatively small heads for their body sizes; spatulate or leaf-shaped, often denticulated marginal teeth; extensive palatal dentition; evidence of at least rudimentary oral processing of food (e.g., robust hyoids suggesting the presence of a powerful tongue and evidence of a palinal [anterior-posterior sliding] motion of the jaw in some taxa); and large ribs that enclose a barrel-shaped trunk (e.g., King 1996, Hotton *et al.* 1997, Sues and Reisz 1998, Reisz and Sues 2000, Reisz 2006). *Eocasea* is noteworthy in this regard because it has simple conical teeth and its trunk is not expanded, implying that it had not yet evolved a herbivorous lifestyle (Reisz and Fröbisch 2014). *Datheosaurus* and *Callibrachion*, potential caseids from the Late Pennsylvanian of Poland and early Permian of France (respectively), retain conical marginal dentitions but have body proportions that resemble later,

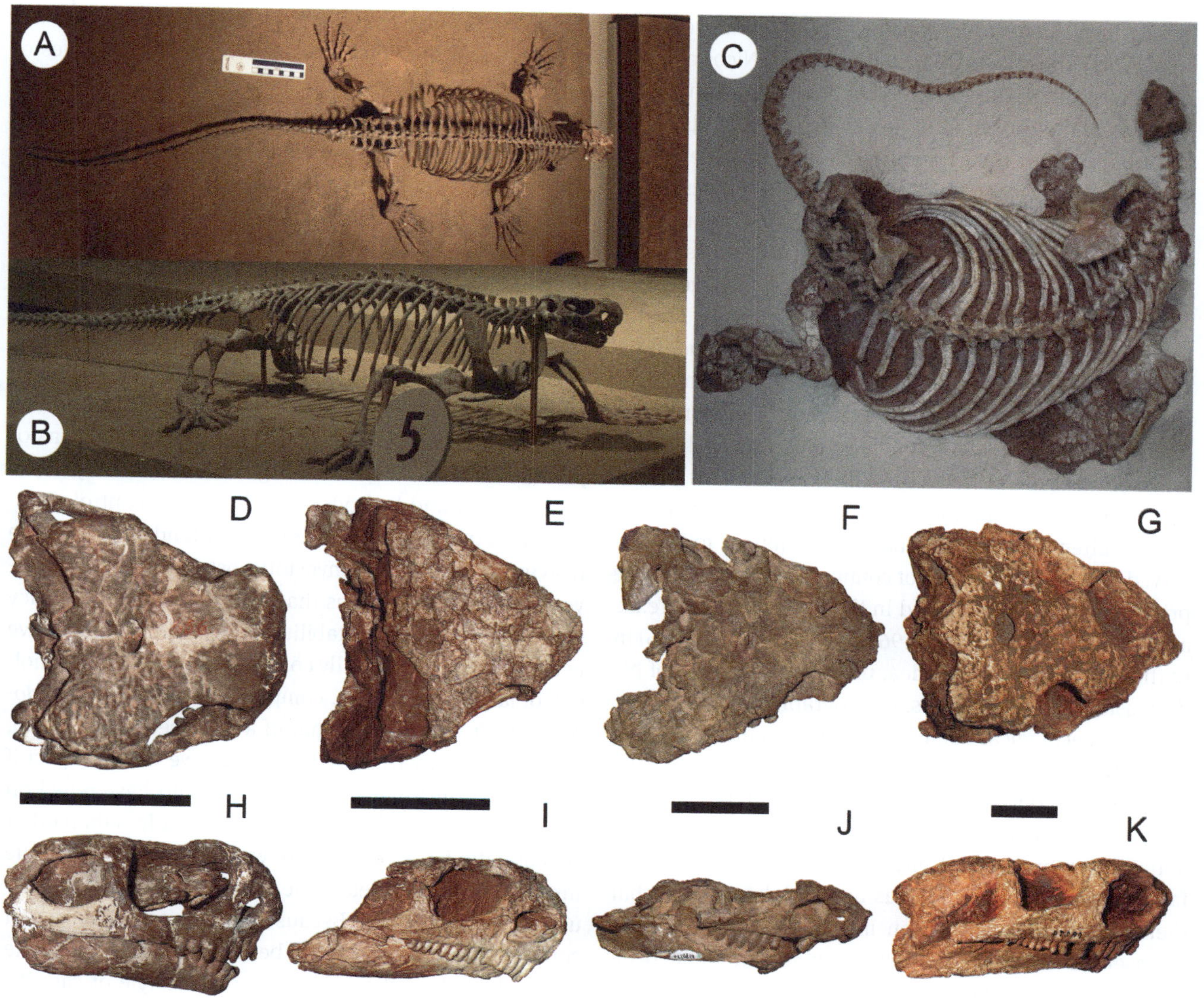

Fig. 5.6: Caseidae. (A, B) Mounted skeleton of *Casea broilii* (FMNH UC 656) in dorsal and right lateral views. (C) Plaque-mounted skeleton of *Cotylorhynchus romeri* (AMNH FARB 7517) in dorsal view. (D, H) Skull of *C. broilii* (FMNH UC 656) in dorsal and lateral views. (E, I) Skull of *Euromycter rutenus* (MNHN.F.MCL-2) in dorsal and lateral views. (F, J) Dorsoventrally crushed skull of *Ennatosaurus tecton* (PIN 1580/24) in dorsal and lateral views. (G, K) Skull of *C. romeri* (FMNH PR 274, cast of OMNH 04329). Specimens in H, I, J, and K mirrored for comparative purposes. Scale bars equal 5 cm.

herbivorous caseids (Spindler *et al.* 2016). The next-oldest definite caseids, *Oromycter dolesorum* from Oklahoma and an undescribed taxon from Germany, both have somewhat spatulate but non-denticulated teeth, although they also show other skeletal evidence of herbivory (Reisz 2005, Reisz and Fröbisch 2014). Together, these observations suggests that caseids acquired their morphological specializations for herbivory in a mosaic fashion. Even within derived members of the group, there is variation in features related to feeding. For example, there is not a simple trend for increased tooth complexity, and the extent of the palatal dentition differs among taxa (Reisz and Sues 2000, Maddin *et al.* 2008).

Caseids show a wide variation in body size. *Eocasea* is known only from a juvenile specimen that is estimated to have had a mass of less than 10 kg (Reisz and Fröbisch 2014). Other early caseids are somewhat larger (e.g., *Oromycter* and *Casea* are estimated to have body masses at the lower end of the 10- to 100-kg range; Reisz and Fröbisch 2014; also see Spindler *et al.* 2016), but some of the later-occurring caseids reach very large sizes and were likely the largest terrestrial animals of their time. For example, *Cotylorhynchus* (Fig. 5.6 C) has been suggested to reach sizes of four to six meters in total body length, and masses of 330 to 500 kg (Romer and Price 1940, Stovall *et al.* 1966, Olson 1968, Reisz and Fröbisch 2014), and the more fragmentary *Alierasaurus* appears to have reached comparable sizes (Ronchi *et al.* 2011, Romano and Nicosia 2014). Based on similar patterns in other Permo-Carboniferous herbivores, the evolution of larger sizes in caseids likely is correlated with their transition to high-fiber herbivory (Reisz and Fröbisch 2014). Romano (2017) considered patterns of limb bone scaling in caseids and found that the bones of the forelimbs increased in robustness early in the clade's history, before the evolution of large sizes, whereas the hindlimbs remained relatively gracile regardless of size. He suggested that although the robust forelimbs might have been exapted to help support the body in large caseids, the initial strengthening of the forelimbs likely was related to a different function, such as digging.

From their first discovery (Williston 1910, 1911), there has been a sense in the literature that early Permian caseids often required unusual circumstances to be preserved and that they were not common components of the primary ecosystems sampled in the fossil record of Texas and Oklahoma. Olson (e.g., 1968, 1971, 1974, 1975, 1986) in particular developed this idea, using it as the basis of his discussion of chronofaunas, or geographically restricted, persistent ecological communities that characterize specific environments (for more general discussions of the chronofauna concept, see Olson 1952, 1983). He hypothesized that caseids were part of a chronofauna that inhabited drier, more upland environments than the typical faunas of early Permian Texas and Oklahoma, and that their occasional preservation in the latter communities represented occurrences where their remains had been transported before being buried. When habitats became drier as the Permian proceeded, caseids were able to expand their geographic range and become more common in the fossil record. The chronofauna concept is not widely applied in vertebrate paleontology today, but the idea of caseids preferring drier, more upland environments has persisted, especially with the discovery of caseid specimens in two localities (the Dolese Quarry near Richards Spur, Oklahoma and the Bromacker Quarry in Germany) that do seem to sample upland environments (Reisz 2005, Reisz *et al.* 2010a, Berman *et al.* 2014).

Lambertz *et al.* (2016) challenged the prevailing view of large, derived caseids such as *Cotylorhynchus* as highly terrestrial animals and instead proposed that they had an aquatic lifestyle. The observations underlying this argument included morphological features such as the extremely short neck of large caseids, which would have limited their feeding range on land, and the bone microstructure of their humeri, femora, and ribs, which have extremely thin cortices, a highly cancellous, osteoporotic-like trabecular structure, and no distinct medullary cavities. They also estimated that the joints between the vertebrae and the dorsal ribs had very limited ranges of motion and proposed that a mammal-like diaphragm must have been present to facilitate respiration, especially in an aquatic setting. Their bone microstructure observations are particularly interesting: although there is variation in the exact characters found in extant and extinct semiaquatic and aquatic tetrapods (e.g., de Ricqlès and de Buffrénil 2001, Houssaye *et al.* 2016), extremely reduced cortices and osteoporotic-like trabecular structures are present in some secondarily aquatic tetrapods and are not typical for terrestrial taxa. However, the degree to which these features are developed in caseids is most like what is seen in highly pelagic taxa such as cetaceans or pinnipeds, which emphasize high maneuverability, rapid acceleration, and hydrodynamic buoyancy control. Semiaquatic taxa and pelagic herbivores tend to have more heavily ossified (i.e., pachyostotic, osteosclerotic, or pachyosteosclerotic) skeletons that provide passive buoyancy control and increased stability against current and wave action. Given the generally conservative skeletal morphology of large caseids (e.g., compare the skeleton of *Cotylorhynchus* in Fig. 5.6 C to that of a pinniped or cetacean), their bone microstructure is not what we would expect if they were semiaquatic. Either they would have relied on passive stability and buoyancy control far less than other semiaquatic tetrapods or they would represent a unique instance of highly pelagic tetrapods retaining a largely unmodified body plan. Because of these issues, we do not consider available data to yet be sufficient to overturn the more widely hypothesized terrestrial lifestyle of caseids. Nevertheless, the unexpected discovery of the unusual bone microstructure of *Cotylorhynchus* by Lambertz *et al.* (2016) shows how little we know about many aspects of the anatomy, paleobiology, and paleoecology of early synapsids and the need for more research on these topics.

5.3.2 Eupelycosauria

Under the prevailing phylogenetic hypothesis for Synapsida (Fig. 5.3), Eupelycosauria is the sister taxon of Caseasauria (Reisz 1980, 1986, Kemp 1982, Gauthier *et al.* 1988, Hopson 1991, 1994, Reisz *et al.* 1992a, Reisz and Fröbisch 2014, Spindler *et al.* 2018). This clade includes five main subgroups: Varanopidae, Ophiacodontidae, Edaphosauridae, Sphenacodontidae, and Therapsida (including mammals as an extant subclade), as well as several taxa whose relationships are less certain (e.g., the "haptodonts"). Diagnostic characters for Eupelycosauria include an antorbital region of the skull that is deeper than wide, preorbital region of the skull longer than the postorbital region, broad frontal contribution to the margin of the orbit, anteroposteriorly short parietal, small parietal foramen located on the posterior half of the parietals, and splintlike supratemporal located in a groove between the parietal and the squamosal (Reisz 1986, Reisz *et al.* 1992a). As noted above, Benson's (2012) analysis did not recover a monophyletic Eupelycosauria when his full cranial and postcranial data sets were taken into account (also see Romano and Nicosia 2015), although his cranial character partition did support the clade.

5.3.2.1 Varanopidae

Varanopidae (frequently called Varanopsidae or Varanopseidae in the older literature; see Reisz and Dilkes 2003) is a cosmopolitan, stratigraphically long-ranging clade of carnivorous synapsids (Figs. 5.2 C and 5.7). *Archaeovenator hamiltonensis*, the oldest and most plesiomorphic varanopid, was found at the Hamilton Quarry in Kansas (the same Late Pennsylvanian locality that produced *Eocasea*; Reisz and Dilkes 2003, Reisz and Fröbisch 2014), but most of the clade's diversity is known from the latter part of the Cisuralian (early Permian) and the Guadalupian (middle Permian) (e.g., Brocklehurst *et al.* 2013; also see below). A number of characters diagnose varanopids (Reisz and Dilkes 2003 found 11 synapomorphies for the clade, and Campione and Reisz 2010 and Spindler *et al.* 2018 each reported seven), including a posterodorsally expanded external naris, gracile subtemporal bar, extension of the parietals over the orbital region of the skull roof, slender mandible with no contribution of the splenial to the symphysis, gracile femur, and two sacral ribs of subequal size (Reisz and Dilkes 2003).

Our understanding of the relationships of varanopids to other synapsids has undergone considerable change over the last century. Initially, they were considered to be close relatives of ophiacodontids (e.g., Williston 1911, 1925; for a modern iteration of this hypothesis, see Benson 2012), but Romer and Price (1940) suggested a closer relationship with sphenacodontids, a hypothesis that was subsequently followed by Reisz (1980) and Kemp (1982). Olson (1965) noted similarities to ophiacodontids and sphenacodontids but prophetically stated that varanopids also had a very primitive appearance in many ways. Brinkman and Eberth (1983) suggested a relationship with caseids, but since Reisz's (1986) study, cladistic analyses that have included a sample of most higher-level "pelycosaur" clades usually recover varanopids as the most stemward clade of Eupelycosauria (Reisz *et al.* 1992a, Fröbisch *et al.* 2011, Reisz and Fröbisch 2014; some trees of Romano and Nicosia 2015, Spindler *et al.* 2018).

At the time of the two most recent monographic treatments of "pelycosaurs" (Romer and Price 1940, Reisz 1986), varanopids were a species-poor group of small carnivores represented by a limited number of specimens. Like caseids, they were thought to require unusual circumstances for preservation (e.g., *Varanops brevirostris*, the first varanopid discovered, was found at the same locality as the first caseid; Williston 1911), and Olson included them as a component of his caseid chronofauna (e.g., Olson 1965, 1971, 1974, 1975). Nevertheless, the stratigraphic and geographic distributions of the known specimens hinted at the group's long duration and cosmopolitanism (Romer and Price 1940, Olson 1965, Reisz 1986; also see Broom 1937, Reisz *et al.* 1998). The last two decades have witnessed a dramatic expansion of our knowledge of varanopids. Several taxa that were initially identified as members of other clades (often sauropsids), or of uncertain affinities, have since been recognized as varanopids (Berman and Reisz 1982, Dilkes and Reisz 1996, Reisz and Dilkes 2003, Reisz and Laurin 2004, Reisz and Modesto 2007, Evans *et al.* 2009, Reisz *et al.* 2010b), and new varanopid species have been described (Anderson and Reisz 2004, Berman *et al.* 2014, Spindler *et al.* 2018). These and other works have confirmed that varanopids were found in both Laurasia and Gondwana and that their stratigraphic range extended until the end of the middle Permian (Reisz and Berman 2001, Modesto *et al.* 2001, 2011; the specimen described by Piñeiro *et al.* 2003 likely is not a varanopid, see Dias-da-Silva *et al.* 2006). As the diversity of varanopids has become better known, phylogenetic analyses have shown that this group includes two main subclades: Varanodontinae and Mycterosaurinae e.g., Reisz and Berman 2001, Modesto *et al.* 2001, Anderson and Reisz 2004, Reisz and Laurin 2004, Maddin *et al.* 2006, Botha-Brink and Modesto 2009, Campione and Reisz 2010, Berman *et al.* 2014; but for a discussion of the appropriate names of these subclades, see Kammerer and Angielczyk 2009. Both clades are of approximately equal richness and have members known from North America and Europe. However, it is unclear if the ranges of both clades extended into Gondwana, given the persistent uncertainty about the phylogenetic placement of the South African *Elliotsmithia longiceps* (e.g., see the alternate phylogenetic topology and taxonomy of Spindler *et al.* 2018). It also has become apparent that not all varanopids were small: at least three varanodontines reached large sizes (>1 m length) and were likely high-level predators in their communities (Olson 1974, Reisz and Laurin 2004, Maddin *et al.* 2006, Campione and Reisz 2010, Berman *et al.* 2014).

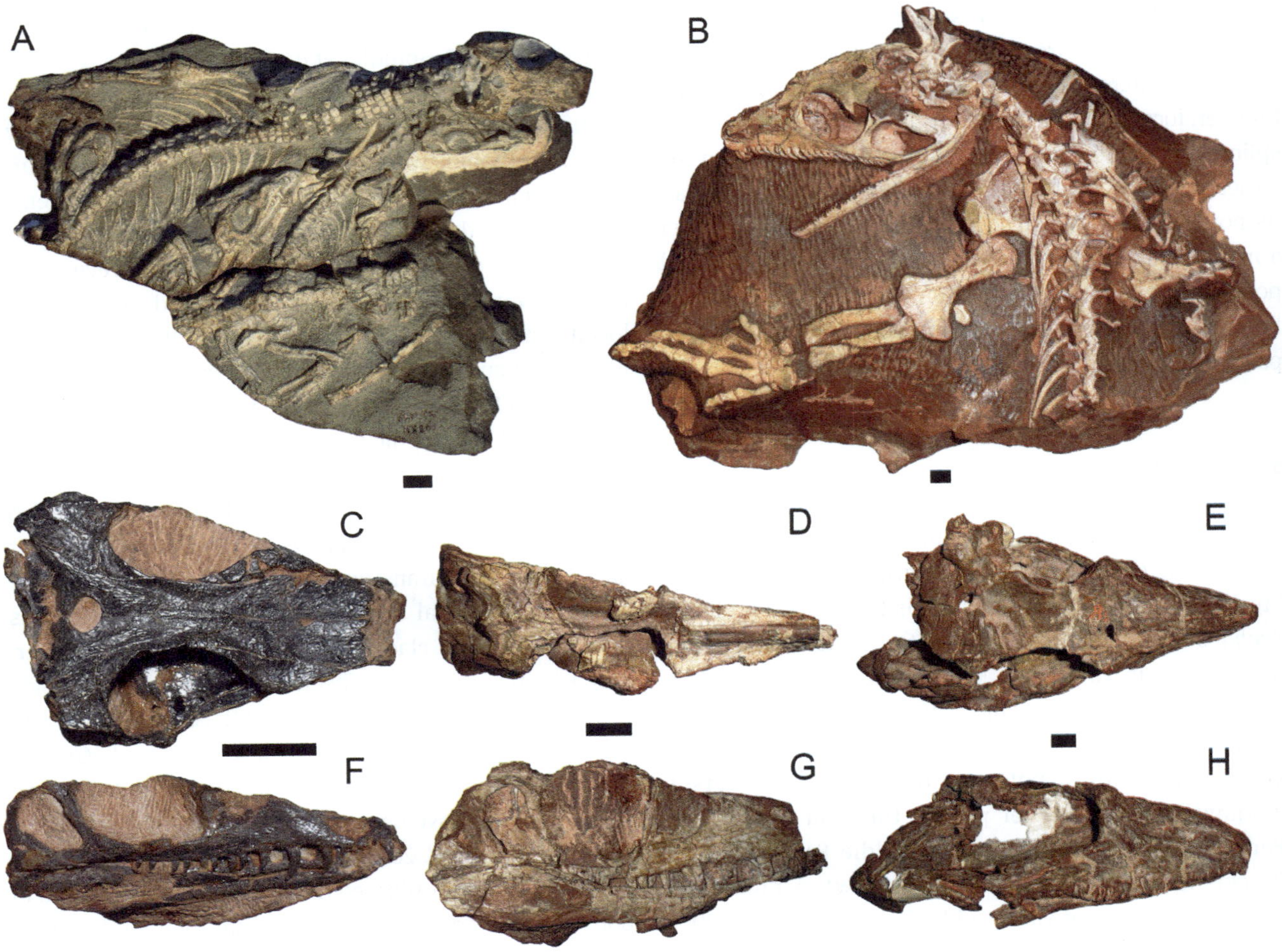

Fig. 5.7: Varanopidae. (A) Social aggregation of five specimens of *Heleosaurus scholtzi* (SAM-PK-K8305) representing two age classes (one large, presumed adult, specimen with four smaller, presumed juvenile or subadult, specimens). (B) Articulated anterior skeleton of *Aerosaurus wellesi* (UCMP 40096) in dorsal view. (C, F) Skull of *Mesenosaurus romeri* (PIN 3706/11) in dorsal and lateral views. (D, G) Skull of *Mycterosaurus longiceps* (FMNH UC 692) in dorsal and lateral views. (E, H) Skull of *Varanops brevirostris* (MCZ 1926) in dorsal and lateral views. Specimen in panel G mirrored for comparative purposes. Scale bars equal 1 cm.

The paleobiology and paleoecology of varanopids has not been studied in great detail, but several observations underscore that these were likely very interesting animals. For example, varanopids are one of the few clades of synapsids that evolved extensive dermal osteoderms (particularly evident in *Heleosaurus scholtzi*; Botha-Brink and Modesto 2009; Fig. 5.7 A). Ecologically, the rarity and unusual preservational circumstances of varanopids in the classic North American Permian localities have led to the hypothesis that they had different environmental or ecological preferences than more common components of the faunas (Romer and Price 1940). More recent observations lend credence to this hypothesis. The greatest diversity of sympatric varanopids is known from the Dolese Quarry, near Richards Spur, Oklahoma (Maddin *et al.* 2006), which has generally been considered to have sampled an upland tetrapod community (e.g., Sullivan and Reisz 1999). Likewise, the large varanopid *Tambacarnifex* was the apex predator in the fauna sampled at the Bromacker locality in Germany (Berman *et al.* 2014), which represents an internally drained upland basin (Eberth *et al.* 2000). A skeleton of *Varanops brevirostris* also preserves evidence of a direct ecological interaction: bite marks and a tooth lodged between the radius and the ulna demonstrate that the individual's carcass was scavenged by a dissorophoid temnospondyl (Reisz and Tsuji 2006). Finally, varanopids are interesting because they show evidence of behavioral patterns that are typically thought of as mammalian. A mixed-age aggregation of *Heleosaurus* individuals from South Africa has been interpreted as evidence of parental care in this early synapsid (Botha-Brink and Modesto 2007). The proportions of the sclerotic rings and orbits of *Heleosaurus* and *Aerosaurus* (Fig. 5.7 A, B) suggest that the eyes of these taxa were optimized for functioning under scotopic (low light) conditions, raising the possibility that they were nocturnal (Angielczyk and Schmitz 2014). Based on the strongly curved claws and elongate phalanges of a series of exceptionally preserved specimens from Germany, Spindler *et al.* (2018) suggested a clinging arboreal lifestyle for the recently described varanopid *Ascendonanus*. If correct, this would represent the oldest known example of arboreality in tetrapods.

As mentioned above, several taxa currently classified as varanopids were originally described as sauropsids (e.g., *Apsisaurus*, *Heleosaurus*, and *Mesenosaurus*, all of which were initially considered diapsid reptiles), suggesting extensive anatomical convergence between these groups. Recent work suggests an alternate possibility, however. CT reconstruction of the unquestioned early Permian diapsid *Orovenator* has revealed that this taxon shares a number of unique features with varanopids (Ford and Benson 2017), bringing the very synapsid nature of Varanopidae into question. Further research untangling the knotty early evolutionary history of Amniota is clearly required to address this issue.

5.3.2.2 Ophiacodontidae

Ophiacodontidae is a clade of small to large carnivorous "pelycosaurs" known primarily from the Late Pennsylvanian to Cisuralian (early Permian) of North America (Figs. 5.2 D and 5.8). The taxonomic history of the clade (frequently called Poliosauridae in the older literature) is somewhat complicated because different authors included in their classifications various combinations of ophiacodontids in the modern sense as well as primitive members of other "pelycosaur" clades (see historical reviews in Case 1907, Romer and Price 1940, Reisz 1986). Romer and Price's (1940) concept of Ophiacodontidae is the foundation for the modern view of the clade, although they grouped ophiacodontids with eothyridids in Ophiacodontia, a plesiomorphic grade of basal synapsids that they considered to represent the ancestral stock of other "pelycosaur" taxa. Subsequent cladistic analyses have not upheld this grouping, and ophiacodontids are generally considered to be a subclade of Eupelycosauria (although see Benson 2012). The oldest known ophiacodontid is *Archaeothyris florensis* from the Late Pennsylvanian of Nova Scotia (Reisz 1972), which is also one of the two oldest known definite synapsid species (see above). Synapomorphies of Ophiacodontidae include an antorbital region of the skull at least twice as long as the postorbital region, nasal longer than frontal, maxillary supracanine buttress with a slender ascending process, short, thickened paroccipital process of the opisthotic, greatly expanded axial neural spine, only the anterior sacral rib directly attaching to the ilium, presence of a dorsal groove on the ilium, and long and slender posterior process of the ilium (Reisz *et al.* 1992a, Berman *et al.* 1995).

Although ophiacodontids were among the first non-mammalian synapsids to be described (Cope 1875, Marsh 1878), they are not a highly diverse clade (seven to eight genera are known), and most ophiacodontids are represented by comparatively fragmentary material (e.g., Gaudry 1880, Romer and Price 1940, Reisz 1972, Brinkman and Eberth 1986, Spielmann and Lucas 2010). *Ophiacodon* and *Varanosaurus* are known from much more complete material (Fig. 5.8 A), however, that provides detailed information about their anatomy and appearance (Williston and Case 1913a, Watson 1914b, Romer and Price 1940, Sumida 1989a, Berman *et al.* 1995). The picture that emerges is one of a generalized, carnivorous basal synapsid, with a large, tall, narrow skull and slender mandibles lined with fairly uniform conical teeth. These features of the skull contributed to the long-standing hypothesis that ophiacodontids, and especially *Ophiacodon*, were semiaquatic piscivores (Case 1907, Romer and Price 1940, Kemp 1982, Reisz 1986, Germain and Laurin 2005). However, Felice and Angielczyk (2014) considered the evidence for a semiaquatic lifestyle in *Ophiacodon* equivocal at best, and they suggested that the null hypothesis for studies going forward should be that it was terrestrial (also see Laurin and de Buffrénil 2015).

Even if ophiacodontids were no less terrestrial than other "pelycosaurs", their skeletons do show some peculiar features. The vertebral morphology of *Varanosaurus* is unique among basal synapsids in having expanded neural arches and neural spines that alternate in height (Sumida 1989a, 1990), although other basal amniotes and early sauropsids exhibit a similar morphology. Given its phylogenetic position, *Varanosaurus* most likely evolved its unusual vertebral morphology independently (i.e., it is not a retained primitive character; Sumida 1989a, 1990, Sumida and Modesto 2001), perhaps as a means to increase dorsoventral and rotational flexibility in the vertebral column. *Ophiacodon* shows unusual patterns of ossification and suture closure compared with other "pelycosaurs". The sutures between the elements of the braincase of other basal synapsids are normally indistinguishable in adults, but they are unfused and discernible in *Ophiacodon* (Reisz 1986). The sutures in the pelvis also remain poorly fused, even in *Ophiacodon major*, the largest species of the genus (Romer and Price 1940, Olson 1941). The joint surfaces of the long bones appear to have remained cartilaginous throughout a longer period of ontogeny in *Ophiacodon* than in the sphenacodont *Dimetrodon*, with the ulna and femur

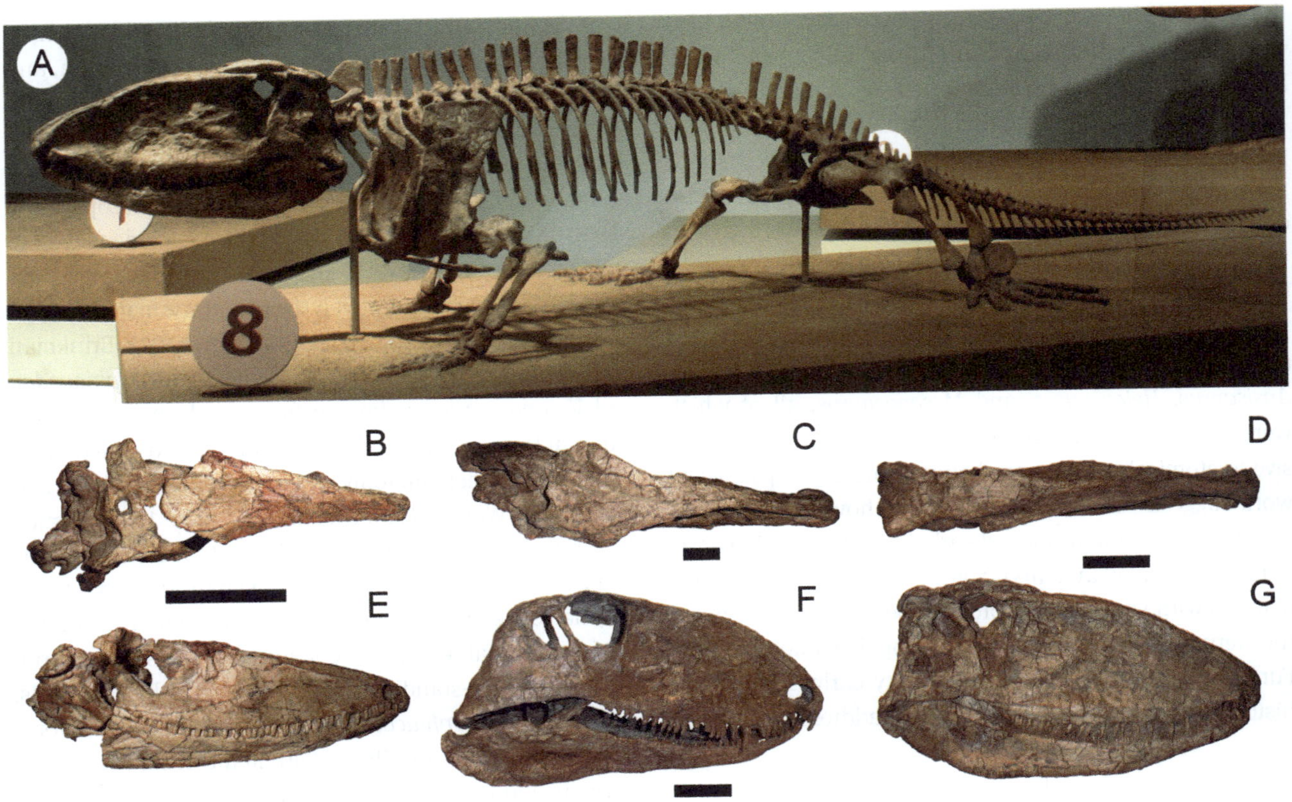

Fig. 5.8: Ophiacodontidae. (A) Mounted skeleton of *Ophiacodon mirus* (FMNH UC 671) in left lateral view. (B, E) Skull of *Varanosaurus acutirostris* (BSPG 1901 XV 20) in dorsal and lateral views. (C) Skull of *Ophiacodon retroversus* (FMNH UC 458) in dorsal view. (D, G) Skull of *Ophiacodon mirus* (FMNH UC 671) in dorsal and lateral views. (F) Skull of *Ophiacodon uniformis* (MCZ 1366) in lateral view. Specimens in panels E and G mirrored for comparative purposes. Scale bars equal 5 cm.

never reaching the degree of ossification seen in adult *Dimetrodon* specimens (Brinkman 1988). Persistent open sutures and reduced ossification of long bone articular surfaces are common in secondarily aquatic tetrapods (de Ricqlès 1989, de Ricqlès and de Buffrénil 2001), and they have been used as evidence to infer a semiaquatic lifestyle in *Ophiacodon*.

Finally, the bone histology of *Ophiacodon* is quite distinctive among "pelycosaur"-grade synapsids. Enlow and Brown (1957; also see Enlow 1969) initially noted that the limb bone tissue of *Ophiacodon* was more highly vascularized than in other "pelycosaurs" and that the lamellar bone seen in the latter was largely absent. They interpreted this as evidence of rapid growth, but noted that it could represent rapid growth at an early ontogenetic stage that was not sustained through ontogeny. de Ricqlès (1974) corroborated these observations and also noted that the longitudinal orientation of vascular canals and lack of distinction between the cortex and the spongiosa resembled the tissue organization of secondarily aquatic tetrapods, consistent with the semiaquatic lifestyle hypothesized for *Ophiacodon*. Germain and Laurin (2005) considered the compactness of the limb bone tissue of *Ophiacodon*, and although it was an outlier in their study, they again concluded that its tissue structure likely indicated a semiaquatic lifestyle. More recently, the bone histologies of *Ophiacodon* and *Clepsydrops* have been considered in more detail (Laurin and de Buffrénil 2015, Shelton and Sander 2015), and these studies have placed greater emphasis on the histological evidence for rapid growth in *Ophiacodon* while remaining more agnostic about its habitat preferences. Shelton and Sander (2015) in particular emphasized the apparently rapid growth of *Ophiacodon* and argued that it pushed the origin of mammal-like growth rates far back in synapsid history.

5.3.2.3 Edaphosauridae

Edaphosauridae is one of two clades of "pelycosaurs" characterized by hyperelongate neural spines on the presacral vertebrae that form a dorsal sail on the back (the other being Sphenacodontidae; see below) (Figs. 5.2 E and 5.9). The clade is also noteworthy because some members show extensive modifications to the skull and dentition for high-fiber herbivory (e.g., Romer and Price 1940,

Modesto and Reisz 1992, Modesto 1995, Hotton *et al.* 1997, Sues and Reisz 1998, Reisz and Sues 2000, Reisz 2006). Similar to Ophiacodontidae, the taxonomic concept of Edaphosauridae has changed over time. The first edaphosaurid vertebral column to be described (Cope 1878b, 1880) did not have an associated skull, and the elongate neural spines led to it being described as a species of the sphenacodontid *Dimetrodon*. Meanwhile, the first skull of *Edaphosaurus* to be described was not associated with postcrania (Cope 1882). It was not until the discovery of associated cranial and postcranial material (Case 1907, Williston and Case 1913b) that the differences between *Edaphosaurus* and *Dimetrodon* were realized and Edaphosauridae began to be treated as a distinct group of sail-backed herbivores. Romer and Price (1940) considered Edaphosauridae to include only members of the genus *Edaphosaurus* but also erected the more expansive taxon Edaphosauria to include the herbivorous caseids, which they thought to be closely related to edaphosaurids, as well as a handful of other taxa such as *Lupeosaurus* and *Mycterosaurus* (the latter subsequently identified as a varanopid; Berman and Reisz 1982). Following Romer and Price's (1940) monograph, a number of authors suggested that caseids were more closely related to eothyridids than to edaphosaurids (Watson 1957, Vaughn 1958, Olson 1962, Langston 1965, Reisz 1980, Brinkman and Eberth 1983, Reisz 1986), leading Reisz (1986) to again restrict Edaphosauridae to include only members of *Edaphosaurus*. Subsequent phylogenetic analyses have upheld this division (Modesto 1994, Berman *et al.* 1995, Benson 2012, Reisz and Fröbisch 2014; although see some trees of Romano and Nicosia 2015), but later research has expanded Edaphosauridae to include several less specialized taxa (Reisz and Berman 1986, Modesto and Reisz 1990, Modesto 1994, Mazierski and Reisz 2010). *Ianthasaurus hardestiorum* (Fig. 5.9 B), from the Late Pennsylvanian Garnett Quarry in Kansas (Reisz and Berman 1986, Kissel and Reisz 2004), is the oldest known edaphosaurid, but most members of the clade are known from the early Permian (Mazierski and Reisz 2010). Although Modesto (1995) found 18 synapomorphies of the skull, dentition, vertebral column, and ribs that diagnosed the better-known species of *Edaphosaurus*, inclusion of more incompletely known *Edaphosaurus* species and non-*Edaphosaurus* edaphosaurids shows that relatively few characters serve to unite the clade (e.g., Modesto and Reisz 1992, Reisz *et al.* 1992a, Mazierski and Reisz 2010, Benson 2012). The most consistent synapomorphy for Edaphosauridae across these analyses is the presence of hyperelongate neural spines that are subcircular in cross section and bear lateral tubercles or "crossbars" (e.g., Fig. 5.9 A, B, H). However, even that character may not be perfectly consistent within the clade, because the neural spines of the putative edaphosaurid *Lupeosaurus* lack tubercles (Sumida 1989b).

The most iconic aspect of the morphology of edaphosaurids is the dorsal sail formed by hyperelongate neural spines. In contrast to Sphenacodontidae, whose members include a mixture of sail-backed and more normally proportioned taxa, all members of Edaphosauridae possessed a sail (except possibly *Glaucosaurus*, for which postcranial material is not known; Modesto 1994). Edaphosaurids also seem to have evolved the sail earlier than sphenacodontids: the presence of *Ianthasaurus* at the Garnett Quarry in Kansas (e.g., Reisz and Berman 1986, Kissel and Reisz 2004, Mazierski and Reisz 2010) as well as edaphosaurid occurrences in the Sangre de Cristo Formation in Colorado (Sumida and Berman 1993) place sailed edaphosaurids earlier in the Late Pennsylvanian than the first occurrence of the sailed sphenacodontid *Dimetrodon* in the Bursum Formation of New Mexico (Harris *et al.* 2004). Several details differentiate the sails of edaphosaurids and sphenacodontids. Most obviously, edaphosaurid neural spines are characterized by crossbars, whereas sphenacodontid neural spines lack these features (crossbars seem to only be present on the more anterior vertebrae of *Ianthasaurus* and are absent in *Lupeosaurus*; Reisz and Berman 1986, Sumida 1989b, Modesto and Reisz 1990). The neural spines of sphenacodontids tend to be nearly vertical for most of the sail, whereas in edaphosaurids the anterior spines angle anteriorly (*Ianthasaurus*) or curve anteriorly (*Edaphosaurus*) and the posterior spines curve posteriorly, sometimes very strongly (Romer and Price 1940, Reisz and Berman 1986, Reisz 1986, Modesto and Reisz 1990). The sails of edaphosaurids and sphenacodontids scale differently with body size: edaphosaurids show negative allometry, whereas sphenacodontids show positive allometry (Romer and Price 1940, Romer 1948, Pivorunas 1970). Finally, the hyperelongate neural spines of edaphosaurids typically are subcircular in shape (Romer and Price 1940, Reisz and Berman 1986, Sumida 1989), although the distal ends of the anterior spines can become expanded and club-shaped in *Edaphosaurus* (e.g., Romer and Price 1940), and the neural spine is laterally compressed in the possible edaphosaurid *Xyrospondylus* (Reisz *et al.* 1982). Cross-sectional shape is more variable in sphenacodontids, ranging from "figure-eight" or "dumbbell"-shaped in *Dimetrodon* and *Secodontosaurus* (e.g., Romer and Price 1940, Reisz *et al.* 1992a, b, Rega *et al.* 2012) to laterally compressed in *Ctenospondylus* (Romer 1936, Romer and Price 1940, Vaughn 1964, Berman 1978).

The function of the "pelycosaur" sail has been a persistent mystery in synapsid paleontology (also see discussion of sphenacodontid sails below). In the case of

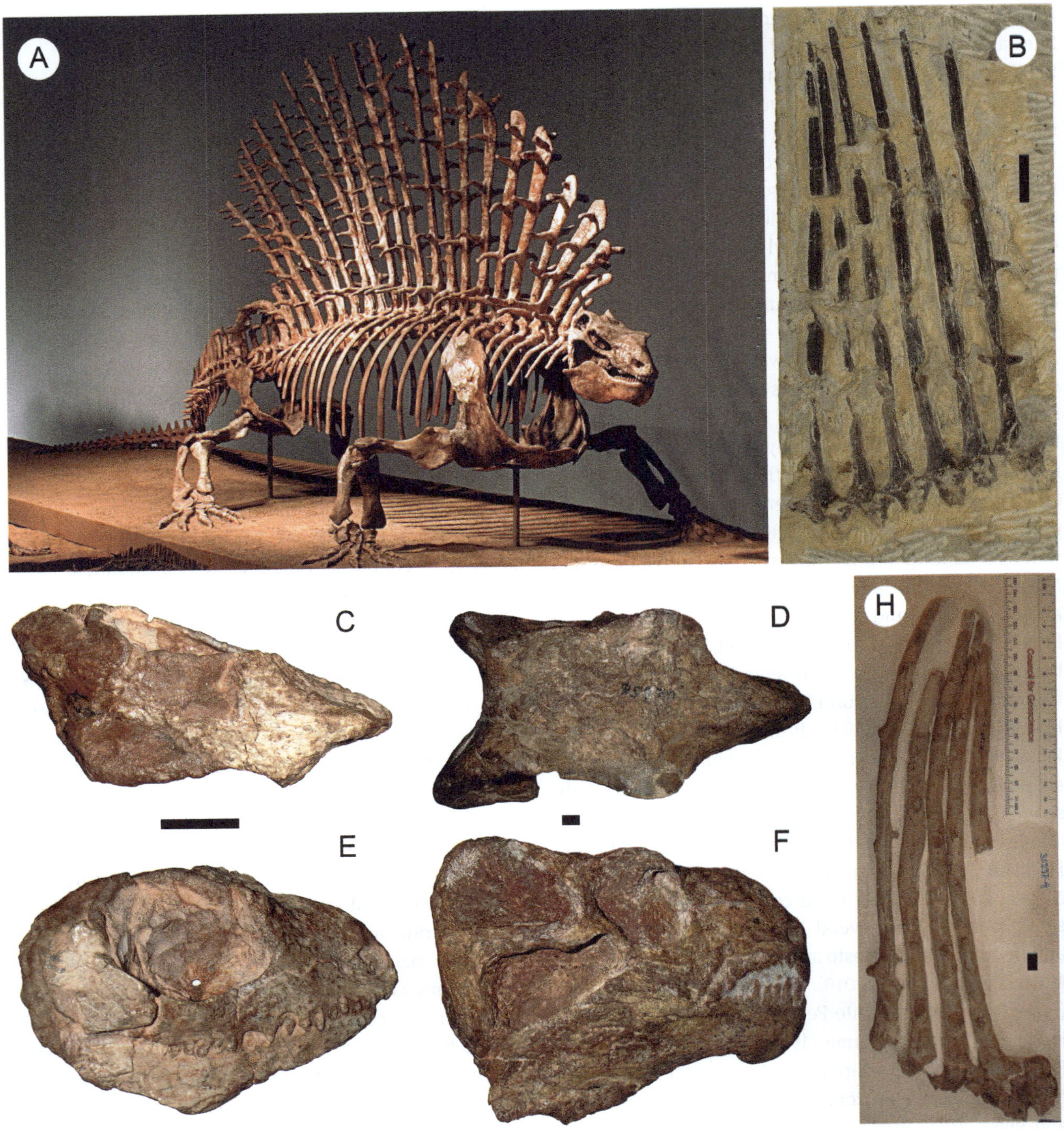

Fig. 5.9: Edaphosauridae. (A) Mounted skeleton of *Edaphosaurus pogonias* (FMNH UC 239) in anterolateral view. (B) Dorsal neural spines of *Ianthasaurus hardestiorum* (CM 70291). (C, E) Skull of *Glaucosaurus megalops* (FMNH UC 691) in dorsal and lateral views. (D, F) Skull of *Edaphosaurus cruciger* (FMNH UC 658) in dorsal and lateral views. (H) Dorsal neural spines of *Edaphosaurus* sp. (TMM 31227-4). Specimens in panels E and F mirrored for comparative purposes. Scale bars equal 1 cm.

edaphosaurids, proposed functions include thermoregulation (de Ricqlès 1974, Bennett 1996), defense (Modesto and Reisz 1990), intraspecific display (Modesto and Reisz 1990, Brocklehurst and Brink 2017), and fat storage (Romer and Price 1940, Pivorunas 1970), with the details of the thermoregulation hypothesis having been developed in the most detail. Using data from the gross anatomy of the spines of *Edaphosaurus* and their bone histology, de Ricqlès (1974) developed a detailed model of the circulatory system in the sail that would facilitate a thermoregulatory function. He noted the large central canal in the neural spine and the longitudinal arrangement of smaller vascular canals in the bone tissue and proposed that a large artery and associated smaller arterial vessels moved blood toward the distal end of the spine. Most of the cortical surface of the neural spine is avascular, so de Ricqlès (1974) suggested that capillaries leading to the soft tissues of the sail emerged from the more vascularized tubercles and crossbars. Blood would be returned to the body core by large veins located in the grooves on the anterior and posterior surfaces of the spines. Bennett (1996) conducted airflow and heatflow experiments with models of *Edaphosaurus* to investigate the potential functional importance of the tubercles and crossbars in the context of thermoregulation. He found that these structures would increase turbulent airflow around the sail, improving its effectiveness for shedding heat but reducing its utility as a means for absorbing heat.

Recently, Huttenlocker *et al.* (2011a; also see Huttenlocker and Rega 2012) reexamined the bone histology of edaphosaurids and questioned the role of the sail in thermoregulation. They found that there was little evidence to support the vascular system model proposed by de Ricqlès (1974), especially in terms of vascular communication between the proposed central artery and the sail via the tubercles. In *Ianthasaurus*, the bone forming the tubercles was essentially avascular, precluding their use as a means to transmit blood to the sail. The tubercles and crossbars of *Edaphosaurus* were more vascularized, but this primarily seemed to reflect higher growth rates associated with their rapid outgrowth from the spine. Instead of a thermoregulatory function, Huttenlocker *et al.* (2011a) proposed that the central canal of edaphosaurid neural spines likely formed through bone remodeling associated with the growth of the spine and the need for a shape that maximized its moment of inertia to resist bending stresses, with the lateral tubercles and crossbars functioning for intraspecific display.

In addition to their highly distinctive dorsal sails, edaphosaurids also are noteworthy because of the extensive modifications of the feeding system presented by *Edaphosaurus*. Although a diet of soft-bodied or hard-shelled invertebrates has been suggested at times for the genus (Williston 1914a, Case 1915, 1918, Munk and Sues 1993), it typically is considered to be a high-fiber herbivore (e.g., Romer and Price 1940, Modesto 1995, King 1996, Hotton *et al.* 1997, Sues and Reisz 1998, Reisz and Sues 2000, Reisz 2006). Modesto (1995) provided a detailed review of the morphological features of *Edaphosaurus* associated with herbivory, including the relatively small skull for its body size, shortened snout, large temporal fenestra, ventral off-set of the jaw joint relative to the tooth row, articular and quadrate bones that suggest the presence of a palinal (anterior-posterior sliding) motion of the lower jaw, modification of the jaw adductor muscles to power the palinal power stroke of the jaws, and the large barrel-shaped body (for a discussion of the evolution of large size associated with herbivory in Edaphosauridae, also see Fröbisch and Reisz 2014, Brocklehurst and Brink 2017). The most striking features, however, concern the dentition. The marginal teeth in the upper and lower jaws are isodont and roughly conical in shape, although they are widest near their midpoint before tapering toward their tips. Each tooth bears an anterior and posterior cutting edge, but these edges are oriented obliquely relative to the long axis of the tooth row. Unworn marginal teeth also bear fine serrations that are emphasized by oblique grooves. The orientation of the teeth also changes along the course of the tooth row. The more anterior teeth in the jaws are oriented approximately vertically, whereas the more posterior teeth in the maxilla face somewhat laterally and the posterior dentary teeth face somewhat medially. Besides the marginal dentition, the palatal dentition of *Edaphosaurus* is highly elaborated compared with its closest relatives. The palatine, ectopterygoid, and pterygoid bones are modified into expanded tooth plates, each of which bears a battery of up to 150 small, conical teeth. The tooth plates in the upper jaw occluded with similar tooth plates on the mandible formed by the anterior and posterior coronoids and the prearticular. Tooth wear features of the palatal dentition provide additional support for a palinal movement of the jaw during chewing (Modesto 1995, Hotton *et al.* 1997). Modesto (1995) posited that these extensive dental modifications allowed *Edaphosaurus* to engage in a two-step system of mastication. In the first step, plant material would be cropped by the anterior marginal teeth, and then would be comminuted by the tooth plates in the second step using a palinal motion of the mandible. Despite the sophistication of its feeding system, *Edaphosaurus* was only moderately successful, at least by comparison with later Permian synapsid herbivores. It is represented by about five species that are limited to the latest Pennsylvanian and early Permian (Huttenlocker *et al.* 2011a), and it tends to be a rare component of the communities in which it is found (Romer 1928).

The edaphosaurids *Ianthasaurus* and *Glaucosaurus* provide some insight into the evolution of herbivory within the clade. *Ianthasaurus* is much smaller than *Edaphosaurus*, with a larger head relative to its body size and a proportionally longer snout (Reisz and Berman 1986, Modesto and Reisz 1990, Mazierski and Reisz 2010, Fröbisch and Reisz 2014). The dentition of *Ianthasaurus* seems to be variable. The initial specimens described preserved pointed, recurved teeth, with an enlarged caniniform tooth located about one third of the way along the maxillary tooth row (Reisz and Berman 1986). This morphology led to *Ianthasaurus* being considered to have had an insectivorous diet. However, a larger, more recently described specimen has conical to slightly bulbous teeth that resemble those of *Edaphosaurus*, although it retains an enlarged caniniform tooth in the maxillary tooth row (Mazierski and Reisz 2010). In their description of the newer specimen, Mazierski and Reisz (2010) considered the possibility that the pointed, recurved teeth initially assigned to *Ianthasaurus* might have actually represented a different taxon due to the disarticulated nature of the original specimens, but they also noted that the apparently different ontogenetic stage of the more recently described specimen could indicate that *Ianthasaurus* was primarily insectivorous early in life with a later switch to herbivory or omnivory near adulthood. The somewhat more derived *Glaucosaurus* is known from a single poorly preserved skull (Modesto 1994) (Fig. 5.9 C, E). It shows great similarities to *Edaphosaurus* in some respects (e.g., isodont dentition, shorted snout, reduction of the transverse flange of the pterygoid similar to that seen in the tooth plates of *Edaphosaurus*), but it also has sharp, laterally compressed teeth. Modesto (1994) interpreted this combination of features as evidence for a diet of hard-bodied arthropods. Based on these observations, it appears that edaphosaurids originated as insectivores and/or omnivores before making the transition to high-fiber herbivory, a scenario that is particularly intriguing in light of the hypothesis that the initial inoculation of the antecessors of early amniote herbivores with crucial gut microbiota occurred by feeding on insects (Reisz and Sues 2000).

5.3.2.4 "Haptodonts"

The "haptodonts" are an assemblage of mostly small carnivorous synapsids known from the Late Pennsylvanian and Cisuralian (early Permian) of Europe and North America (Figs. 5.3 and 5.10 F). Initial assessments of the affinities of these species were quite varied, with authors sometimes placing them outside of Synapsida (e.g., Credner 1888, Osborn 1903, von Huene 1908), and even when identified as "pelycosaurs", they were suggested to be related to various subgroups (e.g., Jaekel 1910, Williston 1914a, Nopcsa 1923, Piveteau 1927). Romer and Price (1940) made a strong argument that they constituted a primitive subgroup of Sphenacodontidae. They also proposed that they might be the sphenacodontids most closely related to therapsids on account of their lack of unusual specializations, such as a dorsal sail, that were not expected in a therapsid ancestor, a hypothesis echoed by several other authors (Olson 1962, 1975, Sigogneau and Chudinov 1972, Currie 1977, 1979). The first cladistic analyses of "pelycosaurs" either did not include any "haptodonts" as terminal taxa (e.g., Reisz 1980, 1986, Brinkman and Eberth 1983, Gauthier *et al.* 1988) or followed Currie's (1979) synonymization of the various "haptodont" species and included only *Haptodus* as a terminal taxon (Reisz *et al.* 1992a, Berman *et al.* 1995). The latter analyses recovered *Haptodus* as the sister taxon of Sphenacodontoidea (i.e., Sphenacodontidae + Therapsida; also see Kemp 1982, 1988b, Hopson 1991), a phylogenetic position suggesting that the haptodonts were not especially close to the origin of therapsids. Laurin (1993, 1994) revised the taxonomy of the "haptodonts", recognizing the genera *Haptodus*, *Palaeohatteria*, *Pantelosaurus*, and *Cutleria* as valid, and found that they comprised a paraphyletic assemblage of taxa falling between edaphosaurids and sphenacodontids. More recent phylogenetic analyses (Sidor and Hopson 1998, Kissel and Reisz 2004, Fröbisch *et al.* 2011, Benson 2012, Reisz and Fröbisch 2014, Spindler *et al.* 2015) have supported this hypothesis, although there is disagreement as to whether *Cutleria* is a "haptodont" or an early sphenacodontid. Spindler (2015) undertook a detailed analysis of the taxonomy and phylogeny of the "haptodonts". Although only parts of this work have been published formally (Spindler *et al.* 2015, 2016), it promises to greatly improve our understanding of the origins of Sphenacodontoidea.

From an evolutionary perspective, perhaps the most interesting aspect of the "haptodont" grade is the insight that it provides on the early evolution of the reflected lamina of the angular. Extant mammals are famously characterized by a highly derived auditory system in which the ancestral quadrate and articular have been incorporated into the impedance-matching system of the middle ear as the malleus and incus (see reviews in Luo 2011, Kemp 2016, Luo *et al.* 2016). In addition to these elements, another bone of the ancestral mandible, the angular, is homologous with the ectotympanic, which provides support for the tympanic membrane and also can be incorporated into the auditory bulla (e.g., Novacek 1993). In all therapsids and sphenacodontids, the angular bears a posteriorly emarginated ventral keel (the reflected lamina), such that a pocket (angular cleft of Allin 1975, 1986) is formed between the medial surface of the reflected lamina and the lateral surface of the mandibular portion of the angular. Comparisons between modern mammals, fossil mammals, and non-mammalian synapsids indicate that the reflected lamina is homologous with the posterior (or ventral) portion of the ectotympanic annulus, and because of this, the reflected lamina of non-mammalian synapsids has long been of interest in understanding the evolution of the

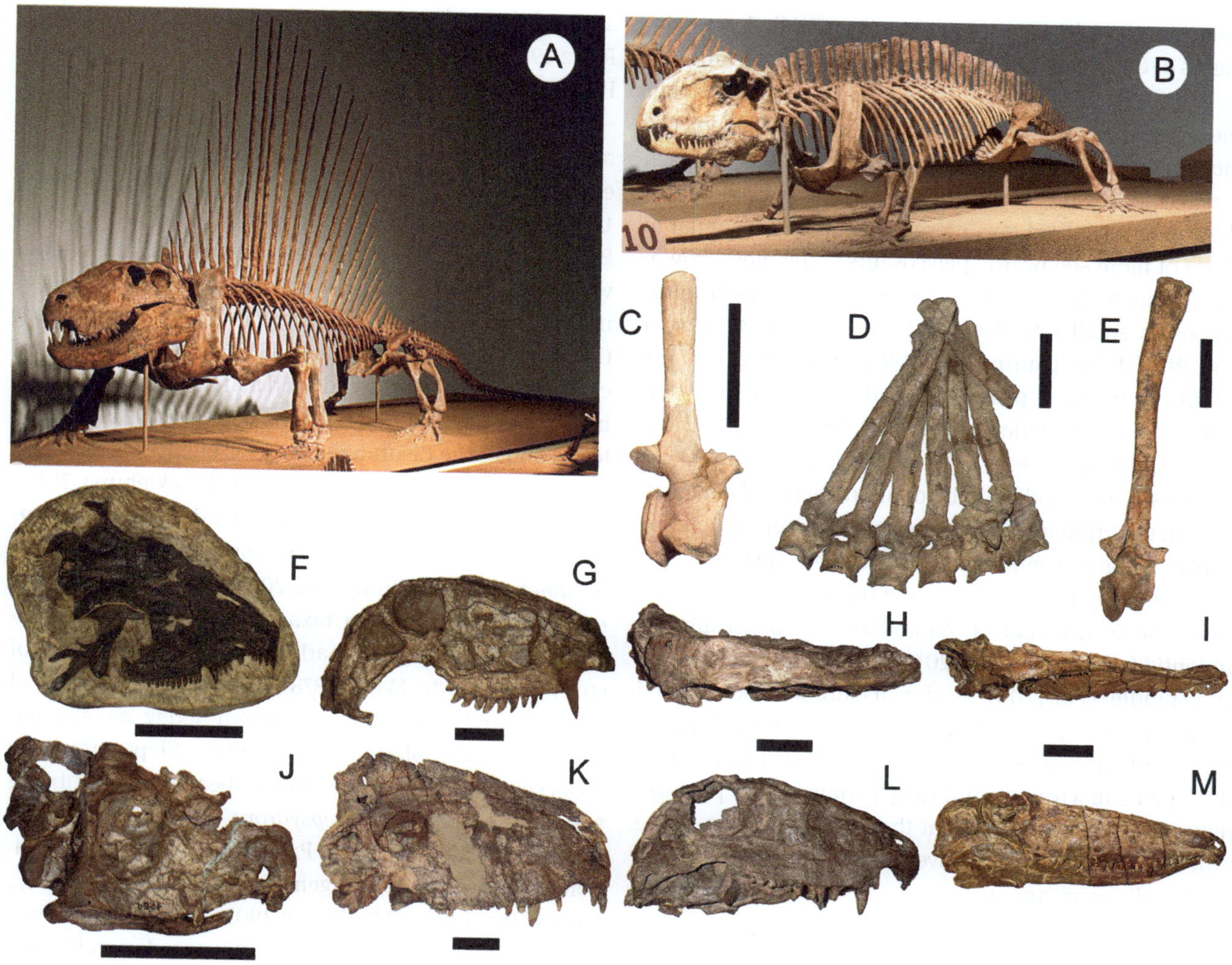

Fig. 5.10: Sphenacodontia. (A) Mounted skeleton of *Dimetrodon grandis* (FMNH UC 1002) in anterolateral view. (B) Mounted skeleton of *Sphenacodon ferox* (FMNH UC 35) in anterolateral view. (C) Caudal vertebra of *Dimetrodon giganhomogenes* (TMM 30966-329) in left lateral view. (D) Articulated series of dorsal vertebrae of *Ctenorhachis jacksoni* (USNM 437711) in left lateral view. (E) Dorsal vertebra of *Ctenospondylus casei* (TMM 40333-217) in left lateral view. (F) Crushed skull of *Haptodus garnettensis* (CM 71536, cast of ROM 43606) in dorsolateral view. (G) Skull of *Ctenospondylus casei* (NTM VP 1001) in lateral view. (H, L) Skull of *Dimetrodon loomisi* (FMNH UC 40) in dorsal and lateral views. (I, M) Skull of *Secodontosaurus obtusidens* (MCZ 1124) in dorsal and lateral views. (J) Skull of *Tetraceratops insignis* (AMNH FARB 4526) in lateral view. (K) Crushed skull of *S. ferox* (NMMNHS P-55367) in dorsolateral view. Specimens in panels J, K, L, and M mirrored for comparative purposes. Scale bars equal 5 cm.

mammalian auditory system (e.g., Broom 1912a, Palmer 1913, Watson 1914a, Sushkin 1927, Westoll 1943, 1945, Olson 1944, Parrington 1955, 1979, Schute 1956, Hopson 1966, Allin 1975, 1986, Kermack 1982, Kermack and Mussett 1983, Allin and Hopson 1992, Luo and Crompton 1994, Ivakhnenko 2003a, b, 2008, Kemp 2007, Takechi and Kuratani 2010, Anthwal *et al.* 2013, Laaß 2014, 2016, Tatarinov 1968, Gaetano and Abdala 2015). There has been substantial debate about when in synapsid history a mammal-like tympanic membrane supported by the angular evolved (see reviews in Allin and Hopson 1992, Meng *et al.* 2011, Gaetano and Abdala 2015, Kemp 2016), but there is a general consensus that the initial evolution of the reflected lamina was not driven by selection related to hearing. Instead, it has been hypothesized that the reflected lamina represents a modification of the ventral keel on the angular that is present in more basal "pelycosaurs", upon which the pterygoideus musculature inserts (Romer and Price 1940, Watson 1948, Crompton 1963a, b, Fox 1964, Barghusen 1968, 1973, Kemp 1972a, 1982, Allin 1975, Allin and Hopson 1992). Under this scenario, the angular cleft between the reflected lamina and the body of the mandible formed a space for the accommodation of the muscle. The lateral surface of the mandibular portion of the angular, and perhaps the reflected lamina itself, provided an expanded area for the attachment of the musculature, although note that the detailed arrangements of the attachments vary somewhat among authors.

The reflected lamina of the angular evolved within the "haptodont" grade. *Haptodus* and *Ianthodon*, the most stemward "haptodonts" (Spindler *et al.* 2015), lack a reflected lamina (Laurin 1993, Spindler 2015). Instead, the angulars of these taxa present a ventral keel similar to those of more stemward "pelycosaurs". The more crownward "haptodonts" *Palaeohatteria* and *Pantelosaurus* possess a small reflected lamina (Credner 1888, Romer and Price 1940, Laurin 1993, 1994, Spindler 2015). The lamina is rounded posteriorly and is not notched dorsally in *Palaeohatteria* (Laurin 1993, Spindler 2015), but its morphology appears to be ontogenetically variable in *Pantelosaurus*: smaller individuals display a *Palaeohatteria*-like morphology, but larger ones present a notched dorsal margin, comparable with (albeit smaller than) that seen in sphenacodontids and therapsids. In *Cutleria*, either the most advanced "haptodont" or a basal sphenacodontid (compare Benson 2012, Reisz and Fröbisch 2014, and Spindler *et al.* 2015), the dorsal margin of the reflected lamina is strongly notched (Laurin 1993, 1994, Spindler 2015). Although these taxa occur at different times during the Late Pennsylvanian and early Permian, their phylogenetic relationships imply that the initial evolution of the reflected lamina occurred over a short period of time in the Late Pennsylvanian.

5.3.2.5 Sphenacodontidae

Sphenacodontidae (sensu Hopson 1991) is doubtlessly the most famous of all non-mammalian synapsid clades, on account of the inclusion of the sphenacodontid *Dimetrodon* (Fig. 5.10 A) in countless books about dinosaurs and in sets of "dinosaur" toys (Figs. 5.2 F and 5.10). Although this fame is understandable because of *Dimetrodon*'s spectacular dorsal sail and ferocious appearance, its association with dinosaurs has caused widespread confusion about what it actually is (a synapsid more closely related to modern mammals than to any reptile or amphibian) and when it lived (more time separates the first specimens of *Dimetrodon* from the first dinosaurs than separates humans from *Tyrannosaurus rex*). The scientific importance of Sphenacodontidae lies in the fact that they are the "pelycosaurs" most closely related to therapsids, and as such they help us to understand the evolution of many of the distinctive characters of therapsids. *Dimetrodon* itself also was an extremely successful animal, with a long stratigraphic range and a cosmopolitan distribution, and it is one of the most common taxa in the Permian strata of Texas and Oklahoma (e.g., Romer 1928, Romer and Price 1940, Reisz 1986, Berman *et al.* 1997, 2001, Harris *et al.* 2004, Brocklehurst 2015).

Sphenacodontids are medium to large carnivores, and most were apex predators in their communities, well equipped to deal with large prey items (Romer and Price 1940, van Valkenburgh and Jenkins 2002, Brink *et al.* 2014, Brink and Reisz 2014) and likely employing a "sit-and-wait" foraging strategy (Hopson 2012). They first appear in the Late Pennsylvanian Sangre de Cristo Formation of Colorado (Sumida and Berman 1993; even older, but less certain potential sphenacodontid material has been found in Nova Scotia, see Reisz 1972), and they survive until close to the end of the Cisuralian (early Permian; e.g., see compilation of specimen occurrences in Brocklehurst 2015). The first "pelycosaur" specimen to be described (*Bathygnathus borealis* Leidy 1854, now considered to be a species of *Dimetrodon*; see Brink *et al.* 2015) was a sphenacodontid, and the well-known taxa *Sphenacodon* and *Dimetrodon* were also named early in the group's history of study (Cope 1878c, Marsh 1878). In early classifications of "pelycosaurs", the taxa recognized today as sphenacodontids were placed in the family Clepsydropsidae on account of some *Dimetrodon* species initially being assigned to the genus *Clepsydrops* (now considered an ophiacodontid; Romer and Price 1940), and a close relationship between the two genera also was assumed (e.g., Cope 1878c, 1880, 1881, Broom 1903a, Case 1907). Lesser known sphenacodontids such as *Sphenacodon* and *Bathygnathus* were also included in Clepsydropsidae on account of their similarities to *Dimetrodon* (Case 1905, 1907, von Huene 1905), and Clepsydropsidae remained in use throughout most of the early 20th century (e.g., Case 1915, Watson 1917a, Williston 1925). Romer and Price (1940) recognized that *Clepsydrops* was an ophiacodontid and presented a revised Sphenacodontidae, the composition of which has largely remained stable in more recent studies (Reisz *et al.* 1992a, Fröbisch *et al.* 2011, Benson 2012, Reisz and Fröbisch 2014, Brink *et al.* 2015, Brocklehurst *et al.* 2016a, b). Romer and Price (1940) also included the "haptodonts" within the group, but Hopson (1991) redefined Sphenacodontidae to exclude these taxa when it became clear that they represented a grade at the base of Sphenacodontoidea (Sphenacodontidae + Therapsida).

At the time of Romer and Price's (1940) monograph, the general relationship between "pelycosaurs" and therapsids was well established, but there was not a clear consensus on which "pelycosaurs" were most closely related to therapsids. Romer and Price (1940) undertook a detailed comparison of "pelycosaur" and therapsid skeletons and concluded that the sphenacodontids were the "pelycosaurs" most closely related to Therapsida. Following Romer and Price's work, Everett Olson argued instead for a polyphyletic origin of therapsids from multiple "pelycosaur" groups (Olson 1944, 1962, 1974, 1986, Olson and Beerbower 1953, also see Barghusen 1976 and more recently Ivakhnenko 2002a, 2003b, 2008), but other workers continued to accept that sphenacodontids were the "pelycosaurs" most closely related to or directly ancestral to therapsids (e.g., Boonstra 1963, 1971, 1972). Early cladistic studies also concluded that sphenacodontids were the sister group of therapsids (e.g., Reisz 1980, Hopson and Barghusen 1986, Gauthier *et al.* 1988, Kemp 1988a, b), but most did not include multiple sphenacodontid (sensu Hopson 1991) terminal taxa. Brinkman and Eberth (1983) is an exception, including two sphenacodontid terminals but no therapsid terminals. Newer phylogenetic analyses with more comprehensive taxon sampling have confirmed that Sphenacodontidae is the monophyletic sister group of Therapsida (Reisz *et al.* 1992a, Laurin 1993, Fröbisch *et al.* 2011, Benson 2012, Reisz and Fröbisch 2014, Brink *et al.* 2015, Brocklehurst *et al.* 2016a, b).

Reisz *et al.* (1992a; also see Reisz *et al.* 1992b) provided the most detailed discussion of characters relevant to sphenacodontid relationships and reported nine sphenacodontid synapomorphies, including a premaxilla lacking a distinct palatal process, narrow anterior process of the frontals, long anterior process of the frontals, cervical centra with ventral keels, dorsal central with ventral keels, and elongated neural spines. They also listed 27 synapomorphies that sphenacodontids share with therapsids, such as a robust premaxilla, nasal longer than frontal, ridge surrounding the pineal foramen, ventral margin of maxilla strongly convex, lacrimal excluded from the external naris, paroccipital process of the opisthotic extending ventrolaterally, retroarticular process of the articular well developed, caniniform teeth present that are more than twice as long as other maxillary teeth, less than four precanine maxillary teeth, scapular blade narrow at base, anterodorsal process present on the iliac blade, and shallow intertrochanteric fossa on the femur. Nevertheless, therapsids themselves are highly distinctive from sphenacodontids, with Sidor and Hopson (1998) and Kemp (2006) reporting as many as 55 therapsid synapomorphies (see below). This number may not represent elevated rates of morphological evolution in early therapsids, however, merely that the long ghost lineage separating the divergence of sphenacodontids and therapsids in the Late Pennsylvanian and the first appearance of therapsids in the middle Permian provided ample time for novel therapsid character states to accumulate. Species-level relationships within Sphenacodontidae have been difficult to reconstruct, and there have been suggestions that many of the species might instead represent parts of ontogenetic series (Bakker 1982; but see Brinkman 1988). However, new data on braincase morphology, tooth structure, and bone histology have improved our understanding of sphenacodontid evolution, demonstrating probable shifts in feeding style within the clade and that there was size differentiation among adults of different *Dimetrodon* species (Brink and Reisz 2012, 2014, Shelton *et al.* 2013, Brink *et al.* 2014, 2015).

Sphenacodontids display a mosaic of primitive and derived (i.e., therapsid-like) characters, reflecting their phylogenetic position just outside of the therapsid radiation. For example, compared with more stemward "pelycosaurs", the jaw joint of sphenacodontids has been shifted ventrally, the jaw has developed a prominent coronoid eminence, and the line of action of M. adductor mandibulae externus is more posteriorly directed. These changes increase the mechanical advantage of the jaw and would help to resist anterior displacement of the jaw by struggling prey, and therapsids show additional modifications of the feeding system that further emphasize these benefits (e.g., expansion and emargination of the temporal fenestra) (Barghusen 1968, 1972, 1973, DeMar and Barghusen 1972). A reflected lamina of the angular is present in sphenacodontids, but it is less extensively developed than in therapsids, and it has been regarded as providing an expanded area for attachment of the pterygoideus musculature instead of being related to the auditory system (Romer and Price 1940, Watson 1948, Crompton 1963a, b, Fox 1964, Barghusen 1968, 1973, Kemp 1972a, 1982, Allin 1975, Allin and Hopson 1992). Indeed, a number of authors have questioned whether a tympanic membrane was present at all in sphenacodontids (and other "pelycosaurs"), implying that they lacked sensitivity to airborne sound (Watson 1953, Tumarkin 1955, Allin 1975, 1986, Lombard and Bolt 1979, Allin and Hopson 1992). The enlarged caniniform teeth of sphenacodontids represent differentiation of the dentition that is carried forward and expanded upon in therapsids. The limbs of sphenacodontids are more elongate than those of other "pelycosaurs" (Fig. 5.10 A, B), indicating optimization for more rapid locomotion (Romer and Price 1940), but they still retain a sprawling posture (Watson 1917b, Romer 1922, Romer and Price 1940, Haines 1942, 1952, Jenkins 1971, 1973, Hopson 2015) and lateral undulation likely continued to play a major role in their locomotory patterns (Sumida and Modesto 2001, Hopson 2012, 2015, contra Kemp 1982, MacDonald 1994, Hunt and Lucas 1998).

Initial surveys of sphenacodontid bone histology reported primarily slow-growing lamellar-zonal tissues divided into annual growth zones by lines of arrested growth (Enlow and Brown 1957, Enlow 1969, de Ricqlès

1974), which contributed to the widespread perception of sphenacodontids and other "pelycosaurs" as having reptile-like ectothermic metabolisms. However, more recent studies have found faster-growing tissues (variously interpreted as fibrolamellar bone or "incipient fibrolamellar bone") in subadults of *Sphenacodon ferocior* (Huttenlocker *et al.* 2006; also see Huttenlocker *et al.* 2010, Huttenlocker and Rega 2012) and throughout ontogeny in *Dimetrodon natalis* (Shelton *et al.* 2013). Fibrolamellar bone is ubiquitous in therapsids (e.g., Ray *et al.* 2004, Botha-Brink and Angielczyk 2010, Chinsamy-Turan and Ray 2012, Ray *et al.* 2012, Botha-Brink *et al.* 2012, 2016, Huttenlocker and Botha-Brink 2014), and its presence in *Sphenacodon* and *Dimetrodon* raises the possibility that increased growth rates were present in the common ancestor of sphenacodontids and therapsids. However, histological sampling within Sphenacodontidae is currently too limited to definitively eliminate the possibility that faster growth rates evolved independently in this clade and in therapsids.

Ever since their first discovery, the hyperelongate neural spines of many sphenacodontids have been a source of wonder and speculation in both the paleontological community and among the general public. Sphenacodontids show two basic configurations of the neural spines. *Sphenacodon*, *Ctenospondylus*, and *Ctenorhachis* have laterally compressed, bladelike neural spines that are somewhat elongated (e.g., Romer 1936, Romer and Price 1940, Vaughn 1964, Berman 1978, Hook and Hotton 1991) (Fig. 5.10 B, D, E), but not to the extreme degree that is seen in the true "sail-backed" sphenacodontids *Dimetrodon* and *Secodontosaurus*. In *Dimetrodon* and *Secodontosaurus*, the neural spines are extremely elongated (18 to 30 times the height of the corresponding vertebral centra; Romer and Price 1940, Huttenlocker *et al.* 2010) (Fig. 5.10 A), and while they are somewhat flattened near their bases, they have a "dumbbell" or "figure-eight" shape in cross section for most of their length (e.g., Romer and Price 1940, Reisz *et al.* 1992a, b, Rega *et al.* 2012). In both morphologies, evidence from muscle scars and bone histology suggest that the epaxial musculature was limited to the lower portions of the spine (i.e., the flattened portions in *Dimetrodon*), with the upper parts of the spine being embedded in a much thinner soft tissue membrane (Olson 1936, Romer and Price 1940, Huttenlocker *et al.* 2010) (the distal tips of the neural spines of at least some *Dimetrodon* individuals may have been free of the membrane as well, although those specimens may be pathological; see Rega *et al.* 2012). Therefore, it seems likely that all sphenacodontids had at least a rudimentary dorsal crest (Huttenlocker *et al.* 2010), and recent phylogenetic analyses suggest that it was a basal feature of the clade (Fröbisch *et al.* 2011, Benson 2012, Brink and Reisz 2014, Brink *et al.* 2015). Surprisingly, the position of *Secodontosaurus* outside of a subclade containing *Ctenospondylus*, *Dimetrodon*, and *Sphenacodon* in most of these trees raises the possibility that shorter-spined sphenacodontids such as *Sphenacodon* and *Ctenorhachis* represent a secondarily shortened morphology. Although edaphosaurids and sphenacodontids are closely related, and both clades possess hyperelongate neural spines, the intermediate phylogenetic position of the "haptodonts" with more normally proportioned vertebrae (e.g., Romer and Price 1940, Currie 1977, 1979, Laurin 1993, Spindler 2015, Brocklehurst and Brink 2017) suggests that the dorsal sail evolved independently in the two clades.

Early writings on the possible function of the dorsal sail range from Cope's (1886) whimsical suggestion that it actually functioned as a nautical sail to Jaekel's (1910) proposal that it was an elaborate defense mechanism and Case's (1915) conclusion that it had no function at all and instead represented "excess vitality" (p. 115) that was not put to more constructive use on account of benign environmental conditions. More recent functions proposed for sphenacodontid sails fall into the categories of biomechanics (Romer 1927, Romer and Price 1940, Rega *et al.* 2012), thermoregulation (Romer 1948, Robard 1949, Pivorunas 1970, Bramwell and Fellgett 1973, de Ricqlès 1974, Haack 1986, Tracy *et al.* 1986, Florides *et al.* 1999, 2001), display (Bakker 1971, Tomkins *et al.* 2010, Brocklehurst and Brink 2017)), and hearing (Tumarkin 1965, 1968). We find Tumarkin's proposal that the hyperelongate neural spines supported an analog of the tympanic membrane to be highly implausible and therefore focus on biomechanical, thermoregulatory, and display-related functions below.

Romer (1948) was the first to suggest a thermoregulatory function for the sail of *Dimetrodon*, although he and Robard (1949) described this function in strictly qualitative terms. The sail was seen as a structure that would greatly increase the surface area of the animal relative to its volume, improving its ability to absorb heat. Moreover, the structure was thought to be well vascularized, with large blood vessels situated in the grooves that extend along most of the anterior and posterior surfaces of the neural spines that comprise the sail (Romer 1927, de Ricqlès 1974). Romer (1948) also noted that the sail displayed positive allometry (also see Romer and Price 1940, Pivorunas 1970), which would allow it to maintain its function as body mass increased in larger individuals and larger species.

Subsequently, a number of authors used modeling approaches with varying assumptions to quantitatively test the effectiveness of the sail of *Dimetrodon* as a heat exchanger (Bramwell and Fellgett 1973, Haack 1986, Tracy *et al.* 1986, Florides *et al.* 1999, 2001). Despite the ubiquity of the idea that the sail functioned in thermoregulation, the results of these quantitative studies vary widely. Bramwell and Fellgett (1973) concluded that the sail of *Dimetrodon* would have been an efficient heat exchanger for both warming and cooling the animal. Tracy *et al.* (1986) largely agreed, but they noted that the sail might have been most important in maintaining a stable body temperature, with the relative significance of its heating and cooling functions varying in relation to the size of the animal in question. Haack (1986) suggested that the sail was likely most useful in its warming function and for maintaining a stable body temperature, not cooling, but also concluded that the effect was not as great as previously hypothesized. Florides *et al.* (1999, 2001) favored a heating function, but noted that the importance of this function would vary depending on environmental conditions. Specifically, they hypothesized that the sail provided the most benefit under cooler conditions, but that it could be a liability under hot conditions because it would promote overheating. Tomkins *et al.* (2010) noted that the extreme dorsal sail morphology in sphenacodontids (and edaphosaurids) initially evolved in animals with relatively small body sizes (e.g., *Dimetrodon milleri* and *Ianthasaurus hardestiorum*), an observation confirmed by recent phylogenies of the groups (e.g., Mazierski and Reisz 2010, Brink and Reisz 2014). Citing previous modeling experiments, Tomkins *et al.* (2010) raised the issue that the small size of these species would have conferred a low thermal inertia regardless of the presence of a dorsal sail, reducing the need for an elaborate heat exchange system. Moreover, the sail could have made these smaller animals liable to overheat if it was a highly vascularized heat exchanger. Finally, Bennett's (1996) results suggesting a cooling function for the sail of *Edaphosaurus* (see above) further complicate matters.

The lingering uncertainty about the manner in which the sail would have functioned in heat exchange and its effectiveness in that role indicates that a thermoregulatory function is not as definite as often portrayed (e.g., Kemp 1982). Other aspects of a thermoregulatory function for the sail also are problematic. For example, Huttenlocker *et al.* (2010) noted that the highly vascularized bone tissue of the neural spines of *Dimetrodon* likely was associated with their rapid growth and need not reflect a thermoregulatory function. Rega *et al.* (2012) stated that the lack of vascular canals in the anterior and posterior grooves of the spines called the presence of blood vessels in the grooves into question (see Huttenlocker *et al.* 2011a, b and discussion above for similar problems with the neural spine anatomy of edaphosaurids). Moreover, the presence of two to three large, vertical blood vessels at each vertebral segment in the sphenacodontid (and edaphosaurid) sail would have made the sail very vulnerable to injury and would have greatly increased resistance to blood flow (Huttenlocker *et al.* 2011a, b). If the blood supply of the sail was instead associated with the superficial skin covering, it would have been much less efficient at raising or lowering the core body temperature of the animal (Haack 1986).

A hypothesized biomechanical function for the sphenacodontid sail predates the thermoregulatory hypothesis. Romer (1927) posited that the elongation of the neural spines in basal synapsids formed part of a system for strengthening an otherwise relatively weak vertebral column in animals with long, slender bodies. Romer and Price (1940) reiterated this hypothesis, but both Romer (1927, 1948) and Romer and Price (1940) were at a loss as to why such a trend was taken to the extreme observed in *Dimetrodon*. When the thermoregulatory hypothesis for the function of the sail came to prominence, the previous suggestions of a biomechanical function for the sail largely were forgotten. Recently, Huttenlocker *et al.* (2010, 2011a, b) and Rega *et al.* (2012) reassessed a possible biomechanical role for the sail. In particular, Rega *et al.* (2012) used finite element analysis, and data on bone remodeling associated with a microfracture that occurred in a neural spine of the large species *Dimetrodon giganhomogenes*, to posit that the "dumbbell" cross-sectional shape of the neural spines (resulting from the anterior and posterior grooves) was strongly indicative of their being specialized for resisting lateral bending stresses. They also noted that similar morphologies can be found in the dorsal fin of the yellowfin tuna. In that animal, the spines are embedded in a dense web of collagen that may function in resisting dorsoventral bending and increasing the efficiency of lateral undulation, and Rega *et al.* (2012) hypothesized that the sail of *Dimetrodon* would have been similar in structure and function. If this is the case, it closely matches Romer's (1927) original conception of the construction and function of the sail.

The third function proposed for the sails of sphenacodontids is intraspecific display (Bakker 1971, Tomkins *et al.* 2010, Brocklehurst and Brink 2017). The information presented in support of this hypothesis largely consists of anecdotal comparisons to various living animals. Bakker (1971) discussed the display function of the similar dorsal crest of *Bos gaurus* (for more information on this display behavior, see Grzimek 1990). Tomkins *et al.* (2010) noted that the scaling exponent of the sail of *Dimetrodon* was similar to that of antler length relative to shoulder height in cervids, and they suggested that this scaling was typical of structures whose function was primarily related to display. In general, a display function for the sail is difficult to test, and there is no reason why the sail could not have also functioned in this capacity even if its primary role was thermoregulatory or biomechanical (as is the case for the expanded ears of the African elephant; Wright 1984).

Given the ubiquity of sphenacodontids, especially *Dimetrodon*, in many of the classic Permian localities in Texas (e.g., Romer 1928), the group has played an important role in considerations of "pelycosaur" paleoecology (see review in Falconnet 2015). In general, *Dimetrodon* and *Secodontosaurus* have been considered to favor coastal lowland habitats, floodplains, and lake margins, whereas *Sphenacodon* was thought to occur in more upland settings (e.g., Olson 1958, 1977, Vaughn 1966, Berman 1978, Sander 1987, 1989). However, the two classic Permian upland localities (the Dolese Quarry near Richards Spur, Oklahoma and the Bromacker Quarry in Germany; see e.g., Reisz 2005, Reisz *et al.* 2010a, Berman *et al.* 2014) show that the situation was more complex than a simple lowland/upland dichotomy. Although the Bromacker locality represents an internally drained upland basin, the fossiliferous Tambach Formation includes sheet flood deposits, alluvial paleochannels, and ephemeral lake deposits (Eberth *et al.* 2000). The sphenacodontid *Dimetrodon teutonis* was able to invade the Bromacker habitat, and despite being rare in comparison with tetrapod herbivores in the environment (see initial tabulation in Eberth *et al.* 2000 and more up-to-date ratios in Berman *et al.* 2014), it is more common than the other apex predator (the varanopid *Tambacarnifex unguifalcatus*) in the system (compare Berman *et al.* 2004, 2014). The Dolese Quarry fossils are preserved as fissure fills in an early Permian cave system, and although the surrounding environmental conditions have been difficult to reconstruct, there has been the suggestion that they were relatively dry (Olson 1991, Sullivan and Reisz 1999). Sphenacodontids are quite rare at the Dolese Quarry, even in comparison with other synapsid carnivores (Evans *et al.* 2009), which could indicate an environment less favorable to the clade than what was present at the Bromacker locality. These observations, combined with data from other European localities, led Falconnet (2015) to conclude that the distribution of sphenacodontids was strongly controlled by the specific nature of a given habitat. A noteworthy feature of the Bromacker and Dolese sphenacodontids is their relatively small body size, which has led to the suggestion that a reduction in body size was important for allowing sphenacodontids to survive in these environments (Berman *et al.* 2001, Evans *et al.* 2009). Cantrell *et al.* (2011) turned this idea on its head and suggested that *Dimetrodon* in particular initially evolved in upland environments and subsequently invaded lowland habitats, increasing its body size in the process. This hypothesis must remain speculative until the phylogenetic position of *D. teutonis* and other upland *Dimetrodon* specimens is better resolved, however. Angielczyk and Schmitz (2014) added an additional detail to the environmental preferences of sphenacodontids when they presented evidence that the eyes of *Dimetrodon* and *Sphenacodon* were optimized for scotopic (low light) settings, raising the possibility that they were nocturnal.

5.3.3 Therapsida

Therapsids have long been recognized as having evolved from sphenacodonts (Romer and Price 1940), but they are so morphologically distinct from their "pelycosaurian" ancestors that they were historically accorded their own taxonomic order (Romer 1956). A vast array of synapomorphies diagnoses Therapsida. Sidor and Hopson (1998) and Kemp (2006) recognized as many as 55 characters diagnosing the clade. Some of the more notable include the posterior extension of the septomaxilla between the maxilla and the nasal; posterior elongation of the dorsal process of the premaxilla; elevation of the pineal foramen on a boss; loss of the supratemporal; enlargement of the canines and establishment of discrete incisor-canine-postcanine heterodonty in both the upper and the lower dentition; expansion of the dentary; movement of the coronoid ventrally, so it occurs entirely on the medial face of the mandible; a pattern of ridges on the lateral surface of the reflected lamina; deepening of the glenoid fossa of the scapula and the acetabulum of the pelvis; inflection of the femoral head to allow a more adducted hindlimb position; narrowing of the scapula; and loss of intercentra in the trunk vertebrae (Rubidge and Sidor 2001). Many of these characters are recognizable as antecedents of the characteristic features of mammals. For example, heterodonty in therapsids presages even greater dental differentiation in mammals, and therapsid limb and girdle characters are indicative of a more upright gait and active lifestyle than their "pelycosaurian" forebears.

Six major therapsid subclades are recognized (Fig. 5.3): Biarmosuchia, Dinocephalia, Anomodontia, Gorgonopsia, Therocephalia, and Cynodontia. For the most part, these clades are highly distinctive, and their reciprocal monophyly is strongly supported (Sidor 2000, 2001). The interrelationships of these major clades, however, have long been controversial and are still not fully resolved. Particularly problematic for sorting out therapsid relationships is the long-branch issue—like the modern mammalian orders, all the major therapsid subclades appear within a short span of time in the fossil record and are already clearly recognizable in their first appearances in the middle Permian (i.e., there are no taxa that bridge the morphological gaps between the major clades—with the possible exception of the enigmatic Chinese taxon *Raranimus*, even the earliest therapsid fossils are each referable to one of these groups) (Rubidge 1991, 1994, Abdala *et al.* 2008, Liu *et al.* 2009, 2010). Hence, therapsids may have experienced an explosive radiation similar to that of mammals following the K-Pg extinction, and it has even been suggested that the major subclades diverged so rapidly as to represent an insoluble polytomy (Kemp 2006, 2009). It seems more likely, however, that the long-branch problem in therapsid phylogeny is an artifact of missing record (see comments on Olson's Gap above).

Despite these problems, many studies of therapsid phylogeny do exist, in both precladistic (e.g., Watson 1921, Broom 1932, Olson 1944, 1962, Watson and Romer 1956, Romer 1966, Boonstra 1972) and modern frameworks (e.g., Hopson and Barghusen 1986, Kemp 1988a, b, Gauthier *et al.* 1988, Rowe 1986, 1988, Hopson 1991, 1994, Sidor and Hopson 1998, Modesto *et al.* 1999, Sidor 2000, 2001, Liu *et al.* 2009, 2010), and numerous hypotheses of relationship between the six major clades have been proposed. Two of these studies have had particularly lasting influence on our understanding of therapsid phylogeny: those of Watson and Romer (1956; refined and canonized in Romer 1966) and Hopson and Barghusen (1986; expanded in Hopson 1991, 1994). Watson and Romer proposed a deep dichotomy within Therapsida, with all known therapsids belonging to two major groups: Theriodontia and Anomodontia. These groups roughly correspond to feeding ecology: theriodonts (sensu Watson and Romer 1956) were primarily carnivores and included "titanosuchian" (anteosaurid + titanosuchid) dinocephalians, gorgonopsians, therocephalians, and cynodonts, whereas anomodonts included the major lineages of herbivorous therapsids (tapinocephalid dinocephalians, venyukovioids, "dromasaurs", and dicynodonts). This idea of a basal split within Therapsida was the dominant paradigm for over 30 years (see, e.g., King 1988, Sigogneau-Russell 1989; but for opposing viewpoints, see Olson 1962, Boonstra 1972).

The advent of cladistic methods for evaluating evolutionary relationships began to call the Watson and Romer topology into question, however, as exemplified by the hand-drawn cladistic analysis of Hopson and Barghusen (1986). This study was important as the first analysis detailing character evolution in all the major therapsid groups (contemporary analyses typically focused on single therapsid clades; e.g., Battail 1982, Cluver and King 1983), and it introduced a number of novel proposals that proved to be of great importance for subsequent work on therapsid evolution. First, Hopson and Barghusen recognized a new clade, Biarmosuchia, representing the most stemward group of therapsids. Second, they separated Dinocephalia from Anomodontia (a proposal later echoed by Hopson 1991 and Grine 1997), and recognized venyukovioids and "dromasaurs" as a grade outside of dicynodonts within Anomodontia. Third, they proposed lower-level interrelationships for most of the major therapsid subclades, including the first such topologies for dinocephalians and therocephalians. In contrast to the symmetric dichotomy of Watson and Romer, Hopson and Barghusen's tree was highly pectinate, with the major therapsid clades representing successive divergences on the line leading up to mammals. Biarmosuchia is the first offshoot, with all other therapsid groups being united in the clade Eutherapsida (Hopson 1994). Hopson and Barghusen (1986) were uncertain about the relative positions of Dinocephalia, Anomodontia, and Theriodontia (in their classification, the latter group was restricted to Gorgonopsia, Therocephalia, and Cynodontia) and restored this node as an unresolved trichotomy, but later iterations of this analysis treated dinocephalians as the most stemward eutherapsids, with anomodonts and theriodonts forming the clade Neotherapsida (Hopson 1999, Sidor 2001).

The monophyly of Theriodontia (sensu Hopson and Barghusen 1986) has been extensively debated (Kemp 1972a, b, 1982, 1988, Rowe 1986, Hopson 1991, 1994, 1999, Sidor and Hopson 1998, Rubidge and Sidor 2001, Sidor 2001). A sister-group relationship between Therocephalia and Cynodontia (forming the clade Eutheriodontia; Kemp 1982) is overwhelmingly supported (synapomorphies include loss of the temporal roof, loss of postorbital-squamosal contact, expansion of the parietal to form a sagittal crest, anteroposterior expansion of the epipterygoid, and thickening of the posteroventral dentary to form a trough for the angular; Hopson 1991) and is perhaps the only uncontroversial aspect of higher therapsid phylogeny. The association of gorgonopsians with eutheriodonts to the exclusion of other therapsids, however, is substantially less secure, and several analyses have instead recovered anomodonts as the sister-group of eutheriodonts (Rowe 1986, Gauthier *et al.* 1988, Modesto *et al.* 1999). The primary character supporting Theriodontia, a free-standing coronoid process of the dentary, has been argued to be convergent between gorgonopsians and therocephalians (Kemp 1988), and the precise morphology of this process does vary extensively between gorgonopsians, therocephalians, and cynodonts. However, the most recent analyses of therapsid relationships, with character and taxon sampling far exceeding that of previous studies, have generally upheld the monophyly of Theriodontia (Sidor and Hopson 1998, Sidor 2000, 2001). In general, the Hopson and Barghusen paradigm remains the best-known and supported therapsid topology, and it is used to structure our discussion of the therapsid subclades here. That said, alternative proposals based on cladistic analysis (e.g., Liu *et al.* 2009, Kammerer *et al.* 2013a) plus recent, highly unorthodox and explicitly non-cladistic therapsid classifications (Ivakhnenko 2003b, 2008) indicate that deep relationships in the Therapsida are not yet settled.

5.3.3.1 Biarmosuchia

Biarmosuchia is the most recently recognized major clade of therapsids (Figs. 5.2 G and 5.11). Hopson and Barghusen (1986) initially suggested that the problematic therapsid families Biarmosuchidae, Hipposauridae, Ictidorhinidae, and Burnetiidae might form a monophyletic group, and Sigogneau-Russell (1989) formally proposed the infraorder Biarmosuchia to contain these taxa. The majority of biarmosuchians were historically considered "primitive gorgonopsians" (Boonstra 1952, Watson and Romer 1956, Sigogneau 1970, Ivakhnenko 2002b, 2003b) or occasionally dinocephalians (Broom 1923, Olson 1962, Chudinov 1964). The late recognition of Biarmosuchia as a clade can be attributed to their generalized "primitive" morphology (burnetiamorphs excepted) and a historical lack of interest due to their rarity (the majority of biarmosuchian species are known from a single specimen) (Sidor 2015). Only *Biarmosuchus* itself (Fig. 5.11 A–C) is known from an extensive sample (Ivakhnenko 2003b), although the proposed conspecificity of all this material has been questioned. Ivakhnenko (1999) argued that *Eotitanosuchus* (Fig. 5.11 C) represents the adult of *Biarmosuchus* (Fig. 5.11 B), but Hopson and Barghusen (1986) and Sigogneau-Russell (1989) considered *Eotitanosuchus* to represent a more crownward therapsid lineage, outside of Biarmosuchia, on account of its more expansive attachment site for temporal musculature. Despite their rarity, biarmosuchians have a long stratigraphic range: they first appear in the Guadalupian (middle Permian) (Rubidge 2005) and survive until the end of the Permian (Viglietti *et al.* 2016). The earliest known biarmosuchian is *Biarmosuchus tener* from the Roadian–Wordian of Russia (Sennikov 1996, Newell *et al.* 2010); African biarmosuchians are not known until the Capitanian (Day *et al.* 2015b). Biarmosuchian fossils have thus far only been found in Russia (Ivakhnenko 2003b) and the major Permian basins of sub-Saharan Africa (Malawi, South Africa, Tanzania, and Zambia) (Sigogneau-Russell 1989, Jacobs *et al.* 2005, Sidor *et al.* 2010, Sidor 2015).

Recently, biarmosuchians have become the focus of extensive research attention (Ivakhnenko 1999, 2003b, 2008, Rubidge and Sidor 2002, Rubidge and Kitching 2003, Sidor and Welman 2003, Sidor *et al.* 2004, Rubidge *et al.* 2006, Sidor and Rubidge 2006, Smith and Sidor 2006, Sidor and Smith 2007, Kruger *et al.* 2015, Sidor 2015, Day *et al.* 2016, 2018, Kammerer 2016b), in part because of their general tractability compared with more taxonomically sprawling groups like Gorgonopsia and Dicynodontia. Indeed, in terms of relative worker effort per specimen, no therapsid clade has been so intensively studied.

Biarmosuchians can be subdivided into two primary groups: a grade of "basal biarmosuchians" and the highly apomorphic, deeply nested subclade Burnetiamorpha. Basal biarmosuchians include the former Biarmosuchidae, Hipposauridae, and Ictidorhinidae (these families are no longer in use, as they are now recognized as either monotypic or paraphyletic), and they represent a morphologically conservative assortment of generalized early therapsids (Sidor and Rubidge 2006). Most basal biarmosuchians (e.g., Fig. 5.11 D, G) are small-bodied animals (skull length 10–15 cm), with the sole exceptions being the South African *Hipposaurus* (maximum skull length ~30 cm in the problematic *Hipposaurus brinki*) and the Russian *Biarmosuchus* (estimated maximum skull length ~60 cm, assuming that *Eotitanosuchus* and the similar *Ivantosaurus* are correctly interpreted as adult specimens of *Biarmosuchus*) (Ivakhnenko 1999).

Burnetiamorphs are a bizarre group of biarmosuchians with byzantine skull ornamentation taking the form of various horns and bosses (Sidor and Welman 2003). In the burnetiamorph subclade Burnetiidae, the skull roof is also heavily pachyostosed, obliterating cranial sutures and in some cases even forming a frontoparietal "dome" similar to tapinocephalid dinocephalians or pachycephalosaurian dinosaurs (Rubidge and Sidor 2002, Kammerer 2016b). Kammerer (2016b) considered Burnetiidae to contain two subclades: Proburnetiinae (including *Lende* from Malawi, *Paraburnetia* [Fig. 5.11 E, H] from South Africa, and *Proburnetia* from Russia) and Burnetiinae (including *Bullacephalus*, *Burnetia* [Fig. 5.11 F, I], and *Pachydectes* from South Africa and *Niuksenitia* from Russia). These

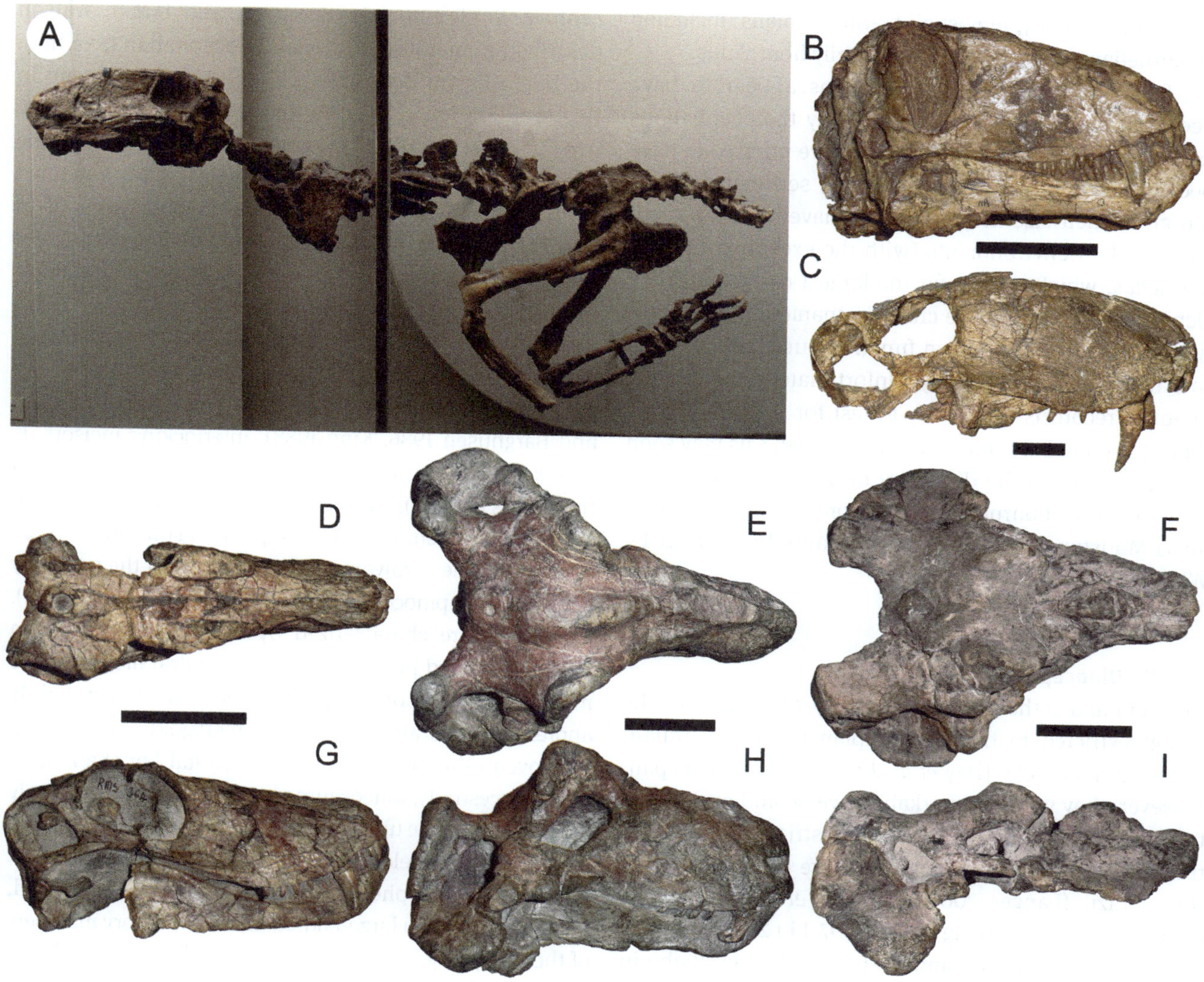

Fig. 5.11: Biarmosuchia. (A) Partial articulated skeleton of subadult *Biarmosuchus tener* (PIN 1758/7, holotype of *Biarmosaurus antecessor*) in left lateral view. (B) Skull of juvenile *B. tener* (PIN 1758/2) in lateral view. (C) Skull of adult *B. tener* (PIN 1758/1, holotype of *Eotitanosuchus olsoni*) in lateral view. (D, G) Skull of *Herpetoskylax hopsoni* (CGP/1/67) in dorsal and lateral views. (E, H) Skull of *Paraburnetia sneeubergensis* (SAM-PK-K10037) in dorsal and lateral views. (F, I) Skull of *Burnetia mirabilis* (NHMUK R5697) in dorsal and lateral views. Specimens in panels B and I mirrored for comparative purposes. Scale bars equal 5 cm.

subfamilies are distinguished by details of cranial boss morphology: proburnetiines are characterized by the presence of a massive interorbital median boss and burnetiines by two pairs of supraorbital bosses (Kammerer 2016b). Recently, however, Day *et al.* (2016, 2018) challenged the monophyly of Burnetiidae by arguing that *Bullacephalus* and *Pachydectes* (which they placed in a new family, Bullacephalidae) fall outside of Burnetiamorpha. Their hypothesis is based on similarity of the palates of these taxa with *Hipposaurus* (interpreted as plesiomorphy), and differences in the details of their cranial ornamentation, which they suggest evolved independently in bullacephalids and burnetiamorphs. The Day *et al.* (2016) phylogeny also better fits the known stratigraphic occurrences of biarmosuchians, although given the extremely poor fossil record of this clade, stratocladistic evidence should be considered somewhat questionable.

Outside of Burnetiamorpha, support for biarmosuchian relationships is relatively low. Only a few synapomorphies support the monophyly of Biarmosuchia, such as a distinct difference in height between the anterior and the posterior portions of the dentary (also present in gorgonopsians, however) and equivalent size of the incisors and postcanines (Sidor and Rubidge 2006). Additional support for biarmosuchian monophyly may be present in the postcranium (as argued by Sidor and Rubidge 2006 and based primarily on the morphologies of *Biarmosuchus* and *Hipposaurus*), but problematically, almost all biarmosuchians (including all burnetiamorphs) are known primarily from skulls.

Paleoecologically, biarmosuchians appear to have been similar to gorgonopsians or early therocephalians (Sennikov 1996). All biarmosuchians are inferred to have been predatory, with well-developed, serrated canines, but given their size they would not have been top predators in their environments (with the exception of *Biarmosuchus*, which was by far the largest carnivore in the Ocher fauna). The bizarre cranial ornamentation of burnetiamorphs is of unknown function but likely played a role in intraspecific display. Unfortunately, the burnetiamorph record is too scanty to test for possible sexual dimorphism in boss morphology, although this may soon change thanks to the discovery of a rich, mostly undescribed fauna of burnetiamorphs from the lower Madumabisa Mudstone Formation of Zambia (Sidor *et al.* 2015, Whitney and Sidor 2016).

5.3.3.2 Dinocephalia

Dinocephalia is the shortest-lived major therapsid clade, being restricted to the Guadalupian (middle Permian) (Figs. 5.2 H and 5.12) (Day *et al.* 2015b). Within that span, however, they were a remarkably diverse and cosmopolitan group, including the largest terrestrial vertebrates of the Permian. Dinocephalian fossils are currently known from Brazil (Langer 2000, Cisneros *et al.* 2012, Boos *et al.* 2015), China (Cheng and Li 1997, Liu 2013), Kazakhstan (Chudinov 1968a, Kammerer 2011), Russia (Riabinin 1932, Ivakhnenko 2003b), South Africa (Rubidge 2005), Tanzania (Simon *et al.* 2010), Zambia (Sidor *et al.* 2014), and Zimbabwe (Boonstra 1946). Dinocephalians include the earliest known therapsids, such as *Parabradysaurus* and *Microsyodon* from the potentially Roadian Golyusherma faunal complex of Russia (Newell *et al.* 2010), and are also abundant components of the earliest therapsid faunas of South Africa (*Eodicynodon* Assemblage Zone [AZ]) and China (Dashankou fauna) (Rubidge 1991, 1994, Li 2001). Dinocephalians were also the earliest therapsids to be described: even before Owen's (1845) description of *Dicynodon* (generally considered the starting point of non-mammalian synapsid research), Kutorga (1838) named three dinocephalian taxa (*Brithopus priscus*, *Orthopus primaevus*, and *Syodon biarmicum*) from the Russian Copper Measures, although he mistakenly believed these taxa to represent edentate mammals. Owen (1876) described the first dinocephalian from South Africa (*Tapinocephalus atherstonei*) (although he considered it a dinosaur at the time), and during the early 20th century, R. Broom, L.D. Boonstra, and D.M.S. Watson described numerous additional dinocephalian taxa from the Karoo Basin. Boonstra (1953, 1969) began to rein in this proliferation of species, but few papers addressing African dinocephalian taxonomy were published in the latter half of the 20th century and the group remained badly oversplit. Only recently have modern revisions of Dinocephalia begun to produce a workable taxonomy for the group (Rubidge and van den Heever 1997, Kammerer 2009, 2011, Güven *et al.* 2013).

Taken as a whole, Dinocephalia is not a particularly well-supported group: the "key character" traditionally used to diagnose dinocephalians (Boonstra 1969, Hopson and Barghusen 1986, King 1988), interlocking incisors, is now known to be more broadly distributed among basal therapsids (Sidor and Welman 2003). Within Dinocephalia, however, two distinctive, well-supported subclades can be recognized: the carnivorous Anteosauria and the primarily herbivorous Tapinocephalia (Hopson and Barghusen 1986). Anteosaurs are characterized by the combination of an upwardly canted premaxilla, convex ventral margin of the maxilla, elongate, overlapping edges of the vomer, tightly appressed quadrate rami of the pterygoids, and strong anteroventral curvature of the postorbital bar (Kammerer 2011). Ancestrally, anteosaurs also possessed an extremely recurved, hooklike upper canine of unknown function (particularly well developed in *Archaeosyodon* and *Syodon*), although this morphology was lost in favor of a serrated, bladelike canine in large-bodied, macropredatory members of the group (e.g., *Anteosaurus*, *Pampaphoneus*, and *Titanophoneus*) (Kammerer 2011, Cisneros *et al.* 2012). Anteosauria contains a single family, Anteosauridae, made up of a few early taxa (*Archaeosyodon* and *Microsyodon*) and two monophyletic subfamilies: Syodontinae and Anteosaurinae. Syodontines (including *Australosyodon* from South Africa, *Notosyodon* from Kazakhstan, *Pampaphoneus* from Brazil, and *Syodon* [Fig. 5.12 C, F] from Russia) are characterized by extremely broad terminal upper postcanines, adjoining palatine bosses, frontal contribution to the pineal boss, and expansion of the attachment site for the adductor musculature onto the pineal boss (Kammerer 2011, Cisneros *et al.* 2012). Anteosaurines (including *Anteosaurus* from South Africa, *Titanophoneus* [Fig. 5.12 A, D, G] from Russia, and possibly *Sinophoneus* from China [although see Liu 2013]) are characterized by enlargement of the postfrontals, a nearly vertical anterior edge of the dentary, and thickening of the skull roof (developed into a pachyostotic dome and massive postfrontal "horns" in *Anteosaurus*) (Kammerer 2011). Anteosaurines include the largest terrestrial predators of the Permian—*Anteosaurus magnificus* reached

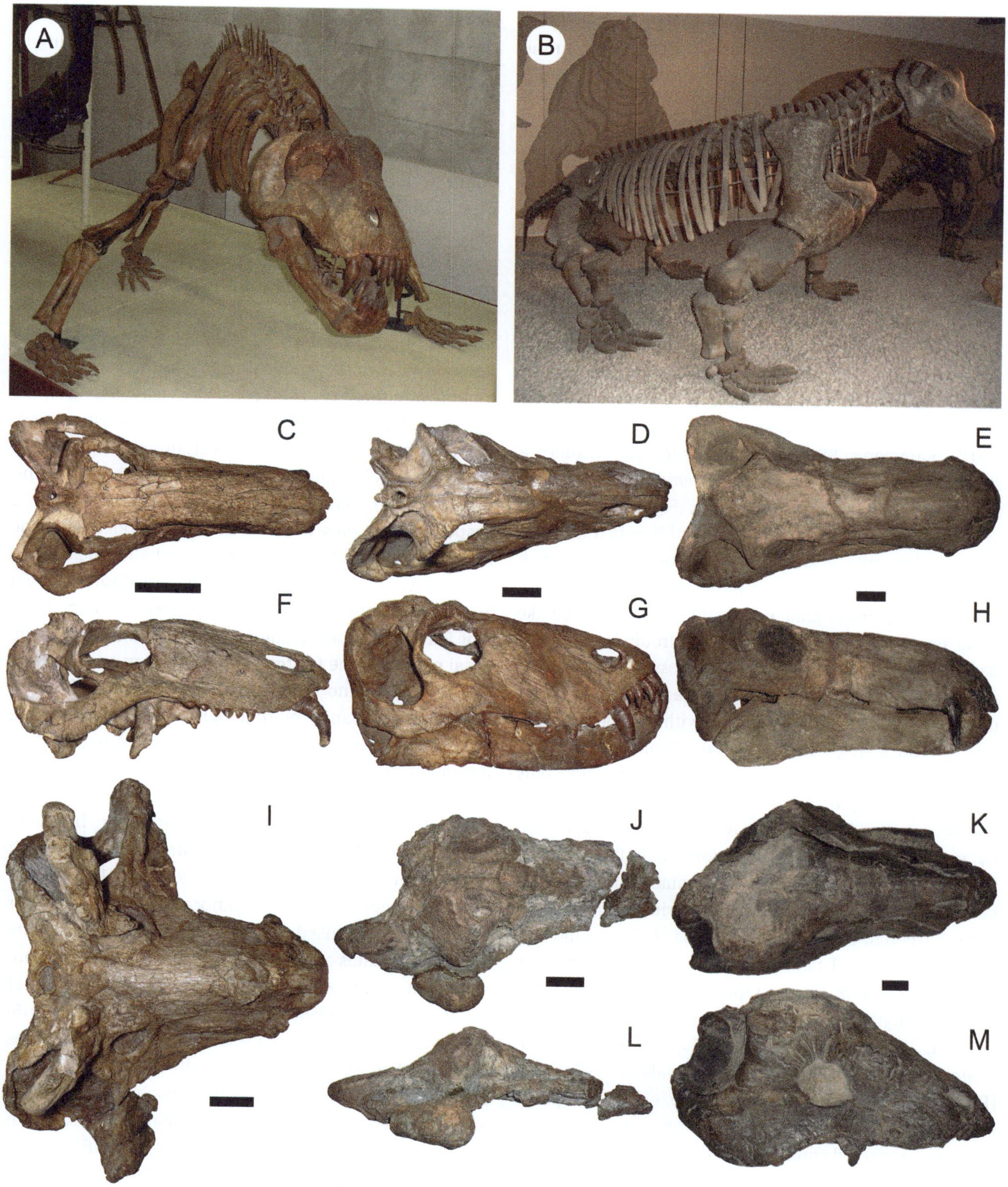

Fig. 5.12: Dinocephalia. (A) Mounted skeleton of adult *Titanophoneus potens* (cast of PIN 157/3, missing elements modeled after PIN 157/1) in anterior view. (B) Mounted skeleton of *Struthiocephalus whaitsi* (GPIT/RE/7102, holotype of *Keratocephalus moloch*) in right lateral view. (C, F) Skull of *Syodon biarmicum* (PIN 157/2) in dorsal and lateral views. (D, G) Skull of subadult *T. potens* (PIN 157/1) in dorsal and lateral views. (E, H) Skull of *Jonkeria truculenta* (TM 212) in dorsal and lateral views. (I) Skull of *Estemmenosuchus mirabilis* (PIN 1758/6) in dorsal view. (J, L) Skull of *Styracocephalus platyrhynchus* (SAM-PK-8936) in dorsal and lateral views. (K, M) Skull of *Tapinocaninus pamelae* (NMQR 2985) in dorsal and lateral views. Specimen in panel F mirrored for comparative purposes. Scale bars equal 5 cm.

a maximum skull length of 80–90 cm (based on the specimen TM 265). They would have been able to prey on the largest herbivores of their time, including pareiasaurs and the giant tapinocephalians.

Tapinocephalians are morphologically diverse compared with anteosaurs, but all possess a unique shelflike posterior process on the transverse process of the pterygoid (Hopson and Barghusen 1986). All tapinocephalians also have leaf-shaped, denticulated postcanines typical of "reptilian" herbivores (Throckmorton 1976), although the anteriormost postcanines are incisiform in tapinocephalids. Tapinocephalia contains four well-circumscribed families, Estemmenosuchidae, Styracocephalidae, Titanosuchidae, and Tapinocephalidae, as well as more enigmatic groups such as Rhopalodontidae and Deuterosauridae (Rubidge and van den Heever 1997, Ivakhnenko 2000, 2003b, Kammerer 2011). Estemmenosuchidae includes a single genus with two species (*Estemmenosuchus uralensis* and *Estemmenosuchus mirabilis*; Fig. 5.12 I) known only from the Wordian Ocher locality of Russia (Chudinov 1960, Ivakhnenko 2003b). Estemmenosuchids are well known for having the most outlandish cranial ornamentation among dinocephalians, combining rhinoceros-like nasal horns, moose-like "antlers", protruding squamosal flanges, and an array of pachyostotic bosses. This group is also unusual in having the postcanine tooth row extending lingual to the canines, uniting with the incisor tooth row to form a single cutting surface (Ivakhnenko 2008). Styracocephalidae is also monogeneric (containing only *Styracocephalus platyrhynchus* [Fig. 5.12 J, L] from the *Tapinocephalus* AZ of South Africa) and characterized by unusual cranial ornamentation, including massive frontal and supraorbital bosses, squamosal flanges, and backward-pointing squamosal horns similar to those of some burnetiamorph biarmosuchians (Rubidge and van den Heever 1997). Titanosuchidae and Tapinocephalidae include the giants among tapinocephalians, huge (4–5 m total length) herbivores with barrel-shaped bodies (Fig. 5.12 B), some of which (e.g., *Tapinocephalus atherstonei*) were the largest terrestrial animals of the Permian (Boonstra 1969). Titanosuchids have elongate, spoon-shaped snouts with numerous (>20) tiny postcanine teeth and large, recurved canines. Two titanosuchid genera are currently recognized based on relative limb proportions (*Jonkeria* [Fig. 5.12 E, H] and *Titanosuchus*) (Boonstra 1969), but associated variation in cranial morphology is uncertain, making it difficult to attribute isolated skulls to genus. Tapinocephalids are the most species-rich and widespread tapinocephalian group, with specimens known from Africa, South America, and Russia (Boonstra 1969, Ivakhnenko 2003b, Boos *et al.* 2015) (tapinocephalids were formerly recorded from China as well, but this material has subsequently been reinterpreted as belonging to pareiasaurs; Liu *et al.* 2014). All tapinocephalids exhibit massively pachyostosed skulls and increased homodonty. The earliest known tapinocephalid, *Tapinocaninus* (Fig. 5.12 K, M) from the South African *Eodicynodon* AZ, retains well-developed canines, and a small canine is also present in the Russian *Ulemosaurus* (Rubidge 1991). In all other tapinocephalids, however, the canine has been lost as a distinct dental morphotype, and the marginal tooth rows consist of "talon and heel"-style incisiform teeth that decrease in size posteriorly but retain the same morphology through what would otherwise be the canine and anterior postcanine regions (the posteriormost part of the tooth row in these taxa still consists of leaf-shaped shearing teeth, however) (Ivakhnenko 2008). Boonstra (1969) divided tapinocephalids into several subfamilies (Moschopinae, Riebeeckosaurinae, Struthiocephalinae, and Tapinocephalinae), but the monophyly of these groups is suspect, as they have not been tested in a modern cladistic framework (Boos *et al.* 2015). Tapinocephalids include the latest-surviving dinocephalians: *Criocephalosaurus* from South Africa extends into the Poortjie Member of the Teekloof Formation, traditionally considered to represent the basal portion of the *Pristerognathus* AZ (Day *et al.* 2015a).

The name Dinocephalia means "terrible heads", and their unusual cranial morphologies have long been the subject of functional discussion (see, e.g., Watson 1914a, Broom 1932). In tapinocephalians, cranial elaborations (horns, bosses, domes) have generally been interpreted as structures used in intraspecific agonistic behavior, similar to that of many extant ungulates. Barghusen (1975; also see Geist 1972, Benoit *et al.* 2016a, 2017a) argued that the domelike skulls of tapinocephalids in particular could have been used in head-butting or flank-butting battles over mates or territory. The development of tapinocephalian cranial elaborations (including pachyostosis of the skull roof) relatively late in ontogeny (Boos *et al.* 2015) supports the idea that they are secondary sexual structures (whether they were used agonistically or just for display). Anteosaurine anteosaurs also have pachyostosed skulls and, in *Anteosaurus* itself, supraorbital "horns", which have been suggested to be useful for intraspecific combat or protection against dense vegetation (Ivakhnenko 2003b). However, Kammerer (2011) noted that similar cranial morphologies (with well-developed supraorbital "horns") are common to all large macropredatory "reptiles", and have been shown to serve as stress sinks during biting in at least thalattosuchian crocodiles (Pierce *et al.* 2009, Young *et al.* 2010). As such, boss development in anteosaurs is likely to be more closely related to protecting the skull from mechanical failure during prey capture than for intraspecific combat (although the possibility of multiple uses cannot be discarded).

5.3.3.3 Anomodontia

In terms of their species richness, numerical abundance, cosmopolitan distribution, and stratigraphic longevity, anomodont therapsids (Figs. 5.2 I and 5.13) (and dicynodont anomodonts in particular) are the most successful clade of synapsids (aside from mammals). There are over 100 well-characterized anomodont species (e.g., see taxon lists for the phylogenetic analyses of Cox and Angielczyk 2015, Kammerer *et al.* 2015a, and Angielczyk *et al.* 2016; Fröbisch 2008 listed 128 species, but that study took place before the major revision of the wastebasket genus *Dicynodon* by Kammerer *et al.* 2011), and in upper Permian strata in the Karoo Basin, dicynodont anomodonts comprise 77%–96% of the specimens discovered (Sidor and Smith 2007, Smith *et al.* 2012). *Biseridens qilianicus*, the oldest and most plesiomorphic known anomodont (Liu *et al.* 2010), co-occurs with *Raranimus dashankouensis* in the Wordian (middle Permian) Dashankou locality of China (Liu *et al.* 2009). The youngest definite anomodont remains are from the Rhaetian (Late Triassic) of Poland (Dzik *et al.* 2008; although see Szulc *et al.* 2015a, 2015b for a suggested older age for the strata in question). Together, this gives Anomodontia a temporal range of about 60 million years, slightly less than the 66 million years of the Cenozoic radiation of therian mammals. Possible dicynodont anomodont remains reported from the Albian (Early Cretaceous) of Australia (Thulborn and Turner 2003) would more than double the temporal range of anomodonts, but the identification of this material is complicated by unexpected morphological features (e.g., presence of a postcanine tooth in addition to a caniniform tusk) and the non-preservation of highly diagnostic elements for dicynodonts such as the quadrate and articular. Ivakhnenko's (2009) hypothesis that monotremes are extant members of Anomodontia is not widely accepted. Anomodont fossils have been found on every continent, with particularly diverse and important records coming from southern and eastern Africa (Fröbisch 2009).

Although fragmentary dinocephalian remains from Russia were described earlier (as mammals; Kutorga 1838), the anomodont *Dicynodon lacerticeps* Owen 1845 was the first non-mammalian synapsid to be recognized as representing a previously unknown lineage of tetrapods. Owen (1860a) erected the name Anomodontia to include *Dicynodon*, *Oudenodon*, *Ptychognathus* (now *Lystrosaurus*; see Seeley 1898), and the rhynchosaur *Rhynchosaurus*. Aside from including rhynchosaurs, which are now recognized as part of the sauropsid clade Archosauromorpha (e.g., Ezcurra 2016), this usage is close to the modern concept of Anomodontia (i.e., dicynodonts and closely related taxa). However, the content of Anomodontia became quite confused during the late 19th century as additional non-mammalian synapsids were discovered and attempts were made to classify them, such that Anomodontia could refer to a group as narrowly circumscribed as Dicynodontia or as wide as all Synapsida plus certain basal amniote and reptile groups (see reviews in Broom 1905a, Kammerer and Angielczyk 2009). Following Broom's (1905a) work, usage of Anomodontia continued to be divided, with some authors using it essentially as a synonym of Synapsida (e.g., Watson 1917a, 1921), whereas others (e.g., Williston 1925, Broom 1932) used it in a much more restricted sense that was basically equivalent to Dicynodontia. Romer and Price (1940) argued that Synapsida was the least ambiguous, and therefore preferred, name to use for the group comprising "pelycosaurs" and therapsids, and the subsequent usage of Anomodontia stabilized around a more restricted concept. Building on previous suggestions of a close relationship between dinocephalians and anomodonts (e.g., Broom 1932), and new data on the basal anomodont *Venyukovia invisa* (now *Ulemica invisa*; see Ivakhnenko 1996) presented by Efremov (1938, 1940a), Watson (1942, 1948) posited that dicynodonts were derived from tapinocephalid dinocephalians by way of an *Ulemica*-like ancestor. Watson and Romer (1956) and Romer (1956) further codified grouping dinocephalians and dicynodonts together within Anomodontia. This arrangement was questioned at times (e.g., Olson 1962, Boonstra 1972, Barghusen 1976) but was followed in the most recent monographic treatment of dinocephalians and dicynodonts (King 1988). Early cladistic studies of therapsids (Hopson and Barghusen 1986, Rowe 1986, Gauthier *et al.* 1988, Hopson 1994, also see Kemp 1988, Hopson 1991, Grine 1997) did not find strong evidence of dinocephalians as a subgroup of anomodonts, however, and this result has been upheld by more recent analyses (e.g., Sidor and Hopson 1998, Sidor 2000, Liu *et al.* 2009, Amson and Laurin 2011, Brink *et al.* 2015). Current usage of Anomodontia restricts the clade to taxa more closely related to the dicynodont *D. lacerticeps* than to dinocephalians, biarmosuchians, gorgonopsians, or therocephalians (Kammerer and Angielczyk 2009). Because of the great morphological disparity between dicynodonts and non-dicynodont anomodonts (Ruta *et al.* 2013b), relatively few synapomorphies diagnose Anomodontia. For example, Kammerer and Angielczyk (2009) included three synapomorphies in their diagnosis

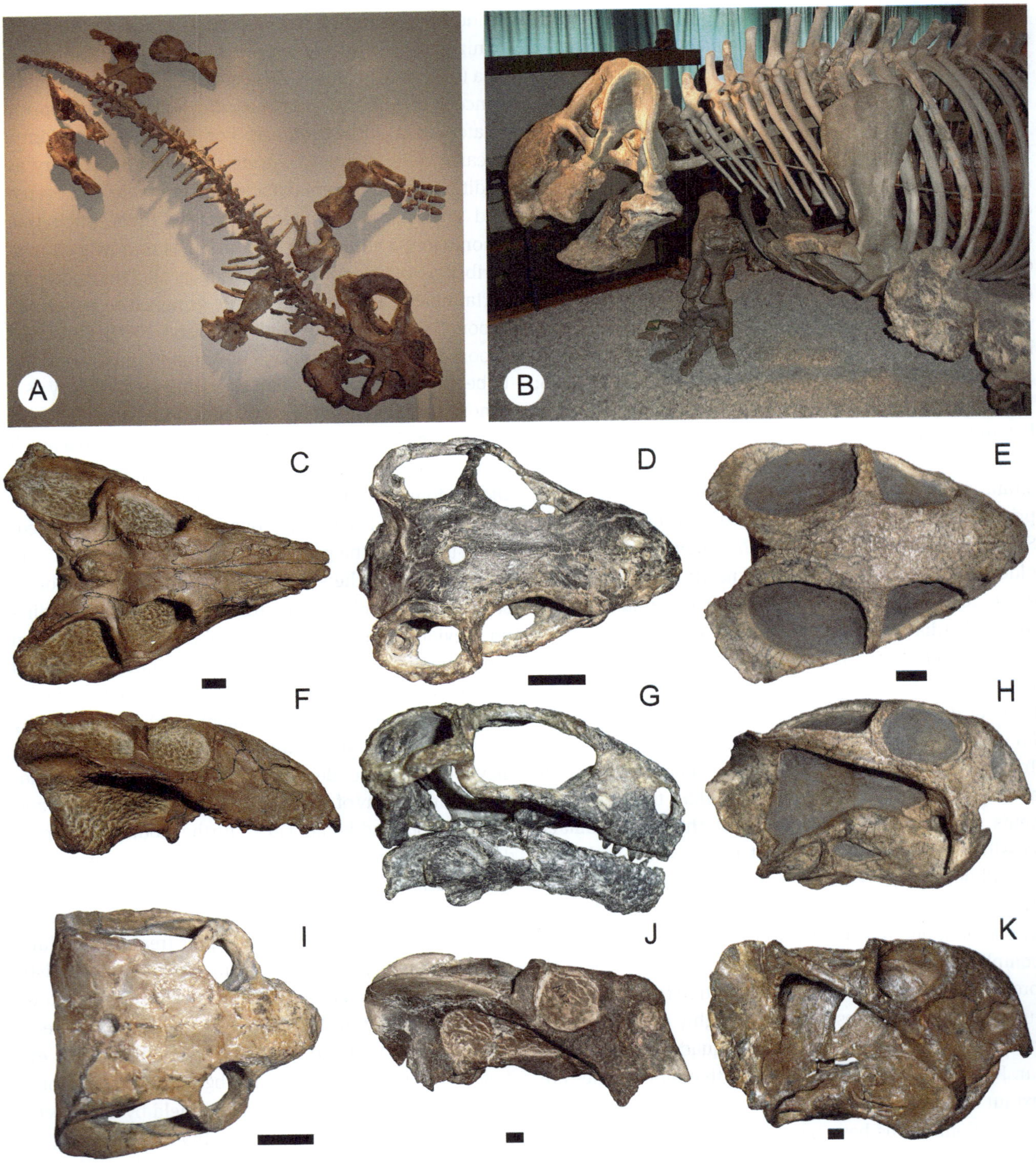

Fig. 5.13: Anomodontia. (A) Mounted skeleton of *Dicynodon huenei* (Museum Korbach, cast of NHMUK R37005) in dorsal view. (B) Mounted skeleton of *Stahleckeria potens* (GPIT/RE/7106) in left lateral view. (C, F) Skull of *Ulemica efremovi* (PIN 2793/1) in dorsal and lateral views. (D, G) Skull of *Patranomodon nyaphulii* (NMQR 3000) in dorsal and lateral views. (E, H) Skull of *Diictodon feliceps* (MB.R.1000) in dorsal and lateral views. (I) Skull of *Cistecephalus microrhinus* (SAM-PK-K7852) in dorsal view. (J) Skull of *Geikia locusticeps* (GPIT/RE/7186) in lateral view. (K) Skull of *Peramodon amalitzkii* (PIN 2005/38) in lateral view. Scale bars equal 1 cm.

of Anomodontia (absence of serrations on marginal dentition; mandibular fenestra present; zygomatic arch bowed dorsally), and Liu *et al.* (2010) listed five (short snout; elevated zygomatic arch; absence of stapedial foramen; dorsal notch of reflected lamina of angular near dentary; absence of serrations on marginal dentition).

The vast majority of anomodont species belong to the subclade Dicynodontia, but there are also 13 species of non-dicynodont anomodonts ("basal anomodonts") that form a paraphyletic assemblage outside of Dicynodontia (e.g., Modesto *et al.* 1999, Modesto and Rybczynski 2000, Angielczyk 2004, Fröbisch 2007, Liu *et al.* 2010, Kammerer *et al.* 2011). The "basal anomodonts" are limited to the middle Permian and the beginning of the late Permian (e.g., Fröbisch 2008) and have been found in Brazil (Cisneros *et al.* 2011, 2015), China (Li and Cheng 1997, Liu *et al.* 2010), Russia (Ivakhnenko 1994, 1996, Rybczynski 2000, Fröbisch and Reisz 2011, Kurkin 2017), and South Africa (Brinkman 1981, Rubidge and Hopson 1996, Modesto *et al.* 1999, Modesto and Rubidge 2000). Despite their low species richness, "basal anomodonts" are noteworthy for a number of reasons. First, they display a great deal of dental disparity. For example, *Biseridens* presents recurved canines that are oval in cross section, paired rows of rounded, heeled marginal teeth, and extensive palatal dentition (Liu *et al.* 2010). *Suminia* possesses enlarged, procumbent incisiform teeth, no distinct canines, and cingulated, coarsely serrated maxillary and dentary teeth that occluded extensively during chewing (Rybczynski 2000, Rybczynski and Reisz 2001). *Tiarajudens* has an extremely distinctive dentition comprised of leaf-shaped incisiform teeth, huge, laterally compressed saber-like canines and expanded, molariform palatal teeth (Cisneros *et al.* 2011, 2015). *Patranomodon* (Fig. 5.13 D, G) has a reduced dentition of simple peglike teeth (Rubidge and Hopson 1990, 1996). Most of this dental disparity appears to represent various experiments in oral processing of food related to high-fiber herbivory (Reisz and Sues 2000, Rybczynski and Reisz 2001, Reisz 2006, Cisneros *et al.* 2011, 2015), which is ironic considering that the most successful clade of anomodonts, the dicynodonts, largely replaced the dentition with a keratinous beak.

A second area of interest for basal anomodonts is the evolution of a palinal (anterior-posterior sliding) motion of the lower jaw (note that this is frequently referred to as propaliny in the anomodont literature, but palinal is a more accurate term because of the posteriorly directed power stroke). Dicynodont anomodonts show extensive modifications of the bony and muscular anatomy of the skull and mandible to emphasize a palinal motion of the jaw when feeding (Watson 1948, Crompton and Hotton 1967, Barghusen 1976, King 1981, 1994, King *et al.* 1989). The jaw joint morphology of some non-dicynodont anomodonts, such as *Patranomodon*, indicates that they were only capable of an orthal (up-and-down) motion of the jaw (King 1994, Rubidge and Hopson 1996), but evidence of at least limited paliny has been noted in *Ulemica* (Fig. 5.13 C, F), *Suminia*, Parasuminia, and *Galeops* (King 1994, Rybczynski 2000, Rybczynski and Reisz 2001, Angielczyk 2004, Kurkin 2017). The phylogenetic position of *Galeops*, just outside of Dicynodontia (e.g., Angielczyk 2004, Fröbisch and Reisz 2011, Cisneros *et al.* 2015, Angielczyk *et al.* 2016), suggests that its features associated with a palinal motion of the jaw likely are homologous with those found in dicynodonts. By contrast, the cranial morphology and distant phylogenetic position of the venyukovioids (*Suminia, Parasuminia, Ulemica, Otsheria, and Venyukovia*) suggest that they evolved a palinal motion of the jaw independently of dicynodonts (Angielczyk 2004). If this is the case, it would suggest that a feeding system characterized by palinal chewing was not the sole reason for the success of the dicynodonts because venyukovioids never underwent a comparable evolutionary radiation (i.e., there are only six venyukovioid species, of which only *Suminia getmanovi* presents evidence of an extensive palinal motion of the jaw).

Finally, "basal anomodonts" make a unique contribution to the ecomorphological diversity of non-mammalian synapsids. Dicynodont species span a wide range of body sizes, but their body forms are morphologically conservative, with robust, stocky limbs, barrel-shaped torsos, and short tails. This contrasts with the majority of basal anomodonts, which have lightly built skeletons with gracile limbs (Watson and Romer 1956, Brinkman 1981, Fröbisch and Reisz 2011) (*Tiarajudens* and likely *Anomocephalus* are larger and more dicynodont-like in appearance; Cisneros *et al.* 2015). The postcranial skeleton of *Suminia getmanovi* is particularly interesting because it presents a number of characters that suggest a clinging arboreal lifestyle, including long limbs; large manus and pes; long, slender digits with particularly elongated penultimate phalanges; divergent first digits on the manus and pes, suggesting grasping abilities; tall, laterally compressed unguals; and a potentially prehensile tail (Fröbisch and Reisz 2009, 2011). *Suminia's* presence in the late Permian makes it not only the oldest arboreal therapsid but also one of the oldest known tetrapods with extensive specializations for this lifestyle

Dicynodontia is the major subclade of Anomodontia. The alpha taxonomy of dicynodonts has a highly convoluted history, with a great deal of superfluous names having been created, and several factors contributed to this problem. The longest-studied and most extensive fossil record of dicynodonts is found in the middle Permian to Middle Triassic rocks of the Beaufort Group in the Karoo Basin of South Africa (e.g., Rubidge 2005). Although abundant, fossils in the Karoo often have undergone varying degrees of plastic deformation and are usually preserved in very hard matrix that does not separate cleanly from the bone when using older preparation techniques. Because of this, the majority of older dicynodont holotypes are poorly preserved and prepared specimens that obscure many important morphological features and patterns of inter- and intraspecific variation. Such taphonomic artifacts and intraspecific variabilities were not widely considered by early workers, with the result that at times nearly every specimen discovered became the holotype of a new species, if not a new genus. Robert Broom is particularly notorious for his profligate naming practices, erecting 162 anomodont species (out of a total of 369 therapsid species he named; Wyllie 2003). This can be excused to some degree because it is clear that Broom thought the Beaufort Group represented a much longer time period than is actually the case, and based on this expected amount of time, he predicted that millions of therapsid species should have existed in South Africa (e.g., Broom 1932). An additional logistical problem facing dicynodont taxonomists was travel. Karoo specimens were deposited in collections spread across Africa, Europe, and North America, with additional specimens collected in basins in North and South America, Africa, and Asia housed in a variety of local and foreign museums. Until recently, few researchers had the resources or time to examine these far-flung collections first-hand, and instead had to rely on often cursory descriptions of species in the literature for comparisons with their material.

Beginning with the work of Toerien (1953), much greater attention has been paid to intraspecific variation and the effects of taphonomy in dicynodont taxonomy. When combined with significant advances in fossil preparation and the greater ease with which geographically dispersed collections can be visited, there have been substantial efforts to revise dicynodont taxonomy (e.g., Cox 1964, Keyser 1973a, b, 1975, 1993, Tollman *et al.* 1981, Cluver and Hotton 1981, Cluver and King 1983, King 1993, King and Rubidge 1993, Renaut 2000, Angielczyk 2002, Sullivan and Reisz 2005, Grine *et al.* 2006, Botha and Angielczyk 2007, Fröbisch and Reisz 2008, Angielczyk *et al.* 2009, 2016, Kammerer *et al.* 2011, 2015a, c, Cox and Angielczyk 2015) that have dramatically reduced the number of valid species and clarified their phylogenetic relationships, as well as their stratigraphic and geographic distributions. However, the majority of this revisionary work has focused on Permian dicynodonts, and there are still important problems in Triassic dicynodont alpha taxonomy that require attention (e.g., the number of valid species and geographic distributions of Russian and Asian kannemeyeriiforms). A full review of the numerous dicynodont species descriptions that exist is beyond the scope of this chapter, but King (1988) provides a useful entry point into this extensive literature.

Early work on the systematics of dicynodonts primarily focused on grouping dicynodont species into higher-level taxa, with relatively little consideration of relationships among the species included in a given taxon or the relationships among the various taxa (e.g., Owen 1860a, 1876, Seeley 1894a, Broom 1903a, 1905a). van Hoepen (1934) presented the first detailed treatment of relationships within Dicynodontia, and he recognized important structural differences in the palates of taxa such as *Endothiodon* and more *Dicynodon*-like forms. Based on these and other distinctions, he proposed that endothiodonts represented a line of descent that was separate from other, more typical dicynodonts (i.e., that Dicynodontia was not monophyletic). Although this view has not been upheld by subsequent phylogenetic analyses, many of the subgroups in van Hoepen's classification are similar to dicynodont clades recovered in more recent cladistic analyses (e.g., Emydopoidea and Pylaecephalidae; Kammerer and Angielczyk 2009).

A second milestone in dicynodont systematics was the work of Toerien (1953). Not only was this one of the first papers to consider intraspecific and taphonomic variation in greater detail than most previous works, it also included a classification whose family- and subfamily-level groupings closely match the results of most recent cladistic analyses (e.g., Kammerer *et al.* 2011, 2013b, 2015a, Castanhinha *et al.* 2013, Angielczyk *et al.* 2016, Olroyd *et al.* 2018). However, the pattern of relationships Toerien proposed for these higher-level clades also differs in many ways from modern phylogenies. For example, he considered all higher taxa that lack "postcanine" teeth to have evolved independently from toothed ancestors, whereas current phylogenies reconstruct most toothed dicynodonts in a series of clades near the base of Dicynodontia. Other important precladistic works on dicynodont classification and phylogeny include Camp (1956), Boonstra (1963, 1972), Cox (1965), Keyser and Cruickshank (1979), and Cox and Li (1983). Aside from Boonstra (1963, 1972), these latter papers focused on relationships among Triassic dicynodonts and their derivation from one or more Permian lineages. Interestingly, Watson and Romer (1956) did not attempt a detailed classification of dicynodonts in their treatment of therapsid taxonomy because they considered the group's alpha taxonomy too confused to allow an accurate assessment of relationships.

The first cladistic analyses of dicynodonts were hand-drawn cladograms that used basic cladistic principles (Cluver and King 1983, King 1988, 1990, Cox 1998). The analyses of Cluver and King (1983) and King (1988, 1990) were more detailed than that of Cox (1998) and listed synapomorphies for several major groups of dicynodonts (using the taxonomy of Kammerer and Angielczyk 2009 and Kammerer *et al.* 2013b: Endothiodontia, Pylaecephalidae, Emydopoidea, Cryptodontia, Dicynodontoidea, and Kannemeyeriiformes). They also proposed a topology in which *Eodicynodon* and Endothiodontia were the sister taxa of all other dicynodonts, and a major dichotomy between Pylaecephalidae + Emydopoidea and Dicynodontoidea (including Kannemeyeriiformes) + Cryptodontia. The first decade of the 21st century saw a number of computerized cladistic analyses of dicynodonts (e.g., Angielczyk 2001, 2002, 2007, Maisch 2001, 2002a, Modesto *et al.* 2002, Angielczyk and Kurkin 2003a, Vega-Dias *et al.* 2004, Fröbisch 2007, Govender and Yates 2009, Angielczyk and Rubidge 2010), but most of these analyses only focused on parts of the dicynodont tree (i.e., primarily Permian taxa with only a handful of Triassic taxa or vice versa) and were based on comparatively few characters. These analyses generally corroborated the main clades proposed by the earlier hand-drawn phylogenies but also found some differences in their relationships (e.g., Pylaecephalidae and Emydopoidea usually were not recovered as sister taxa).

Kammerer *et al.* (2011) presented the first comprehensive cladistic analysis of dicynodonts, with extensive sampling of both Permian and Triassic taxa, and nearly all subsequent analyses have been modifications of this data set (e.g., Castanhinha *et al.* 2013, Kammerer *et al.* 2013b, 2015a, 2016, Cox and Angielczyk 2015, Angielczyk *et al.* 2016, Boos *et al.* 2016, Kammerer and Smith 2017, Olroyd *et al.* 2018, Kammerer 2018). The comprehensive analyses have found support for most of the major clades of Permian dicynodonts noted in previous works (although Boos *et al.* 2016 did not recover a monophyletic Cryptodontia) and have been consistent in a rough three-fold division of Triassic taxa into the clades Shansiodontidae and Stahleckeriidae, which are separated by a paraphyletic group of “kannemeyeriids”. Although the results of all of the most recent cladistic analyses are generally compatible with each other, support values across the trees are uniformly very poor, an issue that has plagued computerized cladistic analyses of dicynodonts from the start (e.g., Angielczyk 2001). The problem seems to arise from a combination of morphological conservatism in the basic dicynodont bauplan and significant homoplasy in various regions of the skeleton, particularly the skull. The addition of taxa from analysis to analysis has not improved the problem, so further refinement and expansion of the character sets underlying the analyses will likely be necessary. Recent reviews of some of the persistent problems in dicynodont phylogenetics can be found in Angielczyk and Kammerer (2017) and Angielcyk *et al.* (2018).

The phylogenies also highlight a significant gap in the dicynodont fossil record. The Middle and Late Triassic radiation of kannemeyeriiform dicynodonts (shansiodontids, stahleckeriids, and “kannemeyeriids”) represents the last major phylogenetic diversification event in dicynodont history. However, almost no kannemeyeriiform fossils are known from the Early Triassic (the recently described *Sungeodon* may be an exception; Maisch and Matzke 2014), resulting in a kannemeyeriiform ghost lineage that extends back into the late Permian. The persistence of the Early Triassic “kannemeyeriiform gap”, despite a well-sampled Early Triassic dicynodont fossil record from several continents (e.g., Fröbisch 2009), implies that the earliest parts of kannenmeyeriiform history occurred in a currently unsampled geographic area. The recent description of fossils from Antarctica that fill a different Triassic dicynodont ghost lineage (that of the genus *Kombuisia*, a relictual Middle Triassic emydopoid; Fröbisch *et al.* 2010) provides reason to hope that the puzzling absence of Early Triassic kannemeyeriiforms may eventually be resolved through continuing exploration of traditionally understudied geographic areas.

The dicynodont skull is highly modified compared with the basic therapsid ground plan, and even to the morphologies found in basal anomodonts (Fig. 5.2 I and 13), to accommodate a palinal motion of the mandible associated with the adoption of high-fiber herbivory (for comparisons of dicynodonts to other early synapsid herbivores, see King 1996, Reisz and Sues 2000, Reisz 2006). These modifications include a significantly shortened snout, a partial bony secondary palate (formed primarily by the fused premaxillae) that supported a keratinous beak, the reduction or complete loss of the dentition, the extreme reduction of the transverse flanges of the pterygoids, extensive emargination of the temporal openings, presence of a large lateral fossa on the squamosal hypothesized to accommodate a neomorphic jaw adductor muscle mass (M. adductor mandibulae externus lateralis; see Crompton and Hotton 1967, Barghusen 1976, King *et al.* 1989, King 1994, Angielczyk 2004), a jaw joint that allows extensive translation of the articular relative to the quadrate, and fusion of the mandibular symphysis. The evolutionary trajectory of the dicynodont skull and jaw deviates sharply from the general trend in Synapsida for the evolution of progressively more mammal-like morphologies (e.g., see Sidor 2003 for an example considering mandible morphology; Benoit *et al.* 2018 describe extensive changes to facial innervation patterns associated with evolution of the beak). As such, dicynodonts represent a striking (and successful) evolutionary experiment within Synapsida that was not paralleled until the evolution of highly disparate groups of placental mammals such as cetaceans. It is also no surprise that Dicynodontia is diagnosed by a large number of synapomorphies in most phylogenetic analyses (Kammerer and Angielczyk 2009 listed 15 commonly cited synapomorphies for Dicynodontia). Although the "basal anomodonts" help to bridge the gap between dicynodonts and other therapsids (e.g., Watson 1948, Barghusen 1976, King *et al.* 1989, King 1994), the earliest known dicynodont (*Eodicynodon* Barry 1974, Rubidge 1984, 1990) shows considerably greater similarity to later dicynodonts than it does to any "basal anomodont". This implies that we are missing the earliest parts of dicynodont history, and the potential exists that additional fossils will help clarify the evolution of many of the distinctive features of the clade. It is interesting to note that despite their extreme commitment to paliny early in their evolutionary history, some Triassic dicynodonts subsequently reemphasized an orthal movement of the mandible (Crompton and Hotton 1967, Cluver 1971, King and Cluver 1991, Renaut 2001).

The function of the dicynodont feeding system has long been an area of interest. Watson (1912a) was the first to describe the dicynodont articular in detail, and he realized that it was modified to allow extensive anteroposterior sliding relative to the quadrate in addition to more traditional orthal movements of the mandible. He later examined the functional morphology and evolutionary history of the dicynodont feeding system in a large comparative study (Watson 1948) that formed the first detailed treatment of the topic. Among the results that Watson (1948) presented are a model of the jaw musculature of dicynodonts; a comparison of the feeding systems of the dicynodonts *Endothiodon*, *Pristerodon* (which he called *Synostocephalus*), *Diictodon* (called *Dicynodon*), *Tropidostoma*, and *Kannemeyeria*; and a hypothesis of how the morphologies present in dicynodonts could have evolved from those of other therapsids by way of a venyukovioid ancestor. Crompton and Hotton (1967) expanded on this work by providing a more quantitative analysis of the dicynodont feeding system and also investigating the functional implications of the distinctive skull morphology of *Lystrosaurus*. Although other authors have considered feeding system function in other dicynodonts (e.g., Cox 1959, 1998, Cluver 1970, 1975, King 1981, Hotton 1986, King and Cluver 1991, Renaut 2001, Maisch 2003, Morato 2006, Jasinoski *et al.* 2009, 2010a, Cox and Angielczyk 2015, Angielczyk *et al.* 2018) and its evolution from "basal anomodont" ancestors (e.g., Barghusen 1976, King *et al.* 1989, King 1994, Rybczynski 2000, Angielczyk 2004), the work of Watson, Crompton, and Hotton still forms the foundation for our understanding of the system and few other studies have approached their level of detail. The latter point is particularly important because it is clear that much of the variation in skull shape and anatomy used in studies of dicynodont taxonomy and phylogeny is also of functional importance, but the quantitative details of these differences are largely unknown. In turn, this lack of firm information obscures how the diversity of dicynodonts present in some late Permian ecosystems partitioned resources in a way that allowed their co-existence (although see Hotton 1986 for an attempt to address the problem). Surkov and Benton (2008) provided some insight into the question of resource partitioning by showing that differences in occipital proportions among dicynodonts could have corresponded to the heights at which different species were feeding and their styles of feeding. Although the palinal feeding system of dicynodonts has been proposed to explain their success in the Permian and Triassic (King *et al.* 1989, King 1990), Angielczyk (2004) questioned whether this was the case because of the diversity difference observed between palinal dicynodont and palinal "basal anomodonts". Ruta *et al.* (2013b) even went so far as to suggest that it might have served as a constraint on their evolution.

Regardless of whether their feeding system facilitated or constrained the diversification of dicynodonts, their status as high-fiber herbivores is noteworthy. Although there is evidence of older communities characterized by high tetrapod herbivore abundances (e.g., the Bromacker quarry; Berman *et al.* 2014), it has long been recognized that the dicynodont-dominated communities of the middle and late Permian represent the first well-documented, widespread instances of terrestrial ecosystems characterized by a high diversity and abundance of tetrapod herbivores (e.g., Olson 1966, 1971, DiMichele *et al.* 1993, Hotton *et al.* 1997, Sues and Reisz 1998, Reisz and Sues 2000, Reisz 2006, Nicolas and Rubidge 2010, Smith *et al.* 2012, Pearson *et al.* 2013). This type of community structure is widely regarded as "modern" in aspect in the paleoecological literature and is contrasted with older tetrapod assemblages in which faunivores are the most diverse and abundant members of the system and herbivore species are few and rare. In this way, the dicynodont radiation stands at a key transition in the evolution of terrestrial ecosystems. Nevertheless, it is important to note that dicynodont-dominated communities appear to have been most characteristic of high-latitude regions, with reptilian herbivores like pareiasaurs and captorhinids playing a more important role in low latitude assemblages (Bernardi *et al.* 2017).

The word dicynodont means "two canine teeth" and refers to the pair of enlarged caniniform tusks found in the maxillae of most members of the clade (Figs. 5.2 I and 5.13) (the informal name "bidental" and the formal name

Bidentalia, which also have been applied to the group, have a similar etymology; see Bain 1845, Owen 1876, Kammerer and Angielczyk 2009). Owen (1845) described the structure of the tusks in the first paper on dicynodonts, reporting that they had an open pulp cavity, were primarily composed of dentine with very thin layers of enamel and cementum, and that the dentine displayed concentric light and dark bands in cross section. Subsequent investigators questioned the presence of enamel in dicynodont tusks, showed that they displayed the cone-in-cone style of growth typical of most mammal tusks, and concluded that the tusks grew throughout life without replacement (e.g., Poole 1956, Camp and Welles 1956, Watson 1960, Cluver 1971, King 1981, Hotton 1986; although Camp 1956 suggested that tusks were periodically replaced in primitive dicynodonts and perhaps some kannemeyeriiforms). The macroscopic banding patterns in dicynodont tusks were suggested to represent periodic growth features (Camp and Welles 1956, Cox 1968, Cluver 1971), and detailed examination of the microstructure of dicynodont tusks has shown that the macroscopic features can be resolved into much finer growth increments whose thicknesses correspond well to Lines of von Ebner (Thackeray 1991, Green 2012, Jasinoski and Chinsamy-Turan 2012), which are daily growth increments in extant vertebrates (Hoffman and Schour 1940, Erickson 1996). A series of dicynodont tusks from the late Permian and Early Triassic of South Africa and Zambia include from about two months to about two (Permian) years of daily growth increments (Green *et al.* 2017), and Thackeray (1991) noted a trend for declining growth increment thicknesses in *Diictodon* tusks across the late Permian, suggesting that dicynodont tusks have the potential to preserve a finely resolved record of the interactions between growth rates and environmental conditions in these therapsids. However, recent work on *Diictodon* contradicts much of the accepted picture of the nature of dicynodont tusks. In a specimen of *Diictodon* that they sampled, Reisz *et al.* (2015) reported the presence of a thin enamel layer, a calcified periodontal ligament that would prevent continuous growth, and a piece of a prior generation of tusk, indicating that tusks were periodically replaced. It remains to be seen how widely these observations apply among other dicynodonts, but they would help to explain rare observations of double-tusked dicynodont specimens (Jinnah and Rubidge 2007, Fröbisch and Reisz 2008), specimens with apparent fragments of tusks in addition to a functional tusk (Camp 1956), and specimens that possess a tusk on one side of the skull but not the other (Angielczyk 2002, Botha and Angielczyk 2007). Although the saber-like canines of the basal anomodont *Tiarajudens* are superficially similar to dicynodont tusks, they differ from the latter by having a reniform instead of round cross section, being nearly straight instead of weakly recurved, and possibly in having an enamel layer (Cisneros *et al.* 2011, 2015). The phylogenetic distribution of enlarged caniniforms in anomodonts also indicates that the saber teeth of *Tiarajudens* evolved separately from the tusks of dicynodonts (Fröbisch 2011).

The function of dicynodont tusks has long been a subject of speculation. Owen (1845) initially suggested that they were used as offensive or defensive weapons (also see Broom's [1932, p. 253] colorful description of a specimen of the archosauriform *Proterosuchus* that he posited was "torn to pieces" by *Lystrosaurus*). As tuskless dicynodont specimens were discovered, there was much discussion of their potential as indicators of sexual dimorphism in the clade (Owen 1860b, 1876, Lydekker 1890, Broom 1902), and Broom's (1912b, 1932) conclusion that some species were always tusked, some always tuskless, and others sexually dimorphic for tusks has proven largely correct (although note that Broom's hypothesis that the genus *Oudenodon* comprised female individuals of *Dicynodon* is in error; see, e.g., Keyser 1975, Cluver and Hotton 1981, Kammerer *et al.* 2011). The analysis of *Diictodon feliceps* by Sullivan *et al.* (2003; also see Angielczyk and Sullivan 2008) is the most statistically rigorous demonstration of tusk dimorphism in a dicynodont. Most other cases of potential dimorphism are based on smaller samples and/or have not been tested statistically (e.g., Kammerer *et al.* 2015c, Angielczyk *et al.* 2016). Wear facets are frequently present on dicynodont tusks (van Hoepen 1934, Watson 1960, Cluver 1971, King 1981, Hotton 1986). A prominent medial (lingual) wear facet is common, presumably formed as the keratinous beak on the mandible slid past the tusk, but wear facets on the anterior, posterior, and lateral surfaces of the apical end of the tusk also may be present. The presence of these wear facets suggests that dicynodont tusks served a function or functions beyond intraspecific display, perhaps grubbing or other foraging behaviors (Hotton 1986; although see Watson and Romer 1956, Green 2009). However, given that some dicynodont species lacked tusks, others were sexually dimorphic for this character, and that the presence and development of wear facets is variable within tusked individuals, it is likely that their use in foraging or similar activities was not universal in the clade.

Like "basal anomodonts", dicynodonts make an important contribution to the ecomorphological diversity present in non-mammalian therapsids. Dicynodonts with very generalized postcranial skeletons, such as *Diictodon* or *Lystrosaurus*, were capable of producing burrow structures (Smith 1987, Groenewald 1991, Retallack *et al.* 2003,

Bordy *et al.* 2011; Botha-Brink 2017; but see Modesto and Botha-Brink 2010), but the late Permian cistecephalid dicynodonts appear to have taken this behavior to an extreme by evolving a scratch-digging fossorial lifestyle. Named by Owen (1876), *Cistecephalus microrhinus* (Fig. 5.13 I) was the first cistecephalid to be described and represents the vast majority of known cistecephalid specimens. *Cistecephalus* was initially suggested to have been aquatic by Broom (1948) and Brink (1950), but these were passing comments based on little hard data other than Brink's observation that the skulls of aquatic turtles were roofed over in a manner somewhat analogous to the skull of *Cistecephalus*. Brink (1952) and Cox (1959) argued for a burrowing lifestyle for *Cistecephalus* on account of the structure of the manus and forelimb (although Brink's interpretation of the manus was subsequently shown to be incorrect; see Cluver 1978), the unusual, boxlike skull, and the absence of a process on the occiput that Cox (1959) thought supported a tympanum (implying reduced auditory capabilities). Keyser (1973b) hypothesized that the forward-facing eyes, small snout, and changes in braincase morphology in the vicinity of the optic lobe were indicative of stereopsis in *Cistecephalus*. He used these observations along with a potentially opposable first digit on the manus to suggest that *Cistecephalus* was a squirrel-like animal that foraged in trees but also spent time in burrows. Cox (1972; also see von Huene 1942) described the skull and forelimb of *Kawingasaurus fossilis* from Tanzania and analyzed the functional morphology of the forelimb. He found that the forelimb was characterized by a number of derived features, including a more laterally facing glenoid than in most dicynodonts, very well ossified joint surfaces, enlarged areas for muscle attachment on the humerus, and a very large olecranon process, leading him to conclude that the limb was highly specialized for powerful digging. Cluver (1974) noted that the rounded, convex occiput of cistecephalids resembled those of extant fossorial mammals such as the golden mole, and both Cox (1972) and Cluver (1974) remarked on the unusually small orbits of *Kawingasaurus* and *Cistecephaloides*. The first detailed description of the postcranial skeleton of *Cistecephalus* (Cluver 1978) reported that the forelimb presented many similarities to that of *Kawingasaurus*, indicating similar functional capabilities for digging. Following Cox (1972) and Cluver (1974, 1978), a specialized fossorial lifestyle has become widely accepted for cistecephalids. Recent work on cistecephalids has shown that their bones are robustly constructed, with very thick cortices (Nasterlack *et al.* 2012). They also have a highly inflated vestibular space in the inner ear, and characters of the skull and mandible suggestive of bone conduction sensing of ground-borne vibrations (Laaß 2014), all features commonly seen in extant fossorial tetrapods.

A hippopotamus-like amphibious ecology has long been proposed for the hyperabundant Early Triassic dicynodont *Lystrosaurus*, based on characters such as its deepened snout, nostrils located relatively high on the snout, supposed flexibility of the vertebral column, posterior position of the coracoids relative to the scapula, apparent lack of ossification of the carpal elements, the notched dorsal margin of the ilium, and potentially pachyostotic bones with nearly completely cancellous medullary cavities (Broom 1903b, 1932, Watson 1912b, 1913, Brink 1951, Camp 1956, Cluver 1971, Kemp 1982, Hotton 1986, Germain and Laurin 2005, Ray *et al.* 2005, Canoville and Laurin 2010; although see Williston 1914b for an early dissenting viewpoint). King (1991) provided a detailed review of the data presented by papers up to that point in support of a semiaquatic lifestyle in *Lystrosaurus* and found most of them to be equivocal at best. Comparisons of the anatomy of *Lystrosaurus* with that of extant semiaquatic taxa (King 1991), aspects of the functional morphology of *Lystrosaurus* (King 1991, King and Cluver 1991, Jasinoski *et al.* 2009, 2010a, b, 2014), the arid habitat in which it lived (King 1991, Botha and Smith 2007), and the almost entirely terrestrial fossil assemblage with which it is associated (King 1991) provide additional strong evidence that it was fully terrestrial. Likewise, the histology and highly cancellous medullary cavities of *Lystrosaurus* bones do not differ substantially from those of other dicynodonts, suggesting that these characters may more strongly reflect its phylogenetic history than its ecology (Botha-Brink and Angielczyk 2010). We therefore urge that an amphibious ecology for *Lystrosaurus* not be assumed as a null hypothesis in paleobiological studies, and we caution against using *Lystrosaurus* as an example of an amphibious ecomorphotype among non-mammalian synapsids.

Dicynodont bone histology has been a subject of extensive inquiry (Gross 1934, Enlow and Brown 1957, de Ricqlès 1972, 1975, 1976, Chinsamy and Rubidge 1993, Botha 2003, Ray and Chinsamy 2004, Ray *et al.* 2004, 2005, 2009a, b, 2012, Germain and Laurin 2005, Botha and Angielczyk 2007, Kriloff *et al.* 2008, Botha-Brink and Angielczyk 2010, Canoville and Laurin 2010, Green *et al.* 2010, Jasinoski *et al.* 2010a, Jasinoski and Chinsamy-Turan 2012, Green 2012, Nasterlack *et al.* 2012) and has provided insight into patterns of growth, life history, and potential lifestyle adaptations in the clade. Fibrolamellar bone tissue predominates in dicynodonts, and although annuli and lines of arrested growth indicate periodic pauses in growth in nearly all dicynodonts (*Eodicynodon* is an exception; Botha-Brink and Angielczyk 2010), most dicynodonts seem to have reached about 50% of their adult size before their first pause in growth. These observations, combined with relatively high densities of vascular channels and the large sizes of individual vascular channels, suggest that dicynodonts were probably some of the most rapidly growing non-mammalian synapsids. Growth rates slow at larger sizes, indicated by a shift to slower-growing parallel-fibered bone near the periphery of skeletal elements, but a true external fundamental system indicating a cessation of growth has only been reported in one specimen of the Late Triassic dicynodont *Placerias* (Green *et al.* 2010). Therefore, it would seem that the majority of dicynodonts had an indeterminate growth strategy. Recent histological work quantifying osteocyte lacuna size in dicynodonts suggests they may have had mammal-like high mass-independent resting metabolic rates (Olivier *et al.* 2017; see Rey *et al.* 2017 for similar results based on stable isotope analyses).

Growth and life history parameters might have played an important role in the survival of some dicynodonts (and other therapsids) during the Permo-Triassic mass extinction. Based on counts of lines of arrested growth and annuli, as well as body size distribution data, Botha-Brink *et al.* (2016) found that nearly all of the Permian dicynodonts that they sampled, including specimens of *Lystrosaurus*, took several years to reach maturity. By contrast, Early Triassic specimens of *Lystrosaurus* had much lower life expectancies and must have been breeding (and dying) at younger ages on average. Demographic simulations they carried out demonstrated that this shift toward earlier breeding would significantly improve the probability of species survivorship in unstable, unpredictable environments, such as are posited in the Early Triassic aftermath of the extinction.

Details of bone microstructure also have played a role in reconstructing the ecology of some dicynodonts. For example, Botha (2003) and Nasterlack *et al.* (2012) used the thick cortices of the limb bones of *Oudenodon* and *Cistecephalus* to argue that digging was an important component of the lifestyles of these animals. Features such as thickened cortices of limb bones and ribs, extensive bone remodeling, lack of a free medullary cavity, and a gradual transition between the cortex and the trabecular bone filling the medullary cavity have all been cited as evidence that *Lystrosaurus* was amphibious (Germain and Laurin 2005, Ray *et al.* 2005, Canoville and Laurin 2010). As noted above, however, most of these features are found in other dicynodonts that are thought to have been fully terrestrial (Kriloff *et al.* 2008, Botha-Brink and Angielczyk 2010).

Finally, dicynodonts have played an important role in terrestrial biostratigraphy for more than a century. Dicynodonts were first used extensively in the subdivision of the middle Permian to Middle Triassic rocks of the Beaufort Group in the Karoo Basin of South Africa (e.g., Seeley 1892, Broom 1906a, b, Watson 1914c, d, Kitching 1977, Keyser and Smith 1977–1978, South African Committee for Stratigraphy 1980, Rubidge 1995; see Day 2013a for a historical review of Karoo biostratigraphy), and work continues on this topic to the present day (e.g., Hancox *et al.* 1995, Hancox and Rubidge 2001, Neveling 2004, Abdala *et al.* 2005, van der Walt *et al.* 2010, Day 2013b, Day *et al.* 2015a, Viglietti *et al.* 2016). Because of the richness and extensive study of those deposits, the Karoo has served as the standard of comparison for other Permo-Triassic assemblages, leading dicynodonts to have a key position in terrestrial Permian and Triassic biostratigraphy globally (e.g., von Huene 1940, Chudinov 1965, Anderson and Cruickshank 1978, Cooper 1982, Cox 1991, Lucas 1993, 1996, 1998a, b, 2006, 2010, Lucas and Wild 1995, Golubev 2005, Rubidge 2005, Lucas *et al.* 2007). Two comparatively recent developments promise to have important implications for dicynodont-based biostratigraphy. High-precision collecting of specimens in measured stratigraphic sections, combined with GIS mapping of historical specimens and improved taxonomic identifications, are allowing further refinement of Karoo biostratigraphy and dicynodonts are still important index fossils in this work (Day 2013b, Viglietti *et al.* 2016). However, taxonomic revisions of dicynodonts have shown that some taxa that were previously thought be highly cosmopolitan, and thus useful for long-distance correlations, actually represent multiple distinct species with much more restricted geographic ranges and variable temporal relationships. *Dicynodon* is a particularly noteworthy example of this (Angielczyk and Kurkin 2003a, b, Kammerer *et al.* 2011, Viglietti *et al.* 2016). This problem is exacerbated by the poorly constrained ages of many dicynodont-bearing assemblages, and the increasing realization that long-assumed age relationships between faunas may be incorrect (e.g., Ottone *et al.* 2014, Marsicano *et al.* 2016; although see Liu *et al.* 2018).

Thus, we seem to be at a turning point in Permo-Triassic tetrapod biostratigraphy. Intrabasinal biostratigraphic schemes are improving in resolution and in their ties to the numerical geological timescale (for examples of the latter from the Karoo, see Rubidge *et al.* 2013, Day *et al.* 2015b), but greater efforts to radiometrically date important fossil assemblages around the globe, especially in the Triassic, are needed to test the accuracy of many previously proposed interbasinal correlations.

5.3.3.4 Gorgonopsia

Among the therapsid groups, gorgonopsians (Figs. 5.2 J and 5.14) are second only to anomodonts in their abundance in Karoo Permian collections (Smith *et al.* 2012). Despite this (or perhaps because of it), they are the most poorly studied major therapsid clade. As for other Karoo vertebrates, there was a surfeit of new gorgonopsian species described in the first half of the 20th century, which has seriously impeded subsequent attempts to understand the diversity and evolution of this group. Unlike other Karoo vertebrate groups, however, little subsequent work has been done on gorgonopsians to revise that earlier taxonomic decadence, and until recently only a single monograph (Sigogneau 1970) seriously addressed this problem. Only very recently has comprehensive revision of gorgonopsian species become a priority in therapsid taxonomy (Gebauer 2007, 2014, Kammerer 2014b, 2015, 2016c, 2017a, b, Kammerer *et al.* 2015b). In addition to their extensive South African record, gorgonopsians are common components of east African (Haughton 1924, 1926, von Huene 1950, Maisch 2002b, Kammerer 2016c) and Russian (Pravoslavlev 1927, Tatarinov 1999, Ivakhnenko 2002b, 2003b, 2008, Kammerer and Masyutin 2018a) Permian tetrapod faunas. Gorgonopsians also appear to have occurred in the arid equatorial belt that covered northern Africa in the late Permian, as indicated by tooth and jaw fragments from the Moradi Formation of Niger (Smiley *et al.* 2008). These elements represent the only therapsid fossils known from this otherwise sauropsid- and amphibian-dominated assemblage (Smith *et al.* 2015).

Gorgonopsians range from the Guadalupian to the Lopingian (middle to late Permian), ending as apparent victims of the end-Permian mass extinction (Ward *et al.* 2005). Gorgonopsians exhibit remarkably low cranial disparity (Fig. 5.14) compared with the other major therapsid clades: all gorgonopsians have broadly similar skulls with serrated, hypertrophied canines (Sigogneau-Russell 1989). Diagnostic features of gorgonopsians include a transverse lamina of the septomaxilla separating the external naris into two compartments; a steep, highly elevated mandibular symphysis; a large, anteriorly curved retroarticular process; and a cruciate pattern of ridges on the surface of the reflected lamina of the angular (Hopson and Barghusen 1986, Kammerer 2016c). A median suture between the palatines (excluding the vomer from contacting the pterygoid) has long been considered a key gorgonopsian synapomorphy (Hopson and Barghusen 1986, Sidor 2000), but new research has shown that a narrow vomerine-pterygoid contact is retained in the Russian taxa *Sauroctonus*, *Suchogorgon*, and *Inostrancevia* (Kammerer and Masyutin 2018a). Despite their overall similarity in skull morphology, gorgonopsians occupied a wide range of body sizes, with adults ranging between ~15 and 60 cm in skull length (Pravoslavlev 1927, Sigogneau 1970). Although historically poorly known, interrelationships among gorgonopsian species are beginning to become better resolved. In studying gorgonopsian phylogeny, Kammerer (2016c) initially focused on the subfamily Rubidgeinae, but more recent analyses have included a wider diversity of non-rubidgeine taxa (Kammerer 2017a, Kammerer and Masyutin 2018a). Intriguingly, these most recent analyses found that African gorgonopsians comprise a subclade distinct from their Russian relatives, and that the most basal gorgonopsians (*Nochnitsa* and *Viatkogorgon*) are from Russia. The biogeographic implications of these results have yet to be fully explored, but they underscore the potential of a renewed research focus on gorgonopsians.

Rubidgeines are the most distinctive group of gorgonopsians and included the top predators in late Permian African terrestrial ecosystems. The largest rubidgeines (e.g., *Dinogorgon* [Fig. 5.14 E, H] and *Rubidgea*) were convergent on anteosaurid dinocephalians and had massive, pachyostosed skulls with prominent bosses (probably also serving as stress sinks during biting). In late Permian Russia, members of a different gorgonopsian subclade (Inostranceviinae) acted as top predators, possibly hunting the coeval giant pareiasaur *Scutosaurus* (Sennikov 1996, Ivakhnenko 2003b). Inostranceviines include the largest known gorgonopsian, *Inostrancevia* itself, with a maximum total skull length of 64 cm (Pravoslavlev 1927). Despite its great size, the skull of *Inostrancevia* is substantially less robust than those of the giant rubidgeines, and it likely employed a different feeding style (Kammerer 2016c).

Gorgonopsians are typically described as "sabertoothed" and are frequently compared with Cenozoic mammalian carnivores with hypertrophied canines (i.e., machairodontine cats, nimravids, and thylacosmilids) (Parrington 1955, van Valkenburgh and Jenkins 2002, Andersson *et al.* 2011). However, those mammals are also strongly convergent in overall skull and forelimb morphology (Slater and van Valkenburgh 2008, Meachen-Samuels 2012) in ways that gorgonopsians are not—no gorgonopsian has a tall, short face comparable with that of *Smilodon* or *Thylacosmilus*, for instance. As such,

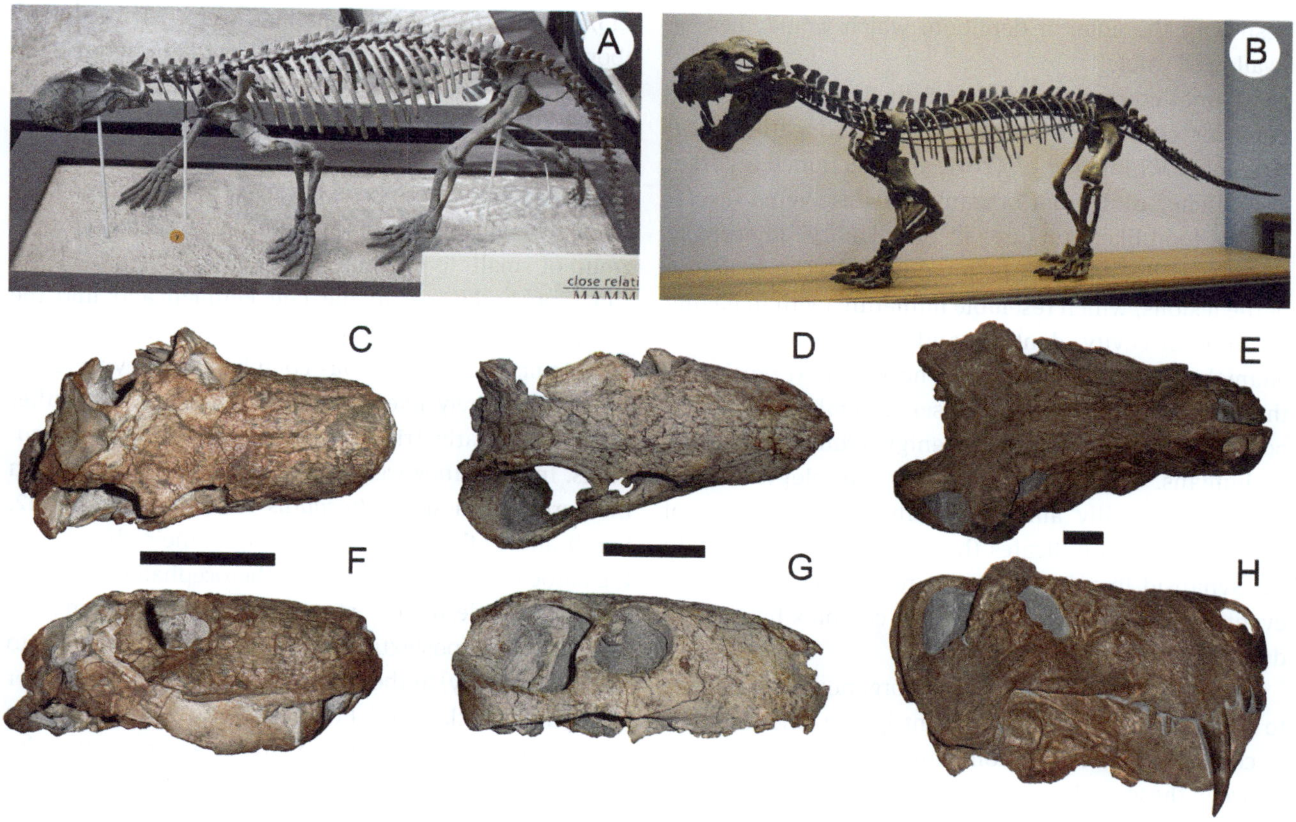

Fig. 5.14: Gorgonopsia. (A) Mounted skeleton of *Lycaenops ornatus* (AMNH FARB 2240) in left lateral view. (B) Mounted skeleton of *"Scymnognathus" parringtoni* (GPIT/RE/7113) in left lateral view. (C, F) Skull of *Eriphostoma microdon* (SAM-PK-K11164) in dorsal and lateral views. (D, G) Skull of *Arctognathus curvimola* (SAM-PK-K11457) in dorsal and lateral views. (E, H) Skull of *Dinogorgon rubidgei* (GPIT/RE/7114) in dorsal and lateral views. Specimen in panel F mirrored for comparative purposes. Scale bars equal 5 cm.

mammalian saber-toothed carnivores may not represent particularly close analogs for gorgonopsians; Komodo dragons have been suggested as a more likely comparison for informing gorgonopsian prey capture behaviors (Cruickshank 1973). Gorgonopsians do appear to converge on saber-toothed carnivores in substantially expanding their gape relative to sister taxa, although the extent of jaw motility in the group is debated. Kemp (1969) argued that the gorgonopsian quadrate was streptostylic, rotating around its contact with the squamosal to permit limited palinal motion of the lower jaw. Laurin (1998) questioned this conclusion, echoing work by Parrington (1955), noting that this motion would require movement of the stapes (which remains tightly associated with the quadrate in gorgonopsians) in a manner otherwise unknown in vertebrates.

Another point of interest for gorgonopsian feeding is the trend toward postcanine loss in several subclades. Independent reduction of the postcanine count occurs in multiple lineages of large gorgonopsians, including the genera *Inostrancevia* (three to four upper postcanines, no lower postcanines), *Smilesaurus* (two to three upper postcanines, one lower postcanine), and *Rubidgea* (one to two upper postcanines, no lower postcanines) (Ivakhnenko 2003b, Kammerer 2016c). In the most extreme example (the South African rubidgeine *Clelandina*), the postcanine teeth were lost entirely and were replaced by a narrow bony ridge similar to that of the coeval whaitsiid therocephalian *Theriognathus* (Cruickshank 1973, Kammerer 2016c). Although total loss of the lower postcanines is only known in the largest gorgonopsians, this tendency is not a direct correlate of body size: *Dinogorgon* (Fig. 5.14 E, H) is comparable in size (~40 cm skull length) with *Rubidgea* and *Inostrancevia* but retained four to five upper and lower postcanines (Kammerer 2016c), whereas some of the smallest gorgonopsians like *Eriphostoma* (10–15 cm skull length; Fig. 5.14 C, F) had only three upper and two lower postcanines (Kammerer 2014b). It should be noted that even in taxa like *Dinogorgon*, however, the postcanines are extremely small compared with the incisors and canine. Taken together, the enlarged canines, expanded gape, and reduced postcanines of gorgonopsians suggest that feeding in this group involved tearing off large chunks of flesh with the anterior dentition, which would then be swallowed whole.

A final noteworthy aspect of the dentition of gorgonopsians concerns a recently discovered pathological specimen. Whitney *et al.* (2017) thin-sectioned a gorgonopsian mandible from Tanzania and discovered several ectopic toothlike structures in the dentary, adjacent to the labial edge of the functional canine. The structure of the lesions, which resemble miniature teeth in having an internal cavity, dentine, and an outer covering of enamel, and the fact that they have caused erosion of the canine root make them closely resemble compound odontoma, a common type of benign odontogenic tumor in humans. The discovery of this pathology in such a phylogenetically and temporally distant relative of extant mammals indicates that it has a deep history in the synapsid lineage and is not related to more recent events such as the evolution of the complex heterodont dentition of mammals.

The evolutionary history of gorgonopsians relative to that of therocephalians is intriguing and suggests a complex history of competition and replacement between these clades. In the middle Permian, gorgonopsians are a rare component of terrestrial ecosystems, and all known middle Permian gorgonopsians are relatively small (10–15 cm skull length), whereas coeval therocephalians were large (30–40 cm skull length) and abundant (Smith *et al.* 2012, Kammerer *et al.* 2015b). These large-bodied predatory therocephalians (lycosuchids and scylacosaurids) went extinct at the end of the middle Permian, however (Day *et al.* 2015b), and late Permian therocephalians are primarily small-bodied (Kemp 2012). The extinction of large-bodied middle Permian therocephalians coincides with a radiation of gorgonopsians—this group became extremely abundant and species-rich in the late Permian, with many gorgonopsian species attaining sizes equal to or exceeding that of the earlier therocephalians (Sigogneau 1970). Given the spottiness of the middle Permian therapsid record, it is impossible to say whether local evidence of this pattern (thus far recorded in South Africa and Russia: Tatarinov 1999, Kammerer *et al.* 2015b, Kammerer and Masyutin 2018b) truly represents clade-wide interactions, but it is certainly suggestive that therocephalian extinction opened up ecospace for gorgonopsians to exploit.

5.3.3.5 Therocephalia

Therocephalians were, ancestrally, superficially similar to gorgonopsians but went on to evolve substantially higher morphological diversity than that group (Figs. 5.2 K and 5.15). The earliest therocephalians were, as mentioned above, large-bodied, saber-toothed predators (van Valkenburgh and Jenkins 2002) that were abundant in middle Permian faunas in southern Africa and Russia (Ivakhnenko 2011, Smith *et al.* 2012, Day *et al.* 2015b). Following the extinction of these taxa, however, the remaining therocephalian lineages radiated into an incredibly diverse group of small- to medium-sized insectivores, predators, and even herbivores in the late Permian and into the Triassic. The response of therocephalians to the Permo-Triassic mass extinction was complicated: they were not wiped out entirely, like gorgonopsians, nor did they suffer the extreme Early Triassic lineage bottleneck of dicynodonts. Instead, they were actually relatively abundant in the earliest Triassic *Lystrosaurus* AZ (Smith *et al.* 2012, Smith and Botha-Brink 2014). Despite their abundance, the Karoo record of Early Triassic therocephalians is indicative of a clade under extreme pressure, exhibiting the Lilliput effect (postextinction miniaturization related to resource scarcity) at the species level and between higher taxa (Huttenlocker and Botha-Brink 2013, Huttenlocker 2014). Therocephalians appear to have gone extinct by the Middle Triassic (Abdala *et al.* 2014a), at the same time when cynodonts were undergoing their major radiation (Eucynodontia) and dicynodonts had an evolutionary "second wind" via their species-rich Triassic subclade Kannemeyeriiformes. Oddly, however, the latest-known therocephalians (Middle Triassic bauriamorphs) were cosmopolitan and species-rich, so therocephalian extinction does not neatly fit the narrative of a "dead clade walking" irreparably damaged by mass extinction (Jablonski 2002).

Therocephalians can be broadly separated into a basal grade of large Guadalupian (middle Permian) predators and the diverse, primarily Lopingian (late Permian) clade Eutherocephalia (Hopson and Barghusen 1986). "Basal therocephalians" consist of two families, Lycosuchidae and Scylacosauridae van den Heever 1980. Lycosuchids are the most plesiomorphic therocephalian group and are characterized by relatively short, straight snouts and reduction of the postcanine dentition (van den Heever 1994). Two genera are known, *Lycosuchus* (Fig. 5.15 C, F) and *Simorhinella*, which are separated by minor features of the palate (Abdala *et al.* 2014b). Scylacosaurids are the most abundant basal therocephalians and numerous species have been described, although their taxonomy requires revision, and it is unlikely that many of the recognized species are actually valid (particularly problematic is the possibly dubious nature of the scylacosaurid *Pristerognathus*, as it is the namesake of the *Pristerognathus* AZ; Keyser and Smith 1977–1978). The best-known taxon is *Glanosuchus macrops* (Fig. 5.15 A), which has been

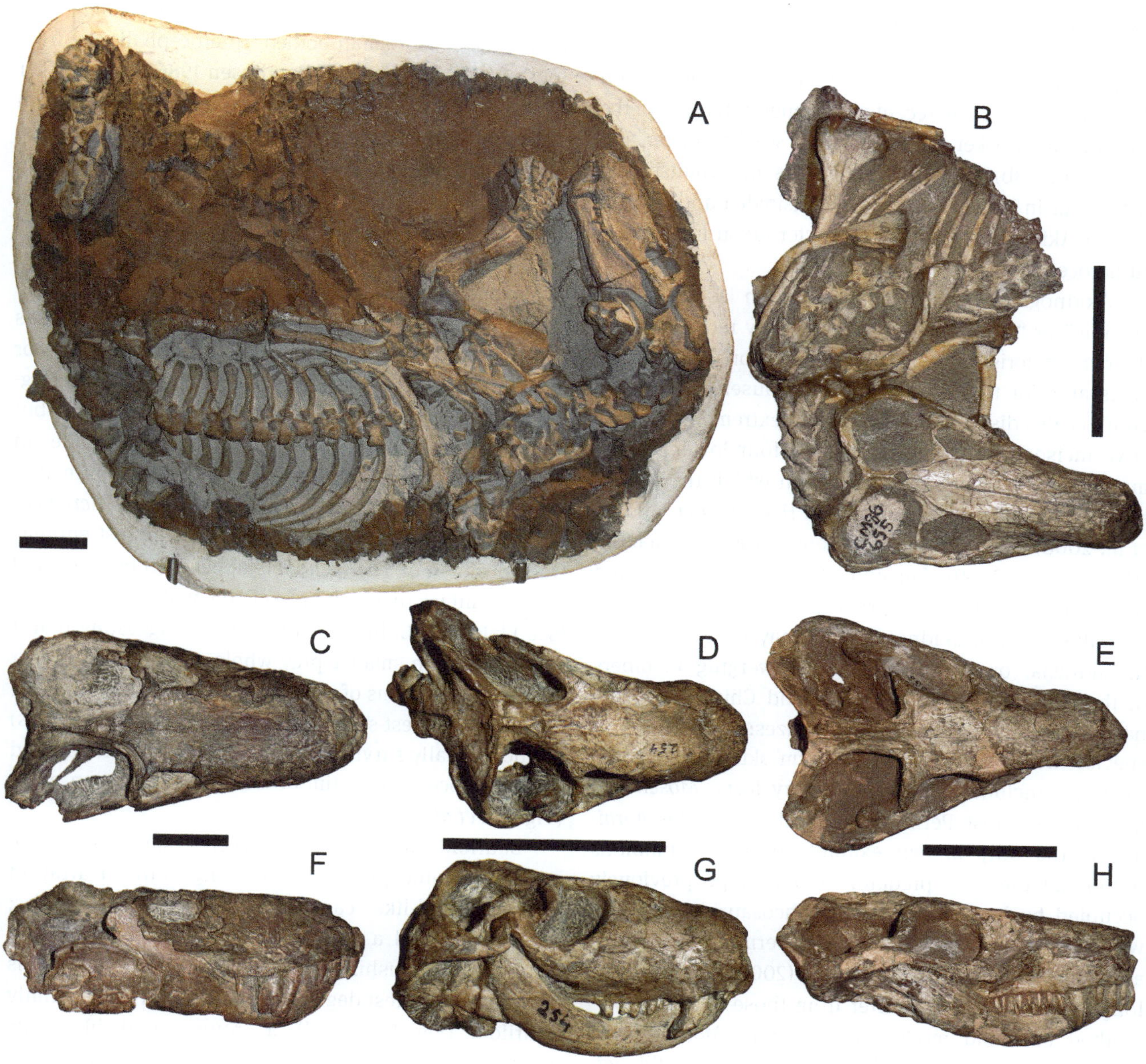

Fig. 5.15: Therocephalia. (A) Articulated partial skeleton of *Glanosuchus macrops* (SAM-PK-K7809) in dorsal view (skull in left lateral view). (B) Articulated anterior skeleton of *Ictidosuchoides longiceps* (CGS CM86-655) in dorsal view. (C, F) Skull of *Lycosuchus vanderrieti* (US D173) in dorsal and lateral views. (D, G) Skull of *Hofmeyria atavus* (TM 254) in dorsal and lateral views. (E, H) Skull of *Bauria cynops* (BP/1/3770) in dorsal and lateral views. Scale bars equal 5 cm.

the subject of extensive cranial (van den Heever 1994) and postcranial (Fourie and Rubidge 2009) description. *Glanosuchus* has also been reported to have had maxillary turbinates, suggesting mammalian-style endothermy (Hillenius 1994), but this report has subsequently been questioned (Sigurdsen 2006). Although these basal therocephalians went extinct shortly after the end of the middle Permian, it is worth noting that their extinction was not synchronous with that of dinocephalians, and both lycosuchids and scylacosaurids survived to the end of the *Pristerognathus* AZ in the Karoo (Day *et al.* 2015b).

Eutherocephalians are diagnosed by a large suite of characters, including the presence of a rostral process of the premaxilla, the absence of the postfrontal, the fusion of the vomers, the presence of a mandibular fenestra, and the smoothly ridged cutting surfaces of the incisors (Hopson and Kitching 1986, Huttenlocker 2009). Hopson and Barghusen (1986) recognized four major lineages of eutherocephalians: Hofmeyriidae, Whaitsiidae, Akidnognathidae, and Baurioidea. They considered hofmeyriids to be the most plesiomorphic group because of their retention of a postfrontal, but Hopson (1991) later placed them in an unresolved tetrachotomy with the other eutherocephalian clades, based on subsequent study of specimens indicating that the postfrontal is lost through fusion during ontogeny. More recent phylogenetic analyses of the group (Huttenlocker 2009, 2014; but see Kammerer and Masyutin 2018b) recognize hofmeyriids and whaitsiids as sister taxa in the larger clade Whaitsioidea and instead recover Akidnognathidae as the sister taxon to all other eutherocephalians.

Akidnognathids (previously known as "euchambersiids" or "moschorhinids") are easily recognized by their characteristic mediolateral expansion of the vomer at its anterior tip (Hopson and Barghusen 1986). This group is also distinctive in having an extremely conservative incisor count (five upper and four lower, unlike most other therocephalian groups in which the incisor number is highly variable; van den Heever 1994, Abdala *et al.* 2008) and facial enlargement of the septomaxilla (Huttenlocker 2009). Liu and Abdala (2017) suggested that unlike most other, primarily Gondwanan therocephalian clades, akidnognathids may have originated in Laurasia, on account of early diverging members of the clade occurring in Russia and China. Akidnognathids occupied a wide range of sizes: the majority of species are relatively small (~10 cm skull length), but they also included a large predatory form, *Moschorhinus*, in the latest Permian (Durand 1991). *Moschorhinus* is the only known example of therocephalians reevolving the "top predator" morphotype previously occupied by lycosuchids and scylacosaurids in otherwise gorgonopsian-dominated late Permian food webs, and van Valkenburgh and Jenkins (2002) noted that its long, spikelike canines differ from those of true sabertoothed cats and modern carnivorans. *Moschorhinus* survived, albeit briefly, where gorgonopsians did not, being one of only a few therapsid taxa known to cross the Permo-Triassic extinction boundary (Huttenlocker and Botha-Brink 2013, Smith and Botha-Brink 2014). This clade also includes the most famous of therocephalians, the late Permian *Euchambersia mirabilis*, which has frequently been hypothesized to have been venomous based on a large, apparently gland-housing fossa on its maxilla (e.g., Nopcsa 1933, Folinsbee *et al.* 2007, Benoit *et al.* 2017b). Akidnognathids in general were relatively abundant and diverse in the Early Triassic (Botha-Brink and Modesto 2011, Huttenlocker *et al.* 2011a, b, Huttenlocker 2014) but did not survive beyond the *Lystrosaurus* AZ, and appear to be victims of the final extinction pulse associated with the end-Permian mass extinction (Smith and Botha-Brink 2014).

Whaitsioidea includes two main subclades, Hofmeyeriidae and Whaitsiidae. Hofmeyriids include three genera (*Hofmeyria* [Fig. 5.15 D, G], *Ictidostoma*, and *Mirotenthes*) of small-bodied therocephalians from South Africa (Hopson and Barghusen 1986, Huttenlocker 2009). They are the first eutherocephalians to appear in the Karoo record, with specimens known from the *Pristerognathus* AZ (Smith *et al.* 2012). Whaitsiids include *Viatkosuchus* and *Moschowhaitsia* from Russia and the highly autapomorphic *Theriognathus* from South and eastern Africa (Tatarinov 1995, Ivakhnenko 2011, Huttenlocker and Abdala 2015). Whaitsioids are most notable for their unusually shaped lower jaws: both hofmeyriids and whaitsiids have a strongly bowed, "boomerang", or "banana"-shaped dentary (Fig. 5.15 G). The functional significance of this mandibular morphology is uncertain, but it appears to be correlated with tooth loss in this clade—in most hofmeyriids, the lower postcanines are confined to a small patch close to the lower canine, and in *Theriognathus* both upper and lower postcanines are lost entirely. One specimen of *Theriognathus* appears to be preserved with an intact dicynodont skeleton within its rib cage (see Plate H of Chinsamy-Turan 2012), suggesting that it may have eaten smaller prey whole. Whaitsioids appear to have been victims of the end-Permian mass extinction, although the latest-surviving Karoo taxon (*Theriognathus*) did not actually survive until the end of the Permian and may have been the victim of an earlier extinction pulse (Viglietti *et al.* 2016).

Baurioids are the most species-rich and diverse group of therocephalians, including both an array of small, weasel-like carnivores (e.g., *Ictidosuchoides*; Fig. 5.15 B), and a Triassic radiation of apparent herbivores with crushing, molariform dentition. This is taken to its greatest degree in the middle Triassic family Bauriidae (Fig. 5.15 E, H), a group remarkable in its degree of convergence with gomphodont cynodonts and mammals. Bauriids developed a complete secondary palate, lost the complete postorbital bar (in at least some taxa, e.g., *Bauria*), and had inset rows of labiolingually expanded postcanines with complex occlusion and cuspulated cingula (Abdala *et al.* 2014a). Indeed, the degree of similarity between bauriids and gomphodont cynodonts is so great that many northern hemisphere bauriids were originally described as cynodonts (e.g., Tatarinov 1973, 2002, Young 1974).

5.3.3.6 Cynodontia

Cynodonts (Figs. 5.2 L and 5.16) are best known for including the only extant synapsids: mammals. Non-mammalian cynodonts were an important group in their own right, however, and represent one of the most abundant and species-rich groups of Triassic tetrapods. It should be noted here that two different definitions have been proposed for Mammalia, one corresponding to an intuitive composition of the clade (i.e., including "Mesozoic mammals" such as *Morganucodon* and docodonts; Luo *et al.* 2002, Kielan-Jaworowska *et al.* 2004) and the other restricting this name to the crown group (Rowe 1988). Both have their advantages: the "traditional" definition offers stability of composition and a robust set of diagnostic osteological characters, whereas the crown definition offers direct knowledge of typically "mammalian" soft tissue and behavioral characters. For the purposes of consistency with our use of the term "non-mammalian synapsids", here we mean the "traditional" definition when we speak of "non-mammalian cynodonts".

Cynodonts are a highly distinctive clade that can be recognized by upward of 27 synapomorphies, including a nasolacrimal contact, secondary palate composed of maxilla and palatine, expansion of the epipterygoid (alisphenoid of mammals) to contact the prootic and frontal, double occipital condyles, dorsoventral orientation of the stapedial foramen, masseteric fossa on the coronoid region of the dentary, reduction of the reflected lamina of the angular into a "spoon"-like process, and complex postcanine morphology, minimally involving the presence of anterior and posterior accessory cusps (Hopson and Kitching 2001).

Alone among the major therapsid clades, cynodonts have a substantially better Triassic than Permian record and thrived in the aftermath of the end-Permian mass extinction. Relatively few Permian cynodont taxa are currently known, most of which were rare components of their faunas (Broom 1937, Kammerer 2016a, Viglietti *et al.* 2016). Cynodontia has no middle Permian fossil record (Kammerer 2014a), despite the fact that its sister group (Therocephalia) is one of the most abundant therapsid clades in the Capitanian (van den Heever 1994, Smith *et al.* 2012), leaving a 10-million-year ghost lineage between its latest possible divergence and first appearance. This lengthy ghost lineage helps to explain the morphological distinctiveness of even the oldest known cynodonts. The earliest known cynodonts appear in the *Tropidostoma* Assemblage Zone (earliest Lopingian) of South Africa and are represented by a single family (Charassognathidae) containing two genera (*Charassognathus* [Fig. 5.16 C, E] and *Abdalodon*, each known from a single specimen) (Botha *et al.* 2007, Kammerer 2016a). The small size (5–6 cm skull length) of charassognathids suggests that sampling bias may be to blame for the poor early record of cynodonts.

By the end of the Permian (*Daptocephalus* AZ of South Africa and equivalents), four additional lineages of cynodonts are known: Dviniidae, Procynosuchidae, Galesauridae, and Thrinaxodontidae. Dviniids and procynosuchids are restricted to the Permian, whereas galesaurids and thrinaxodontids crossed the Permo-Triassic boundary and became important components of Early Triassic ecosystems (Sidor and Smith 2004). Despite being one of the earliest known cynodont clades (Abdala 2007), dviniids are a highly autapomorphic group, characterized by labiolingually expanded postcanine teeth in which the main cusp is surrounded by a ring of numerous cingular cuspules (up to 16 in the posterior lower postcanines) (Tatarinov 1968, Ivakhnenko 2012, 2013). In *Dvinia prima* (the only dviniid known from cranial remains), the skull is also unusual in having anteriorly directed orbits, an enormous, elevated sagittal crest, and a fossa for the lower canine that perforates the dorsolateral margin of the maxilla (Ivakhnenko 2013). Procynosuchidae currently includes a single species, *Procynosuchus delaharpeae* (Fig. 5.16 A), which is by far the most abundant and well-known Permian cynodont—it is represented by dozens of specimens, including multiple complete skeletons (Broom 1948, Kemp 1980c). *Procynosuchus* is also an extremely widespread therapsid taxon, with specimens recorded from southern and eastern Africa, Germany, and Russia (von Huene 1950, Kemp 1979, Sues and Boy 1988, Abdala and Allinson 2005, Weide *et al.* 2009, Ivakhnenko 2012).

Galesaurids include a single late Permian taxon (*Cynosaurus*) and two Early Triassic taxa (*Progalesaurus* and *Galesaurus*) characterized by extremely recurved main cusps on the postcanines overhanging an enlarged, rounded posterior accessory cusp (Sidor and Smith 2004). Thrinaxodontids are best represented by *Thrinaxodon liorhinus* (Fig. 5.16 B), an extremely common animal in Early Triassic Karoo deposits (second only to *Lystrosaurus* in terms of specimen abundance; Smith *et al.* 2012). In addition to its South African record, *Thrinaxodon* is also common in the Lower Fremouw Formation of Antarctica (Hammer 1990; one of the topographical features of the Transantarctic Mountains is officially known as Thrinaxodon Col because of this) and has been reported from the Panchet Formation of India (Satsangi 1987). *Thrinaxodon* is the archetypal "early cynodont" and has been used as the exemplar for this grade in countless studies of mammalian evolution (e.g., Crompton and Jenkins 1968, Hopson *et al.* 1989, Wible 1991, Luo and Crompton 1994, Luo *et al.* 1995, Rowe *et al.* 2011). Because of its abundance and historical importance for understanding mammalian origins, *Thrinaxodon* is one of the most extensively studied non-mammalian synapsids: its cranial anatomy has been described in exquisite detail on the basis of both serial sections (Fourie 1974) and CT data (Rowe *et al.* 1995); its ontogenetic trajectory, tooth replacement

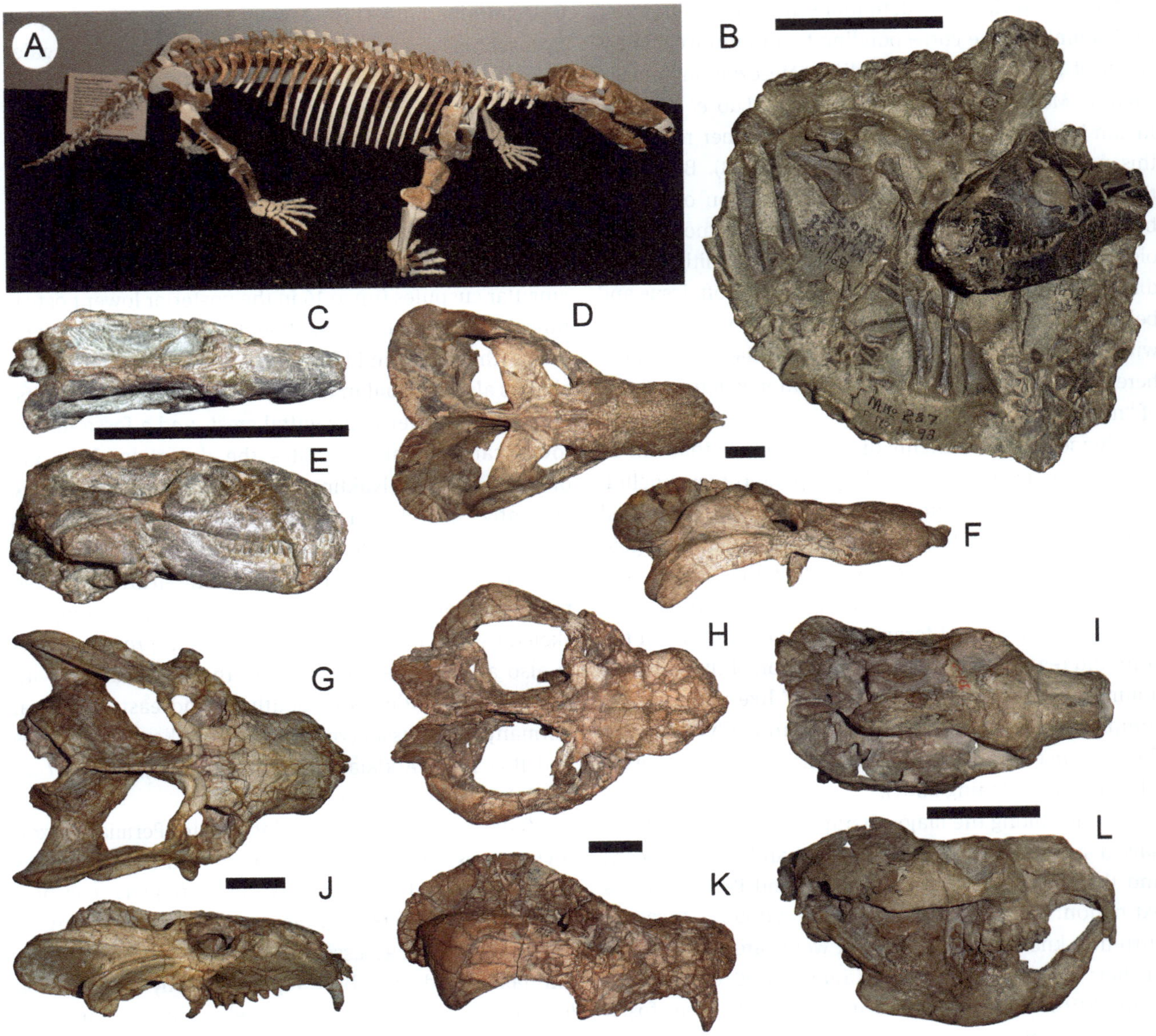

Fig. 5.16: Cynodontia. (A) Mounted skeleton of *Procynosuchus delaharpeae* (Museum Korbach, cast of NHMUK R37054) in right lateral view. (B) Articulated skeleton of *Thrinaxodon liorhinus* (BP/1/1693) preserved curled up in its den. (C, E) Skull of *Charassognathus gracilis* (SAM-PK-K10369) in dorsal and lateral views. (D, F) Skull of *Cynognathus crateronotus* (BSPG 1934 VIII 6) in dorsal and lateral views. (G, J) Skull of *Exaeretodon riograndensis* (MCP-PV-1522T) in dorsal and lateral views. (H, K) Skull of *Chiniquodon theotonicus* (MCZ 1533) in dorsal and lateral views. (I, L) Skull of *Kayentatherium wellesi* (USNM 317210) in dorsal and lateral views. Scale bars equal 5 cm.

pattern, and growth dynamics are well known (Estes 1961, Gow 1985, Botha and Chinsamy 2005, Abdala *et al.* 2013, Jasinoski *et al.* 2015); and even its behavior is reasonably well understood. *Thrinaxodon* appears to have displayed parental care, which is inferred on the basis of several well-preserved mixed-age aggregations of *Thrinaxodon* individuals (Jasinoski and Abdala 2017a). *Thrinaxodon* is also known to be a burrower—multiple specimens have been found preserved inside burrow casts (Damiani *et al.* 2003, Sidor *et al.* 2008, Modesto and Botha-Brink 2010), and burrow casts matching *Thrinaxodon*'s size are common in Early Triassic ichnofacies (Miller *et al.* 2001; although some of these burrows could also belong to coeval cynodonts like *Galesaurus*). Recently, synchotron scanning of the intact living chamber of an Early Triassic Karoo burrow cast revealed an individual of *Thrinaxodon* cohabiting with the temnospondyl *Broomistega* (Fernandez *et al.* 2013), providing evidence for interspecies commensalism similar to that observed in certain extant burrowing tetrapods (e.g., gopher tortoises; Lips 1991). Beyond *Thrinaxodon* itself, the composition of Thrinaxodontidae is uncertain; historically, various early cynodonts were placed in this group (including fragmentary Russian taxa like *Nanocynodon*; Ivakhnenko 2012), but it is uncertain whether they are actually more closely related to *Thrinaxodon* than other cynodonts. The only other taxon that has been recovered as a thrinaxodontid in phylogenetic analyses is *Nanictosaurus kitchingi*, a small late Permian cynodont from South Africa that is important for demonstrating that the evolution of derived cynodonts preceded the end-Permian mass extinction (Abdala 2007).

The major radiation of Triassic cynodonts is the Eucynodontia, a large, diverse clade characterized by fused dentaries, contact between the surangular and a descending flange of the squamosal, loss of the quadrate ramus of the pterygoid, reduction of the postdentary bones to a narrow rod, hook-shaped morphology of the reflected lamina of the angular, and presence of an acromion process of the scapula (Hopson and Kitching 2001). Although the relationships between the major eucynodont taxa have been extensively debated in the past (Battail 1982, Kemp 1982, 1988, Rowe 1986, 1988, Hopson 1991, 1994), there is now overwhelming support for a basal dichotomy in eucynodont phylogeny between the primarily herbivorous Cynognathia and the ancestrally carnivorous Probainognathia (Hopson and Kitching 2001, Abdala 2007, Liu and Olsen 2010), as originally proposed by Hopson and Barghusen (1986).

Cynognathians are characterized by an extremely deep zygomatic arch that extends above the midpoint of the orbit, a suborbital process on the jugal, absence of internal carotid foramina on the basisphenoid, and a skull that is notably triangular in dorsal view, with greatest expansion of the zygomatic arches at their posterior ends (Hopson and Barghusen 1986, Hopson and Kitching 2001). The namesake of Cynognathia, *Cynognathus* (Fig. 5.16 D, F) from the early to middle Triassic of southern Africa, Antarctica, and Argentina (Hammer 1990, Abdala and Ribeiro 2010), is actually a rather aberrant member of its clade, as it was a large, predatory form (indeed, it is the largest known carnivorous non-mammalian cynodont) (Broili and Schröder 1935), whereas the majority of cynognathians were herbivores. The herbivorous cynognathians make up the clade Gomphodontia, all members of which have typically "gomphodont" (molariform, labiolingually expanded) postcanines, and can be broken into three successive families, the species-poor Gomphognathidae (=Diademodontidae) and Trirachodontidae, and the highly diverse, widespread, and abundant Traversodontidae Abdala and Ribeiro 2010, Kammerer *et al.* 2010. Traversodontids include the latest-surviving and largest gomphodonts, gomphodontosuchines such as *Exaeretodon* (Fig. 5.16 G, J) and *Scalenodontoides* from the Norian and possibly even Rhaetian deposits of South America and South Africa (Liu and Abdala 2014).

Probainognathians are characterized by the absence of the pineal foramen, ectopterygoid, and costal plates on the ribs, expansion of the secondary palate, and restriction of the procoracoid so that it does not contribute to the glenoid (Hopson and Kitching 2001). The earliest known probainognathian is *Lumkuia fuzzi*, represented by a single specimen from the Middle Triassic *Cynognathus* AZ of South Africa (Hopson and Kitching 2001), but the group is much more abundant in the Late Triassic. Chiniquodontids (some of which attained large size, e.g., *Chiniquodon theotonicus*; Fig. 5.16 H, K) and probainognathids are common components of the classical Triassic vertebrate localities in the Santa Maria Formation of Brazil and Chañares and Ischigualasto formations of Argentina (Romer 1969, 1973, Abdala and Giannini 2002, Martinelli *et al.* 2016), and more crownward taxa are known from fragmentary but taxonomically diverse remains in coeval and slightly younger beds (e.g., *Prozostrodon* and *Therioherpeton*) (Bonaparte and Barberena 2001, Bonaparte *et al.* 2006).

The most "advanced" cynodonts belong to a set of phylogenetically problematic families best known from the Early Jurassic. Two of these families, Tritheledontidae and Tritylodontidae, have long jockeyed for position as the sister taxon to mammals within Cynodontia (Rowe 1986, 1988, Hopson 1991, Liu and Olsen 2010). Both show suites of extremely mammal-like and, confoundingly, mutually conflicting characters, meaning that competing topologies (including a sister taxon relationship between tritheledontids and tritylodontids, as recovered by Abdala 2007) all incur substantial homoplasy.

Tritylodontids (Fig. 5.16 I, L) are a highly aberrant group that was the first cynodont clade to explore what would become a highly successful "mammalian" morphotype (gnawer), with enlarged, ever-growing incisors and expanded, multicusped postcanine dentition very similar to that later evolved by multituberculates and rodents (Clark and Hopson 1985, Sues 1985). Tritylodontids currently have no definitive Triassic record (which has hindered understanding their relationships), but they are the longest-surviving family of non-mammalian synapsids, with a range extending to the Early Cretaceous (Tatarinov and Matchenko 1999, Matsuoka *et al.* 2016). Tritylodontid postcrania are exceptionally mammal-like (Kemp 1983, Sues and Jenkins 2006), and their massive olecranon process of the ulna suggests burrowing capability. Tritheledontids are morphologically poorly known compared with tritylodontids; most named species are represented by partial jaws and isolated teeth, and very few postcrania are known (Hopson 1991). However, the early evolution of this group is better understood, thanks to an increasingly well-represented Triassic record (Martinelli *et al.* 2005). Tritheledontids are all small, shrewlike faunivores, unusual in their hypertrophy of the lower incisors (like tritylodontids), but not the uppers. The best-known genera are *Pachygenelus* and *Diarthrognathus* from the Upper Elliot Formation (Lower Jurassic) of South Africa and Lesotho (Gow 1980). The holotype of *Diarthrognathus broomi* in particular is notable for being one of the first cynodont specimens recognized as having both the primitive, "reptilian" quadrate-articular jaw joint and the derived, mammalian dentary-squamosal jaw joint (hence the genus name, meaning "two-jointed jaw") (Crompton 1958). The dentary-squamosal contact in *Diarthrognathus* and other tritheledontids is a simple hinge, not the fully developed articulation present in Mesozoic mammals (e.g., *Morganucodon*) that retain a double jaw-joint, making tritheledontids a useful clade for understanding the earliest stages of this morphological transition (Luo 2011).

The ongoing debate on whether tritheledontids or tritylodontids are the sister taxon of mammals has recently been shaken up by the exciting discovery of a new cynodont clade from the Late Triassic of Brazil, Brasilodontidae (Bonaparte *et al.* 2003, 2005, 2010). Brasilodontids possess several classically mammalian characters, such as the presence of a petrosal promontorium and divided postcanine roots, that are not present in other non-mammalian cynodont clades. In the few phylogenetic analyses that have tested their relationships, brasilodontids are strongly supported as the sister taxon of Mammalia (e.g., Liu and Olsen 2010). Numerous brasilodontid specimens have now been recovered (C.F. Kammerer, personal observation), and it seems certain that further description and analysis of this material (e.g., Rodrigues *et al.* 2013, Ruf *et al.* 2014) will significantly advance our understanding of the cynodont-mammal transition.

5.4 Discussion

Non-mammalian synapsids have a history of study stretching back over 150 years. Given this legacy, a fair question to ask is whether there is still interesting and important research that can be carried out on early synapsids. Our answer to this question is an emphatic "yes" for two main reasons. First, their diversity, long temporal range, and their dominance in Permo-Triassic terrestrial ecosystems make a detailed knowledge of the paleobiology and paleoecology of non-mammalian synapsids key for understanding life on land at a time when the first herbivore-dominated ecosystems were emerging, the last major ice-house to greenhouse transition (before the ongoing anthropogenic example) occurred, and the largest mass extinction in Earth history (caused in part by rapid global warming) took place. Yet, the majority of research on non-mammalian synapsids has concerned their taxonomy and systematics and the evolution of a subset of mammalian characters. We are well overdue for a more thorough investigation of synapsid paleobiology in its own right.

Our picture of Mesozoic mammals recently has been transformed by the realization that there was much greater ecomorphological diversity among these taxa than had previously been recognized (e.g., Luo and Wible 2005, Ji *et al.* 2006, Meng *et al.* 2006, 2015, Luo 2007, Zheng *et al.* 2013, Krause *et al.* 2014, Martin *et al.* 2015), and we believe a similar transformation is possible for non-mammalian synapsids if sufficient research effort is applied. This is not to say that there are no taxonomic or phylogenetic problems left (e.g., a comprehensive reassessment of higher-level synapsid phylogenetic relationships is badly needed, and the longstanding alpha taxonomic confusion surrounding gorgonopsians is only beginning to be resolved), but we should also take greater advantage of the considerable progress made in these areas over the past three decades.

Second, the paleontological contribution to much of what we "know" about synapsid paleobiology and how mammalian characters evolved over the course of synapsid history stems from research that was conducted over 50 years ago, and/or that included only a superficial representation of synapsids outside of Mammaliaformes. This work includes many splendid comparative anatomical studies that clearly demonstrate the power and continued relevance of such research. However, they could not take advantage of advances in synapsid taxonomy and systematics, phylogenetic comparative methods, visualization techniques such as high-resolution X-ray computed tomography (CT-scanning), and breakthroughs in evolutionary developmental biology. Geneticists, physiologists, and developmental biologists working on extant mammals have made a great deal of progress in understanding the underpinnings of many components of the mammalian phenotype (e.g., Oftedal 2002, 2012, Grigg *et al.* 2004, Clarke and Pörtner 2010, Takechi and Kuratani 2010, Nespolo *et al.* 2011, Alibardi 2012, Hirasawa and Kuratani 2013, Oftedal and Dhouailly 2013, Buchholtz 2014, Owerkowicz *et al.* 2015, Kim *et al.* 2016, Lovegrove 2017), but the consideration of non-mammalian synapsids in this body of literature varies from extensive to almost non-existent. Like the realization that greater collaboration between paleontologists and systematists was needed to establish rigorous temporal calibrations for molecular phylogenies (e.g., Parham *et al.* 2012), we believe a greater dialog between paleontologists and neontologists is needed to fully realize the potential contributions the synapsid fossil record can make to our picture of how mammalian characters evolved.

5.4.1 The intrinsic interest of synapsid paleobiology

When confronted with early synapsids like the sail-backed *Dimetrodon*, knowing that they are more closely related to extant mammals than to any reptiles or amphibians, one cannot help but wonder what they would have been like when alive. Was *Dimetrodon* basically a Komodo dragon with a sail, or were there hints of mammalian biology beginning to emerge? Part of this interest is tied up with thinking about the evolution of mammals (i.e., how mammal-like were they?), but there is also curiosity about their seemingly unique morphologies. There isn't anything exactly like *Dimetrodon* before or after in the fossil record or in the modern biota, so what are the functional, physiological, and behavioral implications of that unique morphology? The answers to these questions become even more significant when we consider the unqualified success of non-mammalian synapsids during the late Paleozoic and the early Mesozoic. Why were animals like *Dimetrodon* or the dicynodonts able to dominate the terrestrial ecosystems of their time, and why did their fortunes seem to change during the Triassic, as the archosaur radiation began? Answering these questions requires a deep, multifaceted examination of synapsid morphology, paleobiology, and paleoecology, but there are key pitfalls that must be avoided if we are to progress beyond our current level of understanding.

The first pitfall concerns the source of much of the conventional wisdom about various aspects of synapsid paleobiology. Ideas such as *Ophiacodon* and *Lystrosaurus* being semiaquatic, the sails of *Dimetrodon* and *Edaphosaurus* having a thermoregulatory function, or "pelycosaurs" being cold-blooded and slow-growing were proposed long ago (e.g., a semiaquatic lifestyle for *Ophiacodon* and *Lystrosaurus* was first suggested in 1907 and 1903, respectively) and have become entrenched in the synapsid literature. However, the origins of many of these hypotheses can be traced to passing, speculative statements in the older literature that do not stand up well to modern scrutiny (see King 1991 and Felice and Angielczyk 2014 for reevaluations of the semiaquatic lifestyle of *Lystrosaurus* and *Ophiacodon*). Even in cases such as the proposed thermoregulatory function of the sail of *Dimetrodon*, which stem from more rigorous work (Romer 1948, Robard 1949), there is often significantly more disagreement about the details of the idea and potential alternatives than is apparent from textbook treatments of synapsids (see the earlier discussion of "pelycosaurs" for the controversy surrounding the function of their sails). Yet, these hypotheses are often uncritically recycled in the literature and color analytical design and interpretation of results in ways that are not justified. For example, Germain and Laurin (2005) examined the bone microstructure of the radius in *Ophiacodon* and a number of other fossil and extant tetrapods. They found that the low compactness of the bone and the gradual transition between the cortex and the medullary cavity in *Ophiacodon* was reminiscent of the structure of extant semiaquatic tetrapods, but it was an outlier in a linear discriminant analysis they conducted on their data. Despite the implication that the radius structure was not precisely like that of extant terrestrial, semiaquatic, or aquatic tetrapods, they tentatively concluded that it likely reflected a semiaquatic lifestyle because *Ophiacodon* was "known" to have had this ecology. Sometimes conventional wisdom about synapsid paleobiology does stand up to modern scrutiny, however. Geist (1972) and Barghusen (1975) hypothesized that the distinctive, heavily pachyostosed skulls of tapinocephalid dinocephalians evolved for head-butting intraspecific combat. This hypothesis was not seriously tested until Benoit *et al.* (2016a, 2017a) investigated the endocranial structure and paleoneurology of the tapinocephalid *Moschops capensis*. Their documentation of the heavily ossified braincase, reorientation of the brain cavity relative to the long axis of the skull, and the likely presence of thick layers of connective tissue in parts of the brain cavity and pineal tube add to the list of apparent adaptations in tapinocephalids for the use of the head as a weapon, corroborating Geist and Barghusen's hypothesis. Further investigation of other aspects of this unusual morphology, such as biomechanical modeling, will doubtless provide additional insight, but this example demonstrates how rigorous testing of older hypotheses will improve our knowledge of non-mammalian synapsids. Similarly, conventional wisdom about the group should not be uncritically accepted until such testing has occurred.

A second pitfall is assuming that information known about a particular model synapsid will be applicable to all members of its clade, or to non-mammalian synapsids as a whole, without considering the morphological, functional, and physiological diversity involved. A useful example of this problem can be found in dicynodont jaw mechanics. Crompton and Hotton (1967) developed the quantitative model that forms the foundation for our current understanding of the dicynodont feeding system. Their model was primarily derived from the study of a single specimen of the early dicynodont *Pristerodon mackayi* (called *Emydops* in that paper), although they also investigated differences between *Pristerodon* and *Lystrosaurus* that arise from the latter taxon's divergent skull morphology. Aspects of Crompton and Hotton's work were incorporated into studies of the feeding systems of certain other dicynodonts and non-dicynodont anomodonts (e.g., Cluver 1970, Barghusen 1976, King *et al.* 1989, King 1994), but only Renaut (2000, 2001) formally applied their model to another dicynodont species (*Kannemeyeria simocephalus*). It is tempting to assume that *Pristerodon*, *Lystrosaurus*, and *Kannemeyeria* adequately cover the phylogenetic and morphological diversity of dicynodonts, and that we therefore know how the dicynodont feeding system worked. However, a great deal of the morphological disparity of dicynodont skulls and mandibles is related to variations in the sizes and positions of jaw muscle attachments, and there has been very little consideration of the functional implications of this diversity of form (for an exception, see Angielczyk *et al.* 2018). When combined with factors such as the potential for mandibular kinesis in at least some dicynodonts (Cox and Angielczyk 2015), the truth is that we only have a superficial understanding of the dicynodont feeding system. A more detailed consideration has the potential to not only clarify why there are so many variations on the basic dicynodont skull ground plan, but also to serve as a vehicle for addressing issues of broad evolutionary interest, such as whether there are trends for optimization of functional complexes over time, the effects of integration and modularity in skull and jaw evolution, many-to-one relationships between form and function, and niche partitioning in some of the oldest herbivore-dominated terrestrial ecosystems.

It is encouraging that there has been an upswing in recent years in the number of papers dealing with many aspects of synapsid paleobiology and paleoecology. Examples of topics addressed include anatomy (Castanhinha *et al.* 2013, Araújo *et al.* 2017, Benoit *et al.* 2017c, Laaß *et al.* 2017), functional morphology (Kemp 2007, Jasinoski *et al.* 2009, 2010a, Laaß 2014, 2016), neuroanatomy (Laaß 2015, Benoit *et al.* 2016c, 2017d, 2017e, Laaß and Kaestner 2017, Araújo *et al.* 2018), diel activity patterns (Angielczyk and Schmitz 2014), growth and life history (e.g., Botha-Brink and Angielczyk 2010, Botha-Brink *et al.* 2012, 2016, Chinsamy-Turan and Ray 2012, Huttenlocker and Rega 2012, Ray *et al.* 2012, Huttenlocker and Botha-Brink 2013, 2014, Huttenlocker 2014, Jasinoski *et al.* 2015, O'Meara and Asher 2015, Shelton and Sander 2015, Jasinoski and Abdala 2017a, b), tooth attachment (LeBlanc *et al.* 2016), paleopathology (Vega and Maisch 2014, Whitney *et al.* 2017), the evolution of mammal-like elevated metabolic rates (Huttenlocker and Farmer 2017, Olivier *et al.* 2017, Rey *et al.* 2017), lifestyle adaptations (Fröbisch and Reisz 2009, Canoville and Laurin 2010, Nasterlack *et al.* 2012, Brink *et al.* 2014, Felice and Angielczyk 2014), biogeography (Rubidge 2005, Fröbisch 2009, Sidor *et al.* 2013, Angielczyk *et al.* 2014, Huttenlocker *et al.* 2015, Huttenlocker and Sidor 2016, Bernardi *et al.* 2017, Brocklehurst *et al.* 2017), and community and food web structure (Roopnarine *et al.* 2007, 2018, Roopnarine and Angielczyk 2012, 2015, 2016, Smith *et al.* 2012, Codron *et al.* 2017). As the size of available data sets for synapsids have grown, there also has been new interest in addressing macroevolutionary problems concerning the diversity and disparity through time of various synapsid clades, as well as the quality of their fossil records (Nicolas and Rubidge 2009, Abdala and Ribeiro 2010, Irmis and Whiteside 2012, Brocklehurst *et al.* 2013, Fröbisch 2013, 2014, Irmis *et al.* 2013, Pearson *et al.* 2013, Ruta *et al.* 2013a, b, Walther and Fröbisch 2013, Brocklehurst and Fröbisch 2014). Such studies represent only the tip of the iceberg in terms of possible analyses—the non-mammalian synapsid record is so rich and (at least intrabasinally) stratigraphically well resolved that it offers immense opportunities for quantitative analysis.

5.4.2 Non-Mammalian synapsids and the origins of mammalian characters

Modern mammals are extremely distinctive when compared with other extant tetrapods, and it has long been recognized that non-mammalian synapsid fossils can provide unique insights into the evolution of this distinctive phenotype (for a noteworthy early example, see Palmer 1913). Indeed, the evolutionary transformation from very reptile-like basal synapsids to extremely mammal-like early mammaliaforms is so well documented in the fossil record that non-mammalian synapsids are a textbook example of how fossils provide evidence of major macroevolutionary transitions (e.g., Kemp 1982, 2005, Hopson 1987, Cain 1988, Prothero 2007, Angielczyk 2009, Asher 2012). The evolutionary history of the characteristic features of the mammalian cranium has been especially thoroughly studied, revealing the stepwise origins of such important characters as a complete secondary palate, middle ear ossicles, and heterodont dentition with complex occlusion. Here, we will present a brief overview of the evolution of these characters in their mammalian form, as well as discuss the complexity of their distribution across Synapsida as a whole.

Mammals are distinguished from most other tetrapods by the presence of a bony secondary palate (often called the "hard palate") composed of laminar outgrowths of the premaxilla, maxilla, and palatine bones. The most important function of the modern mammalian secondary palate is related to endothermy—hard separation of the nasal and oral cavities improves heat and water retention during breathing, particularly in conjunction with the nasal turbinates (McNab 1978, Hillenius 1992, Owerkowicz *et al.* 2015). This enables mammals to maintain a high metabolic rate without dangerous loss of heat and moisture during respiration. As such, the origin of the secondary palate in deep mammalian evolution has been suggested as a proxy for the origin of endothermy in this lineage (Ruben *et al.* 2012).

The earliest known homologous precursor of the mammalian secondary palate is found in early cynodonts (Fig. 5.17). The Permian cynodonts *Procynosuchus* and *Dvinia* lack a closed secondary palate (i.e., one in which the bony extensions of the marginal elements meet on the midline, covering the vomer) but exhibit medial expansions of the maxillae and anterior palatine relative to the condition in gorgonopsians and early therocephalians (Kemp 1979, Hopson 1991). In these early cynodonts, the medial maxillary-palatine expansions near but do not contact the vomer (indeed, some laterally compressed specimens of *Procynosuchus* were previously believed to represent distinct, more advanced species because of apparent overlap of the maxillae with the vomer due to taphonomic distortion; Abdala and Allinson 2005). The maxillary-palatine expansions are even broader in galesaurids but still do not actually contact the vomer (Sidor and Smith 2004). The first closed secondary palate in Cynodontia occurs in Thrinaxodontidae, and the basic palatal architecture of *Thrinaxodon* is retained throughout non-mammalian cynodont evolution (Fourie 1974, Hopson and Barghusen 1986). The maxillae are the primary contributors to the secondary palate, and they meet on the midline to form a broad palatal plate with a smaller posterior palatine contribution. The premaxilla is also expanded relative to the ancestral theriodont condition, but it is partially separated from the main maxillary-palatine palatal plate by paired incisive foramina. The vomer is largely obscured in ventral view, with only minor exposure between the incisive foramina anteriorly but more extensive exposure between the primary palatal portions of the palatines posteriorly. Subsequent evolution of the secondary palate in cynodonts consists largely of elaboration on the basic *Thrinaxodon* morphology: increased expansion of the premaxilla, with smaller incisive foramina and greater coverage of the anterior vomer, more robust midline suturing of the palatal plate, and most importantly, posterior extension of the palatine portion of the plate, covering the majority of the primary palate with the secondary in ventral view (Hopson 1994). Posterior extension of the palatine contribution to the secondary palate is relatively poorly developed in cynognathians but is substantial in probainognathians, with the mammalian condition essentially being reached in chiniquodontids (in which the bony secondary palate extends to the end of the postcanine tooth row) (Hopson and Kitching 2001, Abdala and Giannini 2002).

As mentioned above, the origin of the bony secondary palate in cynodonts has been argued to be a correlate of the evolution of endothermy along the mammalian stem. However, this idea is complicated by the fact that a bony secondary palate is not unique to mammals among extant tetrapods, also being present in ectothermic crocodylians and a few lizards and turtles (Brock 1941, Langston 1973, Pritchard 1997). Indeed, the bony secondary palate is not even unique to the mammalian lineage within Synapsida, having also evolved in dicynodonts and multiple times within therocephalians (Kemp 1972b, King 1988, Hopson and Barghusen 1986) (Fig. 5.17). In dicynodonts, the secondary palate is formed primarily by an expanded posterior premaxillary plate, with small lateral contributions from the maxilla and palatine. Therocephalians show a variety of non-homologous secondary palate morphologies. The most unusual condition is in whaitsiids, in which the secondary palate is formed by narrow lateral outgrowths of the vomer (giving the vomer a cruciate shape overall) that extend to suture with the maxillae (Mendrez 1975, Huttenlocker and Abdala 2015). A morphology convergent on that of cynodonts is found in the Baurioidea, with early forms (e.g., *Ictidosuchoides*) having expanded medial edges of the maxillae but no contact with the vomer. Later baurioids formed the secondary palate in somewhat different ways: in regisaurids, it is primarily an anterior structure composed of an anteriorly broadened vomer and expanded premaxilla, with only a short anteromedial contribution from the maxilla (Mendrez 1972), whereas in lycideopids the maxillae are broadly sutured to the vomer along their medial edges (but do not cover it ventrally, there is no midline maxillary contact) (Mendrez 1975). Secondary palate development is taken to an extreme in bauriids proper, which achieve a condition comparable with basal eucynodonts, with a complete secondary palate extending roughly halfway the length of the postcanine tooth row (Abdala *et al.* 2014a). The variety of secondary palate morphologies within early synapsids, and cases in which

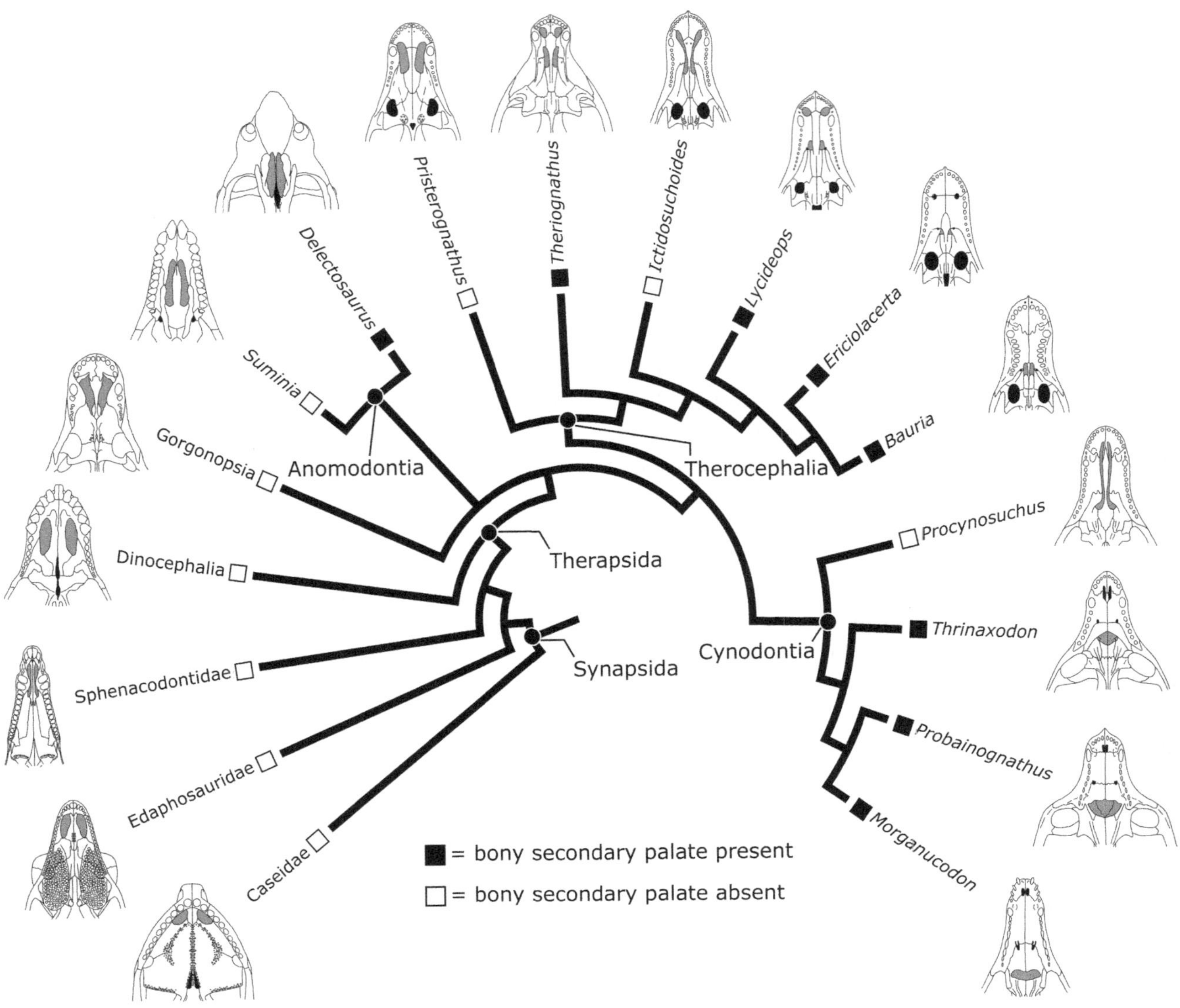

Fig. 5.17: Simplified phylogenetic tree of Synapsida summarizing the distribution of the bony secondary palate. Drawings are ventral views of the snouts of exemplar taxa (not to scale) with the choanae shown in grey; other openings in the palate (e.g., incisive foramina, interpterygoid vacuity, pterygoid fenestrae) are shown in black. Note that a complete secondary palate evolved multiple times within Synapsida (once within Anomodontia, once within Cynodontia, and multiple times within Therocephalia), with the premaxillae, maxillae, vomers, and palatines making differing contributions to the structure in each case. Drawings redrawn and modified from Efremov (1940b), Romer and Price (1940), Kemp (1969), Hopson and Barghusen (1986), Rybczynski (2000), Hopson and Kitching (2001), Maddin *et al.* (2008), Huttenlocker (2013), and Huttenlocker and Abdala (2015).

the formation of the secondary palate would not actually create an osseous barrier between the nasal and the oral capsules (e.g., the vomerine extensions of whaitsiids), suggest that selective pressures other than the development of endothermy underlie the origin of this feature. Thomason and Russell (1986) demonstrated that the secondary palate confers significant torsional strength to the snout in extant mammals and argued that stiffening the rostrum may have been at least as important for the evolution of the secondary palate as energetic concerns. This is an appealing explanation for the initial secondary palate morphologies of cynodonts and therocephalians, which are more of a brace between maxilla and vomer than any real closing-off of the palate. In dicynodonts, development of the secondary palate can be interpreted as an adaptation for herbivory—in many dicynodonts, the palatine surface of the palate is highly rugose, indicating the palate had a keratinous covering used in mastication (Crompton and Hotton 1967, Angielczyk 2004). Thus, although the bony secondary palate undeniably plays an important role in modern mammalian endothermy, this is probably exaptive rather than adaptive.

The transformation of the ancestral amniote jaw bones into the middle ear ossicles of mammals is one of the best-known macroevolutionary changes in the fossil record (Fig. 5.18). “Pelycosaurs” retain the primitive complement of jaw bones: dentary, splenial, anterior and posterior coronoids, prearticular, angular, surangular, and articular (Romer and Price 1940, Sidor 2003). Subsequent evolution of the synapsid jaw shows a trend toward increase in the size of the dentary at the expense of the other jaw bones, culminating in the mammalian condition in which the dentary is the only mandibular element (Allin 1975, Luo 2011, Kemp 2016, Luo *et al.* 2016). The anterior coronoid was the first element to be lost on the lineage leading to mammals, and the posterior coronoid is reduced in size in all therapsids. Compared with “pelycosaurs”, therapsids also exhibit a trend toward restriction of the postdentary bones to the medial face of the jaw: the posterior coronoid goes from making up the dorsal edge of the mandible to being an exclusively medial element, and the splenial goes from making up the ventral edge of the mandible to becoming primarily a medial lamina in therapsids (although it still contributes to the ventral margin of the jaw symphysis in all groups other than eutheriodonts). Loss of the posterior coronoid was once considered a synapomorphy of a dinocephalian + anomodont clade, but it is now clear that this element was lost independently in these two groups, as it is still present in “basal anomodonts” and non-tapinocephalid dinocephalians (Hopson 1991, Grine 1997, Liu *et al.* 2010).

In cynodonts, early taxa such as *Procynosuchus* retain a large set of postdentary bones, although the reflected lamina of the angular is already reduced in size relative to non-cynodont therapsids (Kemp 1979). Moving crownward, however, these elements dramatically reduce in size (Fig. 5.18), with galesaurids and thrinaxodontids exhibiting an expanded coronoid process of the dentary and anteroposteriorly shortened angular (Fourie 1974, Sidor and Smith 2004). By Eucynodontia, the angular, surangular, prearticular, and articular have become narrow, rodlike elements situated in the postdentary trough, an elongate fossa on the medial face of the dentary. Of the non-dentary lower jaw bones, only the splenial and posterior coronoid remain laminar elements in eucynodonts and are isolated from the postdentary rod. This basic configuration is retained through early mammals, albeit with further size reduction of the non-dentary bones and, eventually, separation from direct contact with the dentary (they are still connected to the dentary via an ossified Meckel’s cartilage, however; Meng *et al.* 2011). Complete detachment of the postdentary bones to form the definitive mammalian middle ear (with the reflected lamina of the angular becoming the ectotympanic, prearticular-articular becoming the malleus, and, from the cranium, the quadrate becoming the incus) was previously considered a synapomorphy of crown mammals (Rowe 1988). However, recent research on the Australian stem-monotreme *Teinolophos* confirms that it retained an ossified Meckel’s cartilage connected to the ear ossicles, and that the evolution of the definitive mammalian middle ear must have occurred at least twice (Rich *et al.* 2016).

Mammals exhibit remarkable dental disparity, and their advanced masticatory ability is often cited as a key reason for their success (Crompton 1963b, Turnbull 1970). Heterodonty, complex cusps, and precise crown-to-crown occlusion are not unique to mammals among vertebrates, but they are developed to their greatest extent within the clade. Simple heterodonty is ancestral for synapsids (and probably amniotes in general), with enlarged maxillary caniniforms present in eothyridids (Reisz *et al.* 2009), but mammalian-style heterodonty first appears in therapsids, with the establishment of fixed incisor, canine, and postcanine regions (Rubidge and Sidor 2001). Tooth counts for these regions vary widely among early therapsids—among incisors, for instance, some clades have very conservative counts (five upper and four lower incisors are present in almost all anteosaurs and gorgonopsians), whereas others exhibit extensive variation (between four and seven upper incisors in therocephalians) that can even change during ontogeny (Mendrez 1975, Hopson 1991, van den Heever 1994, Kammerer 2011, 2016c). Almost all therapsids have a single functional canine (with the possible exception of the enigmatic early taxon *Raranimus*; Liu *et al.* 2009), but many taxa have additional precanines, incisiform teeth rooted in the maxilla (rather than the premaxilla like true incisors). Non-mammalian therapsids typically have homodont postcanines, although variation in cusp morphology along the tooth row is present in many cynodonts. Extreme heterodonty of the postcanines is present in cynognathian eucynodonts, in which the tooth row includes both laterally compressed, sectorial postcanines and labiolingually expanded, cingulated gomphodont postcanines (Hopson and Kitching 2001, Abdala and Ribeiro 2003). This heterodonty is partially ontogenetic, with the sectorial dentition of juveniles being replaced by gomphodont dentition in adults, but at least some sectorial teeth are retained at the posterior end of the tooth row even in adults of taxa such as *Diademodon* (Hopson 2005). In tritheledontids, the anterior and posterior upper postcanines differ substantially in complexity, with the posterior postcanines being more complex and in some taxa exhibiting a buccal cingulum, rendering them more “molariform”

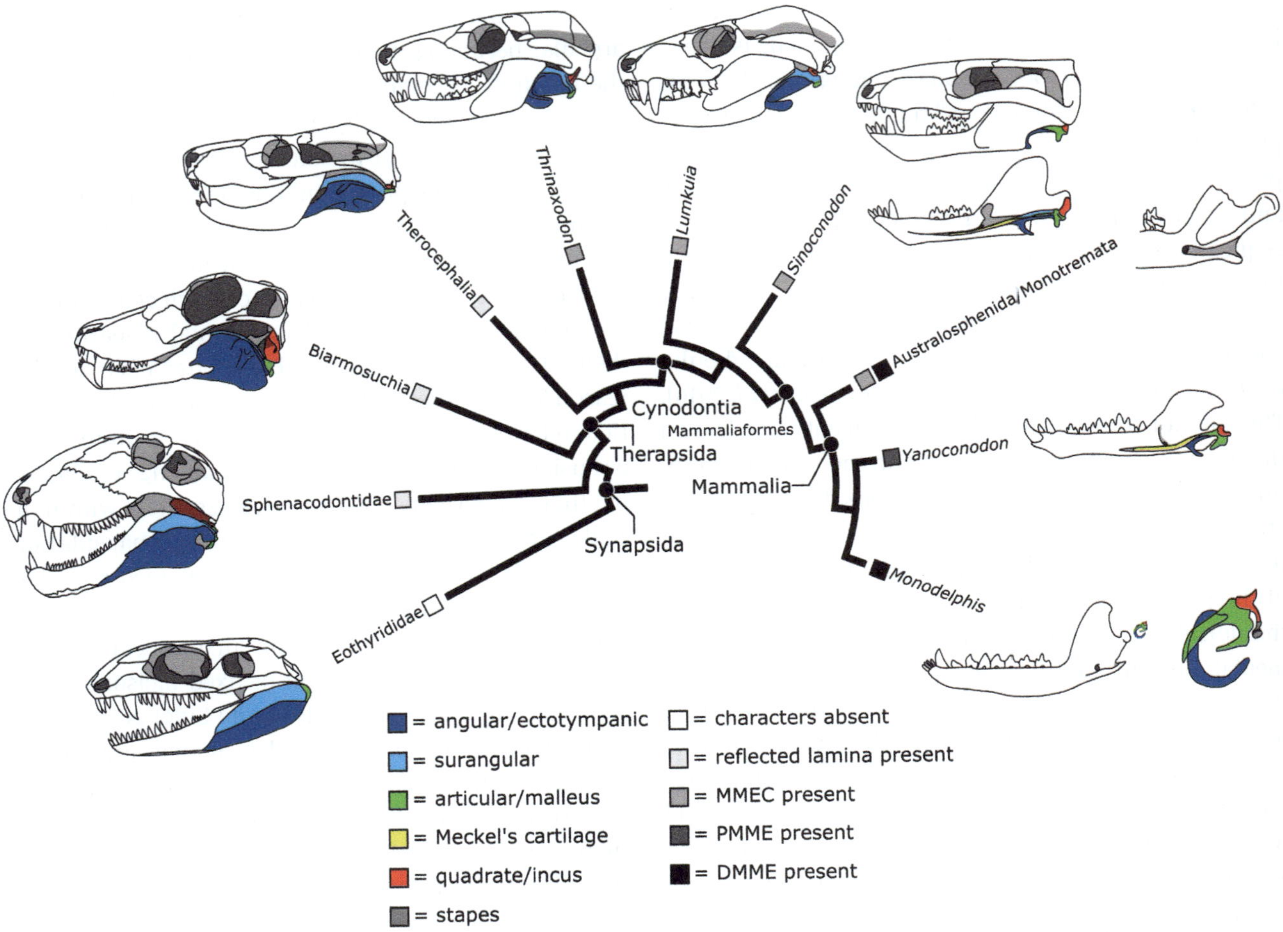

Fig. 5.18: Simplified phylogenetic tree summarizing the evolutionary transformation of the ancestral synapsid postdentary bones into the middle ear bones of the mammalian hearing system. Early synapsids such as *Eothyris* have a full complement of postdentary bones that are structural components of the mandible and only seem to have been involved in feeding. A posteriorly emarginated ventral keel of the angular bone, the reflected lamina, evolved in the "haptodonts" and is present in sphenacodontids and therapsids. The original function of the reflected lamina likely was to provide an expanded space for jaw musculature attachment, not the detection of airborne sound (e.g., Barghusen 1968, Allin 1975). The postdentary bones of non-cynodont therapsids are still structural elements of the mandible, but the reflected lamina has increased in size and often is quite thin. The reflected lamina may have played a role in sound transmission in these taxa, but the details of the associated soft tissue anatomy and the sensitivity of the system to airborne sound and ground-borne vibrations is debated (e.g., Allin 1975, Allin and Hopson 1992, Maier and van den Heever 2002, Laaß 2014, 2016). The postdentary bones and the reflected lamina are reduced in size to varying degrees in non-mammaliaform cynodonts, and the presence of a tympanic membrane supported in part by the reflected lamina has been inferred in members of this grade such as *Thrinaxodon* and *Lumkuia* (e.g., Allin 1975, 1986, Allin and Hopson 1992, Luo and Crompton 1994, although see Kemp 2007 for an alternate view). This morphology has been referred to as the mandibular middle ear of cynodonts (MMEC) in reviews such as Luo (2011) and Luo *et al.* 2016. The postdentary bones of early mammaliforms such as *Sinoconodon* are further reduced in size, and the primary jaw joint is now between the dentary condyle and the squamosal. On the medial surface of the mandible, the postdentary bones rest in a postdentary trough, with a further connection via the straight Meckel's cartilage, which extends into the Meckel's sulcus (e.g., Luo 2011, Luo *et al.* 2016). However, the basic organization of the postdentary bones is essentially the same as in the MMEC. Evolution of the postdentary bones in Mammalia is complex, and the definitive mammalian middle ear (DMME), in which the postdentary bones are highly miniaturized and completely detached from the mandible, likely evolved at least twice (once in monotremes and once in Theriiformes) (see reviews in Luo 2011, Luo *et al.* 2016). Extant monotremes possess a DMME, but early monotremes such as *Teinolophos* and the monotreme stem lineage (Australosphenida) possess a prominent postdentary trough (shown in grey), implying the presence of an MMEC (well-preserved postdentary bones have not been discovered for these taxa). In stem theriiforms, such as the eutriconodont *Yanoconodon*, the postdentary bones are still connected to the mandible by an ossified Meckel's cartilage, but the Meckel's cartilage bends medially, displacing the postdentary bones away from the mandible and positioning them closer to the basicranium. This condition is referred to as the partial mammalian middle ear (PMME; Luo 2011, Luo *et al.* 2016) or transitional mammalian middle ear (Meng *et al.* 2011). A DMME evolved at least once in Theriiformes but may have evolved twice in the clade (once in multituberculates and once in therians) depending on whether the PMME in taxa such as the spalacotherioid *Maotherium* (not shown) represents the true ancestral condition for Theriiformes or a secondary reversal to a primitive state (Luo 2011, Luo *et al.* 2016). Drawings are not to scale and have been redrawn and modified from Romer and Price (1940), Hopson and Kitching (2001), Rich *et al.* (2005), Sidor and Rubidge (2006), Reisz *et al.* (2009), Huttenlocker and Abdala (2015), and Luo *et al.* 2016.

(Gow 1980, Martinelli *et al.* 2005). However, true separation of the postcanine tooth row into distinct premolars and molars is not recognized until Mammaliaformes (Kielan-Jaworowska *et al.* 2004).

Despite our thorough understanding of the assembly of various aspects of the mammalian skeleton based on synapsid fossils, there are still a number of iconic mammalian characters for which the synapsid fossil record has stubbornly refused to yield much information at all. For example, the oldest definitive fossil evidence of hair comes from the Middle Jurassic (Callovian) of China (Ji *et al.* 2006, Meng *et al.* 2006, Zhou *et al.* 2013; for a particularly striking example from the Cretaceous, see Martin *et al.* 2015). However, all of these examples are from taxa within Mammaliaformes, and there is very little direct evidence about the integument of non-mammalian synapsids in the fossil record. Reisz (1975) mentioned the presence of ventral "scales" in Carboniferous "pelycosaurs", but these represent ossified subcutaneous structures homologous with gastralia, not external integumentary scales. Based on preserved skin impressions, Niedźwiedzki and Bojanowski (2012) proposed that "pelycosaurs" had scales at least on the ventral surfaces of the body. This hypothesis seems to be confirmed in a spectacular fashion by the recent discovery of the exceptionally preserved varanopid *Ascendonanus*, the known specimens of which preserve a scaly skin covering that is at least superficially very similar to that of extant lepidosauromorphs (Spindler *et al.* 2018). Chudinov (1968b) described what he interpreted as remains of epidermal tissue in specimens of the dinocephalian therapsid *Estemmenosuchus*, which seems to have had scaleless, glandular skin. Potential impressions of mummified skin preserved in therapsid fossils from the Karoo Basin also have been noted but not described in detail (Smith and Botha-Brink 2014), and none of the known non-mammalian synapsid skin impressions preserve direct evidence of hair. It has been suggested on various occasions that the facial pits found in theriodont therapsids are an osteological correlate for the presence of vibrissae (e.g., Watson 1931, Broili 1941, Attridge 1956, Brink 1956, Tatarinov 1967, Findlay 1968, Lingham-Soliar 2014), which would imply a middle to late Permian origin of hair, but given that similar foramination is associated with the tooth row in reptiles, this is questionable (van Valen 1960). Although the presence of vibrissae and/or hair outside of advanced cynodonts has been debated (Estes 1961, Tatarinov 1967, Hillenius 2000, Ruben and Jones 2000), the recent discovery of hairlike filaments in late Permian coprolites (Smith and Botha-Brink 2011, Bajdek *et al.* 2015) has again raised the possibility that some Paleozoic therapsids had hair.

Extensive biological research has been carried out on the evolution of hair, and a number of scenarios about the early stages of this process, which presumably occurred among non-mammalian synapsids, have been proposed (e.g., Alibardi 2003, 2012, Maderson 2003, Wu *et al.* 2004, 2008, Eckhart *et al.* 2008, Dhouailly 2009, Vandebergh and Bossuyt 2012, Alibardi and Rogers 2015). Given the paucity of information that the fossil record has been able to provide about the evolution of hair, are this and other topics dealing with soft tissue structures and physiology, such as the evolution of mammalian endothermy, best left to purely neontological studies? We argue that the answer is no, and a recent paper by Benoit *et al.* (2016b) demonstrates how a careful, informed consideration of synapsid fossils, when synthesized with neontological data, can yield unique insights into the evolution of mammalian characters.

Benoit *et al.* (2016b) CT scanned a series of theriodont skulls and reconstructed the pathway of the maxillary canal, which transmits the maxillary branch of the trigeminal nerve (cranial nerve V_2), the source of innervation for the vibrissae and soft tissues of the snout. They found that gorgonopsians, therocephalians, and early cynodonts have long, incompletely ossified, ramifying maxillary canals that exit to the surface of the snout through a series of foramina. This is comparable with the condition seen in many extant reptiles, and they interpreted this as indicating that vibrissae and fleshy lips were absent in these taxa. However, basal probainognathian cynodonts had simplified maxillary canals that exited to the surface of the snout through a small number of foramina, and in prozostrodont cynodonts, this trend was carried further, such that they presented essentially the same condition as mammals (i.e., a short, well-ossified, simple maxillary canal that exits through a single, large infraorbital foramen). These osteological features are consistent with thick, mammal-like lips and vibrissae in prozostrodonts, with the evolution of these characters likely taking place among more basal probainognathians in the Middle to Late Triassic. Interestingly, several other significant morphological changes occur near the base of Probainognathia, including loss of the pineal foramen and expansion of the cerebellum. In extant mammals, ossification of the frontoparietal fontanel, which is in a similar position to the pineal foramen, is linked to the homeogene *MSX2*. *MSX2* also has been linked to ossification of the middle ear, development of the mammary glands and cerebellum, and maintenance of hair follicles (see review in Benoit *et al.* 2016b). Benoit *et al.* (2016b) noted that the combined paleontological and neontological evidence suggests that mutation of the *MSX2* gene likely occurred near the base of Probainognathia, in the late Middle Triassic, and that the origin and/or important changes in several characters typical of mammals might have taken place at this time. They correctly stated that because *MSX2* is involved in the maintenance of hair follicles, and not their development, the evolution of hair might have occurred earlier in synapsid history, before the divergence of Probainognathia. However, important changes in aspects of the integument, such as the origin of vibrissae or expansion of the pelage, could have accompanied the inferred mutation of *MSX2*. The insights offered by the study of Benoit *et al.* (2016b) would have been difficult to obtain if they only considered one or two representative synapsid species, did not have access to modern imaging techniques, and if they did not consider data from developmental genetics. Indeed, the long-standing debate about the presence of vibrissae in theriodonts (e.g., Watson 1931, Broili 1941, Attridge 1956, Brink 1956, Estes 1961, Tatarinov 1967, Findlay 1968, Hillenius 2000, Ruben and Jones 2000, Lingham-Soliar 2014) underscores the difficulty of resolving this question with more limited data sets. We hope that the evolution of other mammalian characteristics will be reassessed in similarly synthetic ways, taking full advantage of the extensive synapsid fossil record and the strides that have been made by neontologists.

5.5 Conclusion

Aside from a handful of noteworthy exceptions such as *Dimetrodon*, non-mammalian synapsids are largely unknown to the general public. Even in the paleontological community, they are a relatively obscure group. Yet, they form the foundation of our understanding of the evolution of mammals, and they were the most successful tetrapods of their time by almost every measure. Our goal for this review is to inspire more research on these animals, which we find so fascinating, and also to encourage a more nuanced consideration of non-mammalian synapsids in studies on the evolution of mammalian characters. Synapsid history is our history, and we owe it to these amazing animals to give them their scientific due.

Acknowledgments

We thank F. Zachos and R. Asher for the invitation to write this paper and their patience while we prepared it. We are very grateful to curators, collections managers, and friends at the many museums we have visited for their assistance in accessing specimens. C.F.K.'s work is supported by the Deutsche Forschungsgemeinschaft (KA 4133/1-1). We thank R. Asher, F. Zachos, and an anonymous referee for their helpful reviews of the manuscript.

Literature

Abdala, F. (2007): Redescription of *Platycraniellus elegans* (Therapsida, Cynodontia) from the Lower Triassic of South Africa, and the cladistic relationships of eutheriodonts. *Palaeontology* 50: 591–618.

Abdala, F. & Allinson, M. (2005): The taxonomic status of *Parathrinaxodon proops* (Therapsida: Cynodontia), with comments on the morphology of the palate in basal cynodonts. *Palaeontologia Africana* 41: 45–52.

Abdala, F. & Giannini, N.P. (2002): Chiniquodontid cynodonts: systematic and morphometric considerations. *Palaeontology* 45: 1151–1170.

Abdala, F. & Ribeiro, A.M. (2003): A new traversodontid cynodont from the Santa Maria Formation (Ladinian-Carnian) of southern Brazil, with a phylogenetic analysis of Gondwanan traversodontids. *Zoological Journal of the Linnean Society* 139: 529–545.

Abdala, F. & Ribeiro, A.M. (2010): Distribution and diversity patterns of Triassic cynodonts (Therapsida, Cynodontia) in Gondwana. *Palaeogeography, Palaeoclimatology, Palaeoecology* 286: 202–217.

Abdala, F., Hancox, P.J. & Neveling, J. (2005) Cynodonts from the uppermost Burgersdorp Formation, South Africa, and their bearing on the biostratigraphy and correlation of the Triassic *Cynognathus* Assemblage Zone. *Journal of Vertebrate Paleontology* 25: 192–199.

Abdala, F., Rubidge, B.S. & van den Heever, J. (2008): The oldest therocephalians (Therapsida, Eutheriodontia) and the early diversification of Therapsida. *Palaeontology* 51: 1011–1024.

Abdala, F., Jasinoski, S.C. & Fernandez, V. (2013): Ontogeny of the Early Triassic cynodont *Thrinaxodon liorhinus* (Therapsida): dental morphology and replacement. *Journal of Vertebrate Paleontology* 33: 1408–1431.

Abdala, F., Jashashvili, T., Rubidge, B.S. & van den Heever, J. (2014a): New material of *Microgomphodon oligocynus* (Eutherapsida, Therocephalia) and the taxonomy of southern African Bauriidae. In: Kammerer, C.F., Angielczyk, K.D. & Fröbisch, J. (eds.). *Early Evolutionary History of the Synapsida*. Springer, Dordrecht: 209–231.

Abdala, F., Kammerer, C.F., Day, M.O., Jirah, S. & Rubidge, B.S. (2014b): Adult morphology of the therocephalian *Simorhinella baini* from the middle Permian of South Africa and the taxonomy, paleobiogeography, and temporal distribution of the Lycosuchidae. *Journal of Paleontology* 88: 1139–1153.

Alibardi, L. (2003): Adaptation to the land: the skin of reptiles in comparison to that of amphibians and endotherm amniotes. *Journal of Experimental Zoology Part B: Molecular and Developmental Evolution* 298B: 12–41.

Alibardi, L. (2012): Perspectives on hair evolution based on some comparative studies on vertebrate cornification. *Journal of Experimental Zoology Part B: Molecular and Developmental Evolution* 318B: 325–343.

Alibardi, L. & Rogers, G. (2015): Observations on fur development in echidna (Monotremata, Mammalia) indicate that spines precede hairs in ontogeny. *The Anatomical Record* 298: 761–770.

Allin, E.F. (1975). Evolution of the mammalian middle ear. *Journal of Morphology* 147: 403–438.

Allin, E.F. (1986): The auditory apparatus of advanced mammal-like reptiles and early mammals. In: Hotton, N. III, MacLean, P.D., Roth, J.J. & Roth, E.C. (eds.). *The Ecology and Biology of Mammal-like Reptiles*. Smithsonian Institution Press, Washington DC: 283–294.

Allin, E.F. & Hopson, J.A. (1992): Evolution of the auditory system in Synapsida ("mammal-like reptiles" and primitive mammals) as seen in the fossil record. In: Webster, D.B., Fay, R.R. & Popper, A.N. (eds.). *The Evolutionary Biology of Hearing*. Springer Verlag, New York: 587–614.

Alroy, J. 2013. Online paleogeographic map generator. http://paleodb.org/?a=mapForm

Amson, E. & Laurin, M. (2011): On the affinities of *Tetraceratops insignis*, an Early Permian synapsid. *Acta Palaeontologica Polonica* 56: 301–312.

Anderson, J.M. & Cruickshank, A.R.I. (1978): The biostratigraphy of the Permian and the Triassic. Part 5. A review of the classification and distribution of Permo-Triassic tetrapods. *Palaeontologia Africana* 21: 15–44.

Anderson, J.S. & Reisz, R.R. (2004): *Pyozia mesenensis*, a new small varanopid (Synapsida, Eupelycosauria) from Russia: "pelycosaur" diversity in the Middle Permian. *Journal of Vertebrate Paleontology* 24: 173–179.

Andersson, K., Norman, D. & Wederlin, L. (2011): Sabretoothed carnivores and the killing of large prey. *PLoS One* 6(10): e24971.

Angielczyk, K.D. (2001): Preliminary phylogenetic analysis and stratigraphic congruence of the dicynodont anomodonts (Synapsida: Therapsida). *Palaeontologia Africana* 37: 53–79.

Angielczyk, K.D. (2002): Redescription, phylogenetic position, and stratigraphic significance of the dicynodont genus *Odontocyclops* (Synapsida: Anomodontia). *Journal of Paleontology* 76: 1047–1059.

Angielczyk, K.D. (2004): Phylogenetic evidence for and implications of a dual origin of propaliny in anomodont therapsids (Synapsida). *Paleobiology* 30: 268–296.

Angielczyk, K.D. (2007): New specimens of the Tanzanian dicynodont "*Cryptocynodon*" *parringtoni* von Huene, 1942 (Therapsida, Anomodontia), with an expanded analysis of Permian dicynodont phylogeny. *Journal of Vertebrate Paleontology* 27: 116–131.

Angielczyk, K.D. (2009): *Dimetrodon* is not a dinosaur: using tree thinking to understand the ancient relatives of mammals and their evolution. *Evolution: Education and Outreach* 2: 257–271.

Angielczyk, K.D. & Kammerer, C.F. (2017): The cranial morphology, phylogenetic position and biogeography of the upper Permian dicynodont *Compsodon helmoedi* van Hoepen (Therapsida, Anomodontia). *Papers in Palaeontology* 3: 513–545.

Angielczyk, K.D. & Kurkin, A.A. (2003a): Phylogenetic analysis of Russian Permian dicynodonts (Therapsida: Anomodontia): implications for Permian biostratigraphy and Pangaean biogeography. *Zoological Journal of the Linnean Society* 139: 157–212.

Angielczyk, K.D. & Kurkin, A.A. (2003b): Has the utility of *Dicynodon* for Late Permian terrestrial biostratigraphy been overstated? *Geology* 31: 363–366.

Angielczyk, K.D. & Rubidge, B.S. (2010): A new pylaecephalid dicynodont (Therapsida, Anomodontia) from the *Tapinocephalus* Assemblage Zone, Karoo Basin, Middle Permian of South Africa. *Journal of Vertebrate Paleontology* 30: 1396–1409.

Angielczyk, K.D. & Schmitz, L. (2014): Nocturnality in synapsids predates the origin of mammals by over 100 million years. *Proceedings of the Royal Society of London. Series B* 281: 20141642.

Angielczyk, K.D. & Sullivan, C. (2008): *Diictodon feliceps* (Owen, 1876), a dicynodont (Therapsida, Anomodontia) species with a Pangaean distribution. *Journal of Vertebrate Paleontology* 28: 788–802.

Angielczyk, K.D., Sidor C.A., Nesbitt S.J., Smith R.M.H. & Tsuji, L.A. (2009): Taxonomic revision and new observations on the postcranial skeleton, biogeography, and biostratigraphy of the dicynodont genus *Dicynodontoides*, the senior subjective synonym of *Kingoria* (Therapsida, Anomodontia). *Journal of Vertebrate Paleontology* 29: 1174–1187.

Angielczyk, K.D., Steyer, J.S., Sidor, C.A., Smith, R.M.H., Whatley, R.L. & Tolan, S. (2014): Permian and Triassic dicynodont (Therapsida: Anomodontia) faunas of the Luangwa Basin, Zambia: taxonomic update and implications for dicynodont biogeography and biostratigraphy. In: Kammerer C.F., Angielczyk K.D. & Fröbisch J. (eds.). *Early Evolutionary History of the Synapsida*. Springer, Dordrecht: 93–138.

Angielczyk, K.D., Rubidge, B.S., Day, M.O. & Lin, F. (2016): A reevaluation of *Brachyprosopus broomi* and *Chelydontops altidentalis*, dicynodonts (Therapsida, Anomodontia) from the middle Permian *Tapinocephalus* Assemblage Zone of the Karoo Basin, South Africa. *Journal of Vertebrate Paleontology* 36: e1078342.

Angielczyk, K.D., Hancox, P.J. & Nabavizadeh, A. (2018): A redescription of the Triassic kannemeyeriiform dicynodont *Sangusaurus* (Therapsida, Anomodontia), with an analysis of its feeding system. In: Sidor, C.A. & Nesbitt S. J. (eds.). *Vertebrate and Climatic Evolution in the Triassic Rift Basins of Tanzania and Zambia*. Society of Vertebrate Paleontology Memoir 17: 189–227.

Anthwal, N., Joshi, L. & Tucker, A.S. (2013): Evolution of the mammalian middle ear and jaw: adaptations and novel structures. *Journal of Anatomy* 222: 147–160.

Araújo, R., Fernandez, V., Polcyn, M.J., Fröbisch, J. & Martins, R.M.S. (2017): Aspects of gorgonopsian paleobiology and evolution: insights from the basicranium, occiput, osseus labyrinth, vasculature, and neuroanatomy. *PeerJ* 5:e3119.

Araújo, R., Fernandez, V., Rabbitt, R.D., Ekdale, E.G., Antunes, M.T., Castanhinha, R., Fröbisch, J. & Martins, R.M.S. (2018): *Endothiodon* cf. *bathystoma* (Synapsida: Dicynodontia) bony labyrinth anatomy, variation and body mass estimates. *PLoS One* 13(3): e0189883.

Asher, R.A. (2012): *Evolution & Belief: Confessions of a Religious Paleontologist*. Cambridge University Press, Cambridge.

Attridge, J. (1956): The morphology and relationships of a complete therocephalian skeleton from the *Cistecephlaus* zone of South Africa. *Proceedings of the Royal Society of Edinburgh. Series B* 66: 59–93.

Aulie, R.P. (1974a): The origin of the idea of the mammal-like reptile. *American Biology Teacher* 36: 476–511.

Aulie, R.P. (1974b): The origin of the idea of the mammal-like reptile. II. The evidence accumulates. *American Biology Teacher* 36: 545–553.

Aulie, R.P. (1975): The origin of the idea of the mammal-like reptile. III. The mammal-like reptiles. *American Biology Teacher* 37: 21–32.

Bain, A.G. (1845): On the discovery of fossil remains of bidental and other reptiles in South Africa. *Transactions of the Geological Society of London 2nd series* 7: 53–59.

Bajdek, P., Qvarnström, M., Owocki, K., Sulej, T., Sennikov, A.G., Golubev, V.K. & Niedźwiedzki, G. (2015): Microbiota and food residues including possible evidence of pre-mammalian hair in Upper Permian coprolites from Russia. *Lethaia*. 49: 455–477.

Bakker, R.T. (1971): Dinosaur physiology and the origin of mammals. *Evolution* 25: 636–658.

Bakker, R.T. (1982): Juvenile-adult habitat switching in Permian fossil reptiles and amphibians. *Science* 217: 53–55.

Barghusen, H.R. (1968): The lower jaw of cynodonts (Reptilia, Therapsida) and the evolutionary origin of mammal-like adductor jaw musculature. *Postilla* 116: 1–49.

Barghusen, H.R. (1972): The origin of the mammalian jaw apparatus. In: Schumacher, G.H. (ed.). *Morphology of the Maxillomandibular Apparatus*. VEB Georg Thieme, Leipzig: 26–32.

Barghusen, H.R. (1973): The adductor jaw musculature of *Dimetrodon* (Reptilia, Pelycosauria). *Journal of Paleontology* 47: 823–834.

Barghusen, H.R. (1975): A review of fighting adaptations in dinocephalians (Reptilia, Therapsida). *Paleobiology* 1: 295–311.

Barghusen, H.R. (1976): Notes on the adductor jaw musculature of *Venjukovia*, a primitive anomodont therapsid from the Permian of the U.S.S.R. *Annals of the South African Museum* 69: 249–260.

Barry, T.H. (1974): A new dicynodont ancestor from the Upper Ecca (lower Middle Permian) of South Africa. *Annals of the South African Museum* 64: 117–136.

Battail, B. (1982): Essai de phylogénie des cynodontes (Reptilia: Therapsida). *Geobios Mémoire Spécial* 6: 157–167.

Battail, B. (2000): A comparison of Late Permian Gondwanan and Laurasian amniote faunas. *Journal of African Earth Sciences* 31: 165–174.

Baur, G. (1887): On the phylogenetic arrangement of the Sauropsida. *Journal of Morphology* 1: 93–104.

Baur, G. & Case, E.C. (1897): On the morphology of the skull of the Pelycosauria and the origin of mammals. *Anatomischer Anzeiger* 13: 109–120.

Baur, G. & Case, E.C. (1899): The history of the Pelycosauria, with a description of the genus *Dimetrodon*, Cope. *Transactions of the American Philosophical Society, New Series* 20: 5–62.

Bennett, S.C. (1996): Aerodynamics and thermoregulatory function of the dorsal sail of *Edaphosaurus*. *Paleobiology* 22: 496–506.

Benoit, J., Manger, P.R., Fernandez, V. & Rubidge, B.S. (2016a): Cranial bosses of *Choerosaurus dejageri* (Therapsida, Therocephalia): earliest evidence of cranial display structures in eutheriodonts. *PLoS One* 11 (8): e0161457.

Benoit, J., Manger, P.R. & Rubidge, B.S. (2016b): Palaeoneurological clues to the evolution of defining mammalian soft tissue traits. *Scientific Reports* 6: 25604.

Benoit, J., Abdala, F., Manger, P.R. & Rubidge, B.S. (2016c): The sixth sense in mammalian forerunners: variability of the parietal foramen and the evolution of the pineal eye in South African Permo-Triassic eutheriodont therapsids. *Palaeontologica Polonica* 61: 777–789.

Benoit, J. Manger, P.R., Norton, L., Fernandez, V. & Rubidge B.S. (2017a): Synchotron scanning reveals the palaeoneurology of the head-butting *Moschops capensis* (Therapsida, Dinocephalia). *PeerJ* 5: e3496.

Benoit, J., Norton, L.A., Manger, P.R. & Rubidge, B.S. (2017b): Reappraisal of the envenoming capacity of *Euchambersia mirabilis* (Therapsida, Therocephalia) using µCT-scanning techniques. *PLoS One* 12(2): e0172047.

Benoit, J., Jasinoski S.C., Fernandez V. & Abdala, F. (2017c): The mystery of a missing bone: revealing the orbitosphenoid in basal Epicynodontia (Cynodontia, Therapsida) through computer tomography. *Science of Nature* 104: 66.

Benoit, J., Manger, P.R., Fernandez, V. & Rubidge, B.S. (2017d): The bony labyrinth of late Permian Biarmosuchia: palaeobiology and diversity in non-mammalian Therapsida. *Palaeontologia Africana* 52: 58–77.

Benoit, J., Fernandez, V., Manger, P.R. & Rubidge, B.S. (2017e): Endocranial casts of pre-mammalian therapsids reveal an unexpected neurological diversity at the deep evolutionary root of mammals. *Brain, Behavior, and Evolution* 90: 311–333.

Benoit, J. Angielczyk, K.D., Miyamae, J.A., Manger, P., Fernandez, V. & Rubidge, B. (2018): Evolution of facial innervation in anomodont therapsids (Synapsida): insights from X-ray computerized microtomography. *Journal of Morphology* 279: 673–701.

Benson, R.B.J. (2012): Interrelationships of basal synapsids: cranial and postcranial morphological partitions suggest different topologies. *Journal of Systematic Palaeontology* 10: 601–624.

Benson, R.B.J. & Upchurch, P. (2013): Diversity trends in the establishment of terrestrial vertebrate ecosystems: interactions between spatial and temporal sampling biases. *Geology* 41: 43–46.

Benton, M.J. (2012) No gap in the Middle Permian record of terrestrial vertebrates. *Geology* 40: 339–342.

Benton, M.J. (2013): No gap in the Middle Permian record of terrestrial vertebrates: Reply. *Geology* 41: e294.

Benton, M.J. (2015a): *Vertebrate Palaeontology*. Fourth Edition. John Wiley & Sons, Chichester.

Benton, M.J. (2015b): *When Life Nearly Died: The Greatest Mass Extinction of All Time*. Revised Edition. Thames & Hudson, London.

Benton, M.J. & Newell, A.J. (2014): Impacts of global warming on Permo-Triassic terrestrial ecosystems. *Gondwana Research* 25: 1308–1337.

Benton, M.J., Donoghue, P.C.J., Asher, R.J., Friedman, M., Near, T.J. & Vinther, J. (2015): Constraints on the timescale of animal evolutionary history. *Palaeontologia Electronica* 18.1.1FC: 1–106.

Berman, D.S. (1978): *Ctenospondylus ninevehensis*, a new species (Reptilia, Pelycosauria) from the Lower Permian Dunkard Group of Ohio. *Annals of Carnegie Museum* 47: 493–514.

Berman, D.S. & Reisz, R.R. (1982): Restudy of *Mycterosaurus longiceps* (Reptilia, Pelycosauria) from the Lower Permian of Texas. *Annals of Carnegie Museum* 51: 423–453.

Berman, D.S., Sumida, S.S. & Lombard, R.E. (1997): Biogeography of primitive amniotes. In: Sumida S.S. & Martin K.L.M. (eds.). *Amniote Origins: Completing the Transition to Land*. Academic Press, San Diego: 85–139.

Berman, D.S., Reisz, R.R., Bolt, J.R. & Scott, D. (1995): The cranial anatomy and relationships of the synapsid *Varanosaurus* (Eupelycosauria: Ophiacodontidae) from the Early Permian of Texas and Oklahoma. *Annals of Carnegie Museum* 64: 99–133.

Berman, D.S., Reisz, R.R., Martens, T. & Henrici, A.C. (2001). A new species of *Dimetrodon* (Synapsida: Sphenacodontidae) from the Lower Permian of Germany records first occurrence of genus outside of North America. *Canadian Journal of Earth Sciences* 38: 803–812.

Berman, D.S., Henrici, A.C., Sumida, S.S. & Martens, T. (2004): New materials of *Dimetrodon teutonis* (Synapsida: Sphenacodontidae) from the Lower Permian of Germany. *Annals of Carnegie Museum* 73: 48–56.

Berman, D.S., Henrici, A.C., Sumida, S.S., Martens, T. & Pelletier, V. (2014): First European record of a varanodontine (Synapsida: Varanopidae): member of a unique Early Permian upland paleoecosystem, Tambach Basin, Central Germany. In: Kammerer, C.F., Angielczyk, K.D. & Fröbisch, J. (eds.). *Early Evolutionary History of the Synapsida*. Springer, Dordrecht: 69–86.

Bernardi, M., Petti, F.M., Kustatscher, E., Franz, M., Hartkopf-Fröder, C., Labandeira, C. C., Wappler, T., van Konijnenburg-van Cittert, J.H.A., Peecook, B.R. & Angielczyk, K.D. (2017): Late Permian (Lopingian) terrestrial ecosystems: a global comparison with new data from the low-latitude Bletterbach Biota. *Earth-Science Reviews* 175: 18–43.

Bi, S., Wang, Y., Guan, J. Sheng, X. & Meng, J. (2014): Three new Jurassic euharamiyidan species reinforce early divergence of mammals. *Nature* 514: 579–584.

Bonaparte, J.F. & Barberena, M.C. (2001): On two advanced carnivorous cynodonts from the Late Triassic of southern Brazil. *Bulletin of the Museum of Comparative Zoology* 156: 59–80.

Bonaparte, J.F., Martinelli, A.G., Schultz, C.L. & Rubert, R. (2003): The sister group of mammals: small cynodonts from the Late Triassic of southern Brazil. *Revista Brasileira de Paleontologia* 5: 5–27.

Bonaparte, J.F., Martinelli, A.G. & Schultz, C.L. (2005): New information on *Brasilodon* and *Brasilitherium* (Cynodontia, Probainognathia) from the Late Triassic of southern Brazil. *Revista Brasileira de Paleontologia* 8: 25–46.

Bonaparte, J.F., Soares, M.B. & Schultz, C.L. (2006): A new non-mammalian cynodont from the Middle Triassic of southern Brazil and its implications for the ancestry of mammals. In: Harris J.D., Lucas S.G., Spielmann J.A., Lockley M.G., Milner A.R.C. & Kirkland J.I. (eds.). *The Triassic–Jurassic Transition. New Mexico Museum of Natural History and Science Bulletin* 37: 599–607.

Bonaparte, J.F., Schultz, C.L., Soares, M.B. & Martinelli, A.G. (2010): La fauna local de Faxinal do Soturno, Triásico tardío de Rio Grande do Sul, Brazil. *Revista Brasileira de Paleontologia* 13: 233–246.

Boonstra, L.D. (1946): Report on some reptilian fossils from Gunyanka's Kraal, Busi Valley. *Proceedings and Transactions of the Rhodesian Scientific Association* 41: 46–49.

Boonstra, L.D. (1952): Die gorgonopsiër-geslag, *Hipposaurus*, en die familie, *Ictidorhinidae. Tydskrif vir Wetenskap en Kuns* 12: 142–149.

Boonstra, L.D. (1953): A suggested clarification of the taxonomic status of the South African titanosuchians. *Annals of the South African Museum* 42: 19–28.

Boonstra, L.D. (1963): Early dichotomies in the therapsids. *South African Journal of Science* 59: 176–195.

Boonstra, L.D. (1969): The fauna of the *Tapinocephalus* Zone (Beaufort beds of the Karoo). *Annals of the South African Museum* 56: 1–73.

Boonstra, L.D. (1971): The early therapsids. *Annals of the South African Museum* 59: 17–46.

Boonstra, L.D. (1972): Discard the names Theriodontia and Anomodontia: a new classification of therapsids. *Annals of the South African Museum* 59: 315–338.

Boos, A.D.S., Kammerer, C.F., Schultz, C.L. & Paes Neto, V. (2015): A tapinocephalid dinocephalian (Synapsida, Therapsida) from the Rio do Rasto Formation (Paraná Basin, Brazil): taxonomic, ontogenetic and biostratigraphic considerations. *Journal of South American Earth Sciences* 63: 375–384.

Boos, A.D.S., Kammerer, C.F., Schultz, C.L., Soares, M.B. & Ilha, A.L.R. (2016): A new dicynodont (Therapsida, Anomodontia) from the Permian of southern Brazil and its implications for bidentalian origins. *PloS One* 11(5): e0155000.

Botha, J. (2003): Biological aspects of the Permian dicynodont *Oudenodon* (Therapsida, Dicynodontia), deduced from bone histology and cross-sectional geometry. *Palaeontologia Africana* 39: 7–44.

Botha, J. & Angielczyk, K.D. (2007): An integrative approach to distinguishing the Late Permian dicynodonts *Oudenodon bainii* and *Tropidostoma microtrema* (Therapsida: Anomodontia). *Palaeontology* 50: 1175–1209.

Botha, J. & Chinsamy, A. (2005): Growth patterns of *Thrinaxodon liorhinus*, a non-mammalian cynodont from the Lower Triassic of South Africa. *Palaeontology* 48: 385–394.

Botha, J. & Smith, R.M.H. (2007): *Lystrosaurus* species composition across the Permo-Triassic boundary of South Africa. *Lethaia* 40: 125–137.

Botha, J., Abdala, F. & Smith, R. (2007): The oldest cynodont: new clues on the origin and early diversification of the Cynodontia. *Zoological Journal of the Linnean Society* 149: 477–492.

Botha-Brink, J. & Abdala, F. (2008): A new cynodont record from the *Tropidostoma* Assemblage Zone of the Beaufort Group: implications for the early evolution of cynodonts in South Africa. *Palaeontologia Africana* 43: 1–6.

Botha-Brink, J. & Angielczyk, K.D. (2010): Do extraordinarily high growth rates in Permo-Triassic dicynodonts (Therapsida, Anomodontia) explain their success before and after the end-Permian extinction? *Zoological Journal of the Linnean Society* 160: 341–365.

Botha-Brink, J. & Modesto, S.P. (2007): A mixed-age classed 'pelycosaur' aggregation from South Africa: earliest evidence of parental care in amniotes? *Proceedings of the Royal Society Series B* 274: 2829–2834.

Botha-Brink, J. & Modesto, S.P. (2009): Anatomy and relationships of the Middle Permian varanopid *Heleosaurus scholtzi* based on a social aggregation from the Karoo Basin of South Africa. *Journal of Vertebrate Paleontology* 29: 389–400.

Botha-Brink, J. & Modesto, S.P. (2011): A new skeleton of the therocephalian synapsid *Olivierosuchus parringtoni* from the Lower Triassic South African Karoo Basin. *Palaeontology* 54: 591–606.

Botha-Brink, J., Abdala, F. & Chinsamy-Turan, A. (2012): The radiation and osteohistology of nonmammaliaform cynodonts. In: Chinsamy-Turan, A. (ed.). *Forerunners of Mammals: Radiation, Histology, Biology*. Indiana University Press, Bloomington: 223–246.

Botha-Brink, J., Codron, D., Huttenlocker, A.K., Angielczyk, K.D. & Ruta, M. (2016): Breeding young as a survival strategy during Earth's greatest mass extinction. *Scientific Reports* 6: 24053.

Botha-Brink, J. (2017). Burrowing in Lystrosaurus: preadaptation to a postextinction environment? Journal of Vertebrate Paleontology. DOI: 10.1080/02724634.2017.1365080.

Bordy, E.M., Sztanó, O., Rubidge, B.S. & Bumby, A. (2011): Early Triassic vertebrate burrows from the Katberg Formation of the south-western Karoo Basin, South Africa. *Lethaia* 44: 33–45.

Bramwell, C.D. & Fellgett, P.B. (1973): Thermal regulation in sail lizards. *Nature* 242: 203–205.

Brink, A.S. (1950): On a new species of *Cistecephalus* Owen. *Annals and Magazine of Natural History Series 12* 3: 985–997.

Brink, A.S. (1951): On the genus *Lystrosaurus* Cope. *Transactions of the Royal Society of South Africa* 33: 107–120.

Brink, A.S. (1952): Studies of Karroo Reptiles. III. The manus of *Cistecephalus. South African Journal of Science* 49: 13–15.

Brink, A.S. (1956): Speculations on some advanced mammalian characteristics in the higher mammal-like reptiles. *Palaeontologia Africana* 4: 77–96.

Brink, K.S. & Reisz, R.R. (2012): Morphology of the palate and braincase of *Dimetrodon milleri. Historical Biology* 24: 453–459.

Brink, K.S. & Reisz, R.R. (2014): Hidden dental diversity in the oldest terrestrial apex predator *Dimetrodon*. *Nature Communications* 5: 3269.

Brink, K.S., LeBlanc, A.R.H. & Reisz, R.R. (2014): First record of plicidentine in Synapsida and patterns of tooth root shape change in Early Permian sphenacodontians. *Naturwissenschaften* 101: 883–892.

Brink, K.S., Maddin, H.C., Evans, D.C. & Reisz, R.R. (2015): Re-evaluation of the historic Canadian fossil *Bathygnathus borealis* from the Early Permian of Prince Edward Island. *Canadian Journal of Earth Sciences* 52: 1109–1120.

Brinkman, D. (1981): The structure and relationships of the dromasaurs (Reptilia; Therapsida). *Breviora* 465: 1–34.

Brinkman, D. (1988): Size-independent criteria for estimating relative age in *Ophiacodon* and *Dimetrodon* (Reptilia, Pelycosauria) from the Admiral and lower Belle Plains formations of west-central Texas. *Journal of Vertebrate Paleontology* 8: 172–180.

Brinkman, D. & Eberth, D.A. (1983): The interrelationships of pelycosaurs. *Breviora* 473: 1–35.

Brinkman, D. & Eberth, D.A. (1986): The anatomy and relationships of *Stereophallodon* and *Baldwinonus* (Reptilia, Pelycosauria). *Breviora* 485: 1–34.

Brock, G.T. (1941): The skull of *Acontias meleagris*, with a study of the affinities between lizards and snakes. *Journal of the Linnean Society of London, Zoology*. 41: 71–88.

Brocklehurst, N. (2015): *The Early Evolution of the Synapsida (Vertebrata, Amniota) and the Quality of Their Fossil Record*. Unpublished PhD dissertation, Humbolt-Universität zu Berlin.

Brocklehurst, N. & Brink, K.S. (2017): Selection towards larger body size in both herbivorous and carnivorous synapsids during the Carboniferous. *Facets* 2: 68–84.

Brocklehurst, N. & Fröbisch, J. (2014): Current & historical perspectives on the completeness of the fossil record of pelycosaurian-grade synapsids. *Palaeogeography, Palaeoclimatology, Palaeoecology* 399: 114–126.

Brocklehurst, N. & Fröbisch, J. (2017): A re-examination of the enigmatic Russian tetrapod *Phreatophasma aenigmaticum* and its evolutionary implications. *Fossil Record* 20: 87–93.

Brocklehurst, N., Kammerer, C.F. & Fröbisch, J. (2013): The early evolution of synapsids, and the influence of sampling on their fossil record. *Paleobiology* 39: 470–490.

Brocklehurst, N., Romano, M. & Fröbisch, J. (2016a). Principal component analysis as an alternative treatment for mophometric characters: phylogeny of caseids as a case study. *Palaeontology* 59: 877–886.

Brocklehurst, N., Reisz R., Fernandez V. & Fröbisch, J. (2016b): A re-description of *'Mycterosaurus' smithae*, an early Permian eothyridid, and its impact on the phylogeny of pelycosaurian-grade synapsids. *PLoS One* 11(6): e0156810.

Brocklehurst, N., Day, M.O., Rubidge, B.S. & Fröbisch, J. (2017): Olson's extinction and the latitudinal biodiversity gradient of tetrapods in the Permian. *Proceedings of the Royal Society of London B* 284:20170231.

Broili, F. (1941): Haare bei Reptilien. *Anatomischer Anzeiger* 92: 62–68.

Broili, F. & Schröder, J. (1935): Beobachtungen an Wirbeltieren der Karrooformation. XI. Über den Schädel von *Cynidiognathus* Haughton. *Sitzungsberichte der mathematisch-naturwissenschaftlichen Abteilung der Bayerischen Akademie der Wissenschaften zu München* 1935: 199–221.

Broom, R. (1902): Remarks on certain differences in the skulls of dicynodonts, apparently due to sex. *Proceedings of the Zoological Society of London* 1902: 86–88.

Broom, R. (1903a): On the classification of the theriodonts and their allies. *Report of the South African Association for the Advancement of Science* 1: 286–294.

Broom, R. (1903b): On the structure of the shoulder girdle in *Lystrosaurus*. *Annals of the South African Museum* 4: 139–141.

Broom, R. (1905a): On the use of the term Anomodontia. *Records of the Albany Museum* 1: 266–269.

Broom, R. (1905b): The origin of mammals. *Report of the Seventy-fifth Meeting of the British Association for the Advancement of Science*: 437–438.

Broom, R. (1906a): On the Permian and Triassic faunas of South Africa. *Geological Magazine Decade V* 3: 29–30.

Broom, R. (1906b): The classification of the Karoo Beds of South Africa. *Geological Magazine Decade V* 3: 36.

Broom, R. (1907): On the origin of the mammal-like reptiles. *Proceedings of the Zoological Society of London* 1907: 1047–1061.

Broom, R. (1910): A comparison of the Permian reptiles of North America with those of South Africa. *Bulletin of the American Museum of Natural History* 28: 197–234.

Broom, R. (1912a): On the structure of the internal ear and the relations of the basicranial nerves in *Dicynodon*, and on the homology of the mammalian auditory ossicles. *Proceedings of the Zoological Society of London* 1912: 419–425.

Broom, R. (1912b): On some points in the structure of the dicynodont skull. *Annals of the South African Museum* 7: 337–351.

Broom, R. (1914a): Croonian Lecture: on the origin of mammals. *Philosophical Transactions of the Royal Society Series B* 206: 1–48.

Broom, R. (1914b): A further comparison of the South African dinocephalians with the American pelycosaurs. *Bulletin of the American Museum of Natural History* 33: 135–141.

Broom, R. (1923): On the structure of the skull in the carnivorous dinocephalian reptiles. *Proceedings of the Zoological Society of London* 1923: 661–684.

Broom, R. (1932): *The Mammal-like Reptiles of South Africa and the Origin of Mammals*. H.F. & G. Witherby, London.

Broom, R. (1937): A further contribution to our knowledge of the fossil reptiles of the Karroo. *Proceedings of the Zoological Society of London Series B* 107: 299–318.

Broom, R. (1948): A contribution to our knowledge of the vertebrates of the Karroo Beds of South Africa. *Transactions of the Royal Society of Edinburgh* 61: 577–629.

Buchholtz, E.A. (2014): Crossing the frontier: a hypothesis for the origins of meristic constraint in mammalian axial patterning. *Zoology* 117: 64–69.

Cain, J.A. (1988): Creationism and mammal origins. *Journal of Geological Education* 36: 94–105.

Camp, C.L. (1956): Triassic dicynodont reptiles. Part II. Triassic dicynodonts compared. *Memoirs of the University of California* 13: 305–341.

Camp, C.L. & Welles, S.P. (1956): Triassic dicynodont reptiles. Part I. The North American genus *Placerias*. *Memoirs of the University of California* 13: 255–304.

Campione, N.E. & Reisz, R.R. (2010): *Varanops brevirostris* (Eupelycosauria: Varanopidae) from the Lower Permian of Texas, with discussion of varanopid morphology and in[terrelationships. *Journal of Vertebrate Paleontology* 30: 724–746.

Canoville, A. & Laurin, M. (2010): Evolution of humeral microanatomy and lifestyle in amniotes, and some comments on paleobiological inferences. *Biological Journal of the Linnean Society* 100: 384–406.

Cantrell, A.K., Suazo, T.L., McKeighen, K.L. Jr., McKeighen, H.W., Lucas, S.G., Harris, S.K., Spielmann, J.A. & Rinehart, L.F. (2011): *Dimetrodon* (Eupelycosauria: Sphenacodontidae) from the Lower Permian Abo Formation, Socorro and Torrance counties, New Mexico. In: Sullivan, R.M., Lucas, S.G. & Spielmann, J.A. (eds.). *Fossil Record 3. New Mexico Museum of Natural History and Science Bulletin* 53: 34–37.

Carroll, R.L. (1964): The earliest reptiles. *Journal of the Linnean Society, Zoology* 45: 61–83.

Case, E.C. (1905): *Bathygnathus borealis* Leidy and the Permian of Prince Edward Island. *Science* 22: 52–53.

Case, E.C. (1907): Revision of the Pelycosauria of North America. *Carnegie Institution of Washington Publication* 55: 1–176.

Case, E.C. (1915): The Permo-Carboniferous red beds of North America and their vertebrate fauna. *Carnegie Institution of Washington Publication* 207: 1–176.

Case, E.C. (1918): A mounted skeleton of *Edaphosaurus cruciger* Cope, in the geological collection of the University of Michigan. *Occasional Papers of the Museum of Zoology, University of Michigan* 62: 1–9.

Case, E.C. (1924): A possible explanation of fenestration in the primitive reptilian skull, with notes on the temporal region of the genus *Dimetrodon*. *Contributions from the Museum of Geology, University of Michigan* 2: 1–12.

Castanhinha, R., Araújo, R., Júnior, L.C., Angielczyk, K.D., Martins, G.G., Martins, R.M.S., Chaouiya, C., Beckmann, F. & Wilde, F. (2013): Bringing dicynodonts back to life: paleobiology and anatomy of a new emydopoid genus from the Upper Permian of Mozambique. *PLoS One* 8 (12): e80974.

Chen, Z.-Q. & Benton, M.J. (2012): The timing and pattern of biotic recovery following the end-Permian mass extinction. *Nature Geoscience* 5: 375–383.

Chinsamy, A. & Rubidge, B.S. (1993): Dicynodont (Therapsida) bone histology: phylogenetic and physiological implications. *Palaeontologia Africana* 30: 97–102.

Chinsamy-Turan, A. (ed.): (2012): *Forerunners of Mammals: Radiation, Histology, Biology*. Indiana University Press, Bloomington and Indianapolis.

Chinsamy-Turan, A. & Ray, S. (2012): Bone histology of some therocephalians and gorgonopsians, and evidence of bone degradation by fungi. In: Chinsamy-Turan, A. (ed.). *Forerunners of Mammals: Radiation, Histology, Biology*. Indiana University Press, Bloomington: 199–221.

Chudinov, P.K. (1960): Upper Permian therapsids of the Ezhovo locality [in Russian]. *Paleontologicheskii Zhurnal* 4: 81–94.

Chudinov, P.K. (1964): Contribution to the knowledge of Deinocephalia of the USSR [in Russian]. *Paleontologicheskii Zhurnal* 1964: 85–98.

Chudinov, P.K. (1965): New facts about the fauna of the Upper Permian of the U.S.S.R. *Journal of Geology* 73: 117–130.

Chudinov, P.K. (1968a): A new dinocephalian from the Cisuralian Region (Reptilia, Therapsida; Upper Permian). *Postilla* 121: 1–20.

Chudinov, P.K. (1968b): Structure of the integuments of theromorphs. *Doklady Akademiya Nauk SSSR* [English Translation] 179: 226–229.

Chudinov, P.K. (1983): Early therapsids [in Russian]. *Trudy Paleontologicheskogo Instituta Akademii Nauk SSSR* 202: 1–227.

Cisneros, J.C., Abdala, F., Rubidge, B.S., Dentzien-Dias, P.C. & de Oliveira Bueno, A. (2011): Dental occlusion in a 260-million-year-old therapsid with saber canines from the Permian of Brazil. *Science* 331: 1603–1605.

Cisneros, J.C., Abdala, F., Atayman-Güven, S., Rubidge, B.S., Celâl Sengör, A.M.C. & Schultz, C.L. (2012): Carnivorous dinocephalian from the Middle Permian of Brazil and tetrapod dispersal in Pangaea. *Proceedings of the National Academy of Sciences* 109: 1584–1588.

Cisneros, J.C., Abdala, F., Jashashvili, T., de Oliveira Bueno, A. & Dentzien-Dias, P. (2015): *Tiarajudens eccentricus* and *Anomocephalus africanus*, two bizarre anomodonts (Synapsida, Therapsida) with dental occlusion from the Permian of Gondwana. *Royal Society Open Science* 2(7): 150090.

Clark, J.M. & Hopson, J.A. (1985): Distinctive mammal-like reptile from Mexico and its bearing on the phylogeny of the Tritylodontidae. *Nature* 315: 398–400.

Clarke, A. & Pörtner, H.-O. (2010): Temperature, metabolic power and the evolution of endothermy. *Biological Reviews* 85: 703–727.

Cluver, M.A. (1970): The palate and mandible in some specimens of *Dicynodon testudirostris* Broom & Haughton (Reptilia, Therapsida). *Annals of the South African Museum* 56: 133–153.

Cluver, M.A. (1971): The cranial morphology of the dicynodont genus *Lystrosaurus*. *Annals of the South African Museum* 56: 155–274.

Cluver, M.A. (1974): The skull and mandible of a new cistecephalid dicynodont. *Annals of the South African Museum* 64: 137–155.

Cluver, M.A. (1975): A new dicynodont reptile from the *Tapinocephalus* Zone (Karoo System, Beaufort Series) of South Africa, with evidence of the jaw adductor musculature. *Annals of the South African Museum* 67: 7–23.

Cluver, M.A. (1978): The skeleton of the mammal-like reptile *Cistecephalus* with evidence for a fossorial mode of life. *Annals of the South African Museum* 76: 213–246.

Cluver, M.A. & Hotton, N. III (1981): The genera *Dicynodon* and *Diictodon* and their bearing on the classification of the Dicynodontia (Reptilia, Therapsida). *Annals of the South African Museum* 83: 99–146.

Cluver, M.A. & King, G.M. (1983): A reassessment of the relationships of Permian Dicynodontia (Reptilia, Therapsida) and a new classification of dicynodonts. *Annals of the South African Museum* 91: 195–273.

Codron, J., Botha-Brink, J. Codron, D., Huttenlocker, A.K. & Angielczyk, K.D. (2017): Predator-prey interactions amongst Permo-Triassic terrestrial vertebrates as a deterministic factor influencing faunal collapse and turnover. *Journal of Evolutionary Biology* 30: 40–54.

Cohen, K.M., Finney, S.C., Gibbard, P.L. & Fan, J.-X. (2013): The ICS International Chronostratigraphic Chart. *Episodes* 36: 199–204.

Conrad, J. & Sidor, C.A. (2001): Re-evaluation of *Tetraceratops insignis* (Synapsida: Sphenacodontia). *Journal of Vertebrate Paleontology* 21 (supplement to 3): 42A.

Cooper, M.R. (1982): A mid-Permian to earliest Jurassic tetrapod biostratigraphy and its significance. *Arnoldia* 9: 77–104.

Cope, E.D. (1875): On fossil remains of Reptilia and fishes from Illinois. *Proceedings of the Academy of Natural Sciences of Philadelphia* 27: 404–411.

Cope, E.D. (1878a): Descriptions of the extinct Vertebrata from the Permian and Triassic formations of the United States. *Proceedings of the American Philosophical Society* 17: 182–193.

Cope, E.D. (1878b): The theromorphous Reptilia. *American Naturalist* 12: 829–830.

Cope, E.D. (1878c): Descriptions of extinct Batrachia and Reptilia from the Permian formation of Texas. *Proceedings of the American Philosophical Society* 17: 505–530.

Cope, E.D. (1880): Second contribution to the history of the Vertebrata of the Permian Formation of Texas. *Proceedings of the American Philosophical Society* 19: 38–58.

Cope, E.D. (1881): Catalogue of the Vertebrata of the Permian Formation of the United States. *American Naturalist* 15: 162–164.

Cope, E.D. (1882): Third contribution to the history of the Vertebrata of the Permian Formation of Texas. *Proceedings of the American Philosophical Society* 20: 447–461.

Cope, E.D. (1884a): Fifth contribution to the knowledge of the fauna of the Permian Formation of Texas and the Indian Territory. *Proceedings of the American Philosophical Society* 22: 28–47.

Cope, E.D. (1884b): The relations between the theromorphous reptiles and the monotreme Mammalia. *Proceedings of the American Association for the Advancement of Science* 33: 471–482.

Cope, E.D. (1884c): The origin of Mammalia. *American Naturalist* 11: 1136–1137.

Cope, E.D. (1886): The long-spined Theromorpha of the Permian Epoch. *American Naturalist* 20: 544–545.

Cope, E.D. (1892): On the phylogeny of the Vertebrata. *Proceedings of the American Philosophical Society* 30: 278–281.

Cox, C.B. (1959): On the anatomy of a new dicynodont genus with evidence of the position of the tympanum. *Proceedings of the Zoological Society of London* 132: 321–367.

Cox, C.B. (1964): On the palate, dentition and classification of the fossil reptile *Endothiodon* and related genera. *American Museum Novitates* 2171: 1–25.

Cox, C.B. (1965): New Triassic dicynodonts from South America, their origins and relationships. *Philosophical Transactions of the Royal Society of London Series B* 248: 457–514.

Cox, C.B. (1968): The Chañares (Argentina) reptile fauna IV: the dicynodont fauna. *Breviora* 295: 1–27.

Cox, C.B. (1972): A new digging dicynodont from the upper Permian of Tanzania. In: Joysey, K.A. & Kemp, T.S. (eds.). *Studies in Vertebrate Evolution*. Oliver and Boyd, Edinburgh: 173–189.

Cox, C.B. (1991): The Pangaean dicynodont *Rechnisaurus* and the comparative biostratigraphy of Triassic dicynodont faunas. *Palaeontology* 34: 767–784.

Cox, C.B. (1998): The jaw function and adaptive radiation of the dicynodont mammal-like reptiles of the Karoo basin of South Africa. *Zoological Journal of the Linnean Society* 122: 349–384.

Cox, C.B. & Angielczyk, K.D. (2015): A new endothiodont dicynodont (Therapsida, Anomodontia) from the Permian Ruhuhu Formation (Songea Group) of Tanzania and its feeding system. *Journal of Vertebrate Paleontology* 35: e935388.

Cox, C.B. & Li, J. (1983): A new genus of Triassic dicynodont from east Africa and its classification. *Palaeontology* 26: 389–406.

Crompton, A.W. (1958): The cranial morphology of a new genus and species of ictidosaurian. *Proceedings of the Zoological Society of London* 130: 183–216.

Crompton, A.W. (1963a): On the lower jaw of *Diarthrognathus* and the origin of the mammalian lower jaw. *Proceedings of the Zoological Society of London* 140: 697–748.

Crompton, A.W. (1963b): The evolution of the mammalian jaw. *Evolution* 17: 431–439.

Crompton, A.W. & Hotton, N. III (1967): Functional morphology of the masticatory apparatus of two dicynodonts (Reptilia, Therapsida). *Postilla* 109: 1–51.

Crompton, A.W. & Jenkins, F.A. (1968): Molar occlusion in Late Triassic mammals. *Biological Reviews* 43: 427–458.

Credner, H. (1888): Die Stegocephalen und Saurier aus dem Rothliegenden des Plauen'schen Grundes bei Dresden, vii. Theil: *Palaeohatteria longicaudata* CRED. *Zeitschrift der Deutschen Geologischen Gesellschaft* 40: 490–558.

Cruickshank, A.R.I. (1973): The mode of life of gorgonopsians. *Palaeontologia Africana* 15: 65–67.

Currie, P.J. (1977): A new haptodontine sphenacodont (Reptilia: Pelycosauria) from the Upper Pennsylvanian of North America. *Journal of Paleontology* 51: 927–942.

Currie, P.J. (1979): The osteology of haptodontine sphenacodonts (Reptilia: Pelycosauria). *Palaeontographica Abteilung A* 163: 130–168.

Curtis, N., Jones, M.E.H., Shi, J., O'Higgins, P., Evans, S.E. & Fagan, M.J. (2011): Functional relationship between skull form and feeding mechanics in *Sphenodon*, and implications for diapsid skull development. *PLoS One* 6(12): e29804.

da Silva, R.C., Sedor, F.A. & Fernandes, A.C.S. (2012): Fossil footprints from the Late Permian of Brazil: an example of hidden biodiversity. *Journal of South American Earth Sciences* 38: 31–43.

Damiani, R., Modesto, S., Yates, A. & Neveling, J. (2003): Earliest evidence of cynodont burrowing. *Proceedings of the Royal Society of London* B 270: 1747–1751.

Dawson, J.W. (1860): On a terrestrial mollusk, a chilognathous myriapod, and some new species of reptiles, from the coal-formation of Nova Scotia. *Quarterly Journal of the Geological Society of London* 16: 268–277.

Day, M.O. (2013a): Charting the fossils of the Great Karoo: a history of tetrapod biostratigraphy in the Lower Beaufort Group, South Africa. *Palaeontologia Africana* 48: 41–47.

Day, M.O. (2013b): *Middle Permian continental biodiversity changes as reflected in the Beaufort Group of South Africa: a bio- and lithostratigraphic review of the* Eodicynodon, Tapinocephalus *and* Pristerognathus *Assemblage Zones*. Unpublished PhD dissertation, University of the Witwatersrand, Johannesburg.

Day, M.O., Güven, S., Abdala, F., Jirah, S., Rubidge, B. & Almond, J. (2015a): Youngest dinocephalian fossils extend the *Tapinocephalus* Zone, Karoo Basin, South Africa. *South African Journal of Science* 111: 1–5.

Day, M.O., Ramezani, J., Bowring, S.A., Sadler, P.M., Erwin, D.H., Abdala, F. & Rubidge, B.S. (2015b): When and how did the terrestrial mid-Permian mass extinction occur? Evidence from the tetrapod record of the Karoo Basin, South Africa. *Proceedings of the Royal Society B* 282: 20150834.

Day, M.O., Rubidge, B.S. & Abdala, F. (2016): A new mid-Permian burnetiamorph therapsid from the main Karoo Basin of South Africa and a phylogenetic review of Burnetiamorpha. *Acta Palaeontologica Polonica* 61: 701–719.

Day, M.O., Smith, R.M.H., Benoit, J., Fernandez, V. & Rubidge, B.S. (2018): A new species of burnetiid (Therapsida, Burnetiamorpha) from the early Wuchiapingian of South Africa and implications for the evolutionary ecology of the family Burnetiidae. *Papers in Palaeontology*. DOI: 10.1002/spp2.1114.

DeBraga, M. & Rieppel, O. (1997): Reptile phylogeny and the interrelationships of turtles. *Philosophical Transaction of the Royal Society of London Series B* 353: 501–541.

DeMar, R. & Barghusen, H.R. (1972): Mechanics and the evolution of the synapsid jaw. *Evolution* 26: 622–637.

Dhouailly, D. (2009) A new scenario for the evolutionary origin of hair, feather, and avian scales. *Journal of Anatomy* 214: 587–606.

Dias-da-Silva, S., Modesto, S.P. & Schultz, C.L. (2006): New material of *Procolophon* (Parareptilia, Procolophonoidea) from the Lower Triassic of Brazil, with remarks on the ages of the Sanga do Cabral and Buena Vista formations of South America. *Canadian Journal of Earth Sciences* 43: 1685–1693.

Dilkes, D.W. & Reisz, R.R. (1996): First record of a basal synapsid ('mammal-like reptile') in Gondwana. *Proceedings of the Royal Society Series B* 263: 1165–1170.

DiMichele, W.A., Hook, R.W., Beerbower, R., Boy, J.A., Gastaldo, R.A., Hotton, N. III, Phillips, T.L., Scheckler, S.E., Shear, W.A. & Sues, H.-D. (1993): Paleozoic terrestrial ecosystems. In: Behrensmeyer, A.K., Damuth, J.A., DiMichele, W.A., Potts, R., Sues, H.-D. & Wing, S.L. (eds.). *Terrestrial Ecosystems Through Time: Evolutionary Paleoecology of Terrestrial Plants and Animals*. University of Chicago Press, Chicago: 205–325.

Durand, J.F. (1991): A revised description of the skull of *Moschorhinus* (Therapsida, Therocephalia). *Annals of the South African Museum* 99: 381–413.

Dzik, J., Sulej, T. & Niedźwiedzki, G. (2008): A dicynodont-theropod association in the latest Triassic of Poland. *Acta Palaeontologica Polonica* 53: 733–738.

Eberth, D.A., Berman, D.S., Sumida, S.S. & Hopf, H. (2000): Lower Permian terrestrial paleoenvironments and vertebrate paleoecology of the Tambach Basin (Thuringia, central Germany): the upland holy grail. *Palaios* 15: 293–313.

Eckhart, L., Dalla Valle, L., Jaeger, K., Ballaun, C., Szabo, S., Nardi, A., Buchberger, M., Hermann, M., Alibardi, L. & Tschachler, E. (2008): Identification of reptilian genes encoding hair keratin-like proteins suggests a new scenario for the evolutionary origin of hair. *Proceedings of the National Academy of Sciences* 105: 18419–18423.

Efremov, I.A. (1938): Some new Permian reptiles of the USSR. *Doklady Akademiya Nauk SSSR* 19: 771–776.

Efremov, I.A. (1940a): Preliminary description of new Permian and Triassic Tetrapoda from USSR [in Russian]. *Trudy Paleontologicheskogo Instituta Akademii Nauk SSSR* 10: 5–140.

Efremov, I.A. (1940b): *Ulemosaurus svijagensis* Riab. – ein Dinocephale aus den Ablagerungen des Perm der UdSSR. *Nova Acta Leopoldina* 9: 155–205.

Efremov, I.A. (1956): American elements in the fauna of the USSR [in Russian]. *Doklady Akademiya Nauk SSSR* 111: 1091–1094.

Enlow, D.H. (1969): The bone of reptiles. In: Gans, C., Bellairs, A. & Parsons, T. *The Biology of Reptilia Volume 1*. Academic Press, London: 45–77.

Enlow, D.H. & Brown, S.O. (1957): A comparative histological study of fossil and recent bone tissues. Part II. *Texas Journal of Science* 9: 186–214.

Erickson, G.M. (1996): Daily deposition of dentine in juvenile *Alligator* and assessment of tooth replacement rates using incremental line counts. *Journal of Morphology* 228: 189–194.

Estes, R. (1961): Cranial anatomy of the cynodont reptile *Thrinaxodon liorhinus*. *Bulletin of the Museum of Comparative Zoology* 125: 165–180.

Evans, D.C., Maddin, H.C. & Reisz, R.R. (2009): A re-evaluation of sphenacodontid synapsid material from the Lower Permian fissure fills near Richards Spur, Oklahoma. *Palaeontology* 52: 219–227.

Ezcurra, M.D. (2016): The phylogenetic relationships of basal archosauromorphs, with an emphasis on the systematics of proterosuchian archosauriforms. *PeerJ* 4: e1778.

Falcon-Lang, H.J., Benton, M.J. & Stimson, M. (2007): Ecology of the earliest reptiles inferred from basal Pennsylvanian trackways. *Journal of the Geological Society of London* 164: 1113–1118.

Falconnet, J. (2015): The sphenacodontid synapsid *Neosaurus cynodus*, and related material, from the Permo-Carboniferous of France. *Acta Palaeontologica Polonica* 60: 169–182.

Felice, R.N. & Angielczyk, K.D. (2014): Was *Ophiacodon* (Synapsida, Eupelycosauria) a swimmer? A test using vertebral dimensions. In: Kammerer, C.F., Angielczyk, K.D. & Fröbisch, J. (eds.). *Early Evolutionary History of the Synapsida*. Springer, Dordrecht: 25–51.

Fernandez, V., Abdala, F., Carlson, K.J., Rubidge, B.S., Yates, A. & Tafforeau, P. (2013): Synchrotron reveals Early Triassic odd couple: injured amphibian and aestivating therapsid share burrow. *PLoS One* 8(6): e64978.

Findlay, G.H. (1968): On the scaloposaurid skull of *Olivieria parringtoni*, Brink with a note on the origin of hair. *Palaeontologia Africana* 11: 47–59.

Florides, G.A., Wrobel, L.C., Kalogirou, S.A. & Tassou, S.A. (1999): A thermal model for reptiles and pelycosaurs. *Journal of Thermal Biology* 24: 1–13.

Florides, G.A., Kalogirou, S.A., Tassou, S.A. & Wrobel, L.C. (2001): Natural environment and thermal behaviour of *Dimetrodon limbatus*. *Journal of Thermal Biology* 26: 15–20.

Folinsbee, K.E., Müller, J. & Reisz, R.R. (2007): Canine grooves: morphology, function, and relevance to venom. *Journal of Vertebrate Paleontology* 27: 547–551.

Ford, D. & Benson, R. 2017. A re-examination of *Orovenator mayorum* using µCT data, and its consequences for early amniote phylogeny. *The 65th Symposium on Vertebrate Paleontology and Comparative Anatomy & The 26th Symposium on Palaeontological Preparation and Conservation, University of Birmingham 12th–15th September 2017, Abstract Volume*: 42–43.

Fourie, H. & Rubidge, B.S. (2009): The postcranial skeleton of the basal therocephalian *Glanosuchus macrops* (Scylacosauridae) and comparison of morphological and phylogenetic trends amongst the Theriodontia. *Palaeontologia Africana* 44: 27–39.

Fourie, S. (1974): The cranial morphology of *Thrinaxodon liorhinus* Seeley. *Annals of the South African Museum* 65: 337–400.

Fox, R.C. (1964): The adductor muscles of the jaw in some primitive reptiles. *University of Kansas Publications, Museum of Natural History* 12: 657–680.

Frazetta, T.H. (1968): Adaptive problems and possibilities in the temporal fenestration of tetrapod skulls. *Journal of Morphology* 125: 145–158.

Fröbisch, J. (2007): The cranial anatomy of *Kombuisia frerensis* Hotton (Synapsida, Dicynodontia) and a new phylogeny of anomodont therapsids. *Zoological Journal of the Linnean Society* 150: 117–144.

Fröbisch, J. (2008): Global taxonomic diversity of anomodonts (Tetrapoda, Therapsida) and the terrestrial rock record across the Permian-Triassic Boundary. *PLoS One* 3(11): e3733.

Fröbisch, J. (2009): Composition and similarity of global anomodont-bearing tetrapod faunas. *Earth-Science Reviews* 95: 119–157.

Fröbisch, J. (2011): On dental occlusion and sabre teeth. *Science* 331: 1525–1528.

Fröbisch, J. (2013): Vertebrate diversity across the end-Permian mass extinction—separating biological and geological signals. *Palaeogeography, Palaeoclimatology, Palaeoecology* 372: 50–61.

Fröbisch, J. (2014): Synapsid diversity and the rock record in the Permian-Triassic Beaufort Group (Karoo Supoergroup), South Africa. In: Kammerer, C.F., Angielczyk, K.D. & Fröbisch, J. (eds.). *Early Evolutionary History of the Synapsida*. Springer, Dordrecht: 305–319.

Fröbisch, J. & Reisz, R.R. (2008): A new species of *Emydops* (Synapsida, Anomodontia) and a discussion of dental variability and pathology in dicynodonts. *Journal of Vertebrate Paleontology* 28: 770–787.

Fröbisch, J. & Reisz, R.R. (2009): The Late Permian herbivore *Suminia* and the early evolution of arboreality in vertebrate ecosystems. *Proceedings of the Royal Society Series B* 276: 3611–3618.

Fröbisch, J. & Reisz, R.R. (2011): The postcranial anatomy of *Suminia getmanovi* (Synapsida: Anomodontia), the earliest known arboreal tetrapod. *Zoological Journal of the Linnean Society* 162: 661–698.

Fröbisch, J., Angielczyk, K.D. & Sidor, C.A. (2010): The Triassic dicynodont *Kombuisia* (Synapsida, Anomodontia) from Antarctica, a refuge from the terrestrial Permian-Triassic extinction. *Naturwissenschaften* 97: 187–196.

Fröbisch, J., Schoch, R.R., Müller, J., Schindler, T. & Schweiss, D. (2011): A new basal sphenacodontid synapsid from the Late Carboniferous of the Saar-Nahe Basin, Germany. *Acta Palaeontologia Polonica* 56: 113–120.

Gaetano, L.C. & Abdala, F. (2015): The stapes of gomphodont cynodonts: insights into the middle ear structure of non-mammaliaform cynodonts. *PLoS One* 10 (7): e0131174.

Gaudry, A. (1880): Sur un reptile tre perfectionné trouvé dans le terrain permien. *Comptes rendus hebdomadaires des séances de l'Académie des sciences* 91: 669–672.

Gauthier, J., Kluge, A.J. & Rowe, T. (1988): Amniote phylogeny and the importance of fossils. *Cladistics* 4: 105–209.

Gebauer, E.V.I. (2007): *Phylogeny and evolution of the Gorgonopsia with a special reference to the skull and skeleton of GPIT/RE/7113 (*'Aelurognathus' parringtoni*).* Unpublished PhD Dissertation, Eberhard-Karls-Universität Tübingen, Tübingen.

Gebauer, E.V.I. (2014): Re-assessment of the taxonomic position of the specimen GPIT/RE/7113 (*Sauroctonus parringtoni* comb. nov., Gorgonopsia). In: Kammerer, C.F., Angielczyk, K.D. & Fröbisch, J. (eds.). *Early Evolutionary History of the Synapsida*. Springer, Dordrecht: 185–207.

Geist, V. (1972): An ecological and behavioral explanation of mammalian characteristics, and their implication to therapsid evolution. *Zeitschrift Säugetierkunde* 37: 1–15.

Germain, D. & Laurin, M. (2005): Microanatomy of the radius and lifestyle in amniotes (Vertebrata, Tetrapoda). *Zoologica Scripta* 34: 335–350.

Golubev, V.K. (2005): Permian tetrapod stratigraphy. In: Lucas, S.G. & Ziegler, K.E. (eds.). *The Nonmarine Permian. New Mexico Museum of Natural History and Science Bulletin* 30: 95–99.

Govender, R. & Yates, A. (2009): Dicynodont postcrania from the Triassic of Namibia and their implication for the systematics of Kannemeyeriiforme dicynodonts. *Palaeontologia Africana* 44: 41–57.

Gow, C.E. (1980): The dentitions of the Tritheledontidae (Therapsida: Cynodontia). *Proceedings of the Royal Society of London B* 208: 461–581.

Gow, C.E. (1985): Dentitions of juvenile *Thrinaxodon* (Reptilia: Cynodontia) and the origin of mammalian diphyodonty. *Annals of the Geological Survey of South Africa* 19: 1–17.

Green, J.L. (2009): *Enamel reduction and orthodentine in Dicynodontia (Therapsida) and Xenarthra (Mammalia): an evaluation of the potential ecological signal revealed by dental microwear.* Unpublished PhD dissertation, North Carolina State University, Raleigh.

Green, J.L. (2012): Bone and dental histology of Late Triassic dicynodonts from North America. In: Chinsamy-Turan, A. (ed.). *Forerunners of Mammals: Radiation, Histology, Biology*. Indiana University Press, Bloomington: 179–196.

Green, J.L., Schweitzer, M.H. & Lamm, E.T. (2010): Limb bone histology and growth in *Placerias hesternus* (Therapsida: Anomodontia) from the Upper Triasic of North America. *Palaeontology* 53: 347–364.

Green, J.L., Angielczyk, K.D., Beld, S. & Fisher, D.C. (2017): Hierarchically-organized growth increments in the tusk orthodentin of dicynodonts (Therapsida, Dicynodontia). *Journal of Vertebrate Paleontology*, Program and Abstracts, 2017: 121.

Gregory, W.K. & Adams, L.A. (1915): The temporal fossae of vertebrates in relation to the jaw muscles. *Science* 41: 763–765.

Grigg, G.C., Beard, L.A. & Augee, M.L. (2004): The evolution of endothermy and its diversity in mammals and birds. *Physiological and Biochemical Zoology* 77: 982–997.

Groenewald, G.H. (1991): Burrow casts from the *Lystrosaurus-Procolophon* Assemblage-zone, Karoo Sequence, South Africa. *Koedoe* 34: 13–22.

Gross, W. (1934): Die Typen des mikroskopischen Knochenbaues bei folsen Stegocephalen und Reptilien. *Zeitschrift für Anatomie und Entwicklungsgeschichte* 103: 731–764.

Grine, F.E. (1997): Dinocephalians are not anomodonts. *Journal of Vertebrate Paleontology* 17: 177–183.

Grine, F.E., Forster, C.A., Cluver, M.A. & Georgi, J.A. (2006): Cranial variability, ontogeny, and taxonomy of *Lystrosaurus* from the Karoo Basin of South Africa. In: Carrano, M.T., Gaudin, T.J., Blob, R.W. & Wible, J.R. (eds.). *Amniote Paleobiology: Perspectives on the Evolution of Mammals, Birds, and Reptiles.* University of Chicago Press, Chicago: 432–503.

Grzimek, B. (1990): *Grzimek's Encyclopedia of Mammals. Volume 5.* McGraw-Hill, New York.

Güven, S., Rubidge, B.S. & Abdala, F. (2013). Cranial morphology and taxonomy of South African Tapinocephalidae (Therapsida, Dinocephalia): the case of *Avenantia* and *Riebeeckosaurus*. *Palaeontologia Africana* 48: 24–33.

Haack, S.C. (1986): A thermal model of the sailback pelycosaur. *Paleobiology* 12: 450–458.

Hammer, W.R. (1990): Triassic terrestrial vertebrate faunas of Antarctica. In: Taylor, T.N. & Taylor, E.L. (eds.). *Antarctic Paleobiology: Its Role in the Reconstruction of Gondwana*. Springer, New York: 42–50.

Haines, R.W. (1942): The tetrapod knee joint. *Journal of Anatomy* 76: 270–301.

Haines, R.W. (1952): The shoulder joint of lizards and the primitive reptilian shoulder mechanism. *Journal of Anatomy* 86: 412–422.

Harris, S.K., Lucas, S.G., Berman, D.S. & Henrici, A.C. (2004): Vertebrate fossil assemblage from the Upper Pennsylvanian Red Tanks Member of the Bursum Formation, Lucero Uplift, central New Mexico. In: Lucas, S.G. & Zeigler, K.E. (eds.). *Carboniferous-Permian Transition at Carrizo Arroyo, Central New Mexico. New Mexico Museum of Natural History and Science Bulletin* 25: 267–284.

Haubold, H., Allen, A., Atkinson, T.P., Buta, R., Lacefield, J.A., Minkin, S.C. & Relihan B.A. (2005): Interpretation of the tetrapod footprints from the Early Pennsylvanian of Alabama. In: Rindsberg, A.K. & Kopaska-Merkel, D.C. (eds.). Pennsylvanian footprints in the Black Warrior Basin of Alabama. *Alabama Paleontological Society Monograph* 1: 75–111.

Haughton, S.H. (1924) Investigations in South African Fossil Reptiles and Amphibia. 12. On some gorgonopsian skulls in the collection of the South African Museum. *Annals of the South African Museum* 12: 499–517.

Haughton, S.H. (1926) On Karroo vertebrates from Nyasaland. *Transactions of the Geological Society of South Africa* 29: 69–83.

Hillenius, W.J. (1992): The evolution of nasal turbinates and mammalian endothermy. *Paleobiology* 18: 17–29.

Hillenius, W.J. (1994): Turbinates in therapsids: evidence for Late Permian origins of mammalian endothermy. *Evolution* 1994: 207–229.

Hillenius, W.J. (2000): Septomaxilla of nonmammalian synapsids: soft-tissue correlates and a new functional interpretation. *Journal of Morphology* 245: 29–50.

Hirasawa, T. & Kuratani, S. (2013): A new scenario of the evolutionary derivation of the mammalian diaphragm from shoulder muscles. *Journal of Anatomy* 222: 504–517.

Hoffman, M.M. & Schour, I. (1940): Quantitative studies in the development of the rat molar I. The growth pattern of primary and secondary dentin from birth to 500 days of age. *The Anatomical Record* 78: 233–251.

Hook, R.W. & Hotton, N. III. (1991): A new sphenacodontid pelycosaur (Synapsida) from the Wichita Group, Lower Permian of north-central Texas. *Journal of Vertebrate Paleontology* 11: 37–44.

Hopson, J.A. (1966): The origin of the mammalian middle ear. *American Zoologist* 6: 437–450.

Hopson, J.A. (1969): The origin and adaptive radiation of mammal-like reptiles and nontherian mammals. *Annals of the New York Academy of Sciences* 167: 199–216.

Hopson, J.A. (1987): The mammal-like reptiles: a study of transitional fossils. *American Biology Teacher* 49: 16–26.

Hopson, J.A. (1991): Systematics of the nonmammalian Synapsida and implication for patterns of evolution in synapsids. In: Schultze, H.P. & Trueb, L. (eds.). *Origins of the Higher Groups of Tetrapods: Controversy and Consensus*. Comstock Publishing Associates, Ithaca: 635–693.

Hopson, J.A. (1994): Synapsid evolution and the radiation of non-eutherian mammals. In: Prothero, D.R. & Schoch, R.M. (eds.). *Major Features of Vertebrate Evolution. Short Courses in Paleontology* 7: 190–219.

Hopson, J.A. (1999): Therapsida. In: Singer, R. (ed.). *Encyclopedia of Paleontology*. Fitzroy Dearborn, Chicago: 1256–1266.

Hopson, J.A. (2001): Origin of mammals. In: Briggs, D.E.G. & Crowther, P.R. (eds.). *Palaeobiology II*. Blackwell Science Ltd., Oxford: 88–94.

Hopson, J.A. (2005): A juvenile gomphodont cynodont specimen from the *Cynognathus* Assemblage Zone of South Africa: implications for the origin of gomphodont postcanine morphology. *Palaeontologia Africana* 41: 53–66.

Hopson, J.A. (2012): The role of foraging mode in the origin of therapsids: implications for the origin of mammalian endothermy. *Fieldiana: Life and Earth Sciences* 5: 126–148.

Hopson, J.A. (2015): Fossils, trackways, and transitions in locomotion: a case study of *Dimetrodon*. In: Dial, K.P., Shubin N. & Brainerd, E.L. (eds.). *Great Transformations in Vertebrate Evolution*. University of Chicago Press, Chicago: 125–141.

Hopson, J.A. & Barghusen, H.R. (1986): An analysis of therapsid relationships. In: Hotton, N. III, MacLean, P.D., Roth, J.J. & Roth, E.C. (eds.). *The Ecology and Biology of Mammal-like Reptiles*. Smithsonian Institution Press, Washington D.C.: 83–106.

Hopson, J.A. & Kitching, J.W. (2001): A probainognathian cynodont from South Africa and the phylogeny of nonmammalian cynodonts. *Bulletin of the Museum of Comparative Zoology* 156: 3–35.

Hopson, J.A., Kielan-Jaworowska, Z. & Allin, E.F. (1989): The cryptic jugal of multituberculates. *Journal of Vertebrate Paleontology* 9: 201–209.

Hotton, N. III (1986): Dicynodonts and their roles as primary consumers. In: Hotton, N. III, MacLean, P.D., Roth, J.J. & Roth, E.C. (eds.). *The Ecology and Biology of Mammal-like Reptiles*. Smithsonian Institution Press, Washington D.C.: 71–82.

Hotton, N. III (1991): The nature and diversity of synapsids: prologue to the origin of mammals. In: Schultze H.P. & Trueb L. (eds.). *Origins of the Higher Groups of Tetrapods: Controversy and Consensus*. Comstock Publishing Associates, Ithaca: 598–634.

Hotton, N. III, Olson, E.C. & Beerbower, R. (1997): Amniote origins and the discovery of herbivory. In: Sumida, S. & Martin, K.L.M. (eds.). *Amniote Origins: Completing the Transition to Land*. Academic Press, San Diego: 207–264.

Houssaye, A., Sander, P.M. & Klein, N. (2016): Adaptive patterns in aquatic amniote bone microanatomy—more complex than previously thought. *Integrative and Comparative Biology*. 56: 1349–1369.

Huene, F. von (1905): Pelycosaurier im deutschen Muschelkalk. *Neues Jahrbuch für Mineralogie, Geologie und Paläontologie (Beilage-Band)* 20: 321–353.

Huene, F. von. (1908): Bemerkungen über *Callibrachion*. *Centralblatt für Mineralogie, Geologie und Paläontologie* 1908: 532–534.

Huene, F. von. (1940): Die Saurier der Karroo-, Gondwana- und verwandten Ablagerungen in faunistischer, biologischer, und phylogenetischer Hinsicht. *Neues Jahrbuch für Mineralogie, Geologie und Paläontologie Abteilung B: Geologie und Paläontologie* 83: 246–347.

Huene, F. von. (1942): Die anomodontier des Ruhuhu-Gebietes in der Tübinger Sammlung. *Palaeontographica Abteilung A* 94: 154–184.

Huene, F. von. (1950): Die Theriodontier des ostafrikanischen Ruhuhu-Gebietes in der Tübinger Sammlung. *Neues Jahrbuch für Geologie und Paläontologie Abhandlungen* 92: 47–136.

Hunt, A.P. & Lucas, S.G. (1998): Vertebrate tracks and the myth of the belly-dragging, tail-dragging tetrapods of the late Paleozoic. In: Lucas, S.G., Estep, J.W. & Hoffer, J.M. (eds.). *Permian Stratigraphy and Paleontology of the Robledo Mountain, New Mexico. New Mexico Museum of Natural History and Science Bulletin* 12: 67–69.

Hunt, A.P., Lucas, S.G. & Lockley, M.G. (2004): Large pelycosaur footprints from the Lower Pennsylvanian of Alabama, USA. *Ichnos* 11: 39–45.

Hunt, A.P., Lucas, S.G. & Pyenson, N.D. (2005): The significance of the Union chapel Mine Site: a Lower Pennsylvanian (Westphalian A) ichnological Konzentrat-Laggerstätte, Alabama, USA. In: Rindsberg, A.K. & Kopaska-Merkel, D.C. (eds.). Pennsylvanian footprints in the Black Warrior Basin of Alabama. *Alabama Paleontological Society Monograph* 1: 3–14.

Huttenlocker, A.K. (2009): An investigation into the cladistic relationships and monophyly of therocephalian therapsids (Amniota: Synapsida). *Zoological Journal of the Linnean Society* 157: 865–891.

Huttenlocker, A.K. (2013): *The paleobiology of South African Therocephalian Therapsids (Amniota, Synapsida) and the Effects of the End-Permian Extinction on Size, Growth, and Bone Microstructure*. Unpublished PhD dissertation, University of Washington, Seattle.

Huttenlocker, A.K. (2014): Body size reductions in nonmammalian eutheriodont therapsids (Synapsida) during the end-Permian mass extinction. *PLoS One* 9(2): e87553.

Huttenlocker, A.K. & Abdala, F. (2015): Revision of the first therocephalian, *Theriognathus* Owen (Therapsida: Whaitsiidae), and implications for cranial ontogeny and allometry in nonmammaliaform eutheriodonts. *Journal of Paleontology* 89: 645–664.

Huttenlocker, A.K. & Botha-Brink, J. (2013): Body size and growth patterns in the therocephalian *Moschorhinus kitchingi* (Therapsida: Eutheriodontia) before and after the end-Permian extinction in South Africa. *Paleobiology* 39: 253–277.

Huttenlocker, A.K. & Botha-Brink, J. (2014): Bone microstructure and the evolution of growth patterns in Permo-Triassic therocephalians (Amniota, Therapsida) of South Africa. *PeerJ* 2: e325.

Huttenlocker, A.K. & Farmer, C.G. (2017): Bone microvasculature tracks red blood cell size diminuation in Triassic mammal and dinosaur forerunners. *Current Biology* 27: 48–54.

Huttenlocker, A.K. & Rega, E. (2012): The paleobiology and bone microstructure of pelycosaurian-grade synapsids. In: Chinsamy-Turan, A. (ed.). *Forerunners of Mammals: Radiation, Histology, Biology*. Indiana University Press, Bloomington: 91–119.

Huttenlocker, A.K. & Sidor, C.A. (2016): The first karenitid (Therapsida, Therocephalia) from the upper Permian of Gondwana and the biogeography of Permo-Triassic therocephalians. *Journal of Vertebrate Paleontology* 36(4): e1111897.

Huttenlocker, A.K., Angielczyk, K.D. & Lee, A.S. (2006): Osteohistology of *Sphenacodon* (Synapsida: Sphenacodontidae) and the hidden diversity of growth patterns in basal synapsids. *Journal of Vertebrate Paleontology* 26 (supplement to 3): 79A–80A.

Huttenlocker, A.K., Rega, E. & Sumida, S.S. (2010): Comparative anatomy and osteohistology of hyperelongate neural spines in the sphenacodontids *Sphenacodon* and *Dimetrodon* (Amniota: Synapsida). *Journal of Morphology* 271: 1407–1421.

Huttenlocker, A.K., Mazierski, D. & Reisz, R.R. (2011a): Comparative osteohistology of hyperelongate neural spines in the Edaphosauridae (Amniota: Synapsida). *Palaeontology* 54: 573–590.

Huttenlocker, A.K., Sidor, C.A. & Smith, R.M.H. (2011b): A new specimen of *Promoschorhynchus* (Therapsida: Therocephalia: Akidnognathidae) from the Lower Triassic of South Africa and its implications for theriodont survivorship across the Permo-Triassic boundary. *Journal of Vertebrate Paleontology* 31: 405–421.

Huttenlocker, A.K. Sidor, C.A. & Angielczyk, K.D. (2015): A new eutherocephalian (Therapsida, Therocephalia) from the upper Permian Madumabisa Mudstone Formation (Luangwa Basin) of Zambia. *Journal of Vertebrate Paleontology* 35: e969400.

Huxley, T.H. (1878): On the characters of the pelvis in the Mammalia, and the conclusions respecting the origin of mammals which may be based on them. *Proceedings of the Royal Society of London* 28: 367–372.

Huxley, T.H. (1880): On the application of the laws of evolution to the arrangement of the Vertebrata, and more particularly of the Mammalia. *Proceedings of the Zoological Society of London* 1880: 649–662.

Irmis, R.B. & Whiteside, J.H. (2012): Delayed recovery of non-marine tetrapods after the end-Permian mass extinction tracks global carbon cycle. *Proceedings of the Royal Society Series B* 279: 1310–1318.

Irmis R.B., Whiteside, J.H. & Kammerer, C.F. (2013): Non-biotic controls of observed diversity in the paleontologic record: an example from the Permo-Triassic Karoo Basin of South Africa. *Palaeogeography, Palaeoclimatology, Palaeoecology* 372: 62–77.

Ivakhnenko, M.F. (1994): A new Late Permian dromasaurian (Anomodontia) from eastern Europe. *Paleontological Journal* 28: 96–103.

Ivakhnenko, M.F. (1996): Primitive anomodonts, venyukoviids, from the Late Permian of eastern Europe. *Paleontological Journal* 30: 575–582.

Ivakhnenko, M.F. (1999): Biarmosuches from the Ocher faunal assemblage of eastern Europe. *Paleontological Journal* 33: 289–296.

Ivakhnenko, M.F. (2000): Estemmenosuches and primitive theriodonts from the Late Permian. *Paleontological Journal* 34: 189–197.

Ivakhnenko, M.F. (2002a): The origin and early divergence of therapsids. *Paleontological Journal* 36: 168–175.

Ivakhnenko, M.F. (2002b): Taxonomy of East European gorgonopians (Therapsida). *Paleontological Journal* 36: 283–292.

Ivakhnenko, M.F. (2003a): Angular region of the lower jaw in primitive therapsids. *Paleontological Journal* 37: 287–292.

Ivakhnenko, M.F. (2003b): Eotherapsids from the East European placket. *Paleontological Journal* 37 (supplement 4): S339–S465.

Ivakhnenko, M.F. (2008): Cranial morphology and evolution of Permian Dinomorpha (Eotherapsida) of Eastern Europe. *Paleontological Journal* 42: 859–995.

Ivakhnenko, M.F. (2009): Eotherapsid hypothesis for the origin of Monotremata. *Paleontological Journal* 43: 237–250.

Ivakhnenko, M.F. (2011): Permian and Triassic therocephals (Eutherapsida) of eastern Europe. *Paleontological Journal* 45: 981–1144.

Ivakhnenko, M.F. (2012): Permian Cynodontia (Theromorpha) of Eastern Europe. *Paleontological Journal* 46: 199–207.

Ivakhnenko, M.F. (2013): Cranial morphology of *Dvinia prima* Amalitzky (Cynodontia, Theromorpha). *Paleontological Journal* 47: 210–222.

Jablonski, D. (2002): Survival without recovery after mass extinctions. *Proceedings of the National Academy of Sciences* 99: 8139–8144.

Jacobs, L.L., Winkler, D.A., Newman, K.D., Gomani, E.M. & Deino, A. (2005): Therapsids from the Permian Chiweta Beds and the age of the Karoo Supergroup in Malawi. *Palaeontologia Electronica* 8(28A): 1–23.

Jaekel, O. (1910): *Naosaurus credneri* im Rotliegenden von Sachsen. *Zeitschrift der Deutschen Geologischen Gesellschaft* 62: 526–535.

Jasinoski, S.C. & Abdala, F. (2017a): Aggregations and parental care in the Early Triassic basal cynodonts *Galesaurus planiceps* and *Thrinaxodon liorhinus*. *PeerJ* 5: e2875.

Jasinoski, S.C. & Abdala, F. (2017b): Cranial ontogeny of the Early Triassic basal cynodont *Galesaurus planiceps*. *The Anatomical Record* 300: 353–381.

Jasinoski, S.C. & Chinsamy-Turan, A. (2012): Biological inferences of the cranial microstructure of the dicynodonts *Oudenodon* and *Lystrosaurus*. In: Chinsamy-Turan, A. (ed.). *Forerunners of Mammals: Radiation, Histology, Biology*. Indiana University Press, Bloomington: 149–176.

Jasinoski, S.C., Abdala F. & Fernandez, V. (2015): Ontogeny of the Early Triassic cynodont *Thrinaxodon liorhinus* (Therapsida): cranial morphology. *The Anatomical Record* 298: 1440–1464.

Jasinoski, S.C., Cluver, M.A., Chinsamy, A. & Reddy, B.D. (2014): Anatomical plasticity in the snout of *Lystrosaurus*. In: Kammerer, C.F., Angielczyk, K.D. & Fröbisch, J. (eds.). *Early Evolutionary History of the Synapsida*. Springer, Dordrecht: 139–149.

Jasinoski, S.C., Rayfield, E.J. & Chinsamy, A. (2009). Comparative feeding biomechanics of *Lystrosaurus* and the generalized dicynodont *Oudenodon*. *Anatomical Record* 292: 862–874.

Jasinoski, S.C., Rayfield, E.J. & Chinsamy, A. (2010a). Functional implications of dicynodont cranial suture morphology. *Journal of Morphology* 271: 705–728.

Jasinoski, S.C., Rayfield, E.J. & Chinsamy, A. (2010b). Mechanics of the patent premaxilla-nasal suture in the snout of *Lystrosaurus*. *Journal of Vertebrate Paleontology* 30: 1283–1288.

Jenkins, F.A. (1971): The postcranial skeleton of African cynodonts. *Peabody Museum of Natural History Bulletin* 36: 1–216.

Jenkins, F.A. (1973): The functional anatomy and evolution of the mammalian humero-ulnar articulation. *American Journal of Anatomy* 137: 281–297.

Jinnah, Z.A. & Rubidge, B. (2007): A double-tusked dicynodont and its biostratigraphic significance. *South African Journal of Science* 103: 51–53.

Ji, Q., Luo, Z.-X., Yuan, C.-X. & Tabrum, A.R. (2006): A swimming mammaliaform from the Middle Jurassic and ecomorphological diversification of early mammals. *Science* 311: 1123–1127.

Kammerer, C.F. (2009): *Cranial disparity in the non-mammalian Synapsida*. Unpublished PhD dissertation, University of Chicago, Chicago.

Kammerer, C.F. (2011): Systematics of the Anteosauria (Therapsida: Dinocephalia). *Journal of Systematic Palaeontology* 9: 261–304.

Kammerer, C.F. (2014a). Theriodontia: Introduction. In: Kammerer, C.F., Angielczyk, K.D. & Fröbisch, J. (eds.). *Early Evolutionary History of the Synapsida*. Springer, Dordrecht: 165–169.

Kammerer, C.F. (2014b): A redescription of *Eriphostoma microdon* Broom, 1911 (Therapsida, Gorgonopsia) from the *Tapinocephalus* Assemblage Zone of South Africa and a review of Middle Permian gorgonopsians. In: Kammerer, C.F., Angielczyk, K.D. & Fröbisch, J. (eds.). *Early Evolutionary History of the Synapsida*. Springer, Dordrecht: 171–184.

Kammerer, C.F. (2015): Cranial osteology of *Arctognathus curvimola*, a short-snouted gorgonopsian from the Late Permian of South Africa. *Papers in Palaeontology* 1: 41–58.

Kammerer, C.F. (2016a): A new taxon of cynodont from the *Tropidostoma* Assemblage Zone (upper Permian) of South Africa, and the early evolution of Cynodontia. *Papers in Palaeontology* 2: 387–397.

Kammerer, C.F. (2016b): Two unrecognized burnetiamorph specimens from historical Karoo collections. *Palaeontologia Africana* 50: 64–75.

Kammerer, C.F. (2016c): Systematics of the Rubidgeinae (Therapsida: Gorgonopsia). *PeerJ* 4: e1608.

Kammerer, C.F. (2017a): Anatomy and relationships of the South African gorgonopsian *Arctops* (Therapsida, Theriodontia). *Papers in Palaeontology* 3: 583–611.

Kammerer, C.F. (2017b): Rediscovery of the holotype of *Clelandina major* Broom, 1948 (Gorgonopsia: Rubidgeinae) with implications for the identity of the species. *Palaeontologia Africana* 52: 85–88.

Kammerer, C.F. (2018): The first skeletal evidence of a dicynodont from the lower Elliot Formation of South Africa. *Palaeontologia Africana* 52: 102–128.

Kammerer, C.F. & Angielczyk, K.D. (2009): A proposed higher taxonomy of anomodont therapsids. *Zootaxa* 2018: 1–24.

Kammerer, C.F. & Masyutin, V. (2018a). Gorgonopsian therapsids (Nochnitsa gen. nov. and Viatkogorgon) from the Permian Kotelnich locality of Russia. PeerJ 6:e4954.

Kammerer, C.F. & Masyutin, V. (2018b). A new therocephalian (Gorynychus masyutinae gen. et. sp. nov.) from the Permian Kotelnich locality, Kirov Region, Russia. PeerJ 6:e4933.

Kammerer, C.F. & Smith, R.M.H. (2017): An early geikiid dicynodont from the *Tropidostoma* Assemblage Zone (late Permian) of South Africa. *PeerJ* 5: e2913.

Kammerer, C.F., Flynn, J.J., Ranivoharimanana, L. & Wyss, A.R. (2010): The first record of a probainognathian (Cynodontia: Chiniquodontidae) from the Triassic of Madagascar. *Journal of Vertebrate Paleontology* 30: 1889–1894.

Kammerer, C.F., Angielczyk, K.D. & Fröbisch, J. (2011): A comprehensive taxonomic revision of *Dicynodon* (Therapsida, Anomodontia) and its implications for dicynodont phylogeny, biogeography, and biostratigraphy. *Society of Vertebrate Paleontology Memoir* 11: 1–158.

Kammerer, C., Jansen, M. & Fröbisch, J. (2013a): Therapsid phylogeny revisited. *Journal of Vertebrate Paleontology, Program and Abstracts* 2013: 150.

Kammerer, C.F., Fröbisch, J. & Angielczyk, K.D. (2013b): On the validity and phylogenetic position of *Eubrachiosaurus browni*, a kannemeyeriiform dicynodont (Anomodontia) from Triassic North America. *PLoS One* 8(5): e64203.

Kammerer, C.F., Angielczyk, K.D. & Fröbisch, J. (2015a): Redescription of *Digalodon rubidgei*, an emydopoid dicynodont (Therapsida, Anomodontia) from the Late Permian of South Africa. *Fossil Record* 18: 43–55.

Kammerer, C.F., Smith, R.M.H., Day, M.O. & Rubidge, B.S. (2015b): New information on the morphology and stratigraphic range of the mid-Permian gorgonopsian *Eriphostoma microdon* Broom, 1911. *Papers in Palaeontology* 1: 201–221.

Kammerer, C.F., Angielczyk, K.D. & Fröbisch, J. (2015c): Redescription of the geikiid *Pelanomodon* (Therapsida, Dicynodontia), with a reconsideration of *'Propelanomodon.' Journal of Vertebrate Paleontology* 36: e1030408.

Kammerer, C.F., Bandyopadhyay, S. & Ray, S. (2016): A new taxon of cistecephalid dicynodont from the upper Permian Kundaram Formation of India. *Papers in Palaeontology* 2: 569–584.

Kemp, T.S. (1969): On the functional morphology of the gorgonopsid skull. *Philosophical Transactions of the Royal Society of London, Series B, Biological Sciences* 256: 1–83.

Kemp, T.S. (1972a): The jaw articulation and musculature of the whaitsiid Therocephalia. In: Joysey, K.A. & Kemp, T.S. (eds.). *Studies in Vertebrate Evolution*. Oliver and Boyd, Edinburgh: 213–230.

Kemp, T.S. (1972b): Whaitsiid Therocephalia and the origin of cynodonts. *Philosophical Transactions of the Royal Society of London B* 264: 1–54.

Kemp, T.S. (1979): The primitive cynodont *Procynosuchus*: functional anatomy of the skull and relationships. *Philosophical Transactions of the Royal Society B* 285: 73–122.

Kemp, T.S. (1980a): Origin of the mammal-like reptiles. *Nature* 283: 378–380.

Kemp, T.S. (1980b): Matters arising: Origin of mammal-like reptiles. Kemp replies. *Nature* 288: 193.

Kemp, T.S. (1980c): The primitive cynodont *Procynosuchus*: structure, function and evolution of the postcranial skeleton. *Philosophical Transactions of the Royal Society B* 288: 217–258.

Kemp, T.S. (1982): *Mammal-like Reptiles and the Origin of Mammals*. Academic Press, London.

Kemp, T.S. (1983): The relationships of mammals. *Journal of the Linnean Society of London (Zoology)* 77: 353–384.

Kemp, T.S. (1985): Synapsid reptiles and the origin of higher taxa. *Special Papers in Palaeontology* 33: 175–184.

Kemp, T.S. (1988a): Haemothermia or Archosauria? The interrelationships of mammals, birds, and crocodiles. *Zoological Journal of the Linnean Society* 92: 67–104.

Kemp, T.S. (1988b): Interrelationships of the Synapsida. In: Benton, M.J. (ed). *The Phylogeny and Classification of the Tetrapods*. Volume 2. Systematics Association, Special Volume 35B: 1–22.

Kemp, T.S. (2005): *The Origin and Evolution of Mammals*. Oxford University Press, Oxford.

Kemp, T.S. (2006): The origin and early radiation of the therapsid mammal-like reptiles: a palaeobiological hypothesis. *Journal of Evolutionary Biology* 19: 1231–1247.

Kemp, T.S. (2007): Acoustic transformer function of the postdentary bones and quadrate of a nonmammalian cynodont. *Journal of Vertebrate Paleontology* 27: 431–444.

Kemp, T.S. (2009): Phylogenetic relationships and pattern of evolution of the therapsids: testing for polytomy. *Palaeontologia Africana* 44: 1–12.

Kemp, T.S. (2012): The origin and radiation of therapsids. In: Chinsamy-Turan, A. (ed.). *Forerunners of Mammals: Radiation, Histology, Biology*. Indiana University Press, Bloomington: 3–28.

Kemp, T.S. (2016): Nonmammalian synapsids. In: Clack, J.A., Fay, R.R. & Popper, A.N. (eds.). *Evolution of the Vertebrate Ear—Evidence from the Fossil Record. Springer Handbook of Auditory Research* 59. Springer International Publishing, Switzerland: 107–137.

Kermack, K.A. (1982): The ear in the Theropsida. *Geobios Mémoire Spécial* 6: 151–156.

Kermack, K.A. & Mussett, F. (1983): The ear in mammal-like reptiles and early mammals. *Acta Palaeontologica Polonica* 28: 147–158.

Keyser, A.W. (1973a): A re-evaluation of the genus *Tropidostoma* Seeley. *Palaeontologia Africana* 16: 25–35.

Keyser, A.W. (1973b): A preliminary study of the type area of the *Cistecephalus* Zone of the Beaufort Series, and a revision of the anomodont family Cistecephalidae. *Geological Survey of South Africa Memoir* 62: 1–71.

Keyser, A.W. (1975): A re-evaluation of the cranial morphology and systematics of some tuskless Anomodontia. *Memoirs of the Geological Survey of South Africa* 67: 1–110.

Keyser, A.W. & Cruickshank, A.R.I. (1979): The origins and classification of Triassic dicynodonts. *Transactions of the Geological Society of South Africa* 82: 81–108.

Keyser, A.W. & Smith, R.M.H. (1977–1978): Vertebrate biozonation of the Beaufort Group with special reference to the western Karoo Basin. *Annals of the Geological Survey of South Africa* 12: 1–36.

Kielan-Jaworowska, Z., Cifelli, R.L. & Luo, Z.-X. (2004): *Mammals from the Age of Dinosaurs: Origins, Evolution, and Structure*. Columbia University Press, New York.

Kim, J.-W., Yang, H.-J., Oel, A.P., Brooks, M.J., Plachetzki, D.C., Li W., Allison, W.T. & Swaroop, A. (2016): Recruitment of rod photoreceptors from short-wavelength-sensitive cones during the evolution of nocturnal vision in mammals. *Developmental Cell* 37: 520–532.

King, G.M. (1981): The functional anatomy of a Permian dicynodont. *Philosophical Transactions of the Royal Society Series B* 291: 243–322.

King, G.M. (1988): *Anomodontia. Handbuch der Paläoherpetologie* 17C. Gustav Fischer Verlag, Stuttgart: 1–174.

King, G.M. (1990): *The Dicynodonts: A Study in Palaeobiology*. Chapman and Hall, London.

King, G.M. (1991) The aquatic *Lystrosaurus*: a palaeontological myth. *Historical Biology* 4: 285–322.

King, G.M. (1993): How many species of *Diictodon* were there? *Annals of the South African Museum* 102: 303–325.

King, G.M. (1994): The early anomodont *Venjukovia* and the evolution of the anomodont skull. *Journal of Zoology* 232: 651–673.

King, G.M. (1996): *Reptiles and Herbivory*. Chapman and Hall, London.

King, G.M. & Cluver, M.A. (1991): The aquatic *Lystrosaurus*: an alternative lifestyle. *Historical Biology* 4: 323–342.

King, G.M. & Rubidge, B.S. (1993): A taxonomic revision of small dicynodonts with postcanine teeth. *Zoological Journal of the Linnean Society* 107: 131–154.

King, G.M., Oelofsen, B.W. & Rubidge, B.S. (1989): The evolution of the dicynodont feeding system. *Zoological Journal of the Linnean Society* 96: 185–211.

Kissel, R.A. & Reisz, R.R. (2004): Synapsid fauna of the Upper Pennsylvanian Rock Lake Shale near Garnett, Kansas and the diversity pattern of early amniotes. In: Arratia, G., Wilson, M.V.H. & Cloutier, R. (eds.). *Recent Advances in the Origin and Early Radiation of Vertebrates*. Verlag Dr. Friedrich Pfeil, Munich: 409–428.

Kitching, J.W. (1977): The distribution of the Karroo vertebrate fauna. *Bernard Price Institute for Palaeontological Research Memoir* 1: 1–131.

Koyabu, D., Maier, W. & Sánchez-Villagra, M.R. (2014): Paleontological and developmental evidence resolve the homology and dual embryonic origin of a mammalian skull bone, the interparietal. *Proceedings of the National Academy of Sciences* 109: 14075–14080.

Krause, D.W., Hoffmann, S., Wible, J.R., Kirk, E.C., Schultz, J.A., von Koenigswald, W., Groenke, J.R., Rossie, J.B., O'Connor, P.M., Seiffert, E.R., Dumont, E.R., Holloway, W.L., Rogers, R.R., Rahantarisoa, L.J., Kemp, A.D. & Andriamialison, H. (2014): First cranial remains of a gondwanatherian mammal reveal remarkable mosaicism. *Nature* 515: 512–517.

Kriloff, A., Germain, D., Canoville, A., Vincent, P., Sache, M. & Laurin, M. (2008): Evolution of bone microanatomy of the tetrapod tibia and its use in palaeobiological inference. *Journal of Evolutionary Biology* 21: 807–826.

Kruger, A., Rubidge, B.S., Abdala, F., Gomani Chindebvu, E. & Jacobs, L.L. (2015): *Lende chiweta*, a new therapsid from Malawi, and its influence on burnetiamorph phylogeny and biogeography. *Journal of Vertebrate Paleontology* 35: e1008698.

Kurkin, A.A. (2017). A new galeopid (Anomodontia, Galeopidae) from the Permian of Eastern Europe. Paleontological Journal 51: 308–312.

Kutorga, S.S. (1838): *Beitrag zur Kenntniss der organischen Überreste der Kupfersandsteins am Westlichen Abhange des Urals*. N. Gretsch, Saint Petersburg.

Laaß, M. (2014): Bone-conduction hearing and seismic sensitivity of the late Permian anomodont *Kawingasaurus fossilis*. *Journal of Morphology* 276: 121–143.

Laaß, M. (2015): Virtual reconstruction and description of the cranial endocast of *Pristerodon mackayi* (Therapsida, Anomodontia). *Journal of Morphology* 267: 1089–1099.

Laaß, M. (2016): The origins of the cochlea and impedance matching hearing in synapsids. *Acta Palaeontologica Polonica* 61: 267–280.

Laaß, M. & Kaestner, A. (2017): Evidence for convergent evolution of a neocortex-like structure in a late Permian therapsid. *Journal of Morphology* 278: 1033–1057.

Laaß, M., Schillinger, B. & Kaestner, A. (2017): What did the "Unossified zone" of the non-mammalian therapsid braincase house? *Journal of Morphology* 278: 1020–1032.

Lambertz, M., Shelton, C.D., Spindler, F. & Perry, S.F. (2016): A caseian point for the evolution of a diaphragm homologue among the earliest synapsids. *Annals of the New York Academy of Sciences* 1385: 3–20.

Langer, M.C. (2000): The first record of dinocephalians in South America: late Permian (Rio do Rasto Formation) of the Paraná Basin, Brazil. *Neues Jahrbuch für Geologie und Paläontologie Abhandlungen* 215: 69–95.

Langston, W. Jr. (1965): *Oedaleops campi* (Reptilia: Pelycosauria) new genus and species from the Lower Permian of New Mexico, and the family Eothyrididae. *Bulletin of the Texas Memorial Museum* 9: 5–47.

Langston, W. Jr. (1973): The crocodilian skull in historical perspective. In: Gans, C. & Parsons, T.S. (eds.). *Biology of the Reptilia, Vol. 4*. Academic Press, New York: 263–284.

Laurin, M. (1993): Anatomy and relationships of *Haptodus garnettensis*, a Pennsylvanian synapsid from Kansas. *Journal of Vertebrate Paleontology* 13: 200–229.

Laurin, M. (1994): A re-evaluation of *Cutleria wilmarthi* an Early Permian synapsid from Colorado. *Journal of Vertebrate Paleontology* 14: 134–138.

Laurin, M. (1998): New data on the cranial anatomy of *Lycaenops* (Synapsida, Gorgonopsidae), and reflections on the possible presence of streptostyly in gorgonopsians. *Journal of Vertebrate Paleontology* 18: 765–776.

Laurin, M. & de Buffrénil, V. (2015): Microstructural features of the femur in early ophiacodontids: a reappraisal of ancestral habitat use and lifestyle of amniotes. *Comptes Rendus Palevol* 15: 115–127.

Laurin, M. & Reisz, R.R. (1990): *Tetraceratops* is the oldest known therapsid. *Nature* 345: 249–250.

Laurin, M. & Reisz, R.R. (1995): A reevaluation of early amniote phylogeny. *Zoological Journal of the Linnean Society* 113: 165–223.

Laurin, M. & Reisz, R.R. (1996): The osteology and relationships of *Tetraceratops insignis*, the oldest known theraspid. *Journal of Vertebrate Paleontology* 16: 95–102.

Laurin, M. & Reisz, R.R. (1997): A new perspective on tetrapod phylogeny. In: Sumida, S.S. & Martin, K.L.M. (eds.). *Amniote Origins: Completing the Transition to Land*. Academic Press, San Diego: 9–59.

LeBlanc, A.R.H., Reisz, R.R., Brink, K.S. & Abdala, F. (2016): Mineralized periodontia in extinct relatives of mammals shed light on the evolutionary history of mineral homeostasis in periodontal tissue maintenance. *Journal of Clinical Periodontology* 43: 323–332.

Lee, M.S.Y. (1999): Molecular clock calibrations and metazoan divergence dates. *Journal of Molecular Evolution* 49: 385–391.

Leidy, J. (1854): On *Bathygnathus borealis*, an extinct saurian of the New Red Sandstone of Prince Edward's Island. *Journal of the Academy of Natural Sciences, Philadelphia Series 2* 2: 327–330.

Li, J. (2001): The most primitive lower tetrapod fauna in China. *Science in China (Series D)* 44: 47–51.

Li, J. & Cheng, Z. (1997): First discovery of eotitanosuchian (Therapsida, Synapsida) of China. *Vertebrata PalAsiatica* 35: 268–282.

Lingham-Soliar, T. (2014): *The Vertebrate Integument*. Volume 1. Springer-Verlag, Berlin.

Linnæus, C. (1758): *Systema Naturæ per Regnia Tria Naturæ, Secundum Classes, Ordines, Genera, Species, cum Characteribus, Differentiis, Synonymis, Locis. Tomus I. Editio Decima, Reformata.* Laurentii Salvii, Stockholm.

Lips, K.R. (1991): Vertebrates associated with tortoise (*Gopherus polyphemus*) burrows in four habitats in south-central Florida. *Journal of Herpetology* 25: 477–481.

Liu, J. (2013): Osteology, ontogeny, and phylogenetic position of *Sinophoneus yumenensis* (Therapsida, Dinocephalia) from the Middle Permian Dashankou Fauna of China. *Journal of Vertebrate Paleontology* 33: 1394–1407.

Liu, J. & Abdala, F. (2014): Phylogeny and taxonomy of the Traversodontidae. In: Kammerer, C.F., Angielczyk, K.D. & Fröbisch, J. (eds.). *Early Evolutionary History of the Synapsida*. Springer, Dordrecht: 255–279.

Liu, J. & Abdala, F. (2017): The tetrapod fauna of the upper Permian Naobaoguo Formation of China: 1. *Shiguaignathus wangi* gen et sp. nov., the first akidnognathid therocephalian from China. *PeerJ* 5: e4150.

Liu, J. & Olsen, P.E. (2010): The phylogenetic relationships of Eucynodontia (Amniota: Synapsida). *Journal of Mammalian Evolution* 17: 151–176.

Liu, J., Rubidge, B.S. & Li, J. (2009): New basal synapsid supports Laurasian origin for therapsids. *Acta Palaeontologica Polonica* 54: 393–400.

Liu, J., Rubidge, B.S. & Li, J. (2010): A new specimen of *Biseridens qilianicus* indicates its phylogenetic position as the most basal anomodont. *Proceedings of the Royal Society B* 277: 285–292.

Liu, J., Xu, L., Jia, S.-H. & Liu, X.-L. (2014): The Jiyuan tetrapod fauna of the Upper Permian of China—2. Stratigraphy, taxonomical review, and correlation. *Vertebrata PalAsiatica* 52: 328–339.

Liu, J., Ramezani, J., Li, L., Shang, Q.-H., Xu, G.-H., Wang, Y.-Y. & Yang, J.-S. (2018): High precision temporal calibration of Middle Triassic vertebrate biostratigraphy: U=Pb zircon constraints for the *Sinokannemeyeria* Fauna and *Yonghesuchus*. *Vertebrata Palasiatica* 56: 16–24.

Lombard, R.E. & Bolt, J.R. (1979): Evolution of the tetrapod ear: an analysis and reinterpretation. *Biological Journal of the Linnean Society* 11: 19–76.

Lovegrove, B.G. (2017): A phenology of the evolution of endothermy in birds and mammals. *Biological Reviews* 92:1213–1240.

Lozovsky, V.R. (2005): Olson's Gap or Olson's Bridge, that is the question. In: Lucas, S.G. & Ziegler, K.E. (eds.). *The Nonmarine Permian. New Mexico Museum of Natural History and Science Bulletin* 30: 179–184.

Lucas, S.G. (1993): The *Shansiodon* Biochron, nonmarine Middle Triassic of Pangaea. *Albertiana* 11: 40–42.

Lucas, S.G. (1996): Vertebrate biochronology of the Mesozoic of China. *Memoirs of the Beijing Natural History Museum* 55: 110–148.

Lucas, S.G. (1998a): Global Triassic tetrapod biostratigraphy and biochronology. *Palaeogeography Palaeoclimatology, Palaeoecology* 143: 347–384.

Lucas, S.G. (1998b) Toward a tetrapod biochronology of the Permian. In: Lucas, S.G., Estep, J.W. & Hoffer, J.M. (eds.). *Permian Stratigraphy and Paleontology of the Robledo Mountain, New Mexico. New Mexico Museum of Natural History and Science Bulletin* 12: 71–91.

Lucas, S.G. (2001): A global hiatus in the Middle Permian tetrapod fossil record. *Permophiles* 38: 24–27.

Lucas, S.G. (2002): Discussion and reply: The reptile *Macroleter*: first vertebrate evidence for correlation of Upper Permian continental strata of North America and Russia. Discussion. *Geological Society of America Bulletin* 114: 1174–1175.

Lucas, S.G. (2004): A global hiatus in the Middle Permian tetrapod fossil record. *Stratigraphy* 1: 1–11.

Lucas, S.G. (2005): Olson's Gap of Olson's Bridge: an answer. In: Lucas, S.G. & Ziegler, K.E. (eds.). *The Nonmarine Permian. New Mexico Museum of Natural History and Science Bulletin* 30: 185–186.

Lucas, S.G. (2006): Global Permian tetrapod biostratigraphy and biochronology. In: Lucas, S.G., Cassinis, G. & Schneider, J.W. (eds.). *Non-marine Permian Biostratigraphy and Biochronology. Geological Society of London, Special Publication* 265: 65–93.

Lucas, S.G. (2010): The Triassic timescale based on nonmarine tetrapod biostratigraphy and biogeography. In: Lucas, S.G. (ed.). *The Triassic Timescale. Geological Society of London, Special Publication* 334: 447–500.

Lucas, S.G. (2013): No gap in the Middle Permian record of terrestrial vertebrates: Comment. *Geology* 41: e293.

Lucas, S.G. (2017): Permian tetrapod extinction events. *Earth-Science Reviews* 170: 31–60.

Lucas, S.G. (2018): Permian tetrapod biochronology, correlation and evolutionary events. In: Lucas, S.G. & Shen, S.Z. (eds.). *The Permian Timescale. Geological Society of London, Special Publication* 450: 405–444.

Lucas, S.G. & Heckert, A.B. (2001): Olson's Gap: a global hiatus in the record of Middle Permian tetrapods. *Journal of Vertebrate Paleontology* 22 (supplement to 3): 75A.

Lucas, S.G. & Wild, R. (1995): A Middle Triassic dicynodont from Germany and the biochronology of Triassic dicynodonts. *Stuttgarter Beiträge zur Naturkunde, Serie B Geologie und Paläontologie* 220: 1–16.

Lucas, S.G., Hunt, A.P., Heckert, A.B. & Spielmann, J.A. (2007): Global Triassic tetrapod biostratigraphy and biochronology: 2007 status. In: Lucas, S.G. & Spielmann, J.A. (2007) (eds.). *The Global Triassic. New Mexico Museum of Natural History and Science Bulletin* 41: 229–240.

Luo, Z.-X. (2007): Transformation and diversification in early mammal evolution. *Nature* 450: 1011–1019.

Luo, Z.-X. (2011): Developmental patterns in Mesozoic evolution of mammal ears. *Annual Review of Ecology and Systematics* 42: 355–380.

Luo, Z.-X. & Crompton, A.W. (1994): Transformation of the quadrate (incus) through the transition from non-mammalian cynodonts to mammals. *Journal of Vertebrate Paleontology* 14: 341–374.

Luo, Z.-X. & Wible, J.R. (2005): A Late Jurassic digging mammal and early mammal diversification. *Science* 308: 103–107.

Luo, Z.-X., Crompton, A.W. & Lucas, S.G. (1995): Evolutionary origins of the mammalian promontorium and cochlea. *Journal of Vertebrate Paleontology* 15: 113–121.

Luo, Z.-X., Kielan-Jaworowska, Z. & Cifelli, R.L. (2002): In quest for a phylogeny of Mesozoic mammals. *Acta Palaeontologica Polonica* 47: 1–78.

Luo, Z.-X., Yuan, C.-X., Meng, Q.-J. & Ji, Q. (2011): A Jurassic eutherian mammal and divergence of marsupials and placentals. *Nature* 476: 442–445.

Luo, Z.-X., Gatesy, S.M., Jenkins, F.A. Jr., Amaral, W.W. & Shubin, N.H. (2015): Mandibular and dental characteristics of Late Triassic mammaliaform *Haramiyavia* and their ramifications for basal mammal evolution. *Proceedings of the National Academy of Sciences* 112: E7101–E7109.

Luo, Z.-X., Schultz, J.A. & Ekdale, E.G. (2016): Evolution of the middle and inner ears of mammaliaforms: the approach to mammals. In: Clack, J.A., Fay, R.R. & Popper, A.N. (eds.). *Evolution of the Vertebrate Ear–Evidence from the Fossil Record. Springer Handbook of Auditory Research* 59. Springer International Publishing, Switzerland: 139–174.

Lydekker, R. (1890): *Catalogue of the Fossil Reptilia and Amphibia in the British Museum (Natural History). Part IV. Containing the Orders Anomodontia, Ecaudata, Caudata, and Labyrinthodontia; and Supplement.* Trustees of the British Museum (Natural History), London.

MacDonald, J. (1994): *Earth's First Steps: Tracking Life Before the Dinosaurs.* Johnson Books, Boulder.

Maddin, H.C., Evans, D.C. & Reisz, R.R. (2006): An Early Permian varanodontine varanopid (Synapsida: Eupelycosauria) from the Richard's Spur Locality, Oklahoma. *Journal of Vertebrate Paleontology* 26: 957–966.

Maddin, H.C., Sidor, C.A. & Reisz, R.R. (2008): Cranial anatomy of *Ennatosaurus tecton* (Synapsida: Caseidae) from the Middle Permian of Russia and the evolutionary relationships of Caseidae. *Journal of Vertebrate Paleontology* 28: 160–180.

Maderson, P.F.A. (2003): Mammalian skin evolution: a reevaluation. *Experimental Dermatology* 12: 223–236.

Maier, W. & van den Heever, J. (2002): Middle ear structures in the Permian *Glanosuchus sp.* (Therocephalia, Therapsida) based on thin sections. *Mitteilungen aus dem Museum für Naturkunde in Berlin Geowissenschaftliche Reihe* 5: 309–318.

Maisch, M.W. (2001): Observations on Karoo and Gondwana vertebrates. Part 2: a new skull-reconstruction of *Stahleckeria potens* von Huene, 1935 (Dicynodontia, Middle Triassic) and a reconsideration of kannemeyeriiform phylogeny. *Neues Jahrbuch für Geologie und Paläontologie Abhandlungen* 220: 127–152.

Maisch, M.W. (2002a): A new basal lystrosaurid dicynodont from the Upper Permian of South Africa. *Palaeontology* 45: 343–359.

Maisch, M.W. (2002b): Observations on Karoo and Gondwana vertebrates. Part 3: Notes on the gorgonopsians from the Upper Permian of Tanzania. *Neues Jahrbuch für Geologie und Paläontologie Monatshefte* 2002: 237–251.

Maisch, M.W. (2003): Lower jaw morphology and jaw adductor musculature of the giant Permian dicynodont *Rhachiocephalus* Seeley, 1898 (Therapsida) from the Late Permian of Tanzania. *Geologica et Palaeontologica* 37: 89–106.

Maisch, M.W. & Matzke, A.T. (2014): *Sungeodon kimkraemerae* n.gen. n.sp., the oldest kannemeyeriiform (Therapsida, Dicynodontia) and its implications for the early diversification of large herbivores after the P/T boundary. *Neues Jahrbuch für Geologie und Paläontologie Abhandlungen* 272: 1–12.

Marsh, O.C. (1878): Notice of new fossil reptiles. *American Journal of Science* (Series 3) 15: 409–411.

Marsicano, C.A., Irmis, R.B., Mancuso, A.C., Mundil, R. & Chemale, F. (2016): The precise temporal calibration of dinosaur origins. *Proceedings of the National Academy of Sciences* 113: 509–513.

Martin, T., Marugan-Lobón, J., Vullo, R., Martín-Abad, H., Luo, Z.-X. & Buscalioni, A.D. (2015): A Cretaceous eutriconodont and integument evolution in early mammals. *Nature* 526: 380–384.

Martinelli, A.G., Bonaparte, J.F., Schultz, C.L. & Rubert, R. (2005): A new tritheledontid (Therapsida, Eucynodontia) from the Late Triassic of Rio Grande do Sul (Brazil) and its phylogenetic relationships among carnivorous non-mammalian eucynodonts. *Ameghiniana* 42: 191–208.

Martinelli, A.G., Soares, M.B. & Schwanke, C. (2016): Two new cynodonts (Therapsida) from the middle-early Late Triassic of Brazil and comments on South American probainognathians. *PLoS One* 11(10): e0162945.

Matthew, W.D. (1908): A four-horned pelycosaurian from the Permian of Texas. *Bulletin of the American Museum of Natural History* 24: 183–185.

Matsuoka, H., Kusuhashi, N. & Corfe, I.J. (2016): A new Early Cretaceous tritylodontid (Synapsida, Cynodontia, Mammaliamorpha) from the Kuwajima Formation (Tetori Group) of central Japan. *Journal of Vertebrate Paleontology* 36: e1112289.

Mazierski, D.M. & Reisz, R.R. (2010): Description of a new specimen of *Ianthasaurus hardestiorum* (Eupelycosauria: Edpahosauridae) and a re-evaluation of edaphosaurid phylogeny. *Canadian Journal of Earth Sciences* 47: 901–912.

McNab, B.K. (1978): The evolution of endothermy in the phylogeny of mammals. *American Naturalist* 112: 1–21.

Meachen-Samuels, J.A. (2012): Morphological convergence of the prey-killing arsenal of sabertooth predators. *Paleobiology* 38: 1–14.

Mendrez, C. (1972): On the skull of *Regisaurus jacobi*, a new genus and species of Bauriamorpha Watson and Romer 1956 (=Scaloposauria Boonstra 1953), from the *Lystrosaurus* zone of South Africa. In: Joysey, K.A. & Kemp, T.S. (eds.). *Studies in Vertebrate Evolution*. Oliver & Boyd, Edinburgh: 191–212.

Mendrez, C. (1975): Principales variations du palais chez les Thérocephales sud-africains (Pristerosauria et Scaloposauria) au cours du Permien supérieur et du Trias inférieur. *Problémes Actuels de Paléontologie—Évolution des Vertébrés. Coloques Internationaux de Centre National de la Recherche Scientifique* 218: 379–408.

Meng, J., Hu, Y., Wang, Y., Wang, X. & Li, C. (2006): A Mesozoic gliding mammal from northeatern China. *Nature* 444: 889–893.

Meng, J., Wang, Y. & Li, C. (2011): Transitional mammalian middle ear from a new Cretaceous Jehol eutriconodont. *Nature* 472: 181–185.

Meng, Q.-J., Ji, Q., Zhang, Y.-G., Liu, D., Grossnickle, D.M. & Luo, Z.-X. (2015): An arboreal docodont from the Jurassic and mammaliaform ecological diversification. *Science* 347: 764–768.

Miller, M.F., Hasiotis, S.T., Babcock, L.E., Isbell, J.L. & Collinson, J.W. (2001): Tetrapod and large burrows of uncertain origin in Triassic high paleolatitude floodplain deposits, Antarctica. *Palaios* 16: 218–232.

Modesto, S.P. (1994): The Lower Permian synapsid *Glaucosaurus* from Texas. *Palaeontology* 37: 51–60.

Modesto, S.P. (1995): The skull of the herbivorous synapsid *Edaphosaurus boanerges* from the Lower Permian of Texas. *Palaeontology* 38: 213–239.

Modesto, S.P. & Botha-Brink, J. (2010): A burrow cast with *Lystrosaurus* skeletal remains from the Lower Triassic of South Africa. *Palaios* 25: 274–281.

Modesto, S.P. & Reisz, R.R. (1990): A new skeleton of *Ianthasaurus hardestii*, a primitive edaphosaur (Synapsida: Pelycosauria) from the Upper Pennsylvanian of Kansas. *Canadian Journal of Earth Sciences* 27: 834–844.

Modesto, S.P. & Reisz, R.R. (1992): Restudy of Permo-Carboniferous synapsid *Edaphosaurus novomexicanus* Williston and Case, the oldest known herbivorous amniote. *Canadian Journal of Earth Sciences* 29: 2653–2662.

Modesto, S.P. & Rubidge, B. (2000): A basal anomodont therapsid from the lower Beaufort Group, Upper Permian of South Africa. *Journal of Vertebrate Paleontology* 20: 515–521.

Modesto, S.P. & Rybczynski, N. (2000): The amniote faunas of the Russian Permian: implications for Late Permian terrestrial vertebrate biogeography. In: Benton, M.J., Shishkin, M.A., Unwin, D.M. & Kurochkin, E.N. (eds.). *The Age of Dinosaurs in Russia and Mongolia*. Cambridge University Press, Cambridge: 13–34.

Modesto, S., Rubidge, B. & Welman, J. (1999): The most basal anomodont therapsid and the primacy of Gondwana in the evolution of anomodonts. *Proceedings of the Royal Society of London B* 266: 331–337.

Modesto, S.P., Sidor, C.A., Rubidge, B.S. & Welman, J. (2001): A second varanopseid skull from the Upper Permian of South Africa: implications for Late Permian 'pelycosaur' evolution. *Lethaia* 34: 249–259.

Modesto, S.P., Rubidge, B.S. & Welman, J. (2002): A new dicynodont therapsid from the lowermost Beaufort Group, Upper Permian of South Africa. *Canadian Journal of Earth Sciences* 39: 1755–1765.

Modesto, S.P., Smith, R.M.H., Campione, N.E. & Reisz, R.R. (2011): The last "pelycosaur": a varanopid synapsid from the *Pristerognathus* Assemblage Zone, Middle Permian of South Africa. *Naturwissenschaften* 98: 1027–1034.

Morato, L. (2006): Dinodontosaurus *(Synapsida, Dicynodontia): reconsituições morfológicas e aspectos biomechânicos.* Unpublished master's thesis, Universidade Federal do Rio Grande do Sul, Porto Alegre.
Müller, J. & Reisz, R.R. (2006): The phylogeny of early eureptiles: comparing parsimony and Bayesian approaches in the investigation of a basal fossil clade. *Systematic Biology* 55: 503–511.
Nasterlack, T., Canoville, A. & Chinsamy, A. (2012): New insights into the biology of the Permian genus *Cistecephalus* (Therapsida, Dicynodontia). *Journal of Vertebrate Paleontology* 32: 1396–1410.
Nespolo, R.F., Bacigalupe, L.D., Figueroa, C.C., Koteka, P. & Opazo, J.C. (2011): Using new tools to solve an old problem: the evolution of endothermy in vertebrates. *Trends in Ecology and Evolution* 26: 414–423.
Newell, A.J., Sennikov, A.G., Benton, M.J., Molostovskaya, I.I., Golubev, V.K., Minikh, A.V. & Minikh, M.G. (2010): Disruption of playa-lacustrine depositional systems at the Permo-Triassic boundary: evidence from Vyazniki and Gorokhovets on the Russian Platform. *Journal of the Geological Society of London* 167: 695–716.
Nicolas, M. & Rubidge, B.S. (2009): Assessing content and bias in South African Permo-Triassic Karoo tetrapod fossil collections. *Palaeontologia Africana* 44: 13–20.
Nicolas, M. & Rubidge, B.S. (2010): Changes in Permo-Triassic tetrapod ecological representation in the Beaufort Group (Karoo Supergroup) of South Africa. *Lethaia* 43: 45–59.
Niedźwiedzki, G. & Bojanowski, M. (2012): A supposed eupelycosaur body impression form the Early Permian of the intra-Sudetic Basin, Poland. *Ichnos* 19: 150–155.
Nopcsa, F. (1923): Die Familien der Reptilien. *Fortschritte der Geologie und Paläontologie* 2: 1–210.
Nopcsa, F. (1933): On the biology of the theromorphous reptile *Euchambersia*. *Journal of Natural History* (Series 10) 12: 125–126.
Novacek, M.J. (1993): Patterns of diversity in the mammalian skull. In: Hanken, J. & Hall, B.K. (eds.). *The Skull. Volume 2. Patterns of Structure and Systematic Diversity*. University of Chicago Press, Chicago: 438–545.
Novacek, M.J. & Wyss, A.R. (1986): Higher-level relationships of the recent eutherian orders: morphological evidence. *Cladistics* 2: 257–287.
Oftedal, O.T. (2002): The mammary gland and its origin during synapsid evolution. *Journal of Mammary Gland Biology and Neoplasia* 7: 225–252.
Oftedal, O.T. (2012): The evolution of milk secretion and its ancient origins. *Animal* 6 (special issue 3): 355–368.
Oftedal, O.T. & Dhouailly, D. (2013): Evo-devo of the mammary gland. *Journal of Mammary Gland Biology and Neoplasia* 18: 105–120.
Olivier, C., Houssaye, A., Jalil, N.-E. & Cubo, J. (2017): First palaeohistological inference of resting metabolic rate in an extinct synapsid, *Moghreberia nmachouensis* (Therapsida: Anomodontia). *Biological Journal of the Linnean Society* 121: 409–419.
Olroyd, S.L., Sidor, C.A. & Angielczyk, K.D. (2018): New materials of the enigmatic dicynodont *Abajudon kaayai* (Therapsida, Anomodontia) from the lowerMadumabisa Mudstone Formation, middle Permian of Zambia. *Journal of Vertebrate Paleontology*. DOI: 10.1080/02724634.2017.1403442.
Olson, E.C. (1936): The dorsal axial musculature of certain primitive Permian tetrapods. *Journal of Morphology* 59: 265–311.
Olson, E.C. (1941): New specimens of Permian vertebrates in Walker Museum. *Journal of Geology* 49: 753–763.
Olson, E.C. (1944): Origin of mammals based upon cranial morphology of the therapsid suborders. *Geological Society of America Special Papers* 55: 1–136.
Olson, E.C. (1952): The evolution of a Permian vertebrate chronofauna. *Evolution* 6: 181–196.
Olson, E.C. (1955): Parallelism in the evolution of the Permian reptilian faunas of the old and new worlds. *Fieldiana: Zoology* 37: 385–401.
Olson, E.C. (1957): Catalogue of localities of Permian and Triassic terrestrial vertebrates of the territories of the U.S.S.R. *Journal of Geology* 65: 196–226.
Olson, E.C. (1958): Fauna of the Vale and Choza. 14. Summary, review, and integration of the geology and the fauna. *Fieldiana: Geology* 10: 397–448.
Olson, E.C. (1961): Jaw mechanisms: rhipidistians, amphibians, reptiles. *American Zoologist* 1: 205–215.
Olson, E.C. (1962). Late Permian terrestrial vertebrates, U.S.A. and U.S.S.R. *Transactions of the American Philosophical Society New Series* 52: 3–224.
Olson, E.C. (1965): New Permian vertebrates from the Chickasha Formation in Oklahoma. *Oklahoma Geological Survey Circular* 70: 5–70.
Olson, E.C. (1966): Community evolution and the origin of mammals. *Evolution* 47: 291–302.
Olson, E.C. (1968): The family Caseidae. *Fieldiana: Geology* 17: 225–349.
Olson, E.C. (1971): *Vertebrate Paleozoology*. Wiley Interscience, New York.
Olson, E.C. (1974): On the source of therapsids. *Annals of the South African Museum* 64: 27–46.
Olson, E.C. (1975): Permo-Carboniferous paleoecology and morphotypic series. *American Zoologist* 15: 371–389.
Olson, E.C. (1977): Permian lake faunas: a study in community evolution. *Journal of the Paleontological Society of India* 20: 164–163.
Olson, E.C. (1983): Coevolution or coadaptation: Permo-Carboniferous vertebrate chronofauna. In: Nitecki, M. (ed.). *Coevolution*. University of Chicago Press, Chicago: 301–338.
Olson, E.C. (1986): Relationships and ecology of the early therapsids and their predecessors. In: Hotton, N. III, MacLean, P.D., Roth, J.J. & Roth, E.C. (eds.). *The Ecology and Biology of Mammal-like Reptiles*. Smithsonian Institution Press, Washington DC: 47–60.
Olson, E.C. (1991): An eryopid (Amphibia: Labyrinthodontia) from the Fort Sill fissures, Lower Permian, Oklahoma. *Journal of Vertebrate Paleontology* 11: 130–132.
Olson, E.C. & Beerbower, J.R. (1953): The San Angelo Formation, Permian of Texas, and its vertebrates. *Journal of Geology* 61: 389–423.
Olson, E.C. & Chudinov, P. (1992): Upper Permian terrestrial vertebrates of the USA and Russia: 1991. *International Geology Review* 34: 1143–1160.
O'Meara, R.N. & Asher, R.A. (2015). The evolution of growth patterns in mammalian versus nonmammalian cynodonts. *Paleobiology* 42: 439–464.

Orlov, J.A. (1958): Predatory dinocephalians from the Isheevo Fauna (titanosuchians) [in Russian]. *Trudy Paleontologicheskogo Instituta Akademii Nauk SSSR* 72: 1–114.

Osborn, H.F. (1898): The origin of the Mammalia. *American Naturalist* 32: 309–334.

Osborn, H.F. (1903): The reptilian subclasses Diapsida and Synapsida and the early history of the Diaptosauria. *Memoirs of the American Museum of Natural History* 1: 451–507.

Ottone, E.G., Monti, M., Marsicano, C.A., de la Fuente, M.S., Naipauer, M., Armstrong, R. & Mancuso, A.C. (2014): A new Late Triassic age from the Puesto Viejo Group (San Rafael depocenter, Argentina): SHRIMP U-Pb zircon dating and biostratigraphic correlations across southern Gondwana. *Journal of South American Earth Sciences* 56: 186–199.

Owen, R. (1845): Report on the Reptilian Fossils of South Africa. Part I.—Description of certain fossil crania, discovered by A. G. Bain, Esq., in sandstone rocks at the south-eastern extremity of Africa, referable to different species of an extinct genus of Reptilia (*Dicynodon*), and indicative of a new Tribe or Sub-order of Sauria. *Transactions of the Geological Society of London, 2nd Series* 7: 59–84.

Owen, R. (1860a): On the orders of fossil and recent Reptilia, and their distribution in time. *Report of the 29th Meeting of the British Association for the Advancement of Science*: 153–166.

Owen, R. (1860b): On some reptilian fossils from South Africa. *Quarterly Journal of the Geological Society of London* 16: 49–63.

Owen,. R. (1876): *Descriptive and Illustrated Catalogue of the Fossil Reptilia of South Africa in the Collection of the British Museum*. Taylor and Francis, London.

Owerkowicz, T., Musinksky, C., Middleton, K.M. & Crompton, A.W. (2015): Respiratory turbinates and the evolution of endothermy in mammals and birds. In: Dial, K.P., Shubin, N. & Brainerd, E.L. (eds.). *Great Transformations in Vertebrate Evolution*. University of Chicago Press, Chicago: 143–165.

Palmer, W.R. (1913): Note on the lower jaw and ear ossicles of a foetal *Perameles*. *Anatomischer Anzeiger* 43: 510–515.

Parham, J.F., Donoghue, P.C.J., Bell, C.J., Calway, T. D., Head, J.J., Holroyd, P.A., Inoue, J.G., Irmis, R.B., Joyce, W.G., Ksepka, D.T., Patané, J.S.L., Smith, N.D., Tarver, J.E., van Tuinen, M., Yang, Z., Angielczyk, K.D., Greenwood, J.M. Hipsley, C.A., Jacobs, L., Makovicky, P.J., Müller, J., Smith, K.T., Theodor, J.A., Warnock, R.C.M. & Benton, M.J. (2012): Best practices for justifying fossil calibrations. *Systematic Biology* 61: 346–359.

Parrington, F.R. (1955): On the cranial anatomy of some gorgonopsids and the synapsid middle ear. *Proceedings of the Zoological Society of London* 125: 1–40.

Parrington, F.R. (1979): The evolution of the mammalian middle and outer ears: a personal review. *Biological Reviews* 54: 369–387.

Parrish, J.M., Parrish, J.T. & Ziegler, A.M. (1986): Permian-Triassic paleogeography and paleoclimatology and implications for therapsid distributions. In: Hotton, N. III, MacLean, P.D., Roth, J.J. & Roth, E.C. (eds.). *The Ecology and Biology of Mammal-like Reptiles*. Smithsonian Institution Press, Washington D.C.: 109–131.

Pearson, M.R., Benson, R.B.J., Upchurch, P., Fröbisch, J. & Kammerer, C.F. (2013): Reconstructing the diversity of early terrestrial herbivorous tetrapods. *Palaeogeography, Palaeoclimatology, Palaeoecology* 372: 42–49.

Pierce, S.E., Angielczyk, K.D., & Rayfield, E.R. (2009). Shape and mechanics in thalattosuchian (Crocodylomorpha) skulls: implications for feeding behaviour and niche partitioning. Journal of Anatomy 215: 555–576.

Piñeiro, G., Verde, M., Ubilla, M. & Ferigolo, J. (2003): First basal synapsids ("pelycosaurs") from the Upper Permian-?Lower Triassic of Uruguay, South America. *Journal of Paleontology* 77: 389–392.

Piveteau, J. (1927): Études sur quelques amphibiens et reptiles fossiles. *Annales de Paléontologie* 16: 1–44.

Pivorunas, A. (1970): *Allometry in the limbs and sail of* Dimetrodon. Unpublished master's thesis. University of Illinois, Chicago.

Poole, D.F.G. (1956): The structure of the teeth of some mammal-like reptiles. *Quarterly Journal of Microscopical Science* 97: 303–312.

Pravoslavlev, P.A. (1927): Gorgonopsidae from the North Dvina Expedition of V. P. Amalitzki [in Russian]. *Severo-Dvinskie raskopi Prof. V. P. Amalitskogo* 3: 1–117.

Preuschoft, H. & Witzel, U. (2005): Functional shape of the skull in vertebrates: which forces determine skull morphology in lower Primates and ancestral synapsids. *The Anatomical Record Part A: Advances in Integrative Anatomy and Evolutionary Biology* 283A: 402–413.

Pritchard, P.C.H. (1997): Evolution, phylogeny, and current status. In: Lutz, P.L. & Musick, J.A. (eds.). *The Biology of Sea Turtles*. CRC Press, Boca Raton: 1–28.

Prothero, D.R. (2007): *Evolution: What the Fossils Say and Why It Matters*. Columbia University Press, New York.

Ray, S. & Chinsamy, A. (2004): *Diictodon feliceps* (Therapsida, Dicynodontia): bone histology, growth, and biomechanics. *Journal of Vertebrate Paleontology* 24: 180–194.

Ray, S., Botha, J. & Chinsamy, A. (2004): Bone histology and growth patterns of some nonmammalian therapsids. *Journal of Vertebrate Paleontology* 24: 634–648.

Ray, S., Chinsamy, A. & Bandyopadhyay, S. (2005): *Lystrosaurus murrayi* (Therapsida, Dicynodontia): bone histology, growth and lifestyle adaptations. *Palaeontology* 48: 1169–1185.

Ray, S., Bandyopadhyay, S. & Bhawal, D. (2009a): Growth patterns as deduced from bone microstructure of some selected neotherapsids with special emphasis on dicynodonts: phylogenetic implications. *Palaeoworld* 18: 53–66.

Ray, S., Mukherjee, D. & Bandyopadhyay, S. (2009b): Growth patterns of fossil vertebrates as deduced from bone microstructure: case studies from India. *Journal of Biosciences* 34: 661–672.

Ray, S., Botha-Brink, J. & Chinsamy-Turan, A. (2012): Dicynodont growth dynamics and lifestyle adaptations. In: Chinsamy-Turan, A. (ed.). *Forerunners of Mammals: Radiation, Histology, Biology*. Indiana University Press, Bloomington: 121–146.

Rega, E.R., Noriega, K., Sumida, S.S., Huttenlocker, A., Lee, A. & Kennedy, B. (2012): Healed fractures in the neural spines of an associated skeleton of *Dimetrodon*: implications for dorsal sail morphology and function. *Fieldiana: Life and Earth Sciences* 5: 104–111.

Reisz, R.R. (1972): Pelycosaurian reptiles from the Middle Pennsylvanian of North America. *Bulletin of the Museum of Comparative Zoology* 144: 27–62.

Reisz, R. (1975): Pennsylvanian pelycosaurs from Linton, Ohio and Nýřany, Czechoslovakia. *Journal of Paleontology* 49: 522–527.

Reisz, R.R. (1980): The Pelycosauria: a review of phylogenetic relationships. In: Panchen, A.L. (ed.). *The Terrestrial Environment and the Origin of Land Vertebrates*. Academic Press, London: 553–591.

Reisz, R.R. (1986): *Pelycosauria. Handbuch der Paläoherpetologie 17A:* Gustav Fischer Verlag, Stuttgart.

Reisz, R.R. (2005): *Oromycter*, a new caseid from the Lower Permian of Oklahoma. *Journal of Vertebrate Paleontology* 25: 905–910.

Reisz, R.R. (2006): Origin of dental occlusion in tetrapods: signal for terrestrial vertebrate evolution? *Journal of Experimental Zoology Part B: Molecular and Developmental Evolution* 306B: 261–277.

Reisz, R.R. & Berman, D.S. (1986): *Ianthasaurus hardestii* n. sp., a primitive edaphosaur (Reptilia, Pelycosauria) from the Upper Pennsylvanian Rock Lake Shale near Garnett, Kansas. *Candian Journal of Earth Sciences* 23: 77–91.

Reisz, R.R. & Berman, D.S. (2001): The skull of *Mesenosaurus romeri*, a small varanopseid (Synapsida: Eupelycosauria) from the Upper Permian of the Mezen River Basin, Northern Russia. *Annals of Carnegie Museum* 70: 113–132.

Reisz, R.R. & Dilkes, D.W. (2003): *Archaeovenator hamiltonensis*, a new varanopid (Synapsida: Eupelycosauria) from the Upper Carboniferous of Kansas. *Canadian Journal of Earth Sciences* 40: 667–687.

Reisz, R.R. & Fröbisch, J. (2014): The oldest caseid synapsid from the late Pennsylvanian of Kansas, and the evolution of herbivory in terrestrial vertebrates. *PLoS One* 9: e94518.

Reisz, R.R. & Heaton, M.J. (1980): Matters arising. Origin of mammal-like reptiles. *Nature* 288: 193.

Reisz, R.R. & Laurin, M. (2001): The reptile *Macroleter*: first vertebrate evidence for correlation of Upper Permian continental strata of North America and Russia. *Geological Society of America Bulletin* 113: 1299–1233.

Reisz, R.R. & Laurin, M. (2002): Discussion and reply: the reptile *Macroleter*: first vertebrate evidence for correlation of Upper Permian continental strata of North America and Russia. Reply. *Geological Society of America Bulletin* 114: 1176–1177.

Reisz, R.R. & Laurin, M. (2004): A reevaluation of the enigmatic Permian synapsid *Watongia* and of its stratigraphic significance. *Canadian Journal of Earth Sciences* 41: 377–386.

Reisz, R.R. & Modesto, S.P. (2007): *Heleosaurus scholtzi* from the Permian of South Africa: a varanopid synapsid, not a diapsid reptile. *Journal of Vertebrate Paleontology* 27: 734–739.

Reisz, R.R. & Müller, J. (2004): Molecular timescales and the fossil record: a paleontological perspective. *Trends in Genetics* 20: 237–241.

Reisz, R.R. & Sues, H.-D. (2000): Herbivory in late Paleozoic and Triassic terrestrial vertebrates. In: Sues, H.-D. (ed.). *Evolution of Herbivory in Terrestrial Vertebrates*. Cambridge University Press, Cambridge: 9–41.

Reisz, R.R. & Tsuji, L.A. (2006): An articulated skeleton of *Varanops* with bite marks: the oldest known evidence of scavenging among terrestrial vertebrates. *Journal of Vertebrate Paleontology* 26: 1021–1023.

Reisz, R.R., Heaton, M.J. & Pynn, B.R. (1982): Vertebrate fauna of Late Pennsylvanian Rock Lake Shale near Garnett, Kansas: Pelycosauria. *Journal of Paleontology* 56: 741–750.

Reisz, R.R., Berman, D.S. & Scott, D. (1992a): The cranial anatomy and relationships of *Secodontosaurus*, an unusual mammal-like reptile (Synapsida: Sphenacodontia) from the Early Permian of Texas. *Zoological Journal of the Linnean Society* 104: 127–184.

Reisz, R.R., Scott, D. & van Bendegem, J. (1992b): Atlas-axis complex of *Secodontosaurus*, a sphenacodontid mammal-like reptile (Eupelycosauria: Synapsida) from the Lower Permian of Texas. *Canadian Journal of Earth Sciences* 29: 596–600.

Reisz, R.R., Dilkes, D.W. & Berman, D.S. (1998): Anatomy and relationships of *Elliotsmithia longiceps* Broom, a small synapsid (Eupelycosauria: Varanopseidae) from the Late Permian of South Africa. *Journal of Vertebrate Paleontology* 18: 602–611.

Reisz, R.R., Godfrey, S.J. & Scott, D. (2009): *Eothyris* and *Oedaleops*: do these Early Permian synapsids from Texas and New Mexico form a clade? *Journal of Vertebrate Paleontology* 29: 39–47.

Reisz, R., Fröbisch, J., Berman, D. & Henrici, A. (2010a): New Permo-Carboniferous caseid synapsids from North America and Europe, and their evolutionary significance. *Journal of Vertebrate Paleontology* 30 (supplement to 3): 150A.

Reisz, R.R. Laurin, M. & Marjanović, D. (2010b): *Apsisaurus witteri* from the Lower Permian of Texas: yet another small varanopid synapsid, not a diapsid. *Journal of Vertebrate Paleontology* 30: 1628–1631.

Reisz, R.R., Maddin, H.C., Fröbisch, J. & Falconnet, J. (2011): A new large caseid (Synapsida, Caseasauria) from the Permian of Rodez (France), including a reappraisal of *"Casea" rutena* Sigogneau-Russell & Russell, 1974. *Geodiversitas* 33: 227–246.

Reisz, R.R., Bhasin, M. & LeBlanc, A.R.H. (2015): Was the oldest sexually dimorphic armament of the synapsid *Diictodon feliceps* a tusk or a tooth? Comparative dental histology provides the answer. *Abstracts, Program Schedule of Events, 3rd Annual Meeting, 2015*, Canadian Society of Vertebrate Paleontology, 36.

Renaut, A.J. (2000): *A re-evaluation of the cranial morphology and taxonomy of the Triassic dicynodont genus* Kannemeyeria. Unpublished PhD thesis, University of the Witwatersrand, Johannesburg.

Renaut, A.J. (2001): Dicynodont jaw mechanisms reconsidered: the *Kannemeyeria* (Anomodontia Therapsida) masticatory cycle. *Asociación Palentológica Argentine Publicación Especial* 7: 167–170.

Retallack, G.J., Smith, R.M.H. & Ward, P.D. (2003): Vertebrate extinction across the Permian–Triassic boundary in Karoo Basin, South Africa. *Geological Society of America Bulletin* 115: 1133–1152.

Rey, K., Amiot, R., Fourel, F., Abdala, F., Fluteau, F., Jalil, N.-E., Liu, J., Rubidge, B. S., Smith, R. M. H., Steyer, J. S., Viglietti, P. A., Wang, X. & Lécuyer, C. (2017): Oxygen isotopes suggest elevated thermometabolism within multiple Permo-Triassic therapsid clades. *Elife* 2017; 6: e28589.

Riabinin, A.N. (1932): On the recent finding of dinocephalians in the Upper Permian formations of USSR [in Russian]. *Priroda* 1932: 939–941.

Rich, T.H., Hopson, J.A., Musser, A.M., Flannery, A.M. & Vickers-Rich, P. (2005): Independent origins of middle ear bones in monotremes and therians. *Science* 307: 910–914.

Rich, T.H., Hopson, J.A., Gill, P.G., Trusler, P., Rogers-Davison, S., Morton, S., Cifelli, R.L., Pickering, D., Kool, L., Siu, K., Burgmann, F.A., Senden, T., Evans, A.R., Wagstaff, B.E., Seegets-Villiers, D., Corfe, I.J., Flannery, T.F., Walker, K., Musser, A.M., Archer, M., Pian, R. & Vickers-Rich, P. (2016): The mandible and dentition of the Early Cretaceous monotreme *Teinolophos trusleri*. *Alcheringa* 40: 475–501.

Ricqlès, A. de. (1972): Recherches paléohistologiques sur les os longs des tétrapodes III.-Titanosuchiens, dinocéphales et dicynodontes. *Annales de Paléontologie (Vertébrés)* 58: 1–60.

Ricqlès, A. de. (1974): Recherches paléohistologiques sur les os long des tétrapodes IV: éothériodontes et pélycosaures. *Annales de Paléontologie (Vertébrés)* 60: 3–39.

Ricqlès, A. de. (1975): Recherches paléohistologiques sur les os long des tétrapodes. VII, sur la classification, la signification fonctionelle, et l'histoire des tissues osseux des tétrapodes (première partie: structures). *Annales de Paléontologie (Vertébrés)* 61: 49–121.

Ricqlès, A. de. (1976): On bone histology of fossil and living reptiles, with comments on its functional and evolutionary significance. In: Bellairs, A. d'a, Cox, C.B. (eds.). *Morphology and Biology of Reptiles*. Academic Press, London: 123–150.

Ricqlès, A. de. (1989): Les méchanismes hétérochroniques dans le retour des tétrapodes au Milieu aquatique. *Geobios Mémoir Spécial* 12: 337–348.

Ricqlès, A. de & de Buffrénil, V. (2001): Bone histology, heterochronies and the return of tetrapods to life in water. In: Mazin, J.M. & de Buffrénil, V. (eds.). *Secondary Adaptation of Tetrapods to Life in Water*. Verlag Dr. Friedrich Pfeil, Munich: 289–310.

Robard, S. (1949): On the dorsal sail of *Dimetrodon*. *Copeia* 1949: 224.

Rodrigues, P.G., Ruf, I. & Schultz, C.L. (2013): Digital reconstruction of the otic region and inner ear of the non-mammalian cynodont *Brasilitherium riograndensis* (Late Triassic, Brazil) and its relevance to the evolution of the mammalian ear. *Journal of Mammalian Evolution* 20: 291–307.

Ronchi, A., Sacchi, E., Romano, M. & Nicosia, U. (2011): A huge caseid pelycosaur from north-western Sardinia and its bearing on European Permian stratigraphy and palaeobiogeography. *Acta Palaeontologia Polonica* 56: 723–738.

Romano, M. (2017): Long bone scaling of caseid synapsids: a combined morphometric and cladistic approach. *Lethaia* 50: 511–526.

Romano, M. & Nicosia, U. (2014). *Alierasaurus ronchii*, gen. et sp. nov., a caseid from the Permian of Sardinia, Italy. *Journal of Vertebrate Paleontology* 34: 900–913.

Romano, M. & Nicosia, U. (2015): Cladistic analysis of Caseidae (Caseasauria, Synapsida): using the gap-weighting method to include taxa based on incomplete specimens. *Palaeontology* 58: 1109–1130.

Romano, M., Brocklehurst, N. & Fröbisch, J. (2017): The postcranial skeleton of *Ennatosaurus tecton* (Synapsida, Caseidae). *Journal of Systematic Palaeontology*. DOI: 10.1080/14772019.2017.1367729.

Romer, A.S. (1922): The locomotor apparatus of certain primitive and mammal-like reptiles. *Bulletin of the American Museum of Natural History* 46: 517–606.

Romer, A.S. (1927): Notes on the Permo-Carboniferous reptile *Dimetrodon*. *Journal of Geology* 35: 673–689.

Romer, A.S. (1928): Vertebrate faunal horizons in the Texas Permo-Carboniferous red beds. *University of Texas Bulletin* 2801: 67–108.

Romer, A.S. (1936): Studies on American Permo-Carboniferous tetrapods. *Problems of Paleontology* 1: 85–93.

Romer, A.S. (1937): New genera and species of pelycosaurian reptiles. *Proceedings of the New England Zoological Club* 16: 86–96.

Romer, A.S. (1946): The primitive reptile *Limnoscelis* restudied. *American Journal of Science* 244: 149–188.

Romer, A.S. (1948): Relative growth in pelycosaurian reptiles. In: Du Toit, A.L. (ed.). *Robert Broom Commemorative Volume*. Royal Society of South Africa, Cape Town: 45–55.

Romer, A.S. (1956): *Osteology of the Reptiles*. University of Chicago Press, Chicago.

Romer, A.S. (1966): *Vertebrate Paleontology*. 3rd Edition. University of Chicago Press, Chicago.

Romer, A.S. (1969): The Chañares (Argentina) Triassic reptile fauna. V. A new chiniquodontid cynodont, *Probelesodon lewisi*—cynodont ancestry. *Breviora* 333: 1–24.

Romer, A.S. (1973): The Chañares (Argentina) Triassic reptile fauna. XVIII. *Probelesodon minor*, a new species of carnivorous cynodont; family Probainognathidae nov. *Breviora* 401: 1–4.

Romer, A.S. & Price, L.I. (1940): Review of the Pelycosauria. *Geological Society of America Special Papers* 28: 1–538.

Roopnarine, P.D. & Angielczyk, K.D. (2012): The evolutionary palaeoecology of species and the tragedy of the commons. *Biology Letters* 8: 147–150.

Roopnarine, P.D. & Angielczyk, K.D. (2015): Community stability and selective extinction during the Permian-Triassic mass extinction. *Science* 350: 90–93.

Roopnarine, P.D. & Angielczyk, K.D. (2016): The stability of ecological communities as an agent of evolutionary selection: evidence from the Permian-Triassic mass extinction. In: Eldredge, N., Pievani, T., Serrelli, E. & Temkin, I. *Evolutionary Theory: A Hierarchical Perspective*. University of Chicago Press, Chicago.

Roopnarine, P.D., Angielczyk, K.D., Wang, S.C. & Hertog, R. (2007): Trophic network models explain instability of Early Triassic terrestrial communities. *Proceedings of the Royal Society Series B* 274: 2077–2086.

Roopnarine, P.D., Angielczyk, K.D., Olroyd, S.L., Nesbitt, S.J., Botha-Brink, J., Peecook, B.R., Day, M.O. & Smith, R.M.H. (2018): Comparative ecological dynamics of Permian-Triassic communities from the Karoo, Luangwa, and Ruhuhu basins of southern Africa. In: Sidor, C.A. & Nesbitt S. J. (eds.). *Vertebrate and Climatic Evolution in the Triassic Rift Basins of Tanzania and Zambia*. Society of Vertebrate Paleontology Memoir 17: 254–272.

Rowe, T. (1986): *Osteological diagnosis of Mammalia, L. 1758, and its relationships to extinct Synapsida*. Unpublished PhD thesis, University of California, Berkeley.

Rowe, T. (1988): Definition, diagnosis, and origin of Mammalia. *Journal of Vertebrate Paleontology* 8: 241–264.

Rowe, T., Carlson, W. & Bottorff, W. (1995): Thrinaxodon*: digital atlas of the skull*. CD-ROM (Second Edition for Windows and Macintosh platforms). University of Texas Press, Austin.

Rowe, T.B., Macrini, T.E. & Luo, Z.-X. (2011): Fossil evidence on origin of the mammalian brain. *Science* 332: 955–957.

Ruben, J.A. & Jones, T.D. (2000): Selective factors associated with the origin of fur and feathers. *American Zoologist* 40: 585–596.

Ruben, J.A., Hillenius, W.J., Kemp, T.S. & Quick, D.E. (2012): The evolution of mammalian endothermy. In: Chinsamy-Turan, A. (ed.). *Forerunners of Mammals: Radiation, Histology, Biology*. Indiana University Press, Bloomington: 273–286.

Rubidge, B.S. (1984): The cranial morphology of *Eodicynodon* Barry (Therapsida: Dicynodontia). *Navorsinge van die Nasionale Museum Bloemfontein* 14: 326–402.

Rubidge, B.S. (1990): Redescription of the cranial morphology of *Eodicynodon oosthuizeni* (Therapsida: Dicynodontia). *Navorsinge van die Nasionale Museum Bloemfontein* 7: 1–25.

Rubidge, B.S. (1991): A new primitive dinocephalian mammal-like reptile from the Permian of southern Africa. *Palaeontology* 34: 547–559.

Rubidge, B.S. (1994): *Australosyodon*, the first primitive anteosaurid dinocephalian from the Upper Permian of Gondwana. *Palaeontology* 37: 579–594.

Rubidge, B.S. (ed.). (1995): Biostratigraphy of the Beaufort Group (Karoo Supergroup). *South African Committee for Stratigraphy Biostratigraphic Series* 1: 1–46.

Rubidge, B.S. (2005): Re-uniting lost continents—fossil reptiles form the ancient Karoo and their wanderlust. *South African Journal of Geology* 108: 135–172.

Rubidge, B.S. & Kitching, J.W. (2003): A new burnetiamorph (Therapsida: Biarmosuchia) from the lower Beaufort Group of South Africa. *Palaeontology* 46: 199–210.

Rubidge, B.S. & Hopson, J.A. (1990): A new anomodont therapsid from South Africa and its bearing on the ancestry of Dicynodontia. *South African Journal of Science* 86: 43–45.

Rubidge, B.S. & Hopson, J.A. (1996): A primitive anomodont therapsid from the base of the Beaufort Group (Upper Permian) of South Africa. *Zoological Journal of the Linnean Society* 117: 115–139.

Rubidge, B.S. & Sidor, C.A. (2001): Evolutionary patterns among Permo-Triassic therapsids. *Annual Review of Ecology and Systematics* 32: 449–480.

Rubidge, B.S. & Sidor, C.A. (2002): On the cranial morphology of the basal therapsids *Burnetia* and *Proburnetia* (Therapsida: Burnetiidae). *Journal of Vertebrate Paleontology* 22: 257–267.

Rubidge, B.S. & van den Heever, J.A. (1997): Morphology and systematic position of the dinocephalian *Styracocephalus platyrhynchus*. *Lethaia* 30: 157–168.

Rubidge, B.S., Sidor, C.A. & Modesto, S.P. (2006): A new burnetiamorph (Therapsida: Biarmosuchia) from the Middle Permian of South Africa. *Journal of Paleontology* 80: 740–749.

Rubidge, B.S., Erwin, D.H., Ramezani, J., Bowring, S.A. & de Klerk, W.J. (2013): High-precision temporal calibration of Late Permian vertebrate biostratigraphy: U-Pb zircon constraints from the Karoo Supergroup, South Africa. *Geology* 41: 363–366.

Ruf, I., Maier, W., Rodrigues, P.G. & Schultz, C.L. (2014): Nasal anatomy of the non-mammaliaform cynodont *Brasilitherium riograndensis* (Eucynodontia, Therapsida) reveals new insight into mammalian evolution. *The Anatomical Record* 297: 2018–2030.

Ruta, M. & Benton, M.J. (2008): Calibrated diversity, tree topology, and the mother of all mass extinctions: the lesson of temnospondyls. *Palaeontology* 51: 1261–1288.

Ruta, M., Cisneros, J.C., Liebrecht, T., Tsuji, L. & Müller, J. (2011): Amniotes through major biological crises: faunal turnover among parareptiles and the end-Permian mass extinction. *Palaeontology* 54: 1117–1137.

Ruta, M., Botha-Brink, J., Mitchell, S.A. & Benton, M.J. (2013a): The radiation of cynodonts and the ground plan of mammalian morphological diversity. *Proceedings of the Royal Society Series B* 280: 20131865.

Ruta, M., Angielczyk, K.D., Fröbisch, J. & Benton, M.J. (2013b): Decoupling of morphological disparity and taxic diversity during the adaptive radiation of anomodont therapsids. *Proceedings of the Royal Society Series B* 280: 20131071.

Rybczynski, N. (2000) Cranial anatomy and phylogenetic position of *Suminia getmanovi*, a basal anomodont from the Late Permian of eastern Europe. *Zoological Journal of the Linnean Society* 130: 329–373.

Rybczynski, N. & Reisz, R.R. (2001): Earliest evidence for efficient oral processing in a terrestrial herbivore. *Nature* 411: 684–687.

South African Committee for Stratigraphy (1980): Stratigraphy of South Africa. Part I (Comp L. E. Kent). Lithostratigraphy of the Republic of South Africa, South West Africa/Namibia, and the Republics of Bophuthatswana, Transkei, and Venda. *Handbook of the Geological Survey of South Africa* 8: 1–690.

Sahney, S. & Benton, M.J. (2008): Recovery from the most profound mass extinction of all time. *Proceedings of the Royal Society Series B* 275: 759–765.

Sander, P.M. (1987): Taphonomy of the Lower Permian Geraldine Bonebed in Archer County, Texas. *Palaeogeography, Palaeoclimatology, Palaeoecology* 61: 221–236.

Sander, P.M. (1989): Early Permian depositional environments and pond bonebeds in central Archer County, Texas. *Palaeogeography, Palaeoclimatology, Palaeoecology* 69: 1–21.

Satsangi, P.P. (1987): The vertebrate faunas of the Permian and Lower Triassic sequence of India. *Mesozoic Gondwana Vertebrates. Geological Survey of India Special Publication* 11: 165–178.

Schute, C.C.D. (1956): The evolution of the mammalian eardrum and tympanic cavity. *Journal of Anatomy* 90: 261–281.

Schneider, J.W., Körner, F., Roscher, M. & Kroner, U. (2006): Permian climate development in the northern peri-Tethys area—The Lodève Basin, French Massif Central, compared in a European and global context. *Palaeogeography, Palaeoclimatology, Palaeoecology* 240: 161–183.

Seeley, H.G. (1888): Researches on the structure, organization, and classification of the fossil Reptilia. III. On parts of the skeleton of a mammal from Triassic rocks of Klipfontein, Fraserburg, South Africa (*Theriodesmus phylarchus*, Seeley), illustrating the reptilian inheritance in the mammalian hand. *Philosophical Transactions of the Royal Society Series B* 179: 141–155.

Seeley, H.G. (1889): Researches on the structure, organization, and classification of the fossil Reptilia. VI. On the anomodont Reptilia and their allies. *Philosophical Transactions of the Royal Society Series B* 180: 215–296.

Seeley, H.G. (1892): Researches on the structure, organization, and classification of the fossil Reptilia. VII. Further observations on *Pareiasaurus*. *Philosophical Transactions of the Royal Society Series B* 182: 311–370.

Seeley, H.G. (1894a): Researches on the structure, organisation, and classification of the fossil Reptilia. Part IX., Section 1. On the Therosuchia. *Philosophical Transactions of the Royal Society Series B* 185: 987–1018.

Seeley, H.G. (1894b): Researches on the structure, organization, and classification of the fossil Reptilia. Part IX., Section 2. The reputed mammals from the Karroo Formation of Cape Colony. *Philosophical Transactions of the Royal Society Series B* 185: 1019–1028.

Seeley, H.G. (1895a): Researches on the structure, organization, and classification of the fossil Reptilia. Part IX., Section 3. On *Diademodon*. *Philosophical Transactions of the Royal Society Series B* 185: 1029–1041.

Seeley, H.G. (1895b): Researches on the structure, organization, and classification of the fossil Reptilia. Part IX., Section 4. On the Gomphodontia. *Philosophical Transactions of the Royal Society Series B* 186: 1–57.

Seeley, H.G. (1898): On the skull of *Mochlorhinus platyceps* from Bethulie, Orange Free State, preserved in the Albany Museum, Grahamstown. *Annals and Magazine of Natural History Series 7* 1: 164–176.

Shelton, C.D. & Sander, P.M. (2015): *Ophiacodon* long bone histology: the earliest occurrence of FLB in the mammalian stem lineage. *PeerJ PrePrints*. DOI:10.7287/peerj.preprints.1027v1.

Shelton, C.D., Sander, P.M., Stein, K. & Winkelhorst, H. (2013): Long bone histology indicates sympatric species of *Dimetrodon* (Lower Permian, Sphenacodontidae). *Earth and Environmental Sciences Transactions of the Royal Society of Edinburgh* 103: 1–20.

Sidor, C.A. (2000): *Evolutionary trends and relationships within the Synapsida.* Unpublished PhD thesis, University of Chicago, Chicago.

Sidor, C.A. (2001): Simplification as a trend in synapsid cranial evolution. *Evolution* 55: 1419–1442.

Sidor, C.A. (2003). Evolutionary trends and the origin of the mammalian lower jaw. *Paleobiology* 29: 605–640.

Sidor, C.A. (2015): The first biarmosuchian from the upper Madumabisa Mudstone Formation (Luangwa Basin) of Zambia. *Palaeontologia Africana* 49: 1–7.

Sidor, C.A. & Hopson, J.A. (1995): The taxonomic status of the Upper Permian eotheriodont therapsids of the San Angelo Formation (Guadalupian), Texas. *Journal of Vertebrate Paleontology* 15 (supplement to 3): 53A.

Sidor, C.A. & Hopson, J.A. (1998): Ghost lineages and "mammalness": assessing the temporal pattern of character acquisition in the Synapsida. *Paleobiology* 24: 254–273.

Sidor, C.A. & Rubidge, B.S. (2006): *Herpetoskylax hopsoni*, a new biarmosuchian (Therapsida: Biarmosuchia) from the Beaufort Group of South Africa. In: Carrano, M.T., Gaudin, T.J., Blob, R.W. & Wible, J.R. (eds.). *Amniote Paleobiology: Perspectives on the Evolution of Mammals, Birds, and Reptiles*. University of Chicago Press, Chicago: 76–113.

Sidor, C.A. & Smith, R.M.H. (2004): A new galesaurid (Therapsida: Cynodontia) from the Lower Triassic of South Africa. *Palaeontology* 47: 535–556.

Sidor, C.A. & Smith, R.M.H. (2007): A second burnetiamorph therapsid from the Permian Teekloof Formation of South Africa and its associated fauna. *Journal of Vertebrate Paleontology* 27: 420–430.

Sidor, C.A. & Welman, J. (2003): A second specimen of *Lemurosaurus pricei* (Therapsida: Burnetiamorpha). *Journal of Vertebrate Paleontology* 23: 631–642.

Sidor, C.A., Hopson, J.A. & Keyser, A.W. (2004): A new burnetiamorph therapsid from the Teekloof Formation, Permian, of South Africa. *Journal of Vertebrate Paleontology* 24: 938–950.

Sidor, C.A., Miller, M.F. & Isbell, J.L. (2008): Tetrapod burrows from the Triassic of Antarctica. *Journal of Vertebrate Paleontology* 28: 277–284.

Sidor, C.A., Angielczyk, K.D., Weide, D.M., Smith, R.M.H., Nesbitt, S.J. & Tsuji, L.A. (2010): Tetrapod fauna of the lowermost Usili Formation (Songea Group, Ruhuhu Basin) of southern Tanzania, with a new burnetiid record. *Journal of Vertebrate Paleontology* 30: 696–703.

Sidor, C.A., Vilhena, D.A., Angielczyk, K.D., Huttenlocker, A.K., Nesbitt, S.J., Peecook, B.R., Steyer, J.S., Smith, R.M.H. & Tsuji, L.A. (2013): Provincialization of terrestrial faunas following the end-Permian mass extinction. *Proceedings of the National Acadmey of Sciences* 110: 8129–8133.

Sidor, C.A., Angielczyk, K.D., Smith, R.M.H., Goulding, A.K., Nesbitt, S.J., Peecook, B.R., Steyer, J. S. & Tolan, S. (2014): Tapinocephalids (Therapsida, Dinocephalia) from the Permian Madumabisa Mudstone Formation (Lower Karoo, Mid-Zambezi Basin) of southern Zambia. *Journal of Vertebrate Paleontology* 34: 980–986.

Sidor, C.A., Knaub, C.R., Angielczyk, K.D., Beightol, C.V., Nesbitt, S.J., Smith, R.M.H., Steyer, J.-S., Tabor, N.J. & Tolan, S. (2015): Tanzania and Zambia yield an unprecedented fossil record of burnetiamorph therapsids. *Journal of Vertebrate Paleontology, Program and Abstracts* 2015: 213.

Sigogneau, D. (1970): *Révision systématique des Gorgonopsiens sud-africains. Cahiers de Paléontologie.* Centre National de la Recherche Scientifique, Paris.

Sigogneau, D. & Chudinov, P.K. (1972): Reflections on some Russian eotheriodonts (Reptilia, Synapsida, Therapsida). *Palaeovertebrata* 5: 79–109.

Sigogneau-Russell, D. (1989): *Theriodontia I. Handbuch der Paläoherpetologie 17B.* Gustav Fischer Verlag, Stuttgart.

Sigogneau-Russell, D. & Russell, D.E. (1974): Étude du premier caséidé (Reptilia, Pelycosauria) d'Europe occidentale. *Bulletin du Muséum National d'Histoire Naturelle Serié 3* 3: 145–215.

Sigurdsen, T. (2006): New features of the snout and orbit of a therocephalian therapsid from South Africa. *Acta Palaeontologica Polonica* 51: 63–75.

Simon, R.V., Sidor, C.A., Angielczyk, K.D. & Smith, R.M.H. (2010): First record of a tapinocephalid (Therapsida: Dinocephalia) from the Ruhuhu Formation (Songea Group) of southern Tanzania. *Journal of Vertebrate Paleontology* 30: 1289–1293.

Simpson, G.G. (1959): Mesozoic mammals and the polyphyletic origin of mammals. *Evolution* 13: 405–414.

Slater, G.J. & van Valkenburgh, B. (2008): Long in the tooth: evolution of sabertooth cat cranial shape. *Paleobiology* 34: 403–419.

Smiley, T.M., Sidor, C.A., Maga, A. & Ide, O. (2008): The vertebrate fauna of the Upper Permian of Niger. VI. First evidence of a gorgonopsian therapsid. *Journal of Vertebrate Paleontology* 29: 543–547.

Smith, R.M.H. (1987): Helical burrow casts of therapsid origin from the Beaufort Group (Permian) of South Africa. *Palaeogeography, Palaeoclimatology, Palaeoecology* 60: 155–169.

Smith, R.M.H. & Botha-Brink, J. (2011): Morphology and composition of bone-bearing coprolites from the Late Permian Beaufort Group, Karoo Basin, South Africa. *Palaeogeography, Palaeoclimatology, Palaeoecology* 312: 40–53.

Smith, R.M.H. & Botha-Brink, J. (2014): Anatomy of a mass extinction: sedimentological and taphonomic evidence for drought-induced die-offs at the Permo-Triassic boundary in the main Karoo Basin, South Africa. *Palaeogeography, Palaeoclimatology, Palaeoecology* 15: 99–118.

Smith, R.M.H. & Sidor, C.A. (2006): A new burnetiid (Therapsida: Biarmosuchia) from the Upper Permian of South Africa, and its biogeographic implications. *Journal of Vertebrate Paleontology* 26: 331–343.

Smith, R.M.H., Rubidge, B.S. & van der Walt, M. (2012): Therapsid biodiversity patterns and environments of the Karoo Basin, South Africa. In: Chinsamy-Turan, A. (ed.). *Forerunners of Mammals: Radiation, Histology, Biology*. Indiana University Press, Bloomington and Indianapolis: 223–246.

Smith, R.M.H., Sidor, C.A., Tabor, N.J. & Steyer, J.-S. (2015): Sedimentology and vertebrate taphonomy of the Moradi Formation of northern Niger: a Permian wet desert in the tropics of Pangaea. *Palaeogeography, Palaeoclimatology, Palaeoecology* 440: 128–141.

Spielmann, J.A. & Lucas, S.G. (2010): Re-evaluation of *Ruthiromia elcobriensis* (Eupelycosauria: Ophiacodontidae?) from the Lower Permian (Seymourian?) of Cañon del Cobre, Northern New Mexico. In: Lucas, S.G., Schneider, J.W. & Spielmann, J.A. (eds.). *Carboniferous-Permian Transition in Cañon del Cobre, Northern New Mexico. New Mexico Museum of Natural History and Science Bulletin* 49: 151–158.

Spindler, F. (2015): *The Basal Sphenacodontia — Systematic Revision and Evolutionary Implications*. Unpublished PhD dissertation, Technische Universität Bergakademie Freiberg, Freiberg.

Spindler, F., Scott, D. & Reisz, R.R. (2015): New information on the cranial and postcranial anatomy of the early synapsid *Ianthodon schultzei* (Sphenacomorpha: Sphenacodontia), and its evolutionary significance. *Fossil Record* 18: 17–30.

Spindler, F., Falconnet, J. & Fröbisch, J. (2016): *Callibrachion* and *Datheosaurus*, two historical and previously mistaken basal caseasaurian synapsids from Europe. *Acta Palaeontologica Polonica* 61: 597–616.

Spindler, F., Werenebrug, R., Schneider, J.W., Luthardt, L., Annacker, V. & Rößler, R. (2018). First arboreal 'pelycosaurs' (Synapsida: Varanopidae) from the ealry Permian Chemnitz Fossil Lagerstätte, SE Germany, with a review of varanopid phylogeny. Paläontologische Zeitschrift. DOI: 10.1007/s12542-018-0405-9.

Stovall, J.W., Price, L.I. & Romer, A.S. (1966): The postcranial skeleton of the giant Permian pelycosaur *Cotylorhynchus romeri*. *Bulletin of the Museum of Comparative Zoology* 135: 1–30.

Sues, H.-D. (1985). The relationships of the Tritylodontidae (Synapsida). Zoological Journal of the Linnean Society 85: 205–217.

Sues, H.-D. & Boy, J.A. (1988): A procynosuchid cynodont from central Europe. *Nature* 331: 523–524.

Sues, H.-D. & Jenkins, F.A. (2006): The postcranial skeleton of *Kayentatherium wellesi* from the Lower Jurassic Kayenta Formation of Arizona and the phylogenetic significance of postcranial features. In: Carrano, M. T., Gaudin, T.J., Blob, R.W. & Wible, J.R. (eds.). *Amniote Paleobiology: Perspectives on the Evolution of Mammals, Birds, and Reptiles*. University of Chicago Press, Chicago: 114–152.

Sues, H.-D. & Reisz, R.R. (1998): Origin and early evolution of herbivory in terrestrial tetrapods. *Trends in Ecology and Evolution* 18: 141–145.

Sullivan, C. & Reisz, R.R. (1999): First record of *Seymouria* (Vertebrata: Seymouriamorpha) from Early Permian fissure fills at Richards Spur, Oklahoma. *Canadian Journal of Earth Sciences* 36: 1257–1266.

Sullivan, C. & Reisz, R.R. (2005). Cranial anatomy and taxonomy of the Late Permian dicynodont *Diictodon*. *Annals of Carnegie Museum* 74: 45–75.

Sullivan, C., Reisz, R.R. & Smith, R.M.H. (2003): The Permian mammal-like herbivore *Diictodon*, the oldest known example of sexually dimorphic armament. *Proceedings of the Royal Society Series B* 270: 173–178.

Sumida, S.S. (1989a): Reinterpretation of vertebral structure in the Early Permian pelycosaur *Varanosaurus acutirostris* (Amniota, Synapsida). *Journal of Vertebrate Paleontology* 9: 451–458.

Sumida, S.S. (1989b): New information on the pectoral girdle and vertebral column in *Lupeosaurus* (Reptilia, Pelycosauria). *Canadian Journal of Earth Sciences* 26: 1343–1349.

Sumida, S.S. (1990): Vertebral morphology, alternation of neural spine height, and structure in Permo-Carboniferous tetrapods, and a reappraisal of primitive modes of terrestrial locomotion. *University of California Publications in Zoology* 122: 1–129.

Sumida, S.S. & Berman, D.S. (1993): The pelycosaurian (Amniota: Synapsida) assemblage from the Late Pennsylvanian Sangre de Cristo formation of central Colorado. *Annals of Carnegie Museum* 62: 293–310.

Sumida, S.S. & Modesto, S. (2001): A phylogenetic perspective on locomotory strategies in early amniotes. *American Zoologist* 41: 586–597.

Sumida, S.S., Pelletier, V. & Berman, D.S. (2014): New information on the basal pelycosaurian-grade synapsid *Oedaleops*. In: Kammerer, C.F., Angielczyk, K.D. & Fröbisch, J. (eds.). *Early Evolutionary History of the Synapsida*. Springer, Dordrecht: 7–23.

Surkov, M.V. & Benton, M.J. (2008): Head kinematics and feeding adaptations of the Permian and Triassic dicynodonts. *Journal of Vertebrate Paleontology* 28: 1120–1129.

Sushkin, P.P. (1927): On the modification of the mandibular and hyoid arches and their relations to the brain-case in the early Tetrapoda. *Paläontologische Zeitschrift* 8: 263–321.

Szulc, J., Racki G., & Jewuła, K. (2015a). Key aspects of the stratigraphy of the Upper Silesian middle Keuper, southern Poland. Annales Societatis Geologorum Poloniae 85: 557–586.

Szulc, J., Racki, G., Jewuła, K., and Środoń, J. (2015b). How many Upper Triassic bone-bearing levels are there in Upper Silesia (southern Poland)? A critical overview of stratigraphy and facies. Annales Societatis Geologorum Poloniae 85: 587–626.

Takechi, M. & Kuratani, S. (2010): History of studies on mammalian middle ear evolution: a comparative morphological and developmental biology perspective. *Journal of Experimental Zoology Part B: Molecular and Developmental Evolution* 314B: 417–433.

Tatarinov, L.P. (1967): Development of the system of labial (vibrissal) vessels and nerves in theriodonts. *Paleontological Journal* 1: 1–13.

Tatarinov, L.P. (1968): Morphology and systematics of the northern Dvina cynodonts (Reptilia, Therapsida; Upper Permian). *Postilla* 126: 1–51.

Tatarinov, L.P. (1973): Cynodonts of Gondwana habit in the Middle Triassic of the USSR. *Paleontological Journal* 1973: 200–205.

Tatarinov, L.P. (1995): *Viatkosuchus sumini*, a new Late Permian therocephalian from the Kirov region. *Paleontological Journal* 29: 111–128.

Tatarinov, L.P. (1999): New theriodonts (Reptilia) from the Late Permian fauna of the Kotelnich locality, Kirov Region. *Paleontological Journal* 33: 550–556.

Tatarinov, L.P. (2002): Gomphodont cynodonts (Reptilia, Theriodontia) from the Middle Triassic of the Orenburg Region. *Paleontological Journal* 36: 176–179.

Tatarinov, L.P. & Matchenko, E.N. (1999): A find of an aberrant tritylodont (Reptilia, Cynodontia) in the Lower Cretaceous of the Kemerovo Region. *Paleontological Journal* 33: 422–428.

Thackeray, J.F. (1991): Growth increments in the teeth of *Diictodon* (Therapsida). *Koedoe* 34: 7–11.

Thomason, J.J. & Russell, A.P. (1986): Mechanical factors in the evolution of the mammalian secondary palate: a theoretical analysis. *Journal of Morphology* 189: 199–213.

Throckmorton, G.S. (1976): Oral food processing in two herbivorous lizards, *Iguana iguana* (Iguanidae) and *Uromastix aegyptius* (Agamidae). *Journal of Morphology* 148: 363–390.

Thulborn, T. & Turner, S. (2003): The last dicynodont: an Australian Cretaceous relict. *Proceedings of the Royal Society Series B* 270: 985–993.

Toerien, M.J. (1953): The evolution of the palate in South African Anomodontia and its classificatory significance. *Palaeontologia Africana* 1: 49–117.

Tollman, S.M., Grine, F.E. & Hahn, B.D. (1981): Ontogeny and sexual dimorphism in *Aulacephalodon* (Reptilia, Anomodontia). *Annals of the South African Museum* 81: 159–186.

Tomkins, J.L., LeBas, N.R., Witton, M.P., Martill, D.M. & Humphries, S. (2010): Positive allometry and the prehistory of sexual selection. *American Naturalist* 176: 141–148.

Tracy, C.R., Turner, J.S. & Huey, R.B. (1986): A biophysical analysis of possible thermoregulatory adaptations in sailed pelycosaurs. In: Hotton, N. III, MacLean, P.D., Roth J.J. & Roth, E.C. (eds.). *The Ecology and Biology of Mammal-like Reptiles.* Smithsonian Institution Press, Washington D.C.: 195–206.

Tumarkin, A. (1955): On the evolution of the auditory conducting apparatus: a new theory based on functional considerations. *Evolution* 9: 221–242.

Tumarkin, A. (1965): The pelycosaur, the whale and the golden mole. Part I. *Journal of Laryngology and Otology* 79: 667–694.

Tumarkin, A. (1968): Evolution of the auditory conducting apparatus in terrestrial vertebrates. In: de Reuck, A.V.S. & Knight, J. (eds.). *Ciba Foundation Symposium-Hearing Mechanisms in Vertebrates.* John Wiley & Sons, Chichester: 18–40.

Turnbull, W.D. (1970): Mammalian masticatory apparatus. *Fieldiana: Geology* 18: 147–356.

van den Heever, J.A. (1980): On the validity of the therocephalian family Lycosuchidae (Reptilia, Therapsida). *Annals of the South African Museum* 81: 111–125

van den Heever, J.A. (1994): The cranial anatomy of the early Therocephalia (Amniota: Therapsida). *Universiteit van Stellenbosch Annale* 1: 1–59.

van Hoepen, E.C.N. (1934): Oor die indeling van die Dicynodontidae na aanleidung van nuew vorme. *Palaeontologiese Navorsing van die Nasionale Museum* 2: 67–101.

van Tuinen, M. & Hadly, E.A. (2004): Error in estimation of rate and time inferred from the early amniote fossil record and avian molecular clocks. *Journal of Molecular Evolution* 59: 267–276.

van Valen, L. (1960): Therapsids as mammals. *Evolution* 14: 304–313.

van Valkenburgh, B. & Jenkins, I. (2002): Evolutionary patterns in the history of Permo-Triassic and Cenozoic synapsid predators. In: Kowalewski, M. & Kelley, P.H. (eds). *The Fossil Record of Predation. Paleontological Society Papers* 8: 267–288.

Vandebergh, W. & Bossuyt, F. (2012): Radiation and functional diversification of alpha keratins during early vertebrate evolution. *Molecular Biology and Evolution* 29: 995–1004.

Vaughn, P.P. (1958): On a new pelycosaur from the Lower Permian of Oklahoma, and on the origin of the family Caseidae. *Journal of Paleontology* 32: 981–991.

Vaughn, P.P. (1964): Vertebrates from the Organ Rock Shale of the Cutler Group, Permian of Monument Valley and vicinity, Utah and Arizona. *Journal of Paleontology* 38: 567–583.

Vaughn, P.P. (1966): Comparison of the Early Permian vertebrate faunas of the four corners region and north-central Texas. *Los Angeles County Museum Contributions in Science* 105: 1–13.

Vega, C.S. & Maisch, M.W. (2014): Pathological features in Upper Permian and Middle Triassic dicynodonts (Synapsida, Therapsida). In: Kammerer, C.F., Angielczyk, K.D. & Fröbisch, J. (eds.). *Early Evolutionary History of the Synapsida.* Springer, Dordrecht: 151–161.

Vega-Dias, C., Maisch, M.W. & Schultz, C.L. (2004): A new phylogenetic analysis of Triassic dicynodonts (Therapsida) and the systematic position of *Jachaleria candelariensis* from the Upper Triassic of Brazil. *Neues Jahrbuch für Geologie und Paläontologie Abhandlungen* 231: 145–166.

Viglietti, P.A., Smith, R.M.H., Angielczyk, K.D., Kammerer, C.F., Fröbisch, J. & Rubidge, B.S. (2016): The *Daptocephalus* Assemblage Zone (Lopingian), South Africa: A proposed biostratigraphy based on a new compilation of stratigraphic ranges. *Journal of African Earth Sciences* 113: 153–164.

Voigt, S. & Ganzelewski, M. (2010): Toward the origin of amniotes: diadectomorph and synapsid footprints from the early Late Carboniferous of Germany. *Acta Palaeontologica Polonica* 55: 57–72.

Walther, M. & Fröbisch, J. (2013): The quality of the fossil record of anomodonts (Synapsida, Therapsida). *Comptes Rendus Palevol* 12: 495–504.

Ward, P.D., Botha, J., Buick, R., De Kock, M.O., Erwin, D.H., Garrison, G.H., Kirschvink, J.L. & Smith, R. (2005): Abrupt and gradual extinction among Late Permian land vertebrates in the Karoo Basin, South Africa. *Science* 307: 709–714.

Watson, D.M.S. (1912a): On some reptilian lower jaws. *Annals and Magazine of Natural History Series 8* 10: 573–587.

Watson, D.M.S. (1912b): The skeleton of *Lystrosaurus*. *Records of the Albany Museum* 2: 287–299.

Watson, D.M.S. (1913): The limbs of *Lystrosaurus*. *Geology Magazine* 10: 256–258.

Watson, D.M.S. (1914a): The Deinocephalia, an order of mammal-like reptiles. *Proceedings of the Zoological Society of London* 1914: 749–786.

Watson, D.M.S. (1914b): Notes on *Varanosaurus acutirostris* Broili. *Annals and Magazine of Natural History Series 8* 13: 297–310.

Watson, D.M.S. (1914c): The zones of the Beaufort Beds of the Karoo System in South Africa. *Geological Magazine Decade VI* 1: 203–208.

Watson, D.M.S. (1914d): On the nomenclature of the South-African pareiasaurians. *Annals and Magazine of Natural History Series 8* 14: 98–102.

Watson, D.M.S. (1917a): A sketch classification of the pre-Jurassic tetrapod vertebrates. *Proceedings of the Zoological Society of London* 1917: 167–186.

Watson, D.M.S. (1917b): The evolution of the tetrapod shoulder girdle and forelimb. *Journal of Anatomy* 52: 1–63.

Watson, D.M.S. (1921): The bases of classification of the Theriodontia. *Proceedings of the Zoological Society of London* 1921: 35–98.

Watson, D.M.S. (1931): On the skeleton of a bauriamorph reptile. *Proceedings of the Zoological Society of London* 1931: 1163–1205.

Watson, D.M.S. (1942): On Permian and Triassic tetrapods. *Geological Magazine* 79: 81–116.

Watson, D.M.S. (1948): *Dicynodon* and its allies. *Proceedings of the Zoological Society of London* 118: 823–877.

Watson, D.M.S. (1953): The evolution of the mammalian ear. *Evolution* 7: 159–177.

Watson, D.M.S. (1954): On *Bolosaurus* and the origin and classification of reptiles. *Bulletin of the Museum of Comparative Zoology* 111: 299–449.

Watson, D.M.S. (1957): On *Millerosaurus* and the early history of the sauropsid reptiles. *Philosophical Transactions of the Royal Society Series B* 240: 325–400.
Watson, D.M.S. (1960): The anomodont skeleton. *Transactions of the Zoological Society of London* 29: 131–208.
Watson, D.M.S. & Romer, A.S. (1956). A classification of therapsid reptiles. *Bulletin of the Museum of Comparative Zoology* 114: 37–89.
Weide, D.M., Sidor, C.A., Angielczyk, K.D. & Smith, R.M.H. (2009). A new record of *Procynosuchus delaharpeae* (Therapsida: Cynodontia) from the Upper Permian Usili Formation, Tanzania. *Palaeontologia Africana* 44: 21–26.
Westoll, T.S. (1943): The hyomandibular of *Eusthenopteron* and the tetrapod middle ear. *Proceeding of the Royal Society Series B* 131: 393–414.
Westoll, T.S. (1945): The mammalian middle ear. *Nature* 155: 114–115.
Whitney, M.R. & Sidor, C.A. (2016): A new therapsid from the Permian Madumabisa Mudstone Formation (Mid-Zambezi Basin) of southern Zambia. *Journal of Vertebrate Paleontology* 36: e1150767.
Whitney, M.R., Mose, L. & Sidor, C.A. (2017): Odontoma in a 255-million-year-old mammalian forebear. *JAMA Oncology* 3: 998–1000.
Wible, J.R. (1987): The eutherian stapedial artery: character analysis and implications for superordinal relationships. *Zoological Journal of the Linnean Society* 91: 107–135.
Wible, J.R. (1991): Origin of Mammalia: the craniodental evidence reexamined. *Journal of Vertebrate Paleontology* 11: 1–28.
Wible, J.R., Miao, D. & Hopson, J.A. (1990): The septomaxilla of fossil and recent synapsids and the problem of the septomaxilla of monotremes and armadillos. *Zoological Journal of the Linnean Society* 98: 203–228.
Williston, S.W. (1910): *Cacops, Desmospondylus*: new genera of Permian vertebrates. *Geological Society of America Bulletin* 21: 249–284.
Williston, S.W. (1911): *American Permian Vertebrates*. University of Chicago Press, Chicago.
Williston, S.W. (1914a): The osteology of some American Permian vertebrates. *Contributions from the Walker Museum* 1: 107–162.
Williston, S.W. (1914b): *Water Reptiles of the Past and Present*. University of Chicago Press, Chicago.
Williston, S.W. (1925): *The Osteology of the Reptiles*. Harvard University Press, Cambridge.
Williston, S.W. & Case, E.C. (1913a): Description of a nearly complete skeleton of *Ophiacodon* Marsh. In: Case, E.C., Williston, S.W. & Mehl, M.G. (eds.). *Permo-Carboniferous Vertebrates from New Mexico. Carnegie Institution of Washington Publication* 181: 37–59.
Williston, S.W. & Case, E.C. (1913b): A description of *Edaphosaurus* Cope: In: Case, E.C., Williston S.W. & Mehl, M.G. (eds.). *Permo-Carboniferous Vertebrates from New Mexico. Carnegie Institution of Washington Publication* 181: 71–81.
Wright, P.G. (1984): Why do elephants flap their ears? *South African Journal of Zoology* 19: 266–269.
Wu, P., Hou, L., Plikus, M., Hughes, M., Scehnet, J., Suksaweang, S., Widelitz, R.B., Jiang, T. & Chuong, C.-M. (2004): Evo-devo of amniote integuments and appendages. *International Journal of Developmental Biology* 48: 249–270.
Wu, D.-D., Irwin, D.M. & Zhang, Y.-P. (2008): Molecular evolution of the keratin associated protein gene family in mammals, role in the evolution of mammalian hair. *BMC Evolutionary Biology* 8: 241.
Wyllie, A. (2003): A review of Robert Broom's therapsid holotypes: have they survived the test of time? *Palaeontologia Africana* 39: 1–19.
Young, C.C. (1974): A new traversodon from Jiyuan, Henan [in Chinese]. *Vertebrata PalAsiatica* 12: 203–206.
Young, M.T., Brusatte, S.L., Ruta, M. & Andrade, M.B. (2010): The evolution of Metriorhynchoidea (Mesoeucrocodylia, Thalattosuchia): an integrated approach using geometric morphometrics, analysis of disparity, and biomechanics. *Zoological Journal of the Linnean Society* 158: 801–859.
Zheng, X., Bi, S., Wang, X. & Meng, J. (2013): A new arboreal haramiyid shows the diversity of crown mammals in the Jurassic Period. *Nature* 500: 199–201.
Zhou, C.-F., Wu, S., Martin, T. & Luo, Z.-X. (2013): A Jurassic mammaliaform and the earliest mammalian evolutionary adaptations. *Nature* 500: 163–167.

Thomas Martin

6 Mesozoic mammals – early mammalian diversity and ecomorphological adaptations

6.1 Origin of mammals

There exist three key clades relevant to living mammals (Fig. 6.1). The first is Synapsida–based on the total group, i.e., any species more closely related to an opossum than to an iguana. The second is Mammaliaformes based on an apomorphy, i.e., the presence of a secondary or squamosal-dentary jaw joint. The third is Mammalia based on the crown group, i.e., any species descended from the last common ancestor of monotremes, marsupials, and placentals. Here, I use the stem lineage concept (*Stammlinien-Konzept*) of Ax (1984, 1985). Stem taxa can occur on branches leading up to any clade, whether defined by extinction (crown) or by a specific character (apomorphy). According to the apomorphy-based definition, Late Triassic (Rhaetian, 205 Myr) *Morganucodon* and Early Jurassic (Sinemurian, 193 Myr) *Sinoconodon* are the oldest known Mammaliaformes. In both, the primary jaw joint between quadrate and articular still persisted besides the secondary squamosal-dentary jaw joint (Kermack *et al.* 1973, 1981, Crompton and Luo 1993) (Fig. 6.2).

Sinoconodon from Lufeng, China, is the least derived known mammaliaform that is reasonably well represented by cranial remains (Patterson and Olson 1961, Crompton and Sun 1985, Crompton and Luo 1993) (Fig. 6.3). Somewhat older *Adelobasileus* from the Late Triassic (Carnian; about 225 Myr) of Texas is represented by more fragmentary material (Lucas and Hunt 1990, Lucas and Luo 1993) that is difficult to interpret. *Sinoconodon* is characterized by plesiomorphic features such as a continuous growth during lifetime, multiple replacements in the anterior dentition (at least four replacements of the canines), replacement of molariforms (Luo *et al.* 2004), and missing occlusion of postcanine teeth without consistent pattern in the positions of upper and lower postcanines (Crompton and Sun 1985, Crompton and Luo 1993). The incisor and canine replacement is similar to that of gomphodont cynodonts (Hopson 1971, Crompton 1972b), with a loss of anterior postcanines and progressive addition at the end of the postcanine row, where a small tooth is replaced by a larger one (Crompton and Luo 1993). *Sinoconodon* shares with morganucodontids a number of derived characters such as a well-developed squamosal-dentary articulation, double-rooted postcanines, presence of a promontorium (slightly smaller than that of *Morganucodon* and *Dinnetherium*), enlargement of the occipital condyles, and expansion of the brain vault in the parietal region (Crompton and Luo 1993). The growth range of *Sinoconodon* is between 13 g in small and 517 g in large specimens after body mass estimates based on skull length (Luo *et al.* 2004).

The crown group-based Mammalia comprises the last common ancestor of extant mammals (monotremes, marsupials, and placentals) and all of its descendants (Rowe 1988) (Fig. 6.1). This definition (like the stem or total group concept of Synapsida) is character independent. Stem mammals with a squamosal-dentary jaw joint that split prior to the monotreme-therian clade are called Mammaliaformes according to this definition. Taxa without a squamosal-dentary jaw joint, but that are otherwise reconstructed on the stem to Mammaliaformes, belong to the pelycosaur, therapsid, and cynodont grades of total-group Synapsida.

The origin of mammals is connected to a variety of important skeletal and dental derived characters. Besides the secondary jaw joint between squamosal and dentary condyle, these include an enlargement of the brain, a single bone (the petrosal) that houses the inner ear and development of a promontorium for the elongated cochlear canal, separation of the angular (ectotympanic) and articular (malleus) from the mandible and integration of the middle ear into the basicranium, diphyodont dental replacement and precise molar occlusion, which are in connection with determinate skull growth, and development of the squamosal-dentary temporomandibular joint (Brink 1956, Kermack and Kermack 1984, Gow 1985, Crompton and Hylander 1986, Luo *et al.* 2001a, 2004, 2016, Kielan-Jaworowska *et al.* 2004). These mammalian characters were accumulated step by step over an extended period on the synapsid mammalian stem lineage (Kemp 1982, Hopson and Barghusen 1986, Rowe 1993, Luo 1994, Sidor and Hopson 1998, Hopson and Kitching 2001, Sidor 2001, Ruta *et al.* 2013, Martinelli *et al.* 2016).

The synapsid lineage separated from the sauropsid lineage already in the Upper Carboniferous (about 310 Myr ago). Progressively less inclusive synapsid clades leading to Mammalia include Therapsida, Cynodontia (*Thrinaxodon*, about 250 Myr), Mammaliamorpha (tritylodontid *Yunnanodon*, tritheledontid *Pachygenelus*, and *Brasilitherium*,

https://doi.org/10.1515/9783110341553-006

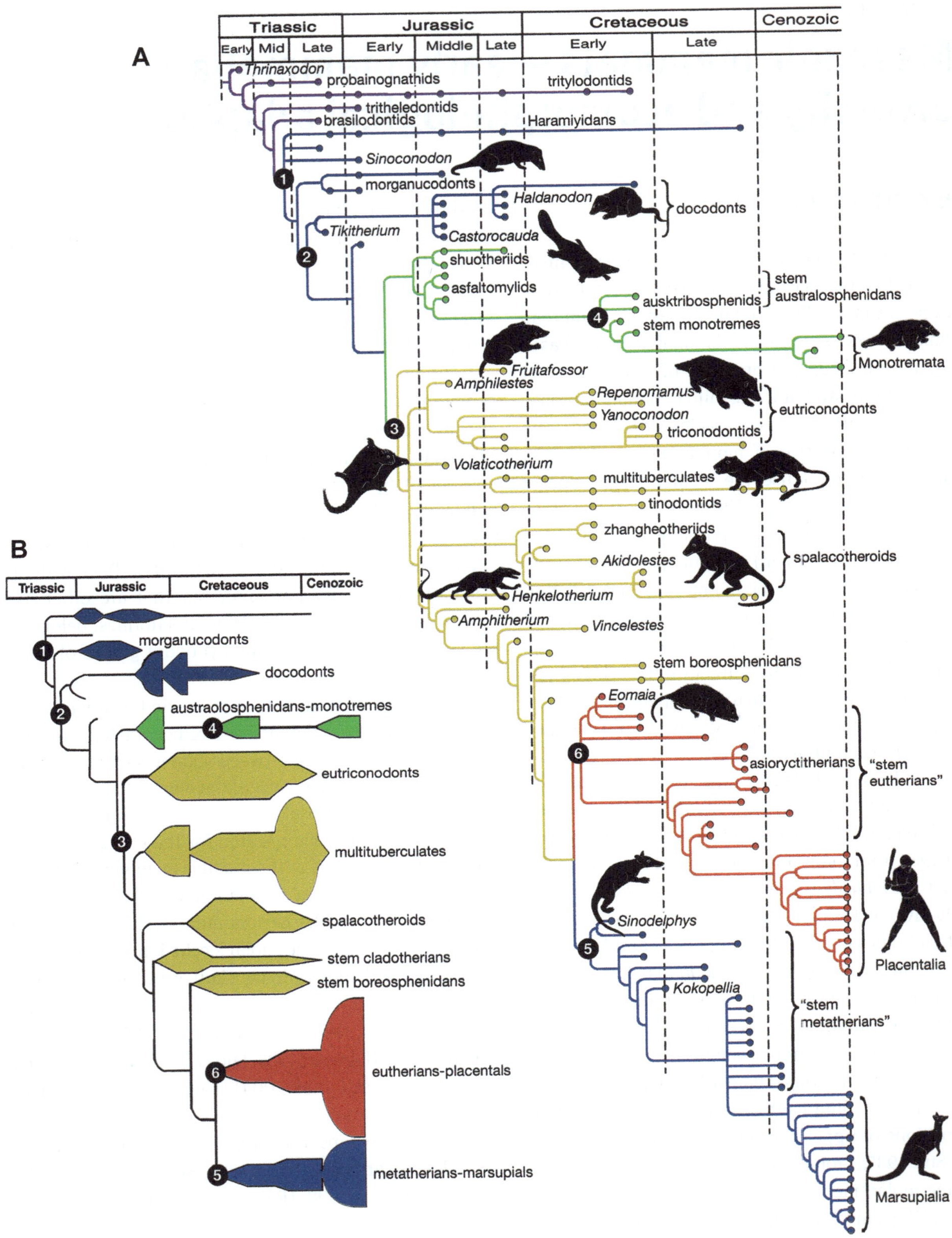

Fig. 6.1: Phylogeny and diversification of Mesozoic mammals in relation to major extant mammal groups. (A) Mesozoic mammalian macroevolution by waves of diversificaton of relatively short-lived clades in succession or by replacement: node 1, the Late Triassic-Early Jurassic diversification of mammaliaform stem clades (blue branches and dots); node 2, diversification of docodontans and splits of several extinct groups in Mammalia (green and yellow); node 3: the Late Jurassic diversification of eutriconodontans, multituberculates, and cladotherians; node 4: Early Cretaceous origins of character-based monotremes; node 5 origins of stem-based metatherians (including marsupials); node 6: origins of stem-based eutherians (including placentals). (B) Diversity patterns of Mesozoic mammaliaforms on the order or family level. Modified after Luo (2007a), reprinted by permission from Springer Nature.

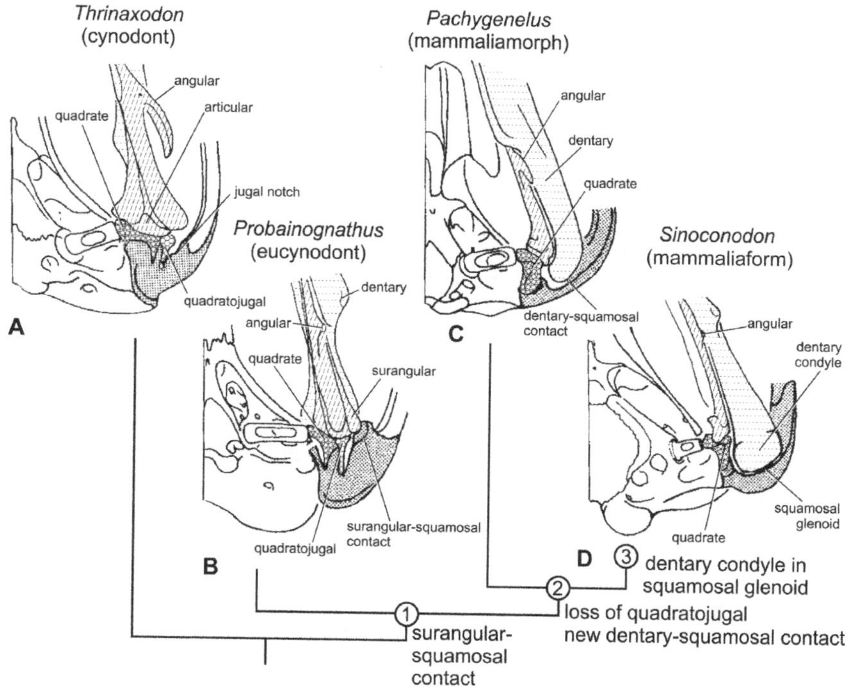

Fig. 6.2: Comparison of craniomandibular joints in cynodonts and mammals in ventral view. From Kielan-Jaworowska *et al.* (2004) compiled from Crompton (1972a), Crompton and Hylander (1986), Allin and Hopson (1992), Crompton and Luo (1993), and Luo *et al.* (2001a). Copyright © 2004 Columbia University Press. Reprinted with permission from the publisher.

about 230 Myr; Mammaliamorpha comprises the last common ancestor of mammals and Tritylodontidae and all of its descendants (this clade is diagnosed by cranial, dental, and postcranial characters; Rowe 1988), Mammaliaformes (*Morganucodon* and *Sinoconodon* about 205 and 193 Myr), and Mammalia (separation of monotremes from therian stem lineage about 165 Myr ago according to paleontological evidence) (Fig. 6.1).

6.2 Dental nomenclature

Tooth enamel is the hardest tissue that occurs in vertebrates, and therefore the dentition is the most durable organ of the vertebrate body. Most Mesozoic mammaliaforms and mammals are exclusively or mainly known from their dentition, and systematics and taxonomy of fossil mammals are largely based on dental characters. Teeth play an important role in the discussion of Mesozoic mammals, and this chapter will make the reader familiar with the special terminology of mammaliaform and mammalian dentitions.

Mammaliaforms and mammals have heterodont dentitions, with incisors, canine, premolars, and molars. In placental mammals, usually two tooth generations occur, with a single replacement of the antemolars (incisors, canine, and premolars). Molars belong to the first generation of teeth and are not replaced by definition. However, in mammaliaforms, multiple replacements of antemolars and even molariforms can occur. The term "molar" is restricted to the "true molars" of mammals that are not replaced; distal cheek teeth of mammaliaforms that are molar shaped but that are replaced are called "molariforms". The tooth categories are abbreviated as follows (upper case letters for upper teeth, lower case letters for lower teeth): I/i, incisor; C/c, canine; P/p, premolar; M/m, molar. Deciduous teeth are designated with D/d, for deciduus (Latin), shed (e.g., dp = lower deciduous premolar).

The cusps on the tooth crown are called cones (singular cone), and the connecting crests are called cristae (singular crista). For the nomenclature of structures on a tribosphenic molar, see Fig. 6.4. The tribosphenic molar is the derived mammalian type of cheek tooth that combines a cutting with a grinding function. It represents the starting point of a remarkable diversity of Cenozoic and modern marsupial and placental molar types. The upper tribosphenic molars originally had a triangular shape, with three cusps forming the trigon: lingual protocone, labiomesial paracone, and labiodistal metacone. The cusps are connected by crests and contour the trigon (Fig. 6.4). The lower molars consist of a mesially situated triangle, the trigonid, formed by the labial protoconid, the linguomesial paraconid, and the linguodistal metaconid, and the distally situated talonid basin, which is bordered by three cusps, the hypoconid, hypoconulid, and entoconid. The talonid has several crests that connect the cusps (Fig. 6.4). The lower molar cusps and other structures bear the suffix "-id" in order to distinguish them from the upper molar structures, e.g., a crest on the upper molar is called "crista", and on the lower molar

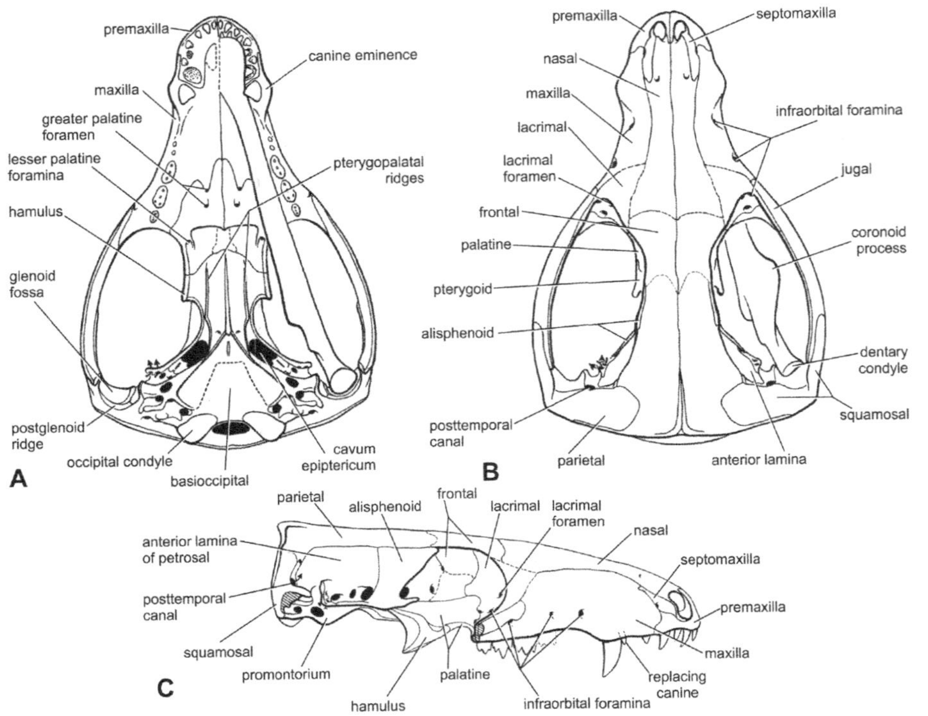

Fig. 6.3: Restoration of the skull of Early Jurassic *Sinoconodon* in palatal (A), dorsal (B), and lateral views (C). No scale; length of skulls ranges from about 23 mm to more than 60 mm. From Kielan-Jaworowska *et al.* (2004) modified after Crompton and Luo (1993). Copyright © 2004 Columbia University Press. Reprinted with permission from the publisher.

"cristid". The often used term "cristid obliqua" (Szalay 1969) for the lower molar talonid crest extending in mesiolingual direction from the hypoconid, replacing "crista obliqua" of earlier authors (Gregory 1920, MacIntyre 1966, van Valen 1966), is linguistically incorrect because "cristid" is not a Latin noun. Therefore, either the vernacular "oblique cristid" (which is used here) or the construction "cristidum obliquum" should be used (Kay and Cartmill 1977).

In cynodonts and stem mammals such as morganucodontids and others, the cusps of the teeth are designated with letters (upper teeth, upper case letters; lower teeth, lower case letters) (Fig. 6.5). In the lower teeth, central main cusp a is homologized with the protoconid, mesial main cusp b with the paraconid, and distal main cusp c with the metaconid (Crompton and Jenkins 1968, Crompton 1971, 1974). The distal cuspule d originally has been homologized with the hypoconid (Osborn 1888a). Later it was considered to represent the hypoconulid (Butler 1939, Patterson 1956, Kermack *et al.* 1968, Crompton 1971), a view that was adopted by Kielan-Jaworowska *et al.* (2004). Later studies (Butler 1990, Martin 2002, Lopatin and Averianov 2006a), including a recent review (Davis 2011), support the original interpretation as hypoconulid. In the upper teeth, central main cusp A is homologized with the paracone, mesial main cusp B with the stylocone, and distal main cusp C with the metacone; distal accessorial cuspule D probably corresponds to the metastyle (Kielan-Jaworowska *et al.* 2004). The protocone of tribosphenic mammals is a neomorphic structure not present in pretribosphenic mammaliaforms (Crompton 1971).

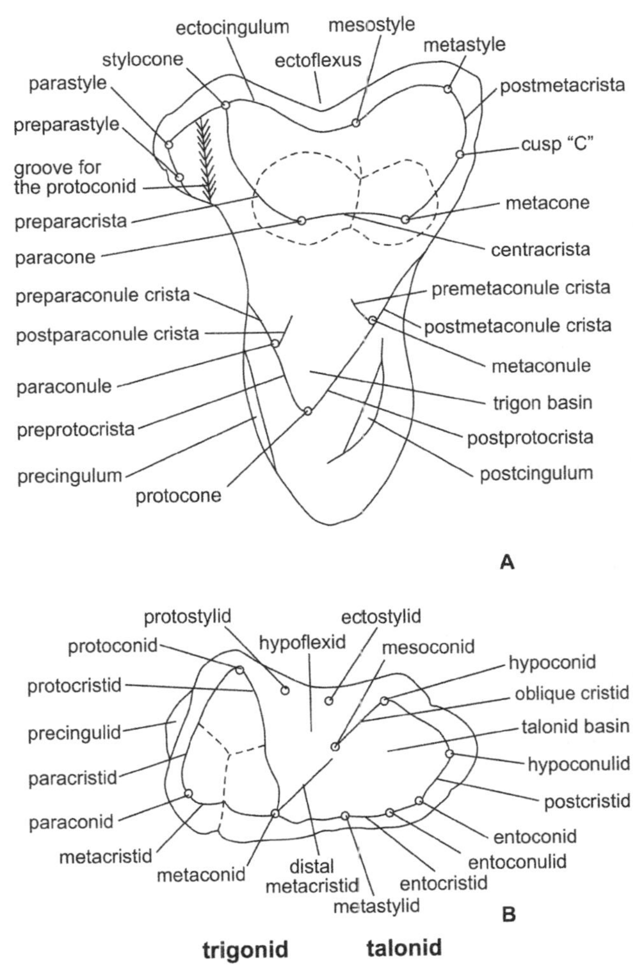

Fig. 6.4: Schematic drawings of upper (A) and lower (B) tribosphenic molars with terminology of cusps and crests. Modified from Kielan-Jaworowska *et al.* (2004). Copyright © 2004 Columbia University Press. Reprinted with permission from the publisher.

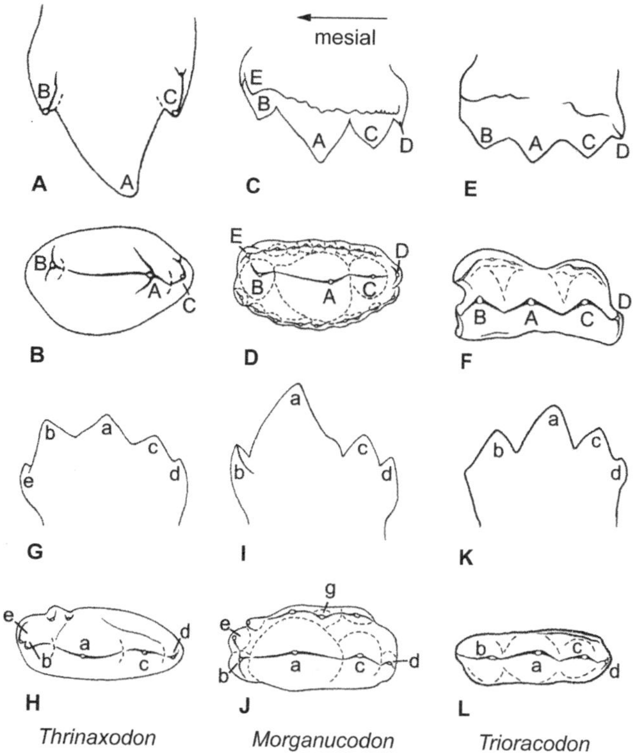

Fig. 6.5: The designation of mammaliaform molariform cusps by letters as exemplified at the triconodont molarifrom pattern, which was already present in advanced cynodonts (e.g., *Thrinaxodon*). No scale, teeth brought to same size. From Kielan-Jaworowska *et al.* (2004). Copyright © 2004 Columbia University Press. Reprinted with permission from the publisher.

6.3 Morganucodonta

Morganucodonta (Kermack *et al.* 1973; Fig. 6.1) represents one of the first radiations of Mammaliaformes, in the Late Triassic/Early Jurassic. They have a typical triconodont molar cusp arrangement with three main cusps in a row (A/a, B/b, and C/c) plus one additional smaller distal cusp (D/d) (Fig. 6.6), which represents the mammaliaform plesiomorphic condition. Morganucodontans cluster as a clade above *Sinoconodon* at the base of Mammaliaformes in several comprehensive cladistic analyses of Mesozoic mammals (Luo *et al.* 2002, Luo 2007a, Rougier *et al.* 2007c). However, in the analysis by Montellano *et al.* (2008), based on the matrix of Rougier *et al.* (2007c) and with the inclusion of two new fossils from Mexico, they represent a paraphyletic array of taxa at the base of Mammaliaformes. Morganucodontans have a worldwide distribution and have been reported from Europe (Britain, France, and Germany), North America, South Africa, and Asia (China), which reflects the Pangaean paleogeographic situation. *Morganucodon* has been thoroughly studied and is well known from its dentition, skull, and postcranial skeleton, and it is one of the most important reference taxa for the study of early mammalian evolution (Kermack 1963, Crompton 1971, 1974, Mills 1971, Kermack *et al.* 1973, 1981, Parrington 1973, 1978, Jenkins and Parrington 1976, Crompton and Luo 1993, Luo 1994, Rowe *et al.* 2011) (Figs. 6.7 and 6.8). The largest number (thousands) of *Morganucodon* teeth and bone elements, almost entirely disarticulated, derive from the Welsh fissure fillings of Early Jurassic age (mammaliamorph-bearing fissures of St. Brides: Hettangian to possibly early Sinemurian; Whiteside *et al.* 2016). The genus *Morganucodon* was coined by Walter Georg Kühne in 1949 after a well-preserved lower molar from Duchy Quarry in Glamorgan, Wales. Specimens collected by Kühne at Holwell Quarry (England) were studied by Parrington (1941), who named the genus *Eozostrodon* based on an upper premolar. Subsequently, there was disagreement on the validity of the genus *Eozostrodon*. Although Parrington (1967, 1971) considered *Morganucodon* a junior synonym of *Eozostrodon*, Kermack *et al.* (1968, 1973) argued that *Eozostrodon* is an indeterminate taxon because it was originally based on a premolar lacking distinctive characters allowing separation from *Morganucodon*. Clemens (1979a) identified some size and morphological differences and suggested restricting the genus *Eozostrodon* to the teeth from Holwell Quarry and retaining the genus *Morganucodon* for the teeth from Wales. Other occurrences of diverse

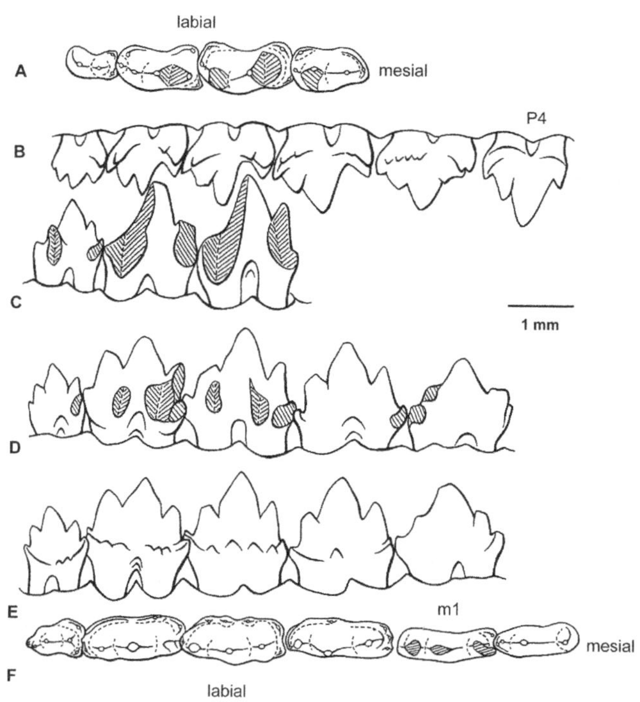

Fig. 6.6: Upper and lower tooth rows of *Dinnetherium nezorum*. (A) Crown view of M2-5. (B) Upper tooth row (P4-M5). (C) m3–m5 with wear facets placed in occlusional relationship with upper tooth row (B). (D–F) Lower tooth row with p4–m5 in labial (D), lingual (E), and occlusal (F) views. Modified after Crompton and Luo (1993), with permission of Springer.

species of morganucodontans in Europe, all isolated teeth, are in the Rhaeto-Liassic of Hallau in Switzerland (*Hallautherium*, *Helvetiodon*; Peyer 1956, Clemens 1980), Baden-Württemberg in Germany (Clemens and Martin 2014), St. Nicolas-de-Port in France (*Brachyzostrodon*, *Rosierodon*; Sigogneau-Russell 1983a, c, Hahn *et al.* 1991, Debuysschere *et al.* 2014), the Early Jurassic of Wales (*Bridetherium*, *Paceyodon*; Clemens 2011), the Middle Jurassic of Kirtlington in Oxfordshire (*Wareolestes*; Freeman 1979, Butler and Sigogneau-Russell 2016), and

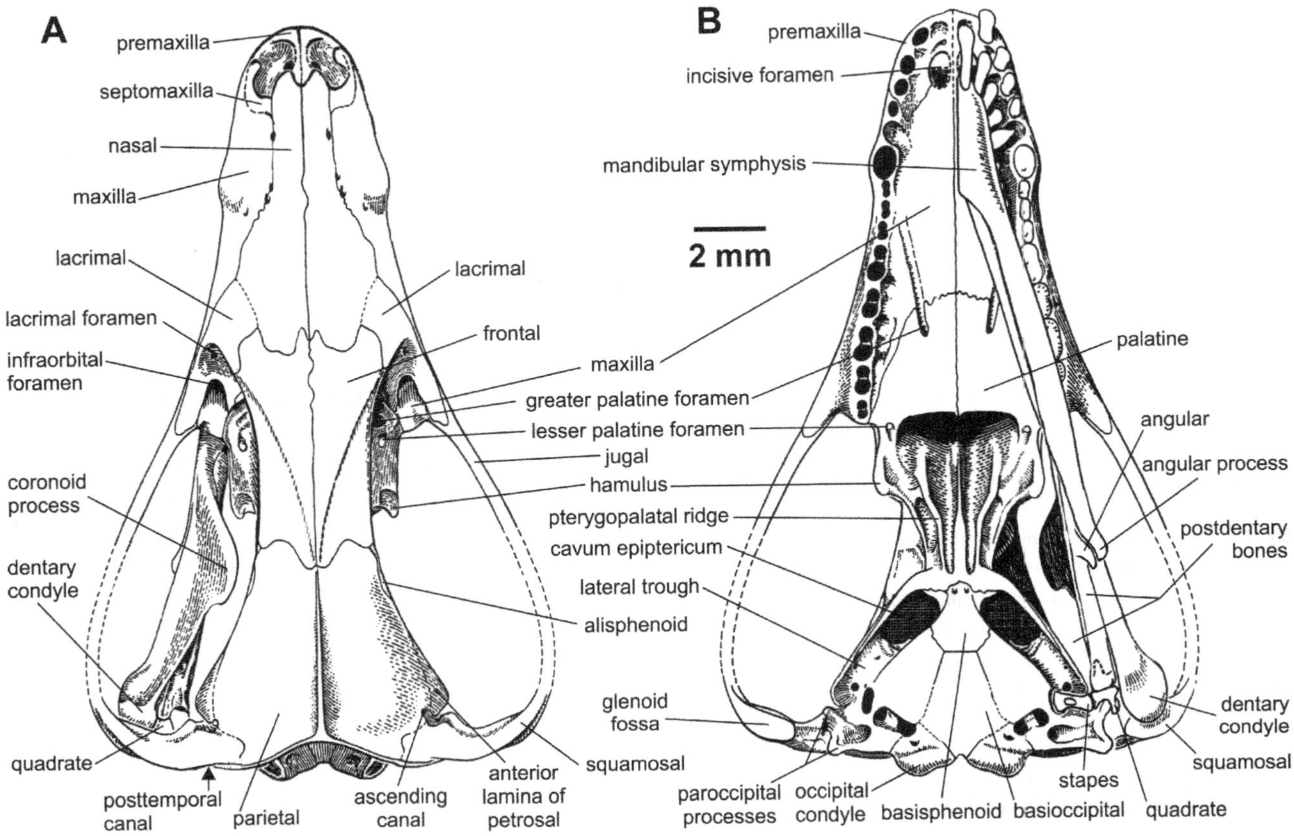

Fig. 6.7: Skull of *Morganucodon watsoni* in dorsal (A) and palatinal (B) views. From Kielan-Jaworowska *et al.* (2004), modified after Kermack *et al.* (1981). Copyright © 2004 Columbia University Press. Reprinted with permission from the publisher.

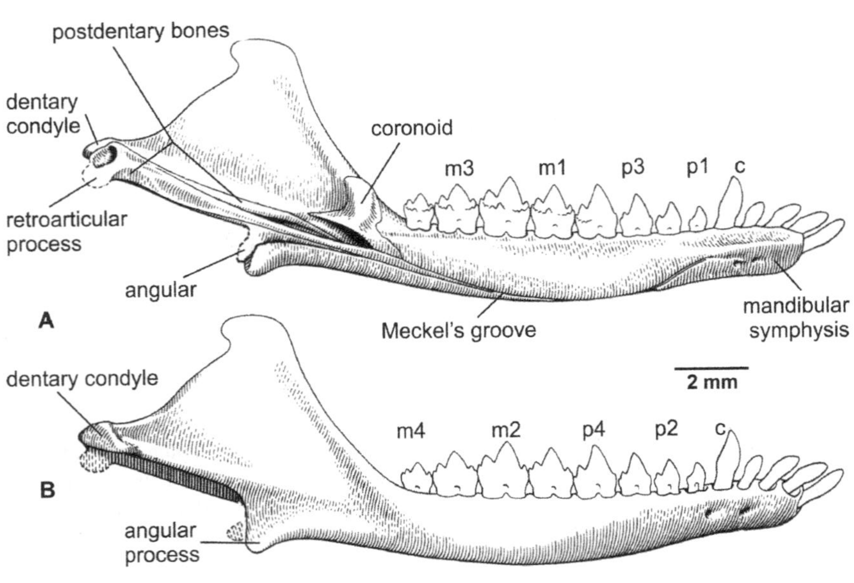

Fig. 6.8: Mandible of *Morganucodon watsoni* in lingual (A) and labial (B) views. From Kielan-Jaworowska *et al.* (2004), modified after Kermack *et al.* (1973). Copyright © 2004 Columbia University Press. Reprinted with permission from the publisher.

possibly the Early Cretaceous Purbeck Limestone Group of southern England (*Purbeckodon*; Butler *et al.* 2012). Panciroli *et al.* (2017) recently described a partial left dentary of *Wareolestes* with two erupted molars, one unerupted molar, and two unerupted premolars from the Middle Jurassic (Bathonian) Kilmaluag Formation of the Isle of Skye, Scotland. The dentary suggests a diphyodont replacement mode of the premolar dentition.

In the Early Jurassic of Lufeng in Yunnan, China, complete skulls of *Morganucodon* were discovered (Young 1978, 1982, Zhang 1984) that yielded important information on cranial anatomy and dentition (Crompton and Luo 1993, Luo 1994). In North America, *Morganucodon*-grade mammaliaforms were first described by Jenkins *et al.* (1983). *Dinnetherium* from the Early Jurassic of Arizona is represented by large parts of the dentition and additional skull materials, and isolated teeth of *Morganucodon* sp. were also described by Jenkins *et al.* (1983) from Arizona. In Greenland, Jenkins *et al.* (1994) discovered fragmentary morganucodontan material, probably belonging to the genus *Brachyzostrodon*. Montellano *et al.* (2008) described a fragmentary mandible with three teeth of a new morganucodontan-grade mammaliaform *Bocaconodon* from the Early Jurassic (Pliensbachian) of Mexico. The molariforms of *Bocaconodon* exhibit similarities to that of morganucodontans (*Morganucodon*, *Megazostrodon*, and *Erythrotherium*) such as cusp b being much smaller than cusp c and sitting at the front margin of the crown. Characters that are more similar to eutriconodontans are a narrower internal cingulum without distinct cusps (Montellano *et al.* 2008). For the southern continents (Gondwana), two morganucodontans were described from the Early Jurassic of South Africa, *Erythrotherium* and *Megazostrodon*, both known from cranial and postcranial skeletons (Crompton 1964, Crompton and Jenkins 1968, Gow 1986). Datta and Das (1996, 2001) reported isolated teeth of *Gondwanadon* and *Indozostrodon* from the Kota Formation of India (originally considered Early Jurassic in age, now late Middle Jurassic to Early Cretaceous according to Prasad and Manhas 2007), which were referred to Morganucodontidae. *Indotherium* from the same formation (Yadagiri 1984, Prasad and Manhas 1997) has been described based on two isolated upper molars and is probably a junior synonym of *Indozostrodon* (Prasad and Manhas 2002).

Originally, Morganucodonta was regarded as ancestral to docodontans because of similarities in the cingulid structure of the lower molars (Patterson 1956, Kermack and Musset 1958). Later it became evident that the molar wear facets of *Morganucodon* are more similar to those of eutriconodontans (Crompton and Jenkins 1968, Crompton 1971, 1974, Mills 1971, Kermack *et al.* 1973), and the morganucodontans were placed within the order "Triconodonta" (Mills 1971, Kermack *et al.* 1973, Parrington 1973, Jenkins and Crompton 1979, Clemens 1980, Jenkins *et al.* 1983, Sigogneau-Russell 1983a). According to later cladistic analyses (e.g., Rowe 1988, Wible and Hopson 1993, Rougier *et al.* 1996, Luo *et al.* 2002), morganucodontans cluster far distantly from eutriconodontans at the base of Mammaliaformes mainly based on non-dental, mandibular, and cranial features, and Triconodonta turned out to be polyphyletic. Currently, morganucodontans are regarded as stem mammals only distantly related to Eutricondonta with which they share the symplesiomorphic "triconodont" molariform pattern.

Although morganucodontans share with *Sinoconodon* the triconodont molariform pattern, they are more similar to crown Mammalia by virtue of the one-to-one occlusal relationship of upper and lower molariforms and the development of precise wear facets, reduced tooth replacement, several characters of the craniomandibular joint (Jenkins and Parrington 1976, Crompton and Luo 1993, Luo 1994) as well as a longer cochlear canal (Luo *et al.* 1995), and somewhat larger cranial capacity. Morganucodont mammaliaforms possess the plesiomorphic postdentary trough at the medial side of the dentary that houses the middle ear bones. The jaw hinge is formed by both the primary quadrato-articular jaw joint and the secondary dentary-squamosal joint. Their ontogenetic size variation is more limited than that of *Sinoconodon* (Luo *et al.* 2001a). The body mass has been estimated between 20 and 30 g for *Megazostrodon rudnerae* (Jenkins and Parrington 1976) and between 27 and 80 g for *Morganucodon oehleri*, according to skull size (Luo *et al.* 2001a). Based on the postcranial evidence, morganucodontans were lightly built locomotorial generalists with scansorial lifestyle and correspond well to the long-held picture of Mesozoic mammals as small shrewlike creatures (Jenkins and Parrington 1976). Gill *et al.* (2014) analyzed the function of the *Morganucodon* mandible and dentition and found evidence for an adaptation toward hard-shelled insects, such as coleopterans, as their main prey.

6.4 Kuehneotheriidae

The Late Triassic-Early Jurassic Kuehneotheriidae Kermack *et al.* 1968 are stem mammals that differ from the morganucodontans by having a slight angulation of the cusps of their molars (Fig. 6.9). Formerly, they were lumped together with tinodontids and spalacotheroids as "Symmetrodonta" in allusion to the symmetric shape of their molars (Simpson 1928, 1929, Patterson 1956). Subsequently, it became clear that "Symmetrodonta" is a

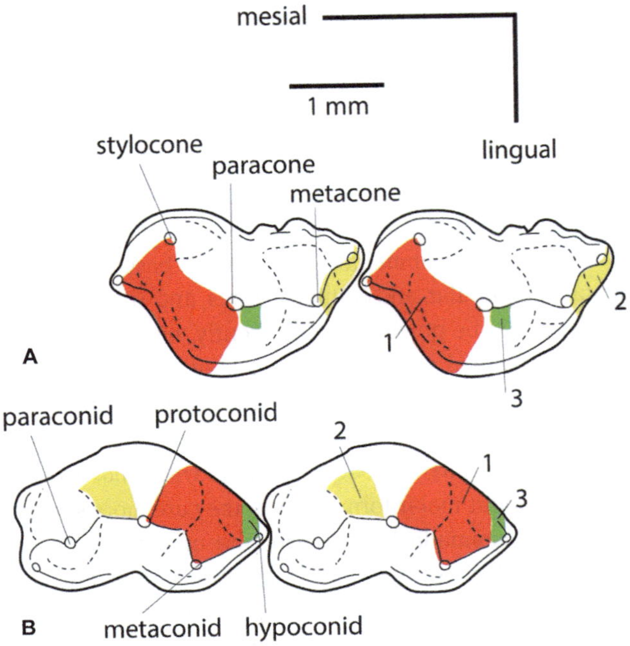

Fig. 6.9: *Kuehneotherium praecursoris*, upper (A) and lower (B) molars in occlusal views with designation of cusps and wear facets. From Williamson *et al.* (2014).

polyphyletic assemblage of stem mammals (Kuehneotheriidae) and crown mammalian Trechnotheria (Spalacotherioidea). *Kuehneotherium* from the Liassic of the Welsh fissure fillings and various Rhaeto-Liassic localities is known from a large number of isolated teeth and mostly edentulous mandible fragments (Kermack *et al.* 1968, Gill 2004, Gill *et al.* 2014, Debuysschere 2016a). Unlike morganucodontans, intact skulls and skeletal fossils remain unknown. The slender dentary has a Meckel's groove and a well-developed postdentary trough that still housed the plesiomorphic postdentary bones. This indicates that *Kuehneotherium* possessed the plesiomorphic primary jaw joint besides the secondary mammalian jaw joint formed by the dentary condyle and the squamosal. The mandible is characterized by a shallow masseteric fossa, a low coronoid process, and the lack of an angular process. *Kuehneotherium* has five or six lower premolars (the first four were single rooted) and up to six molars (Kermack *et al.* 1968, Gill 1974, 2004) (Fig. 6.10). Gill *et al.* (2014) analyzed the wear pattern and mandible structure of *Kuehneotherium* and concluded that it differs from contemporaneous *Morganucodon* by preying mainly on soft-bodied insects.

Debuysschere (2016a) studied the kuehneotheriid teeth from the Rhaetian Saint-Nicolas-de-Port locality in Lorraine (France) and erected a new species of *Kuehneotherium* (*K. stanilavi*) and a new kuehneotheriid genus and species, *Fluctuodon necmergor*. Comparisons with kuehneotheriids from other Upper Triassic sites in Luxembourg, England, and Greenland suggest two distinct kuehneotheriid assemblages west and east of the London-Brabant Massif, and little impact of the Triassic/Jurassic extinction event on *Kuehneotherium* (Debuysschere 2016a).

Kotatherium from the Kota Formation in India is based on one isolated upper molar (Datta 1981) and is attributed to Kuehneotheriidae (Kielan-Jaworowska *et al.* 2004). *Kuehneon* was based on an isolated molar from the Early Jurassic of the Welsh fissure fillings. It had been illustrated and described by Kühne (1949, 1958) but was formally named only in 1960 by Kretzoi. Originally regarded as a lower molar, it later was suggested that it might be an upper (Kermack *et al.* 1968). Because the specimen is no longer available, *Kuehneon* is regarded as a *nomen dubium*. *Woutersia*, based on isolated teeth from the Rhaetic (Late Triassic) of Saint-Nicolas-de-Port, was originally attributed to Kuehneotheriidae (Sigogneau-Russell 1983a) but later was assigned to a family of its own (Woutersiidae) by Sigogneau-Russell and Hahn (1995). *Delsatia* from the same locality was described at the base of a lower molar and an attributed lower premolar and was originally interpreted as a "primitive docodont" (Sigogoneau-Russell and Godefroit 1997). Because of their overall similarity, it cannot be ruled out that the teeth attributed to *Delsatia* represent differing tooth positions of *Woutersia* (Kielan-Jaworowska *et al.* 2004). *Thereuodon* from the Early Cretaceous (Berriasian) of Morocco (Sigogneau-Russell 1989) and Britain, based on isolated upper cheek teeth, has been attributed

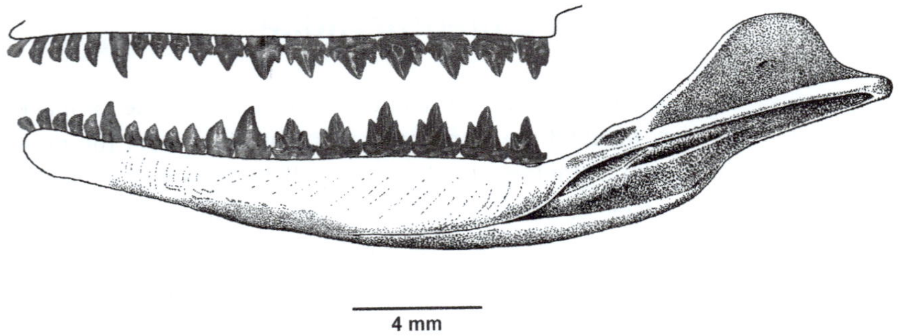

Fig. 6.10: Restoration of mandible with dentition and upper tooth row of *Kuehneotherium praecursoris* in lingual view. Lower tooth row is based on an edentulous mandible reconstruction, with isolated teeth (reconstructed dental formula 6I/6i.1C/1c.6P/6p.6M/6m; exact number of incisors is not known. From Gill (2004), with permission of the author.

to the family "Thereuodontidae" by Sigogneau-Russell and Ensom (1998). The teeth attributed to *Thereuodon* have elongated crowns and widely separated roots as is typical for deciduous premolars of cladotherians (Martin 1999a). Therefore, *Thereuodon* is considered as a *nomen dubium*.

6.5 Docodonta

Docodontans (δοκός (Greek), beam and ὀδούς (Greek), tooth, after the labio-lingually extended upper molars) are mammaliaforms (Fig. 6.1) with more complex molars than morganucodontans and kuehneotheriids (Fig. 6.11). For a long time, they were known by mainly fragmentary remains from western North America (Morrison Formation) and Western Europe (Great Britain and Portugal, e.g., Simpson 1928, 1929, Krusat 1980, Martin and Nowotny 2000), but recently it became evident that they were diverse and widespread in Asia where they often dominate the mammalian assemblages (Martin and Averianov 2001, Averianov *et al.* 2005, Averianov and Lopatin 2006, Martin *et al.* 2010a, 2011). Several virtually complete skeletons have been reported from the Jurassic of northeastern China (Ji *et al.* 2006, Luo *et al.* 2015b, Meng *et al.* 2015). Their geographic range is in northern Laurasia, except for one questionable taxon from India (Prasad and Manhas 2001, 2007). The oldest unambiguous docodontans are known from the Middle Jurassic. They reach their highest diversity in the Middle-Late Jurassic (Kielan-Jaworowska *et al.* 2004, Martin and Averianov 2004, Lopatin and Averianov 2005, Pfretzschner *et al.* 2005, Rougier *et al.* 2014), and their last occurrence is in the late Early Cretaceous (Aptian-Albian) with *Sibirotherium* from Siberia (Maschenko *et al.* 2002). The South American *Reigitherium*, originally described as a docodontan (Pascual *et al.* 2000), is now interpreted as a representative of the Southern Hemisphere Meridiolestida (Rougier and Apesteguia 2004, Averianov *et al.* 2013, Harper and Rougier 2017).

The docodontan molariform dentition is rather complex and includes a crushing and even grinding function in

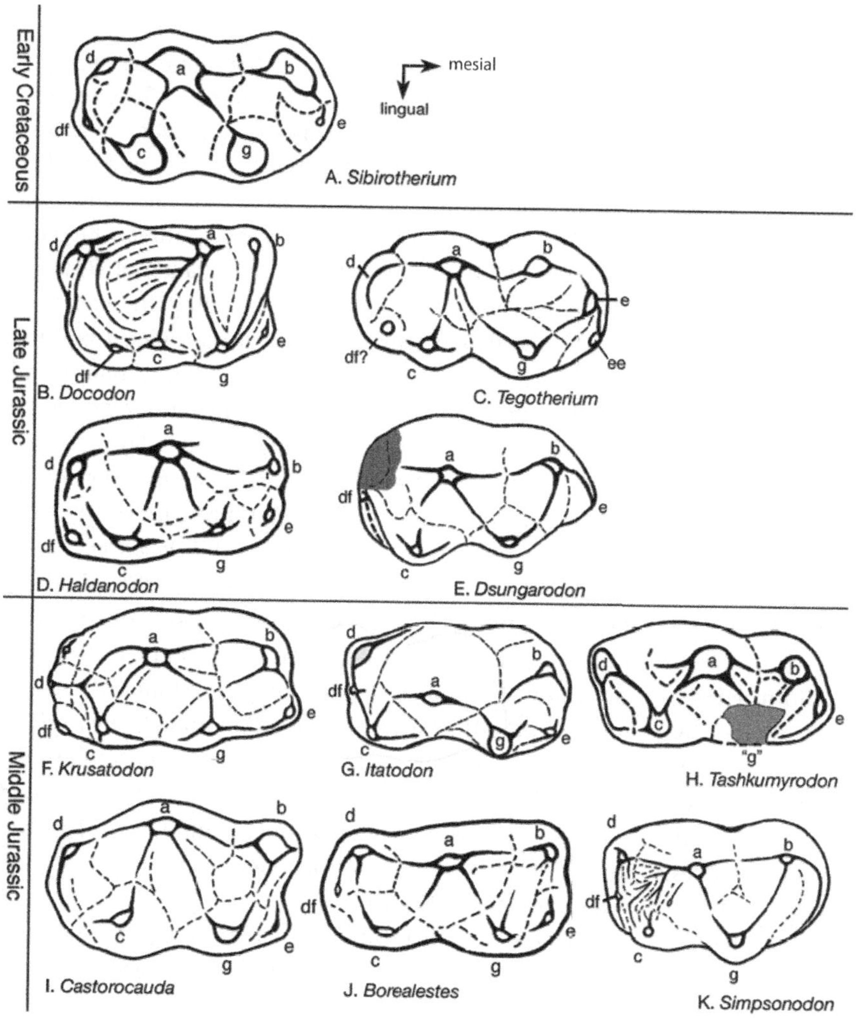

Fig. 6.11: Complexity and systematic character variation of lower molars among selected docodontans. No scale, teeth brought to same size. From Luo and Martin (2007).

some taxa (Gingerich 1973, Butler 1988, Pfretzschner *et al.* 2005, Brinkkötter and Martin 2013). However, docodontan molars are not in the evolutionary lineage leading to the tribosphenic molars of therian mammals but represent an independent early version of a dentition with a grinding and crushing function, indicating a more omnivorous diet. The dentition of the large docodontan *Itatodon* (considered as a shuotheriid by Wang and Li 2016) suggests that it was carnivorously adapted (Lopatin and Averianov 2005, Averianov and Lopatin 2006). The origin of the docodontan dentition is disputed, and there is no consensus on the homologies of the molariform cusps (Luo and Martin 2007). Several hypotheses for the origin of the docodontan molar have been put forward. The docodontan molar main cusps are either homologized with the main cusp row of *Morganucodon* (Patterson 1956) or with that of Late Triassic "symmetrodontans", a polyphyletic group of mammaliaforms with a triangular arrangement of molar cusps (Butler 1997, Sigogneau-Russell and Godefroit 1997). Accordingly, docodontans have been either considered as related to morganucodontans (Hopson and Crompton 1969, Kermack *et al.* 1973, Lillegraven and Krusat 1991, Averianov and Lopatin 2006) or, more recently, as close relatives of "symmetrodont-like" mammaliaforms (Sigogneau-Russell and Hahn 1995, Butler 1997). However, because the straight alignment of the main cusps in "triconodont-like" mammals is primitive, it is uninformative for inferring phylogenetic relationships among mammaliaforms. According to cladistic analyses, docodontans and *Morganucodon* belong to different nodes of the mammalian phylogenetic tree, and therefore the traditional docodontan-morganucodontan grouping should be abandoned. According to Luo and Martin (2007), the sister taxon of docodontans is probably a taxon with a triangulated cusp pattern on the lower molars and transversely wide upper molars; in their parsimony-analysis, Luo and Martin (2007) found the mammaliaform *Tikitherium* from the Late Triassic of India (Datta 2005) as the sister taxon of docodontans, and the Late Triassic mammaliaform *Woutersia* from Europe (Sigogneau-Russell and Hahn 1995, Butler 1997, Sigogneau-Russell and Godefroit 1997) as the next closest relative.

The Middle Jurassic docodontans do not form a monophyletic group, but Luo and Martin (2007) found that two pairs of taxa are closely related: *Krusatodon* from the Middle Jurassic (Bathonian) of Kirtlington (Sigogneau-Russell 2003a) and *Itatodon* from the Middle Jurassic (Bathonian) of Berezovsk, western Siberia (Lopatin and Averianov 2005), as well as *Borealestes* from the Bathonian of Scotland (Waldman and Savage 1972) and *Tashkumyrodon* from the Bathonian Balabansai Formation in Kyrgyzstan (Martin and Averianov 2004). Other Middle Jurassic taxa are *Simpsonodon* (synonym *Cyrtlatherium*; Freeman 1979) from Kirtlington (Kermack *et al.* 1987), *Paritatodon* from Kirtlington and the Callovian Balabansai Formation of Kyrgyzstan (Martin and Averianov 2004), *Castorocauda* from the Jiulongshan Formation of China (Ji *et al.* 2006), and *Hutegotherium* from Berezovsk (Averianov *et al.* 2010b).

The Late Jurassic docodontans of Western Europe (Great Britain and Portugal) and the Western Interior of North America are very similar. Cladistic analysis has found that *Docodon* and *Haldanodon* from Euroamerica form a clade (Martin and Averianov 2004). This grouping was corroborated by Luo and Martin (2007), and *Dsungarodon* (synonym *Acuodulodon*; Hu *et al.* 2007) from the Late Jurassic of the Junggar Basin, northwestern China (Pfretzschner *et al.* 2005), was found as sister taxon of this clade. In the cladistic analysis of Luo and Martin (2007), the central Asian taxa *Sibirotherium* from the Barremian-Aptian of western Siberia (Maschenko *et al.* 2002) and *Tegotherium* from the latest Jurassic-earliest Cretaceous of Shar Teg in Mongolia (Tatarinov 1994, Hopson 1995, Kielan-Jaworowska *et al.* 2004) form a clade. Luo and Martin (2007) corroborated the inclusion of *Gondtherium*, based on an incomplete lower molariform from the late Middle Jurassic to Lower Cretaceous Kota Formation of India, into Docodonta as proposed by Prasad and Manhas (2001, 2007). *Peraiocynodon* from the Purbeck of Durdlestone Bay in southern England (Simpson 1928) was regarded as sister taxon of *Docodon* by Sigogneau-Russell (2003a). According to Averianov (2004), the two species of this taxon are based on milk dentitions of *Docodon* (and probably *Krusatodon*) and therefore are invalid, a view that had been forwarded by previous authors (Butler 1939, Patterson 1956, Kermack *et al.* 1987). Schultz *et al.* (2017) who studied the deciduous lower premolars of *Docodon* advocate again for a generic separation of *Peraiocynodon* and *Docodon*.

Averianov *et al.* (2010a) reanalyzed the phylogenetic relationships of docodontans and recognized three families: Docodontidae Simpson 1929 with *Docodon* and *Haldanodon*; Simpsonodontidae Averianov *et al.* 2010a with *Simpsonodon* and *Dsungarodon*; and Tegotheriidae Tatarinov 1994 with *Tegotherium*, *Sibirotherium*, *Krusatodon*, and *Hutegotherium*. *Borealestes* is considered a stem docodontan, and *Tashkumyrodon* and *Itatodon* as Docodonta indeterminate by Averianov *et al.* (2010a).

Averianov *et al.* (2018) reported a new tegotheriid docodontan, *Khorotherium*, from the Early Cretaceous Sangar

Series of Teete ("Kempendyay") in northern Yakutia, Russia. The paleolatitude of Teete locality is estimated at 63–70° N (Rich *et al.* 2002) which makes it the northernmost Mesozoic mammal locality in Asia.

6.5.1 *Haldanodon exspectatus*

The first docodontan of which skull remains and parts of the postcranial skeleton became known was *Haldanodon exspectatus* from the Late Jurassic (Kimmeridgian) of the Guimarota coal mine in Portugal (Henkel and Krusat 1980, Krusat 1991, Lillegraven and Krusat 1991, Martin 2005, Ruf *et al.* 2013). Lillegraven and Krusat (1991) reconstructed the mandible of *Haldanodon*. In *Haldanodon* and other docodontans, the middle ear bones were still attached to the mandible, and both primary and secondary jaw joints were present (Lillegraven and Krusat 1991). The attachment of the middle ear bones is indicated by the well-developed postdentary trough on the inner side of the mandible (Fig. 6.12). Accordingly, in the middle ear of *Haldanodon*, only the stapes was present. It has long and slender crura, a very large stapedial foramen and a rounded, disklike footplate (Ruf *et al.* 2013) (Fig. 6.13).

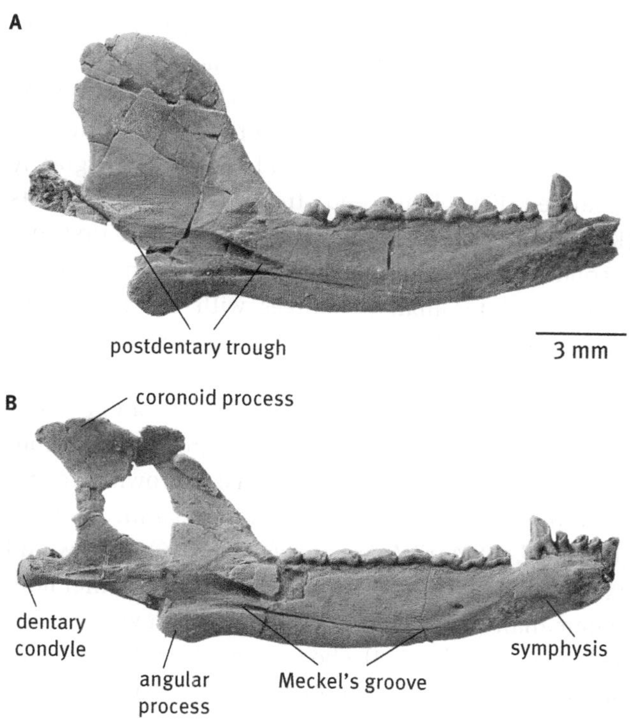

Fig. 6.12: Mandibles of *Haldanodon exspectatus* from the Guimarota coal mine in lingual view. (A) Gui Mam 81/79 with c, p1-3, m1-5. (B) Gui Mam 35/75 with i1-4, c, p1-3, m1-5 posterior premolars and molars are strongly worn. Photographs by Georg Oleschinski.

The skull of *Haldanodon* (Fig. 6.14) is wedge shaped and dorsoventrally flattened, similar as in extant fossorial small mammals (Krusat 1991, Lillegraven and Krusat 1991, Ruf *et al.* 2013). The frontals and the anterior portions of the parietals exhibit a rugose surface (Fig. 6.14), which is interpreted as insertion area for a protective horny shield (Fig. 6.16). In general anatomy, the skull of *Haldanodon* is somewhat more derived than those of *Morganucodon* and *Sinoconodon*. This includes an elongated and curved cochlear canal (nearly 180°)(Fig. 6.15), a single foramen in the lateral flange of the petrosal, the absence of an anterior paroccipital process, and a squamosal constriction between the glenoid fossa and the cranial moiety (Lillegraven and Krusat 1991, Ruf *et al.* 2013). The secondary crus commune of the bony labyrinth is a plesiomorphic character of mammals. The apical inflation of the cochlear canal, which is connected to a distinct sulcus, and the presence of a separate notch in the internal acoustic meatus support the existence of a lagenar nerve and macula, as in monotremes (Ruf *et al.* 2013). *Haldanodon* has a hypertrophied paroccipital region with large tympanic pneumatic recesses, which are connected with the extensive porous (and probably also pneumatic) internal structures of the surrounding bones. This pneumatization of the middle ear region is unique for known Mesozoic mammaliaforms and, together with the curved cochlear canal, indicates that *Haldanodon* had good low-frequency hearing, connected to a fossorial mode of life (Ruf *et al.* 2013).

The dental formula of *Haldanodon* is six incisors (6I), one canine (1C), three premolars (3P), five molars (5M) in the maxilla and four incisors (4i), one canine (1c), three premolars (3p), and five to six molars (5–6m) in the mandible (Krusat 1980, Martin and Nowotny 2000, Nowotny *et al.* 2001). A large sample of juvenile dentitions from the Guimarota mine demonstrates that *Haldanodon* had diphyodont replacement of incisors and premolars (Martin and Nowotny 2000), but most likely more than one replacement of the upper canine occurred (Martin *et al.* 2010b). Upper and lower premolars were replaced sequentially from front to back. In the mandible, four deciduous premolar positions are present, but only three permanent premolars erupted, with loss of the position of dp2. Although the diphyodont replacement of antemolars of crown mammals was largely established in *Haldanodon*, it had retained the plesiomorphic multiple replacement of at least the upper canine, as seen in stem mammals such as *Sinoconodon* (Crompton and Luo 1993, Luo *et al.* 2004). In a number of maxillae and mandibles of *Haldanodon* from the Guimarota coal mine, the molars are worn down to almost featureless knoblike structures (Fig. 6.12 B). This strong wear was probably

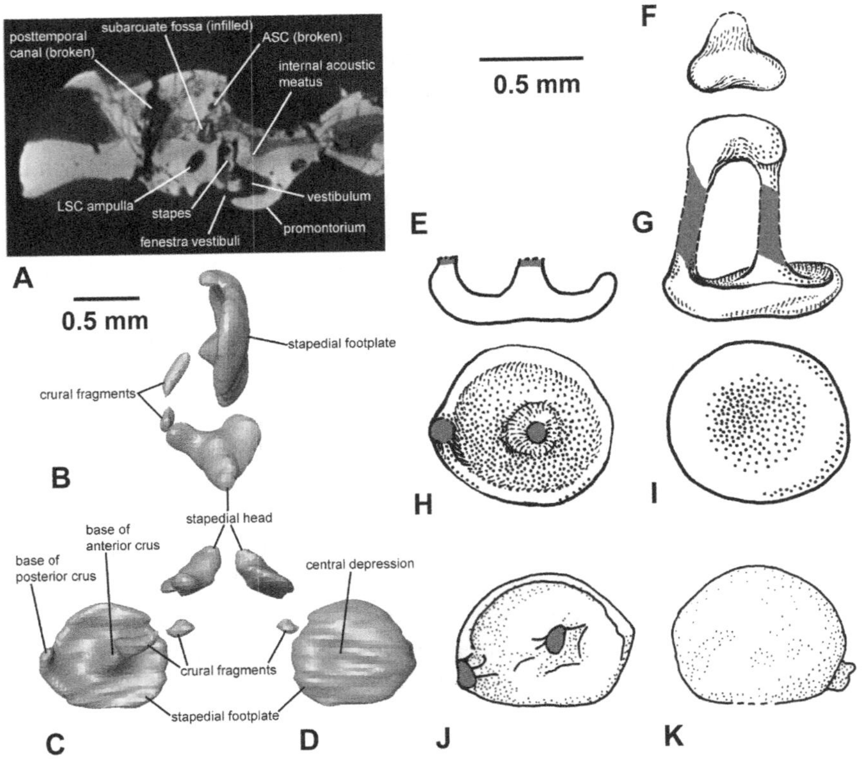

Fig. 6.13: The right stapes of *Haldanodon exspectatus*. (A) Transverse µCT image of ear region showing the displaced stapes (specimen 6722, slice 519). (B–D) 3D model of stapes (specimen 6722): (B) dorsal view of the broken stapes in same orientation as in A; (C) lateral and (D) medial aspects. (E–I) Restoration of stapes of specimen 6722 (lengths of stapedial crura are hypothetical): (E) schematic section of the bullate footplate; (F) stapedial head in lateral view; (G) stapes in ventral view; (H) footplate in lateral view; (I) footplate in medial view. (J, K) Stapedial footplate of specimen 6723 (reproduced from Lillegraven and Krusat 1991 for comparison) in lateral (J) and medial (K) views. Abbreviation: ASC, anterior semicircular canal; LSC, lateral semicircular canal. Modified after Ruf *et al.* (2013), reprinted by permission of the Society of Vertebrate Paleontology, www.vertpaleo.org.

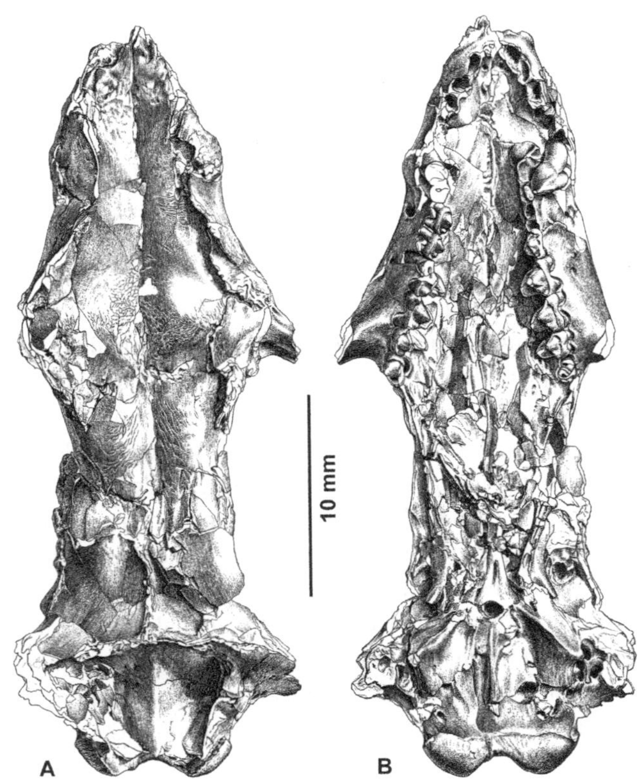

Fig. 6.14: Skull of *Haldanodon exspectatus* (Museu Geológico 6721) in dorsal (A) and palatinal (B) views. Drawings by Peter Berndt.

caused by earth particles adhering to insect larvae and other small subterranean invertebrates (Silcox and Teaford 2002) that were the likely prey of *Haldanodon*.

The postcranial skeleton of *Haldanodon* exhibits its strong adaptations for a fossorial lifestyle such as stout and short limb bones and humeri with greatly expanded distal joints and strong deltopectoral crests (Krusat 1991, Martin 2005) (Fig. 6.16). Short first and second phalanges and moderately curved and laterally compressed terminal phalanges with lateral grooves suggest that *Haldanodon* was a scratch digger. The wide size range of humeri and femora that lack separate epiphyses and the polyphyodont canine locus indicate that *Haldanodon* may have grown throughout life, like *Sinoconodon* but in contrast to crown Mammalia and even early mammaliaform morganucodontans (Luo *et al.* 2004, O'Meara and Asher 2016). A slow and extended lifelong growth in *Haldanodon* and other nonmammalian mammaliaforms is concordant with a study by Newham *et al.* (2018) who estimated for Middle Jurassic docodontans considerably longer lifespans (up to 12 years) than for crown mammals by counting cementum increments in the molar roots. The shoulder girdle exhibits a number of plesiomorphic features. The scapula is triangular and dorsoventrally elongated, with a convex transverse profile, and strongly laterally reflected anterior and posterior scapula margins that enclose a deep troughlike "infraspinous fossa". A supraspinous fossa is not present. The glenoid facet is saddle shaped and mainly formed by the coracoid bone; it is oriented anteroventrally, which suggests a sprawling gait. However, the shoulder girdle of *Haldanodon* is more derived than that of *Morganucodon* in its loss of the procoracoid, absence of the procoracoid foramen, and a peglike coracoid (Martin 2005). *Haldanodon* can be reconstructed as a fossorial small mammaliaform with a subterraneous mode of life similar to modern moles (Fig. 6.17).

6.5.2 *Castorocauda lutrasimilis*

Recent discoveries of remarkably well-preserved docodontan skeletons from northeastern China have revealed further unexpected adaptations. *Castorocauda lutrasimilis* from the Middle Jurassic Jiulongshan Formation in Liaoning Province was a large docodontan with semiaquatic adaptations (Ji *et al.* 2006), such as a beaverlike broad tail covered by horny scales and webbing at the hind feet (Fig. 6.18). The caudal vertebrae have dorsoventrally compressed centra, and caudals 5 to 15 have bifurcate transverse processes, both typical features of mammals with tails specialized for swimming. Similar to *Haldanodon* and the modern monotreme *Ornithorhynchus*, *Castorocauda* has specializations for digging and swimming at the forelimb, including a distally wide humerus with hypertrophied epicondyles, a supinator process, and massive, clearly separated ulnar and radial condyles. The olecranon of the ulna is massive and asymmetrical, the radius is robust, and the hands are massive with blocklike carpals, as well as robust and wide metacarpals and proximal phalanges (Ji *et al.* 2006). *Castorocauda* differs from other docodontans in its mediolaterally compressed tooth crowns of first and second molars, each with five cusps in straight alignment, three of which are slightly recurved. They converge to those of certain eutriconodontans (e.g., *Ichthyoconodon*; Sigogneau-Russell 1995), Eocene whales, and extant seals, and they are interpreted as a specialization for preying fish and other aquatic organisms.

Castorocauda is the largest Jurassic mammaliaform that is known to date. According to the scaling relation of the skull (length about 60 mm), the body mass of the holotype specimen is estimated as at least at 500 g, with an upper limit of 800 g (Ji *et al.* 2006).

6.5.3 *Docofossor brachydactylus*

Two other recently discovered docodontan skeletons from the Middle and Late Jurassic of northeastern China exhibit fossorial (*Docofossor*) and, unexpectedly, climbing (*Agilodocodon*) adaptations. *Docofossor* (Docodontidae) from the Oxfordian of the Tiaojishan Formation is a small docodontan (body mass 13–17 g), with striking adaptations for scratch digging and a subterranean

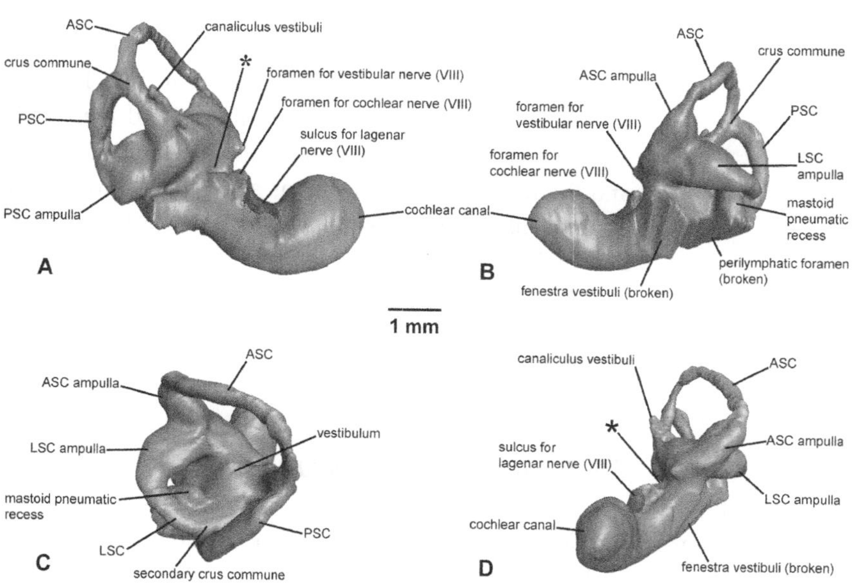

Fig. 6.15: Virtual endocast of the left bony labyrinth of *Haldanodon exspectatus*. Museu Geológico specimen 6721 in caudomedial (A), rostrolateral (B), caudodorsal (C), and anterolateral (D) aspects. Asterisks (*) indicate sulcus between vestibulum and cochlear canal on the endocast. Abbreviations: ASC, anterior semicircular canal; LSC, lateral semicircular canal; PSC, posterior semicircular canal. Modified after Ruf *et al.* (2013), reprinted by permission of the Society of Vertebrate Paleontology, www.vertpaleo.org.

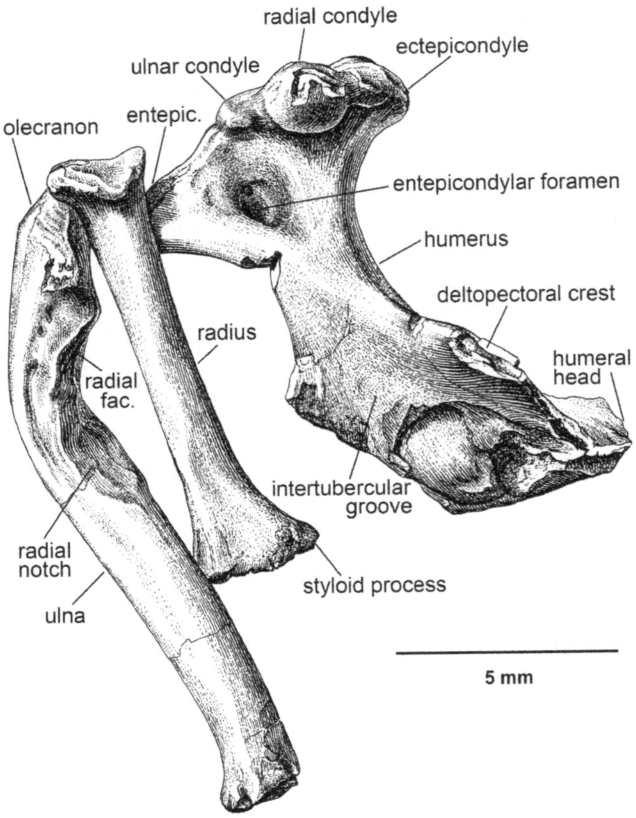

Fig. 6.16: Right forelimb of *Haldanodon exspectatus* (Gui Mam 30/79) as originally preserved with (from right to left) humerus in anterior view (distal end pointing upward), radius in anterior view, and ulna in lateral aspect. Abbreviations: entepic., entepicondyle; fac., facet. From Martin (2005), by permission of the Linnean Society.

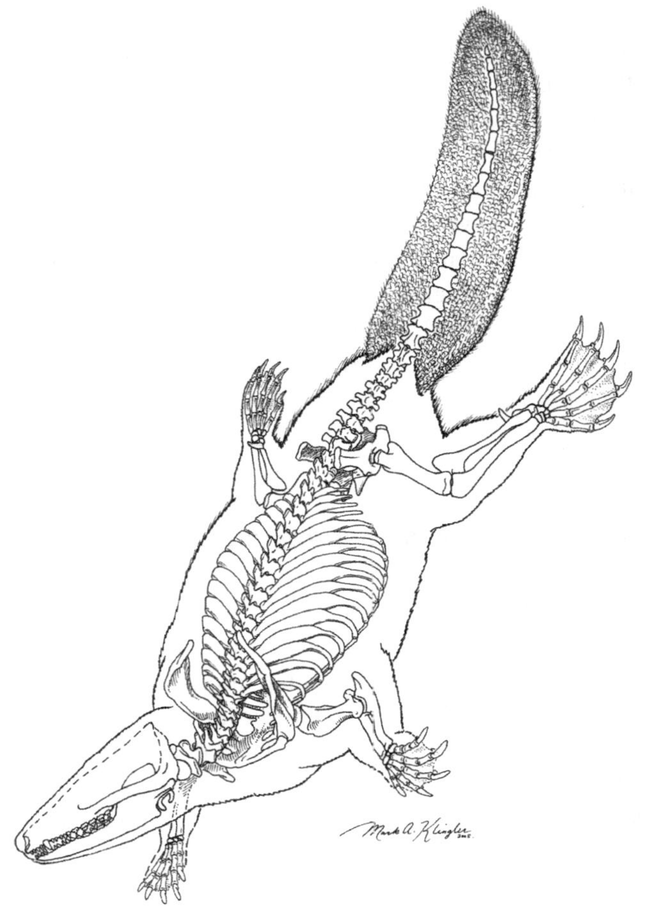

Fig. 6.18: Semiaquatic docodont *Castorocauda lutrasimilis*. Skeletal reconstruction with body outline of holotype specimen. From Ji *et al.* (2006), reprinted with permission from AAAS.

Fig. 6.17: Life reconstruction of *Haldanodon exspectatus*. From Martin and Nowotny (2000).

lifestyle (Fig. 6.19) such as hypertrophied olecranon processes and sprawling posture of fore- and hindlimbs (Luo *et al.* 2015b). The upper molars of *Docofossor* are mesiodistally shortened and labiolingually expanded, a phenomenon known as zalambdodonty and convergently present in certain fossorial mammals such as marsupial moles, golden moles, and the extinct meridiolestidan *Necrolestes* (Asher and Sánchez-Villagra 2005, Asher *et al.* 2007, Archer *et al.* 2011, O'Meara and Thompson 2014). *Docofossor* has robust hands and feet, with a reduced number of phalanges for the hand (2-2-2-2-2) and the foot (1-2-2-2-2), as well as shovel-like, broadened terminal phalanges. A similar reduction of phalanges is known in the African Golden Moles (Chrysochloridae) with a manual formula of 2-2-1-2-0 for *Eremitalpa* and *Chrysospalax* and a pedal formula of 2-2-2-2-2 for *Eremitalpa* and *Chrysochloris*.

6.5.4 *Agilodocodon scansorius*

Agilodocodon scansorius (Docodontidae) from the Middle Jurassic part of the Tiaojishan Formation represents a different adaptational type (Fig. 6.20) and demonstrates that docodontans were ecomorphologically much more diverse than previously thought (Meng *et al.* 2015). With a head-tail length of about 140 mm and a body mass of 27 to 40 g, *Agilodocodon* was slightly larger than *Docofossor*. The lower incisors have incipiently divided roots and spatulate crowns that are convex on the labial side and concave on the lingual side. Meng *et al.* (2015) observed similarities to incisors of small New World primates such as spider monkeys (*Ateles*) and howler monkeys (*Alouatta*) that feed on plant exudates such as gum and sap and assumed a similar diet for *Agilodocodon*. This was contested by Wible and Burrows (2016), who observed no particular adaptation to gummivory in the lower mesial dentition of *Agilodocodon*. They found the mesial dentition of *Agilodocodon* to be most similar to that of some elephant shrews and South American marsupials, which are primarily insectivorous. The crest pattern of the upper molars resembles that of galagid and some lorisid primates that have a mixed diet of insects, other small invertebrates, fruits, tree gums, and sap. Based on its postcranial skeletal characters, *Agilodocodon* was a fully arboreal mammaliaform (Meng *et al.* 2015). *Agilodocodon* has elongated phalanges, and the terminal manual phalanges are laterally compressed and have an arched dorsal margin. The distal humerus is gracile and narrow, and the ankle of the foot had a greater range of mobility than in other mammaliaforms. The caudal vertebrae indicate a highly mobile and well-muscularized tail, which is consistent with the arboreal locomotory adaptation.

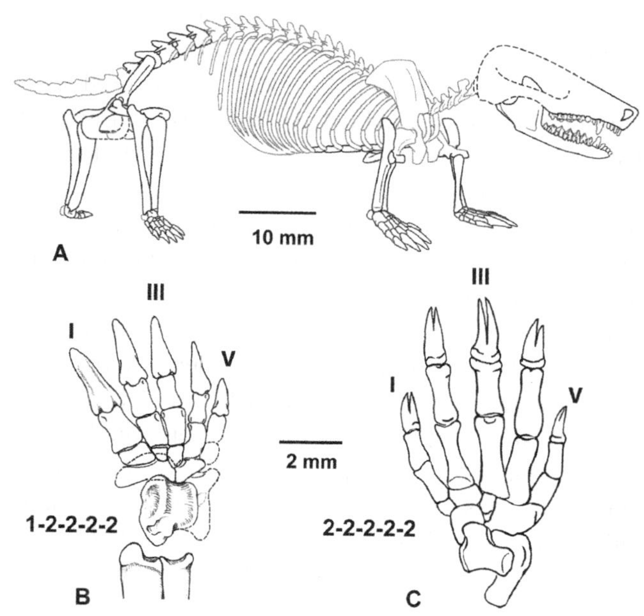

Fig. 6.19: Fossorial docodontan *Docofossor brachydactylus*. Skeletal restoration (A) and right foot (B) of holotype specimen in comparison with right foot of golden mole *Chrysochloris* (C). Modified after Luo *et al.* (2015a), reprinted with permission from AAAS.

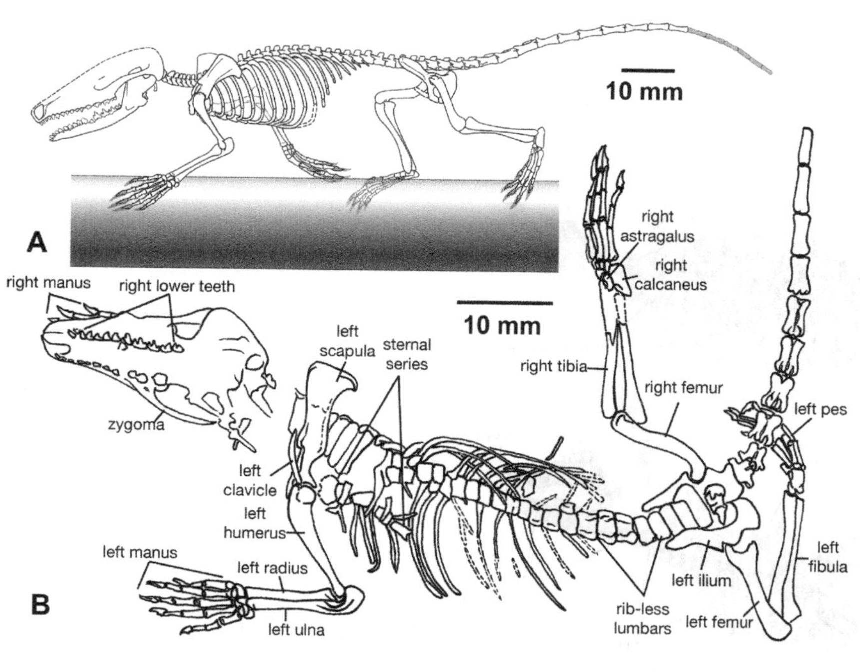

Fig. 6.20: Arboreal docodontan *Agilodocodon scansorius*. Skeletal restoration (A) and drawing of holotype specimen in original position (B). From Meng *et al.* (2015), reprinted with permission from AAAS.

6.6 Haramiyida

For more than 100 years, haramiyidans (Fig. 6.1) were known only by isolated teeth, mainly from Late Triassic strata from southern Germany, eastern France, Belgium, and Luxembourg. Their multicusped molariforms resemble those of Multituberculata, and consequently both groups were placed in the taxon "Allotheria" Marsh 1880 (Hahn *et al.* 1989). Subsequently, it has been argued that Haramiyida are stem mammals that are not closely related to multituberculates, which belong to crown Mammalia (Zhou *et al.* 2013, Luo *et al.* 2015b, Puttick *et al.* 2017). A multituberculate-haramiyid clade ("Allotheria") is upheld by Meng *et al.* (2014) based on disputed dental and mandibular characters in the haramiyidan *Arboroharamiya*, a view that has been supported by Bi *et al.* (2014) and Krause *et al.* (2014a). The dental similarities between Haramiyida and Multituberculata may be the result of convergent evolution. Haramiyidan molariforms are characterized by two rows (on lower molariforms) or two to three (on upper molariforms) rows of cusps which are designated as buccal row A (upper)/a (lower) and lingual row B (upper)/b (lower), with the cusps numbered consecutively from mesial to distal. The molariform cusps are of differing size, with the mesiobuccal cusp A1/a1 being the largest. Striation analysis of the enamel surface demonstrates that the jaw movements of Haramiyida were mainly orthal (vertical) and not palinal (anteroposterior) as in Multituberculata. Butler (2000) has proposed a scenario for the evolution of palinal movement in multituberculates from orthal movement in haramiyidans. Haramiyida is separated into two suborders: Theroteinida Hahn *et al.* 1989 with the family Theroteinidae Sigogneau-Russell *et al.* 1986 and Haramiyoidea Hahn 1973 with families Haramiyaviidae Butler 2000, Haramiyidae Simpson 1947 and Eleutherodontidae Kermack *et al.* 1998.

The first information on skull and mandible of haramiyidans was provided by *Haramiyavia* from the Late Triassic (Norian-Rhaetian) of Greenland (Jenkins *et al.* 1997). *Haramiyavia* revealed striking differences between haramiyidans and multituberculates, such as presence of a postdentary trough and lack of a pterygoid fossa, as well as specializations in the mesial dentition that would appear to preclude *Haramiyavia* from multituberculate ancestry (Fig. 6.21). The dental formula of *Haramiyavia* has been described by Luo *et al.* (2015b) as 4I.?C.?P.3M/3i.1c.4p.3m. The roots of upper and lower molariforms have multiple, partially divided roots similar to those of the Middle-Late Jurassic haramiyidans (Zheng *et al.* 2013, Zhou *et al.* 2013, Bi *et al.* 2014). Luo *et al.* (2015b) confirmed in a kinematic functional analysis that *Haramiyavia* had primarily orthal occlusion, with the tallest lingual cusp of the lower molars occluding with the lingual embrasure of the upper molars, followed by a short palinal movement along the cusp rows. The body mass of *Haramiyavia* is estimated at 50 to 70 g, and the multicusped molariforms indicate an adaptation to an omnivorous-herbivorous diet (Luo *et al.* 2015b).

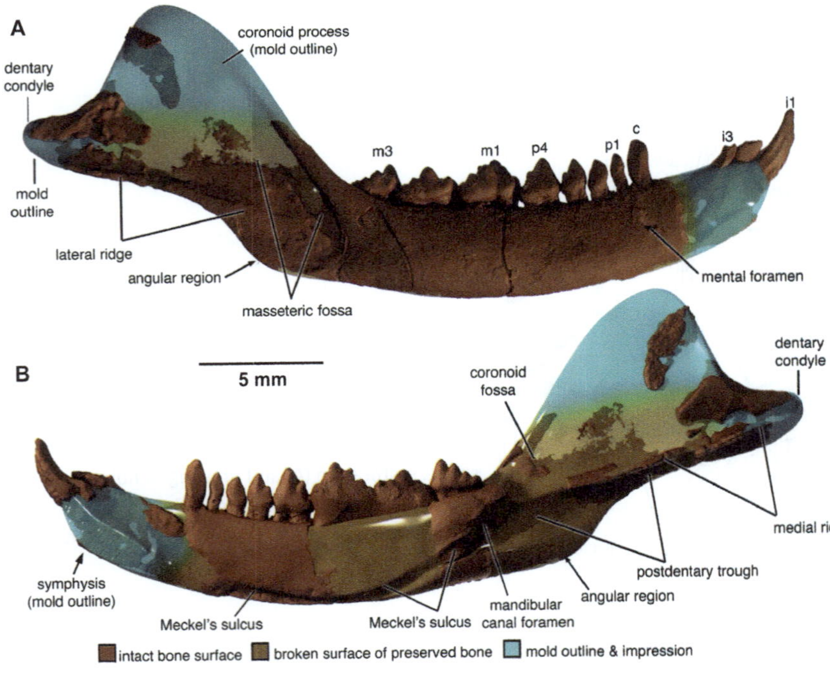

Fig. 6.21: Composite restoration of *Haramiyavia clemmenseni* right mandible in lateral (A) and medial (B) views. From Luo et al. (2015b).

Recently, the skull of a putative large haramiyidan (estimated body mass 0.91–1.27 kg) has been described from the Early Cretaceous Cedar Mountain Formation of Utah (USA) (Huttenlocker *et al.* 2018). The skull of *Cifelliodon* is 70 mm long and exhibits a mix of stem mammaliaform plesiomorphies and crown mammalian apomorphies. Based on the alveoli, the dental formula comprises two incisors, one canine, and four postcanines (PC). The preserved four postcanines closely resemble isolated molars of *Hahnodon* from the Early Cretaceous of Anoual (Morocco), which originally had been considered a paulchoffatiid multituberculate (Sigogneau-Russell 1991a, Hahn and Hahn 2003). Huttenlocker *et al.* (2018) demonstrated the haramiyidan affinities of these Gondwanan hahnodontid teeth and suggested that Hahnodontidae Sigogneau-Russell 1991a had possibly a Pangaean distribution in the Late Jurassic/Early Cretaceous.

6.6.1 *Megaconus mammaliaformis*

The eleutherodontid *Megaconus* from the Middle Jurassic (Callovian) of Inner Mongolia (Fig. 6.22) is the first haramiyidan for which the postcranial skeleton became known, and it was reconstructed in a phylogenetic analysis outside of crown Mammalia, apart from multituberculates (Zhou *et al.* 2013). *Megaconus* has a dental formula of 2I.0C.2P.3M/1i.0c.2p.3m, with a procumbent, enlarged i1 and a rhomboidal I2. The molariforms exhibit root hypsodonty, and their multiple roots are fused proximally but divided distally (Fig. 6.23), a character that is shared with *Eleutherodon* and *Sineleutherus*. *Megaconus* has a postdentary trough, which indicates that the middle ear bones were still attached to the mandible and the primary quadrate-articular jaw joint was still present (Fig. 6.23). The mesial upper molariforms (M1 and M2) of *Megaconus* have a lingual offset with a third row of cusps, whereas a lingual offset occurs on the last upper molar in multituberculates. *Megaconus* has a gracile postcranial skeleton, and its body mass is estimated at about 250 g (Zhou *et al.* 2013). An important difference to multituberculates is the thoracolumbar transition: in *Megaconus*, it is gradational from dorsal (= thoracolumbar) vertebrae D16 to D20, with the anticlinal vertebra positioned at D21 (Zhou *et al.* 2013), whereas multituberculates have a distinct boundary at thoracic vertebra D13, and the anticlinal vertebra at position D10 (Kielan-Jaworowska and Gambaryan 1994). *Megaconus* is terrestrially adaptated, with generalized manual and pedal phalanges. The proximally fused tibia and fibula (similar to *Dasypus* [armadillo], *Orycteropus* [aardvark],

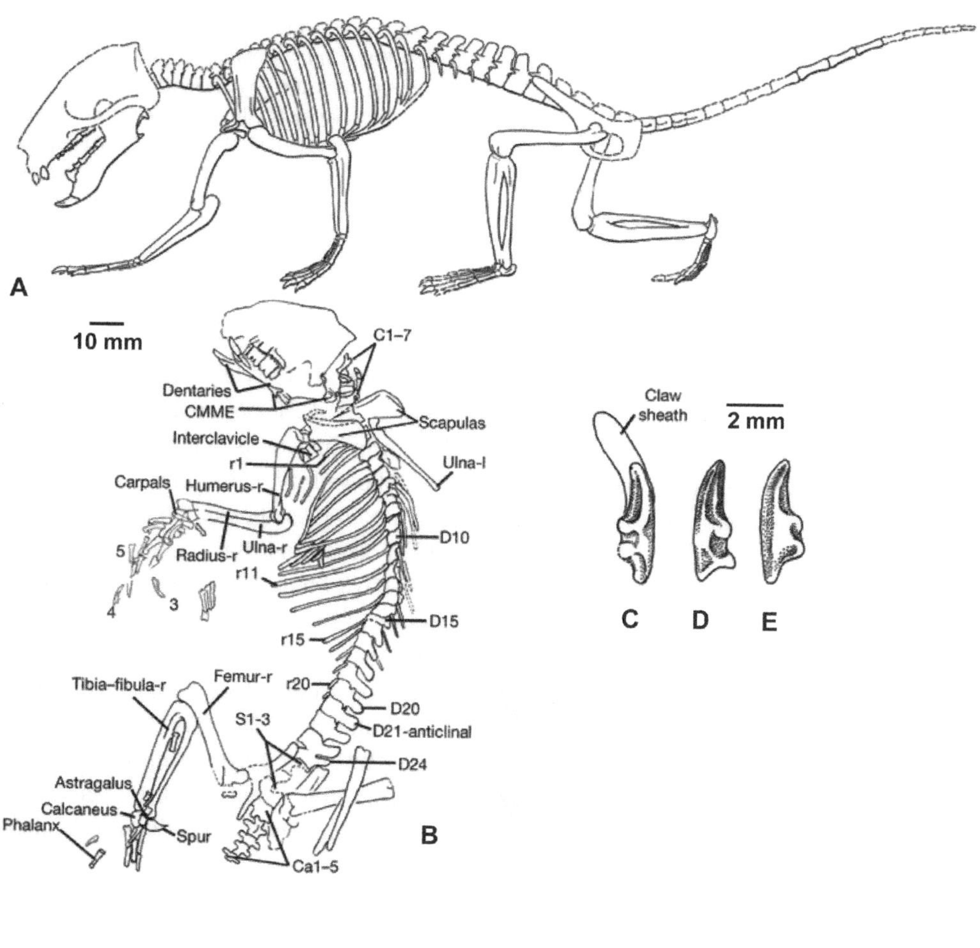

Fig. 6.22: *Megaconus mammaliaformis*. (A) Skeletal restoration; (B) holotype counterpart; and (C–E) manual terminal phalanges. From Zhou *et al.* (2013).

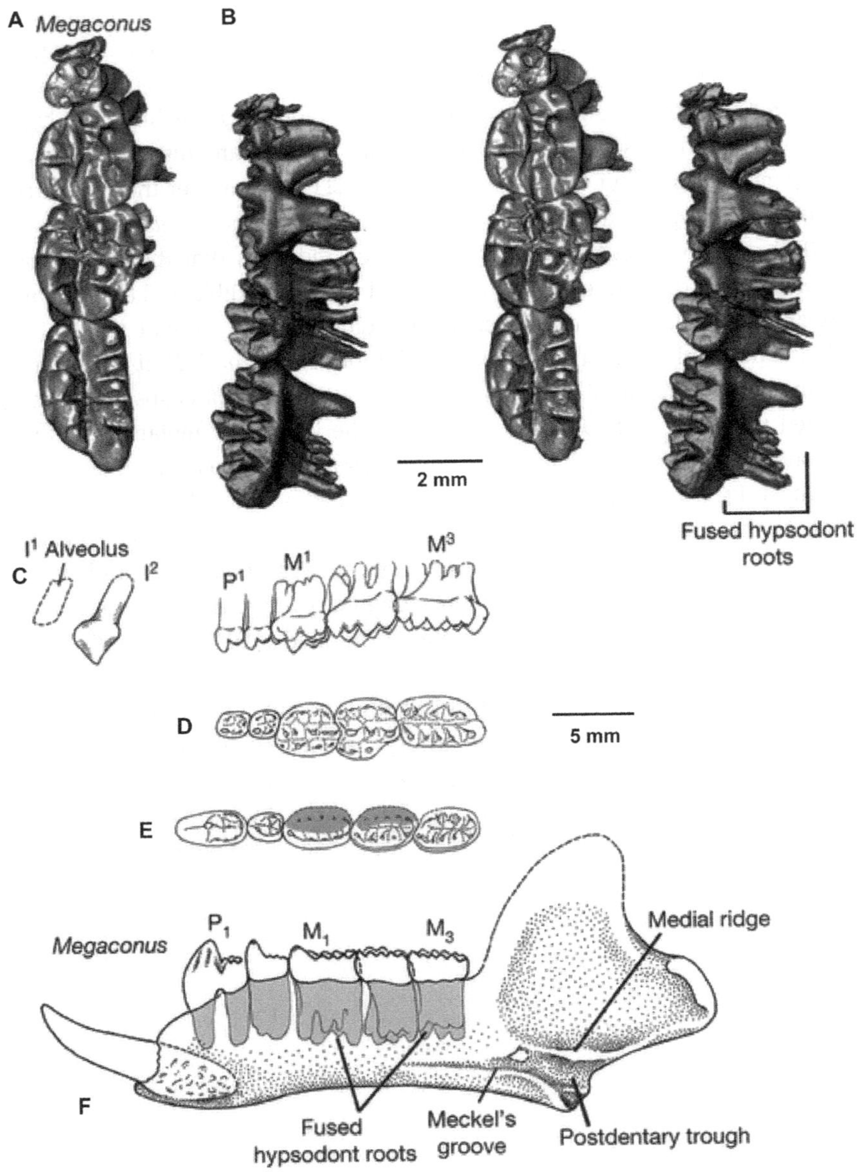

Fig. 6.23: Dental and mandibular structures of *Megaconus mammaliaformis*. (A and B) Occlusal and lingual views of P2-M3 (stereo-pairs); (C) lingual view restoration of I1 (alveolus) and I2-M3; (D) occlusal view of P1-M3; (E) occlusal view of p1-m3; (F) lingual view of lower teeth and dentary. From Zhou *et al.* (2013).

Erinaceus [hedgehog], and others) indicate an ambulatory mode of locomotion. The ankle joint bears an extratarsal (poisonous) spur as a protective device, similar to the platypus. Impressions of guard hairs in the halo around the skeleton demonstrate that mammalian fur originated earlier in evolutionary history than the last common ancestor of extant mammals (Zhou *et al.* 2013).

6.6.2 *Arboroharamiya, Shenshou,* and *Xianshou*

Arboroharamiya jenkinsi from the Middle to Late Jurassic of Hubei Province (China) is the largest known haramiyidan (body mass estimate 354 g), and it exhibits postcranial features adapted for an arboreal lifestyle (Zheng *et al.* 2013). The dental formula is 1I.0C.2P.2M/1i.0c.1p.2m. Similar to *Megaconus*, the lower incisor is enlarged and fully covered by enamel, followed by a diastema and a large lower premolar with a high mesial cusp. Three additional arboreal haramiyidan skeletons have been described from the Late Jurassic (Oxfordian) of Liaoning Province in northeastern China (Meng *et al.* 2014), for which the new clade Euharamiyida Meng *et al.* 2014 has been coined. *Shenshou lui* (family indeterminate) is a large-sized haramiyidan (body mass estimate 300 g) (Fig. 6.24) with the dental formula 1I.0C.2P.2M/4i.0c.1p.2m. *Xianshou* (family Eleutherodontidae) from the same locality has fewer teeth (1I.0C.2P.2M/1i.0c.1p.2m) and is considerably smaller

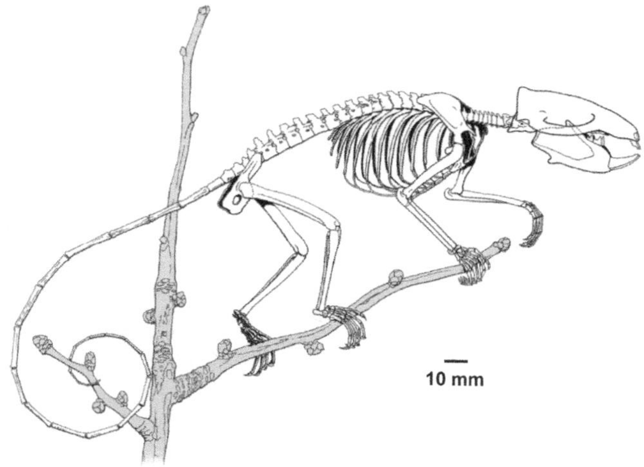

Fig. 6.24: Composite skeletal restoration of arboreal haramiyidan *Shenshou lui*. From Bi *et al.* (2014), reprinted by permission from Springer Nature.

(*Xianshou linglong* with a body mass of 83 g and *Xianshou songae* with 40 g). All three skeletons have proximal and intermediate manual and pedal phalanges that are long relative to the metapodials, as is typical for arboreal species (Meng *et al.* 2014). A second species of *Arboroharamiya*, *A. allinhopsoni* from the Oxfordian Tiaojishan Formation, has gliding membranes and suggests that gliding locomotion was probably common in euharamiyidans (Han *et al.* 2017); see also following paragraph. According to Han *et al.* (2017), *A. allinhopsoni* possessed a five-boned auditory apparatus consisting of the stapes, incus, malleus, ectotympanic, and surangular, whereupon the situation is not unambiguosly clear from the published figures.

6.6.3 *Maiopatagium furculiferum* and *Vilevolodon diplomylos*

Recently, two new eleutherodontids with well-preserved patagial membranes, indicating a gliding and possibly roosting behavior similar to modern dermopterans and bats have been described from the Late Jurassic part of the Tiaojishan Formation in Liaoning (Meng *et al.* 2017, Luo *et al.* 2017). *Maiopatagium furculiferum* is a mid-sized mammaliaform with a head-body-length of 140 mm and body mass estimates between 130 g (after femur length) and 178 g (after humerus length). In overall habitus, *Maiopatagium* is most comparable with the sciuroid gliding rodents with similar proportions of the propatagium, plagiopatagium, and uropatagium (Fig. 6.25). Meng *et al.* (2017) hypothesized a roosting behavior after the proportions of hand and feet with strikingly elongated proximal and intermediate phalanges, similar to the feet of bats. As the other eleutherodontids, *Maiopatagium* has enlarged incisors, lacks canines, and has a reduced number of premolars and molars (1I.0C.2P.2M; lower tooth count unknown). *Vilevolodon diplomylos* from the same formation possesses similar gliding membranes and has skull and dentition more completely preserved (Luo *et al.* 2017). It exhibits a hitherto unknown dual mortar-pestle occlusional pattern of opposing molars where large anterior cusp a1 of a lower molar occludes into the anterior basin of the upper antagonist and simultaneously large posterior cusp A1 of an upper molar into the posterior basin of the lower antagonist. These cheek tooth morphology and occlusion indicate both crushing and grinding and suggest a herbivorous diet, with possible granivory or feeding on soft plant tissues, and represent a remarkable convergence with gliding therian mammals (Luo *et al.* 2017). With a body mass estimate between 35 and 55 g, *Vilevolodon* is smaller than *Maiopatagium*. The dental formula 1I.0C.2P.2M/1i.0c.1p.2m is identical to that of *Arboroharamiya*, *Xianshou*, and *Shenshou*. In *Vilevolodon*, a reduced postdentary trough has been detected, which is also present in *Megaconus*, *Arboroharamiya*, *Shenshou*, *Xianshou linglong*, and *X. songae* (Luo *et al.* 2017), corroborating the mammaliaform relationship of eleutherodontids.

Excepting the Greenland mandibular fossils of *Haramiyavia* (Jenkins *et al.* 1997), other haramiyidans are known only from isolated teeth. The first haramiyidan to be described was *Thomasia* (family Haramiyidae) from the Late Triassic (Rhaetian) *Grenzbonebed* at Degerloch, near Stuttgart, Germany (Plieninger 1847, Poche 1908). Later *Thomasia* was reported from other Late Triassic localities in Germany, France, Belgium, Switzerland, and Britain, as well as from the Early Jurassic (Sinemurian) of Wales. *Allostaffia* from the Late Jurassic (Kimmeridgian-Tithonian) Tendaguru Beds in Tanzania (Heinrich 1999, 2004) and an isolated molar of uncertain affinities (*Avashishta*) from the Maastrichtian of India (Anantharaman *et al.* 2006) were the only haramiyidan records from the southern continents before the reinterpretation of *Hahnodon* as haramiyidan (Huttenlocker *et al.* 2018). The two lower cheek teeth referred to *Allostaffia* differ from *Thomasia* in the central position of cusp a1 (Heinrich 1999).

The family Eleutherodontidae (Kermack *et al.* 1998) is characterized by wide upper molars with three rows of cusps. *Eleutherodon* from the late Middle Jurassic (late Bathonian) Forest Marble at Kirtlington (southern England) is characterized by numerous marginal cusps and enamel fluting (Kermack *et al.* 1998, Butler 2000).

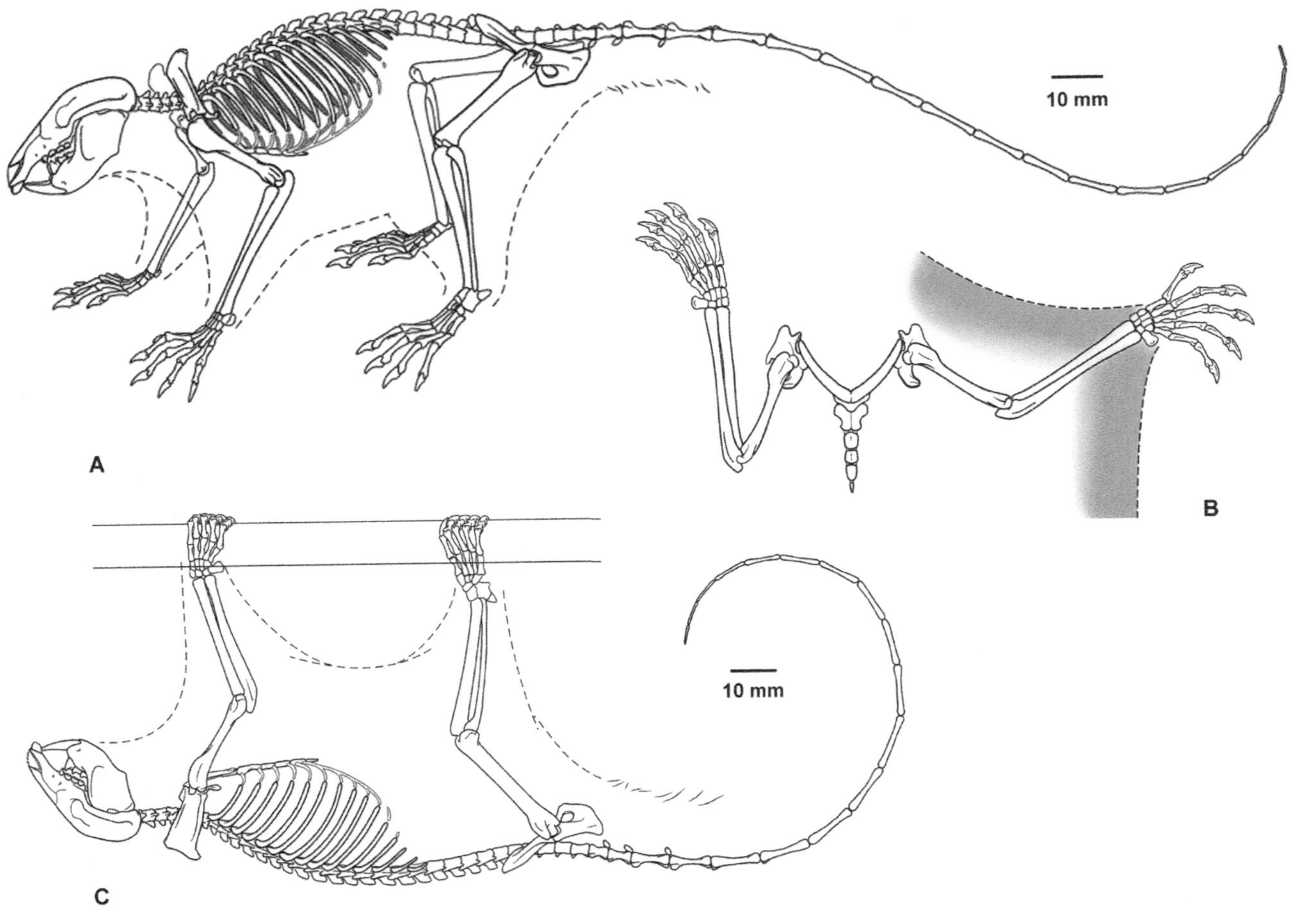

Fig. 6.25: *Maiopatagium furculiferum*, skeletal and patagial membrane restoration. (A) Standing posture; (B) furcula-like shoulder girdle in ventral aspect and the propatagium and plagiopatagium associated with forelimb; (C) four-limbed suspending roosting posture as indicated by hand and foot structures. Modified from Meng *et al.* (2017), adapted with permission from Springer Nature.

Sineleutherus from the Middle Jurassic (Bathonian) Berezovsk coal mine in western Siberia (Averianov *et al.* 2011) and the Late Jurassic (Oxfordian) of the Junggar Basin in northwestern China (Martin *et al.* 2010a) is represented by isolated incisors, premolars, and molars without enamel fluting. Recently, Averianov *et al.* (2018) reported an upper eleutherodontid molariform tooth from the Early Cretaceous Teete locality in northern Yakutia, Russia and tentatively assigned it to *Sineleutherus*. This tooth currently represents the northernmost (paleolatitude estimated at 63–70° N, Rich *et al.* 2002) and the geologically youngest record for haramiyidans. Haramiyidans of uncertain familial affiliation are *Millsodon* (a lower molar taxon) and *Kirtlingtonia* (an upper molar taxon) from Kirtlington (Butler and Hooker 2005).

Suborder Theroteinida Hahn *et al.* 1989 with the monotypic family Theroteinidae (Sigogneau-Russell *et al.* 1986), comprises haramiyidans with fully orthal occlusion and alternating upper and lower molars, so that each lower molar occludes against two upper molars (Butler 2000). *Theroteinus* from the Late Triassic (lower Rhaetian) of Saint-Nicolas-de-Port (eastern France) is represented by two species based on isolated molars (Debuysschere 2016b).

6.7 Australosphenida

Australosphenida Luo *et al.* 2001b (Fig. 6.1) is a clade of mammals from the southern continents that includes Monotremata (Luo *et al.* 2001b, Rougier *et al.* 2007a). The first australosphenidans to be recognized as such came from Madagascar and Australia: *Ambondro* from the Middle Jurassic of Madagascar (Fig. 6.26), represented by a mandibular fragment with the ultimate premolar and the first two molars with fully developed talonid basins (Flynn *et al.* 1999), and *Ausktribosphenos* (Fig. 6.27) and *Bishops* (Fig. 6.28), by largely complete mandibles from the Early

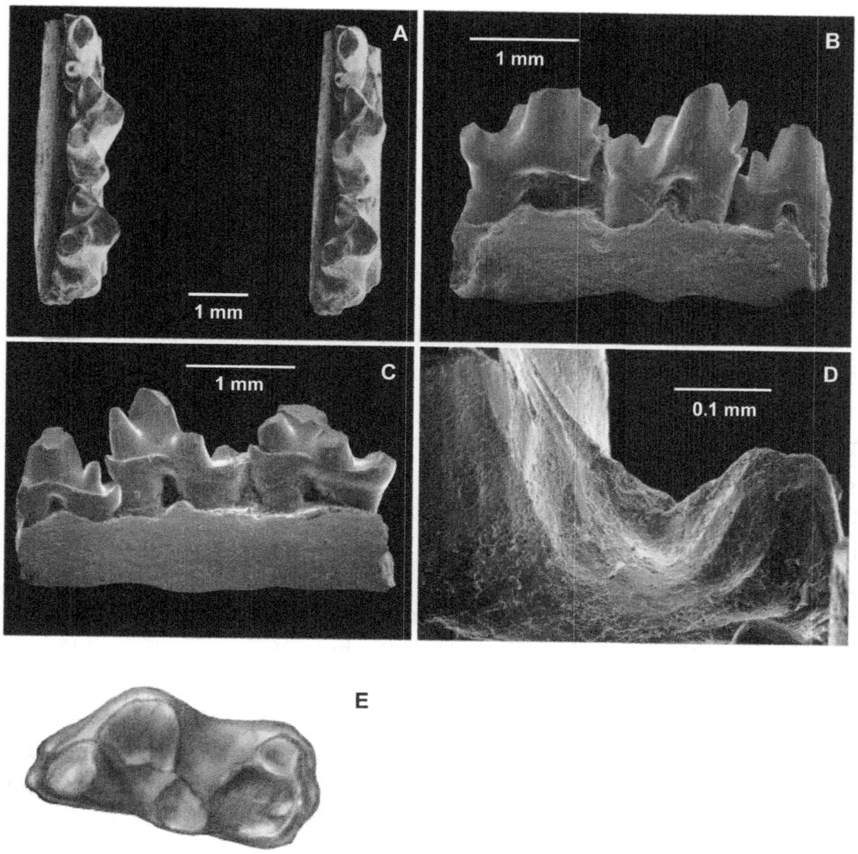

Fig. 6.26: Holotype specimen of *Ambondro mahabo* with ultimate p and m1-2. (A) Occlusal view (stereo-pair), (B) medial view, (C) lingual view, (D) detail of m1 talonid in lingual view. (E) posterior lower molar (m2) in occlusal view (anterior to the left). From Flynn *et al.* (1999), adapted with permission from Springer Nature.

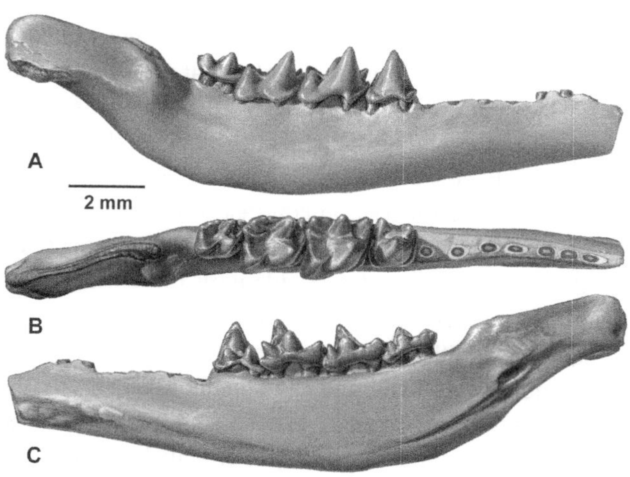

Fig. 6.27: *Ausktribosphenos nyctos*, incomplete right dentary with ultimate premolar and three molars (holotype specimen). (A) Lateral, (B) dorsal, and (C) medial views. Modified after Rich *et al.* (1999), courtesy of Tom Rich.

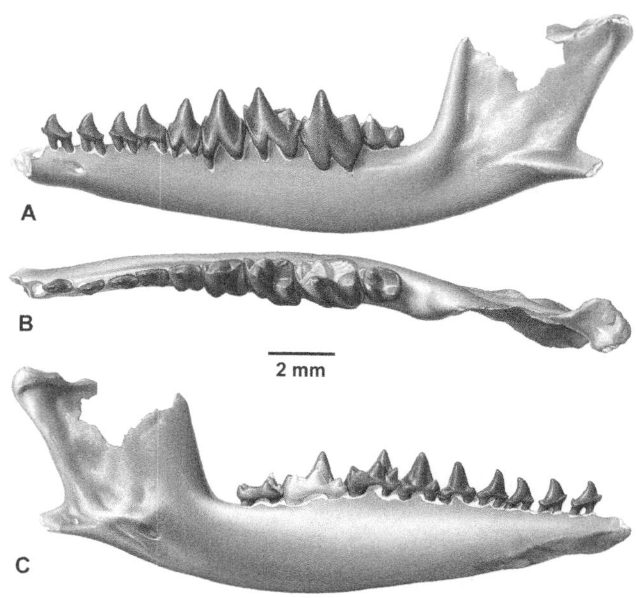

Fig. 6.28: *Bishops whitmorei*, left dentary with nine teeth (probably p1-6 and m1-3). (A) Lateral, (B) dorsal, and (C) medial views. Modified after Rich *et al.* (2001), courtesy of Tom Rich.

Cretacous (Aptian) of the Flat Rocks locality in Victoria, Australia (Rich *et al.* 1997, 1999, 2001). *Ausktribosphenos* and *Bishops* are also characterized by lower molars with fully developed talonids and were assigned to placental erinaceids in their original descriptions (Rich *et al.* 1997, 2001;

see also Woodburne *et al.* 2003). The talonid of the molars of boreosphenidans (see chapter 16) is a basin, which is surrounded by three cusps, the lingual entoconid, the labial hypoconid, and the distal hypoconulid, of which either the hypoconid or the hypoconulid are homologized with distal cuspule d of stem mammals (see chapter 2).

Subsequently, the South American taxa *Asfaltomylos* and *Henosferus* were identified as australosphenidans (Rauhut *et al.* 2002, Rougier *et al.* 2007a). *Asfaltomylos* from the Middle (possibly Early) Jurassic Cañadon Asfalto Formation in Chubut Province of Argentina is represented by a tiny mandible (Fig. 6.29) and was the first Jurassic mammal to be described from South America (Rauhut *et al.* 2002). The mandible has a postdentary trough for accommodation of the middle ear bones and three molars with fully developed talonid basins (Martin and Rauhut 2005). *Henosferus* from the same locality near Cerro Condor is similar to *Asfaltomylos* but about two times larger and has large diastemata between the premolars (Rougier *et al.* 2007a). The dental formula of *Henosferus* is 4i.1c.5p.3m, and, as in *Asfaltomylos*, the molars have fully developed talonid basins with wear facets restricted to the rim of the basins (apical wear). The mandible has a well-developed Meckel's groove and postdentary trough for accommodation of the middle ear bones (Rougier *et al.* 2007a).

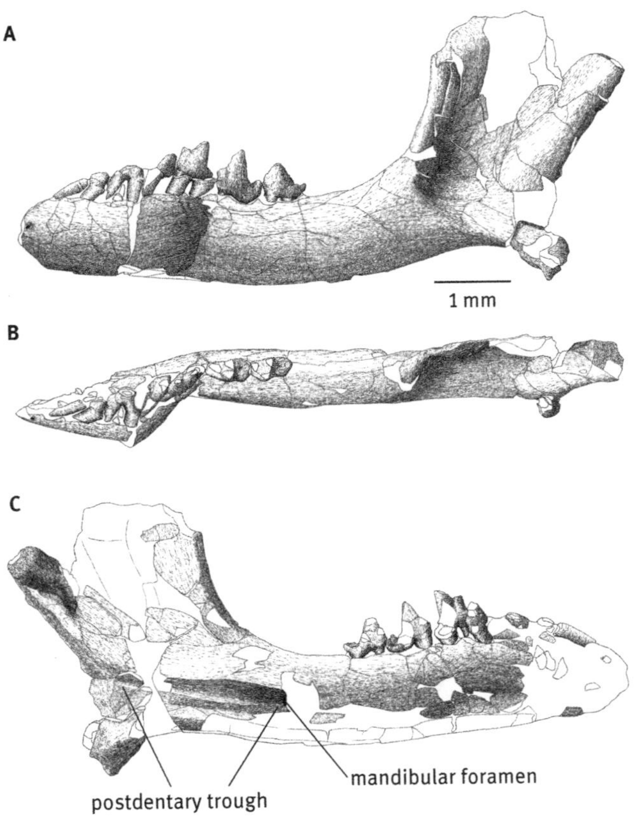

Fig. 6.29: *Asfaltomylos patagonicus*, left mandible (holotype) with p3-5, m1-3. (A) Lateral, (B) dorsal, and (C) medial views. Modified from Martin and Rauhut (2005), reprinted by permission of the Society of Vertebrate Paleontology, www.vertpaleo.org.

Based on the discoveries from Madagascar and Australia, Luo *et al.* (2001b) coined the hypothesis of a dual origin of tribosphenic mammals, independently in Australosphenida on Gondwana and Boreosphenida Luo *et al.* 2001b on the northern continents. Detailed analysis of the molar dentition (Fig. 6.30) suggests the convergent evolution of the tribosphenic pattern. In Australosphenida, the wear facets at the talonid are located on the apices of the talonid cusps, whereas they occur within the talonid basin in boreosphenidan molars, because the protocone of the upper molar occludes into the talonid basin like the pestle into a mortar. In australosphenidans, the upper molar protocone apparently did not completely occlude in the basin, but lingually to the metaconid-hypoconid (Martin and Rauhut 2005). Given this situation, the tribosphenic molars of boreosphenidans and australosphenidans are homoplastic. So far, only lower dentitions of Southern Hemisphere australosphenidans have been described, but unpublished upper australosphenidan molars from the Jurassic of Argentina exhibit a hooklike protocone (Guillermo Rougier, personal communication) that concurs with this occlusional scenario. An autapomorphic character of the australosphenidan lower molars is the presence of a lingual cingulid that wraps onto the mesial side (Luo *et al.* 2001b) (Fig. 6.30). Besides the plesiomorphic postdentary trough for attachment of the middle ear bones, the australosphenidan mandible possesses an elevated angular process, in contrast to the downturned angular process of early boreosphenidans (Luo *et al.* 2001b). The presence of a postdentary trough indicates that in australosphenidans the primary jaw joint was still present, in addition to the secondary jaw joint formed by the dentary condyle and the squamosal. The unique combination of superficially tribosphenic molars with a plesiomorphic mandible clearly distinguishes australosphenidans from the northern continent boreosphenidans.

There are two genera of Northern Hemisphere pseudotribosphenic mammals, *Shuotherium* and *Pseudotribos*, that appear in cladistic analyses as australosphenidans. They are called "pseudotribosphenic" because they have a basinlike structure (talonid) at the mesial side of the lower molars, instead of at the distal side as in boreosphenidans (Chow and Rich 1982, Luo *et al.* 2007a).

The mandible of the enigmatic *Shuotherium dongi* (Fig. 6.31) from the Late Jurassic of Sichuan, China, bears a peculiar dentition with four premolars and three molars of "symmetrodont" shape with a talonid-like structure

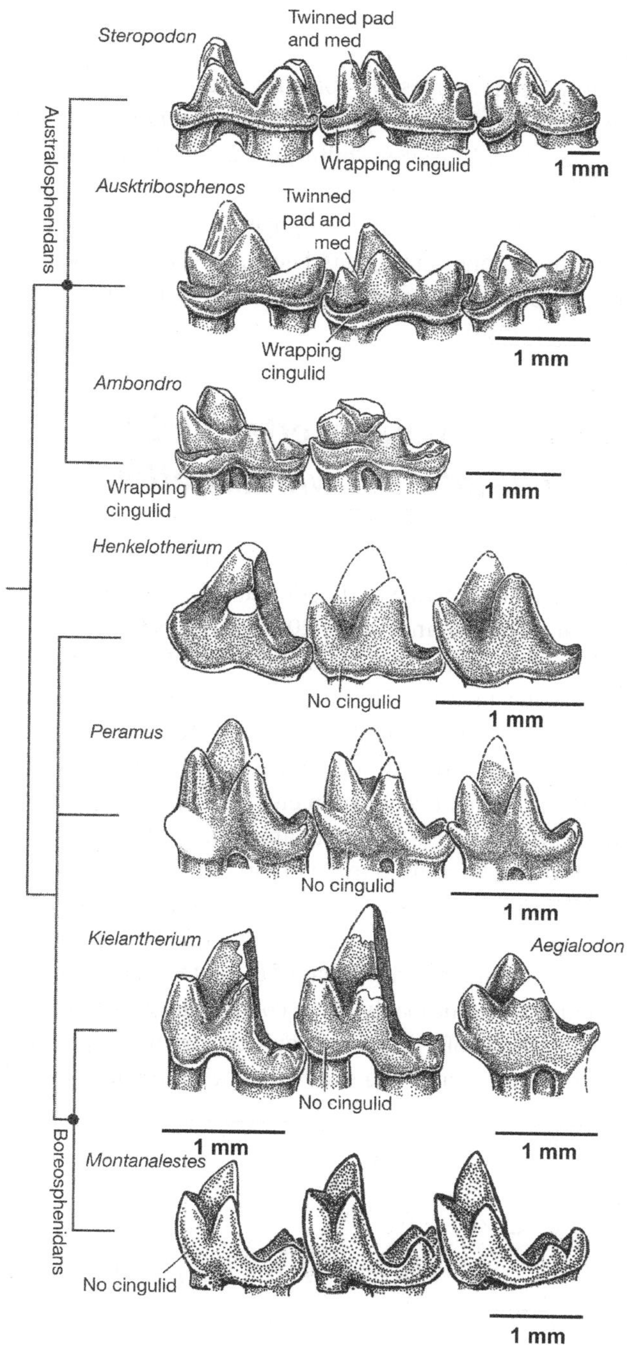

Fig. 6.30: Comparison of molars of australosphenidans, pretribosphenids *Henkelotherium* and *Peramus*, and boreosphenidans, in lingual view. Lower molars of australosphenidans are characterized by a mesial cingulid wrapping around the lingual side of the paraconid (pad), and the reduced height of the trigonid relative to the talonid. From Luo *et al.* (2001a), reprinted with permission from Springer Nature.

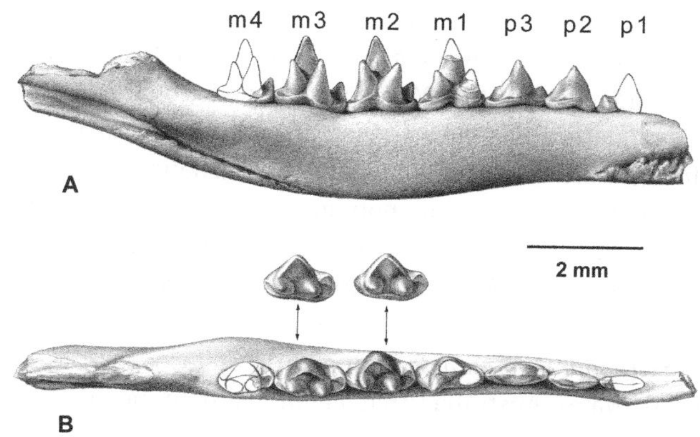

Fig. 6.31: Mandible of *Shuotherium dongi*. (A) Lingual and (B) dorsal view (m2 and m3 separately depicted in exact occlusal view). Modified after Chow and Rich (1982), courtesy of Tom Rich.

("pseudo-talonid") at the mesial side of the molars, in front of the trigonid (Chow and Rich 1982). Because of its peculiar morphology, Chow and Rich (1982) assigned the mandible to a new higher taxon, Yinotheria. Isolated upper and lower molars of *Shuotherium* have been described from the Middle Jurassic (Bathonian) of England and were assigned to "Symmetrodonta" (Sigogneau-Russell 1998). From the type locality of *S. dongi*, Wang *et al.* (1998) described an isolated upper molar of *Shuotherium shilongi*, which is too large to belong to *S. dongi*. In their original description, Chow and Rich (1982) regarded Yinotheria as sister taxon of Cladotheria McKenna 1975. Kermack *et al.* (1987) proposed an assignment to Docodonta because of the mesial pseudo-talonid, which also occurs in some docodontans. The cladistic analysis of Luo *et al.* (2002) placed *Shuotherium* as sister taxon of Australosphenida.

Pseudotribos from the Middle Jurassic Jiulongshan Formation of Daohugou in Inner Mongolia, China is represented by an incomplete skeleton with fossorial adaptations. These are a massive shoulder girdle with the lateral processes of interclavicle and clavicles forming an expanded area for attachment of the sternocleidomastoid muscle, as well as expanded deltopectoral crest, teres major tubercle, and expanded distal end of the humerus (Luo *et al.* 2007). *Pseudotribos* has a reversed tribosphenic molar pattern with mesially placed "pseudo-talonids" and "pseudo-protocones" preserved in occlusion, demonstrating a pestle-to-mortar crushing function. At the mandible a postdentary trough is present, and shallow Meckel´s groove extends anteriorly to below M1. The shoulder girdle of *Pseudotribos* exhibits plesiomorphies for crown Mammalia such as a large interclavicle with posterior club-foot and expanded end of the manubrium. These plesiomorphic structures differ from the pivotal claviculo-interclavicle joint and mobile shoulder girdle of theriimorphs and indicate that *Pseudotribos* and shuotheriids are early diverging mammals not closely related to "symmetrodontans" or to the clade of *Peramus* and more derived lineages. According to the cladistic analysis by Luo *et al.* (2007) they are more closely related to australosphenidans than other Mesozoic mammal groups.

Modern adult monotremes are toothless, but tooth *Anlagen* are developed in early ontogenetic stages of *Ornithorhynchus*, and monotremes with persistent teeth are known in the fossil record (Archer *et al.* 1985, 1992, 1993). The first Mesozoic monotreme to be described was an opalized mandible fragment with three molars of *Steropodon galmani* from the Early Cretaceous of New South Wales, Australia (Archer *et al.* 1985). In the original description of the dentition, the close resemblance to tribosphenic molars was noticed and the same dental terminology was applied, but Kielan-Jaworowska *et al.* (1987) stated that the dentition was not fully tribosphenic. Later additional toothed monotremes were discovered, such as *Teinolophos* (Rich *et al.* 1999, 2016) and *Kollikodon*, a monotreme or close monotreme relative with four bunodont molars (Flannery *et al.* 1995, Pian *et al.* 2016). The only monotreme described from outside Australia is *Monotrematum sudamericanum*, represented by several isolated teeth (Pascual *et al.* 1992a, b, 2002) and two partial femora (Forasiepi and Martinelli 2003) from the early Paleocene of Patagonia (Argentina). *Monotrematum* has been assigned to ornithorhynchids. Paleogene and Neogene toothed ornithorhynchids are *Obdurodon insignis*, based on two teeth of Late Oligocene age in southern Australia (Woodburne and Tedford 1975, Archer *et al.* 1992); *Obdurodon dicksoni*, based on a skull (Archer *et al.* 1992, 1993, Musser and Archer 1998); and the giant *Obdurodon tharalkooschild*, based on a lower first molar (Pian *et al.* 2013), both from the Middle Miocene of Riversleigh, Queensland.

In its original description, *Teinlophos* was regarded as a stem monotreme, diverging before the split between ornithorhynchids and tachyglossids (Rich *et al.* 1999). In the holotype specimen, a postdentary trough for accommodation of the middle ear bones was identified (Rich *et al.* 2005). The presence of a postdentary trough in *Teinolophos* corroborates a separate origin of the definitive mammalian middle ear with three ear ossicles in monotremes, independently from therian mammals (Rich *et al.* 2005, Martin and Luo 2005). Rowe *et al.* (2008) reanalyzed several mandible specimens of early Cretaceous *Teinolophos* by µCT and assigned this taxon based on a cladistic analysis to the monotreme crown group, as the oldest known representative of the Ornithorhynchidae (platypuses). The slightly younger *Steropodon* was also assigned to Ornithorhynchidae Gray 1825. Rowe *et al.* (2008) and Rich *et al.* (2016) rejected the presence of a postdentary trough in *Teinolophos* but recognized an inflated mandibular canal instead, a shared character with the ornithorhynchids. As a consequence, they regarded Australosphenida as a polyphyletic assemblage of crown monotremes (*Teinolophos* and *Steropodon*) and Theriiformes (*Ausktribosphenos*, *Bishops*, *Ambondro*, and *Asfaltomylos*). However, Phillips *et al.* (2009) argued that *Teinolophos* and *Steropodon* are stem monotremes, not crown monotremes, and monophyly of Australosphenida (including monotremes) has been recovered in most, if not all, phylogenetic analyses published recently (e.g., Pian *et al.* 2016).

6.8 Mammalia incertae sedis: *Fruitafossor windscheffeli*

Fruitafossor windscheffeli is a mammal of uncertain systematic position with striking dental and postcranial specializations (Luo and Wible 2005) (Figs. 6.32 and 6.33). In the cladogram of Luo and Wible (2005), *Fruitafossor* is the sister taxon of Theriimorpha Rowe 1993 (common ancestor of Eutriconodonta and crown Theria; Luo 2007b, node 3 in Fig. 1a) within Mammalia. *Fruitafossor* is represented by relatively complete mandible, incomplete cranium, and half of the postcranium from the Late Jurassic Morrison Formation in Colorado (USA). It differs from all other known Mesozoic mammaliaforms in its tubular and single-rooted molariforms with open-ended roots. These teeth are similar to that of the extant aardvark and armadillos and probably had a continuous growth in life. The forelimbs of *Fruitafossor* exhibit strong adaptations for scratch digging. The saddle-shaped scapular glenoid

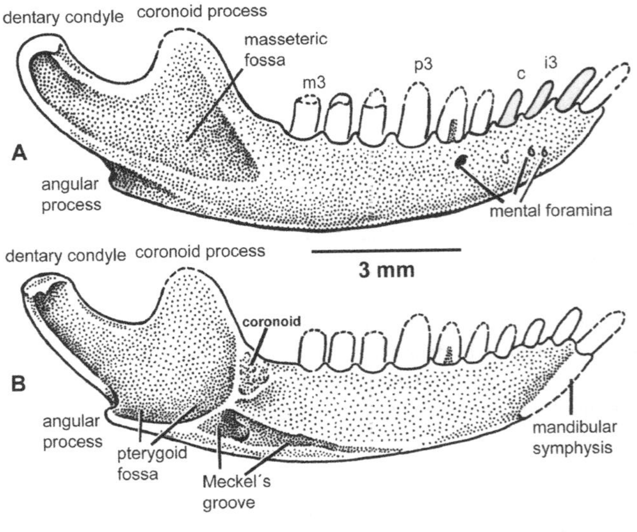

Fig. 6.32: *Fruitafossor windscheffeli*, restoration of left mandible in lateral (A) and medial (B) views. Modified after Luo and Wible (2005), reprinted with permission from AAAS.

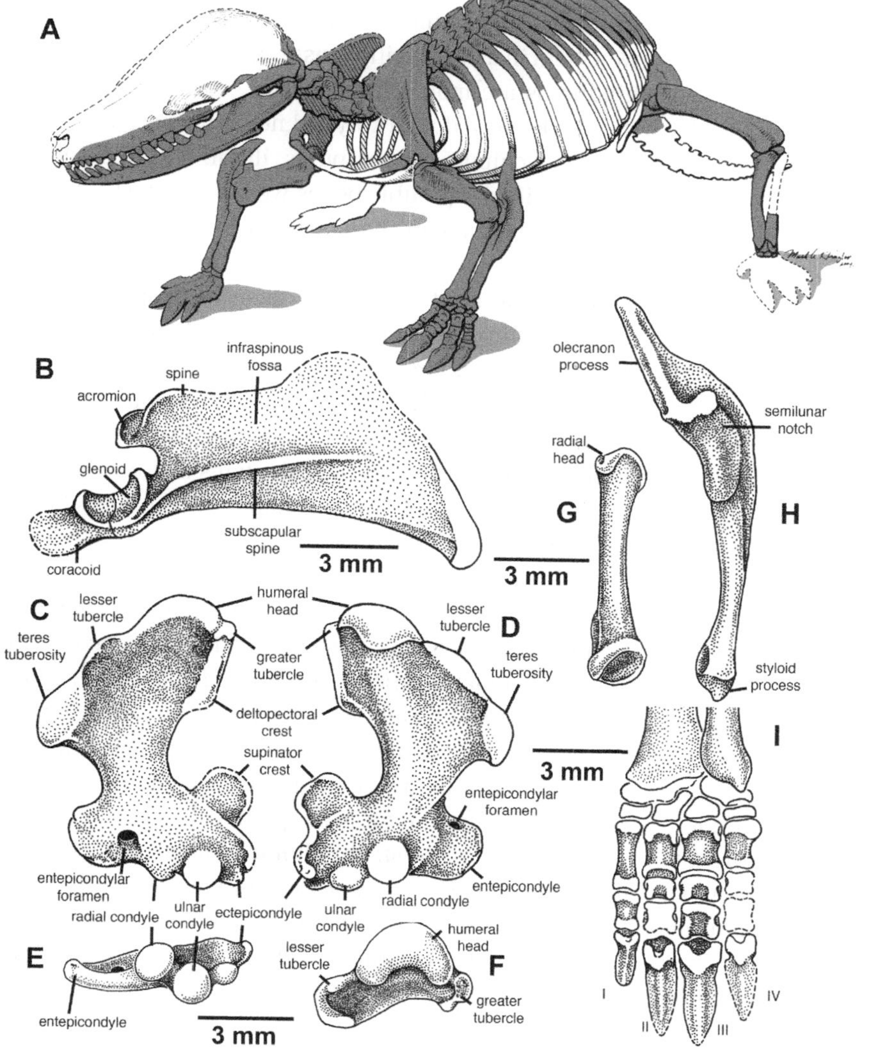

Fig. 6.33: *Fruitafossor windscheffeli*. (A) Restoration of skeleton; (B) left scapula in ventrolateral view; (C–F) left humerus in ventral (C), dorsal (D), distal (E), and anteromedial (F) views; (G) left radius in posteromedial view; (H) left ulna in anterior view; (I) plantar view of left manus. Modified after Luo and Wible (2005), reprinted with permission from AAAS.

is formed by the scapula and a separate coracoid, suggesting that the range of mobility of the shoulder joint was similar to that of monotremes (Luo and Wible 2005). The infraspinous fossa is large and an incipient small supraspinous fossa is present on the cranial border of the scapula. The scapula is similar to that of monotremes, *Morganucodon*, and *Haldanodon*. The humerus lacks a distinctive humeral head but exhibits a number of characters typical for fossorial mammals such as a large deltopectoral crest, hypertrophied tuberosity for insertion of a large teres muscle, and very wide distal portion with prominent epicondyles. The manus has only four digits, and the carpals are all proximodistally shortened. The terminal phalanges are the longest and are distally dorsoventrally flattened. The lumbar vertebrae of *Fruitafossor* have accessorial articulations, which arose independently from those of Xenarthra. The presence of a broad Meckel's groove indicates that the middle ear bones were still connected with the dentary. According to the parsimony analysis of Luo and Wible (2005), *Fruitafossor* is not closely related to eutherians or placental xenarthrans. *Fruitafossor* is a vivid example of unexpected ecomorphological specialization within Mesozoic mammals.

6.9 Eutriconodonta

Eutriconodonta Kermack *et al.* 1973 (Fig. 6.1) is characterized by the triconodont molariform pattern, with three laterally compressed main cusps arranged in a row. This distinctive character was recognized early on and led to the introduction of a formal clade "Triconodonta" (Marsh 1887, Osborn 1888b, c). Later the concept of Triconodonta was expanded, and the newly discovered Late Triassic/Early Jurassic morganucodontans (Parrington 1941, 1947, Kühne 1949, 1958) were

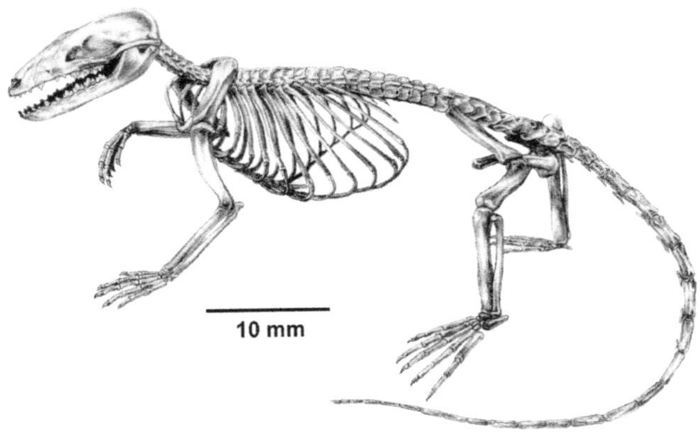

Fig. 6.34: *Jeholodens jenkinsi*, skeletal restoration. From Ji *et al.* (1999), reprinted with permission from Springer Nature.

incorporated into Triconodonta. However, soon considerable morphological disparities within the expanded Triconodonta became evident. Kermack *et al.* (1973) divided Triconodonta into two suborders, the Eutriconodonta for the taxa treated by Simpson (1928, 1929) and the Morganucodonta for the Late Triassic/Early Jurassic taxa. It is now generally accepted that Triconodonta is a polyphyletic assemblage characterized by the symplesiomorphic triconodont molariform pattern. Eutriconodonta has the middle ear bones fully detached from the mandible and is nested within crown Mammalia, whereas Morganucodonta represents mammaliaforms outside the mammalian crown group.

Eutriconodontans differ in a number of apomorphic characters from morganucodontans and other stem mammals. They have a well-developed pterygoid fossa on the lingual side of the dentary, and the dentary lacks an angular process. The postdentary trough is lost, which indicates that the middle ear bones were largely detached from the mandible and hence that only the secondary jaw joint was present. In the eutriconodontan *Yanoconodon*, the postdentary bones were still attached to the dentary via Meckel's cartilage (Luo *et al.* 2007a). The skeleton of *Jeholodens jenkinsi* (Fig. 6.34) was the first eutriconodontan postcranium to be described (Ji *et al.* 1999). The shoulder girdle largely resembles that of therians, except for the presence of an interclavicle. The scapula is plate-like and has a supraspinous and an infraspinous fossa. The scapula spine ends in a large acromion. The coracoid is small and fused with the scapula, and a procoracoid is not present. The glenoid facet is concave and not saddle shaped as in stem mammals. *Jeholodens* has a slender, weakly bent scapula and a small interclavicle. The forelimbs of *Jeholodens* are morphologically intermediate between those of morganucodontans and therians in terms of posture. The humerus has a plesiomorphic proximodistal torsion, weakly developed ent- and ectepicondyles, and an incipient ulnar trochlea. The forelimbs were held in a parasagittal posture. The pelvic girdle and hindlimbs are more plesiomorphic. The acetabulum of the pelvis has a dorsal cotyloid notch between the ilium and the ischium, and the pubis, ilium, and ischium are not coossified. Epipubic bones were present as a plesiomorphic mammalian character. The femoral head is dorsomedially oriented, without a femoral neck, and lacks a fovea. The hind limbs have been reconstructed in a semisprawling posture in *Jeholodens*, whereas the more derived pectoral girdle indicates a parasagittal arrangement of the forelimbs (Ji *et al.* 1999). The mobile shoulder girdle with well-demarcated triangular fossa that forms a large area for attachment of the teres major muscle (Ji *et al.* 1999) suggests arboreal habits (Chen and Wilson 2015). *Yanoconodon* is interpreted as a generalized terrestrial mammal with the capability of swimming (Chen *et al.* 2017). In their multivariate analysis of locomotor adaptations, Chen and Wilson (2015) found *Liaoconodon* to be semiaquatic, whereas *Gobiconodon ostromi* exhibits ground-dwelling and *Repenomamus* burrowing adaptations. *Volaticotherium* possessed a patagium for gliding (Meng *et al.* 2006), which impressively demonstrates the wide range of locomotorial adaptations in eutriconodontans. In *Yanoconodon*, the thoracolumbar transition is gradational, whereas *Jeholodens* (Chen *et al.* 2017) and *Chaoyangodens* (Hou and Meng 2014) lack lumbar ribs and have a very distinctive thoracolumbar transition (Chen *et al.* 2017).

The dentitions of eutriconodontans, with large canines and mainly trenchant molariforms, suggest a faunivorous diet of insects and invertebrates in the smaller taxa, and vertebrates in the larger taxa. It has also been suggested that the larger taxa such as *Repenomamus* and *Gobiconodon* were scavengers. For taxa with recurved molariform cusps (resembling the molars of seals), such as *Ichthyoconodon* (Sigogneau-Russell 1995), a piscivorous diet was suggested (Slaughter 1969). However, Jenkins and Crompton (1979) pointed out that such similarities can be misleading because the molariform function was mainly shearing in eutriconodontans, and it is grasping and piercing in seals. Molariforms of the *Ichthyoconodon*-type have been found in the maxilla and mandible of *Volaticotherium* from the Middle Jurassic of China (Meng *et al.* 2006). Originally, *Volaticotherium* was placed outside of Eutriconodonta, within stem mammals (Meng *et al.* 2006), but in more recent cladistic analyses it clusters as a member of Alticonodontinae within Eutriconodonta (Gaetano and Rougier 2011, 2012, Martin *et al.* 2015) (Fig. 6.35), which had already been suggested by Luo (2007a). *Repenomamus giganticus* is the largest Mesozoic mammal that is known

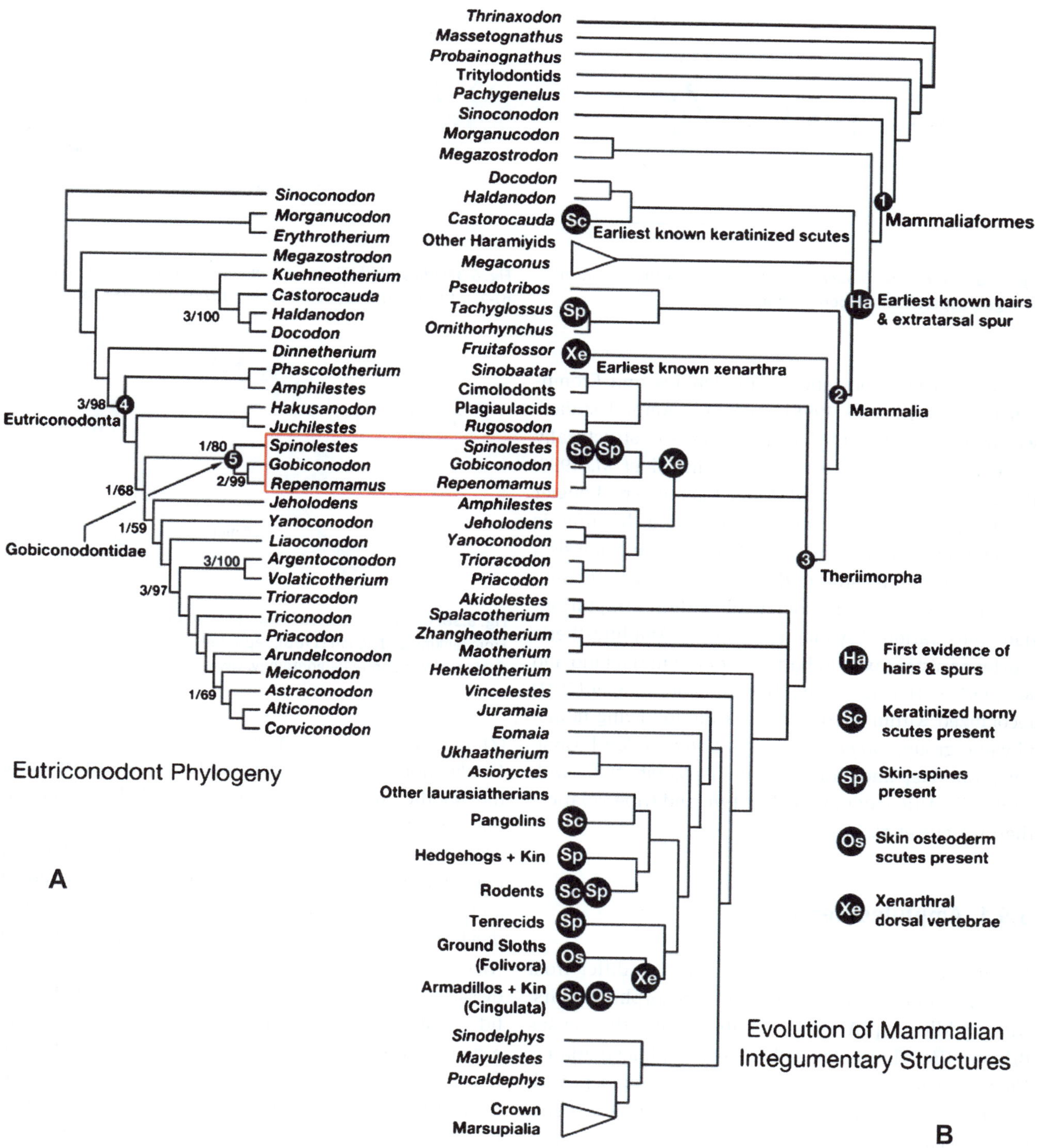

Fig. 6.35: Eutriconodont phylogeny (A) and evolution of mammalian integumentary structures (B). From Martin *et al.* (2015).

and certainly belonged to the top predators in its paleocommunity. Hu *et al.* (2005b) reported a *Repenomamus* specimen with a small dinosaur in its stomach. At the other extreme, the body mass of *Jeholodens* is estimated at less than 30 g, which demonstrates that eutriconodontans spanned a remarkable size range.

The mode of tooth replacement is variable in eutriconodontans. In *Triconodon*, a diphyodont replacement of antemolars has been observed, whereas in *Repenomamus* a single replacement of molariforms and multiple replacement of incisors and canines occurs (Kielan-Jaworowska *et al.* 2004).

Within Eutriconodonta, two groups have been traditionally recognized, Triconodontidae Marsh 1887 and "Amphilestidae" (Osborn 1888c; Simpson 1928, 1929). Subsequently, it became evident that Triconodontidae

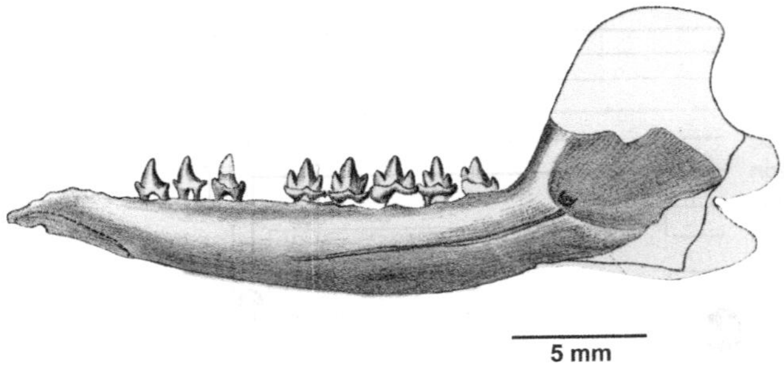

Fig. 6.36: *Amphilestes broderipii*, right mandible with p1-3 and m1-5 in lingual aspect. From Osborn (1888b).

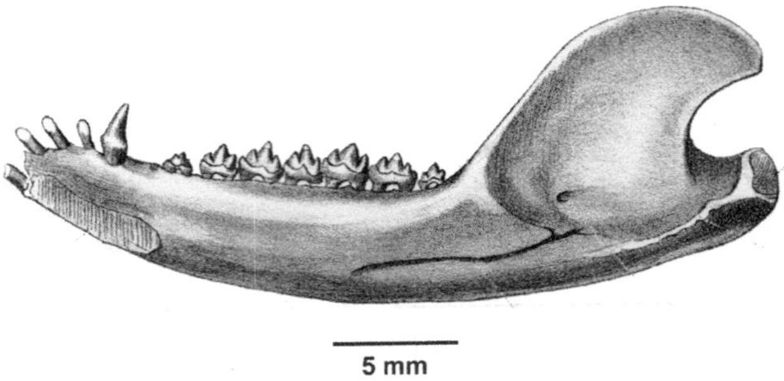

Fig. 6.37: *Phascolotherium bucklandi*, right mandible with i1-4, c, p1-2, m1-5 in medial aspect. Modified after Osborn (1888b).

is indeed a monophyletic group (Jenkins and Crompton 1979, Cifelli *et al.* 1998), whereas '"Amphilestidae" was treated as a paraphyletic assemblage of structurally more plesiomorphic eutriconodontans (Ji *et al.* 1999, Rougier *et al.* 1999, Luo *et al.* 2002). Gaetano and Rougier (2011) described a new triconodontid from the Jurassic of Argentina and recognized in their cladistic analysis a new clade Amphilestheria that comprises the "triconodontan" taxa formerly combined within "Amphilestidae" plus various "symmetrodontans". Amphilestheria cluster as sister taxon to Eutriconodonta in Gaetano and Rougier's (2011) cladogram. Gaetano and Rougier (2011) recognized within Eutriconodonta the following monophyletic groups: *Repenomamus* + *Gobiconodon* [=Gobiconodontidae] and Triconodontidae Marsh 1887, with subfamily Alticonodontinae Fox 1976 and tribe Volaticotherini Meng *et al.* 2006.

6.9.1 Amphilestheria

Amphilestheria Gaetano and Rougier 2011 differ from eutriconodontans in their more plesiomorphic molars, with middle cusp A/a considerably higher than cusps B/b and C/c. The geologically oldest amphilestherian is *Huasteconodon*, represented by a maxilla fragment with two molariforms from the late Early Jurassic (Pliensbachian) La Boca Formation of Mexico (Montellano *et al.* 2008). The next oldest record is from the Middle Jurassic (Bathonian) Stonesfield slate in Britain. Among these are the first Mesozoic mammal to be discovered (*Amphilestes broderipii*) (Fig. 6.36) and the first Mesozoic mammal species to be named (*Phascolotherium bucklandi*, Fig. 6.37) (Broderip 1828). Slightly younger (Callovian) is a molariform of the amphilestherian *Ferganodon* from the Balabansai Formation in Kyrgyzstan (Martin and Averianov 2007). Another Middle Jurassic occurrence is the dentary of the amphilestherian *Liaotherium* from the Haifanggou Formation in Liaoning, northeastern China

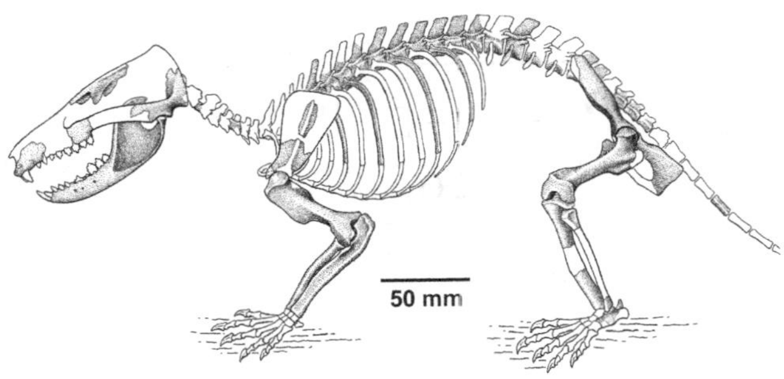

Fig. 6.38: Composite skeletal restoration of *Gobiconodon ostromi*. From Jenkins and Schaff (1988), reprinted by permission of the Society of Vertebrate Paleontology, www.vertpaleo.org.

(Zhou *et al.* 1991). Gaetano and Rougier (2012) described *Condorodon* based on an isolated lower molariform from the Middle Jurassic of the Queso Rallado locality in Chubut Province (Patagonia, Argentina) that clustered as sister taxon of *Tendagurodon* in their cladistic analysis. *Condorodon* is one of the oldest amphilestherians known and the oldest representative from Gondwana. *Tendagurodon* from the Late Jurassic Tendaguru locality in Tanzania (Heinrich 1998) was tentatively placed by Kielan-Jaworowska *et al.* (2004) among eutriconodontans of "amphilestid" grade. *Aploconodon* and *Comodon* are each known by a dentary fragment with the last two, respectively, last four molariforms from the Late Jurassic (late Kimmeridgian-early Tithonian) of Quarry 9 in the Morrison Formation of Wyoming (USA) (Simpson 1925b, c). *Triconolestes* from the Morrison Formation of Utah is poorly known and represented by a lower molariform fragment (Engelmann and Callison 1998). The recent discoveries in the Late Jurassic-Early Cretaceous of northeastern China have provided a number of complete eutriconodontan skeletons. Based on the skeletons of *Jeholodens jenkinsi* from the Barremian Yixian Formation of Liaoning, China (Ji *et al.* 1999), and *Yanoconodon allini* from the same formation in Hubei Province, China, the new eutriconodontan family Jeholodentidae Luo *et al.* 2007b was erected. In the cladistic analyses by Gaetano and Rougier (2011, 2012),

Jeholodens clusters within Amphilestheria. *Juchilestes* from the Barremian Yixian Formation is represented by a partial skull with mandible, and it is the first representative of the former "amphilestids" in which the upper dentition is known (Gao *et al.* 2010). It has a dental formula of 4I.1C.3p.5m/4i.1c.3p.6m. *Paikasigudodon* and *Dyskritodon indicus*, based on single teeth from the late Middle Jurassic to Lower Cretaceous Kota Formation of India (Prasad and Manhas 2002), are highly questionable in terms of stratigraphical age and referral to Amphilestheria.

6.9.2 Gobiconodontidae

Gobiconodontidae Jenkins and Schaff 1988 is a well-supported monophyletic group near the base of Eutriconodonta (Gaetano and Rougier 2011, 2012, Martin *et al.* 2015). Six genera of Gobiconodontidae have been described: *Hangjinia*, of which a single dentary is known (Godefroit and Guo 1999); *Gobiconodon* and *Repenomamus*, both of which are represented by skulls and postcranial material (Li *et al.* 2001, Wang *et al.* 2001); *Meemannodon* from northeastern China (Meng *et al.* 2005); *Acinacodus* from Siberia (Lopatin *et al.* 2010); and *Spinolestes* from Spain (Martin *et al.* 2015). *Gobiconodon* (Fig. 6.38) is remarkable because it has a wide geographic and stratigraphic distribution. It was first described from the Aptian-Albian of Mongolia (Trofimov 1978, Kielan-Jaworowska and Dashzeveg 1998), but *Gobiconodon* is now known from the Valanginian of Mongolia (Rougier *et al.* 2001), Barremian of China (Wang *et al.* 2001), Spain (Cuenca-Bescós and Canudo 1999), and Britain (Sweetman 2006, Butler and Sigogneau-Russell 2016), the Albian of Russia (Maschenko and Lopatin 1998) and China's Gansu Province (Tang *et al.* 2001), and the Aptian-Albian of Montana (Jenkins and Schaff 1988). It is well represented by skulls and most of the postcranium. *Gobiconodon* species vary greatly in size and range from small, insectivorous taxa to large, probably carnivorous representatives (e.g., *Gobiconodon ostromi*). *Repenomamus giganticus* was a carnivore and the largest Mesozoic mammal known that reached a body mass of about 10 kg (Hu *et al.* 2005b). *Meemannodon* is represented by a left mandible with two incisors, canine, two premolars, and four molars from the late Early Cretaceous (Aptian) of the Yixian Formation at Lujiatum village, Beipiao, China (Meng *et al.* 2005). *Acinacodus* is represented by a mandible with preserved canine, two premolars, and four molariforms from the Early Cretaceous Shestakovo locality in western Siberia (Lopatin *et al.* 2010). A gobiconodontid skeleton with spectacular preservation of integument and inner organs, *Spinolestes xenarthrosus*, has recently been described from the Early Cretaceous (Barremian) of the Las Hoyas Fossillagerstätte in eastern central Spain (Martin *et al.* 2015).

6.9.3 *Spinolestes xenarthrosus*

Spinolestes xenarthrosus (Fig. 6.39) provides exceptional insights into early mammalian integument and soft part evolution (Martin *et al.* 2015) (Fig. 6.40). A dense mane of long guard hairs 3–5 mm long covers the neck region (Fig. 6.40B). Along the dorsal region, there are long and fine hairs, forming a median crest, and most of the tail is also covered by this type of hair. The other parts of the body are covered in a short and soft underfur (Fig. 6.40J). Some pieces of skin are preserved with microstructural details of hair (Fig. 6.40G), which is unique for Mesozoic mammals. There are discernible primary and smaller secondary hairs (Fig. 6.40C, D, I), most probably the result of phosphatized preservation. In some parts of the body, the hairs are truncated and have dark-colored distal ends. They resemble the so-called block hairs, which are caused by a fungal infection (dermatophytosis) that is widespread among modern mammals. On its lower back, *Spinolestes* has small spines with a diameter of 80 to 130 μm (Fig. 6.40E, F). These protospines are formed by longitudinally fused tubules that are derived from primary and secondary hairs, similar to extant mammals. In addition to the protospines, a number of oval dermal scutes up to 4 mm long are present in the lower back region (Fig. 6.40H). *Spinolestes* demonstrates that integument differentiation (different types of hair, spines, and horny scutes) in mammaliaforms was already present in the Mesozoic. The evolution of spines has occurred several times independently in mammalian history, namely, in eutriconodontans (*Spinolestes*), hedgehogs, spiny mice, porcupines, and others. *Spinolestes* is a further example that similar ecomorphological adaptational types evolved several times independently in mammalian history (in Mesozoic stem lineage representatives of modern mammals, in monotremes (e.g., tachyglossids), marsupials, and placental mammals). The selective pressure "generated" a certain array of adaptational types independently at various evolutionary levels in the mammalian tree. *Spinolestes* raises the question of whether integumental differentiation was especially diverse in eutriconodontans, or if it was common in Mesozoic mammals in general.

Besides integumentary stuctures, the *Spinolestes* specimen also shows a remarkable preservation of inner

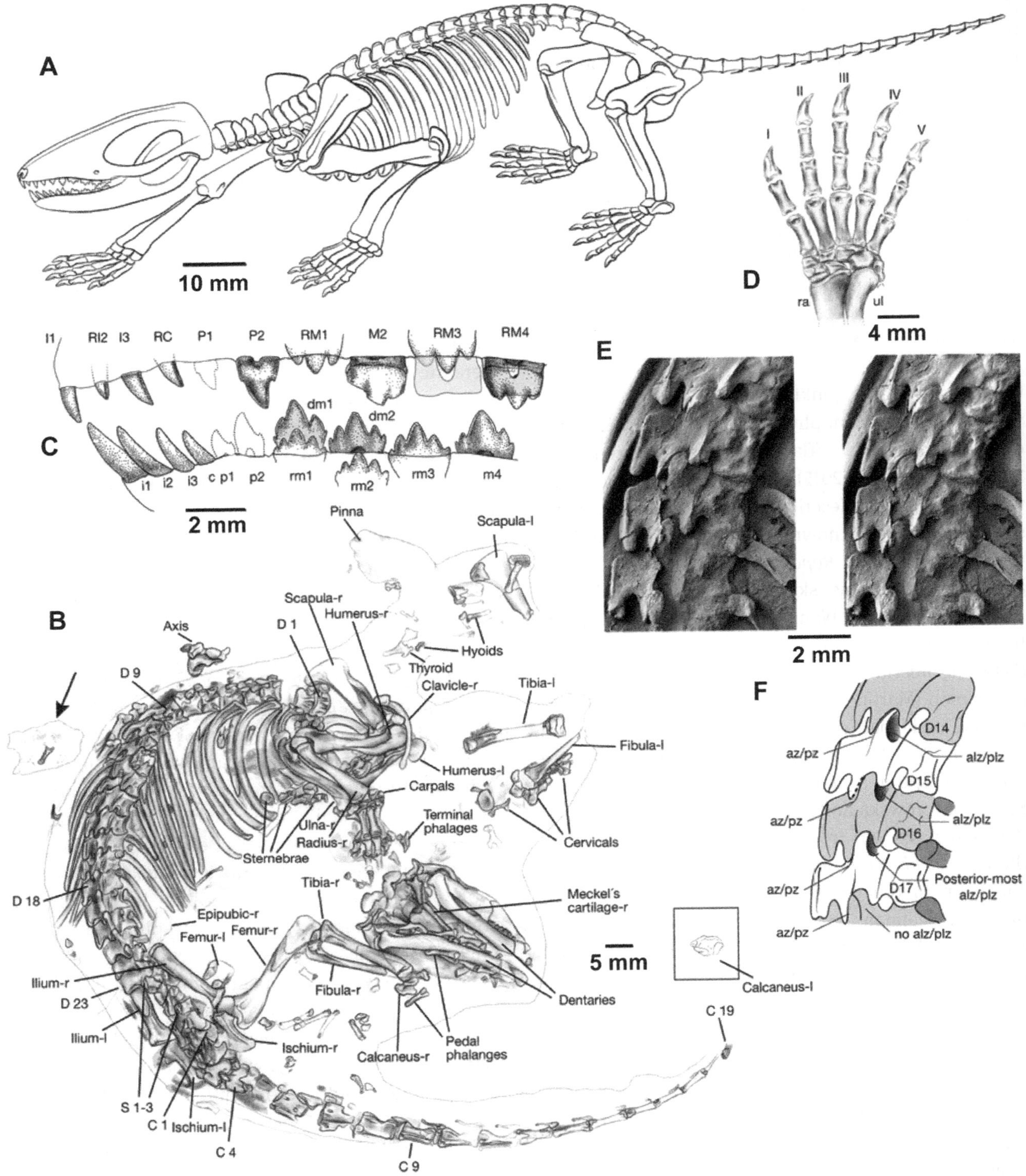

Fig. 6.39: *Spinolestes xenarthrosus*. (A) Skeletal restoration; (B) holotype specimen in original position; (C) restoration of dentition; (D) restoration of manual skeleton; (E) SEM photographs of dorsal vertebrae D14-D17; (F) drawing of D14-D17 with xenarthral articulations. RM, replacing molariform; RI, replacing incisor. From Martin *et al.* (2015).

organs. In the thoracal region, a patch of lung tissue with tubular structures and a branching pattern is preserved. The structures of this fossilized tissue closely resemble the bronchioles of extant small mammals. Right behind the fossilized lung tissue, an area of reddish-brown soft tissue represents residues of the liver (Fig. 6.40A). Because liver tissue is rich in iron, it produces brownish colors when fossilized, as has been described from the late Early

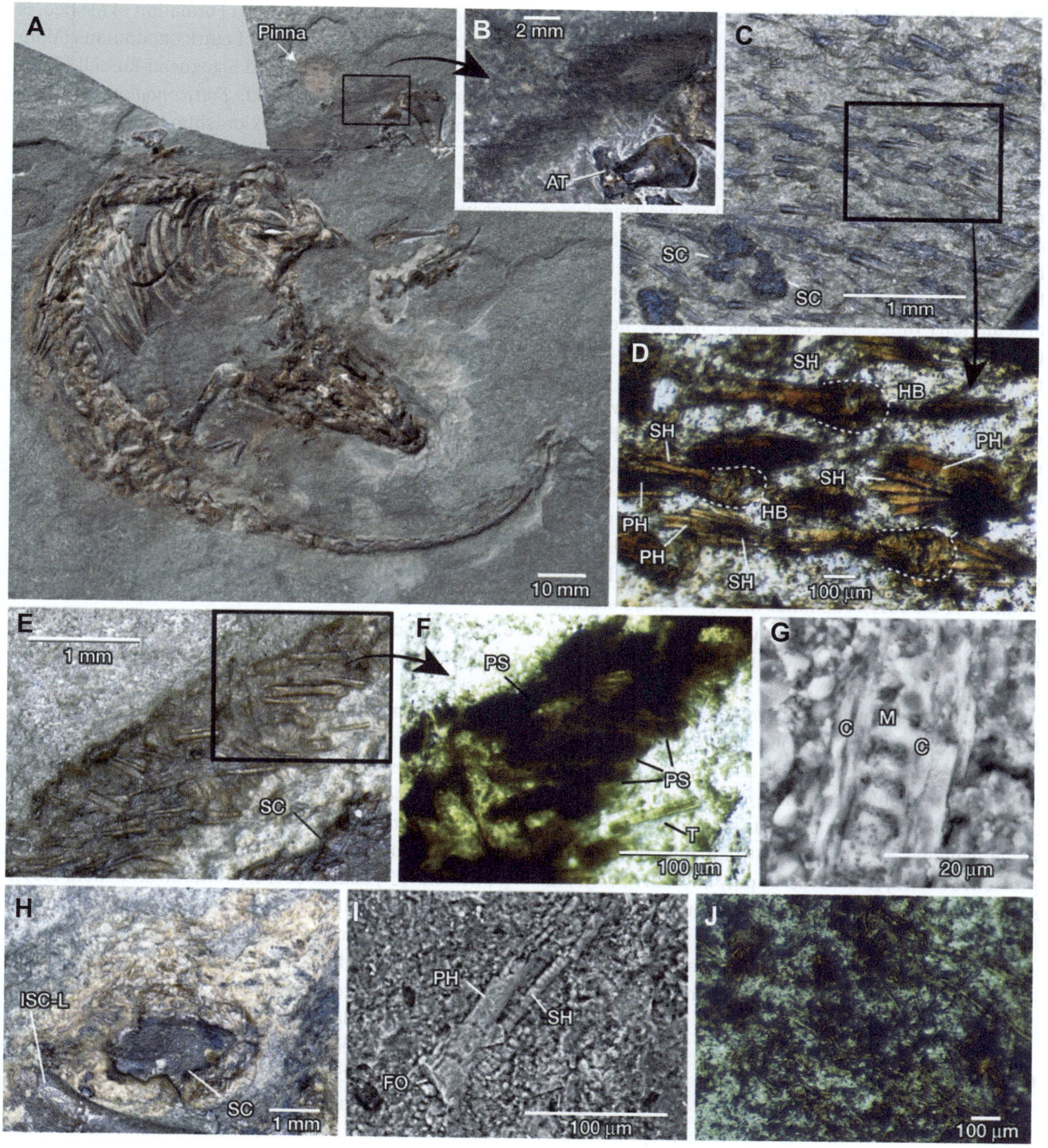

Fig. 6.40: *Spinolestes xenarthrosus*, holotype counter slab, and integumentary structures. (A) Holotype counter slab on rock matrix; (B) guard hair of dense fur in the cervical region (At, atlas); (C) hairs and keratinous dermal scutes (SC); (D) compound hair follicles with hair bulbs (HB), primary hairs (PH), and secondary hairs (SH); (E) area with protospines and dermal scute (SC); (F) protospines (PS) and isolated tubules (T) under translucent light; (G) SEM image of a longitudinal section of a primary hair shaft with cortex (C) and medulla of discontinuous type (M); (H) oval keratinous dermal scute (SC) dorsally of left ischium (ISC-L) with tubules and the homogeneous matrix; (I) SEM image of compound follicle (FO) with primary hair of mosaic cuticular pattern and secondary hairs with coronary cuticular pattern; (J) underfur hairs with annular cuticular pattern in the abdominal region under translucent light. From Martin *et al.* (2015).

Cretaceous small theropod dinosaur *Scipionyx* from Italy (Dal Sasso and Signore 1998). The border between lung and liver tissue in *Spinolestes* is exactly in the position where the muscular diaphragm is located in extant mammals. This is evidence that the complex mammalian respiratory apparatus was already well established in the Early Cretaceous.

6.9.4 Lifestyle of *Spinolestes*

According to body mass estimates based on long bone measurements (Campione and Evans 2012), the weight of *Spinolestes* was between 52 and 72 g. It had a head-body length of 130 mm and a tail length of 110 mm (complete length of *Spinolestes* about 240 mm). This is similar to the modern marsupial *Monodelphis* and makes *Spinolestes* a medium-sized Early Cretaceous mammal. *Spinolestes* has a very strong spine, with additional articulations between the vertebrae (Fig. 6.39E, F). These additional articulations differ somewhat from that of Xenarthra (sloths, armadillos, and anteaters) in that they lack the typical metapophysis and the ventral part of the two accessorial articulation facets but still give the vertebral column an extraordinary strength. A similar phenomenon is known from the living armored shrews (*Scutisorex*). Armored shrews have strongly fused vertebrae, and they use their strong vertebral column for forcing away the bases of palm leaves from the trunk of palm trees, in order to recover insect larvae hidden underneath (Stanley *et al.* 2013). A similar lifestyle can be assumed for *Spinolestes*. *Spinolestes* lived in a swampy environment, where it searched for insects and other small animals in the undergrowth and used its strong vertebral column for forcing apart bark and other plant material for finding larvae. It also had strong anterior legs, which were used for scratch digging. With its spiny fur, *Spinolestes* (Fig. 6.41) resembles a modern spiny mouse (*Acomys*), although it is not closely related to spiny mice or any other living mammal. The spines most probably functioned as protective device, as in modern spiny mice. In *Acomys*, the spines are easily shed, and if a spiny mouse is bitten in the back, the predator is left with a mouth full of spines and *Acomys* can escape. The spines regrow later easily, and the animal is not seriously hurt (Montandon *et al.* 2014).

6.9.5 Triconodontidae

The stem-triconodontid *Victoriaconodon*, represented by an anterior dentary fragment with p2-m1 from the late Early Jurassic (Pliensbachian) La Boca Formation of Mexico, is the geologically oldest record of eutriconodontans (Montellano *et al.* 2008). Butler and Sigogneau-Russell (2016) reported a new triconodontid, *Eotriconodon*, from the Bathonian of Kirtlington in Oxfordshire based on a lower molar. Triconodontidae reached their peak in terms of abundance and diversity in the Late Jurassic-earliest Cretaceous, when they occurred in North America, Western Europe, Asia, and Africa. *Priacodon* and *Trioracodon* (Fig. 6.42) from the Late Jurassic Morrison Formation in western North America are represented by cranial and postcranial (*Priacodon*) remains (Simpson 1929, Rasmussen and Callison 1981, Rougier *et al.* 1996). *Trioracodon* also occurs in the basal Cretaceous of the Purbeck in southern Britain (Simpson 1928), and Krusat (1989) referred a single molariform from strata of the same age at Porto Pinheiro (or Dinheiro) (Portugal) to *Priacodon*. Within the British Purbeck, *Trioracodon* and *Triconodon* are represented by numerous mandibles and teeth, and *Triconodon* also by cranial material (Kermack 1963).

6.9.5.1 Alticonodontinae

Alticonodontines are the dentally most derived triconodontids, with a well-developed cusp d, main cusps of the lower molariforms almost equal in height and distally oriented, and strongly interlocked molariforms. Meckel's groove is vestigial or even missing in some taxa. *Arundelconodon* has been described from the Aptian of Maryland (Cifelli *et al.* 1999, Rose *et al.* 2001), and *Astroconodon* has been reported from the Aptian-Albian of Texas and Oklahoma (Patterson 1951, Slaughter 1969, Turnbull and Cifelli 1999). The Cloverly Formation of the same age has produced two alticonodontine genera, *Astroconodon* and *Corviconodon*, plus an unidentified triconodontid (Cifelli *et al.* 1998). From the slightly younger Cedar Mountain Formation, three genera have been described: *Astroconodon*, *Corviconodon*, and *Jugulator* (Cifelli and Madsen 1998). The last North American occurrence of triconodontids is that of *Alticonodon* in the early Campanian of Alberta, Canada (Fox 1969, 1976). A South American occurrence of Triconodontidae is *Argentoconodon* represented by an isolated upper molariform

Fig. 6.41: Life reconstruction of *Spinolestes xenarthrosus*. From Martin *et al.* (2015).

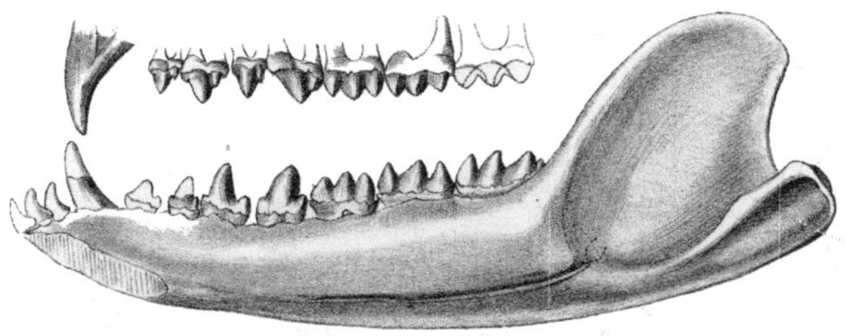

Fig. 6.42: *Trioracodon ferox*, left upper tooth row (C, P1-4, M1-3) and mandible with c, p1-4, and m1-3 in lingual aspect. Modified after Osborn (1888b).

(holotype), a fragmentary maxilla, and a partial disarticulated skeleton from the Jurassic of Chubut Province, Argentina (Gaetano and Rougier 2011). The eutriconodontan affinity for another Gondwanan taxon, *Austrotriconodon*, represented by isolated teeth from the Campanian-Maastrichtian Los Alamitos Formation in Argentina (Chubut Province) as proposed by Bonaparte (1990, 1992), has been disputed by Rougier *et al.* (2007b, 2011). Gaetano *et al.* (2013) suggested that the *Austrotriconodon* specimens should be assigned to Meridiolestida and Mesungulatoidea. *Volaticotherium antiquum*, with its spectacular gliding adaptations (Meng *et al.* 2006), has been recognized as an alticonodontine triconodontid (Gaetano and Rougier 2011). The molariforms of *Volaticotherium* closely resemble those of *Ichthyoconodon* from Morocco (for which a piscivorous diet had been suggested; Sigogneau-Russell 1995), and both cluster together with *Argentoconodon* from Argentina (Rougier *et al.* 2007b) in the clade Volaticotherini (Gaetano and Rougier 2011). Alticonodontinae has also been reported from the lower Cretaceous of northeastern China (*Meiconodon*), suggesting a faunal exchange between North America and Asia before the late Early Cretaceous (Aptian/Albian) (Kusuhashi *et al.* 2009).

6.9.6 Eutriconodonta incertae sedis

A small eutriconodontan of uncertain familial affiliation is *Chaoyangodens* from the Early Cretaceous (Barremian) Yixian Formation at Lingyan in western Liaoning Province of China (Hou and Meng 2014). Slightly younger (Aptian) is the skeleton of *Liaoconodon* from the Juifotang Formation in Liaoning Province, which exhibits a transitional mammalian ear (Meng *et al.* 2011, Weil 2011). Other eutriconodontan remains from Asia comprise dentaries and fragmentary lower dentitions (Chow and Rich 1984, Zhou *et al.* 1991, Rougier *et al.* 1999).

Averianov *et al.* (2018) recently reported a new eutriconodontan of uncertain familial affiliation from the Early Cretaceous Teete locality in northern Yakutia, Russia. *Sangarotherium* is based on a dentary fragment with erupted penultimate molariform and cryptic ultimate molariform. Together with the docodontan *Khorotherium* and the haramiyid cf. *Sineleutherus* it represents the northermost record of Mesozoic mammals in Asia (estimated paleolatitude of the Teete locality is 63–70° N; Rich *et al.* 2002).

6.10 Gondwanatheria

Gondwanatheria Mones 1987 is a distinctive but enigmatic clade of non-therian mammals that has been recorded exclusively from southern landmasses (South America, Antarctica, Madagascar, and India) with the possible exception of one isolated tooth from the Campanian of Texas that resembles *Ferugliotherium* (Brink 2015). Before the discovery of the skull of *Vintana sertichi* from the Late Cretaceous (Maastrichtian) of Madagascar (Krause *et al.* 2014a) (Fig. 6.43), gondwanatherians were known from isolated teeth and a few mandibular fragments only. Although the monophyly of Gondwanatheria is generally accepted, their phylogenetic relationship is under dispute. Early on, they were assigned to Xenarthra (Scillato-Yané and Pascual 1984, 1985, Mones 1987) or to the ancestry of Xenarthra (e.g., Bonaparte 1986a, 1990). Later they were attributed to Multituberculata (e.g., Krause and Bonaparte 1990, 1993, Kielan-Jaworowska and Bonaparte 1996) or considered as closely related to Multituberculata (Krause *et al.* 1997, Goin *et al.* 2012). The presence of four molariform positions (of which the first one possibly is a modified premolar; Gurovich and Beck 2009) in the mandible of the Paleocene gondwanatherian *Sudamerica* (Pascual *et al.* 1999) contradicts an assignment to multituberculates because only two molar positions occur in Multituberculata. Accordingly, Pascual *et al.* (1999) doubted a multituberculate assignment and considered Gondwanatheria as Mammalia incertae sedis, a view that was followed by Koenigswald *et al.* (1999) and Kielan-Jaworowska *et al.* (2004). Cladistic analyses by Gurovich and Beck (2009) and Krause *et al.* (2014a), the latter including the skull of *Vintana*, supported multituberculate affinities of Gondwanatheria. In a recent cladistic analysis, Huttenlocker *et al.* (2018) found *Vintana* in a polytomy with the newly discovered Early Cretaceous putative haramiyidan *Cifelliodon* and with *Hahnodon*. Because the position of gondwanatherians within haramiyidans is only weakly supported, Gondwanatheria is here considered as Mammaliaformes incertae sedis.

Eight gondwanatherian genera are known, which were assigned to the Ferugliotheriidae Bonaparte 1986a with low-crowned molariforms, and the high-crowned Sudamericidae Scillato-Yané and Pascual 1984: *Bharattherium*,

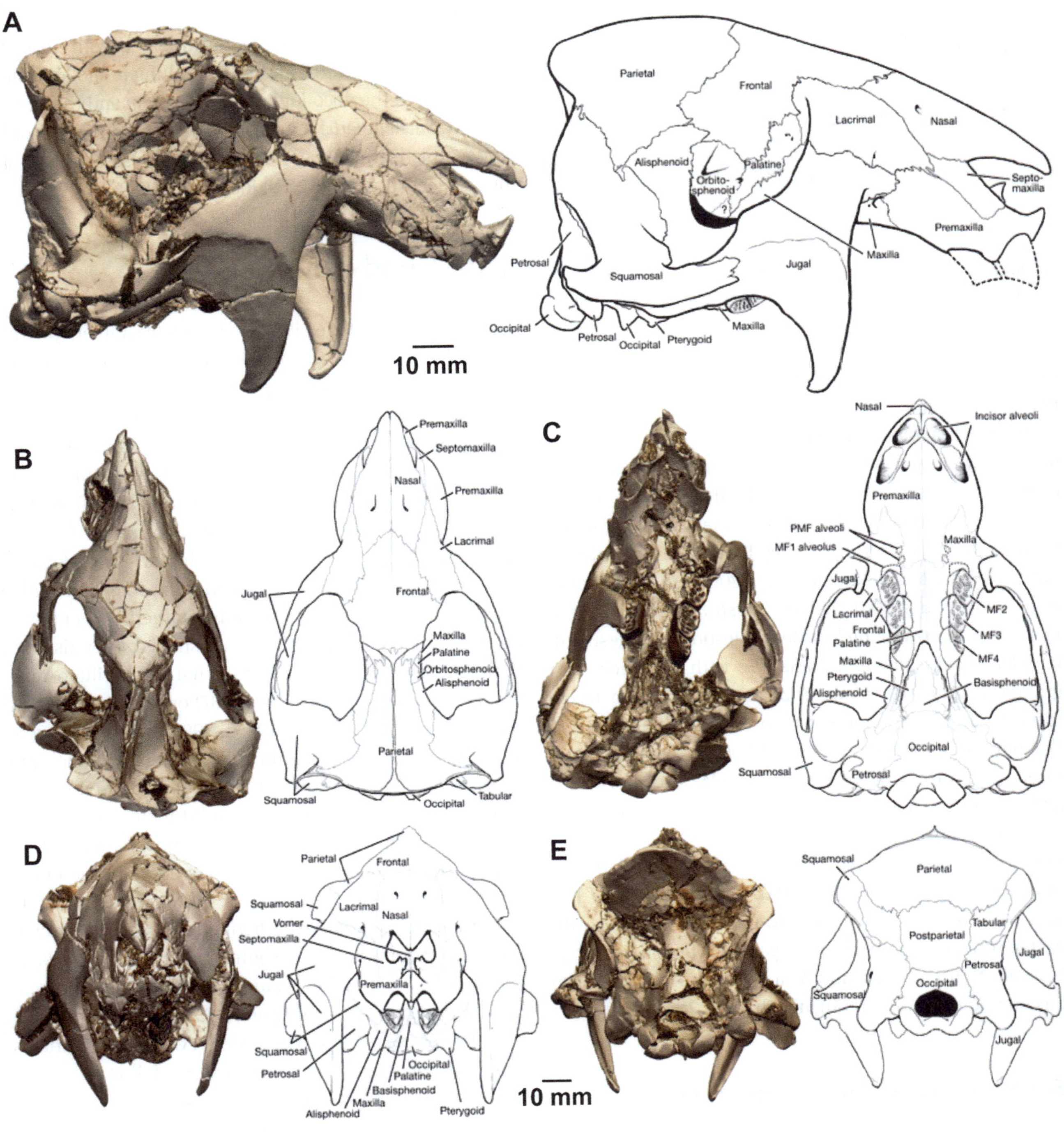

Fig. 6.43: Skull of the Cretaceous gondwanatherian mammal *Vintana sertichi* from Madagascar. (A) From lateral, (B) from dorsal, (C) from palatinal, (D) from frontal, and (E) from caudal. From Krause *et al.* (2014a, b), reprinted by permission from Springer Nature.

Gondwanatherium, *Greniodon*, *Lavanify*, *Sudamerica*, and *Vintana* (Scillato-Yané and Pascual 1984, Bonaparte 1986a, Krause *et al.* 1997, 2014a, Prasad *et al.* 2007a, b, Goin *et al.* 2012, Weil 2014). Ferugliotheriidae comprises *Ferugliotherium* from the Late Cretaceous (Campanian-Maastrichtian) Los Alamitos Formation in Argentina and *Trapalcotherium* from the Campanian-Maastrichtian Allen Formation of Argentina (Bonaparte 1986a, Rougier *et al.* 2009a).

6.10.1 *Vintana sertichi*

With a skull length of 124 mm and an estimated body mass of 9 kg, *Vintana* is the largest known Late Cretaceous mammal from Gondwana (Krause *et al.* 2014a, b). The skull (Fig. 6.43) exhibits unusual features such as elongate, wing-like jugal flanges, very large orbits, strong klinorhynchy, and a vaulted nuchal region (Krause *et al.* 2014b). *Vintana* has a relatively small endocranial cast, with an encephalization quotient of 0.28 to 0.56 (depending on body mass estimates), which is similar to that of stem mammals (Hoffmann *et al.* 2014). The olfactory bulbs are very large, and the cerebral hemispheres are only slightly expanded, which is more similar to the situation in *Morganucodon* than in multituberculates. The cochlear canal is short and only slightly curved (Hoffmann *et al.* 2014). The upper dental formula of *Vintana* is 2I.0C.1P.4M, but the lower dentition is unknown. Incisors are represented only by their alveoli, which indicate that they were enlarged, buccolingually compressed, procumbent, and separated from the cheek teeth by a wide diastema. No evidence for a canine is present, and the single premolariform tooth was apparently small and two-rooted. The molariform cheek teeth are large, with a quadrangular outline and hypsodont crowns. The occlusal surfaces are worn almost flat and exhibit many cementum-filled infundibula that partially invaginate from the buccal side (Krause 2014). The roots have multiple short apices. The enamel of the molariforms of *Vintana* consists of modified radial enamel with thin interrow sheets and resembles that of the sudamericids *Lavanify* from the Late Cretaceous of Madagascar and *Bharattherium* from the Late Cretaceous of India (Koenigswald and Krause 2014). Functional analysis of the dentition of *Vintana* indicates a primarily palinal (distally directed) power stroke, with a significant buccally directed component that is absent in multituberculates and haramiyidans. Schultz *et al.* (2014) suggest that *Vintana* was a herbivorous mixed feeder, with a diet that included relatively hard food items such as roots and twigs. A similar diet had been suggested for the Paleocene *Sudamerica* from Patagonia (Gurovich 2008).

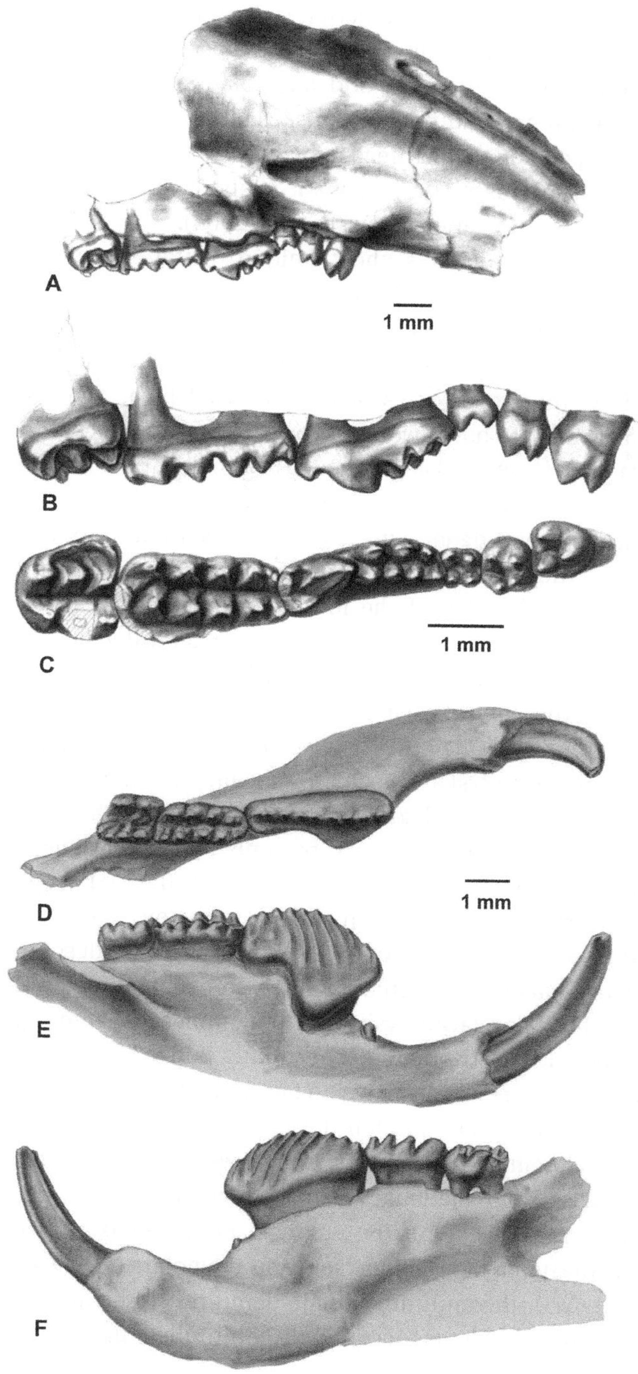

Fig. 6.44: *Cimexomys judithae*. (A) Right maxilla and premaxilla in lateral view; (B, C) right upper dentition (P1-4, M1-2) in labial (B) and occlusal (C) views; (C–F) right mandible with i1, peglike p3, bladelike p4, and m1-2 in dorsal (D), lateral (E), and medial (F) views. Modified after Montellano *et al.* (2008), reprinted by permission of the Society of Vertebrate Paleontology, www.vertpaleo.org.

6.11 Multituberculata

Multituberculata Cope 1884 (multi [Latin], many and tuberculum [Latin], cuspule) (Fig. 6.1) have one of the geologically most extensive records for any mammalian group and are known from the Middle Jurassic to the Eocene. Formerly, they were lumped together with the haramiyidans into the "Allotheria", based on the multicusped molars with two, respectively, three rows of longitudinally arranged cusps (Fig. 6.44). However, these similarities in tooth shape are possibly due to convergent evolution (Jenkins *et al.* 1997, Zhou *et al.* 2013, Luo *et al.* 2015b, Puttick *et al.* 2017; but see Meng *et al.* 2016 as summarized above). An important functional difference is the palinal (powerstroke from front to back) chewing movement in multituberculates, whereas the haramiyidans had mainly an orthal jaw movement. Although haramiyidans are mammaliaforms with the middle ear bones still attached to the dentary in the postdentary trough, multituberculates appear to be crown Mammalia, with three ear ossicles and loss of both the postdentary trough and Meckel's groove. The cheek tooth pattern of multituberculates differs from all other crown Mammalia, and it is still unclear from which ancestral molar pattern it is derived. Together with the fact that multituberculates have a peculiar mosaic of plesiomorphic, derived, and autapomorphic characters in the cranial and postcranial skeleton, their origin and exact position within the mammalian phylogenetic tree is still unclear, as are the systematic relationships within the group (e.g., Kielan-Jaworowska *et al.* 2004).

Multituberculates show unique dental specializations among Mesozoic Mammalia. They have multicusped premolars and molars, with the cusps arranged in longitudinal rows. The upper molars have two, respectively, three cusp rows in more derived taxa, and the lower molars have two cusp rows with an open central basin in between. The third cusp row in the upper molars of some taxa (e.g., *Kryptobaatar* and *Cimexomys*) occurs on a posterolingual wing. In less derived multituberculates, the upper molar lingual cusp row occludes with the central valley of the lower molars, and the labial row of the lower molars into the upper central valley. In more derived multituberculates with three upper cusp rows, the central cusp row occludes with the central longitudinal valley of the lower molars (Simpson 1929, Kielan-Jaworowska *et al.* 2004). The number of molars is reduced to two in all known multituberculates. There are three to two upper incisors, but only one lower incisor. The lower incisor (and to a lesser extent, the first upper incisor) is enlarged, with enamel only on the labial surface: it is hypselodont (evergrowing and rootless) in derived taxa (e.g., Taeniolabididae Granger and Simpson 1929) (Kielan-Jaworowska 1980). Canines are present only in less derived multituberculates and are otherwise absent. There may be between one and five upper premolars present and up to four lower premolars. The mesial lower premolars are mostly small, whereas p4 is often enlarged and bladelike with an arcuate cutting edge (Fig. 6.44D–F). With the enlarged mesial incisors and the diastema between incisors and premolars (Figs. 6.44 and 6.45), the multituberculate dentition closely resembles that of rodents. Accordingly, multituberculates have

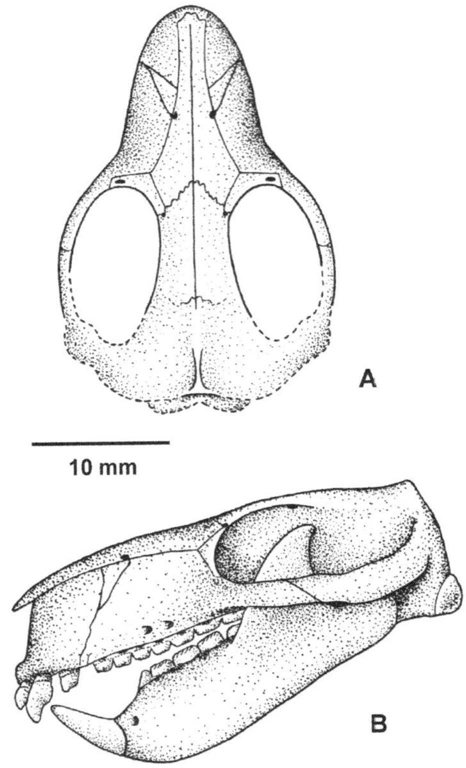

Fig. 6.46: Skull restoration of the multituberculate *Pseudobolodon* in dorsal (A) and lateral (B) views. From Hahn and Hahn (1994).

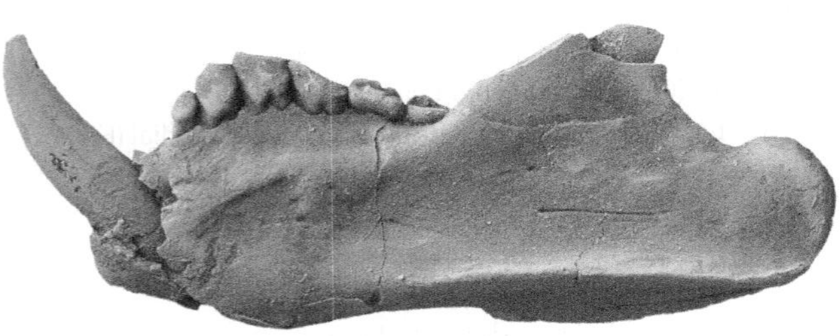

Fig. 6.45: *Meketibolodon robustus*, left mandible in lateral view (Gui Mam 89/76) with i1, p1-4, m1-2; coated with ammonium chloride. Photograph by Georg Oleschinski.

been dubbed as the "rodents of the Mesozoic", and a mainly herbivorous-omnivorous diet has been suggested. However, in contrast to muroid rodents, multituberculates had a palinal chewing movement, which means that the power stroke of the masticatory cycle was oriented from front to back, whereas in muroids it is oriented from back to front (Lazzari *et al.* 2010).

The multituberculate skull (Fig. 6.46) is rodentlike, dorsoventrally flattened and wide with well-developed zygomatic arches. The nasal bone is perforated by numerous foramina, which indicates a high degree of innervation at the snout. The glenoid facet is large and flat for the anterior to posterior chewing movement (e.g., Kielan-Jaworowska *et al.* 2004). Multituberculates possess three ear ossicles, which are fully detached from the mandible (Fox and Meng 1997). Schultz *et al.* (2018) suggested that an asymmetric bicrural stapes as observed in the Late Jurassic paulchoffatiid multituberculate *Pseudobolodon* represents the ancestral morphotype of the mammalian symmetric bicrural stapes. The brain case is largely therian-like, with participation of the squamosal and alisphenoid at the lateral skull wall. Differing from therians but similar to monotremes, the petrosal has a large anterior lamina. Although a well-developed anterior lamina is plesiomorphic for Mammaliaformes, the very large anterior lamina of multituberculates and monotremes is presumably derived.

The mandible is also rodentlike, with the condyle sitting above the occlusal plane of the tooth row and oriented in a posterior, rather than dorsal, direction. The postdentary trough and Meckel's groove are absent; the only postdentary jawbone that has been found is the coronoid in the Late Jurassic *Kuehneodon* (Hahn 1969). The multituberculate mandible lacks an angular process and has a large pterygoid fossa and well-developed pterygoid shelf.

The shoulder girdle consists of the scapulocoracoid, interclavicle, and clavicle (Sereno and McKenna 1995). The scapulocoracoid is narrower than that of therians and has a deep infraspinous fossa and a prominent scapula spine. There is no supraspinous fossa, but acromion and scapular spine are positioned lateral to the glenoid (interpreted as "incipient supraspinous fossa" by Sereno and McKenna 1995). The coracoid does not comprise a separate ossification, unlike as seen in monotremes, haramiyidans, and other taxa stemward of Mammalia. The acromion is large and surrounds the humeral head craniolaterally. The clavicle is long and arched. The humerus has a ball-shaped head but separate radial and ulnar condyles without formation of a trochlea. There is a humeral torsion of 15°–30°. The pelvis is narrow, with ventrally fused pubes and ischia, and a ventral keel. The angle between the ilium and sacrum is 33°–36°, which is considerably larger than in modern therians (9°–19°). The acetabulum is dorsally open, as in stem mammals (e.g., *Morganucodon*). The multituberculate femur is robust and, as in therians, has a well-developed femoral head that sits on a long neck oriented at an angle of 50°–60° to the shaft. The greater trochanter is large and rises above the femoral head. The lesser trochanter is also prominent and, unlike the therian situation, oriented ventrally. At the origin of greater trochanter and neck sits a subtrochanteric tubercle, an autapomorphic character of multituberculates. A third trochanter is not present. The foot bones of multituberculates show a peculiarity in common with cynodonts that distinguishes them from therians. In multituberculates and the so-called Manda cynodont (right cynodont foot from the Middle Triassic Manda Formation, Tanzania [BMNH TR.8]), the metatarsal V articulated via a cartilaginous remnant of a fifth tarsal with the calcaneus, whereas it articulates with the cuboid in therians (Szalay 1993, Kielan-Jaworowska 1997, Kielan-Jaworowska *et al.* 2004).

Multituberculates are generally divided into the less derived "Plagiaulacida" (Ameghino 1889) or "Plagiaulacoidea", and the more derived Cimolodonta McKenna 1975. "Plagiaulacoidea" is a paraphyletic assemblage, whereas Cimolodonta is commonly regarded as monophyletic (Simmons 1993). The resolution of the phylogenetic tree of multituberculates is poor (e.g., Rougier *et al.* 1997, Kielan-Jaworowska and Hurum 2001). Multituberculates have their first occurrence in the Middle Jurassic (late Bathonian) Forest Marble of Kirtlington in Oxfordshire (England) an in the Berezovsk coal mine (Bathonian) of western Siberia (Martin *et al.* 2011, Averianov *et al.* 2016). The two families described from Kirtlington lie outside the morphological range of the "Plagiaulacida" and are regarded as earlier offshoots from the multituberculate stem (Butler and Hooker 2005). Families Kermackodontidae Butler and Hooker 2005 (*Kermackodon*) and Hahnotheriidae Butler and Hooker 2005 (*Hahnotherium*) are based on isolated upper molars; referred specimens are premolars and lower molars. *Kermackodon* has sharp cusps with strong ridges, connected by longitudinal crests. *Hahnotherium* has blunt cusps, the buccal row lacks a longitudinal crest, and the cusps are separated by transverse valleys (Butler and Hooker 2005).

6.11.1 "Plagiaulacoidans"

The Middle Jurassic to Early Cretaceous multituberculates are combined into "plagiaulacoidans", a paraphyletic assemblage of less derived members of the group. The "plagiaulacoidans" are characterized by a number of plesiomorphic dental characters, such as three upper incisors, lower incisors that are completely covered by enamel, five upper premolars, and four or three lower bladelike premolars. "Plagiaulacoidans" comprise the oldest known multituberculates (Middle Jurassic, Bathonian). They have been reported from the Middle Jurassic to Early Cretaceous (Barremian) of Europe, the Late Jurassic to Early-Late Cretaceous boundary of North America, the Middle Jurassic to Early Cretaceous (Aptian-Albian) of Asia, and the Early Cretaceous (Berriasian) of Africa (Morocco). Within "plagiaulacoidans", three (partially paraphyletic) informal lineages are distinguished, namely, allodontid, pauchoffatiid, and "plagiaulacid" lineages (Kielan-Jaworowska and Hurum 2001).

6.11.2 Allodontid lineage

The allodontid lineage differs from the paulchoffatiid and "plagiaulacid" lineages in the presence of a two-cusped I2 and small, single-cusped I3. It differs from the "plagiaulacid" lineage in the presence of well-developed cusps on its lower molars and a smooth enamel surface. In contrast to the paulchoffatiid lineage, it has two rows of cusps on m2, and a p3 without labial cusps. Allodontids have only been reported from the Late Jurassic Morrison Formation of the western USA. Their dental formula is 3I.1C.5P.2M/1i.0c.4p.2m, and they possess an upper canine and an incipient distolingual ridge on M1 (the latter is absent in Paulchoffatiidae). Allodontidae Marsh 1889 consists of the genera *Ctenacodon* and *Psalodon*. Zofiabaataridae Bakker 1992 is monotypic with the genus *Zofiabaatar* represented by a mandible. The dental formula is 1i.0c.4p.2m, with the p4 distinctly longer than the p3. Of uncertain familial affiliation is *Glirodon* from the Late Jurassic Morrison Formation in Utah (Dinosaur National Monument) and Colorado (Fruita). *Glirodon* has the complete plesiomorphic multituberculate dental formula of 3I.1C.5P.2M/1i.0c.4p.2m and gigantoprismatic enamel (Engelmann and Callison 1999).

6.11.3 Paulchoffatiid lineage

The paulchoffatiid lineage was first recognized in the Late Jurassic of the Guimarota coal mine in Portugal (e.g., Hahn 1969, Hahn and Hahn 2000), where Paulchoffatiidae Hahn 1969 are represented by partial skulls, jaws, and teeth (Figs. 6.45 and 6.46). The Guimarota beds have also yielded the highest diversity, although this is clearly an overestimate because most taxa are known only by upper or lower dentitions, and association of upper and lower dentitions is not possible in most cases (Hahn and Hahn 2000). Yuan *et al.* (2013) tentatively associated the lower dentition of *Plesiochoffatia* with the upper dentition of "paulchoffatiid genus D" from the Guimarota locality in the supplementary information for the description of *Rugosodon*. Representatives of the paulchoffatiid lineage are the least derived multituberculates and differ from members of the other lineages in having molars with cusps of differing heights, very large, four-cusped I2 and I3, a labial cusp row at p3, and a basin-shaped m2 with only one cusp. The dental formula of paulchoffatiids is 3I.1-0C.5-4P.2M/1m.0c.4-3p.2m, with the upper canine still present in some taxa (e.g., *Kuehneodon*) (Hahn 1969). Because of the fact that most dentitions from the Guimarota coal mine have not been found in association, separate taxonomies for lower and upper dentitions have been established for the subfamily Paulchoffatiinae Hahn 1969 (lower dentition taxa: *Paulchoffatia*, *Guimarotodon*, *Meketibolodon*, *Plesiochoffatia*, and *Xenachoffatia*; upper dentition taxa: *Bathmochoffatia*, *Henkelodon*, *Kielanodon*, *Meketichoffatia*, *Pseudobolodon*, and *Renatodon;* Hahn 1977, 1987, 1993, 2001, Hahn and Hahn 1998, 1999). *Kuehneodon* (subfamily Kuehneodontinae Hahn 1971) is the only paulchoffatiid of which upper and lower dentitions have been found in association. *Kuehneodon* has the dental formula 3I.1C.4P.2M./1i.0c.4-3p.2m and has, besides the Guimarota coal mine, also been found in the Late Kimmeridgian-Tithonian Lourinhã Formation at Pai Mongo and Porto das Barcas (Portugal). Isolated teeth of Paulchoffatiidae have also been reported from the Lourinhã Formation at Porto Pinheiro (or Dinheiro) (Hahn and Hahn 1999, 2001a) and the Lower Cretaceous (Barremian) of the Spanish localities Galve and Uña (*Galveodon*) (Hahn and Hahn 1992, 2001b). Recently, a largely complete paulchoffatiid skeleton has been reported from the Late Jurassic Tiaojishan Formation in Inner Mongolia (Yuan *et al.* 2013; see next paragraph). The Pinheirodontidae Hahn and Hahn 1999 are known only from isolated teeth (about 250). They have lower molar cusps of differing heights and a basin-like m2, and they differ from all other multituberculates in the shape of I3, which has an obliquely elongated crown and main ridge having up to five cusps, plus accessorial distolabial and mesiolabial cusps. Pinheirodontidae comprise the taxa *Pinheirodon*, *Bernardodon*, *Ecprepaulax*, *Gerhardodon*, *Iberodon*, and *Lavocatia*. They have been reported from the Early Cretaceous (Berriasian) Purbeck Limestone Group of southern England, the Berriasian of Porto Pinheiro (or Dinheiro) at Lourinhã (Portugal), and the Early Cretaceous (Barremian) of Galve (Spain) (Kielan-Jaworowska and Ensom 1992, Canudo and Cuenca-Bescós 1996, Hahn and Hahn 1999).

Sunnyodon, based on a P5 from the Early Cretaceous (Berriasian) of the Purbeck Limestone Group of Dorset (southern England), is tentatively assigned to Paulchoffatiidae (Kielan-Jaworowska and Ensom 1992). *Mojo*, known from a single incomplete upper premolar of the Late Triassic (early Rhaetian) of Gaume (Belgium) and originally tentatively assigned to Paulchoffatiidae (Hahn 1987), is not a multituberculate but a haramiyidan (Butler 2000, Butler and Hooker 2005). A plagiaulacoidan of uncertain familial affiliation but with resemblances to paulchoffatiids, *Argillodon*, has been described based on a M2 with enamel ornamentation from the Early Cretaceous (Aptian) of Maryland in the eastern USA (Cifelli *et al.* 2013).

6.11.3.1 *Rugosodon eurasiaticus*

Rugosodon eurasiaticus is known from a largely complete skeleton from the Late Jurassic (Oxfordian) part of the Tiaojishan Formation of northeastern China and provides, apart from an isolated ulna from the Guimarota coal mine (Martin 2013), the only information on paulchoffatiid postcranial anatomy (Yuan *et al.* 2013). It had a body mass of 65 to 80 g and a terrestrial mode of locomotion. The vertebral column has a clear thoracic-lumbar boundary, and the 10th thoracal is the anticlinal vertebra. According to its dentition, with a bladelike p4 and trenchant mesiolingual cusps at m1-2, plus the multicusped rows of M1/m1, an omnivorous diet has been suggested for *Rugosodon*, including arthropods and worms as well as tough plant material (Yuan *et al.* 2013). The foot bone arrangement

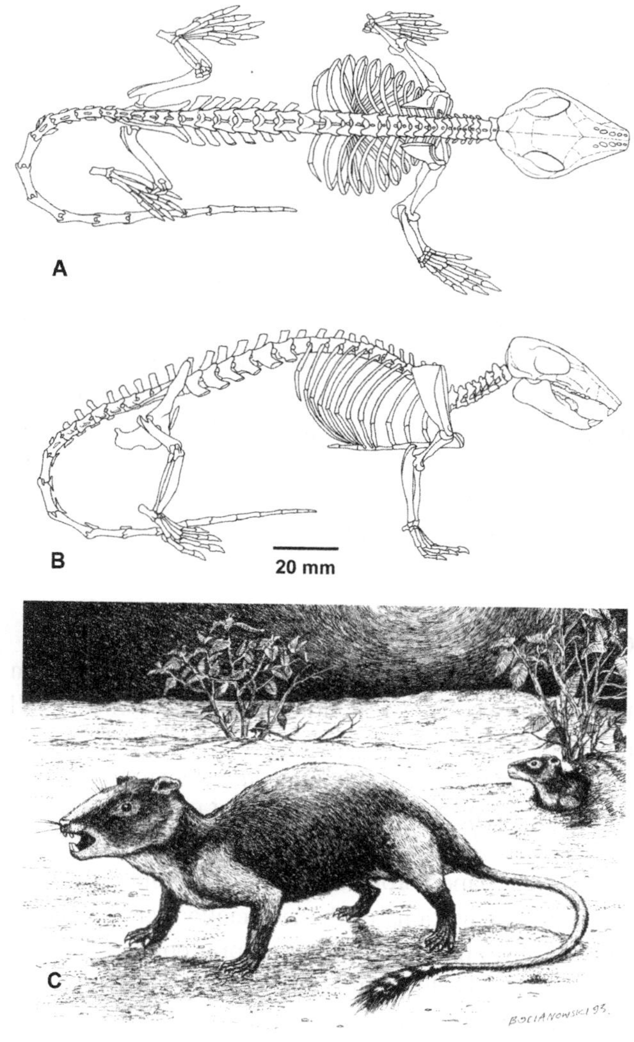

Fig. 6.47: *Nemegtbaatar gobiensis*, skeletal restoration in (A) dorsal and (B) lateral views; (C), life reconstruction. Modified after Kielan-Jaworowska and Gambaryan (1994), reprinted with permission from John Wiley and Sons.

of *Rugosodon* allows a high mobility of metatarsal 1 and therefore allows a greater dorsoventral excursion of the hallux (pedal digit 1). The high mobility of metatarsal 1 is unique among Mesozoic mammals and is a preadaptation for later arboreality (Krause and Jenkins 1983), but it is even retained in fossorial (Kielan-Jaworowska *et al.* 2004) (Fig. 6.47) and saltatorial (Kielan-Jaworowska and Gambarian 1994) multituberculates.

6.11.4 "Plagiaulacid" lineage

The paraphyletic "plagiaulacid" lineage comprises more derived taxa than the paulchoffatiid lineage with a blade-like p4 that is longer than p3 and has more ridges. The I3 is still large but bears fewer cusps than that of paulchoffatiids. The taxa of the "plagiaulacid" lineage differ from those of the allodontid lineage in the presence of ornamentation of the enamel surface. In contrast to representatives of the paulchoffatiid lineage, the m2 is not basinlike, and the M1 has an incipient third row of cusps at the lingual side. Representatives of the "plagiaulacid" lineage are known exclusively from teeth and gnathic remains. The family Plagiaulacidae Gill 1872 (*Bolodon* and *Plagiaulax*) has been reported from the Early Cretaceous (Berriasian) Purbeck Limestone Group of southern England and from the Late Jurassic Morrison Formation of Wyoming, where a P5 of possible plagiaulacid affinity has been found (Bakker 1998). The family Eobaataridae Kielan-Jaworowska *et al.* 1987 (*Eobaatar*, *Loxaulax*, *Monobaatar*, and *Parendotherium*) has been reported from the Early Cretaceous (Barremian) of China (Liaoning Province) and Spain (Galve and Uña), as well as the Early Cretaceous (Aptian or Albian) of the Gobi Desert, Mongolia. The family Albionbaataridae Kielan-Jaworowska and Ensom 1994 (*Albionbaatar* and *Proalbionbaatar*) occurs in Western Europe (Late Jurassic Guimarota Beds of Portugal and Early Cretaceous Purbeck Limestone Group of southern England) and the Early Cretaceous of northeastern China (Wang *et al.* 1995). The family Arginbaataridae Hahn and Hahn 1983, with the monotypic *Arginbaatar dmitrievae*, is of uncertain subordinal affiliation and has been reported from the Early Cretaceous (Aptian or Albian) of the Gobi Desert in Mongolia (Trofimov 1980). *Arginbaatar* differs from all other multituberculates by its very large p4 with limited enamel, and which rotates during ontogeny over the worn p3 and p2, which are then lost. A new taxon of the "plagiaulacid" lineage (family indeterminate), *Teutonodon langenbergensis*, has recently been described based on an M1 from the Late Jurassic of the Langenberg Quarry near Goslar, which is the first Jurassic mammal from Germany (Martin *et al.* 2016).

6.11.5 Cimolodonta

The more derived Late Cretaceous to Paleogene multituberculates are members of the apparently monophyletic (Kielan-Jaworowska and Hurum 2001) Cimolodonta McKenna 1975. Cimolodontans are characterized by (giganto-)prismatic (= prisms with unusually large diameter; Fosse *et al.* 1978) enamel and differ from "plagiaulacoids" in at least five additional apomorphic characters: loss of I1, loss of P4, loss of p1-2, a vestigial p3, and an arched p4 (Clemens 1963, Kielan-Jaworowska and Hurum 2001). Based on the size of p4, two evolutionary lineages can be distinguished within Cimolodonta: (1) one in which there is progressive enlargement of p4 and increase of number of crests and (2) one in which there is progressive reduction in the size of p4 and reduction in the number of upper premolars

(Kielan-Jaworowska and Hurum 2001). The dental formula of Cimolodonta is the most reduced within multituberculates with 2I.0C.1-4P.2M/1i.0p.1-2p.2m. Although Cimolodonta seems to be monophyletic, the phylogeny and the taxonomy within the group remain unclear. The four least derived genera of Cimolodonta have been provisionally assigned to the apparently paraphyletic "*Paracimexomys* group" (Kielan-Jaworowska and Hurum 2001): *Paracimexomys*, *Bryceomys*, *Cedaromys*, and *Dakotamys*. The members of this informal group are mostly known from isolated teeth except for a partial skull with dentition and lower jaw of *Cimexomys judithae* (Fig. 6.44) from the Campanian of Montana (Montellano *et al.* 2000). They differ from most "plagiaulacoids" in the presence of gigantoprismatic tooth enamel and an arcuate p4. Tentatively attributed to the *Paracimexomys* group are *Ameribaatar*, *Barbatodon*, and *Cimexomys*. Representatives of the *Paracimexomys* group sensu *lato* have been reported from the Early Cretaceous (Aptian-Albian) to Late Cretaceous (Maastrichtian) of North America and the Late Cretaceous (Maastrichtian) of Europe (Romania). The range of *Cimexomys* extends to the Early Paleocene in North America (Kielan-Jaworowska *et al.* 2004).

Late Cretaceous *Yubaatar* from Luanchuan County, Henan Province, China is the largest multituberculate from Eurasia (Xu *et al.* 2015). It is represented by a three dimensionally preserved partial skeleton with nearly complete cranium and associated mandibles. The dentary is 56.6 mm long and the estimated skull length is 70 mm. The dental formula of *Yubaatar* is ?I.0C.4P.2M/1i.0c.1p.2m. According to the phylogenetic analysis by Xu *et al.* (2015), *Yubaatar* represents the immediate outgroup of Taeniolabidoidea, that comprises a North American and an Asian clade, and therefore indicates a faunal interchange of multituberculates at least before the K-Pg transition.

6.11.5.1 Djadochtatherioidea

The superfamily Djadochtatherioidea Kielan-Jaworowska and Hurum 2001 represents relatively large cimolodontans that are known by abundant skull material (skull length 19–70 mm). Their dental formula is 2I.0C.3-4P.2M/1i.0c.2p.2m., with single-cusped I2 and I3 and double-rooted upper premolars. A striking synapomorphy is the presence of large frontals that are deeply inserted between the nasals and pointed anteriorly medially. The superfamily comprises Djadochtatheriidae Kielan-Jaworowska and Hurum 1997 (*Djadochtatherium*, *Catopsbaatar*, *Kryptobaatar*, and *Tombaatar*), Sloanbaataridae Kielan-Jaworowska 1974 (*Sloanbaatar*, *Nessovbaatar*, and, tentatively, *Kamptobaatar*), as well as three Asian genera of uncertain familial affiliation (*Chulsanbaatar*, *Nemegtbaatar* [Fig. 6.47] and *Bulganbaatar*). The djadochtatherioids have been reported from the Late Cretaceous (Campanian) of Mongolia (Gobi Desert) (e.g., Simpson 1925a) and China, as well as of Kazakhstan (*Bulganbaatar*).

6.11.5.2 Cimolomyidae

Cimolomyidae Sloan and Van Valen 1965 is a cimolodontan multituberculate family of uncertain superfamilial affiliation from the Late Cretaceous of North America. The dental formula of cimolomyids is 1I.0C.4P.2M/1i.0c.2p.2m, and they share with taeniolabidoids an M1 with three rows of cusps. *Meniscoessus* is a large cimolomyid with a skull length of about 70 mm (Cope 1882, Weil and Tomida 2001). *Cimolomys* is based on isolated upper and lower premolars and molars; it differs from *Meniscoessus* in its long and low (rather than high and short) P4 (Marsh 1889, Sahni 1972, Lillegraven 1969).

6.11.5.3 Taeniolabidoidea

The superfamily Taeniolabidoidea Sloan and van Valen 1965 comprises the largest multituberculates, with a skull length up to 160 mm and a body weight possibly exceeding 100 kg in the Paleocene *Taeniolabis taoensis* (Williamson *et al.* 2015). The dental formula of taeniolabidoids is 1I.0C.1-2P.2M/1i.0c.1p.2m, and they are characterized by a number of synapomorphies such as a short and wide snout with anterior portions of the zygomatic arches oriented transversely, and short and posteriorly pointed frontals. Taeniolabidoids differ from all other multituberculates in possessing only one upper premolar, a long diastema between I3 and the premolars, strongly reduced P4 and p4 relative to the enlarged molars, and a p4 without oblique ridges. Like most Djadochtatheroidea and Eucosmodontidae Jepsen 1940, they have self-sharpening incisors, with enamel limited to the outer surface. The superfamily comprises two families, the primarily North American Taeniolabididae Granger and Simpson 1929 (*Taeniolabis*, *Kimbetopsalis*, *"Catopsalis"* [probably paraphyletic], and *Bubodens*) and the Asian Lambdopsalidae Chow and Qi 1978 (*Lambdopsalis*, *Sphenopsalis*, and *Prionessus*). *Viridomys*, represented by several isolated P4s from the Late Cretaceous (early Campanian) of Alberta (Canada), was originally assigned to Taeniolabidoidea Fox 1971, but it is now considered as superfamily and family incertae sedis (Kielan-Jaworowska *et al.* 2004). Taeniolabidoids have been reported from the Late Cretaceous to Paleocene of North America and the Paleocene of Asia (Mongolia, China).

The European Kogaionidae Rădulescu and Samson 1996 is of uncertain superfamilial affiliation. *Kogaionon* from the Late Cretaceous (late Maastrichtian) of the Haţeg Basin in Romania is represented by a largely complete skull without mandibles. Kogaionids differ from other cimolodontans in their strongly elongated upper premolars, making the premolar row twice as long as the molar row (Rădulescu and Samson 1996). *Hainina*, originally attributed to Cimolomyidae Vianey-Liaud 1979 and now assigned to Kogaionidae (Kielan-Jaworowska and Hurum 2001), is represented by isolated premolars and molars and has been reported from the Late Cretaceous (late Maastrichtian) of Romania and the Paleocene of Belgium and Spain (Vianey-Liaud 1979).

Of uncertain superfamiliar and familiar status is *Uzbekbaatar* from the Late Cretaceous (Turonian-Coniacian) of Dzharakuduk in Uzbekistan. The taxon was based on a single isolated p4 that is arcuate, but relatively low, with nine serrations and eight ridges (Kielan-Jaworowska and Nessov 1992), and subsequently two additional p4s, an edentulous maxilla, and dentary fragments have been described (Averianov 1999, Averianov and Archibald 2006).

6.11.5.4 Ptilodontoidea

The superfamily Ptilodontoidea Sloan and van Valen 1965 comprises several Late Cretaceous and Paleocene-Eocene families. The ptilodontoid dental formula is 2I.0C.4P.2M/1i.0c.2p.2m, with a very large p4 that is arcuate and protrudes strongly over the occlusal level of the molars. The enamel has small prism diameters (2–5 µm), which is similar to the condition in most other mammals, except for *Cimolodon* and *Boffius* that have gigantoprismatic enamel with prism diameters of about 7–13 µm (Carlson and Krause 1985). The family Ptilodontidae Gregory and Simpson 1926, from the Late Cretaceous to the Late Paleocene of North America, comprises the genera *Ptilodus*, *Baiotomeus*, *Gobiabaatar*, and *Kimbetohia*. The family Neoplagiaulacidae Ameghino 1890 occurs in the Late Cretaceous-Late Eocene of North America, Late Paleocene-Early Eocene of Europe, and Early Eocene of Asia and comprises small ptilodontoids with relatively low p4s, such as *Neoplagiaulax*, *Mesodma*, *Parectypodus*, *Cernaysia*, *Krauseia*, *Mesodmops*, *Mimetodon*, *Nidimys*, *Parikimys*, *Xanclomys*, and *Xyronomys*. The Family Cimolodontidae Marsh 1889 is characterized by an extremely high-crowned and strongly vaulted p4 with up to 15 serrations. It comprises the genera *Anconodon*, *Cimolodon*, and *Liotomus*. Cimolodontids have been reported from the Late Cretaceous (Cenomanian) to Late Paleocene of North America (Kielan-Jaworowska *et al.* 2004).

Kielan-Jaworowska *et al.* (2007) described *Argentodites coloniensis* from the Late Cretaceous (Campanian or Maastrichtian) La Colonia Formation in Chubut Province (Argentina) based on an isolated p4 and tentatively assigned it to Cimolodonta. The presence of normal prismatic enamel of the tooth suggests an affiliation with Ptilodontoidea. A previously described p4 from the Late Cretaceous Los Alamitos Formation in Patagonia suggested a "plagiaulacidan" affinity (Kielan-Jaworowska and Bonaparte 1996, Kielan-Jaworowska *et al.* 2004). Gurovich and Beck (2009) argued that this p4 from the Los Alamitos Formation, which was originally referred to the ferugliotheriid gondwanatherian *Ferugliotherium*, represents the same taxon as *Argentodites*. Gurovich and Beck (2009) further argued that the Los Alamitos p4 does indeed belong to *Ferugliotherium*, that ferugliotheriids had bladelike p4s, and that gondwanatherians are either multituberculates or close multituberculate relatives.

6.12 "Symmetrodontans"

"Symmetrodonta" Simpson 1925d is a polyphyletic assemblage of mammals that are characterized by an angulation of the three molariform main cusps, giving the tooth crown a symmetric triangular shape. The upper molars are formed by the primary trigon, and the lower molars by the trigonid with a very weakly developed talonid cusp. The "symmetrodontan" dentition suggests an insectivorous diet. "Symmetrodontans" are mostly small, with slender mandibles that uniformly lack an angular process (Figs. 6.48 and 6.49). They are mainly known from teeth and dentaries, but the skull has been described for *Anebodon* (Bi *et al.* 2016), and complete skeletons are known for *Zhangheotherium*, *Maotherium*, and *Akidolestes* from the Early Cretaceous of Liaoning (northeastern China) (Hu *et al.* 1997, Li and Luo 2006, Ji *et al.* 2009). Traditionally, two groups were distinguished according to their molar shape, the "obtuse-angled symmetrodontans" (Kuehneotheriidae Kermack *et al.* 1968, Amphidontidae Simpson 1925d, and Tinodontidae Marsh 1887) and the "acute-angled symmetrodontans" (Spalacotherioidea Prothero 1981) (Kielan-Jaworowska *et al.* 2004).

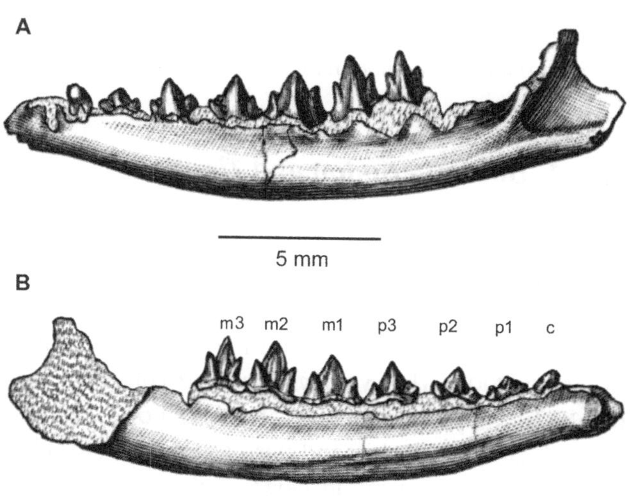

Fig. 6.48: *Tinodon bellus*, left mandible in lateral (A) and medial (B) views. Modified after Marsh (1887).

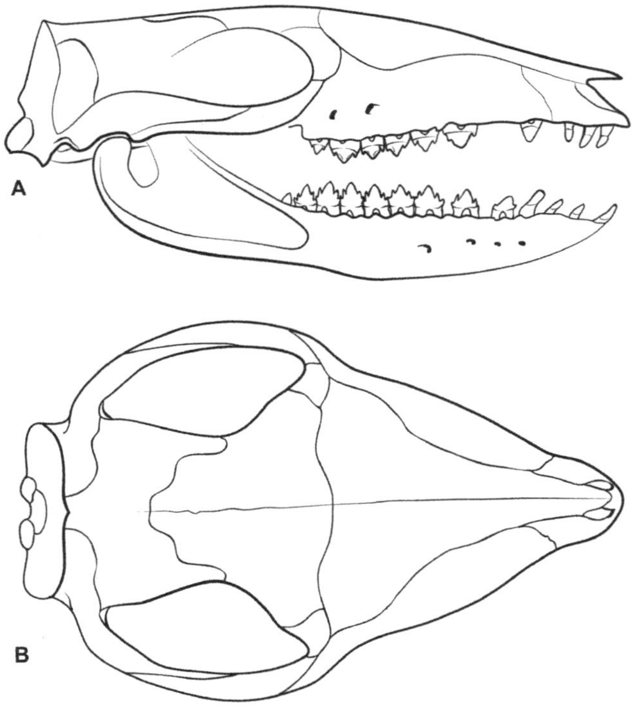

Fig. 6.49: Restoration of skull and mandible of *Maotherium sinense*. (A) Lateral view; (B) dorsal view. No scale. Redrawn from Rougier *et al.* (2003b).

Kuehneotheriids, which previously had been lumped with "Symmetrodonta", are now recognized as mammaliaforms outside crown Mammalia. In recent cladistic analyses (Bi *et al.* 2014, Krause *et al.* 2014a, Luo *et al.* 2015b), the "acute-angled symmetrodontans" form a monophyletic group within Trechnotheria. Trechnotheria McKenna 1975 (Fig. 6.50, node A) is the clade that includes the last common ancestor of *Zhangheotherium* and living therians plus all of its descendants and is characterized by a strong postvallum/prevallid shear between upper and lower molars (Luo *et al.* 2002). Li and Luo (2006) and Bi *et al.* (2016) found that Spalacotherioidea is monophyletic, with *Gobiotheriodon* and *Tinodon* as successive outgroups.

6.12.1 *Gobiotheriodon* and *Tinodon*

Gobiotheriodon from the Early Cretaceous (Aptian or Albian) of Mongolia is based on a mandible that lacks the coronoid and condylar processes, but which preserves m2-4 and alveoli for i1-3, c, p1-3, and m1 (Trofimov 1980, 1997, Averianov 2002). The symphysis is short compared with that of *Tinodon* and terminates at the level of the canine. Meckel's groove is distinct and starts as a narrow slit below m1, gradually increasing in height and becoming flatter posteriorly. There is no postdentary trough and no clear evidence for attachment of postdentary bones (Averianov 2002). *Gobiotheriodon* is characterized by a hypertrophied cusp e and an extreme development of the prevallid shearing surface that extends on the labial surface of cusp e. The trigonid angle is ~85° on m2 and ~95° on m3, becoming more obtuse on m4 (~133°) (Averianov 2002).

Tinodon (Tinodontidae Marsh 1887) is based on a partial dentary with c, p1-3, and m1-3 (Fig. 6.48) from the Late Jurassic Morrison Formation in Wyoming (Marsh 1879a, Simpson 1929). An upper molar from the Morrison Formation was originally assigned to the genus *Eurylambda* (Simpson 1929), which is now regarded as synonymous with *Tinodon* (Crompton and Jenkins 1967, Prothero 1981, Fox 1985). Rougier *et al.* (2003) described an additional upper molar (M1) from the Morrison Formation (Como Bluff) and discussed its relevance for cusp homologies in therian upper molars. Isolated upper and lower molars of the genus have recently been reported from the basal Cretaceous (Berriasian) of the Purbeck Limestone Group in Britain (Ensom and Sigogneau-Russell 2000), and a single lower molar that putatively belongs to *Tinodon* has been described from the Lourinhã Formation in Portugal (Krusat 1989).

6.12.2 *Amphidon* and "obtuse-angled symmetrodontans" of uncertain affinities

Amphidon (Amphidontidae Simpson 1925d) is based on a single dentary fragment from the Late Jurassic Morrison Formation, which preserves the last premolar and damaged m1-4 (Simpson 1925d). The molars are characterized by very small para- and metaconids. Averianov (2002) excluded *Amphidon* from "Symmetrodonta", and in a recent phylogenetic analysis (Gaetano and Rougier 2011), *Amphidon* clustered with triconodontids as member of Amphilestheria. "Obtuse-angled symmetrodontans" of uncertain affinities and based on isolated molars are *Atlasodon* and *Microderson* from the Early Cretaceous (Berriasian) of the Anoual Syncline in Morocco (Sigogneau-Russell 1991b). The phylogenetic relationships of *Nakunodon* from the late Middle Jurassic to Lower Cretaceous Kota Formation of India (Yadagiri 1985) and *Manchurodon* from the Middle Jurassic Wafangdian Formation of eastern China (Yabe and Shikama 1938) that have been assigned to Amphidontidae, are questionable due to poor preservation (Averianov 2002).

6.12.3 Spalacotherioidea

According to the phylogenetic analyses by Gaetano and Rougier (2011, 2012), the "acute-angled symmetrodontans"

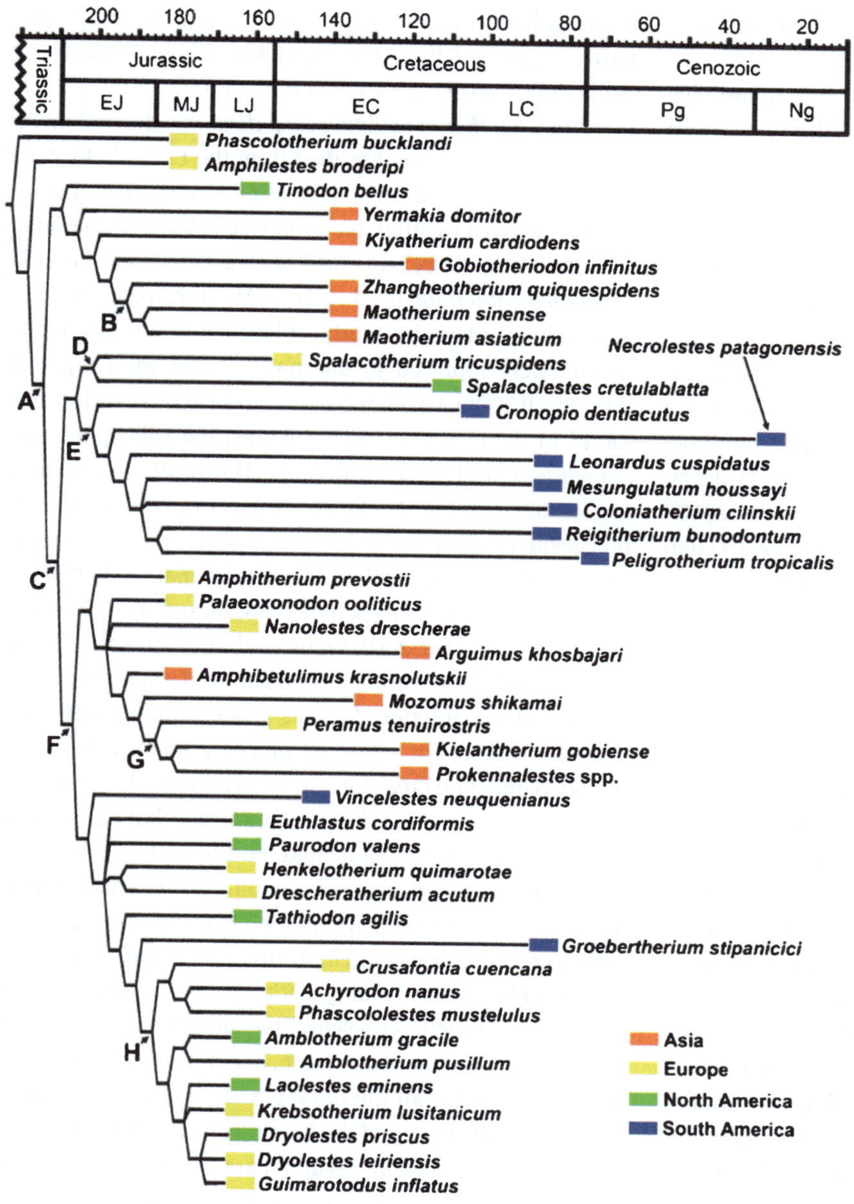

Fig. 6.50: Phylogeny of Trechnotheria (node A) with monophyletic Meridiolestida (node E). (A) Trechnotheria, (B) Zhangheotheriidae, (C) Alethinotheria, (D) Spalacotheriidae, (E) Meridiolestida, (F) Cladotheria, (G) Zatheria, (H) Dryolestidae. EC, Early Cretaceous; EJ, Early Jurassic; LC, Late Cretaceous; LJ, Late Jurassic; MJ, Middle Jurassic; Ng, Neogene; Pg, Paleogene. Strict consensus tree of five most parsimonious trees produced by TNT ratchet based on 44 taxa and 137 characters from the dentition, skull, and postcranium (CI 0.42, RI 0.76) (data matrix available from supplementary information of original publication). From Averianov *et al.* (2013), with permission of Springer.

Spalacotheriidae and Zhangheotheriidae are sister taxa within Spalacotherioidea (Fig. 6.1). Bi *et al.* (2016) supported monophyly of spalacotheriids but found zhangheotheriids to be a paraphyletic assemblage of less derived representatives of this clade.

6.12.4 Zhangheotheriidae

Zhangheotheriidae Rougier *et al.* 2003 (Fig. 6.50, node B) is the least derived spalacotheroid family and is well documented by complete skeletons and skulls (e.g., *Zhangheotherium* and *Maotherium*). Zhangheotheriids share a peculiar shape of the posterior part on the mandible, with a deep notch separating the slender and reclined coronoid process from the robust, upturned condylar process. As in other "symmetrodontans", an angular process is missing. The molars are morphologically intermediate between that of tinodontids and spalacotheriids with the primary cusps somewhat more acutely triangulated than in *Tinodon* (Bi *et al.* 2016). Other probable plesiomorphic characters are the lack of tall sharp crests connecting the cusps and relatively low crowns. The skull of *Anebodon* is relatively low-vaulted and generally plesiomorphic, resembling that of *Maotherium*. The lateral margin of the nasal cavity is formed by the septomaxilla, a plesiomorphic character found also in other zhangheotheriids (Bi *et al.* 2016). The dental formula of *Anebodon* is 4I.1C.5P.3M/3i.1c.4p.4m, and the molars are more acutely triangulated than those of *Tinodon*. The postcanine count of *Anebodon* comprises more premolars and fewer molars than that of *Zhangheotherium* (2P.5M/3p.5-6m) and *Maotherium* (1-2P.4-5M./1-3p.5-6m).

Kiyatherium is based on maxillae and mandibles from the Early Cretaceous Shestakvo 3 locality in western Siberia (Maschenko and Lopatin 2002) and appears as sister taxon of *Anebobon* in the cladistic analysis of Bi *et al.* (2016). The plesiomorphic condition of *Kiyatherium* within Spalacotherioidea is indicated by the relatively large trigon angles (comparable with *Tinodon*) of the molars. The Meckel's groove is wide and deep and the pterygoid fossa is small. As is characteristic of Zhangheotheriidae, the coronoid process of the mandible is reduced and strongly inclined posteriorly. The holotype mandible of *Kiyatherium* exhibits replacement of the first molariform, which is unique within Trechnotheria (Lopatin *et al.* 2010).

Maotherium sinense from the Early Cretaceous Yixian Formation in Liaoning Province (northeastern China) is known from a fully articulated, flattened skeleton with hair and body contour impressions (Rougier *et al.* 2003), and *Maotherium asiaticum* by a three-dimensionally preserved skull with large parts of the postcranium (Ji *et al.* 2009). *Maotherium* is a mid-sized "symmetrodontan" with terrestrial adaptations. The skull of *Maotherium* is conservative and resembles that of stem mammals such as *Morganucodon* in possessing a substantial petrosal with a lateral flange and anterior lamina, septomaxilla, broad nasals, and anteriorly projected parietals (Rougier *et al.* 2003). More derived are the larger promontorium and the elongated glenoid cavity. The dental formula of *Maotherium sinense* is 3I.1C.2P.4M/3i.1c.3p.5-6m (Rougier *et al.* 2005), whereupon m6 is variably present (Plogschties and Martin 2014, in review). *Maotherium asiaticum* (Fig. 6.49) differs in minor dental and skeletal features from *M. sinense* and has a different dental formula: 3I.1C.1P.5M/3i.1c.1p.6m. The match of upper and lower wear facets occurred after eruption and after substantial occlusal contact between upper and lower molars (Ji *et al.* 2009). The postcranium, with its short proximal and intermediate phalanges, indicates that *M. asiaticum* was a generalized terrestrial mammal. Its body length has been estimated between 150 and 155 mm, and its body mass between 72 and 83 g based on mandible, respectively, skull length. *Maotherium asiaticum* has an ossified Meckel's cartilage similar to that of eutriconodontans. In the eutriconodontan *Yanoconodon*, the ossified Meckel's cartilage is connected to the ectotympanic and malleus, and a similar, ossified connection between the mandible and the middle ear is inferred for *M. asiaticum*, although it lacks a postdentary trough (Ji *et al.* 2009). The curved shape of Meckel's groove indicates that the middle ear was oriented at an angle to the mandible in *M. asiaticum*, which suggests that before its full separation from the mandible, the middle ear was already mediolaterally partly separated from the mandible (Maier 1993, Meng *et al.* 2003, Luo *et al.* 2007, Ji *et al.* 2009).

Zhangheotherium quinquecuspidens was the first mammalian skeleton to be described from the Early Cretaceous Jehol Biota (Hu *et al.* 1997). Like other known zhangheotheriids, it was a terrestrial, medium-sized Mesozoic mammal (Hu *et al.* 1997, 1998, Luo and Ji 2005). The dental formula of *Zhangheotherium* is 3I.1C.2P.5-6M./3i.1c.3p.5m, with a diphyodont antemolar tooth replacement. The lower premolars were replaced in alternating positions (p1-p3-p2) (Luo and Ji 2005), which represents a derived condition also found in the cladotherian *Dryolestes*, the tribosphenidan *Slaughteria*, and some eutherians (Martin 1997, Kobayashi *et al.* 2002, Luo *et al.* 2004), and differs from the plesiomorphic sequential replacement pattern seen in eutriconodontans, most multituberculates, and stem mammals (Greenwald 1988, Jenkins and Schaff 1988, Martin and Nowotny 2000, Nowotny *et al.* 2001, Wang *et al.* 2001, Luo *et al.* 2004, Schultz *et al.* 2018). Hu *et al.* (1997) noticed a slightly rugose area along the anterior Meckel's groove and a rough area near the base of the coronoid process, which were interpreted as insertion areas for a splenial and poorly developed coronoid, respectively. The promontorium that houses the cochlea is cylindrical, elongated, and fingerlike, similar to that of *Sinoconodon*, morganucodontans, triconodontans, and multituberculates (Hu *et al.* 1997). A cylindrical and fingerlike promontorium is an indicator of an uncoiled cochlea. The shoulder girdle of *Zhangheotherium* is plesiomorphic in that an interclavicle is present. The scapula has a very high spine with a prominent acromion, and a supraspinous fossa (much narrower than the infraspinous fossa) is developed along the entire length of the blade. The articulations of the interclavicle, clavicle, and scapula are mobile, allowing the clavicle to move and act as a pivotlike strut for a wider rotation of the scapula (Hu *et al.* 1997).

The humerus exhibits a torsion of the proximal and distal ends of about 30°. At the distal end, the humerus has an incipient trochlea for articulation with the ulna which represents a therian apomorphy. The ulnar condyle is weakly developed and the radial condyle is large and spherical, both plesiomorphic features of non-therian mammals. The pelvis possesses large epipubic bones. On the femur, the spherical head is set off from the shaft by a well-defined neck. The calcaneus and astragalus in the ankle joint of *Zhangheotherium* lack superposition and have retained the primitive condition of mammaliaforms in which the astragalus is side-by-side with the calcaneus (Luo and Ji 2005). *Zhangheotherium* has an external pedal spur, similar to that of the extant male platypus, associated with a poisonous gland (Hu *et al.* 1997).

6.12.5 Spalacotheriidae

Spalacotheriidae Marsh 1887 (Fig. 6.50, node D) represents the most derived “symmetrodontans”. Spalacotheriids are characterized by molars with an acute angulation of the principal cusps and a higher molar count than in less derived “symmetrodontans”. They have five or more molars, and the single talonid cusp, the hypoconid or hypoconulid, is small and situated at the lingual margin of the tooth. Spalacotheriids have been reported from the Early Cretaceous (Berriasian to Barremian) of Western Europe (Britain and Spain) and from the Barremian of China. Early to Late Cretaceous occurrences comprise the Aptian-Albian to early Campanian of North America and the Turonian of Uzbekistan. The type genus *Spalacotherium* from the Early Cretaceous Purbeck Limestone Group of Britain and the Early Cretaceous (Barremian) Camarillas Formation of Uña (Spain) differs from members of Spalacolestinae in the presence of a Meckel’s groove and less acutely angled molars (both plesiomorphic features) as well as a lower number of premolars (which is apomorphic) (Cifelli and Madsen 1999). *Infernolestes* from the Early Cretaceous (upper Berriasian-Valanginian) Lakota Formation (Wyoming), based on a lower molar (m1), is a spalacotheriid, with an obtuse angle of the trigonid and relatively poor transverse development of the trigonid. It has been placed within Spalacotheriidae based on derived characters, such as the marked height difference of the alveolar margin of the tooth crown, and a metaconid that is much higher than the paraconid (Cifelli *et al.* 2014).

Akidolestes cifellii (Fig. 6.51) from the Early Cretaceous Yixian Formation of Liaoning, northeastern China, provides information on the skull and postcranium of spalacotheriids (Li and Luo 2006). *Akidolestes* has a dental formula of 4I.1C.?5P.?5M/4i.1c.5p.6m and acute-angled molars (less than 50° in the distal molars), which distinguishes it from zhangheotheriids. The mandible of *Akidolestes* closely resembles those of *Zhangheotherium* and *Maotherium*, with elongate and gracile coronoid processes and mediolaterally compressed dentary condyles. The anterior portion of the mandible is more gracile, concomitant with the slender rostrum and the upper jaws. Shoulder girdle and forelimbs are similar to those of zhangheotheriids (Li and Luo 2006). Differing from zhangheotheriids, *Akidolestes* possesses mobile lumbar ribs, which also occur in the eutriconodontan *Repenomamus*

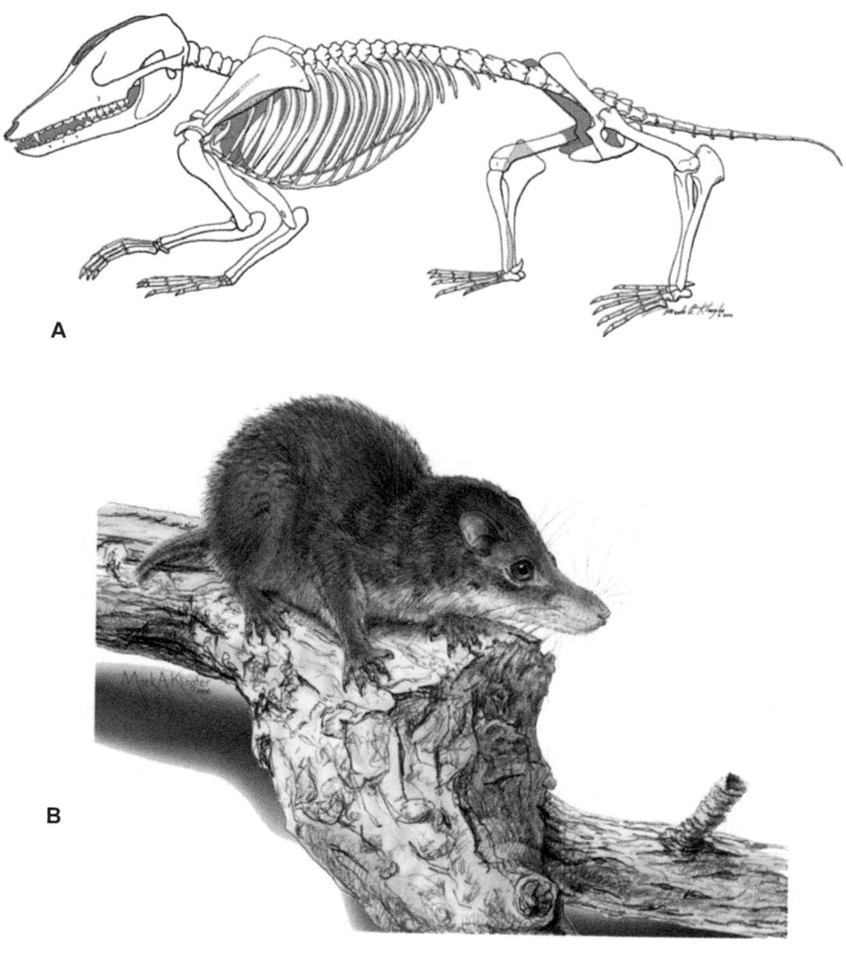

Fig. 6.51: *Akidolestes cifellii.* (A) Skeletal restoration; (B) life reconstruction. No scale. From Chen and Luo (2013), with permission of Springer.

(Hu *et al.* 2005a), in *Fruitafossor* (Luo and Wible 2005), and in many premammalian cynodonts (Jenkins 1971). Li and Luo (2006) found a number of similarities between the postcranium of *Akidolestes* and that of monotremes, such as the presence of a prominent tubercle for the psoas minor muscle and a hypertrophied parafibular process of the fibula. Based on the extremely enlarged parafibular process, a sprawling hindlimb posture similar to that of monotremes has been inferred (Li and Luo 2006). Based on phalangeal length ratios, Li and Luo (2006) inferred a generalized terrestrial adaptation for *Akidolestes*, and Chen and Wilson (2015) in their quantitative multivariate analysis found it to be semifossorial.

Symmetrolestes from the Early Cretaceous (likely Barremian) of Central Japan is based on a fragmentary mandible with the first incisor and p5-m4. It differs from other spalacotheriids in the lower number of molariforms, higher number of premolariforms, and the more gradual transition between premolariforms and molariforms. A poorly developed Meckel's groove is present (Tsubamoto *et al.* 2004).

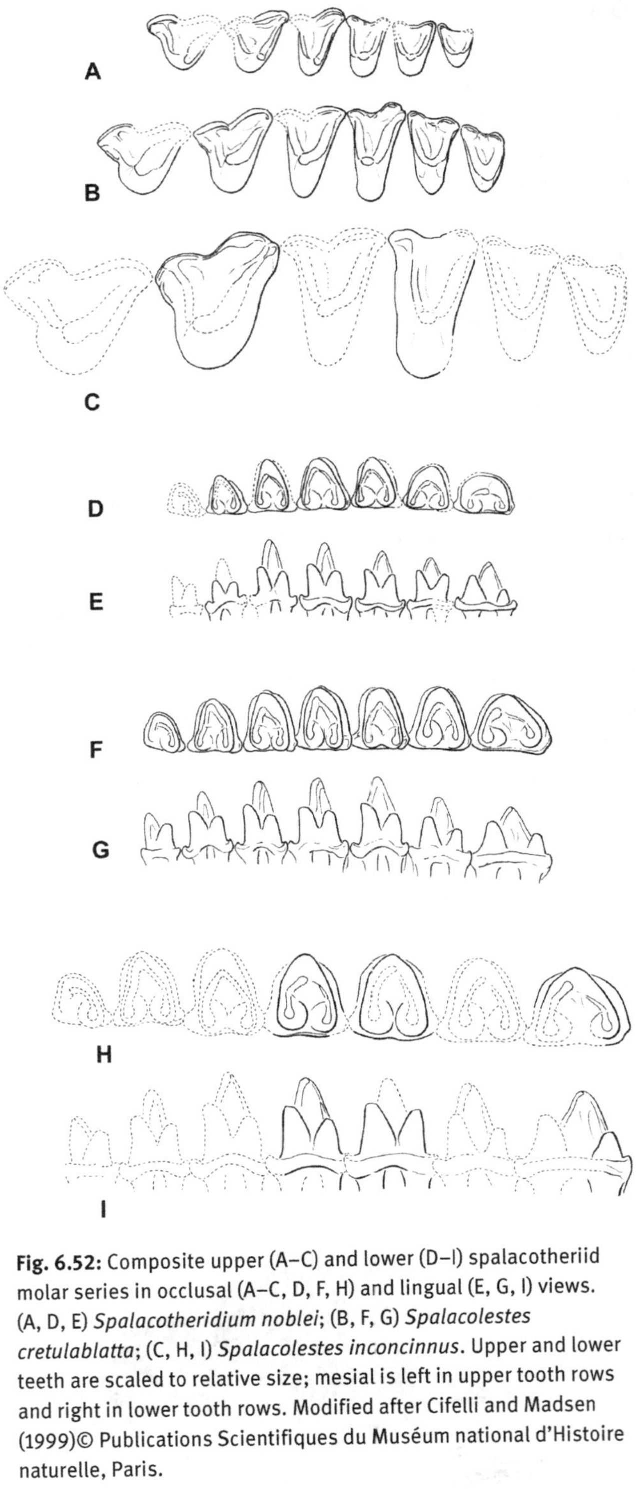

Fig. 6.52: Composite upper (A–C) and lower (D–I) spalacotheriid molar series in occlusal (A–C, D, F, H) and lingual (E, G, I) views. (A, D, E) *Spalacotheridium noblei*; (B, F, G) *Spalacolestes cretulablatta*; (C, H, I) *Spalacolestes inconcinnus*. Upper and lower teeth are scaled to relative size; mesial is left in upper tooth rows and right in lower tooth rows. Modified after Cifelli and Madsen (1999)© Publications Scientifiques du Muséum national d'Histoire naturelle, Paris.

6.12.4.1 Spalacolestinae

The remaining spalacotheriid taxa have been combined in the subfamily Spalacolestinae Cifelli and Madsen 1999. Spalacolestines are more derived than *Spalacotherium* based on their lack of Meckel's groove, a more anteriorly expanded pterygoid fossa, more premolar positions, and prolonged retention of deciduous canine and premolars. The molar triangulation is more acute, and molar crowns are higher in Spalacolestinae (Fig. 6.52). Taxa of Spalacolestinae have been reported from the Early Cretaceous of Britain (Barremian) and China (Aptian), the late Early to Late Cretaceous (Aptian-Albian to early Campanian) of North America, and the Late Cretaceous (Turonian) of Uzbekistan. *Yaverlestes* from the Early Cretaceous (Barremian) of the Isle of Wight, southern England, is represented by an incomplete dentary and isolated upper and lower molariforms. The mandible of *Yaverlestes* is very slender and lacks Meckel's groove, and the dental formula is ?i.1c.3p.5m. (Sweetman 2008). *Heishanlestes* is based on a nearly complete dentary and two mandibular fragments from the Early Cretaceous (Albian) of Badaohao, Liaoning (northeastern China). The dental formula is ?i.1c.4p.6m and the dentary is small (only 17 mm long), slender, and dorsoventrally shallow. A Meckel's groove is absent (Hu *et al.* 2005a). The premolars of *Heishanlestes* are closely spaced, overlap each other, and show an ovate worn area on the distolingual side of the main cusp. This shallowly to deeply concave area suggests a crushing action with the cusps of the upper premolars (Hu *et al.* 2005a). Recently described *Lactodens* from the Aptian Jiufotang Formation in Shangheshou area (Liaoning Province, China) is represented by a partial skeleton with mandible and upper and lower teeth (Han and Meng 2016). The dental formula of *Lactodens* is 3I.1C.3P.6M/3i.1c.5p.6m, with double-rooted canines, extremely low-crowned and transversely narrow premolars, and acute-angled molars. The dental morphologies of molars and deciduous premolars are similar to those of *Spalacolestes* from North America, and phylogenetic analysis indicates a close relationship to North American spalacolestines suggesting faunal interchanges between Eurasia and North America (Han and Meng 2016).

Spalacotheridium from the Albian-Cenomanian to Turonian of Utah has even more symmetrical and acutely angled molars than *Spalacotherium*, and lower molar crowns than *Spalacolestes*; molars have a complete labial cingulid (Cifelli 1990a). This taxon is based on isolated teeth, but for *Spalacotheridium noblei* most of the molar series could be reconstructed (Cifelli and Gordon 1999, Cifelli and Madsen 1999) (Fig. 6.52). The molars of *Spalacotheroides* from the Early Cretaceous (Aptian-Albian) of Texas differ from those of most other spalacotheriids in their incomplete labial cingulid. *Spalacotheroides* was the first spalacotheriid to be described from North America and is based on a dentary fragment with a single lower molar (Patterson 1955). Later, a complete upper molar and two fragments were discovered (Patterson 1956). Although the position of the single preserved tooth of the holotype is debated and its taxonomic state is now (in the light of newly discovered specimens) questionable, the referred upper molars are more distinctive (Kielan-Jaworowska *et al.* 2004). The genus *Spalacolestes* is the best known North American spalacotheriid, represented by several dentaries and many isolated upper and lower teeth, and by deciduous premolars (Cifelli and Madsen 1999). Although the dental formula is unknown, Cifelli and Madsen (1999) suggested a reasonable approximation of 4p and 7m for the lower dentition. The genus *Symmetrodontoides* from the Late Cretaceous (Turonian-early Campanian) of Canada and the western United States (Fox 1976) clusters as sister to *Spalacolestes* in the cladogram presented by Bi *et al.* (2016). It is most similar to *Spalacolestes* but differs in its labiolingually wider distal molars, greater height differential between paraconid and metaconid, and few other morphological details. A possible spalacolestine is *Shalbaatar* from the Late Cretaceous (Cenomanian-Turonian) Bissekty Formation in Uzbekistan (Archibald and Averianov 2005), which is based on an edentulous partial dentary and originally was thought to be a multituberculate (Nessov 1997).

6.13 Meridiolestida

Meridiolestida Rougier *et al.* 2011 is an exclusively South American clade that originally was interpreted as "symmetrodontans" and dryolestoids (Bonaparte 1986b, 1990, 2002), but later usually as Dryolestoidea (Rougier *et al.* 2011 and references therein). According to the recent cladistic analysis by Averianov *et al.* (2013), Meridiolestida are non-cladotherian trechnotherians related to spalacotheriids (Fig. 6.50, node E). This is in contrast to the phylogenetic analysis by Rougier *et al.* (2011), in which the South American *Groebertherium* is linked with the Laurasian dryolestids, and the remaining taxa are united into a monophyletic Meridiolestida within Dryolestoidea. *Groebertherium* (Fig. 6.53 A, B) is a dryolestid-like taxon with an incipient talonid and possibly a reduced distal root of the lower molars. The molar wear pattern resembles that of other dryolestoids (Crompton *et al.* 1994, Schultz and Martin 2010). The other meridiolestidans lack the typical dryolestoid characters. Unlike dryolestoids, the distal lower molar root is not reduced in comparison with the mesial root in meridiolestidans. The cross sections of the molar roots are mesiodistally compressed, similar to the condition in spalacotheriids, and not oval as is the case in dryolestoids. The lower molars often have prominent mesial and distal cingulids, as in spalacotheroids, but they lack the distinct talonid found in dryolestoids (Averianov *et al.* 2013). The dentary of *Cronopio*, the oldest and best known meridiolestidan possesses according to Rougier *et al.* (2011) a small and inflected angular process. The partially broken structure interpreted as angular process by Rougier *et al.* (2011: fig. 2) is, however, quite different from the well-developed and backwards pointing angular process of dryolestoids. According to Averianov *et al.* (2013), the dentary of *Cronopio* resembles in shape and general appearance the dentaries of zhangheotheriids and spalacotheriids rather than those of dryolestoids. In all known meridiolestidans, Meckel's groove is missing, as is also the case in spalacolestines, whereas it is pronounced in Dryolestoida. In more recent analyses by Rougier *et al.* (2012) and O'Meara and Thompson (2014), Meridiolestida is no longer immediate sister taxon to Dryolestidae but comprises part of a paraphyletic Dryolestoida on the stem leading toward Theria.

Cronopio dentiacutus, known by two incomplete skulls (Fig. 6.54) from the Late Cretaceous (Cenomanian) Candeleros Formation of the Neuquén Basin, Argentina (Rougier *et al.* 2011), is characterized by an extremely elongated rostrum and very long canines. The skull of *Cronopio* shares a number of plesiomorphic characters with mammaliaforms, such as the presence of a septomaxilla, an anterior lamina of the petrosal, and a lateral flange (Rougier *et al.* 2011). The mesiodistally highly compressed molariforms indicate an insectivorous diet. The dental formula of *Cronopio* was originally interpreted as 2I.1C.4P.3M/i.1c.3+p.3m, but Averianov *et al.* (2013) reinterpreted it as three premolars and four

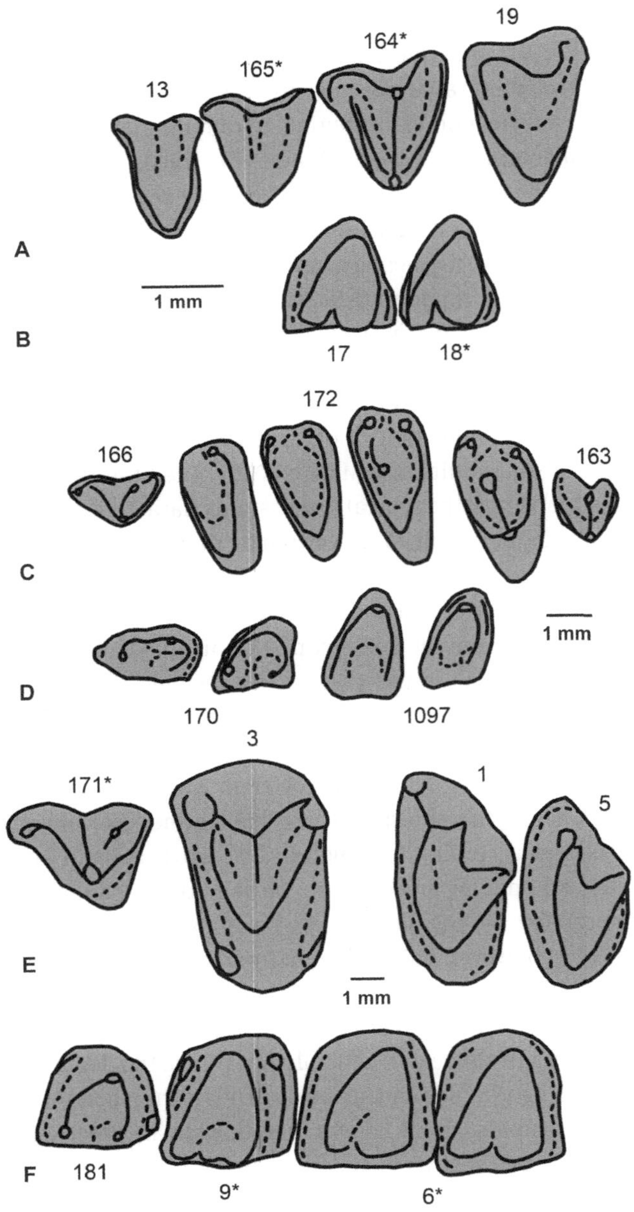

Fig. 6.53: Interpretation of cheek teeth of *Groebertherium* and other mammals from the Campanian Los Alamitos Formation (Argentina). Upper teeth are left, lower teeth are right, anterior side is to the left. (A and B) *Groebertherium stipanici*, 13: possible M1; 164 and 165: anterior upper molars; 17 and 18: lower molars. (C and D) *Leonardus cuspidatus*, 166: possible upper ultimate deciduos premolar; 172: four upper molars from maxilla fragment; 163: upper ultimate molar; 170: lower penultimate premolar and possible m1; 1097: lower molars. (E and F) *Mesungulatum houssayi*, 171: possible upper ultimate deciduous premolar; 3: upper anterior molar; 1: upper penultimate molar; 5: upper ultimate molar; 181: possible m1; 9: lower anterior molar; 6: lower posterior molars. From Averianov *et al.* (2013), with permission of Springer

molars based on the sharp morphological break in the cheek tooth row (P4 of original interpretation is fully molariform). *Leonardus* is known from a maxilla with four molariform teeth and a referred dentary fragment with two molariform teeth from the Late Cretaceous (Campanian) Los Alamitos Formation of Argentina (Bonaparte 1990, Chornogubsky 2011). The upper molars of *Leonardus* are strongly mesiodistally compressed and have a very large median stylar cusp (Fig. 6.53 C, D). The parastyle is absent, and the stylocone and metastyle are subequal in size. *Barberenia* and *Casamiquelia* from the Los Alamitos Formation (Bonaparte 1990) were synonymized with *Leonardus* by Averianov *et al.* (2013).

Mesungulatidae Bonaparte 1986a comprises the relatively large taxa *Mesungulatum* and *Coloniatherium*. Mesungulatids have a strong pre- and postcingulum on the upper molars, and a mesial cingulid on the lower molars, but no distinct talonid. These characters are not found in dryolestoids (see above). *Mesungulatum* is based on isolated upper and lower molars (Fig. 6.53 E, F) from the Campanian-Maastrichtian Los Alamitos and La Colonia Formations (Bonaparte 1986a, 1990), as well as from the Late Cretaceous Allen Formation of Patagonia, Argentina (Rougier *et al.* 2009a). The holotype of *Quirogatherium* was considered to be the last deciduous premolar of *Mesungulatum* by Averianov *et al.* (2013). *Coloniatherium* is based on fragmentary dentaries, isolated teeth, and a number of petrosals from the La Colonia Formation (Rougier *et al.* 2009b). The morphology of the petrosals suggests a phylogenetic position similar to *Vincelestes*, and derived features such as a fully coiled cochlea (one and a half turns) possibly have been attained convergently with therians (Rougier *et al.* 2009b).

Reigitherium is a poorly known taxon based on the holotype lower molar from the Los Alamitos Formation and referred dentary fragment with a premolar and two molars from the La Colonia Formation. Originally, it was attributed to Dryolestoidea (Bonaparte 1990) and later considered a docodontan (Pascual *et al.* 2000). According to the cladistic analyses of Rougier *et al.* (2011) and Averianov *et al.* (2013), *Reigitherium* belongs to Meridiolestida. A peculiar feature of *Reigitherium* is that the lower molars strongly decrease in length and increase in width distally. The molars have an extensive lingual cingulid with three cusps on m2; a talonid is not present (Kielan-Jaworowska *et al.* 2004). Based on newly discovered mandibular fragments and teeth, Harper and Rougier (2017) corroborated the meridiolestidan nature of *Reigitherium* and its placement as sister taxon of *Peligrotherium*.

Peligrotherium tropicalis from the early Paleocene Salamanca Formation of Argentina (Bonaparte *et al.* 1993) is a Cenozoic survivor of Meridiolestida. It is a large taxon (skull length about 200 mm) and known from cranial, mandibular, and dental remains. The teeth closely resemble those of mesungulatids. The other, much younger, Cenozoic meridiolestidan survivor is *Necrolestes patagoniensis*

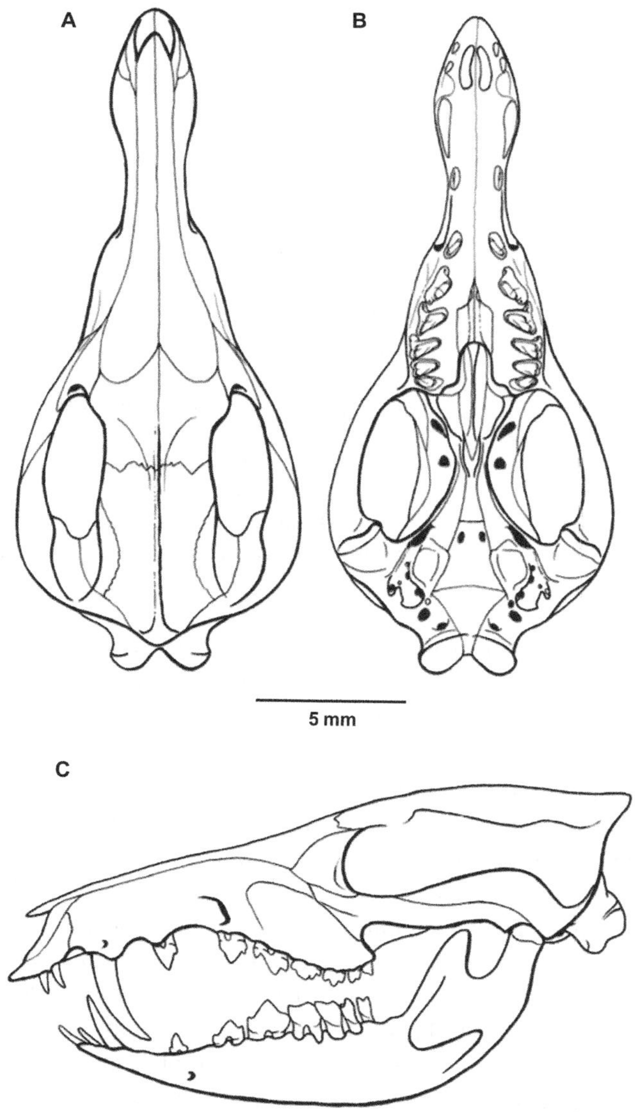

Fig. 6.54: Skull of *Cronopio dentiacutus*. (A) Dorsal; (B) palatinal; (C) lateral views. From Rougier *et al.* (2011), reprinted by permission from Springer Nature.

from the Early Miocene Santa Cruz beds (Ameghino 1891) of Argentina. The phylogenetic relationships of *Necrolestes* were enigmatic for a long time (e.g., Asher *et al.* 2007), but recent cladistic analyses based on new anatomical observations (Chimento *et al.* 2012, Rougier *et al.* 2012), including a study that did not assume cusp homologies between Mesozoic and Miocene taxa and applied both Bayesian and parsimony tree reconstruction techniques (O'Meara and Thompson 2014), found it to be a meridiolestidan.

6.14 Cladotheria

Formerly lumped within the paraphyletic "eupantotherians", taxa such as dryolestoids are within the monophyletic Cladotheria (Kielan-Jaworowska *et al.* 2004). Cladotheria McKenna 1975 (Fig. 6.50, node F) comprises the common ancestor of Dryolestoidea Butler 1939 and Boreosphenida Luo *et al.* 2001b (Northern Hemisphere tribosphenic mammals, including Tribosphenida McKenna 1975 plus all of its descendants). Cladotherians outside of Boreosphenida do not form a natural group by themselves (Kielan-Jaworowska *et al.* 2004). In the dentition, cladotherians are characterized (and differ from the non-cladotherian "symmetrodontans") by the presence of a larger talonid cusp d, which is situated above the cingulid and which has a well-developed wear facet from occlusion with the paracone. Ascending from node F toward Zatheria McKenna 1975 in Fig. 6.50, the dentition of cladotherians becomes more and more boreosphenidan-like. In Dryolestida Prothero 1981, the talonid is short and comprises a single cusp d (hypoconulid or hypoconid; see chapter 2) (Fig. 6.55). In Amphitheriida Prothero 1981, the talonid is still unicusped but somewhat expanded in comparison with dryolestidans. In the stem lineage of Zatheria (Fig. 6.50, node G; defined as common ancestor of *Peramus* and living marsupials and placentals plus all of its descendants by Luo *et al.* 2002), the talonid expansion is continued, and accessory talonid cuspules may occur (not homologous with the boreosphenidan regular talonid cusps). In *Peramus* and close relatives (traditionally combined in the family "Peramuridae" Kretzoi 1946, although they represent a paraphyletic assemblage of stem boreosphenidans), the talonid is incipiently basined and possesses more than one regular talonid cusp (hypoconulid and hypoconid in *Peramus*). A fully developed entoconid is never present on the talonid of non-zatherian cladotherians. A functional protocone is not yet developed, although a bulgelike ledge lingually of paracone and metacone may be present. However, this bulge does not occlude into the developing talonid basin, which does not exhibit any wear facets inside. Wear facets within the talonid basin occur only in Boreosphenida in the next step of dental evolution, when a fully functional protocone is present that occludes with the fully developed talonid basin, which is surrounded by a distinct hypoconulid, hypoconid, and entoconid.

Similarly to "symmetrodontans", the molars of cladotherians are arranged as reversed and interlocking triangles, but the upper molars are significantly wider than the lowers and have three roots. The triangular structure (trigonid) of the lower molars is formed by the lingually placed protoconid, the mesiolabially situated paraconid, and the distolabially placed metaconid, which are homologous to the trigonid cusps of the tribosphenic molar. In the upper molars, the triangular structure (primary trigon) is not fully homologous with the trigon of the tribosphenic

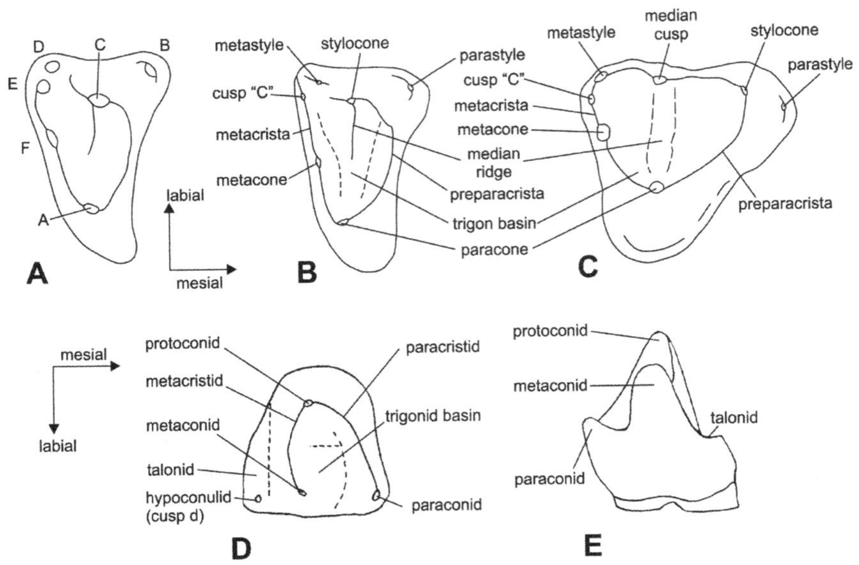

Fig. 6.55: Molar morphology and terminology in Dryolestidae. (A–C) Upper molars in occlusal view; (D and E) lower molars in occlusal (D) and lingual (E) views. No scale. From Kielan-Jaworowska *et al.* (2004) after Simpson (1961) (A) and Martin (1999a) (B, D, E), and courtesy of Sigogneau-Russell (C). Copyright © 2004 Columbia University Press. Reprinted with permission from the publisher.

molar. The primary trigon does not possess a protocone, and the lingual cusp of the primary trigon is the paracone. The metacone is situated on the distal crest. On the labial side, three stylar cusps are developed: the mesiolabial parastyle, the distolabial metastyle, and the stylocone in between. The stylocone is often quite prominent, with a median ridge running in lingual direction into the primary trigon basin. For dryolestoid cusp terminology, see Fig. 6.55.

Cladotheria share with boreosphenidans an angular process of the mandible that is missing in "symmetrodontans". Most non-zatherian cladotherians are known by mandibular and maxillary remains or isolated teeth besides few postcranial bones. Up to now, only one virtually complete skeleton has been found: *Henkelotherium guimarotae* from the Late Jurassic of the Guimarota coal mine in Portugal (Krebs 1991, 2000). The skull has been described and published in detail for *Vincelestes neuquenianus* from the Early Cretaceous of Argentina (Bonaparte and Rougier 1987, Rougier *et al.* 1992, Macrini *et al.* 2007), and some information on the skull of *Dryolestes* is available (Martin 1999a). The description of disassociated upper and lower dentitions led to an artificial inflation of taxonomic diversity (e.g., Simpson 1928) that could be resolved by the detailed study of more extensive fossil material (Martin 1999a).

6.14.1 Dryolestida

Dryolestidans comprise the cladotherians with a short unicusped talonid. Traditionally, two major families have been distinguished within Dryolestida Prothero 1981, Dryolestidae Marsh 1879b, and "Paurodontidae" Marsh 1887. A recent phylogenetic analysis has demonstrated that "paurodontids" are not a monophyletic group but represent early diverging dryolestidans (Averianov *et al.* 2013).

Dryolestidans are important faunal elements in Late Jurassic and Early Cretaceous strata of the Western Hemisphere. Dryolestidans are diverse and common in the Late Jurassic Morrison Formation of the western USA (Simpson 1929, Prothero 1981, Martin 1999a) and the Purbeck of southern England (Simpson 1928). In the Late Jurassic (Kimmeridgian) of the Guimarota coal mine in Portugal, dryolestidan teeth represent 49% of all mammaliaform teeth (Martin 1999b, 2001). Rare findings of dryolestidans have been reported from the Bathonian of Central Asia, with at least one (unnamed) taxon of non-dryolestid dryolestidan from the Balabansai Formation in Kyrgyzstan (Martin and Averianov 2010) and the dryolestid *Anthracolestes* from the Berezovsk coal mine near Krasnoyarsk in western Siberia (Averianov *et al.* 2014). Together with the dryolestids from the Bathonian Forest Marble of Kirtlington in England (Freeman 1976, 1979), these represent the geologically oldest dryolestidans. So far, no dryolestidans have been reported from Eastern Asia despite the wealth of newly discovered mammalian specimens in the Jurassic and Cretaceous of Liaoning and adjacent northeastern Chinese provinces. Dryolestidans apparently were replaced early on by tribosphenidans in Asia (Martin *et al.* 2011, Averianov *et al.* 2013). Dryolestidans survived in western North America (Lakota Formation, upper Berriasian-Valanginian; Cifelli *et al.* 2014) and Western Europe (southern England, Iberian Peninsula) with a diminished diversity into the Lower Cretaceous and have their last occurrence (*Crusafontia*) in the Barremian (Henkel and Krebs 1969). In South America, they survived with *Groebertherium* into the Late Cretaceous (Bonaparte 1986b, Rougier *et al.* 2011, Averianov *et al.* 2013), and a fragmentary lower molar of a possible dryolestid has been reported from North America (Late Cretaceous Mesaverde Formation of Wyoming; Lillegraven and McKenna 1986). The late survival in South America may be linked to the apparent absence of Late Cretaceous tribosphenidans on the South American continent.

6.14.2 Skull and Mandible

Information of the dryolestidan skull is limited because only fragmentary remains are known. A rostral fragment of *Dryolestes leiriensis* shows an elongated and narrow snout, and the skull widens at the position of M1 (Martin 1999a). The nasals are posteriorly widened and rounded. The zygomatic arch is largely formed by the jugal, with participation of the maxilla in its anterior part. The lacrimal participates in the orbital margin. For *Dryolestes leiriensis* and *Henkelotherium guimarotae*, the petrosal and inner ear structures are known (Ruf *et al.* 2009, Luo *et al.* 2011a, 2012). Hughes *et al.* (2015) described a partial petrosal from the Morrison Formation and assigned it to Dryolestoidea based on is resemblance to the petrosal of *Dryolestes*. The inner ear of *Henkelotherium* exhibits a division between the utricle and the saccule, a coiling of the cochlear canal of at least 270°, a distinctive primary bony lamina for the basilar membrane, and a secondary bony lamina. The occurrence of these structures in Late Jurassic cladotherians suggests a more ancient origination of high-frequency hearing than previously assumed (Ruf *et al.* 2009). In the inner ear of *Dryolestes leiriensis* (Fig. 6.56), the secondary lamina is also present, but it is less developed than in *Henkelotherium* (Luo *et al.* 2011a, 2012). The presence of a bony canal for the cochlear ganglion that is embedded in the base of the primary bony lamina in *Dryolestes* demonstrates that the cochlear innervation characteristic of modern therians originated in stem taxa of the cladotherian clade (Luo *et al.* 2012).

The dryolestidan mandible is relatively robust and the horizontal ramus is high. It clearly differs from the zatherian mandible, which is much more slender and gracile.

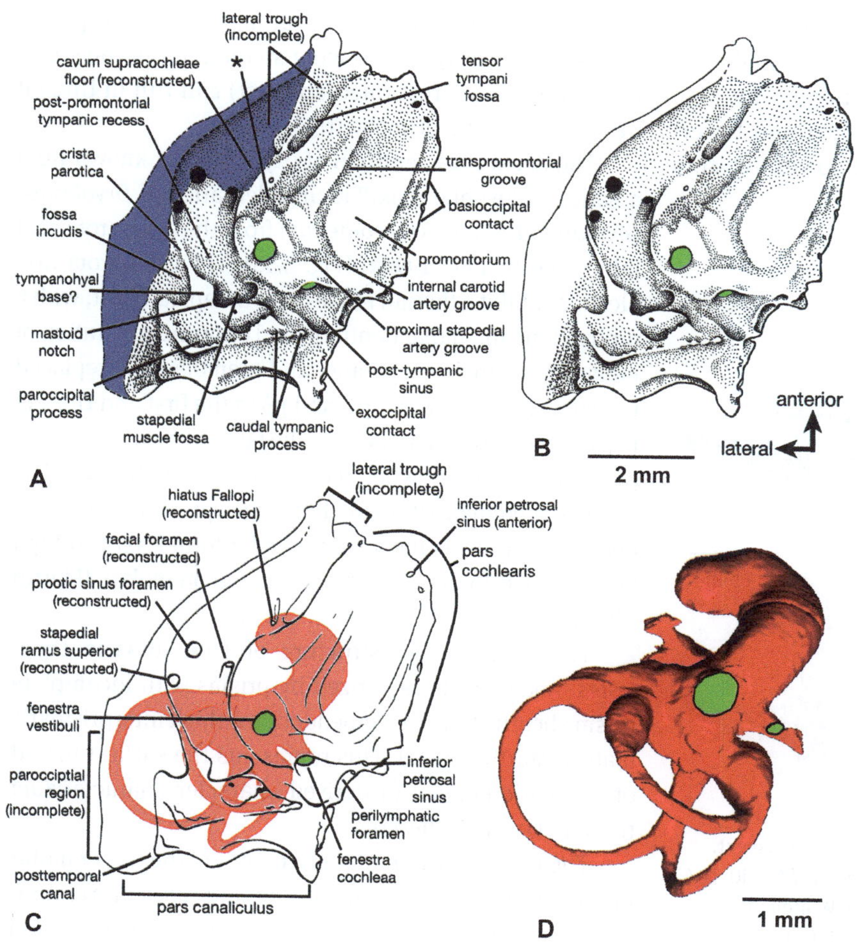

Fig. 6.56: Petrosal and inner ear of *Dryolestes leiriensis*. (A–C) Ventral surface petrosal structures; (C) location of the inner ear in the petrosal; (D) virtual endocast of inner ear from visualization of μCT-data. From Luo *et al.* (2012), by permission of the Linnean Society.

The high horizontal ramus houses the long roots of the dryolestidan dentition (Fig. 6.57D, E). The rim of the mandible is lower labially than lingually. A well-developed, posteriorly oriented angular process is present. The coronoid process is steeply inclined and is separated by a deep indentation from the articular process. The articular process has a well-developed, transversely oriented articular condyle. There is a deep masseteric fossa on the external side of the mandible and the pterygoid fossa on the internal side is also deep, with a sharp ventral border (medial pterygoid shelf). Of the accessorial jawbones, rudiments of the coronoid and splenial were still present in Jurassic dryolestidans (Krebs 1969, 1971, Martin 1995, 1999a) but had disappeared in the Early Cretaceous *Crusafontia* (Henkel and Krebs 1969, 1971). Meckel's groove is present in all known dryolestidan dentaries.

6.14.3 Dentition and dental function

The dryolestidan dentition is characterized by a high number of molars that are compressed mesiodistally. The dental formula comprises five upper and four lower incisors, the canine, four premolars, and up to eight or nine molars. The last molar is often greatly reduced in size, especially in the mandible, and the number of lower molars may exceed that of the uppers by one. The shape of the incisors is peglike (upper and lower) to spatulate (lower), and the mesial lower incisors are procumbent. The canine and premolars are double-rooted and have a large main cusp and a small distal cuspule (mesial cuspule may be present or absent). The lower molars (Fig. 6.55D, E) consist of the trigonid, formed by the proto-, para-, and metaconid, and the talonid cusp (hypoconulid or hypoconid). The lower molars are double-rooted, and the mesial root is much larger than the distal in derived dryolestids, whereas there is little size difference in stem dryolestids ("paurodontids"). In Dryolestidae, the upper molars (Fig. 6.55A–C) have a triangular shape and are strongly compressed mesiodistally and widened labiolingually (less pronounced in stem dryolestids). The crown is bordered by mesial and distal cutting edges (cristae), namely, the paracrista and the metacrista. Because a protocone is not yet developed in the pretribosphenic dryolestidan upper molar, the triangular structure enclosed by paracrista and metacrista is named the primary trigon to distinguish it from the trigon of the tribosphenic molar. The lingual cusp is the paracone, and the metacone sits about half-way between paracone and metastyle in labial direction on the metacrista. The hypotenuse of the triangle of the crown is represented by the stylar region, with the mesial parastyle, the medial stylocone, and the distal metastyle. The parastyle is often enlarged and wing shaped. From the stylocone, a bulge or crest (median crest) can run in lingual direction into the primary trigon basin.

A

B

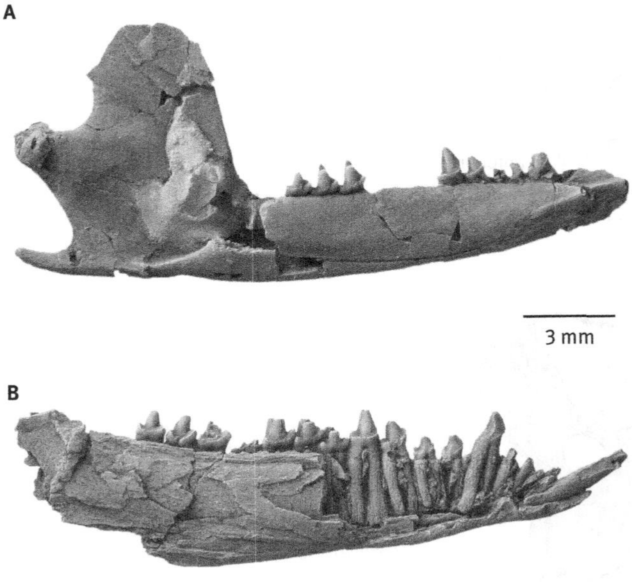

Fig. 6.57: *Dryolestes leiriensis*, left mandibles in medial aspect. (A) Gui Mam 41/79 with c, p1-4, m4-6; (B) Gui Mam 49/79 with i1-4, c, p1-4, m1-7, roots of anterior dentition exposed. Whitened with ammonium chloride. Photographs by Georg Oleschinski.

The deciduous dentition of dryolestids is known from a series of juvenile maxillae and dentaries of *Dryolestes*, *Guimarotodus*, and *Krebsotherium* from the Guimarota coal mine (Martin 1997, 1999a) (Fig. 6.58). The Guimarota dryolestids exhibit a therian mode of tooth replacement, with a diphyodont antemolar dentition and only one generation of molariforms. Incisors, canine, and premolars are replaced in two waves, with i2, i4, p1, and p3 in the first and i1, i3, c, p2, and p4 in the second wave. The last premolar to erupt is p4, and it is present when m6 starts to break through. The first two premolars (P/p1 and P/p2) have premolariform predecessors, whereas the large premolariform P/p3 and P/p4 are preceded by molariform deciduous premolars (Martin 1997, 1999a).

The enamel microstructure of Late Jurassic *Dryolestes* molars is characterized by prisms with incomplete prism sheaths that are embedded in a thick interprismatic matrix (Martin 1999a). This represents the earliest record of prismatic enamel (a derived character characterizing therians) in cladotherians.

The reversed triangular alignment of the upper molar primary trigon and lower molar trigonid creates an embrasure shearing arrangement. This type of mastication

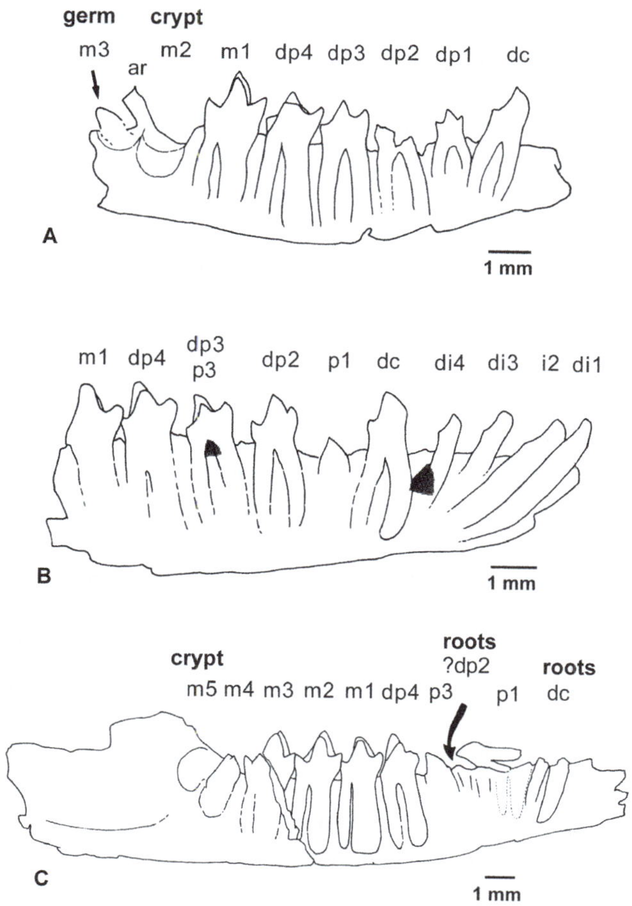

Fig. 6.58: Juvenile left mandibles of *Dryolestes leiriensis* (A, B) and *Guimarotodus inflatus* (C) from the Guimarota coal mine in various stages of tooth replacement. (A) Gui Mam 37/76; (B) Gui Mam 112/76; (C) Gui Mam 129/75. Modified after Martin (1997), with permission of Springer.

is typical of an insectivorous diet. Insect cuticulae are pierced with the pointed cusps and subsequently cut by the crests. Accordingly, the cusps exhibit pitted apical wear facets on their tips, and polished wear facets with striations along protocrista and paracrista of the upper and protocristid and paracristid of the lower molars. These striations have been used to reconstruct the orientation of the chewing movements. The unicusped talonid does not perform a crushing function as in tribosphenic molars but acts together with the hypoflexid as a guide for the paracone of the corresponding upper molar (Schultz and Martin 2011, 2014). The upper molar paracone performs a certain grinding function when it moves along the hypoflexid groove in a ventrolabial direction. The chewing cycle of pretribospenic dryolestidans corresponds to phase I of the tribosphenidan chewing cycle. Phase II, which ends with central occlusion of the protocone within the talonid basin in tribosphenidans, does not occur in the dryolestidan chewing cycle (Schultz and Martin 2014).

6.14.4 Postcranial skeleton and locomotory adaptation

Despite the wealth of newly discovered Mesozoic mammalian skeletons from northeastern China, the skeleton of *Henkelotherium guimarotae* (Fig. 6.59) and a few isolated postcranial bones from the Guimarota coal mine are the only sources of information on dryolestidan postcranial anatomy (Krebs 1991, 2000, Vasquez-Molinero *et al.* 2001, Martin 2013, Jäger *et al.* 2013, in review). *Henkelotherium* was a small mammal (head-body length about 70 mm), with a gracile skeleton adapted for a scansorial lifestyle. The vertebral column is incomplete, with the last two thoracics, five lumbars, and two sacral vertebrae preserved together. The five or six basal tail vertebrae have similar proportions to the presacrals, but subsequently the caudals become increasingly elongated, indicating that *Henkelotherium* used its tail for balancing and steering during leaping. The tail length considerably exceeds

Fig. 6.59: Skeleton of *Henkelotherium guimarotae* holotype on artificial polyester resin matrix. Photograph by Sven Tränkner.

the head-body length (Krebs 1991). In the pectoral girdle, the scapula, clavicle, and interclavicle are present (Krebs 1991, Jäger *et al.* 2013, in review). The coracoid process of the scapula is large, with a suture to the scapula, indicating a separate ossification, and forms part of the glenoid cavity; a procoracoid is not present. The scapula is largely therianlike, with a well-developed spine separating a larger supraspinous from a smaller infraspinous fossa. The dryolestidan humerus is a relatively robust bone with a large bulbous radial and a smaller ulnar condyle at the distal articulation. Both condyles are separated by an intercondylar groove but form an incipient trochlea in posterodistal aspect, as seen at an isolated humerus of *Dryolestes leiriensis* from the Guimarota mine (Martin 2013). The *Dryolestes* humerus also exhibits a distinct medial keel in distal view, which is a character of crown therians (Boyer *et al.* 2000, Chester *et al.* 2010). The distal articulation of the humerus is twisted 15°–20° counterclockwise in relation to the mediolateral plane of the shaft in *Dryolestes*. The radius and the ulna of Henkelotherium are long and slender. The incompletely preserved hand has elongated phalanges and laterally compressed, curved claws, both indicators of an arboreal lifestyle. The pelvis largely resembles that of therians. The ilium is elongate and articulates on its inner side with the two sacrals. The ischium and pubis enclose the large obturator foramen, but only the ischium and ilium contribute to the acetabulum. *Henkelotherium* possesses epipubic bones, a plesiomorphic character present in a number of Mesozoic mammaliaform taxa. The femur (Fig. 6.60) has the head offset from the shaft, with the femoral neck inclined at 30°. The strong greater trochanter reaches the height of the femoral head, and the lesser trochanter is wide. The knee joint appears therianlike, with the two femoral condyles articulating with tibia and fibula. Similarly to the hand, in the foot the metapodials and phalanges are elongated and the claws are laterally compressed and curved (Fig. 6.60), both characters indicating an arboreal lifestyle (Krebs 1991, Jäger *et al.* 2013, in review) (Fig. 6.61).

6.14.5 "Paurodontidae" (stem dryolestids)

Until recently, the following taxa were classified within "Paurodontidae" Marsh 1887. However, there are no unambiguous synapomorphic characters for "paurodontids", and a comprehensive recent cladistic analysis (Averianov *et al.* 2013) demonstrated that they represent a paraphyletic assemblage of stem dryolestids (Fig. 50, taxa below node H). "Paurodontids" are less derived than dryolestids in a number of plesiomorphic dental characters, such as the lower number of molars, weaker mesiodistal compression of molars, and absence of enlargement of the mesial root of the lower molars. Stem dryolestids ("paurodontids") are known from the Late Jurassic to Early Cretaceous of Europe, the Late Jurassic of North America, and possibly the Late Jurassic of Africa. Several taxa of stem dryolestids have been described based on isolated upper or lower dentitions of various ontogenetic ages, which lead to an artificial proliferation of taxa. A recent revision of stem dryolestids by Averianov *et al.* (2013) and Averianov and Martin (2015) lead to a reduction of taxonomic diversity, which is followed here. *Paurodon* (Fig. 6.62) from the Late Jurassic of the Morrison Formation in the western USA was the first stem dryolestid to be described (Marsh 1887), based on a left dentary fragment with c, p2-3, and m1-4. Averianov and Martin (2015) have synonymized the following taxa with *Paurodon*: *Archaeotrigon*, *Araeodon*, and *Foxraptor* (all based on lower dentitions; Simpson 1927b, 1937, Bakker and Carpenter 1990) as well as the upper dentition taxon *Pelicopsis* (Simpson 1927a).

Tathiodon from Quarry 9 of the Morrison Formation is a poorly known stem dryolestid based on a dentary fragment with two molariform teeth (Simpson 1927b), likely dp4 and m1 according to Averianov *et al.* (2013). The upper dentition of *Tathiodon* is known from a maxillary fragment with three molars and an isolated upper

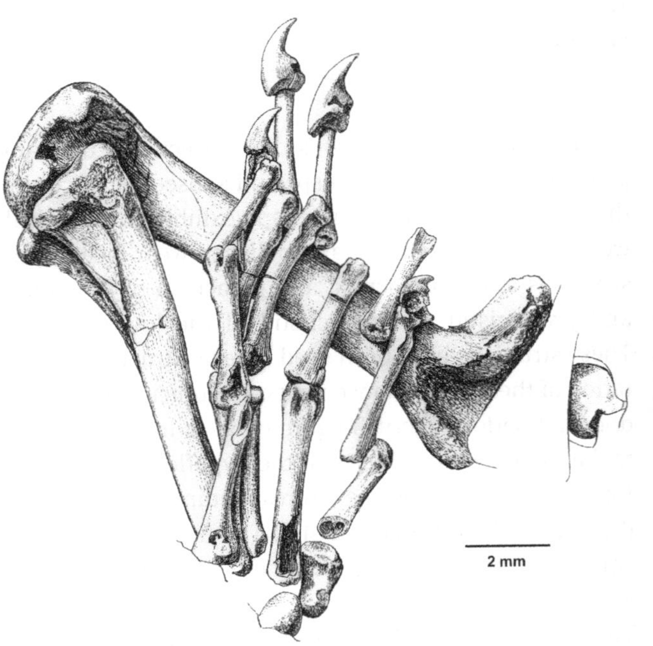

Fig. 6.60: *Henkelotherium guimarotae*, right femur, tibia, and partial foot. Proximal part of femur partially covered by a tail vertebra; femur head partially visible on the right side. From Krebs (1991).

Fig. 6.61: Life reconstruction of *Henkelotherium guimarotae*. From Krebs (1991).

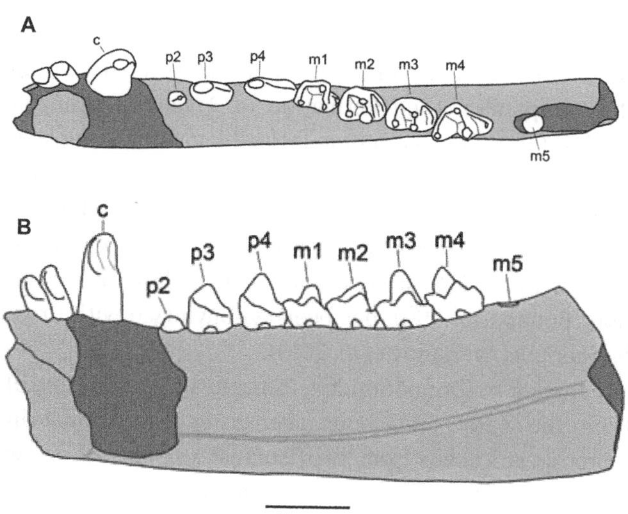

Fig. 6.62: Right mandible of *Paurodon valens* in occlusal (A) and lingual (B) aspects and new interpretation of tooth loci. From Averianov and Martin (2015).

molar, originally described as *Comotherium* by Prothero (1981) and synonymized with *Tathiodon* by Averianov et al. (2013).

Euthlastus is represented by a poorly preserved maxillary fragment with four molars from Quarry 9 (Simpson 1927b), and Engelmann and Callison (1998) referred a well-preserved upper molar from Dinosaur National Monument in Utah (USA) to this taxon. *Euthlastus* is one of the smallest cladotherians from the Morrison Formation, and no lower dentitions are associated with it. The upper dentition of an undescribed stem dryolestid skull from Fruita, Colorado (Morrison Formation) closely resembles the holotype of *Euthlastus*, but is larger. According to Martin (1999a), the similarities are so close that the undescribed skull (dental formula: 4I.1C.4P.4M/4i.1c.4p.5m) from Fruita can be attributed to *Euthlastus*.

Dorsetodon from the Early Cretaceous (Berriasian) Purbeck Limestone Group in southern England is based on several lower molars (Ensom and Sigogneau-Russell 1998) that are similar to the molars of *Paurodon* but are somewhat smaller.

Drescheratherium from the Late Jurassic (Kimmeridgian) of the Guimarota coal mine in Portugal is based on a maxillary fragment with teeth (Fig. 6.63), indicating a dental formula of 5I.1C.3P.5M (Krebs 1998). *Drescheratherium* differs from *Henkelotherium* in its very long and sharp canine and the presence of only three premolars. *Henkelotherium* is the best known cladotherian, and details of its cranial and postcranial anatomy have been given in the introductory paragraph. The dental formula of *Henkelotherium* is 5I.1C.4P.6M/4i.1c.4p.7m, with the last lower molar greatly reduced in size. *Henkelotherium* has more postcanines than other stem dryolestids. In addition to the type specimen described in the monograph by Krebs (1991), there are several dozens of undescribed maxillary fragments and mandibles from the Guimarota coal mine which are currently under study.

Vincelestes neuquenianus from the Hauterivian-Barremian La Amarga Formation in Argentina (Bonaparte 1986b) is one of the best known Early Cretaceous mammals, represented by several skulls (Fig. 6.64) and numerous postcranial elements. The cranial anatomy has been described in a number of publications (Bonaparte and Rougier 1987, Rougier *et al.* 1992, Macrini *et al.* 2007), but detailed descriptions of the dentition and postcranium remain unpublished. Previously, *Vincelestes* was almost universally considered as a pretribosphenic mammal (e.g., Rowe 1993, Rougier *et al.* 1996, Novacek *et al.* 1997, Macrini *et al.* 2007) (Bonaparte 2008 interpreted it as an australosphenidan), but according to the cladistic analysis by Averianov *et al.* (2013), *Vincelestes* is a member of Dryolestida. The molars of *Vincelestes* have no functional protocone and lack an incipiently basined, multicusped talonid, as evident from the only published detailed picture of its dentition (Sigogneau-Russell 1999: Fig. 6.7). According to Averianov *et al.* (2013), *Vincelestes* is dentally very similar to *Paurodon* and not close to tribosphenidans.

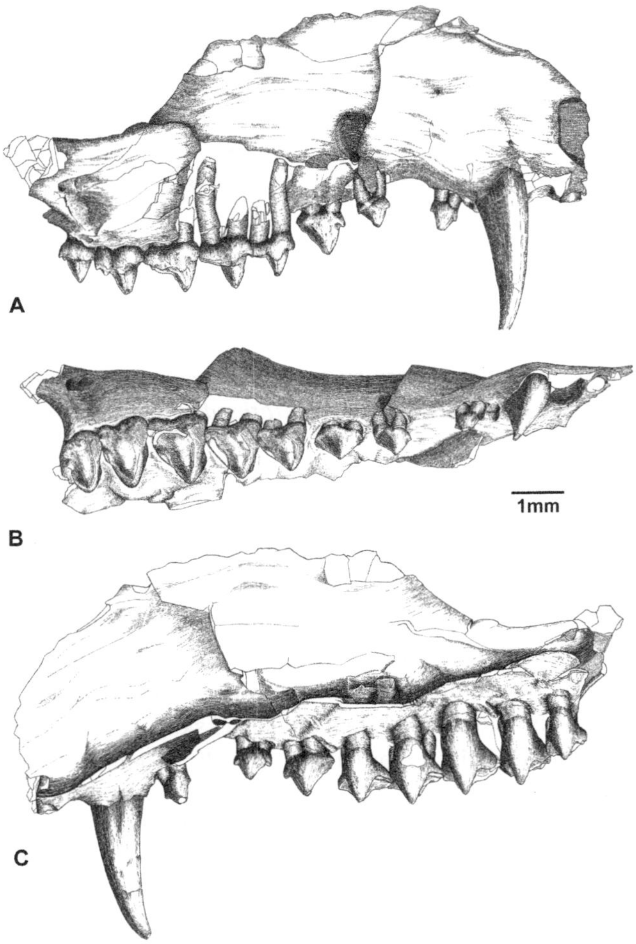

Fig. 6.63: *Drescheratherium acutum*, right maxilla (Gui Mam 4/73, holotype) with C, P1, P3, P4, and M1-5 in lateral (A), occlusal (B), and medial (C) views. Modified after Krebs (1998).

Brancatherulum from the Late Jurassic (Kimmeridgian-Tithonian) Tendaguru Beds in Tanzania is based on a fragmentary, edentulous dentary (Dietrich 1927, Heinrich 1991). According to Averianov and Martin (2015), the dental formula can be interpreted as at least two incisors, double-rooted canine, three double-rooted premolars, and three molars, with the last molar reduced in size. At the penultimate and ante-penultimate tooth, the mesial root is wider than the distal root, but the size difference is not so pronounced as in dryolestids. Based on the limited information available, *Brancatherulum* could be either a stem dryolestidan or a stem zatherian.

Groebertherium (Fig. 6.53 A, B) was interpreted as a member of Dryolestidae by Bonaparte (1986b). Originally based on isolated upper and lower molariform teeth from the Campanian Los Alamitos Formation of Argentina, *Groebertherium* has more recently also been reported from the Campanian Allen Formation of Argentina (Rougier *et al.* 2009a). *Groebertherium* lacks the most striking synapomorphy of Dryolestidae, the unequal roots of lower molars, and clusters outside Dryolestidae in the cladistic analysis of Averianov *et al.* (2013). *Brandonia*, based on isolated upper molariforms from the Los Alamitos Formation (Bonaparte 1990), has been synonymized with *Groebertherium* (Averianov *et al.* 2013).

Donodon (Donodontidae, Sigogneau-Russell 1991b) from the Early Cretaceous (Berriasian) of Anoual in Morocco is known from two isolated upper and a few referred lower molars. The upper molars differ from that of dryolestidans in the presence of a high labial cingulum and the lower molars in having a roughly circular outline in occlusal view (Sigogneau-Russell 1991b).

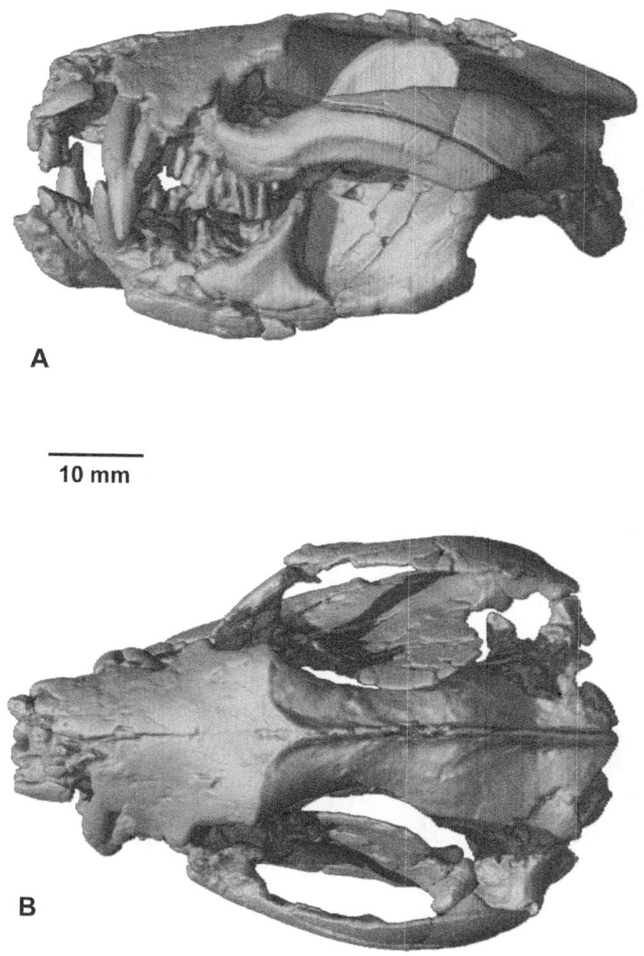

Fig. 6.64: *Vincelestes neuquenianus*, digital rendering of skull and mandible from μCt scan in lateral (A) and dorsal (B) views. From Macrini (2006), with permission of the author.

6.14.6 Dryolestidae

Dryolestidae Marsh 1879b (Fig. 6.50, node H) is clearly distinct from other cladotherians in a number of autapomorphic dental characters. The upper and lower molars are strongly compressed mesiodistally and the uppers widened labiolingually. The mesial root of the lower molars is very strong and enlarged, and the distal root is weak and thin. The number of molars is increased in comparison with other dryolestidans, up to eight or nine, and the alveolar border on the lingual side of the mandible is higher than on the labial side. Dryolestids have their first appearance in the fossil record in the Middle Jurassic of Britain and western Siberia. They are most abundant in the Late Jurassic of Portugal (Kimmeridgian) and of the United States (late Kimmeridgian-early Tithonian Morrison Formation). In Europe, they have been recorded from the Jurassic-Cretaceous boundary of Porto Pinheiro (or Dinheiro) (Portugal), the Early Cretaceous (Berriasian) Purbeck Limestone Group of Britain, the Early Cretaceous (Valanginian) of the Wealden Supergroup on the Isle of Oxney (Britain), and the Early Cretaceous (Barremian) of Spain (Uña and Galve). From the Late Cretaceous Mesaverde Formation of Wyoming (USA), a fragmentary lower molar has been reported and assigned to Dryolestidae indeterminate (Lillegraven and McKenna 1986).

The type genus of Dryolestidae is *Dryolestes* from the Morrison Formation of the western USA (Marsh 1878) and the Guimarota coal mine in Portugal. *Dryolestes* is a large representative of Dryolestidae with a mandible length of about 35 mm (Martin 1999a). *Dryolestes leiriensis* from Guimarota is the best known dryolestid with numerous, partially well-preserved dentaries (Fig. 6.57), a partial skull rostrum, a petrosal preserving the inner ear structure, and isolated postcranial bones (Martin 1999a, 2000, 2013, Ruf *et al.* 2009). *Dryolestes* has a dental formula of 4I.1C.4P.8M/4i.1c.4p.8-9m (Figs. 6.65 and 6.66), and it differs from other dryolestids in its large metacone that projects into the primary trigon basin and its enlarged procumbent i1 in the mandible.

Amblotherium from the Late Jurassic (late Kimmeridgian-early Tithonian) Morrison Formation of the western USA and the Early Cretaceous (Berriasian) of the Purbeck Limestone Group in southern England (Lulworth Formation) is the smallest known dryolestid (Owen 1871, Simpson 1928, 1929). *Amblotherium* is represented by numerous mandibles and mandibular and maxillary fragments. The estimated mandible length is 18 mm and the lower dental formula 4i.1c.4p.7-9m. Lower molars differ from that of other dryolestids in their erect paraconid, which is separated from the simple and pointed metaconid by a deep V-shaped notch. The roots of m1 and m2 are of subequal size and the coronoid process rises at a lower angle (45°) than in other dryolestids.

Achyrodon nanus from the Lulworth Formation (Owen 1871) was referred by Simpson (1928) to the genus *Amblotherium* (*A. nanum*) but was recognized as generically separate in the cladistic analysis by Averianov *et al.* (2013). It differs from *Amblotherium pusillum* Owen 1866 in its somewhat smaller size, double-rooted lower canine, a p2 that is smaller than p1, molar talonids without a shelflike labial extension, and upper molars without cusps along the metacrista (Averianov *et al.* 2013).

Crusafontia from the Early Cretaceous (Barremian) of Galve and Uña in Spain is represented by several mandibular fragments, isolated upper and lower premolars, and molars (Henkel and Krebs 1969, Krebs 1993, Martin 1998). It differs

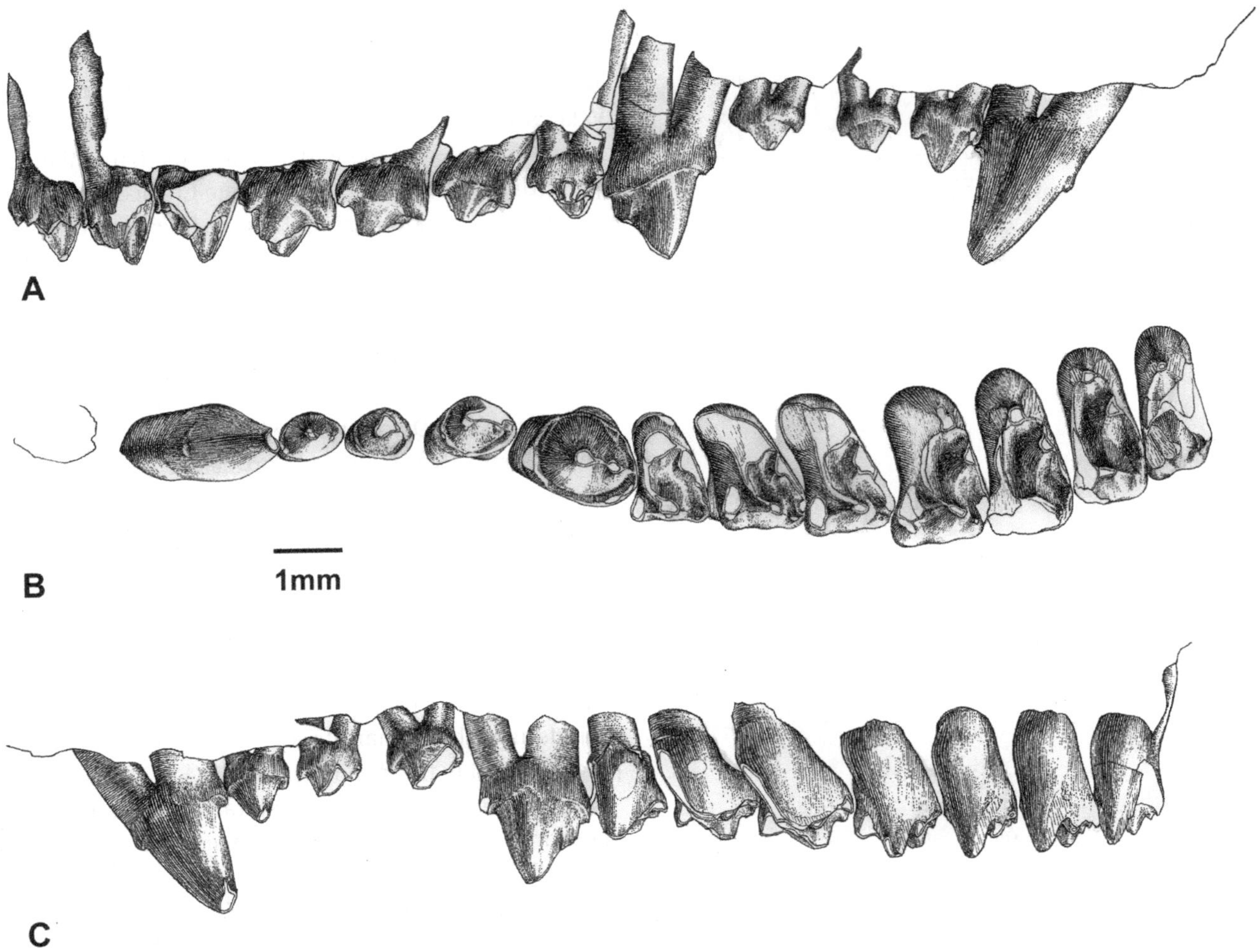

Fig. 6.65: *Dryolestes leiriensis* (Gui Mam 51/75), right upper dentition with C, P1-4, M1-7 in labial (A), occlusal (B), and lingual (C) views. Modified after Martin (1999a).

from the Jurassic dryolestids in the absence of the coronoid and splenial bones at the mandible (Fig. 6.67) and a more reduced Meckel's groove. It is different from other dryolestids in the absence of a median stylar cusp and median ridge, as well as the lack of a metacone at the upper molars.

Guimarotodus from the Late Jurassic (Kimmeridgian) Guimarota coal mine is represented by about a dozen mandibular fragments with teeth. The mandible is robust, and the molars differ from all other dryolestids in the presence of an inflated metaconid. The lower dental formula is ?i.1c.4p.8-9m. Upper teeth of *Guimarotodus* have not been identified so far (Martin 1999a).

Krebsotherium is well represented by about 50 dentaries and about 20 maxillary fragments from the Guimarota coal mine. *Krebsotherium* is smaller than *Dryolestes*, and only slightly larger than *Amblotherium*. Differing from *Amblotherium*, the paraconid is not erect and much smaller than the metaconid. The dental formula of *Krebsotherium* is 5I.1C.4P.8-9M/4i.1c.4p.8-9m (Martin 1999a).

Lakotalestes is a small dryolestid from the late Berriasian-Valanginian Lakota Formation in South Dakota, and it is based on an isolated upper molar with a large and mesially placed stylocone (Cifelli *et al.* 2014). It represents the geologically youngest named dryolestid from North America.

Laolestes differs from all other dryolestids in the bifid metaconid of its lower molars. The lower dentition dental formula is 4i.1c.4p.8m. Upper dentitions had originally been described as *Melanodon* (Simpson 1927b), which was synonymized with the lower dentition taxon *Laolestes* by Martin (1999a). *Laolestes* has been reported from the Late Jurassic (late Kimmeridgian-early Tithonian) of the Morrison Formation in the western USA, the latest Jurassic/earliest Cretaceous of Porto Pinheiro (or Dinheiro) in Portugal, and the Early Cretaceous (Valanginian) of the Cliff End bonebed of Hastings in southern England.

Phascolestes is known from dentary and maxillary fragments with teeth from the Early Cretaceous (Berriasian) of the Purbeck Limestone Group (Lulworth Formation)

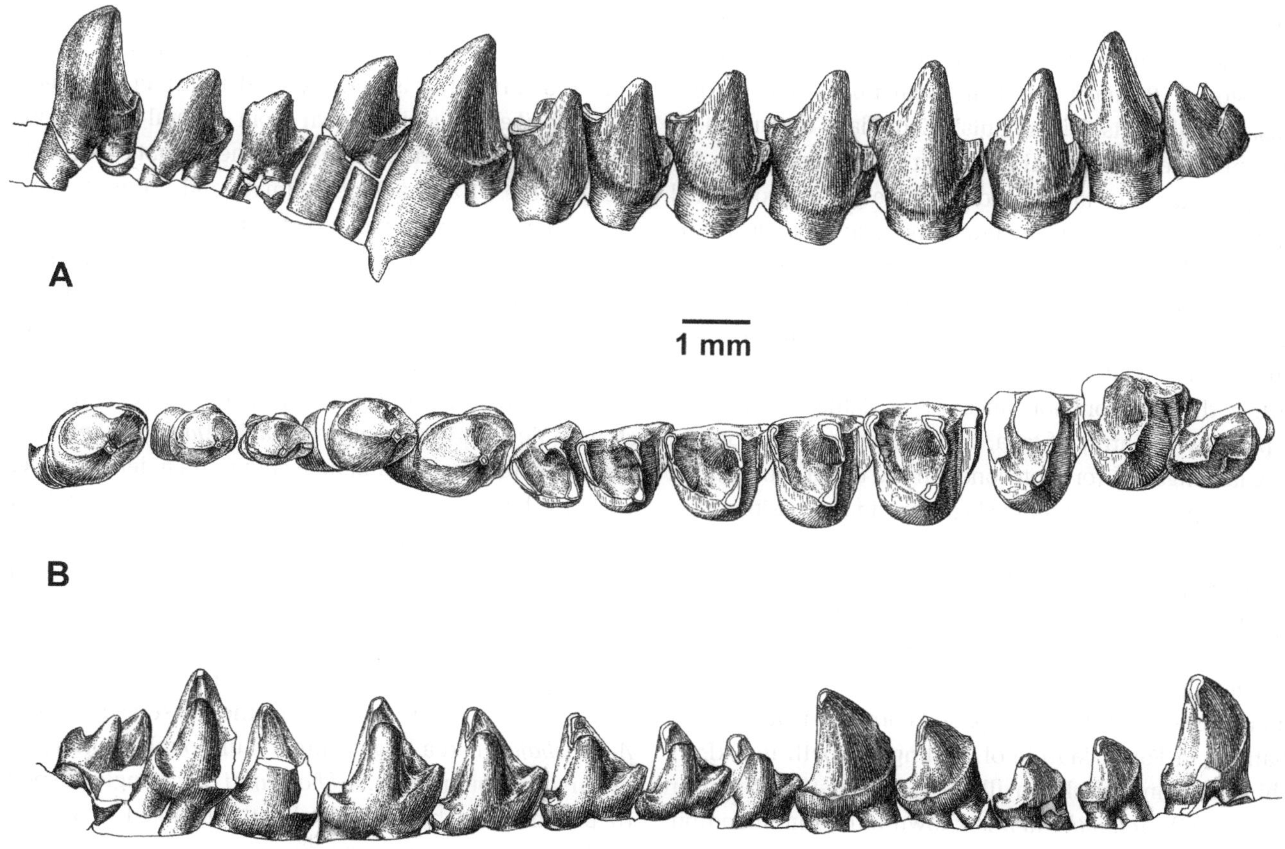

Fig. 6.66: *Dryolestes leiriensis* (holotype, Gui Mam 130/74), left lower dentition with c, p1-4, m1-8 in labial (A), occlusal (B), and lingual (C) views. Modified after Martin (1999a).

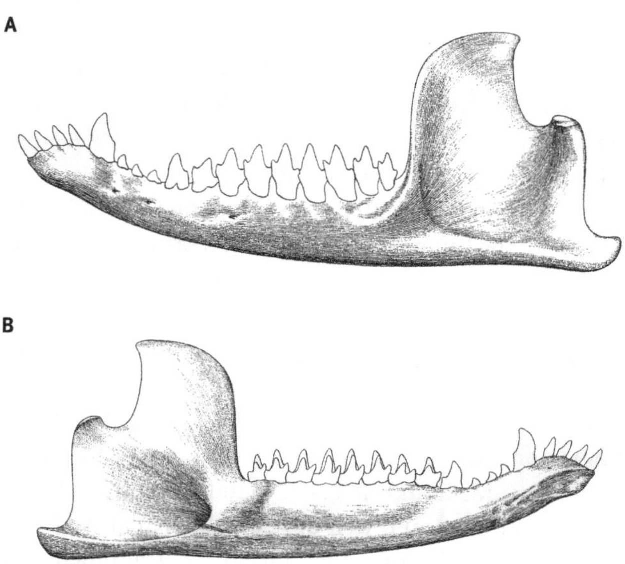

Fig. 6.67: *Crusafontia cuencana*, restoration of left mandible with i1-4, c, p1-4, m1-8. (A) Lateral and (B) medial aspect. Length about 20 mm. Modified after Krebs (1971), from original drawings.

of southern England (Owen 1871). Averianov *et al.* (2013) synonymized the lower dentition taxon *Peraspalax* (Owen 1871) and the upper dentition taxon *Kurtodon* (Osborn 1887) from the same formation with *Phascolestes*, as already suggested by Simpson (1928) and Sigogneau-Russell and Kielan-Jaworowska (2002).

Portopinheirodon from the Jurassic-Cretaceous boundary of Porto Pinheiro (or Dinheiro) (Portugal) is based on a strongly asymmetrical upper molar with an enlarged and protruding metastylar region, small parstyle, and high stylocone connected by a median ridge to the paracone (Martin 1999a).

6.15 Zatheria (including stem Zatheria)

Zatheria McKenna 1975 was defined as the clade including "Peramura" and Tribosphenida. Because "Peramura" is paraphyletic (Sigogneau-Russell 1999, Martin 2002,

Kielan-Jaworowska *et al.* 2004), Luo *et al.* (2002) defined Zatheria as a node-based taxon "comprising the common ancestor of *Peramus* and living marsupials and placentals, plus all of its descendants". Zatheria is diagnosed by the presence of a lingual cingulum (or protocone) on the upper molars, a metacone and paracone that are positioned approximately at the same level, and a cusplike metacone comparable in size with the paracone (Averianov *et al.* 2013). The talonid of the lower molars is basined and has more than one cusp (hypoconid and hypoconulid in *Peramus*, plus entoconid in more derived zatherians). The stem lineage below the last common ancestor of Zatheria comprises a paraphyletic array of taxa, that had not yet developed a lingual cingulum (or protocone) on the upper molars, that have the paracone remaining in a more lingual position than the metacone, in which the metacone is still much smaller than the paracone, and that have only one talonid cusp (hypoconulid or hypoconid) (Averianov *et al.* 2013).

The first stem zatherian to become known is *Amphitherium* (Fig. 6.68) from the Middle Jurassic (middle Bathonian) Stonesfield slate (Sharps Hill Formation) and the late Bathonian Forest Marble of Kirtlington, both in Oxfordshire, England (de Blainville 1838, Owen 1871, Simpson 1928, 1929). *Amphitherium* is known only from mandibular remains and differs from dryolestidans in the presence of five premolars (Butler and Clemens 2001). The talonid on the lower molars is larger than that of dryolestidans and overlaps the following trigonid labially. The number of molars (six or seven) is lower than that of Dryolestidae but higher than that of Zatheria (up to five) and corresponds to that of stem dryolestids such as *Henkelotherium* (seven). In contrast to dryolestidans, the angular process of the mandible is downturned in *Amphitherium*.

Chunnelodon is based on isolated lower molars from the Early Cretaceous (Berriasian) of the Purbeck Limestone Group of Dorset in southern England (Ensom and Sigogneau-Russell 1998). The lower molar trigonid is very narrow labiolingually, the paraconid is small and recurved, and the talonid is a sharp and relatively high cusp. The systematic position of *Chunnelodon* is uncertain, and currently it is best regarded as a stem zatherian.

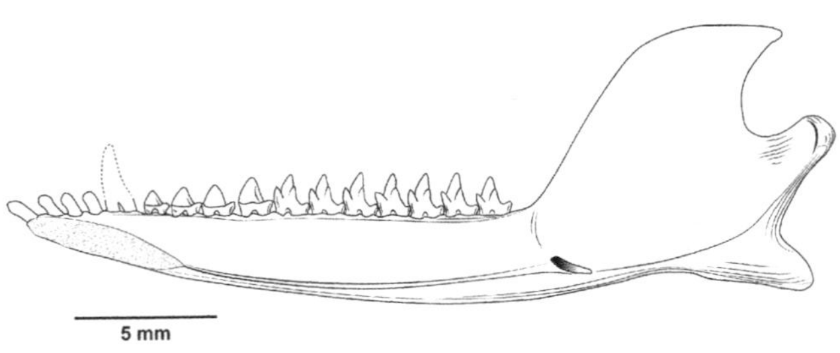

Fig. 6.68: *Amphitherium prevostii*, restoration of right mandible with four incisors, canine, four premolars, and seven molars in medial aspect. Redrawn from Simpson (1928). The number of premolars in Simpson's restoration is four, but Butler and Clemens (2001) argued for the presence of five premolars in *Amphitherium*.

Amphibetulimus is known from several edentulous and three tooth-bearing dentary fragments that collectively preserve the p1 and antepenultimate and ultimate lower molars, and one upper molar (Lopatin and Averianov 2007, Averianov *et al.* 2015). It is unique among stem zatherians in its lingually widely open trigonids on the distal lower molars. *Amphibetulimus* differs from more derived stem zatherians in its unicusped talonid without an incipient talonid basin. The lack of an ectotympanic facet and the long straight Meckel's groove suggest that in *Amphibetulimus* the transitional mammalian middle ear was in a derived state, whereby the ear ossicles were connected to the dentary by a narrow Meckel's cartilage (Averianov *et al.* 2015).

Arguimus khosbajari from the Early Cretaceous (Aptian or Albian) of Höövör in the Gobi Desert (Mongolia) is known from a dentary fragment with two premolars (p3 and p4) and three molars (m1-3) (Dashzeveg 1979, Butler and Clemens 2001). Judging by the preserved alveoli, *A. khosbajari* had a large canine, four double-rooted premolars, and four molars. *A. khosbajari* has a single-cusped (hypoconulid or hypoconid) slightly enlarged talonid at the lower molars (Sigogneau-Russell 1999).

Arguitherium cromptoni, also from Höövör (Dashzeveg 1994), is represented by a dentary fragment with p3-5 (Sigogneau-Russell 1999, Martin 2002). The p5 was originally interpreted as m1 by Dashzeveg (1994) and has a single-cusped elongated talonid (Sigogneau-Russell 1999, Martin 2002). Lopatin and Averianov (2006a) described additional dentary fragments of *Arguimus khosbajari* and interpreted the teeth preserved in the type specimen as p4-5 and m1-3. They confirmed Dashzeveg's (1994) original interpretation of the teeth preserved in the holotype of *Arguitherium cromptoni* as p4-5 and m1 but considered *A. cromptoni* to be conspecific with *A. khosbajari*.

Nanolestes (Fig. 6.69) has been reported from the Late Jurassic of the Guimarota coal mine in Portugal, the earliest Cretaceous (Berriasian) of Porto Pinheiro (or Dinheiro) (Portugal), and the Late Jurassic (Oxfordian) Qigu Formation of Liuhuanggou in the Junggar Basin of Xinjiang, China (Martin 2002, Martin *et al.* 2010a). It is one of the best known stem zatherians and is represented by mandibular remains and isolated upper and lower teeth. *Nanolestes* has a gracile and elongated mandible, with a distinct Meckel's groove and well-developed angular process. The lower dental formula is 4i.1c.5p.5m. The unicusped talonid of the lower molars is elongated and has no basin (Krusat 1969, Martin 2002). On the oblique cristid, there sits a small additional cuspule that does not correspond to any of the standard talonid cusps. The upper molars have

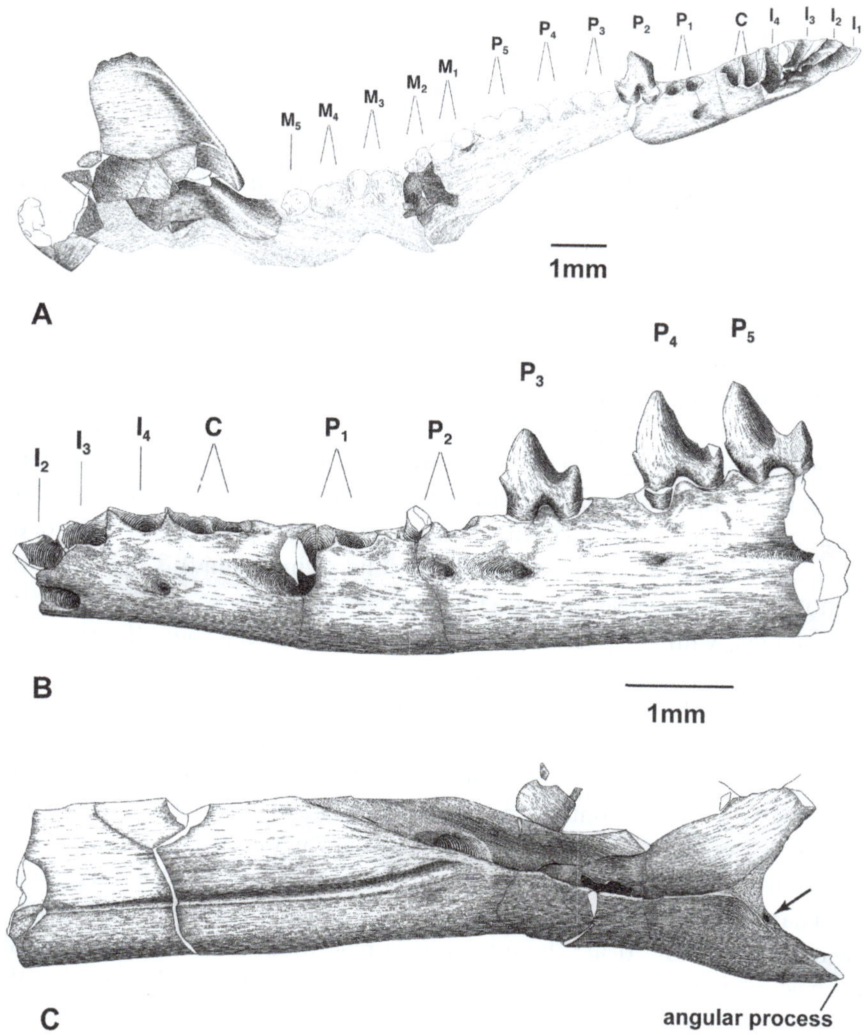

Fig. 6.69: *Nanolestes drescherae*, mandibular fragments in lateral (A, B) and medial (C) aspects. (A) Right mandible partially preserved as plastic cast of natural mold (Guimarota 19). (B) Anterior fragment of left dentary (Gui Mam 66/79). (C) Posterior part of right dentary (Guimarota 19). Modified after Martin (2002), reprinted by permission of the Society of Vertebrate Paleontology, www.vertpaleo.org.

a small extoflexus, and the paracone is the lingualmost cusp; it is lower than the stylar cusps. The metacone sits in the middle of the metacrista, and the paracrista bears additional cusps (Martin 2002).

Palaeoxonodon (Fig. 6.70) is represented by abundant isolated upper and lower molars from the Middle Jurassic (Bathonian) Forest Marble Formation at Kirtlington, England (Freeman 1976, 1979, Sigogneau-Russell 2003). Recently, a largely complete mandible has been described from the late Bathonian of the Isle of Skye (Close *et al.* 2015). According to Close *et al.* (2015), the dental formula of this mandible is 4i.1c.5p.5m, which is intermediate between the primitively higher number of postcanines (5p.6-7m) in *Amphitherium* and the reduced postcanine count (5p.3m) of *Peramus* and tribosphenidans. The mandible has a well-developed Meckel's groove and a possible insertion scar for a rudimentary coronoid. *Palaeoxonodon* has a short and very narrow, but distinctly basined, single-cusped talonid with hypoconulid (hypoconid according to Davis 2011) and entocristid (Kielan-Jaworowska *et al.* 2004,

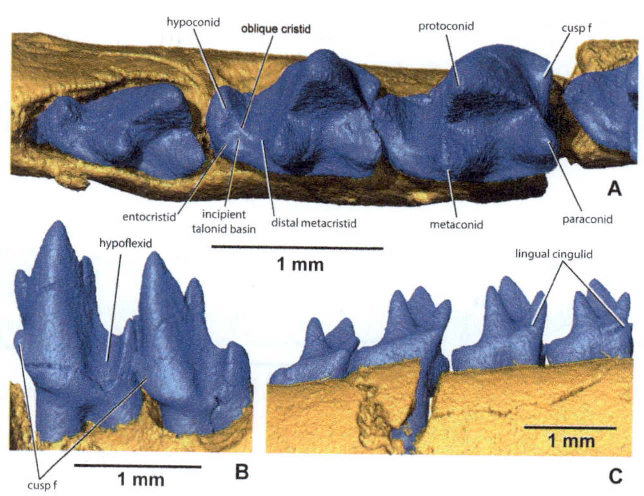

Fig. 6.70: Left lower postcanine dentition of *Palaeoxonodon oolithicus*. (A) m3-5 in occlusal view; (B) m3-4 in labial view; (C) m1-4 in oblique lingual view. Modified after Close *et al.* (2015), reprinted with permission from John Wiley and Sons.

Close *et al.* 2015). On the upper molars, the metacone is shifted lingually and sits only slightly more labially than the larger paracone. It differs from *Peramus* in the absence of a lingual cingulum on the upper molars. Close *et al.* (2015) synoymized *Kennetheridium* which had been based on isolated molars (Sigogneau-Russell 2003b) with *Palaeoxonodon*. Panciroli *et al.* (2018) reported two additional partial dentaries from Scotland that exhibit a deep masseteric fossa with prominent anterior crest. The mandibular foramen is offset from the Meckel´s groove and positioned below the alveolar plane. Panciroli et al´s (2018) phylogenetic analysis confirms the sister taxon relationship between *Paleoxonodon* and *Amphitherium*.

Mozomus is known from a single fragmentary dentary with seven teeth, interpreted as two premolars and m1-5 (Averianov *et al.* 2010b) from the Early Cretaceous (Aptian) Shahai Formation of Liaoning, China (Li *et al.* 2005). According to the description by Li *et al.* (2005), the lower molars of *Mozomus* have relatively large talonids, with one regular talonid cusp and additional irregular cuspules at the lingual side; a basin is not present. From the published photographs (Li *et al.* 2005: Fig. 6.2), a judgment on the number of talonid cusps is difficult, and it is possible that more than one regular talonid cusp is present.

Tendagurutherium from the Late Jurassic (Kimmeridgian-Tithonian) of Tendaguru, Tanzania, is based on a posterior dentary fragment with a damaged last molar (Heinrich 1998). The talonid has been broken and partially displaced postmortally, and it is unclear if it was basined or not. The dentary has a sharp and posteriorly projecting angular process. A coronoid bone and Meckel's groove are present.

Abelodon from the Early Cretaceous (Barremian-Aptian) of Cameroon is based on a single upper molar (holotype) and an attributed lower molar (Brunet *et al.* 1990). The upper molar of *Abelodon* is characterized by a deep ectoflexus. The paracone sits more lingually than the metacone, and a lingual cingulum is not present.

Magnimus is based on isolated upper and lower molars from the Early Cretaceous (Berriasian) of the Purbeck Limestone Group in Dorset, southern England. The upper molars lack an ectoflexus, and the paracone is in a more lingual position than the metacone (Sigogneau-Russell 1999). On the lower molars, the talonid is relatively small compared with the large trigonid, and it is distinctly basined with a single cusp; the presence of only one talonid cusp (hypoconulid or hypoconid) is more primitive than the situation in *Peramus*.

Minimus from the Early Cretaceous (Berriasian) of Anoual, Morocco, is represented by isolated lower molars. *Minimus* is a very small taxon (lower molars less than 1 mm long) and differs from *Magnimus* in its labiolingually relatively wider trigonid and a smaller, unicusped talonid with an incipient basin (Sigogneau-Russell 1999).

Afriquiamus, also from Anoual, is based on an isolated upper molar. The upper molar has a very deep indentation (ectoflexus) and appears almost symmetrical. The metacone is placed relatively lingually but is lower than the paracone; a lingual cingulum is absent (Sigogneau-Russell 1999).

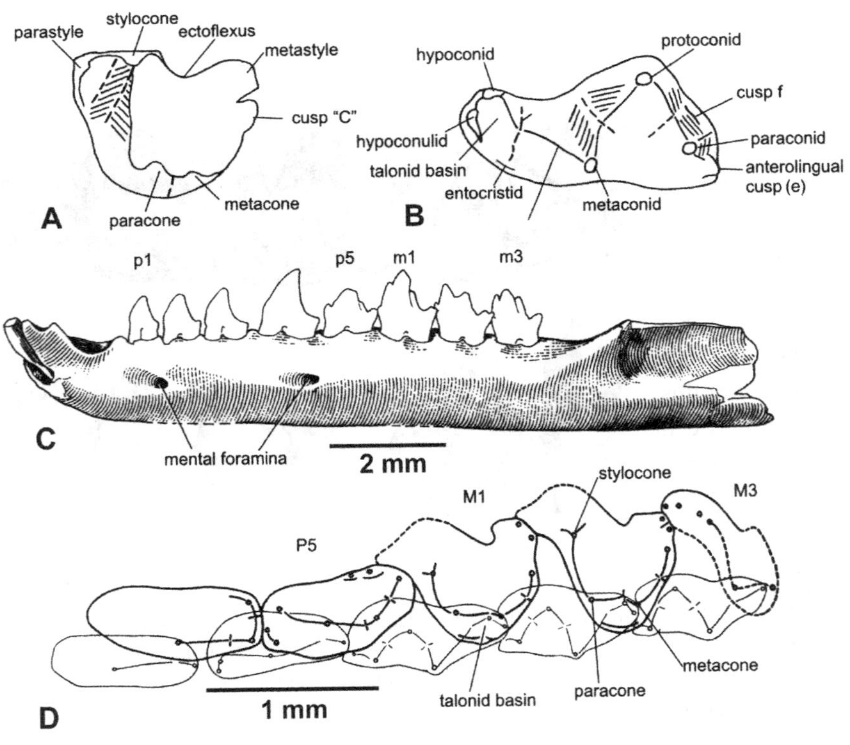

Fig. 6.71: Teeth and mandible of *Peramus*. (A) Upper and (B) lower molar schematic drawing; (C) left dentary in lateral view; (D) restoration of part of the upper and lower postcanine teeth in superposition, in occlusal view. From Kielan-Jaworowska *et al.* (2004) after Sigogneau-Russell (1999) (A) and Clemens and Mills (1971) (B).

Peramus (Fig. 6.71) from the Early Cretaceous (Berriasian) Lulworth Formation of the Purbeck Limestone Group of Dorset in southern England (Owen 1871) has played a crucial role in the understanding of the evolution of the tribosphenic molar. *Peramus* is represented by upper and lower dentitions, which were not found in association but regarded as conspecific (Clemens and Mills 1971). Davis (2012) restudied the hypodigm (nine specimens) of *Peramus tenuirostris* housed in the Natural History Museum London by μCT. He observed minor differences in size and morphology and established three additional taxa: *Peramus dubius*, *Kouriogenys minor*, and *Peramuroides tenuiscus*. In the large sample of the relatively closely related Late Jurassic dryolestids *Dryolestes leiriensis* (87 maxillae and dentaries) and *Krebsotherium lusitanicum* (77 maxillae and dentaries) from the Guimarota coal mine, a considerable intraspecific variation in size and morphology has been observed (Martin 1999a), which casts some doubt on this new interpretation of the *Peramus* hypodigm. The dental formula of *Peramus* has originally been reconstructed as ?I.1C.4P.4M/4i.1c.4p.4m (Simpson 1928, Mills 1964, Clemens and Mills 1971). However, the tooth identified as M1 is narrower than M2, has only two roots, and is semimolariform like m1. Therefore, McKenna (1975) considered the dental formula of *Peramus* comprising five premolars and three molars, an interpretation adopted by most subsequent authors (e.g., Prothero 1981, Novacek 1986, Butler and Clemens 2001, Martin 2002, Kielan-Jaworowska *et al.* 2004, Davis 2012). *Peramus* has at least two talonid cusps on the lower molars, hypoconid and hypoconulid, which are connected by the hypocristid (Fig. 6.72). At the lingual side of the talonid, an entocristid or small isolated entoconid (without connecting cristid) is variably present (Clemens and Mills 1971, Davis 2012). Kielan-Jaworowska *et al.* (2004) stated the presence of a small talonid basin in *Peramus*, but according to Davis (2011), it is not basined. On the upper molars, paracone and metacone are sitting lingually side by side, and a small lingual cingulum (as structural predecessor of the protocone) is present (Fig. 6.72). According to the reconstruction of the occlusal relationships of *Peramus* by Clemens and Mills (1971), the lingual portion of the upper molar (with the lingual cingulum) occludes with the talonid of the lower molars. Because of the absence of a protocone on the upper molars and a functional talonid basin at the lowers, the molars of *Peramus* are not yet tribosphenic.

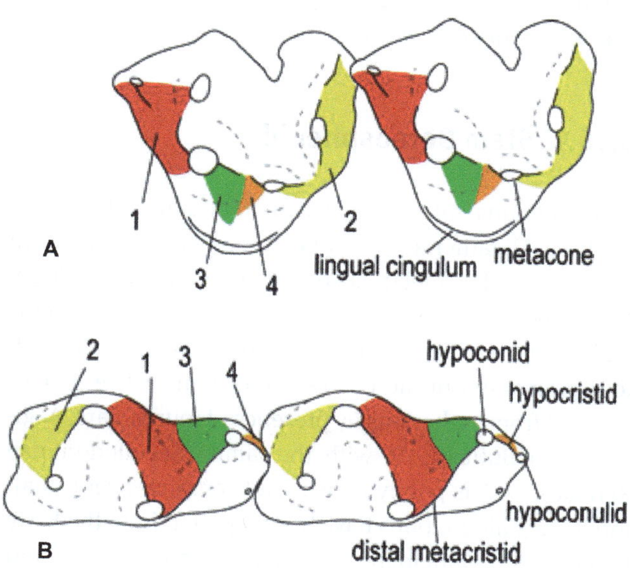

Fig. 6.72: Upper (A) and lower molars (B) of *Peramus* in occlusal view with structures and wear facets. Modified after Davis (2011), with permission of Springer.

6.16 Boreosphenida

Boreosphenida Luo *et al.* 2001b possesses fully tribosphenic molars and comprises the common ancestor of *Kielantherium* and living marsupials and placentals plus all of its descendants (Luo *et al.* 2002). The term tribosphenic molar (from Greek τρίβειν, to grind and σφήν, wedge) was introduced by Simpson (1936; see also Osborn 1907) in allusion to its combined grinding and cutting functions. The tribosphenic molar has three main cusps that enclose the trigon basin: the protocone in the lingual position, the paracone (cusp A of stem mammals) in mesiolabial position, and the metacone (cusp B) in distolabial position. The protocone is a neomorphic cusp of the tribosphenic molar that did not exist in pretribosphenic mammalian molars, but homoplastic protocone-like structures ("pseudoprotocone") occur in several mammaliaform lineages (Luo 2007a). It may be surrounded by cingula, the mesial precingulum, and the distal postcingulum. In more derived forms, a new cusp, the "hypocone", can be present on the distal side giving the tooth a roughly rectangular shape. The "hypocone" can have different origins—it may develop from a cingulum distal to the protocone (the usual condition in placentals that have a "hypocone"), or it may be a distolingually displaced metaconule (the usual condition in marsupials that have a "hypocone"). Beck *et al.* (2008) distinguished between the two by calling the first a "cingular hypocone" and the second a "metaconular hypocone". The term "pseudohypocone" has also been used for the distolingually displaced metaconule. The area labial to the paracone and metacone is the stylar shelf, which is frequently wide in early forms and can have an ectoflexus. The main cusps on the stylar shelf are (from mesial to distal) the parastyle, stylocone, mesostyle, and metastyle. In metatherians, the stylar cusps are conventionally referred to as A–E, of which stylar cusp B is generally considered to be homologous to the stylocone of eutherians, whereas the homology of the other cusps is less clear (Kielan-Jaworowska *et al.* 2004). Given that early metatherians (e.g., *Kokopellia*) only have a stylocone, the stylar cusps of metatherians presumably evolved independently of those of eutherians. The parastylar wing is separated from the trigon by a deep groove for occlusion with the protoconid. The mesial region of the lower tribosphenic molar consists of three cusps in triangular arrangement: the protoconid (homologous to cusp a of stem mammals) in the labial position, the paraconid (cusp b) in the mesiolingual position, and the metaconid (cusp c) in the distolingual position. The distal region is formed by the talonid basin that is surrounded by three cusps, the lingual entoconid, the labial hypoconid, and the distal hypoconulid. As pointed out before, there is an ongoing discussion if the hypoconulid or the hypoconid is homologous to cusp d of stem mammals (Davis 2011). A distinct crest, the oblique cristid, runs from the hypoconid to the distal wall of the trigonid, and an additional cusp, the mesoconid, may be present on the oblique cristid. The indentation on the lingual side between trigonid and oblique cristid is the hypoflexid, which is where the main shearing occurred in pretribosphenic molars (Schultz and Martin 2014). The crest extending distally from the metaconid in early tribosphenidans is the distal metacristid. For detailed anatomical description of tribospheny, see Kielan-Jaworowska *et al.* (2004).

During the masticatory cycle, the alternating trigon and trigonid perform a cutting function, whereas the upper molar protocone performs a grinding function in the talonid basin, similar to a pestle in a mortar (Crompton and Hiiemae 1969). The combination of a cutting and grinding function is an important innovation in mammalian evolutionary history and was the precondition for the great diversity of Cenozoic mammalian dentitions. Although protocone-like structures have evolved in several mammaliaform lineages (e.g., docodontans, shuotheriids, and australosphenidans), their dentitions never attained a comparable diversity. This innovation enabled the mammals to feed on all kind of tough plant materials and greatly contributed to their evolutionary success. The occlusion of the protocone with the talonid basin creates characteristic wear facets inside the talonid basin and on the mesio- and distolingual flanks of the protocone (Crompton 1971). The development of a protocone and a fully basined talonid caused a fundamental functional shift in the talonid region. The main shearing function of the pretribosphenic molar occurred at the hypoflexid on the labial side of the talonid, when the paracone of the upper molar traveled along the hypoflexid groove in a ventrolabial direction during the one-phased masticatory cycle (Crompton and Sita-Lumsden 1970, Crompton and Kielan-Jaworowska 1978, Schultz and Martin 2014). In contrast to pretribosphenic mammals, tribosphenic forms have a two-phased chewing cycle, with a second phase of tooth contacts in which the protocone moves out of the talonid basin after centric occlusion (Crompton and Hiiemae 1969, Schultz and Martin 2014).

Boreosphenida (Greek Βορέας, mythological personification of northern wind, and σφήν, wedge), the Northern Hemisphere tribosphenic mammals, are separate from the Australosphenida, the Southern Hemisphere tribosphenic mammals. Boreosphenida differ from australosphenidans in the placement of the mandibular angle, that is not elevated above the level of the postcanine alveoli, but is either at the level of the ventral border of the mandible (metatherians) or projects below the level of the horizontal ramus (majority of eutherians) (Luo *et al.* 2002). They differ from mammaliaforms and australosphenidans by the absence of the plesiomorphic postdentary trough for accommodation of the postdentary bones, and they differ from all mammals except Australosphenida by the presence of tribosphenic molars. They differ from *Shuotherium* in the lack of the postdentary trough and by having the talonid placed distally (rather than mesially) to the trigonid of the lower molars, and from Australosphenida by the lack of the continuous mesiolingual cingulid on the lower molars.

Boreosphenida comprises the "tribotherians" (i.e., taxa on the stem leading to Boreosphenida), Metatheria Huxley 1880, and Eutheria Gill 1872. Boreosphenida originated in the Northern Hemisphere and were restricted to it in the Cretaceous. They are present from the early Cenozoic to Recent of the World.

6.16.1 Stem Boreosphenida

"Tribotherians" of the Cretaceous that cannot be attributed to Metatheria or Eutheria are treated as stem Boreosphenida (Kielan-Jaworowska *et al.* 2004). They are known from isolated teeth or fragmentary dentitions and differ from metatherians (except for *Sinodelphys* with four premolars) in the possession of four or five premolars and from eutherians by presence of four molars. Aegialodontia Butler 1978, with the family Aegialodontidae Kermack 1967, is poorly known and comprises very small forms. Aegialodontians have been reported from the Early Cretaceous (Valanginian to Aptian or Albian) of England and Mongolia (Gobi Desert). *Aegialodon* (Fig. 6.73) from the Early Cretaceous (Valanginian) Cliff End Bonebed of

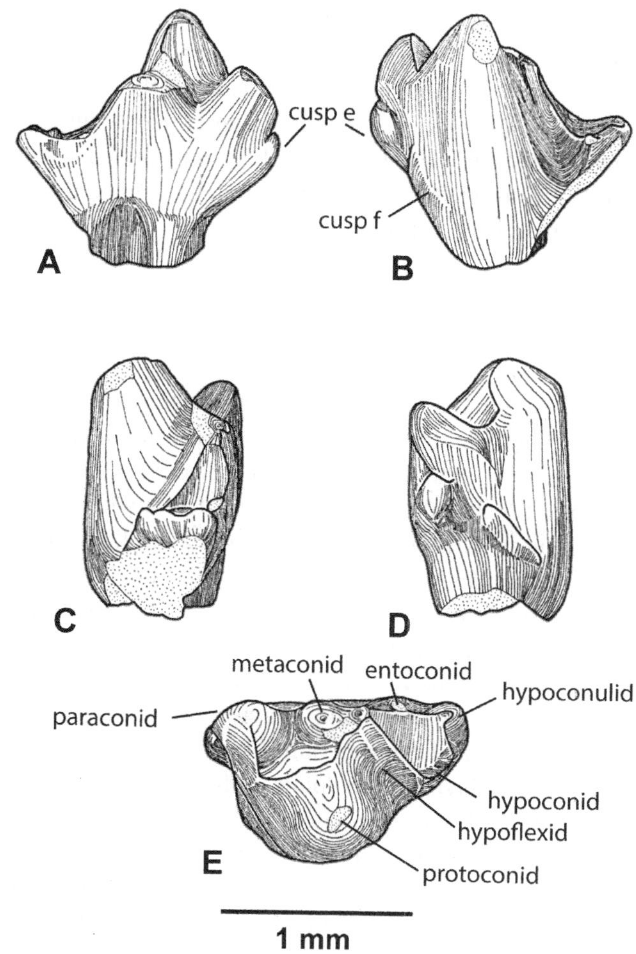

Fig. 6.73: *Aegialodon dawsoni*, lower left molar (holotype) in lingual (A), labial (B), distal (C), mesial (D), and occusal (E) views. Modified after Kermack *et al.* (1965), with permission of the Royal Society.

the Wadhurst Formation in Britain is based on a single lower molar with a basined talonid and possibly three talonid cusps (the presence of an entoconid is uncertain) (Kermack *et al.* 1965).

Kielantherium from the Aptian-Albian of Hövöör in Mongolia is based on a lower molar (holotype) and a dentary with four molars and possibly four double-rooted alveoli for premolars (Dashzeveg 1975); the presence of five premolars cannot be excluded. The trigonid is higher than in *Aegialodon*, and the talonid is more elongated and rectangular and has only two cusps. *Kielantherium* has a distinct Meckel's groove and a facet for the coronoid bone. Lopatin and Averianov (2006b) described an upper molar of *Kielantherium* from Hövöör, which is the first known upper aegialodontid molar. The molar has a small and very low, but distinct protocone as previously hypothetically reconstructed for *Aegialodon* by Crompton (1971). Averianov *et al.* (2010b) proposed that *Kielantherium* might be a stem metatherian with a distinct Meckel's groove, diverging prior to the loss of a postcanine position. An independent separation of the postdentary bones (and loss of Meckel's groove) in eutherians and metatherians has recently been suggested based on developmental studies on extant *Monodelphis* (Urban *et al.* 2017), which would gain additional support if *Kielantherium* is indeed a stem metatherian.

Tribactonodon from the Early Cretaceous (Berriasian) of the Purbeck Limestone Group of Dorset, southern England (Sigogneau-Russell *et al.* 2001), has been tentatively assigned to Aegialodontia. Before the discovery of the possible eutherian *Juramaia* from the Middle Jurassic of northeastern China (Luo *et al.* 2011b), *Tribactonodon* was the geologically oldest boreosphenidan. The *Tribactonodon*-type specimen (right lower molar) has a relatively long talonid, with a distinct hypoconid and entoconid well separated from the taller hypoconulid (Sigogneau-Russell *et al.* 2001, Kielan-Jaworowska *et al.* 2004).

The "Trinity Theria" or "Trinity Tribosphenida" from the Early Cretaceous (Aptian-Albian) Trinity Sands of Texas (Slaughter 1968a, b, 1969, 1971, 1981) have played an important role for analysis of early boreosphenidan dental evolution. In a recent revision, Davis and Cifelli (2011) attributed *Pappotherium* to metatherians and *Holoclemensia* to eutherians, which is the opposite of the original view, which had *Holoclemensia* as metatherian (Slaughter 1968a, Fox 1975) and *Pappotherium* as eutherian (Slaughter 1965, Fox 1975). Two Trinity taxa cannot be allied with either of the living groups of tribosphenidan mammals. *Kermackia*, the single genus of Kermackiidae Butler 1978, has been described based on an isolated lower molar (Slaughter 1971). *Kermackia* is a very small boreosphenidan and differs from *Aegialodon* in its relatively larger talonid with three cusps. Davis and Cifelli (2011) synonymized *Trinitherium*, which is based on a left dentary fragment with m3 (Butler 1978), with *Kermackia*. *Slaughteria*, from the upper Antlers Formation in Texas (Butler 1978), is known from a left dentary fragment with p2, p3, dp4, and dp5 with unerupted replacement teeth at the p4-5 positions (Davis and Cifelli 2011).

Hypomylos and *Tribotherium* are stem boreosphenidans from the Early Cretaceous (Berriasian) of Anoual in Morocco (Sigogneau-Russell 1991c, 1992). *Hypomylos* is known from lower molars that have a strongly reduced paraconid and a very long talonid with two cusps (the entoconid is absent) (Fig. 6.74). *Tribotherium* is known from upper molars and tentatively attributed lower molars. The upper molars lack an ectoflexus, and they have a narrow stylar area. The talonid is shorter than

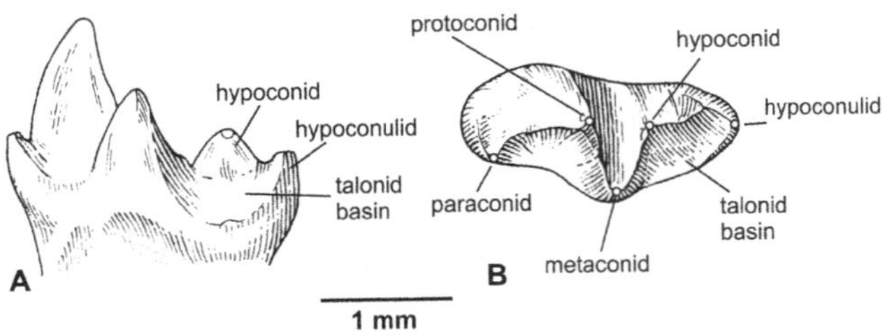

Fig. 6.74: *Hypomylos phelizoni*, right lower molar in lingual (A) and occusal (B) views. From Kielan-Jaworowska *et al.* (2004) after Sigogneau-Russell (1992).

that of *Hypomylos* and has also only two cusps (Kielan-Jaworowska *et al.* 2004).

A number of Early to mostly Late Cretaceous "tribotherians" has been described from North America. Picopsidae Fox 1980 is based on upper molars that lack an ectoflexus and have, as autapomorphic characters, a partially or completely reduced mesial stylar shelf and a large metastylar region (*Picopsis*, *Comanchea*, and *Falepetrus*). *Picopsis* from the upper part of the Milk River Formation (early Campanian) of Alberta (Canada) and the Late Cretaceous (Santonian) of Utah is characterized by upper molars that are longer mesiodistally than transversely. *Comanchea* is known only from an incomplete upper molar from the Early Cretaceous (Albian) Paluxy Formation of Texas (Jacobs *et al.* 1989). *Comanchea* is similar to *Picopsis*, but the molar is labiolingually wider than long. *Falepetrus* is represented by two upper molars from the Late Cretaceous (middle Campanian) of Wyoming and Montana. *Falepetrus* is more than two-times larger than *Picopsis* and *Comanchea* and has a large, but low protocone (Clemens and Lillegraven 1986).

The remaining "tribotherians" currently cannot be assigned to particular families. *Palaeomolops*, from the Late Cretaceous (Campanian) Aguja Formation in Texas, is represented by several lower molars and two tentatively assigned last upper molars (Cifelli 1994). The labial margin of the lower molars is concave in occlusal view and the talonid bears three cusps. *Potamotelses* from the Late Cretaceous (early Campanian) Milk River Formation of Alberta, Canada, is represented by isolated upper and lower molars. The upper molar crown is triangular in occlusal view, with a long labial border. The talonid of the lower molars is short and narrow and bears an additional fourth cusp between hypoconulid and entoconid (Fox 1972). Davis and Cifelli (2011) stated that *Potamotelses* likely could be referred to metatherians, if the hypothesized molar count of four (Fox 1975) would be corroborated. *Zygiocuspis* from the Late Cretaceous (early Campanian) of Utah is known from one incomplete upper molar. The molar is large and has a straight labial margin and a large protocone (Cifelli 1990b). From the Cenomanian Cedar Mountain Formation of Utah, *Dakotadens* and remarkably large *Culicolestes* have been reported based on upper and lower fragmentary molars (Eaton 1993, Cifelli 2004, Cifelli *et al.* 2016).

6.17 Metatheria

Metatheria Huxley 1880 comprises Marsupialia Illiger 1811 (crown-Metatheria) plus their fossil stem lineage (Rougier *et al.* 1998). According to molecular and fossil evidence (Beck 2008, Meredith *et al.* 2011, Mitchell *et al.* 2014), the metatherian crown group (Marsupialia) originated about 65 Million years ago at the beginning of the Cenozoic. Accordingly, all Mesozoic metatherians are outside Marsupialia (Horovitz *et al.* 2009). For a recent comprehensive cladistic analysis of Cretaceous-Paleogene metatherians and a review of the origin and early evolution of metatherians, see Williamson *et al.* (2012, 2014). Living marsupials differ in the mode of reproduction from their sister group, the placentals. Both have life birth, in contrast to the egg-laying monotremes, but marsupials lack the trophoblast (Hubrecht 1889) with an "inner cell mass" (ICM) that is present in placentals. The development of an ICM from which derive all tissues that will form definitive parts of the embryo is a derived character unique to placental mammals (Lillegraven 2004). The ICM also contributes to formation of extra-embryonic membranes such as the amnion, yolk sac, and allantois. It is invested against the inner wall of the trophoblast, which is involved in formation of extra-embryonic membranes that mechanically, nutritionally, and physiologically support and protect the embryo (Loke and Whyte 1983, Cedard and Firth 1992). Effects of the early pregnancy factor (Morton *et al.* 1980) serve as a protective immunosuppressant and as a growth factor during the earliest stages of the developing embryo in placentals (Morton 1989), and subsequent immunological protection of the embryo is accomplished by morphological and physiological elaborations of the enveloping trophoblastic layers (e.g., Edwards *et al.* 1975, Zeller 1999). The absence of the placental immunological protective capabilities in marsupials, first via early pregnancy factor and then by trophoblastic tissues (under participation of ICM) (Lillegraven 2004), prevents a prolonged intrauterine development due to immunological rejection of the fetus by the mother. However, these differences from soft tissue anatomy cannot be recognized directly in the fossil record.

6.17.1 Skeleton

The metatherian skeleton is generally very similar to that of eutherians. The presence of epipubic bones in metatherians is plesiomorphic because epipubics are also present in early eutherians (Kielan-Jaworowska 1975a, Novacek *et al.* 1997, Ji *et al.* 2002), monotremes, and various extinct Mesozoic taxa. According to Szalay (1984), the metatherian ankle is not much derived relative to the common therian ancestor. In contrast to placentals, the metatherian tarsus does not have the restricted mobility of the astragalus by the medial and lateral malleoli at the upper ankle joint (Szalay 1984). Metatherians, including *Sinodelphys* (Fig. 6.75), have a number of features in the wrist and ankle that distinguish them from eutherians. In the manus, the hamate is hypertrophied, and the triquetrum and the scaphoid are enlarged (Szalay 1994, Szalay and Trofimov 1996, Horovitz and Sánchez-Villagra 2003, Luo *et al.* 2003). The trapezium is small and bean shaped, whereas it is large and oblong in eutherians (Kielan-Jaworowska 1978, Ji *et al.* 2002, Luo *et al.* 2003). The tarsals have a transversely broad and anteroposteriorly short navicular, and the navicular facet of the astragalar head is spread medially along the length of the neck (Szalay 1994, Luo *et al.* 2003). By contrast, the navicular of Cretaceous eutherians is transversely narrow and anteroposteriorly elongate, with the navicular facet restricted anteriorly to the astragalar head (Kielan-Jaworowska 1978, Szalay 1994, Ji *et al.* 2002, Luo *et al.* 2003). At the calcaneus, the calcaneocuboid facet is obliquely oriented and buttressed by a large anteroventral tubercle in *Sinodelphys* and other metatherians (Luo *et al.* 2003). In Cretaceous eutherians, the calcaneocuboid facet is oriented anteriorly and lacks a well-defined anteroventral tubercle, representing the primitive condition, which also occurs in taxa outside Theria such as *Vincelestes* and *Zhangheotherium* (Luo *et al.* 2003).

6.17.2 Skull and mandible

The metatherian skull is generally characterized by an alisphenoid bulla. In the ear region, metatherians lack a groove on the promontorium for the stapedial artery, which is possibly apomorphic (Kielan-Jaworowska *et al.* 2004). Further apomorphies include the lack of a foramen for the superior ramus of the stapedial artery and the absence of the related ascending canal, the separation of the jugular foramen from the inferior petrosal opening, and the posterolateral placement of the transverse sinus with regard to the subarcuate fossa (Rougier *et al.* 1998).

The known metatherian mandibles lack postdentary bones, but Urban *et al.* (2017) reported the presence of

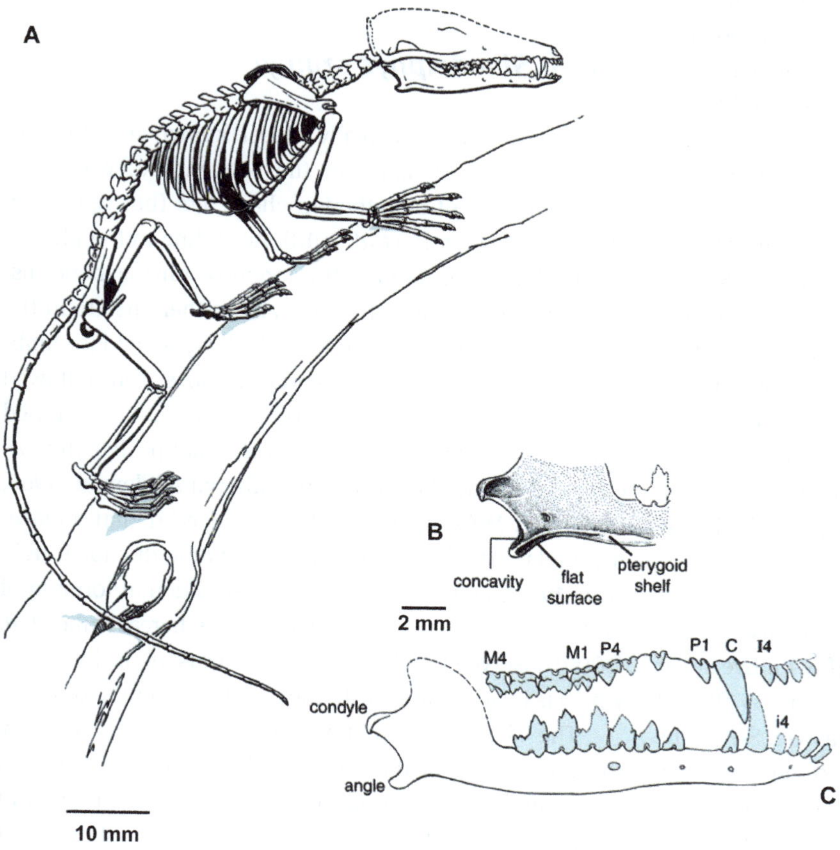

Fig. 6.75: *Sinodelphys szalayi.* (A) Skeletal restoration; (B) posterior part of mandible in medial aspect; (C) restoration of upper and lower dentition with mandible in labial aspect. Modified after Luo *et al.* (2003), reprinted with permission from AAAS.

Meckel's groove in *Kokopellia* and use this and developmental evidence to argue for independent development of the definitive mammalian middle ear in metatherians and eutherians. The most striking feature of the metatherian dentary is the shelflike angular process, which is medially inflected in known Cretaceous and Cenozoic metatherian mandibles (Sánchez-Villagra and Smith 1997) with one exception: *Sinodelphys*, the possibly geologically oldest metatherian from the Early Cretaceous (Barremian) of Liaoning, China, lacks this medial inflection of the angular process but has the posterior shelf of the masseteric fossa (Fig. 6.75 B), another metatherian character (Luo *et al.* 2003).

The primitive dental formula for marsupials is 5I.1C.3P.4M/4i.1c.3p.4m, which represents the plesiomorphic boreosphenidan condition except for the reduction of the number of premolars from five or four to three. Three upper and lower premolars and four upper and lower molars are present in all Mesozoic metatherians for which the postcanine dental formula is known (Kielan-Jaworowska *et al.* 2004) except for *Sinodelphys* from the Early Cretaceous of China, which has four upper and four lower premolars (Vullo *et al.* 2009). Generally, the metatherian postcanine dentition is referred to as P/p1-3 and M/m1-4, although homology of the tooth positions is debated (Archer 1978, Luckett 1993). Averianov *et al.* (2010b) and O'Leary *et al.* (2013) hypothesized that the first molar of metatherians is in fact an unreplaced deciduous premolar (dp5). Marsupials have an apomorphic mode of tooth replacement, with only the third premolar position being replaced postnatally (Luckett 1993). The geologically oldest evidence for this derived mode of tooth replacement is a mandible of *Alphadon* from the early Late Cretaceous of North America in which only at the third premolar position a germ of the succeeding p3 has been detected below dp3 (Cifelli *et al.* 1996). There is circumstantial evidence for marsupial-type replacement in *Deltatheridium* (Rougier *et al.* 1998) and *Lotheridium* (Bi *et al.* 2015), but the significance of this for inferring a marsupial-type replacement pattern was questioned by Fox and Naylor (2006). The suppression of diphyodonty in marsupials has been linked to their specialized mode of reproduction with the birth of very altricial young and the subsequent fixation to the teats during extended maternal lactation (Luckett 1993), which has been disputed by van Nievelt and Smith (2005).

In contrast to eutherians, metatherians (except Pediomyidae where it is apparently secondarily reduced; Davis 2007) retain a primitively wide stylar shelf, with five stylar cusps (StA–StE) on the upper molars (Clemens 1979b). Of these, StA is generally referred to the parastyle, StB to the stylocone, StC to the mesostyle, and StE to the metastyle, although the homology to the stylar cusps of eutherians is not clear (Marshall 1987, Kielan-Jaworowska *et al.* 2004). On the lower molars, metatherians are characterized by an approximation of the entoconid to the hypoconulid, a character that is already present in *Sinodelphys* and other early metatherians such as *Asiatherium* (Szalay and Trofimov 1996) and *Sulestes* (Averianov and Kielan-Jaworowska 1999, Averianov *et al.* 2010b), and a labial postcingulid (Cifelli and Muizon 1997). In *Kokopellia*, entoconid and hypoconulid are not approximated but connected by a strong labial postcingulid, which gives them an integrated appearance (Cifelli and Muizon 1997). In Late Cretaceous metatherians, entoconid and hypoconulid are even more closely approached (twinned) (Clemens 1966). This approximation or twinning appears to be functionally linked with enlargement of the metacone on the upper molars (Robin Beck, personal communication 2017). The dentition of early metatherians such as *Sinodelphys*, *Kokopellia*, small alphadontids, and pediomyids remained conservative and indicates an insectivorous diet. Some larger taxa (e.g., some Stagodontidae) exhibit dental specializations (Clemens 1966, Fox 1981, Fox and Naylor 1986) and suggest a carnivorous diet (e.g., *Didelphodon*). Based on dental-shape disparity and morphospace occupancy via geometric morphometric studies, Wilson (2013) found the dentition of the stagodontid *Didelphodon vorax* and the pediomyid *Leptalestes krejcii* adapted for a carnivorous diet. It was further suggested that *Didelphodon vorax* is the earliest known therian to invade a durophagous predator-scanvenger niche (Wilson *et al.* 2016).

6.17.3 *Sinodelphys szalayi*

As the possibly geologically oldest known metatherian, *Sinodelphys* sheds light on the sequence of evolutionary acquisition of metatherian characters (but see below opposing view by Bi *et al.* 2018). According to the phylogeny by Luo *et al.* (2003), the foremost phylogenetic distinctions between metatherians and eutherians are in the anatomy of the wrist and ankle, which are related to climbing specializations in metatherians. Those were followed by metatherian dental apomorphies such as the twinned entoconid and hypoconulid and the labial postcingulid as observed in Late Cretaceous metatherians (Clemens 1966, Cifelli and Muizon 1997) and the reduced dental replacement related to the marsupial mode of reproduction (Cifelli *et al.* 1996). Luo *et al.* (2003) reported a dental formula of 4I.1C.4P.4M/4i.1c.4p.3m for *Sinodelphys*, whereas Vullo *et al.* (2009) observed four lower molars in a cast of the holotype, which is in accordance with the marsupial dental formula.

The skeletal adaptations of *Sinodelphys*, especially the proportions of the hands and the laterally flattened claws, indicate scansorial and arboreal adaptations and suggest that *Sinodelphys* was an agile scansorial mammal capable of grasping and branch walking (Luo *et al.* 2003) (Fig. 6.75 A). This makes a very ancient evolutionary origin of the arboreal adaptations of didelphid marsupials possible, although some later metatherians such as *Pucadelphys* and *Herpetotherium* that are more closely related to the marsupial crown clade show more terrestrial adaptations (Robin Beck, personal communication 2017). The hands of *Sinodelphys* had a strong capacity to flex the digits, as indicated by bony protuberances for the fibrous tendon sheaths of the flexor digitorum (Luo *et al.* 2003). The wide navicular and expanded navicular facets present in *Sinodelphys* are associated with an effective grasping of the medial pedal digits, as in modern didelphids, and a greater flexibility of the distal pedal bones, for inversion capabilities associated with reversal of the foot, again as in modern didelphids (Jenkins and McLearn 1984). The body mass of *Sinodelphys* is estimated between 25 and 40 g (Luo *et al.* 2003).

The metatherian affinities of *Sinodelphys* have recently been challenged by Bi *et al.* (2018) based on a new eutherian skeleton (*Ambolestes zhoui*) from the Lower Cretaceous Yixian Formation. With an estimated body size of 34–44 g, *Ambolestes* is similar in size to the other Jehol eutherians, and it was either arboreal or scansorial (Bi *et al.* 2018). The dental formula of *Ambolestes* is ?I.1C.5P.3M/?i.1c.4p.3m and differs with eight upper postcanine teeth from that of metatherians which have seven. According to the phylogenetic analysis by Bi *et al.* (2018) under inclusion of the new evidence from *Ambolestes*, the four Jehol therians (*Ambolestes, Sinodelphys, Acristatherium* and *Eomaia*) are placed within Eutheria.

6.17.4 Origin and early dispersal of metatherians

Sinodelphys supports the evolutionary scenario that metatherians originated in Asia and subsequently spread to North America where they appeared about 110 million (late Early Cretaceous, Albian) years ago (Cifelli and Davis 2003). However, the recently forwarded re-interpretation of *Sinodelphys* as a eutherian (Bi *et al.* 2018) would make the Early Cretaceous North American metatherians the oldest known (Cifelli and Davis 2003) and would imply a 50-million-year ghost lineage for Metatheria. The Early Cretaceous diversification of metatherians mainly occurred in North America, whereas metatherians appear to have been rare and little diversified in Asia. The oldest occurrence of metatherians in Europe is *Arcantiodelphys* from the lower Cenomanian of Charente, France (Vullo *et al.* 2009). The most probable metatherian dispersal route was from North America via the Canadian and European Subarctic (North Atlantic Route), although a dispersal directly from Asia cannot be completely excluded (Vullo *et al.* 2009). Before the discovery of *Arcantiodelphys*, an upper molar of *Maastrichtidelphys* from the Late Cretaceous (Maastrichtian) of Maastricht (The Netherlands) was the oldest evidence for metatherians in Europe (Martin *et al.* 2005). Metatherians existed in North America and Europe until the Early Neogene and became extinct in the Lower Miocene (Mammal Neogene Zone 6 in Europe). The oldest record of metatherians in South America are the Early Paleocene *Cocatherium* from the Grenier Farm locality of Chubut Province in Patagonia (Goin *et al.* 2006) and *Mayulestes* from Tiupampa in Bolivia (Muizon 1998), although the latter is less clearly constrained in terms of age. In Australia, metatherians (*Chulpasia*) have been recorded from the Early Eocene Tingamarra Fauna (Godthelp *et al.* 1992, Beck *et al.* 2008, Sigé *et al.* 2009).

Despite the fact that their earliest appearance (*Sinodelphys*) might have been in northeastern Asia, metatherians were rare and little diversified on that continent during the Cretaceous (Cifelli and Davis 2003, Kielan-Jaworowska *et al.* 2004). Most of the Asian metatherians are deltatheroidans; one of the rare exceptions is the

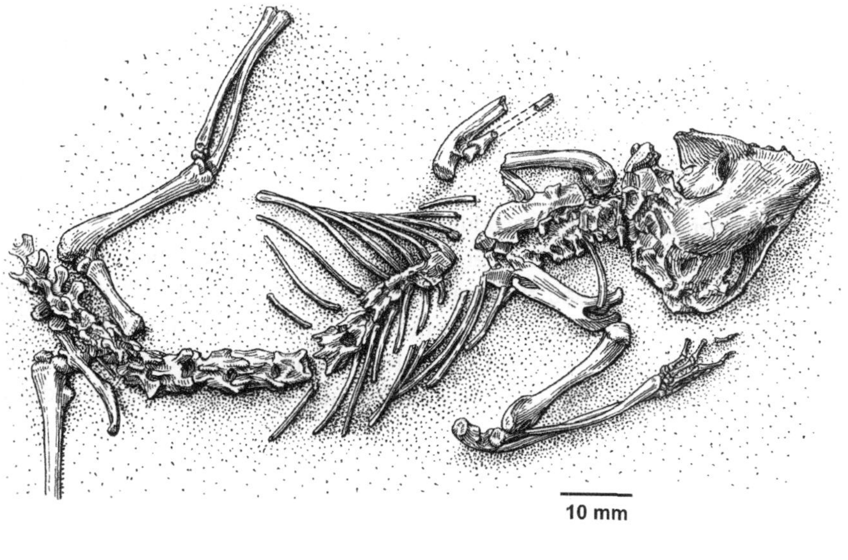

Fig. 6.76: Skeleton of *Asiatherium reshetovi* (holotype). From Szalay and Trofimov (1996), reprinted with permission of the Society of Vertebrate Paleontology, www.vertpaleo.org.

skeleton of *Asiatherium* (Fig. 6.76) from the Late Cretaceous (late Campanian) of Mongolia (Trofimov and Szalay 1994). *Asiatherium* exhibits typical metatherian characters such as three premolars and four molars, closely twinned hypoconulid and entoconid, and an alisphenoid component to the bulla (Szalay and Trofimov 1996). The head-body length is about 70 mm, and the postcranium suggests terrestrial locomotion (Trofimov and Szalay 1994). The skull and skeleton do not exhibit any particular specializations and resemble that of generalized therians. *Marsasia*, based on two edentulous mandibles and a dentary fragment with one molar from the Late Cretaceous (Coniacian) of Uzbekistan (Nessov 1997), has been synonymized with *Sulestes* by Averianov *et al.* (2010b). The "Gurlin Tsav skull" from the Maastrichtian of Mongolia (Kielan-Jaworowska and Nessov 1990, Szalay and Trofimov 1996) remains undescribed up to now.

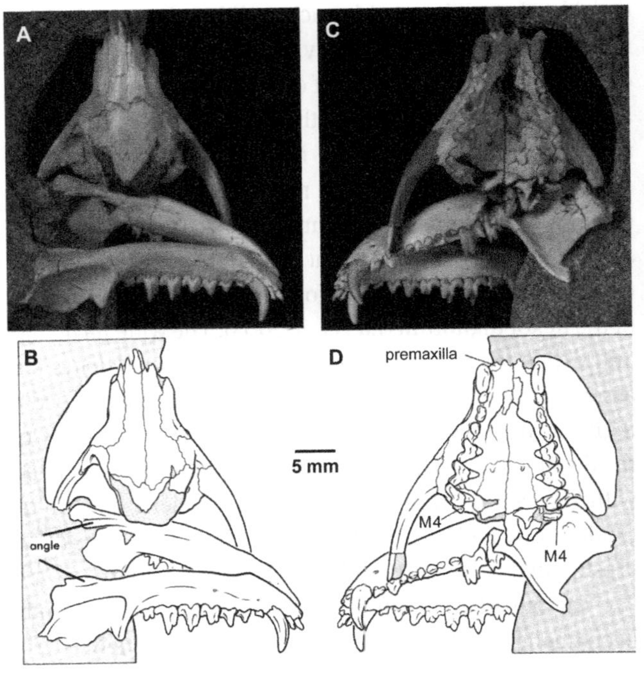

Fig. 6.77: *Deltatheridium* skull in dorsal (A, B) and palatinal (C, D) views, with associated mandible. Modified after Rougier *et al.* (1998), adapted with permission from Springer Nature.

6.17.5 Deltatheroida

Deltatheroida Kielan-Jaworowska 1982 is the sister taxon of Marsupialiformes (= Marsupialia plus stem metatherians more closely related to Marsupialia than to Deltatheroida; Vullo *et al.* 2009). Deltatheroidans are known from the Early-Late Cretaceous (Aptian or Albian through latest Maastrichtian) of North America and the Late Cretaceous (Turonian-Late Campanian) of Asia. Deltatheroidans were first recognized from skulls collected by the American Museum of Natural History's Central Asiatic Expeditions in the 1920s and were originally regarded as eutherians by Gregory and Simpson (1926). On the basis of newly collected specimens, Butler and Kielan-Jaworowska (1973) reinterpreted the dental formula of *Deltatheridium* as including three premolars and four molars and regarded it as a therian of metatherian-eutherian grade. Rougier *et al.* (1998) listed a number of synapomorphies for Deltatheroida and metatherians (of which the presence of three premolars is the most compelling), and the cladistic analysis by Rougier *et al.* (2015) corroborated a monophyletic Deltatheroidea occupying a position closer to the crown than *Sinodelphys* within metatherians. The dentition of deltatheroidans shows trends toward carnivorous specializations, such as a strongly developed postmetacrista, pronounced postvallum-prevallid shear, and reduction of the ultimate molar (Muizon and Lange-Badré 1997, Rougier *et al.* 2015). Deltatheroidans were the first therian lineage that modified their dentition from an insectivorous toward a carnivorous adaptation (Rougier *et al.* 2015).

The type genus *Deltatheridium* (Fig. 6.77), from the Late Cretaceous (Campanian) Djadokhta Formation in the Gobi Desert (Mongolia), is relatively large and is known from partial skulls and mandibles (Gregory and Simpson 1926, Kielan-Jaworowska 1975b, Rougier *et al.* 1998). This taxon has also been reported from the early Campanian of Kazhakstan (Averianov 1997). The angular process is metatherian-like, that is to say medially inflected and shelflike. The dental formula of *Deltatheridium* is 4I.1C.3P.4M/3i.1c.3p.4m (Rougier *et al.* 1998), and the same metatherian postcanine dental formula has been found in *Deltatheroides* (Kielan-Jaworowska 1975b).

Deltatheroidans were important faunal elements in the Cretaceous of Asia. *Sulestes* (including *Marsasia*, see Averianov *et al.* 2010b) has unspecialized molars, with a large talonid and an unreduced last molar (Rougier *et al.* 2015) and is well represented in the Turonian of Uzbekistan (Nessov 1985, Averianov *et al.* 2010b). The Campanian deltatheroidans, which are known by partial skulls and jaws from the Djadokhta and Barun Goyot Formations in Mongolia, are more derived. *Deltatheridium* and *Deltatheroides* both have a small talonid and small protocone, and *Deltatheridium* additionally has a strong reduction of the last molar (Rougier *et al.* 2015). The most complete deltatheroidan skull has recently been described for *Lotheridium* from the Upper Cretaceous of Henna in China (Bi *et al.* 2015). The skull and associated mandibles have the full adult dentition and provide additional evidence that deltatheroidans already possessed the metatherian dental formula.

In the North American fossil record, only a few fragmentary specimens of deltatheroidans from Alberta,

Utah, and Wyoming have been reported until the Late Maastrichtian (Fox 1974, Cifelli 1990c). Highly specialized *Nanocuris* from the Late Maastrichtian of Saskatchewan and Wyoming has a deep dentary, stout canine, and evidence of powerful jaw musculature (Fox *et al.* 2007, Wilson and Riedel 2010, Wilson 2013). Carnivorous adaptations apparently evolved independently in the Asian and North American lineages (Rougier *et al.* 2015).

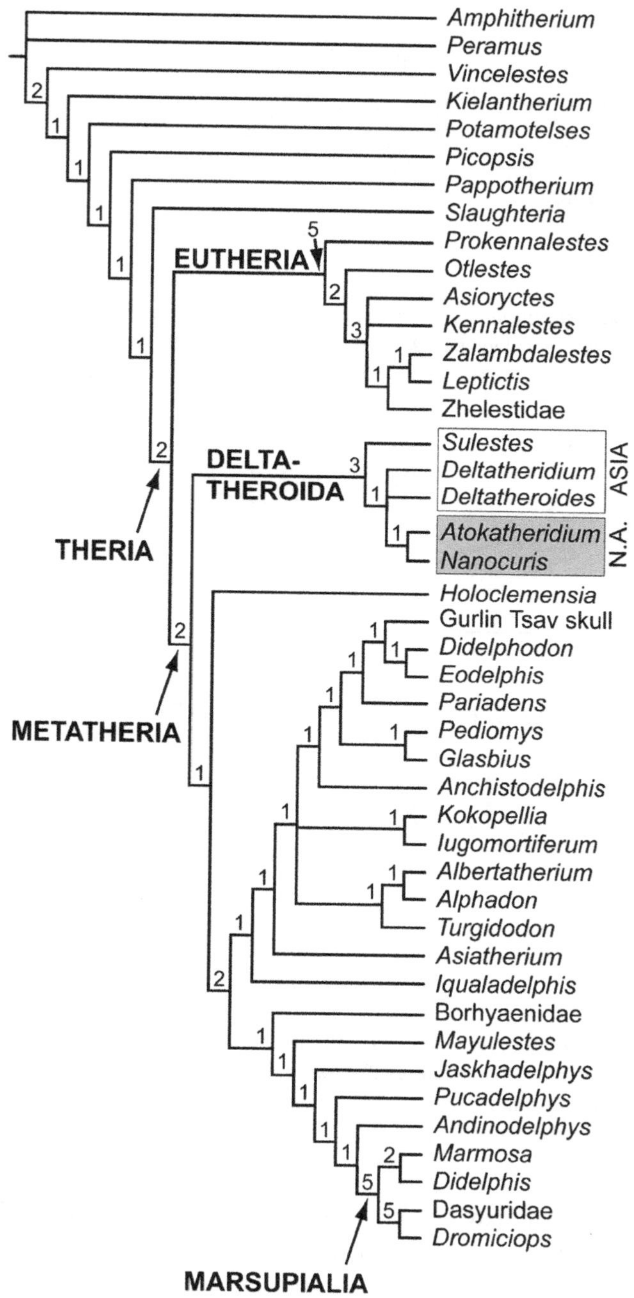

Fig. 6.78: Phylogentic analysis of Cretaceous-Paleogene metatherians and selected eutherians. From Wilson and Riedel (2010), reprinted by permission of the Society of Vertebrate Paleontology, www.vertpaleo.org.

Traditionally, Deltatheroida was regarded as a group of Asian origin, but with the discovery of North American deltatheroidans, this became less evident. In the most recent cladograms presented by Wilson and Riedel (2010) and Rougier *et al.* (2015), the North American *Atokatheridium* and *Nanocuris* cluster with the Asian deltatheroidans (Fig. 6.78). *Sulestes* from the Late Cretaceous of Asia and *Oklatheridium* from the Early Cretaceous of North America form a clade within deltatheroidans (Rougier *et al.* 2015). *Pappotherium*, one of the "Trinity therians" (Slaughter 1965) and hitherto placed outside crown therians (Cifelli 1993b, Rougier *et al.* 1998, Luo *et al.* 2002), has been regarded as a possible metatherian by Davis and Cifelli (2011) and appears as the sister taxon of all other Deltatheroida in the cladogram presented by Rougier *et al.* (2015). Recently, it was proposed that at least one deltatheroidan lineage survived the K-Pg mass extinction in Asia (Ni *et al.* 2016).

6.17.6 Marsupialiformes

Marsupialiformes Vullo *et al.* 2009 were the most diverse mammalian clade in North America during the Late Cretaceous. *Kokopellia* (Fig. 6.79) from the Early-Late Cretaceous (Albian-Cenomanian boundary) of the Cedar Mountain Formation of Utah (Cifelli 1993a) is the best represented pre-Campanian marsupialiform from North America, being

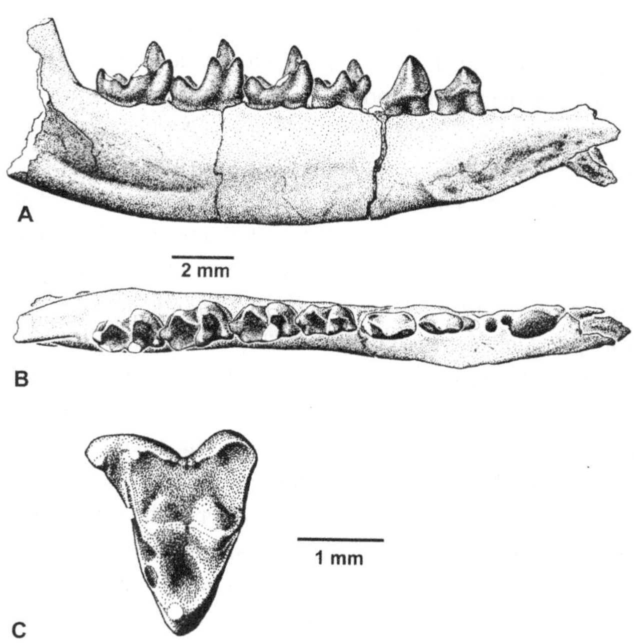

Fig. 6.79: *Kokopellia juddi*. (A, B) Left dentary (holotype) with p2-3, m1-4 in medial (A) and dorsal views (B); (C) left M3 in occlusal view. (A, B) From Cifelli (1993a), copyright (1993) National Academy of Sciences, U.S.A.; (C) from Kielan-Jaworowska *et al.* (2004), copyright © 2004 Columbia University Press. Reprinted with permission from the publisher.

known from most of the dentary, lower dentition, and upper molar series (Cifelli and Muizon 1997). *Kokopellia* is more plesiomorphic than most other Cretaceous marsupialiforms by the more centrally placed hypoconulid that is not fully twinned with the entoconid on the lower molars, and upper molars with a proportionately broader stylar shelf (Kielan-Jaworowska *et al.* 2004). Other pre-Campanian North American metatherians such as *Anchistodelphys*, *Iugomortiferum*, *Adelodelphys*, and *Sinbadelphys* are known only by isolated upper and lower teeth (Cifelli 1990d, 2004).

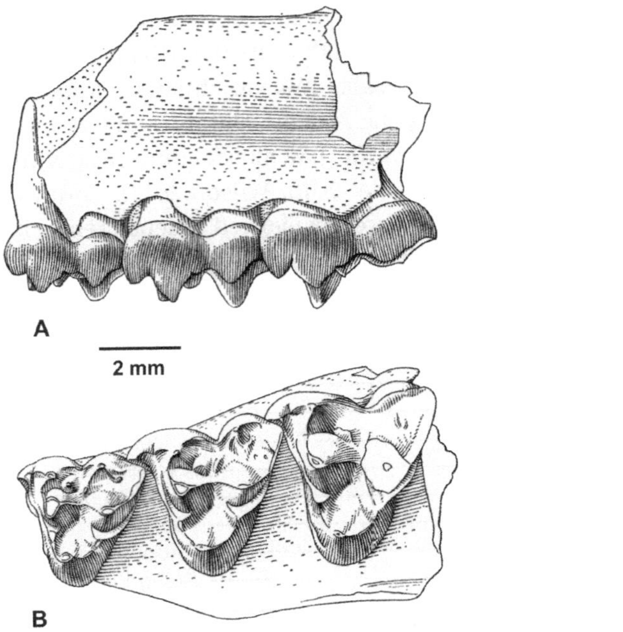

Fig. 6.80: *Turgidodon rhaister*, left M1-3 in labial (A) and occlusal (B) views. From Clemens (1966), © 1966 by the Regents of the University of California. Published by the University of California Press.

6.17.7 "Alphadontidae"

The systematics of Late Cretaceous North American marsupialiforms is confusing and continues to be debated (Williamson *et al.* 2012, 2014). "Alphadontidae" Marshall *et al.* 1990 traditionally comprises an array of Late Cretaceous (Cenomanian-Lancian) North American marsupialiforms that are dentally more derived than *Kokopellia*. They are structurally similar taxa in that their upper molars have well-developed conules, placed near the bases of paracone and metacone, have a proportionally narrower stylar shelf with at least stylar cusp D and often stylar cusp C present, and in that their lower molars have the paraconid placed at the lingual margin of the crown (Kielan-Jaworowska *et al.* 2004). Traditionally, the genus *Alphadon* has been broadly conceived (e.g., Clemens 1966, Lillegraven 1969). Cifelli (1990a, b) removed several species to *Turgidodon* and *Protalphadon* (Figs. 6.80 and 6.81), Johanson (1996) removed two species to *Varalphadon*, and Eaton (2009) recognized a further genus, *Eoalphadon*. Cifelli and Johanson (1994) added *Aenigmadelphys*, based on upper and lower isolated molars, to "Alphadontidae". Davis (2007) restricted "Alphadontidae" to *Alphadon* and *Turgidodon* in his cladistic analysis, whereas *Aenigmadelphys*, *Iqualadelphys*, and *Varalphadon* were excluded from "Alphadontidae". In their description of mammal teeth from the Late Cretaceous (Santonian) of northern Montana, Davis *et al.* (2016) listed *Albertatherium*, *Alphadon*, *Varalphadon*, and *Turgidodon* among "Alphadontidae". In the cladogram presented by Averianov *et al.* (2010b), *Turgidodon* does not cluster in the same clade as *Alphadon*. According to the large-scale cladistic analysis by Williamson *et al.* (2012, 2014), "Alphadontidae" grouping of taxa closely related to *Alphadon marshi* is paraphyletic.

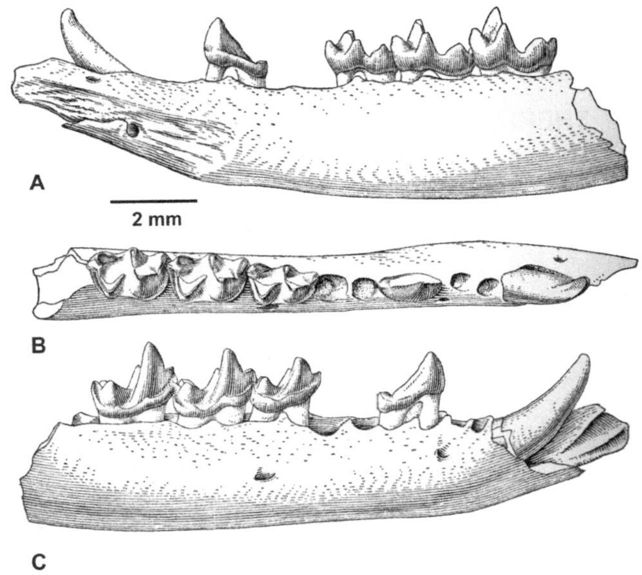

Fig. 6.81: *Protalphadon lulli*, right dentary with c, p2, m1-3 in medial (A), dorsal (B), and lateral (C) views. From Clemens (1966), © 1966 by the Regents of the University of California. Published by the University of California Press.

6.17.8 Pediomyidae

According to Williamson *et al.* (2014), Pediomyidae Simpson 1927c is one of the few marsupialiform clades that is strongly supported by explicit synapomorphies and high tree-support in phylogenetic analyses (Davis 2007, Williamson *et al.* 2012). Pediomyids are characterized by a reduction of the mesial stylar shelf and reduction of the stylocone at the upper molars and a labial shift in the attachment of the oblique cristid in the lower molars. Larger-bodied taxa such as *Aquiladelphis* and *Protolambda* were found to resemble stagodontids with explicit molar crushing function and postvallum/prevallid shear (Williamson 2014). In the cladistic analysis of Williamson *et al.* (2012), Pediomyidae contains *Iqualadelphis*, *Leptalestes*, *Pediomys*, *Protolambda*, and *Aquiladelphis*, whereas the formerly included (or closely related) *Glasbius* (Rougier *et al.* 1998, 2004, Davis 2007, Averianov *et al.* 2010b, Wilson and Riedel 2010) was not found to be closely related to Pediomyidae. The cladistic analysis of Williamson *et al.* (2012) placed *Glasbius* as the sister taxon to the South American Paleogene *Roberthoffstetteria*, supporting the close relationship between both genera noted previously by several authors (Case *et al.* 2005, Goin *et al.* 2003, Marshall *et al.* 1990). If this is correct, the *Glasbius* + *Roberthoffstetteria* clade would have crossed the K-Pg boundary, as one of the few metatherian lineages to do so (Williamson *et al.* 2012).

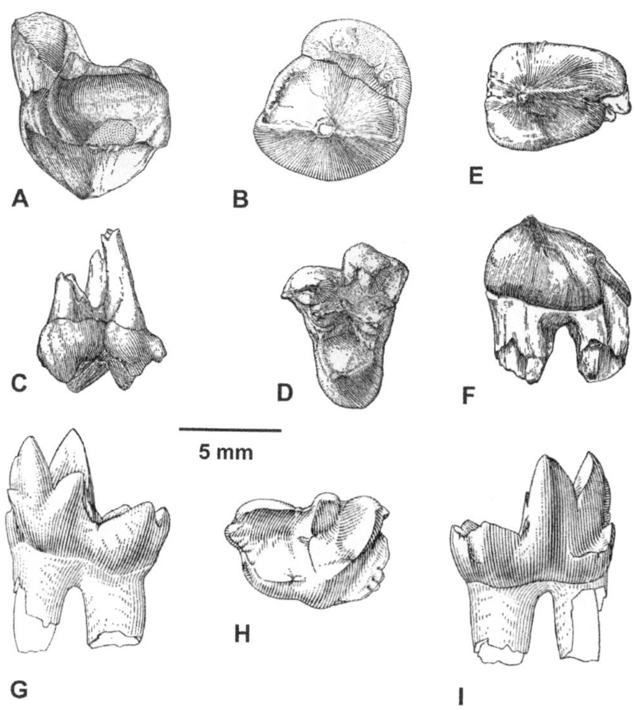

Fig. 6.82: *Didelphodon vorax* upper (A–D) and lower (E–I) teeth. (A and B) Left P3 in labial (A) and occlusal (B) views; (C and D) left M1 in labial (C) and occlusal (D) views; (E and F) left p3 in occlusal (E) and labial (F) views; (G–I) right m4 in lingual (G), occlusal (H), and labial (I) views. Modified after Clemens (1966), © 1966 by the Regents of the University of California. Published by the University of California Press.

6.17.9 Stagodontidae

Stagodontidae Marsh 1889 includes relatively large carnivorous Late Cretaceous (Cenomanian-Lancian [=Maastrichtian]) metatherians with an estimated body size between 400 and 2000 g (Gordon 2003). Stagodontids are the only North American Cretaceous metatherians that are known by partial skulls (Wilson *et al.* 2016). Apomorphic characters include squamosals with large epitympanic sinuses, and upper molars with a marked ectoflexus and strongly developed metastylar region. Upper and lower third premolars are large and bulbous, suggesting a durophagous adaptation (Clemens 1968, Wilson *et al.* 2016, Cohen 2017) (Fig. 6.82). A monophyletic clade comprising the stagodontid genera *Didelphodon*, *Eodelphis*, and *Pariadens* (Marsh 1889, Matthew 1916, Cifelli and Eaton 1987) was recognized in the cladograms presented by Averianov *et al.* (2010b) and Williamson *et al.* (2012). *Pariadens* later was regarded as the immediate outgroup of Stagodontidae by Rougier *et al.* (2015) and Bi *et al.* (2015), whereas Cohen (2017) corroborated the grouping of Stagodontidae by Averianov *et al.* (2010a, b) under inclusion of the newly described *Hoodotherium* and *Fumodelphodon* from the Turonian of Utah. The "Gurlin Tsav skull" (Kielan-Jaworowska and Nessov 1990, Szalay and Trofimov 1996: Fig. 6.22) from Mongolia, that was placed with Stagodontidae in the cladogram of Rougier *et al.* (1998), appears as outgroup in the analysis of Averianov *et al.* (2010b).

6.18 Eutheria

Eutheria Huxley 1880 is the most diverse living mammalian group and has dominated all continents since the Cenozoic except for Australia. Eutheria is the sister taxon of Metatheria, and so its origin was (by definition) simultaneous with that of metatherians. The eutherian *Eomaia scansoria* from the Early Cretaceous (Barremian) of Liaoning in northeastern China (Ji *et al.* 2002) is of the same age as the possible oldest known metatherian, *Sinodelphys szalayi*, which is from the same deposits. The metatherian affinity of *Sinodelphys* has recently been challenged based on a new eutherian skeleton (*Ambolestes zuoi*) from the Yixian Formation (Bi *et al.* 2018) (see chapter 6.17.3). *Eomaia* and *Sinodelphys* indicate a minimum age for the metatherian-eutherian split of 125 million years. The age of this split has been shifted backward by about 35 million years by the discovery of a partial skeleton of *Juramaia sinensis* (Fig. 6.83) from the Late Jurassic (Oxfordian; 161–156 million years old) Tiaojishan Formation of Liaoning, interpreted to be a eutherian (Luo *et al.* 2011b). The interpretation of *Juramaia* as a eutherian would reduce the length of the ghost lineage between *Eomaia* and Jurassic estimates for the eutherian-metatherian divergence based on applications of the molecular clock informed by the fossil record (Benton *et al.* 2009, Meredith *et al.* 2011, Dos Reis *et al.* 2012, Tarver *et al.* 2016). However, not all recent phylogenetic studies support the placement of

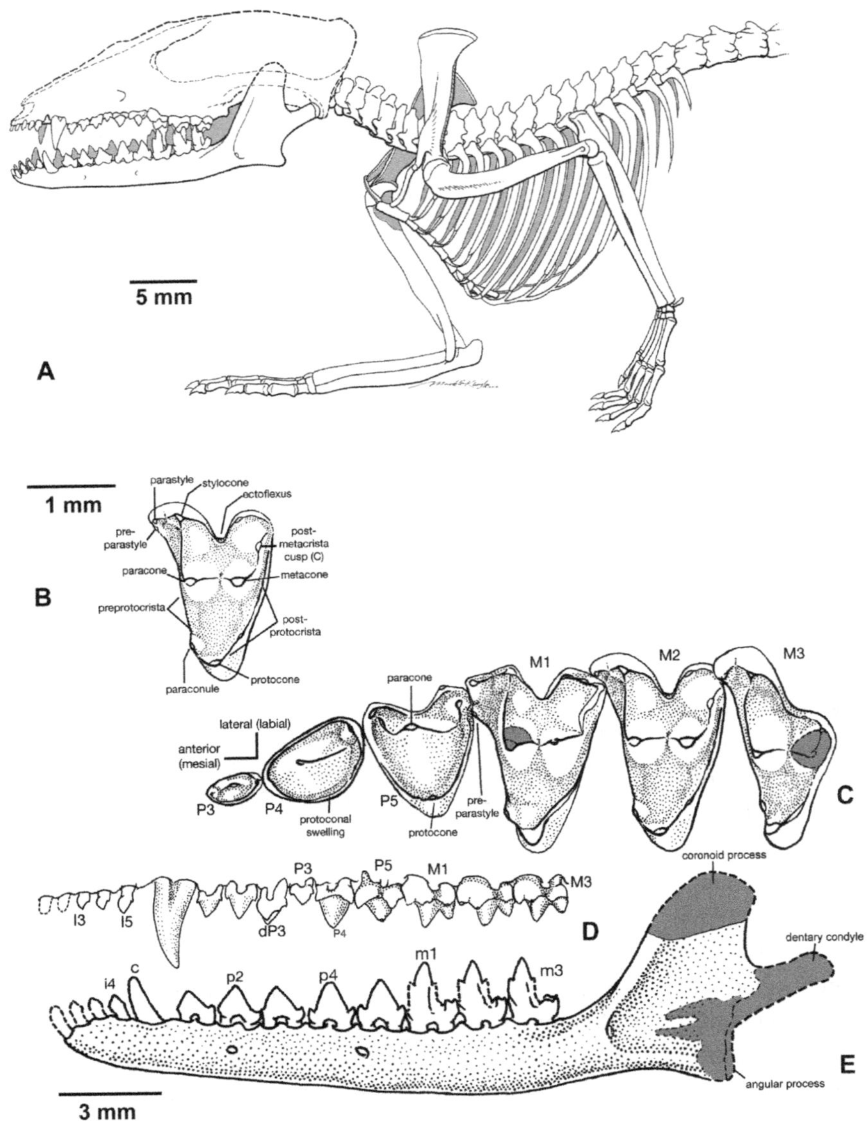

Fig. 6.83: *Juramaia sinensis* (holotype). (A) Restoration of partly preserved skeleton and skull; (B) right M2 in occlusal view (composite restoration); (C) right P3-M3 in occlusal view; (D) left upper dentition restoration in labial view; (E) left lower dentition (restoration) and mandible in labial view. Modified after Luo *et al.* (2011b), reprinted with permission from Springer Nature.

Juramaia as a stem eutherian (e.g., Krause *et al.* 2014a, b). Recently, Sweetman *et al.* (2017) described two isolated upper molars with derived eutherian characters from the earliest Cretaceous (Berriasian) Purbeck Group of Durlston Bay in southern England and assigned it to two new genera, *Durlstotherium* and *Durlstodon*. The molars resemble that of Late Cretaceous eutherians such as asioryctitheres, cimolestids, and gypsonictopsids in height and expansion of the protocone relative to labial cusps and the possession of conules adjacent to the paracone and metacone with sharp internal cristae.

Apart from *Eomaia* (and *Juramaia* if it is indeed eutherian), most Cretaceous eutherians are known by jaws or isolated teeth, and only few by more complete material such as skulls and postcrania. The known Cretaceous eutherians are dentally conservative, which makes it difficult to elucidate their relationships. The taxa which are known by postcranial remains also retain generalized and plesiomorphic skeletons, and only in the Late Cretaceous some skeletal specializations (e.g., fused tibia and fibula in *Zalambdalestes*; Kielan-Jaworowska 1978) become evident.

A problem that has garnered much attention in recent years is the timing of origin and diversification of crown eutherians (the placentals). Some earlier molecular studies (e.g., Kumar and Hedges 1998, Bininda-Emonds *et al.* 2007) suggested an origin of major eutherian crown clades and even most individual orders deep in the Cretaceous. More recent analyses accounting for rate heterogeneity, utilizing soft maximum and hard minimum paleontological calibrations (Benton *et al.* 2009), and applying genomic-scale data sets support substantially younger divergence estimates, with some crown placental clades originating prior to the K-Pg boundary, but others doing so thereafter (Kitazoe *et al.* 2007, Meredith *et al.* 2011, Dos Reis *et al.* 2012, Tarver *et al.* 2016). Morphological evidence from the fossil record does not yet demonstrate the presence of any crown placental clade prior to the K-Pg boundary (Asher *et al.* 2005, Wible *et al.* 2007, Goswami *et al.* 2011). However, such paleontological estimates are widely recognized as minima (Benton *et al.* 2009, 2015), not synonymous with cladistic divergence events, and actual divergences are expected to at least slightly predate such paleontological first appearances.

6.18.1 Skull and skeleton

Cretaceous eutherians resemble early metatherians in general skeletal anatomy. The known early eutherian skulls are small (*Daulestes* about 20 mm, other taxa between 25 and 30 mm, the larger zalambdalestids between 40 and 50 mm) and have an elongated snout with slender mandibles. The therian cochlea is fully coiled: for *Daulestes*, a full turn has been demonstrated (McKenna *et al.* 2000), and Wible *et al.* (2001) estimate a minimum of about 450° for a petrosal referred to *Prokennalestes*. Brain size was similar to extant small placental insectivorans and marsupials, and the large infraorbital foramen suggests that vibrissae were present (Kay and Cartmill 1977). Attached postdentary bones have been lost in eutherians, although some taxa retain a coronoid and a remnant of Meckel's groove (Kielan-Jaworowska *et al.* 2004); based on the presence of a rudimentary Meckel's groove and ontogenetic evidence, an independent evolution of the definitive mammalian middle ear in the eutherian and metatherian lineage has been suggested (Urban *et al.* 2017). Cenozoic and extant eutherians differ from metatherians in lacking epipubic bones, but it has been demonstrated that these were primitively present in Late Cretaceous eutherians from Mongolia (Kielan-Jaworowska 1975a, Novacek *et al.* 1997). *Eomaia* and other early eutherians retain the plesiomorphic mammalian condition of the wrist, with a small scaphoid and triquetrum compared with other carpals, whereas the scaphoid and triquetrum are very large in metatherians (Ji *et al.* 2002, Kielan-Jaworowska *et al.* 2004). The ilium, ischium, and pubis of the eutherian pelvis are fused. A patella is present, which is a derived eutherian feature absent in most early metatherians. The ankle is more plesiomorphic in eutherians than in metatherians, as evident from the condition in stem therians (e.g., *Zhangheotherium*, *Vincelestes*). On the astragalus, the navicular facet is anteriorly restricted in eutherians, whereas it is medially spread in metatherians, and at the calcaneus, the navicular facet is transverse in eutherians and oblique in metatherians (Luo *et al.* 2003).

6.18.2 Dentition

Like metatherians, eutherians are characterized by tribosphenic molars, but they differ both in the dental formula and mode of tooth replacement. The plesiomorphic dental formula of modern placentals is 3I.1C.4P.3M/3i.1c.4p.3m. The earliest eutherians may have up to five upper and four lower incisors and five premolars (Kielan-Jaworowska 1981, Nessov 1985, Kielan-Jaworowska and Daszheveg 1989, Sigogneau-Russell *et al.* 1992, Novacek *et al.* 1997, Cifelli 2000, Ji *et al.* 2002). The presence of four to five incisors is a symplesiomorphy shared with metatherians, whereas the higher number of premolars in eutherians is a retained plesiomorphy. The primitive number of five premolars in eutherians (McKenna 1975, Novacek 1986) is evident from the oldest known eutherian jaws of *Juramaia*, *Eomaia*, *Prokennalestes*, and *Otlestes*. For *Kulbeckia* from the Late Cretaceous of Uzbekistan, a transitional stage of reduction from five to four premolars, with resorption and loss of the dp3 position, has been reported (Averianov and Archibald 2015). Eutherians do not have the clear morphological break between premolars and molars, as seen in metatherians. Eutherian premolars are increasingly molarized in the distal direction, which probably represents a derived condition. It is generally agreed upon that the primitive number of molars in eutherians is three, which is plesiomorphic compared with metatherians which have four molars: it has been suggested that the first molar in metatherians is in fact a retained DP5/dp5 (Averianov *et al.* 2010b, O'Leary *et al.* 2013). The eutherian dentition is characterized by a single replacement (diphyodonty) of the antemolars, which represents the primitive condition for therians (Martin 1997). The metatherian mode of tooth replacement, with only the last premolar (P3/p3) being replaced postnatally, is derived (Luckett 1993, Kobayashi *et al.* 2002). Most extant and Cenozoic eutherians have single-rooted canines, but in Cretaceous taxa double-rooted canines have a broad distribution (Clemens and Lillegraven 1986, Cifelli and Madsen 1999). The canine is consistently single-rooted in metatherians, and double-rooted canines probably represent the plesiomorphic condition, as observed in *Juramaia* (Luo *et al.* 2011b). Early Cretaceous *Eomaia* (Barremian) (Figs. 6.84 and 6.85) and *Acristatherium* (Early Aptian) from northeastern China have single-rooted canines (Ji *et al.* 2002, Hu *et al.* 2010). Early eutherian

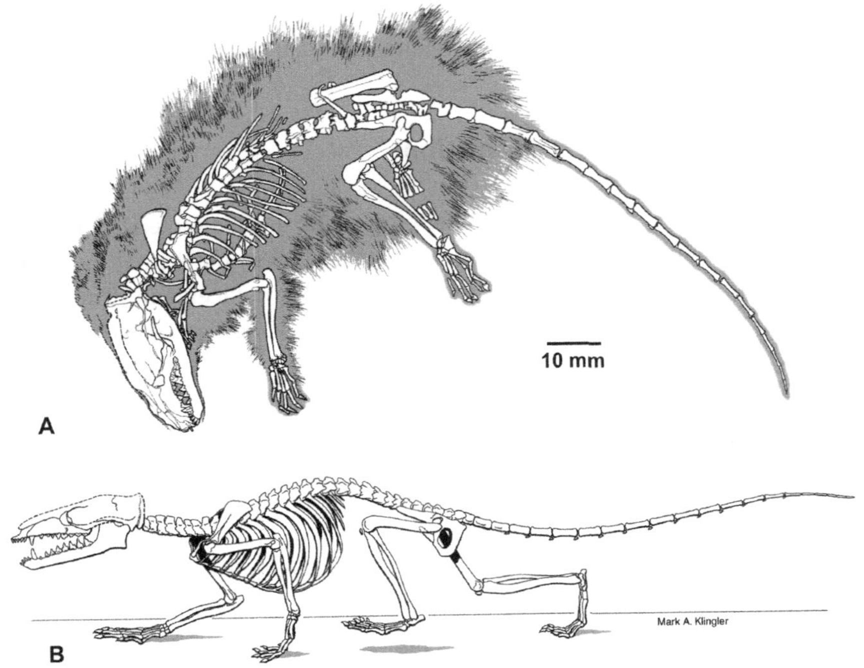

Fig. 6.84: *Eomaia scansoria* (holotype). (A) Skeleton in original position with fur halo around; (B) skeletal restoration in life pose. Modified after Ji *et al.* (2002), adapted with permission from Springer Nature.

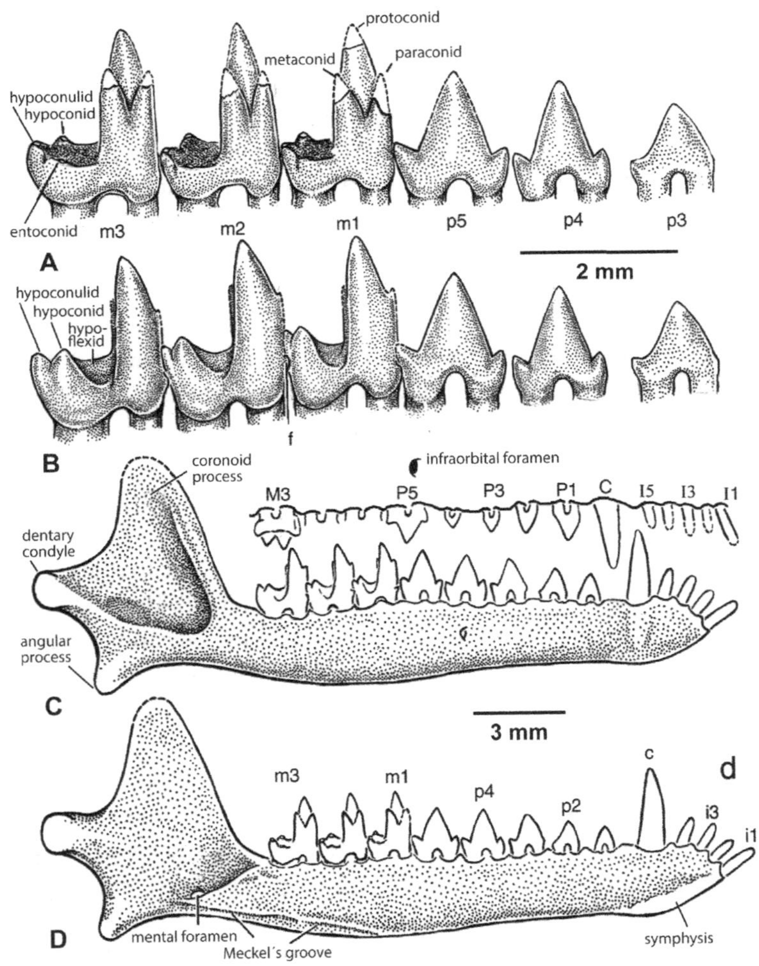

Fig. 6.85: *Eomaia scansoria*. Composite restoration of dentition and mandible. Right lower p3-m3 in lingual (A) and labial (B) views; (C) incomplete right upper dentition and right mandible in lateral view; (D) left mandible in medial view. Modified after Ji *et al.* (2002), adapted with permission from Springer Nature.

molars are characterized by a narrowing of the stylar shelf and reduction of the stylocone, strengthening of the conules with distinct conular cristae, and development of pre- and postcingula. Lower molars are more primitive than those of early metatherians because they lack specializations such as twinning of hypoconulid and entoconid. Eutherian lower molars tend to have paraconids that are reduced in size, and the height difference between trigonid and talonid is larger than in metatherian molars.

6.18.3 Locomotorial adaptations and paleobiology

Juramaia (Fig. 6.83) and *Eomaia* (Fig. 6.84) both exhibit scansorial adaptations in the postcranial skeleton, such as elongated intermediate phalanges and laterally flattened and curved claws (Ji *et al.* 2002, Luo *et al.* 2011b). The feet of *Eomaia* have similar phalangeal proportions and curvature to the grasping feet of extant arboreal mammals. This supports the view that eutherians were primarily arboreal or scansorial, as had been suggested by early authors (e.g., Dollo 1899). For Late Cretaceous eutherians from Mongolia, such as *Asioryctes* and *Kennalestes*, the locomotorial adaptations are less clear, but Kielan-Jaworowska (1977) concluded that they were probably not arboreal because they lived in a semidesert paleoenvironment without any evidence of woodlands. The locomotory adaptations of the zalambdalestids *Zalambdalestes* and *Barunlestes* from the Late Cretaceous of the Gobi Desert, for which the postcranial skeleton is known, are clearer. In *Zalambdalestes*, the tibia and fibula are fused over three quarters of their length, and Kielan-Jaworowska (1978) compared the proportions of their hindlimbs with those of extant macroscelidids. She concluded that the locomotion of zalambdalestids was similar to that of modern macroscelidids, which are facultative bipedal runners. The dentition of early eutherians has a piercing and cutting function, indicating an insectivorous diet (including all kinds of small invertebrates). Early eutherian dentitions mostly remain conservative and do not exhibit strong carnivorous or omnivorous-herbivorous adaptations. Based on dental-shape disparity and morphospace occupancy analyses of the dentition, Wilson (2013) suggested that the large-bodied *Cimolestes magnus* had a highly carnivorous diet (comprising small vertebrates) and that few archaic ungulates (*Baiconodon nordicum* and *Mimatuta minuial*) probably consumed substantially more plant matter. In zalambdalestids, a certain differentiation of the mesial dentition occurs with loss of tooth positions and procumbent lower incisors that are also found in some extant insectivorous placentals (e.g., shrews).

6.18.4 Eutherian origins and early dispersal

The oldest putative eutherians, *Juramaia* (Middle-Late Jurassic), *Eomaia* (Early Cretaceous, Barremain), and *Acristatherium* (Early Cretaceous, early Aptian; Hu *et al.* 2010) are from northeastern Asia, and both are known by partial or complete flattened skeletons. Other Early Cretaceous eutherians from Asia are *Prokennalestes* (Kielan-Jaworowska and Daszheveg 1989, Sigogneau-Russell *et al.* 1992) from the Aptian or Albian of Mongolia and *Murtoilestes* from possibly slightly older strata of Transbaikalia (Averianov and Skutschas 2001). Recently, an Early Cretaceous (early Albian) eutherian, *Sasayamamylos*, has been described from Japan (Kusuhashi *et al.* 2013). The North American record of Early Cretaceous eutherians is poor and younger than that of Asia. Based on a recent revision (Davis and Cifelli 2011), *Holoclemensia* (one of the Aptian-Albian "Trinity therians" from Texas and Oklahoma

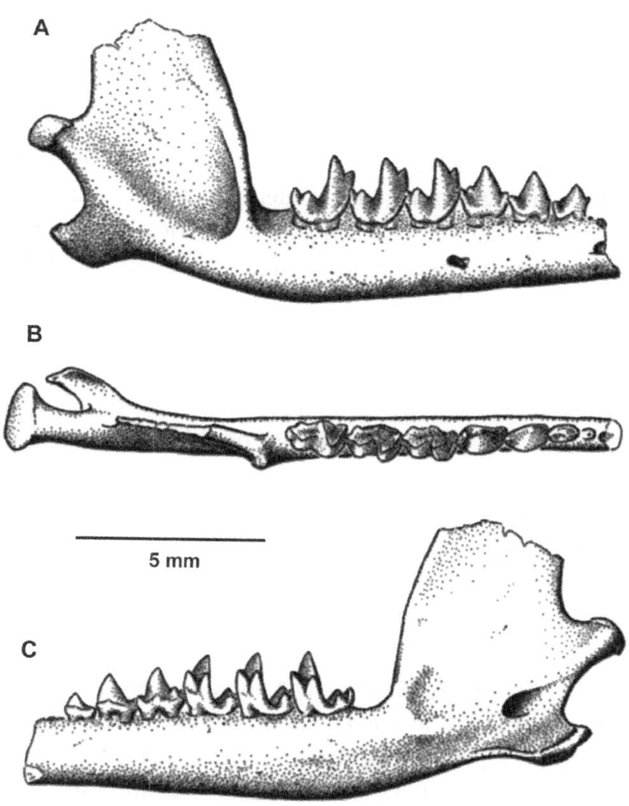

Fig. 6.86: Restored incomplete mandible of *Montanlestes keeblerorum* (holotype) with p3-m3 in lateral (A), dorsal (B), and medial (C) views. From Cifelli (1999), adapted with permission from Springer Nature.

represented by isolated teeth) is now considered as a eutherian, contrary to the original interpretation as a metatherian (Slaughter 1968a). *Montanalestes* (Fig. 6.86) from the late Early Cretaceous (Albian) Cloverly Formation of Montana is an uncontested eutherian, represented by a partial mandible (Cifelli 1999). This scenario suggests that eutherians originated in Asia in the Middle-Late Jurassic and subsequently spread to North America in the Early Cretaceous. Eutherian mammals are the dominant therians in the Cretaceous of Asia. A number of taxa is known from the early Late Cretaceous of Uzbekistan, which are attributed to Zhelestidae Nessov 1985 and Zalambdalestidae Nessov *et al.* 1998, Averianov and Archibald 2005, Archibald and Averianov 2012. Eutherians are also well represented in the Late Cretaceous of Asia, with three families (Asioryctidae, Kennalestidae, and Zalambdalestidae) in Mongolia. In Europe, early eutherians are known from the Campanian and Maastrichtian of northern Spain and southern France from a few isolated teeth and tooth fragments (Ledoux *et al.* 1966, Gheerbrant and Astibia 1999). The best known taxa are *Lainodon* and *Labes*, which are referred to the subfamily Lainodontinae Gheerbrant and Astibia 2012 within Zhelestidae, indicating a certain European endemism. For *Labes*, a mandibular fragment with three damaged molars and an upper molar have been recently reported (Martin *et al.* 2015).

In North America, there is a peculiar gap of about 30 million years without eutherian fossils between the Aptian-Albian and the early Campanian. It is not clear if this is an artifact of the fossil record or if eutherians had become extinct during that period and later immigrated again from Asia. Eutherians reappear in North America in the early Campanian of Alberta, Canada, and the western USA with *Paranyctoides* (Fox 1979, 1984, Lillegraven and McKenna 1986, Rigby and Wolberg 1987, Cifelli 1990e, Montellano 1992). *Paranyctoides* is the only North American Cretaceous eutherian that also has been reported from Asia, from the Turonian and possible Coniacian of Uzbekistan (Nessov 1993, Archibald and Averianov 2001, 2005, Averianov and Archibald 2003, Averianov and Archibald 2013a, b). The slightly older Asian occurrence suggests an immigration from Asia to North America. Montellano-Ballesteros *et al.* (2013) contested the presence of *Paranyctoides* in Central Asia (for discussion, see below). Other eutherian groups such as Zhelestidae *Avitotherium* and *Gallolestes*, Cimolestidae (*Cimolestes* and *Batodon*), and *Gypsonictops* (which, together with Paleogene *Leptictis* may be the sister taxon of Placentalia) (Wible *et al.* 2009) appear in North America by the late Campanian-Maastrichtian (Nessov *et al.* 1998, Lillegraven and McKenna 1986), but they remain minor elements in North American Cretaceous mammal faunas.

There are three genera of eutherian mammals known from the Maastrichtian of the Indian subcontinent, *Deccanolestes*, *Sahnitherium*, and *Kharmerungulatum* (Prasad and Sahni 1988, Prasad *et al.* 1994, Rana and Wilson 2003), that represent the only undisputed pre-Tertiary Gondwanan eutherians. The phylogenetic analysis by Goswami *et al.* (2011) supports a robust relationship between *Deccanolestes* and Paleocene adapisoriculids, but no placental (Euarchonta) affinity as suggested by Boyer *et al.* (2000) and Smith *et al.* (2010). The study of Goswami *et al.* (2011) further demonstrates that eutherian mammals dispersed between India, Africa, and Europe in the Early Cretaceous-Early Paleocene.

6.18.5 Eutherian phylogeny

Recent cladistic analyses have dramatically changed the systematic arrangement of early eutherians. Whereas in earlier classifications (e.g., Kielan-Jaworowska *et al.* 2004), many Cretaceous eutherians were attributed to crown Eutheria (Placentalia), most paleontologists now agree upon that all known Cretaceous eutherians fall outside the crown clade (Asher *et al.* 2005, Wible *et al.* 2007, 2009, Goswami *et al.* 2011, O'Leary *et al.* 2013). However, Wible *et al.* (2009) stated that "latest Cretaceous placentals likely existed, but we have yet to uncover them" and recent genomically based clock analyses (Dos Reis *et al.* 2012, Tarver *et al.* 2016) support this interpretation.

As discussed above, interpretation of the geologically oldest and most plesiomorphic eutherians (*Juramaia*, *Eomaia*, and *Acristatherium*) is not uncontroversial as some authors (e.g., Krause *et al.* 2014a, Averianov and Archibald 2015) consider them to be stem therians. *Eomaia* from the Early Cretaceous (Barremian) of northeastern China (Ji *et al.* 2002), *Prokennalestes* from the Aptian-Albian of Mongolia (Kielan-Jaworowska and Daszheveg 1989), and *Murtoilestes* from the Early Cretaceous (late Barremian-middle Aptian) of Transbaikalia (Averianov and Skutschas 2001) form an unnamed clade at the base of eutherians (Wible *et al.* 2009) (Fig. 6.87). *Bobolestes* from the Early Cretaceous (late Albian) of Uzbekistan (Nessov 1985) and *Montanalestes* from the Early Cretaceous (Aptian-Albian) of Montana (Cifelli 1999) are successive sister taxa of the remaining eutherians.

The next more crownward clade, Zhelestidae Nessov 1985, has a geographical wide distribution and comprises a number of Late Cretaceous taxa from Asia, Europe, and North America which are mainly known by isolated teeth and jaw fragments. Zhelestidae has been interpreted as a paraphyletic stem lineage leading to placental "ungulates", within "Ungulatomorpha" (Archibald 1996, Nessov *et al.* 1998, Archibald *et al.* 2001, Kielan-Jaworowska *et al.* 2004). The broad-scale cladistic analyses by Wible *et al.* (2007,

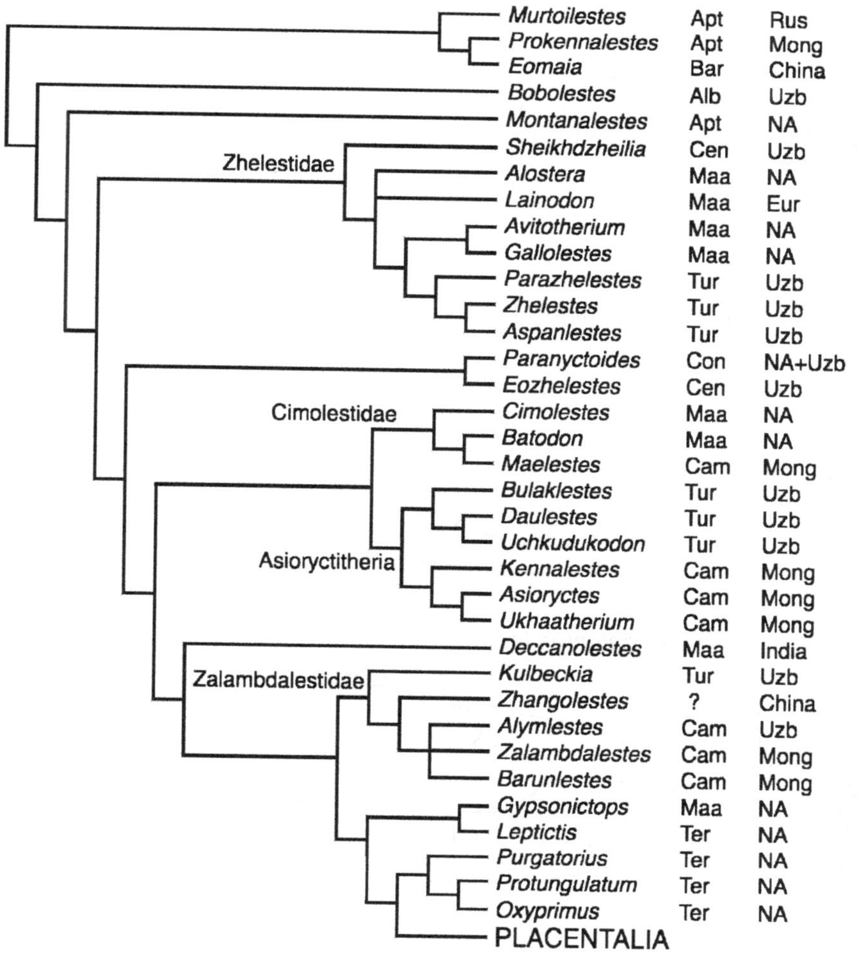

Fig. 6.87: Phylogeny of stem placentals. Abbreviations: Alb, Albian; Apt, Aptian; Bar, Barremian; Cam, Campanian; Cen, Cenomanian; Con, Coniacian; Eur; Europe; Maa, Maastrichtian; Mong, Mongolia; NA, North America; Rus, Russia; Ter, Tertiary; Tur, Turonian; Uzb, Uzbekistan. From Wible *et al.* (2009).

2009) identified zhelestids as stem placentals and sister taxon to the other Late Cretaceous (and Cenozoic) eutherians (Fig. 6.87). *Eozhelestes* from the early Cenomanian of Uzbekistan (Nessov 1997), regarded as zhelestid by Averianov and Archibald (2005), is excluded from Zhelestidae by the analyses of Wible *et al.* (2007, 2009). In a subsequent phylogenetic analysis, Archibald and Averianov (2012) considered *Eozhelestes* as a questionable zhelestid, possibly stem to Zhelestidae. According to Wible *et al.* (2009), monophyly of Zhelestidae is supported by five molar synapomorphies. The Asian taxa *Sheikhdzheilia*, *Parazhelestes*, *Zhelestes*, and *Aspanlestes* are all from the Cenomanian and Turonian of Uzbekistan (Nessov 1985, 1993, Averianov and Archibald 2005). European zhelestids are *Labes* and *Lainodon*, both from the Maastrichtian (Pol *et al.* 1992, Gheerbrant and Astibia 1994, Martin *et al.* 2015). *Alostera*, *Avitotherium*, and *Gallolestes* are Maastrichtian taxa from North America (Fox 1989, Cifelli 1990e, Lillegraven 1976).

Paranyctoides, the only eutherian genus that has been reported from the Coniacian of Asia (Uzbekistan) and North America (Fox 1979, Archibald and Averianov 2005, Averianov and Archibald 2013a), and *Eozhelestes* from the Cenomanian of Uzbekistan (Nessov 1997) form the next higher unnamed clade. It follows a clade consisting of Cimolestidae + Asioryctitheria. Cimolestidae Marsh 1889 comprises two Maastrichtian North American taxa, *Cimolestes* and *Batodon* (Marsh 1889, 1892) plus *Maelestes* from the Campanian of Mongolia (Wible *et al.* 2007). *Maelestes* (Fig. 6.88) is the only Late Cretaceous eutherian that has five premolars in the adult (Wible *et al.* 2009). The taxonomic status of *Paranyctoides* and its occurrence in Central Asia are debated: Montellano-Ballesteros *et al.* (2013) recently contested the presence of *Paranyctoides* in Central Asia and restricted the genus to the North American taxa *Paranyctoides sternbergi*, *Paranyctoides maleficus*, and four unnamed species. Averianov and Archibald (2013b), based on a parsimony analysis of 408 characters in 73 taxa of Cretaceous eutherians, reconfirmed the presence of *Paranyctoides* in North America and Central Asia and its position as sister taxon of Zhelestidae. They recognized only two species, the North American *P. sternbergi* and the Central Asian *Paranyctoides quadrans* that differ by minor features (sharp lingual ridge on p5 and m3 longer than m2 in *P. sternbergi*).

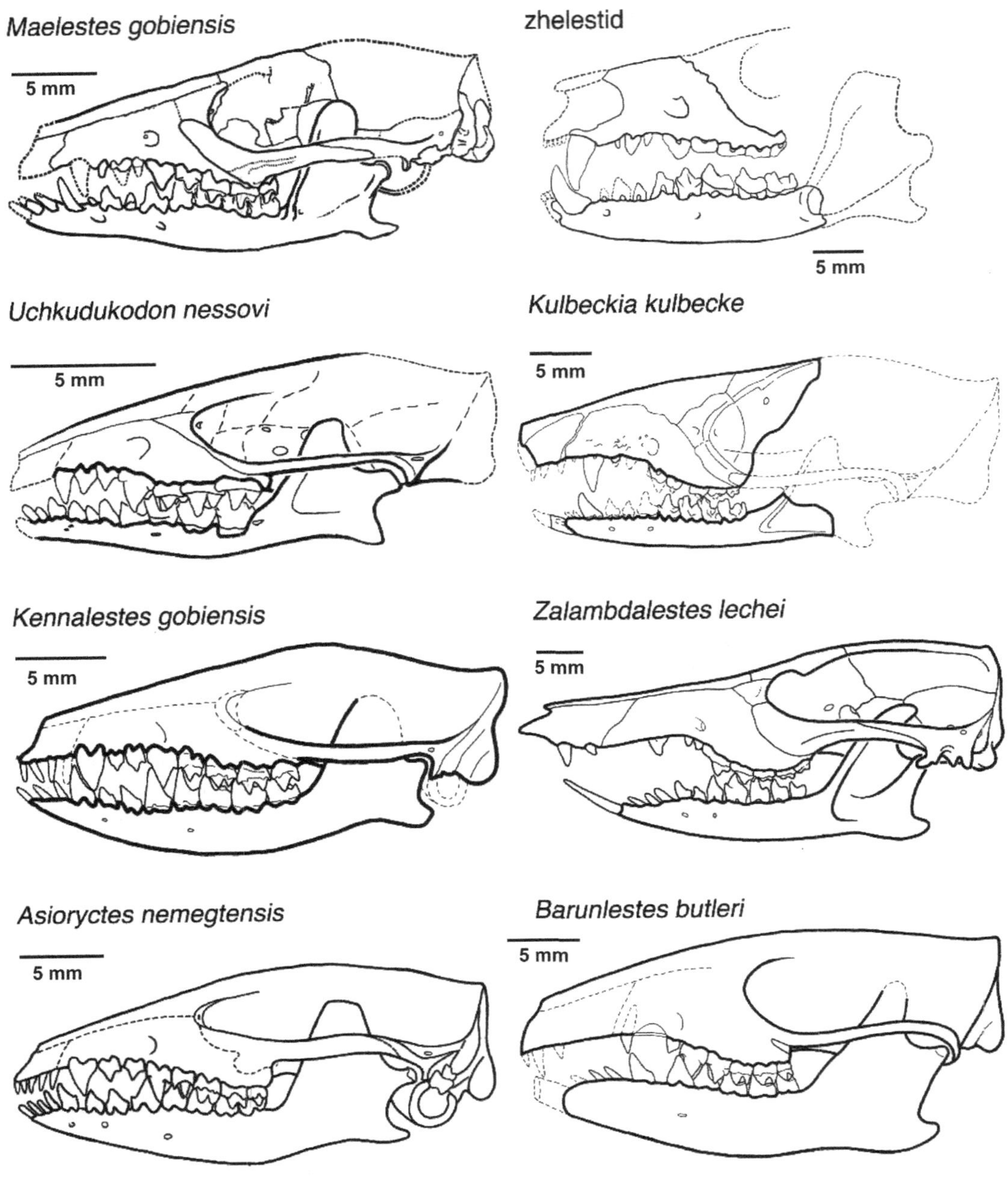

Fig. 6.88: Late Cretaceous eutherian skulls and left lower jaws in lateral view. Modified after Wible *et al.* (2009).

Cimolestidae Marsh 1889 is characterized by an extreme height and width differential between the tall and broad trigonid and the low and narrow talonid of the lower molars (Kielan-Jaworowska *et al.* 2004). The sister taxon of Cimolestidae, Asioryctitheria Novacek *et al.* 1997, consists exclusively of Asian taxa: *Bulaklestes*, *Daulestes*, and *Uchkudukodon* come from the Turonian of Uzbekistan, and *Kennalestes*, *Asioryctes*, and *Ukhaatherium* from the Campanian of Mongolia. Asioryctitheres are small and characterized by a long mesocranial region. They have up to five upper incisors (a plesiomorphic feature) and four premolars, of which the penultimate is the tallest (Kielan-Jaworowska *et al.* 2004). Of the Uzbek taxa, *Daulestes* and *Uchkudukodon*, the skull is known (Nessov and Trofimov 1979, McKenna *et al.* 2000, Archibald and Averianov 2006). The Mongolian asioryctitheres are all known by skulls, and *Ukhaatherium* and *Asioryctes* additionally by postcranial remains (Novacek *et al.* 1997, Horovitz 2003). The skulls of *Kennalestes*, *Asioryctes*, and *Ukhaatherium* (Fig. 6.88) are rather generalized and resemble each other (Wible *et al.* 2009). The postcranial skeleton of *Ukhaatherium* resembles that of generalized insectivores, and both *Ukhaatherium* and *Asioryctes* possess epipubic bones (Novacek *et al.* 1997, Horovitz 2003). Postcranial material of *Deccanolestes* from the Maastrichtian of India indicates an arboreal mode of life for this taxon and suggests that scansoreality and arboreality were prevalent early in eutherian evolution (Goswami *et al.* 2011).

Zalambdalestidae Gregory and Simpson 1926 is a group of Asian Turonian to Campanian taxa from Uzbekistan, Mongolia, and China. They include *Kulbeckia* from the Turonian of Uzbekistan, *Zhangolestes* from the Cretaceous of China, and three Campanian taxa, *Alymlestes* from Uzbekistan, and *Zalambdalestes*, *Barunlestes* (Fig. 6.88), and *Zofialestes* from Mongolia (Gregory and Simpson 1926, Kielan-Jaworowska 1975c, Nessov 1993, Averianov and Nessov 1995, Zan *et al.* 2006, Fostovicz-Frelik 2016). Zalambdalestids are larger than other Late Cretaceous eutherians (body mass estimate for *Zalambdalestes lechei* around 80 g; Rowe *et al.* 2011); have an elongated, narrow snout; and are characterized by certain dental and postcranial specializations (Fig. 6.89). The dental formula is 3I.1C.3-4P.3M/3i.1c.3-4p.3m. The second upper incisor is enlarged and caniniform. Zalambdalestids have a large diastema between the small I3 and large C, a small or fully reduced P1, a small P2, a tall P3, and submolariform P3 and P4. M3 is strongly reduced in size. The first lower incisor is very large and procumbent, with an open root, whereas i2 and i3 are small. Lower molars have very small trigonids, and high talonids that are larger than the trigonids (Archibald *et al.* 2001, Fostowicz-Frelik and Kielan-Jaworowska 2002). On the hindlimb, tibia and fibula are largely fused (Fig. 6.89), and the tibial trochlea at the calcaneus is well developed. The metatarsals of the hind feet are very long (Kielan-Jaworowska *et al.* 2004). The hypothesis that zalambdalestids share close common ancestry with Glires (i.e., rodents and lagomorphs; see van Valen 1964, Archibald *et al.* 2001) has been refuted by combined phylogenetic analyses of anatomical and genetic data, which instead place zalambdalestids outside of crown placentals (Asher *et al.* 2005).

Gypsonictops is known from numerous Maastrichtian localities in Canada and the USA and was the first eutherian to be described from North America (Simpson 1927a). Formerly considered a placental insectivoran within Leptictida McKenna 1975 (e.g., Kielan-Jaworowska *et al.* 2004), *Gypsonictops* is now excluded from Placentalia (Wible *et al.* 2007, 2009). It forms an unnamed clade with sister taxon *Leptictis* in the cladograms presented by Wible *et al.* (2007, 2009). The dentition of *Gypsonictops* is well known: it has molarized premolars and upper molars with strong cingula (Kielan-Jaworowska *et al.* 2004).

6.19 Epilogue

The last two decades witnessed exciting new discoveries of Mesozoic mammals, which have dramatically changed our picture of early mammalian evolution. The increase of knowledge on Mesozoic mammals is impressively demonstrated by the amount of publications that appeared since the publication of the epochal volume *Mammals from the Age of Dinosaurs* by Kielan-Jaworowska *et al.* in 2004. In particular, the discoveries of skeletons of early mammaliaforms and crown mammals in China that until recently had only been known from isolated teeth (e.g., haramiyidans, Zhou *et al.* 2013, Zheng *et al.* 2013, Bi *et al.* 2014) or fragmentary skeletal remains have provided a formerly unconceivable wealth of paleobiologic information. Discovery of a hitherto unknown mammalian clade with tribosphenic molars in the Southern Hemisphere, the australosphenidans (Luo *et al.* 2001a), shook the mammalian tree and initiated vivid and controversial discussions, and there is no end at sight for these exciting times of early mammal research. Farish Jenkins' statement (New York Times, 18 September 1981) that "finding of fossils of early mammals in the world is so rare that all of them would fit in half a shoe box", that was valid for almost 150 years, no longer holds true. Mesozoic mammal research is an extremely dynamic field, and the chapter aimed to provide a concise up-to-date overview on these fascinating animals that stand at the beginning of the evolutionary history of modern day Mammalia, including, of course, ourselves.

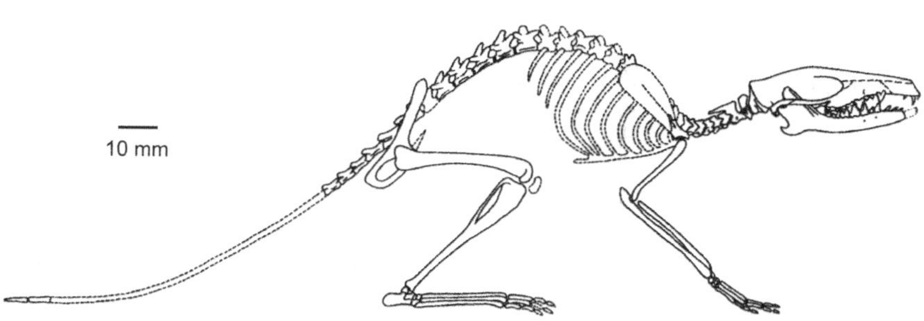

Fig. 6.89: Skeletal restoration of *Zalambdalestes lechei*. From Kielan-Jaworowska (1978).

Acknowledgments

I thank the editors, Frank Zachos (Vienna) and Robert Asher (Cambridge), for the invitation to contribute to the Handbook of Zoology. The reviewers Robin Beck (Manchester) and Augustin Martinelli (Porto Alegre) as well as both editors provided constructive criticism and valuable comments that greatly improved the manuscript. I thank Zhe-Xi Luo (Chicago), Rich Cifelli (Norman), and Alexander Averianov (St. Petersburg) for discussions. Dorothea Kranz adapted the figures, Georg Oleschinski is thanked for photography, and Benjamin Peters (all Bonn) proofread an earlier version of the manuscript. Brian Davis (Louisville), Rich Cifelli, Roger Close (Oxford), Pam Gill (Bristol), Zhe-Xi Luo, Ted Macrini (San Antonio), and Tom Rich (Melbourne) is thanked for providing original files of figures. Financial support was provided by Deutsche Forschungsgemeinschaft (DFG) grant MA 1643/14.

Literature

Allin, E.F. & Hopson, J.A. (1992): Evolution of the auditory system in Synapsida ("mammal-like reptiles" and primitive mammals) as seen in the fossil record. In: Webster, D.B., Fay, R.R. & Popper, A.N. (eds.). *The Evolutionary Biology of Hearing*, Springer-Verlag, New York: 587–614.

Ameghino, F. (1889): Contribución al conocimiento de los mamíferos fósiles de la República Argentina. *Actas de la Academía Nacional de Ciencias de Córdoba* 6: 1–1027.

Ameghino, F. (1890): Los plagiaulácidos Argentinos y sus relaciones zoológicas, geológicas y geográficas. *Boletín del Instituto Geográfico Argentino* 11: 143–208.

Ameghino, F. (1891): Nuevos restos de mamíferos fósiles descubiertos por Carlos Ameghino en el Eocene inferior de la Patagonia austral. Especies nuevas adiciones y correcciones. *Revista Argentina de Historia Natural* 1: 289–328.

Archer, M. (1978): The nature of the premolar-molar boundary of marsupials and a reinterpretation of the homology of marsupial cheekteeth. *Memoirs of the Queensland Museum* 18: 157–164.

Archer, M., Flannery, T.F., Ritchie, A. & Molnar, R. (1985): First Mesozoic mammal from Australia—an Early Cretaceous monotreme. *Nature* 318: 363–366.

Archer, M., Jenkins, F.A., Jr., Hand, S.J., Murray, P. & Godthelp, H. (1992): Description of the skull and non-vestigial dentition of a Miocene platypus (*Obdurodon dicksoni* n. sp.) from Riversleigh, Australia, and the problem of monotreme origins. In: Augee, M.L. (ed.). *Platypus and Echidnas*. The Royal Zoological Society of New South Wales, Sydney: 15–27.

Archer, M., Murray, P., Hand, S.J. & Godthelp, H. (1993): Reconsideration of monotreme relationships based on the skull and dentition of the Miocene *Obdurodon dicksoni*. In: Szalay, F.S., Novacek, M.J. & McKenna, M.C. (eds.). *Mammal Phylogeny: Mesozoic Differentiation, Multituberculates, Monotremes, Early Therians, and Marsupials*. Springer-Verlag, New York: 75–94.

Archer, M., Beck, R., Gott, M., Hand, S., Godthelp, H. & Black, K. (2011): Australia's first fossil marsupial mole (Notoryctemorphia) resolves controversies about their evolution and palaeoenvironmental origins. *Proceedings of the Royal Society of London B: Biological Sciences* 278: 1498–1506.

Archibald, J.D. (1996): Fossil evidence for a Late Cretaceous origin of "hoofed" mammals. *Science* 272: 1150–1153.

Archibald, J.D. & Averianov, A.O. (2001): *Paranyctoides* and allies from the Late Cretaceous of North America and Asia. *Acta Palaeontologica Polonica* 46: 533–551.

Archibald, J.D. & Averianov, A.O. (2005): Mammalian faunal succession in the Cretaceous of the Kyzylkum Desert. *Journal of Mammalian Evolution* 12: 9–22.

Archibald, J.D. & Averianov, A.O. (2006): Late Cretaceous asioryctitherian eutherian mammals from Uzbekistan and phylogenetic analysis of Asioryctitheria. *Acta Paleontologica Polonica* 51: 351–376.

Archibald, J.D. & Averianov, A.O. (2012): Phylogenetic analysis, taxonomic revision, and dental ontogeny of the Cretaceous Zhelestidae (Mammalia: Eutheria). *Zoological Journal of the Linnean Society (Zoology)* 164: 361–426.

Archibald, J.D., Averianov, A.O. & Ekdale, E.G. (2001): Late Cretaceous relatives of rabbits, rodents, and other extant eutherian mammals. *Nature* 414: 62–65.

Asher, R.J. & Sánchez-Villagra, M.R. (2005): Locking yourself out: zalambdodonty as a key adaptation. In: Polly, D. & Luo, Z.-X. (eds.). *Festschrift in honor of Dr. W. A. Clemens. Journal of Mammalian Evolution* 12: 265–282.

Asher, R.J., Horovitz, I., Martin, T. & Sánchez-Villagra, M.R. (2007): Neither a rodent nor a platypus: a reexamination of *Necrolestes patagoniensis* Ameghino. *American Museum Novitates* 3546: 1–40.

Asher, R.J., Meng, J., Wible, J.R., McKenna, M.C., Rougier, G.W., Dashzeveg, D. & Novacek, M.J. (2005): Stem Lagomorpha and the antiquity of Glires. *Science* 307: 1091–1094.

Averianov, A.O. (1997): New Late Cretaceous mammals of southern Kazakhstan. *Acta Palaeontologica Polonica* 46: 243–256.

Averianov, A.O. (1999): A new species of multituberculate mammal *Uzbekbaatar* from the Late Cretaceous of Uzbekistan. *Acta Paleontologica Polonica* 44: 301–304.

Averianov, A.O. (2002) Early Cretaceous "symmetrodont" mammal *Gobiotheriodon* from Mongolia and the classification of "Symmetrodonta". *Acta Palaeontologica Polonica* 47: 705–716.

Averianov, A.O. (2004): Interpretation of the Early Cretaceous mammal *Peraiocynodon* (Docodonta) and taxonomy of some British Mesozoic docodonts. *Russian Journal of Theriology* 3: 1–4.

Averianov, A.O. & Archibald, J.D. (2003): Mammals from the Upper Cretaceous Aitym Formation, Kyzylkum Desert, Uzbekistan. *Cretaceous Research* 24: 171–191.

Averianov, A.O. & Archibald, J.D. (2005): Mammals from the mid-Cretaceous Khodzhakul Formation, Kyzylkum Desert, Uzbekistan. *Cretaceous Research* 26: 593–608.

Averianov, A.O. & Archibald, J.D. (2006): New specimens of the multituberculate mammal *Uzbekbaatar* from the Late Cretaceous of Uzbekistan. *Acta Palaeontologica Polonica* 51: 377–380.

Averianov, A.O. & Archibald, J.D. (2013a). New material and reinterpretation of the Late Cretaceous eutherian mammal *Paranyctoides* from Uzbekistan. *Acta Palaeontologica Polonica* 58: 17–23.

Averianov, A.O. & Archibald, J.D. (2013b). Variation and taxonomy of Asiamerican eutherian mammal *Paranyctoides*. *Canadian Journal of Earth Sciences* 50: 895–903.

Averianov, A.O. & Archibald, J.D. (2015): Evolutionary transition of dental formula in Late Cretaceous eutherian mammals. *Science of Nature* 102: 56. 10.1007/s00114-015-1308-1

Averianov, A.O. & Kielan-Jaworowska, Z. (1999): Marsupials from the Late Createaceous of Uzbekistan. *Acta Palaeontologica Polonica* 44: 71–81.

Averianov, A.O. & Lopatin, A.V. (2006): *Itatodon tatarinovi* (Tegotheriidae, Mammalia), a docodont from the Middle Jurassic of Western Siberia and phylogentic analysis of Docodonta. *Paleontological Journal* 40: 668–677.

Averianov, A.O. & Martin, T. (2015): Ontogeny and taxonomy of *Paurodon valens* (Mammalia, Cladotheria) from the Upper Jurassic Morrison Formation of USA. *Proceedings of the Zoological Institute RAS* 319: 326–340.

Averianov, A.O. & Nessov, L.A. (1995): A new Cretaceous mammal from the Campanian of Kazakhstan. *Neues Jahrbuch für Geologie und Paläontologie, Monatshefte* 1995: 65–74.

Averianov, A.O. & Skutschas, P.P. (2001): A new genus of eutherian mammal from the Early Cretaceous of Transbaikalia, Russia. *Acta Palaeontologica Polonica* 46: 431–436.

Averianov, A.O., Lopatin, A.V., Krasnolutzkii, S.A. & Ivantsov, S.V. (2010a): New docodontans from the Middle Jurassic of Siberia and reanalysis of Docodonta interrelationships. *Proceedings of the Zoological Institute RAS* 314: 121–148.

Averianov, A.O., Archibald, J.D. & Ekdale, E.G. (2010b): New material of the Late Cretaceous deltatheroidan mammal *Sulestes* from Uzbekistan and phylogenetic reassessement of the metatherian-eutherian dichotomy. *Journal of Systematic Palaeontology* 8: 301–330.

Averianov, A.O., Lopatin, A.V. & Krasnolutskii, S.A. (2011): The first haramiyid (Mammalia, Allotheria) from the Jurassic of Russia. *Doklady Biological Sciences* 437: 103–106.

Averianov, A.O., Martin, T. & Lopatin, A.V. (2013): A new phylogeny for basal Trechnotheria and Cladotheria and affinities of South American endemic Late Cretaceous mammals. *Naturwissenschaften* 100: 311–326.

Averianov, A.O., Martin, T. & Lopatin, A.V. (2014): The oldest dryolestid mammal from the Middle Jurassic of Siberia. *Journal of Vertebrate Paleontology* 34: 924–931.

Averianov, A.O., Martin, T., Lopatin, A.V. & Krasnolutzkii, S. (2015): Stem therian mammal *Amphibetulimus* from the Middle Jurassic of Siberia. *Paläontologische Zeitschrift* 89: 197–206.

Averianov, A.O., Lopatin, A.V., Skutschas, P.P., Martynovich, N.V., Leshchinskiy, S.V., Rezvyi, A.S., Krasnolutskii, S.A. & Fayngertz, A.V. (2005): Discovery of Middle Jurassic mammals from Siberia. *Acta Palaeontologica Polonica* 50: 789–797.

Averianov, A., Martin, T., Lopatin, A., Skutschas, P., Schellhorn, R., Kolosov, P. & Vitenko, D. (2018): A high-latitude fauna of mid-Mesooic mammals from Yakutia, Russia. *PLoS ONE* 13(7):e0199983. https://doi.org/10.1371/journal.pone.0199983

Averianov, A.O., Martin, T., Skutschas, P.P., Danilov, I.G., Schultz, J.A., Schellhorn, R., Obraztsova, E., Lopatin, A.V., Sytchevskaya, E.K., Kuzmin, I.T., Krasnolutskii, S.A. & Ivantsov, S.V. (2016): Middle Jurassic vertebrate assemblage of Berezovsk coal mine in Western Siberia (Russia). *Global Geology* 19: 187–204.

Ax, P. (1984): *Das Phylogenetische System*. Gustav Fischer Verlag, Stuttgart.

Ax, P. (1985): Stem species and the stem lineage concept. *Cladistics* 1: 279–287.

Bakker, R.T. (1992): Zofiabaataridae, a new family of multituberculate mammals from the Breakfast Bench fauna at Como Bluff. *Hunteria* 2: 24.

Bakker, R.T. (1998): Dinosaur mid-life crisis: The Jurassic-Cretaceous transition in Wyoming and Colorado. In: Lucas, S.G., Kirkland, J.I. & Estep, J.W. (eds.). *Lower and Middle Cretaceous Terrestrial Ecosystems. New Mexico Museum of Natural History and Science Bulletin* 14: 67–77.

Bakker, R.T. & Carpenter, K. (1990): A new latest Jurassic vertebrate fauna from the highest levels of the Morrison Formation at Como Bluff, Wyoming, with comments on Morrison biochronology. Part III. The mammals: a new multituberculate and a new paurodont. *Hunteria* 2: 2–8.

Beck, R.M.D. (2008): A dated phylogeny of marsupials using a molecular supermatrix and multiple fossil constraints. *Journal of Mammalogy* 89: 175–189.

Beck, R.M.D., Godthelp, H., Weisbecker, V., Archer, M., & Hand, S.J. (2008). Australia's oldest marsupial fossils and their biogeographical implications. *PLoS ONE* 3(3): e1858. https://doi.org/10.1371/journal.pone.0001858.

Benton, M.J., Donoghue, P.C.J. & Asher, R.A. (2009): Calibrating and constraining molecular clocks. In: Hedges, S.B. & Kumar, S. (eds.). *The Time-tree of Life*. Oxford University Press, Oxford: 35–86.

Benton, M.J., Donoghue, P.C.J., Asher, R.J., Friedman, M., Near, T.J. & Vinther, J. (2015): Constraints on the timescale of animal evolutionary history. *Palaeontologia Electronica* 18.1.1.FC: 1–106; palaeo-electronica.org/content/fc-1

Bi, S., Wang, Y., Guan, J. & Meng, J. (2014): Three new Jurassic euharamiyidan species reinforce early divergence of mammals. *Nature* 514: 579–584.

Bi, S., Jin, X., Li, S. & Du, T. (2015): A new Cretaceous metatherian mammal from Henan, China. *PeerJ* 3:e896. https://doi.org/10.7717/peerj.896.

Bi, S., Zhen, X., Meng, J., Wang, X., Robinson, N. & Davis, B. (2016). A new symmetrodont mammal (Trechnotheria: Zhangheotheriidae) from the Early Cretaceous of China and trechnotherian character evolution. *Scientific Reports*. doi: 10.1038/srep26668.

Bi, S., Zheng, X., Wang, X., Cignetti, N.E., Yang, S. & Wible, J.R. (2018): An Early Cretaceous eutherian and the placental-marsupial dichotomy. *Nature*. doi.org/10.1038/s41586-018-0210-3

Bininda-Emonds, O.R., Cardillo, M., Jones, K.E., MacPhee, R.D., Beck, R.M., Grenyer, R., Price, S.A., Vos, R.A., Gittleman, J.L. & Purvis, A. (2007): The delayed rise of present-day mammals. *Nature* 446: 507–512.

Blainville, H.M.D. de (1838): Doutes sur le prétendu Didelphe fossile de Stonefield. *Comptes Rendus hebdomadaires des Séances de l´Académie des Sciences Paris* 7: 402–418.

Bonaparte, J.F. (1986a): A new and unusual Late Cretaceous mammal from Patagonia. *Journal of Vertebrate Paleontology* 6: 264–270.

Bonaparte, J.F. (1986b): Sobre *Mesungulatum houssayi* y nuevos mamíferos cretacicos de Patagonia. *Actas IV Congreso Argentino de Paleontología y Bioestratigrafía* 2: 48–61.

Bonaparte, J.F. (1990): New Late Cretaceous mammals from the Los Alamitos Formation, Northern Patagonia. *National Geographic Research* 6: 63–93.

Bonaparte, J.F. (1992): Una nueva especie de Triconodonta (Mammalia) de la Formación Los Alamitos, Provincia de Río Negro y comentarios sobre su fauna de mamíferos. *Ameghiniana* 29: 99–110.

Bonaparte, J.F. (2002): New Dryolestidae (Theria) from the Late Cretaceous of Los Alamitos, Argentina, and paleogeographical comments. *Neues Jahrbuch für Geologie und Paläontologie, Abhandlungen* 224: 229–271.

Bonaparte, J.F. (2008): On the phylogenetic relationships of *Vincelestes neuquenianus*. *Historical Biology* 20: 81–86.

Bonaparte, J.F. & Rougier, G.W. (1987): Mamíferos del Cretácico Inferior de Patagonia, Argentina. *IV Congreso Latinoamericano de Paleontologia* 1: 343–359.

Bonaparte, J.F., Van Valen, L. & Kramartz, A.G. (1993): La fauna local de Punta Peligro, Paleoceno inferior, de la Provincia del Chubut, Patagonia, Argentina. *Evolutionary Monographs* 14: 1–61.

Boyer, D.M., Prasad, G.V.R., Krause, D.W., Godinot, M., Goswami, A., Verma, O. & Flynn, J.J. (2000): New postcrania of *Deccanolestes* from the Late Cretaceous of India and their bearing on the evolutionary and biogeographic history of euarchontan mammals. *Naturwissenschaften* 97: 365–377.

Brink, A.A. (1956): Speculations on some advanced mammalian characteristics in higher mammal-like reptiles. *Palaeontologica Africana* 4: 77–95.

Brink, A.A. (2015): An Early Campanian mammalian fauna from the Big Bend region of Texas. *Society of Vertebrate Paleontology Annual Meeting October 2015, Program and Abstracts*: 96.

Brinkkötter, J.J. & Martin, T. (2013): Functional aspects of Haldanodon (Docodonta, Mammaliaformes) molar dentition. In: Reitner, J., Yang, Q., Wang, Y. & Reich, M. (eds.). *Palaeobiology and Geobiology of Fossil Lagerstätten through Earth History*, Universitätsverlag Göttingen, Göttingen: 23.

Broderip, W.J. (1828): Observations on the jaw of a fossil mammiferous animal found in the Stonesfield Slate. *Zoological Journal of London* 3: 408–412.

Brunet, M., Coppens, Y., Dejax, J., Flynn, L.J., Heintz, E., Hell, J., Jacobs, L.L., Jehenne, Y., Mouchelin, G., Pilbeam, D. & Sudre, J. (1990): Nouveaus mammifères du Crétacé inférieur du Cameroun, Afrique de l´Oueste. *Comptes Rendus de l´Académie des Sciences Paris, Série II* 310: 1139–1146.

Butler, P.M. (1939): The teeth of the Jurassic mammals. *Proceedings of the Zoological Society of London B* 109: 329–356.

Butler, P.M. (1978): A new interpretation of the mammalian teeth of tribosphenic pattern from the Albian of Texas. *Breviora* 446: 1–27.

Butler, P.M. (1988): Docodont molars as tribosphenic analogues (Mammalia, Jurassic). In: Russell, D.E., Santoro, J.-P. & Sigogneau-Russell, D. (eds.). *Teeth Revisited: Proceedings of the VIIth International Symposium on Dental Morphology. Mémoirs du Muséum national d'Histoire naturelle Paris C* 53: 329–340.

Butler, P.M. (1990): Early trends in the evolution of tribosphenic molars. *Biological Reviews* 65: 529–552.

Butler, P.M. (1997): An alternative hypothesis on the origin of docodont molar teeth. *Journal of Vertebrate Paleontology* 17: 435–439.

Butler, P.M. (2000): Review of the early allotherian mammals. *Acta Palaeontologica Polonica* 45: 317–342.

Butler, P.M. & Clemens, W.A. (2001): Dental morphology of the Jurassic holotherian mammal, *Amphitherium* with a discussion of the evolution of mammalian post-canine dental formulae. *Palaeontology* 44: 1–20.

Butler, P.M. & Hooker, J.J. (2005). New teeth of allotherian mammals from the English Bathonian, including the earliest multituberculates. *Acta Palaeontologica Polonica* 50: 185–207.

Butler, P.M. & Kielan-Jaworowska, Z. (1973): Is *Deltatheridium* a marsupial? *Nature* 245: 105–106.

Butler, P.M. & Sigogneau-Russell, D. (2016): Diversity of triconodonts in the Middle Jurassic of Great Britain. *Palaeontologia Polonica* 67: 35–65.

Butler, P.M., Sigogneau-Russel, D. & Ensom, P.C. (2012): Possible persistence of the morganucodontans in the Lower Cretaceous Purbeck Limestone Group (Dorset, England). *Cretaceous Research* 33: 135–145.

Campione, N.E. & Evans, C.C. (2012): A universal scaling relationship between body mass and proximal limb bone dimensions in quadrupedal terrestrial tetrapods. *BMC Biology* 10: 60. doi: 10.1186/1741-7007-10-60.

Canudo, J.I. & Cuenca-Bescós, G. (1996): Two new mammalian teeth (Multituberculata and Peramura) from the Lower Cretaceous of Spain. *Cretaceous Research* 17: 215–228.

Carlson, S.J. & Krause, D.W. (1985): Enamel ultrastructure of multituberculate mammals: an investigation of variability. *Contributions from the Museum of Paleontology, University of Michigan* 27: 1–50.

Case, J.A., Goin, F.J. & Woodburne, M.O. (2005): "South American" marsupials from the Late Cretaceous of North America and the origin of marsupial cohorts. *Journal of Mammalian Evolution* 11: 223–255. doi: 10.1023/b:jomm.0000047339.39630.82

Cedard, L. & Firth, A. (eds.) (1992). *Placental Signals: Autocrine and Paracrine Control of Pregnancy*. Trophoblast Research Vol. 6, University of Rochester Press, Rochester, NY.

Chen, M. & Luo, Z.-X. (2013): Postcranial skeleton of the Cretaceous mammal *Akidolestes cifellii* and its locomotor adaptations. *Journal of Mammalian Evolution* 20: 159–189.

Chen, M. & Wilson, G.P. (2015): A multivariate approach to infer locomotor modes in Mesozoic mammals. *Paleobiology* 41: 280–312.

Chen, M. Luo, Z.-X. & Wilson, G.P. (2017). The postcranial skeleton of *Yanoconodon allini* from the Early Cretaceous of Hebei, China, and its implications for locomotor adaptation in eutriconodontan mammals. *Journal of Vertebrate Paleontology*. doi: 10.1080/02724634.2017.1315425.

Chester, S.G.B., Sargis, E.J., Szalay, F.S., Archibald, J.D. & Averianov, A.O. (2010): Mammalian distal humeri from the Late Cretaceous of Uzbekistan. *Acta Palaeontologica Polonica* 55: 199–211.

Chimento, N.R., Agnolin, F.L. & Novas, F.E. (2012): The Patagonian fossil mammal *Necrolestes*: a Neogene survivor of Dryolestoidea. *La Revista del Museo Argentino de Ciencias Naturales* 14: 261–306.

Chornogubsky, L. (2011): New remains of the dryolestoid mammal *Leonardus cuspidatus* from the Los Alamitos Formation (Late Cretaceous, Argentina). *Paläontologische Zeitschrift* 85: 343–350.

Chow, M. & Qi, T. (1978): Paleocene mammalian fossils from Nomogen Formation of Inner Mongolia. *Vertebrata PalAsiatica* 16: 77–85.

Chow, M. & Rich, T.H. (1982): *Shuotherium dongi*, n. gen. and sp., a therian with pseudo-tribosphenic molars from the Jurassic of Sichuan, China. *Australian Mammalogy* 5: 127–142.

Chow, M. & Rich, T.H. (1984): A new triconodontan (Mammalia) from the Jurassic of China. *Journal of Vertebrate Paleontology* 4: 226–231.

Cifelli, R.L. (1990a): Cretaceous mammals of southern Utah. III. Therian mammals from the Turonian (early Late Cretaceous). *Journal of Vertebrate Paleontology* 10: 332–345.

Cifelli, R.L. (1990b): A primitive higher mammal from the Late Cretaceous of southern Utah. *Journal of Mammalogy* 71: 343–350.

Cifelli, R.L. (1990c): Cretaceous mammals of southern Utah. I. Marsupial mammals from the Kaiparowitz Formation (Judithian). *Journal of Vertebrate Paleontology* 10: 295–319.

Cifelli, R.L. (1990d): Cretaceous mammals of southern Utah. II. Marsupials and marsupial-like mammals from the Wahweap Formation (Early Campanian). *Journal of Vertebrate Paleontology* 10: 320–331.

Cifelli, R.L. (1990e): Cretaceous mammals of southern Utah. IV. Eutherian mammals from the Wahweap (Aquilan) and Kaiparowits (Judithian) formations. *Journal of Vertebrate Paleontology* 10: 346–360.

Cifelli, R.L. (1993a): Early Cretaceous mammal from North America and the evolution of marsupial dental characters. *Proceedings of the National Academy of Sciences, USA* 90: 9413–9416.

Cifelli, R.L. (1993b): Theria of metatherian–eutherian grade and the origin of marsupials. In: Szalay, F.S., Novacek, M.J. & McKenna, M.C. (eds.). *Mammal Phylogeny: Mesozoic Differentiation, Multituberculates, Monotremes, Early Therians, and Marsupials*. Springer-Verlag, New York: 205–215.

Cifelli, R.L. (1994): Therian mammals of the Terlingua Local Fauna (Judithian), Aguja Formation, Big Bend of the Rio Grande, Texas. *Contributions to Geology, University of Wyoming* 30: 117–136.

Cifelli, R.L. (1999): Tribosphenic mammal from the North American Early Cretaceous. *Nature* 401: 363–366.

Cifelli, R.L. (2000): Counting premolars in early eutherian mammals. *Acta Palaeontologica Polonica* 45: 195–198.

Cifelli, R.L. (2004): Marsupial mammals from the Albian-Cenomanian (Early-Late Cretaceous) boundary, Utah. *Bulletin of the American Museum of Natural History* 285: 62–79.

Cifelli, R.L. & Davis, B.M. (2003): Marsupial origins. *Science* 302: 1899–1890.

Cifelli, R.L., Davis, B.M. & Sames, B. (2014): Earliest Cretaceous mammals from the western United States. *Acta Palaeontologica Polonica* 59: 31–52.

Cifelli, R.L. & Eaton, J.G. (1987): Marsupial from the earliest Late Cretaceous of western US. *Nature* 325: 520–522.

Cifelli, R.L. & Gordon, C.I. (1999): Symmetrodonts from the Late Cretaceous of southern Utah and distribution of archaic mammals in the Cretaceous of North America. *Geology Studies, Bringham Young University* 44: 1–16.

Cifelli, R.L. & Johanson, Z. (1994): New marsupial from the Upper Cretaceous of Utah. *Journal of Vertebrate Paleontology* 14: 292–295.

Cifelli, R.L. & Madsen, S.K. (1998): Triconodont mammals from the medial Cretaceous of Utah. *Journal of Vertebrate Paleontology* 18: 403–411.

Cifelli, R.L. & Madsen, S.K. (1999): Spalacotheriid symmetrodonts (Mammalia) from the medial Cretaceous (upper Albian or lower Cenomanian) Mussentuchit local fauna, Cedar Mountain Formation, Utah, USA. *Geodiversitas* 21: 167–214.

Cifelli, R.L. & Muizon, C. de (1997): Dentition and jaw of *Kokopellia juddi*, a primitive marsupial or near marsupial from the medial Cretaceous of Utah. *Journal of Mammalian Evolution* 4: 241–258.

Cifelli, R.L., Wible, J.R. & Jenkins, F.A. Jr. (1998): Triconodont mammals from the Cloverly Formation (Lower Cretaceous), Montana and Wyoming. *Journal of Vertebrate Paleontology* 18: 237–241.

Cifelli, R.L., Gordon, C.L. & Lipka, T.R. (2013): New multituberculate mammal from the Early Cretaceous of eastern North America. *Canadian Journal of Earth Sciences* 50: 315–323.

Cifelli, R.L., Cohen, J.E. & Davis, B.M. (2016): New tribosphenic mammals from the Mussentuchit local fauna (Cedar Mountain Formation, Cenomanian), Utah, USA. *Palaeontologia Polonica* 67: 67–81.

Cifelli, R.L., Lipka, T.R., Schaff, C.R. & Rowe, T.B. (1999): First Early Cretaceous mammal from the eastern seaboard of the United States. *Journal of Vertebrate Paleontology* 19: 199–203.

Cifelli, R.L., Rowe, T.B., Luckett, W.P., Banta, J., Reyes, R. & Howes, R.I. (1996): Origin of marsupial pattern of tooth replacement: fossil evidence revealed by high resolution X-ray CT. *Nature* 379: 715–718.

Clemens, W.A. (1963): Fossil mammals from the type Lance Formation, Wyoming. Part I. Introduction and Multituberculata. *University of California Publications in Geological Sciences* 48: 1–105.

Clemens, W.A. (1966): Fossil mammals from the type Lance Formation Wyoming. Part II. Marsupialia. *University of California Publications in Geological Sciences* 62: 1–122.

Clemens, W.A. (1968): A mandible of *Didelphodon vorax* (Marsupialia, Mammalia). *Los Angeles County Museum Contributions in Science* 133: 1–11.

Clemens, W.A. (1979a): A problem in morganucodontid taxonomy. *Zoological Journal of the Linnean Society* 66: 1–14.

Clemens, W.A. (1979b): Marsupialia. In: Lillegraven, J.A., Kielan-Jaworowska, Z. & Clemens, W.A. (eds.). *Mesozoic Mammals: The First Two-thirds of Mammalian History*. University of California Press, Berkeley: 192–220.

Clemens, W.A. (1980): Rhaeto-Liassic mammals from Switzerland and West Germany. *Zitteliana* 5: 51–92.

Clemens, W.A. (2011): New morganucodontans from an Early Jurassic fissure filling in Wales (United Kingdom). *Palaeontology* 54: 1139–1156.

Clemens, W.A. & Lillegraven, J.A. (1986): New Late Cretaceous, North American advanced therian mammals that fit neither the marsupial nor eutherian molds. *Contributions to Geology, University of Wyoming, Special Paper* 3: 55–85.

Clemens, W.A. & Martin, T. (2014): Review of the non-tritylodontid synapsids from bone beds in the Rhaetic Sandstone, southern Germany. *Paläontologische Zeitschrift* 88: 461–479.

Clemens, W.A. & Mills, J.R.E. (1971): Review of *Peramus tenuirostris*. *Bulletin of the British Museum (Natural History), Geology* 20: 89–113.

Close, R.A., Davis, B.M., Walsh, S., Wolniewicz, A.S., Friedman, M. & Benson, R.B.J. (2015): A lower jaw of *Palaeoxonodon* from the Middle Jurassic of the Isle of Skye, Scotland, sheds new light on the diversity of British stem therians. *Palaeontology*. doi: 10.1111/pala.12218.

Cohen, J.E. (2017): Earliest divergence of stagodontid (Mammalia: Marsupialiaformes) feeding strategies from the Late

Cretaceous (Turonian) of North America. *Journal of Mammalian Evolution*. doi: 10.1007/s10914-017-9382-0.

Cope, E.D. (1882): Mammalia in the Laramie Formation. *American Naturalist* 16: 830–831.

Cope, E.D. (1884): The Tertiary Marsupialia. *American Naturalist* 18: 686–697.

Crompton, A.W. (1964): A preliminary description of a new mammal from the Upper Triassic of South Africa. *Proceedings of the Zoological Society of London* 142: 441–452.

Crompton, A.W. (1971): The origin of the tribosphenic molar. In: Kermack, D.M. & Kermack, K.A. (eds.). *Early Mammals. Zoological Journal of the Linnean Society* 50, Supplement 1: 65–87.

Crompton, A.W. (1972a): Evolution of the jaw articulation in cynodonts. In: Joysey, K.A. & Kemp, T.S. (eds.). *Studies in Vertebrate Evolution*. Oliver & Boyd, Edinburgh: 231–353.

Crompton, A.W. (1972b): Postcanine occlusion in cynodonts and tritylodonts. *Bulletin of the British Museum (Natural History)* 21: 27–71.

Crompton, A.W. (1974): The dentition and relationships of the southern African Triassic mammals, *Erythrotherium parringtoni* and *Megazostrodon rudnerae*. *Bulletin of the British Museum (Natural History), Geology* 24: 397–437.

Crompton, A.W. & Hiiemae, K.M. (1969): Functional occlusion in tribosphenic molars. *Nature* 222: 678–679.

Crompton, A.W. & Hylander, W.L. (1986): Changes in mandibular function following the acquisition of a dentary-squamosal joint. In: Hotton, N., III, MacLean, P.D., Roth, J.J. & Roth, E.C. (eds.). *The Ecology and Biology of Mammal-like Reptiles*. Smithsonian Institution Press, Washington, DC: 263–282.

Crompton, A.W. & Jenkins, F.A. (1968): Molar occlusion in Late Triassic mammals. *Biological Reviews* 43: 427–458.

Crompton, A.W. & Kielan-Jaworowska, Z. (1978): Molar structure and occlusion in Cretaceous therian mammals. In: Butler, P.M. & Joysey K.A. (eds.). *Studies in the Development, Function and Evolution of Teeth*. Academic Press, London: 249–287.

Crompton, A.W. & Luo, Z.-X. (1993): Relationships of the Liassic mammals *Sinoconodon, Morganucodon,* and *Dinnetherium*. In: Szalay, F.S., Novacek, M.J. & McKenna, M.C. (eds.). *Mammal Phylogeny: Mesozoic Differentiation, Multituberculates, Monotremes, Early Therians, and Marsupials*. Springer-Verlag, New York: 30–44.

Crompton, A.W. & Sita-Lumdsen, A. (1970): Functional significance of the therian molar pattern. *Nature* 227: 197–199.

Crompton, A.W. & Sun, A.L. (1985): Cranial structure and relationships of the Liassic mammal *Sinoconodon*. *Zoological Journal of the Linnean Society* 85: 99–119.

Crompton, A.W., Wood, C.B. & Stern, D.N. (1994): Differential wear of enamel: a mechanism for maintaining sharp cutting edges. *Advances in Comparative and Environmental Physiology* 18: 321–346.

Cuenca-Bescós, G. & Canudo, J.I. (1999): A Lower Cretaceous traveler: *Gobiconodon* ("triconodont", Mammalia) from Vallipón (upper Barremian, Teruel, Spain), *IV European Workshop on Vertebrate Paleontology, Albarracin (Teruel, Spain). Programme and Abstracts Field Guide*. Servicio Publicaciones Universidad de Zaragoza: 41.

Dal Sasso, C. & Signore, M. (1998): Exceptional soft tissue preservation in a theropod dinosaur from Italy. *Nature* 392: 383–387.

Dashzeveg, D. (1975): New primitive therian from the Early Cretaceous of Mongolia. *Nature* 256: 402–403.

Dashzeveg, D. (1979): *Arguimus khosbajari* gen. n., sp. n. (Peramuridae, Eupantotheria) from the Lower Cretaceous of Mongolia. *Acta Palaeontologica Polonica* 24: 199–204.

Dashzeveg, D. (1994): Two previously unknown eupantotheres (Mammalia, Eupantotheria). *American Museum Novitates* 3107: 1–11.

Datta, P.M. (1981): The first Jurassic mammal from India. *Zoological Journal of the Linnean Society* 73: 307–312.

Datta, P.M. (2005): Earliest mammal with transversely expanded upper molar from the Late Triassic (Carnian) Tiki Formation. South Rewa Gondwana Basin, India. *Journal of Vertebrate Paleontology* 25: 200–207.

Datta, P.M. & Das, D.P. (1996): Discovery of the oldest fossil mammal from India. *India Minerals* 50: 217–222.

Datta, P.M. & Das, D.P. (2001): *Indozostrodon simpsoni*, gen. et sp. nov., an Early Jurassic megazostrodontid mammal from India. *Journal of Vertebrate Paleontology* 21: 528–534.

Davis, B.M. (2007): A revision of "pediomyid" marsupials from the Late Cretaceous of North America. *Acta Palaeontologica Polonica* 52: 217–256.

Davis, B.M. (2011): Evolution of the tribosphenic molar pattern in early mammals, with comments on the "dual-origin" hypothesis. *Journal of Mammalian Evolution*. doi: 10.1007/s10914-011-9168-8.

Davis, B.M. (2012): Micro-computed tomography reveals a diversity of peramuran mammals from the Purbeck Group (Berriasian) of England. *Palaeontology* 55: 789–817.

Davis, B.M. & Cifelli, R.C. (2011): Reappraisal of the tribosphenidan mammals from the Trinity Group (Aptian-Albian) of Texas and Oklahoma. *Acta Palaeontologica Polonica* 56: 441–462.

Davis, B.M., Cifelli, R.L. & Cohen, J.E. (2016): First fossil mammals from the Upper Cretaceous Eagle Formation (Santonian, northern Montana, USA), and mammal diversity during the Aquilan North American Land Mammal Age. *Palaeontologia Polonica* 67: 101–126.

Debuysschere, M. (2016a): The Kuehneotheriidae (Mammaliaformes) from Saint-Nicolas-de-Port (Upper Triassic, France): a systematic review. *Journal of Mammalian Evolution*. doi: 10.1007/s10914-016-9335-z.

Debuysschere, M. (2016b): A reappraisal of *Theroteinus* (Haramiyida, Mammaliaformes) from the Upper Triassic of Saint-Nicolas-de-Port (France). *PeerJ* 4: e2592. doi: 10.7717/peerj.2592.

Debuysschere, M., Gheerbrant, E. & Allain, R. (2014): Earliest known European mammals: a review of the Morganucodonta from Saint-Nicolas-de-Port (Upper Triassic, France). *Journal of Systematic Palaeontology*. doi: 10.1080/14772019.2014.960486.

Dietrich, W.O. (1927): *Brancatherulum* n. g., ein Proplacentalier aus dem obersten Jura des Tendaguru in Deutsch-Ostafrika. *Centralblatt für Mineralogie, Geologie und Paläontologie* 1927: 423–426.

Dollo, L. (1899): Les ancêtres des Marsupiaux étaient-ils arboricoles? *Traveaux de la Station Zoologique de Wimereux* 7: 188–203.

Dos Reis, M., Inoue, J., Hasegawa, M., Asher, R.J., Donoghue, P.C.J. & Yang, Z. (2012): Phylogenomic datasets provide both precision and accuracy in estimating the timescale of placental mammal phylogeny. *Proceedings of the Royal Society of London B* 279: 3491–3500.

Eaton, J.G. (1993): Therian mammals from the Cenomanian (Upper Cretaceous) Dakota Formation, southwestern Utah. *Journal of Vertebrate Paleontology* 13: 105–124.

Eaton, J.G. (2009). Cenomanian (Late Cretaceous) mammals from Cedar Canyon, southwestern Utah, and a revision of Cenomanian *Alphadon*-like marsupials. *Museum of Northern Arizona Bulletin* 65: 97–110.

Edwards, R.G., Howe, C.W.S. & Johnson, M.H. (1975): *Immunobiology of Trophoblast*. Cambridge University Press, Cambridge, UK.

Engelmann, G.F. & Callison, G. (1998): Mammalian fauna of the Morrison Formation. *Modern Geology* 23: 343–379.

Engelmann, G.F. & Callison G. (1999): *Glirodon grandis*, a new multituberculate mammal from the Upper Jurassic Morrison Formation. In: Gillette, D.D. (ed.). *Vertebrate Paleontology in Utah*. Utah Geological Survey, Miscellaneous Publication 99–1: 161–177.

Ensom, P.C. & Sigogneau-Russell, D. (1998): New dryolestoid mammals from the basal Cretaceous Purbeck Limestone group of southern England. *Palaeontology* 41: 35–55.

Ensom, P.C. & Sigogneau-Russell, D. (2000): New symmetrodonts (Mammalia, Theria) from the Purbeck Limestone Group, Lower Cretaceous, southern England. *Cretaceous Research* 21: 767–779.

Flannery, T.F., Archer, M., Rich, T.H. & Jones, R. (1995): A new family of monotremes from the Cretaceous of Australia. *Nature* 377: 418–420.

Flynn, J.J., Parrish, J.M., Rakotosamimanana, B., Simpson, W.F. & Wyss, A.E. (1999): A Middle Jurassic mammal from Madagascar. *Nature* 401: 57–60.

Forasiepi, A.M. & Martinelli, A.G. (2003): Femur of a monotreme (Mammalia, Monotremata) from the Early Paleocene Salamanca Formation of Patagonia, Argentina. *Ameghiniana* 40: 625–630.

Fosse, G., Eskildsen, O., Risnes, S. & Sloan, R.E. (1978): Prism size in tooth enamel in some Late Cretaceous mammals and its value in multituberculate taxonomy. *Zoologica Scripta* 7: 7–61.

Fostovicz-Frelik, L. (2016): A new zalambdalestid (Eutheria) from the Late Cretaceous of Mongolia and its implications for the origin of Glires. *Palaeontologia Polonica* 67: 127–136.

Fostovicz-Frelik, L. & Kielan-Jaworowska, Z. (2002): Lower incisor in zalambdalestid mammals (Eutheria) and its phylogenetic implications. *Acta Palaeontologica Polonica* 47: 177–180.

Fox, R.C. (1969): Studies of Late Cretaceous vertebrates. III. Atriconodont mammal from Alberta. *Canadian Journal of Zoology* 47: 1253–1256.

Fox, R.C. (1971): Marsupial mammals from the early Campanian Milk River Formation, Alberta, Canada. In: Kermack, D.M. & Kermack, K.A. (eds.): *Early Mammals. Zoological Journal of the Linnean Society* 50, Supplement 1: 145–164.

Fox, R.C. (1972): A primitive therian mammal from the Upper Cretaceous of Alberta. *Canadian Journal of Earth Sciences* 9: 1479–1494.

Fox, R.C. (1974): *Deltatheroides*-like mammals from the Upper Cretaceous of North America. *Nature* 249: 392.

Fox, R.C. (1975): Molar structure and function in the Early Cretaceous mammal *Pappotherium*: evolutionary implications for Mesozoic Theria. *Canadian Journal of Earth Sciences* 12: 412–442.

Fox, R.C. (1976): Additions to the mammalian local fauna from the upper Milk River Formation (Upper Cretaceous), Alberta. *Canadian Journal of Earth Sciences* 13: 1105–1118.

Fox, R.C. (1979): Mammals from the Upper Cretaceous Oldman Formation, Alberta. III. Eutheria. *Canadian Journal of Earth Sciences* 16: 114–125.

Fox, R.C. (1980): *Picopsis pattersoni*, n. gen. and sp., an unusual therian from the Upper Cretaceous of Alberta, and the classification of primitive tribosphenic mammals. *Canadian Journal of Earth Sciences* 17: 1489–1498.

Fox, R.C. (1981): Mammals from the Upper Cretaceous Oldman Formation, Alberta. V. *Eodelphis* Matthew, and the evolution of the Stagodontidae. *Canadian Journal of Earth Sciences* 18: 350–365.

Fox, R.C. (1984): *Paranyctoides maleficus* (new species), an early eutherian mammal from the Cretaceous of Alberta. *Special Publication, Carnegie Museum of Natural History* 9: 9–20.

Fox, R.C. (1985): Upper molar structure in the Late Cretaceous symmetrodont *Symmetrodontoides* Fox, and a classification of the Symmetrodonta (Mammalia). *Journal of Paleontology* 59: 21–26.

Fox, R.C. (1989): The Wounded Knee local fauna and mammalian evolution near the Cretaceous-Tertiary boundary, Saskatchewan, Canada. *Palaeontographica A* 208: 11–59.

Fox, R.C. & Meng, J. (1997): An X-radiographic and SEM study of the osseous inner ear of multituberculates and monotremes (Mammalia): implications for mammalian phylogeny and evolution of hearing. *Zoological Journal of the Linnean Society* 121: 249–291.

Fox, R.C. & Naylor, B.G. (1986): A new species of *Didelphodon* Marsh (Marsupialia) from the Upper Cretaceous of Alberta, Canada: paleobiology and phylogeny. *Neues Jahrbuch für Geologie und Paläontologie, Abhandlungen* 172: 357–380.

Fox, R.C. & Naylor, B.G. (2006): Stagodontid marsupials from the Late Cretaceous of Canada and their systematic and functional implications. *Acta Palaeontologica Polonica* 51: 13–36.

Fox, R.C., Scott, C.S. & Bryant, H.N. (2007): A new, unusual therian mammal from the Upper Cretaceous of Saskatchewan, Canada. *Cretaceous Research* 28: 821–829.

Freeman, E.F. (1976): A mammalian fossil from the Forest Marble (Middle Jurassic) of Dorset. *Proceedings of the Geological Association* 87: 231–236.

Freeman, E.F. (1979): A Middle Jurassic mammal bed from Oxfordshire. *Palaeontology* 22: 135–166.

Gaetano, L.C. & Rougier, G.W. (2011): New materials of *Argentoconodon fariasorum* (Mammaliaformes, Triconodontidae) from the Jurassic of Argentina and its bearing on triconodont phylogeny. *Journal of Vertebrate Paleontology* 31: 829–843.

Gaetano, L.C. & Rougier, G.W. (2012): First amphilestid from South America. A molariform from the Jurassic Cañadon Asfalto Formation, Patagonia, Argentina. *Journal of Mammalian Evolution* 19: 235–248.

Gaetano, L.C., Marsicano, C.A. & Rougier, G.W. (2013): A revision of the putative Late Cretaceous triconodonts from South America. *Cretaceous Research* 46: 90–100.

Gao, C.-L., Wilson, G. P., Luo, Z.-X., Maga, A.M., Meng, Q. & Wang, X. (2010): A new mammal skull from the Lower Cretaceous of China with implications for the evolution of obtuse-angled molars and 'amphilestid' eutriconodonts. *Proceedings of the Royal Society of London B* 277: 237–246.

Gheerbrant, E. & Astibia, H. (1994): Un nouveau mammifère du Maastrichtien de Laño (Pays basque espagnol). *Comptes* Rendus *de l'Académie des Sciences Paris, Séries II* 318: 1125–1131.

Gheerbrant, E. & Astibia, H. (1999): The Upper Cretaceous mammals from Laño (Spanish Basque Country). *Estudios del Museo de Ciencias naturales de Alava* 14: 295–323.

Gheerbrant, E. & Astibia, H. (2012): Addition to the Late Cretaceous Laño mammal faunule (Spain), and to the knowledge of European "Zhelestidae" (Lainodontinae nov.). *Bulletin de la Société géologique de la France* 183: 537–546.

Gill, P.G. (1974): Resorption of premolars in the early mammal *Kuehne-otherium praecursoris. Archives of Oral Biology* 19: 327–328.

Gill, P.G. (2004): *Kuehneotherium* from the Mesozoic fissure fillings of South Wales. *Dissertation, University of Bristol*. http://research-information.bristol.ac.uk

Gill, P.G., Purnell, M.A., Crumpton, N., Brown, K.R., Gostling, N.J., Stampanoni, M. & Rayfield, E.J. (2014): Dietary specializations and diversity in feeding ecology of the earliest stem mammals. *Nature* 512: 303–305.

Gill, T.N. (1872): Arrangement of the families of mammals. With analytical tables. *Smithonian Miscellaneous Collections* 11: 1–98.

Gingerich, P.D. (1973): Molar occlusion and function in the Jurassic mammal *Docodon*. *Journal of Mammalogy* 254: 1008–1013.

Godefroit, P. & Guo, D.-Y. (1999): A new amphilestid mammal from the Early Cretaceous of China. *Bulletin d´Institut royal des Sciences naturelles de Belgique* 69, Supplement B: 7–16.

Godthelp, H., Archer, M., Cifelli, R.L., Hand, S.J. & Gilkeson, C.F. (1992). Earliest known Australian Tertiary mammal fauna. *Nature* 356: 514–516.

Goin, F.J., Candela, A.M. & Muizon, C. de (2003): The affinities of *Roberthoffstetteria nationalgeograhica* (Marsupialia) and the origin of the polydolopine molar pattern. *Journal of Vertebrate Paleontology* 23: 869–876. doi: 10.1671/2383-11

Goin, F.J., Pascual, R., Tejedor, M.F., Gelfo, J.N., Woodburne, M.O., Case, J.A., Reguero, M.A., Bond, M., López, G.M., Cione, A.L., Sauthier, D.U., Balarino, L., Scasso, R.A., Medina, F.A. & Ubaldon, M.C. (2006): The earliest therian mammal from South America. *Journal of Vertebrate Paleontology* 26: 505–510.

Goin, F.J., Tejedor, M.F., Chornogubsky, L., López, G.M., Gelfo, J.N., Bond, M., Woodburne, M.O., Gurovich, Y. & Reguero, M. (2012): Persistence of a Mesozoic, non-therian mammalian lineage (Gondwanatheria) in the mid-Paleogene of Patagonia. *Naturwissenschaften* 99: 449–463.

Gordon, C.L. (2003): A first look at estimating body size in dentally conservative marsupials. *Journal of Mammalian Evolution* 10: 1–21.

Goswami, A., Prasad, G.V.R., Upchurch, P., Boyer, D.M., Seiffert, E.R., Verma, O., Gheerbrant, E. & Flynn, J.J. (2011). A radiation of arboreal basal eutherian mammals beginning in the Late Cretaceous of India. *Proceedings of the National Academy of Sciences, USA* 108: 16333–16338.

Gow, C.E. (1985): Apomorphies of the Mammalia. *South African Journal of Science* 81: 558–560.

Gow, C.E. (1986): A new skull of *Megazostrodon* (Mammalia: Triconodonta) from the Elliot Formation (Lower Jurassic) of southern Africa. *Palaeontologia Africana* 26: 13–23.

Granger, W. & Simpson, G.G. (1929): A revision of the Tertiary Multituberculata. *Bulletin of the American Museum of Natural History* 56: 601–676.

Gray, K.E. (1825): An outline of an attempt at the disposition of the Mammalia into tribes and families with a list of the genera apparently appertaining to each tribe. *Annals of Philosophy (New Series)* 10: 337–344.

Greenwald, N.S. (1988): Patterns of tooth eruption and replacement in multituberculate mammals. *Journal of Vertebrate Paleontology* 8: 265–277.

Gregory, W.K. (1920): On the structure and relationships of *Notharctus*, an American Eocene primate. *Memoirs of the American Museum of Natural History (New Series)* 3: 49–243.

Gregory, W.K. & Simpson, G.G. (1926): Cretaceous mammal skulls from Mongolia. *American Museum Novitates* 225: 1–20.

Gurovich, Y. (2008): Additional specimens of sudamericid (Gondwanatheria) mammals from the early Paleocene of Argentina. *Palaeontology* 51: 1069–1089.

Gurovich, Y. & Beck, R. (2009): The phylogenetic affinities of the enigmatic mammalian clade Gondwanatheria. *Journal of Mammalian Evolution* 16: 25–49.

Hahn, G. (1969): Beiträge zur Fauna der Grube Guimarota Nr. 3. Die Multituberculata. *Palaeontographica A* 133: 1–100.

Hahn, G. (1971): The dentition of Paulchoffatiidae (Multituberculata, Upper Jurassic). *Memória dos Serviços Geológicos de Portugal* 17: 7–39.

Hahn, G. (1973): Neue Zähne von Haramiyiden aus der deutschen Ober-Trias und ihre Beziehungen zu den Multituberculaten. *Palaeontographica A* 142: 1–15.

Hahn, G. (1977): Neue Schädel–Reste von Multituberculaten (Mamm.) aus dem Malm Portugals. *Geologica et Paleontologica* 11: 161–186.

Hahn, G. (1987): Neue Beobachtungen zum Schädel- und Gebiss-Bau der Paulchoffatiidae (Multituberculata, Ober-Jura). *Palaeovertebrata* 17: 155–196.

Hahn, G. (1993): The systematic arrangement of the Paulcho-ffatiidae (Multituberculata) revisited. *Geologica et Palaeontologica* 27: 201–214.

Hahn, G. (2001): Neue Beobachtungen an Schädel-Resten von Paulchoffatiidae (Multituberculata; Ober-Jura). *Geologica et Palaeontologica* 35: 121–143.

Hahn, G. & Hahn, R. (1983): Multituberculata. In: Westphal, F. (ed.). *Fossilium Catalogus, I: Animalia, Pars 127,* Kugler Publications, Amsterdan: 1–409.

Hahn, G. & Hahn, R. (1992): Neue Multituberculaten-Zähne aus der Unter-Kreide (Barremium) von Spanien (Galve und Uña). *Geologica et Palaeontologica* 26: 143–162.

Hahn, G. & Hahn, R. (1994): Nachweis des Septomaxillare bei *Pseudobolodon krebsi* n. sp. (Multituberculata) aus dem Malm Portugals. *Berliner geowissenschaftliche Abhandlungen E* 13: 9–29.

Hahn, G. & Hahn, R. (1998): Neue Beobachtungen an Plagiaulacoidea (Multituberculata) des Ober-Juras 3. Der Bau der Molaren bei den Paulchoffatiidae. *Berliner geowissen-schaftliche Abhandlungen E* 28: 39–84.

Hahn, G. & Hahn, R. (1999): Pinheirodontidae n. fam. (Multitu-berculata) (Mammalia) aus der tiefen Unter-Kreide Portugals. *Palaeontographica A* 253: 7–222.

Hahn, G. & Hahn, R. (2000): Multituberculates from the Guimarota mine. In: Martin, T. & Krebs, B. (eds.): *Guimarota—A Jurassic Ecosystem*. Verlag Dr. Friedrich Pfeil, München: 97–108.

Hahn, G. & Hahn, R. (2001a): Multituberculaten-Zähne aus dem Ober-Jura von Porto das Barcas (Portugal). *Paläontologische Zeitschrift* 74: 583–586.

Hahn, G. & Hahn, R. (2001b): Multituberculaten-Zähne aus der Unter-Kreide von Pié Pajarón (Prov. Cuenca, Spanien). *Paläontologische Zeitschrift* 74: 587–589.

Hahn, G. & Hahn, R. (2003): New multituberculate teeth from the Early Cretaceous of Morocco. *Acta Palaeontologica Polonica* 48: 349–356.

Hahn, G., Sigogneau-Russell, D. & Wouters, G. (1989): New data on Theroteinidae—their relations with Paulchoffatiidae and Haramiyidae. *Geologica et Palaeontologica* 23: 205–215.

Hahn, G. Sigogneau-Russel, D., Godefroit, P. (1991): New data on *Brachyzostrodon* (Mammalia; Upper Triassic). *Geologica et Palaeontologica* 25: 237–249.

Han, G. & Meng, J. (2016): A new spalacolestine mammal from the Early Cretaceous Jehol Biota and implications for the morphology, phylogeny, and palaeobiology of Laurasian 'symmetrodontans'. *Zoological Journal of the Linnean Society* 178: 343–380.

Han, G., Mao, F.-Y., Bi, S.-D., Wang, Y.-Q. & Meng, J. (2017): A Jurassic gliding euharamiyidan mammal with an ear of five auditory bones. *Nature* 551: 451–456.

Harper, T. & Rougier, G. (2017): Systematic and functional implications of new material from the Late Cretaceous mammal *Reigitherium*. *Society of Vertebrate Paleontology Annual Meeting August 2017, Program and Abstracts*: 125.

Heinrich, W.-D. (1991): Über *Brancatherulum tendagurense* Dietrich 1927 (Mammalia: Eupantotheria) aus dem Oberjura von Tendaguru, Tansania. *Mitteilungen aus dem Zoologischen Museum Berlin* 67: 97–104.

Heinrich, W.-D. (1998): Late Jurassic mammals from Tendaguru, Tanzania, East Africa. *Journal of Mammalian Evolution* 5: 269–290.

Heinrich, W.-D. (1999): First haramiyid (Mammalia, Allotheria) from the Mesozoic of Gondwana. *Mitteilungen aus dem Museum für Naturkunde Berlin, Geowissenschaftliche Reihe* 2: 159–170.

Heinrich, W.D. (2004): *Allostaffia*, a new genus name for *Staffia* Heinrich, 1999 (Allotheria, Haramiyida) preoccupied by *Staffia* Schubert, 1911 (Protista, Foraminifera). *Mitteilungen aus dem Museum für Naturkunde Berlin, Geowissenschaftliche Reihe* 7: 153.

Henkel, S. & Krebs, B. (1969): Zwei Säugetier-Unterkiefer aus der Unteren Kreide von Uña (Prov. Cuenca, Spanien). *Neues Jahrbuch für Geologie und Paläontologie, Monatshefte* 1969: 449–463.

Henkel, S. & Krusat, G. (1980): Die Fossil-Lagerstätte in der Kohlengrube Guimarota (Portugal) und der erste Fund eines Docodontiden-Skelettes. *Berliner geowissenschaftliche Abhandlungen A* 20: 209–214.

Hoffmann, S., O´Connor, P.M., Kirk, E.C., Wible, J.R. & Krause, D.W. (2014): Endocranial and inner ear morphology of *Vintana sertichi* (Mammalia, Gondwanatheria) from the Late Cretaceous of Madagascar. *Society of Vertebrate Paleontology Memoir* 14. *Journal of Vertebrate Paleontology* 34 (6, Supplement): 110–136.

Hopson, J.A. (1971): Postcanine replacement in the gomphodont cynodont Diademodon. In: Kermack, D.M. & Kermack, K.A. (eds.). *Early Mammals. Zoological Journal of the Linnean Society* 50, Supplement 1: 1–21.

Hopson, J.A. (1995): The Jurassic mammal *Shuotherium dongi*: "pseudo-tribosphenic therian", docodontid, or neither? *Journal of Vertebrate Paleontology* 15 (3, Supplement): 36A.

Hopson, J.A. & Barghusen, H. (1986): An analysis of therapsid relationships. In: Hotton, N. III, MacLean, P.D., Roth, J.J. & Roth, E.C. (eds.). *The Ecology and Biology of Mammal-like Reptiles*. Smithsonian Institution Press, Washington, DC: 83–106.

Hopson, J.A. & Crompton, A.W. (1969): Origin of mammals. In: Dobzhansky, T., Hecht, M.K. & Steere, W.C. (eds.). *Evolutionary Biology*, Vol. 3. Appleton–Century–Crofts, New York: 15–72.

Hopson, J.A. & Kitching, J.W. (2001): A probainognathian cynodont from South Africa and the phylogeny of nonmammalian cynodonts. *Bulletin of the Museum of Comparative Zoology* 156: 5–35.

Horovitz, I. (2003): Postcranial skeletal morphology of *Ukhaatherium nessovi* (Eutheria, Mammalia) from the Late Cretaceous of Mongolia. *Journal of Vertebrate Paleontology* 23: 857–868.

Horovitz, I. & Sánchez-Villagra, M.R. (2003): A morphological analysis of marsupial mammal higher-level phylogenetic relationships. *Cladistics* 19: 181–212.

Horovitz, I., Martin, T., Bloch, J., Ladevèze, S., Kurz, C. & Sánchez-Villagra, M.R. (2009): Cranial anatomy of the earliest marsupials and the origin of opossums. *PLoS One* 4(12): eB278. doi: 10.1371/journal.pone.0008278.

Hou, S. & Meng, J. (2014): A new eutriconodont mammal from the early Cretaceous Jehol Biota of Liaoning, China. *Chinese Science Bulletin* 59: 546–553.

Hu, Y.-M., Fox, R.C., Wang, Y. & Li, C.-K. (2005a): A new spalacotheriid symmetrodont from the Early Cretaceous of northeastern China. *American Museum Novitates* 3475: 1–20.

Hu, Y.-M., Meng, J., Wang, Y.-Q. & Li C.-K. (2005b): Large Mesozoic mammals fed on young dinosaurs. *Nature* 433: 149–152.

Hu, Y.-M., Meng, J. & Clark, J.M. (2007): A new Late Jurassic docodont (Mammalia) from northeastern Xinjiang, China. *Vertebrata PalAsiatica* 45: 173–194.

Hu, Y.-M., Meng, J., Li, C.-K. & Wang, Y.-Q. (2010): New basal eutherian mammal from the Early Cretaceous Jehol biota, Liaoning, China. *Proceedings of the Royal Society of London B* 277: 229–236.

Hu, Y.-M., Wang, Y.-Q., Li, C.-K. & Luo, Z.-X. (1998): Morphology of dentition and forelimb of *Zhangheotherium*. *Vertebrata PalAsiatica* 36: 102–125.

Hu, Y.-M., Wang, Y.-Q., Luo, Z.-X. & Li, C.-K. (1997): A new symmetrodont mammal from China and its implications for mammalian evolution. *Nature* 390: 137–142.

Hughes, E.M., Wible, J.R., Spaulding, M. & Luo, Z.-X. (2015): Mammalian petrosal from the Upper Jurassic Morrison Formation of Fruita, Colorado. *Annals of Carnegie Museum* 83: 1–17.

Huttenlocker, A.K., Grossnickle, D.M., Kirkland J.I., Schultz, J.A. & Luo, Z.-X. (2018): Late-surviving stem mammal links the lowermost Cretaceous of North America and Gondwana. *Nature*. https://doi.org/10.1038/s41586-018-0126-y

Huxley, H.T. (1880): On the application of the laws of evolution to the arrangement of the Vertebrata and more particularly of the Mammalia. *Proceedings of the Zoological Society of London* 43: 649–662.

Illiger, C. (1811): *Prodromus systematis mammalium et avium aditits terminis zoographicis utriusque classis*. C. Salfeld, Berlin.

Jäger, K.R.G., Luo, Z.-X. & Martin, T. (2013): CT scanning and 3D image analysis of the postcranial skeleton of *Henkelotherium guimarotae* (Cladotheria, Mammalia) from the Late Jurassic of Portugal and its locomotor adaptations. *Society of Vertebrate Paleontology 73rd Annual Meeting, Program and Abstracts*: 147.

Jäger, K.R.G., Luo, Z.-X. & Martin, T. (in review): Postcranial skeleton of *Henkelotherium guimarotae* (Cladotheria, Mammalia) and locomotor adaptation. *Journal of Mammalian Evolution.*

Jacobs, L.L., Winkler, D.A. & Murry, P.A. (1989): Modern mammal origins: evolutionary grades in the Early Cretaceous of North America. *Proceedings of the National Academy of Sciences, USA* 86: 4992–4995.

Jenkins, F.A. Jr. (1971): The postcranial skeleton of African cynodonts. *Peabody Museum of Natural History Bulletin* 36: 1–216.

Jenkins, F.A. Jr. & Crompton, A.W. (1979): Triconodonta. In: Lillegraven, J.A., Kielan-Jaworowska, Z. & Clemens, W.A., Jr. (eds.). *Mesozoic Mammals: The First Two-thirds of Mammalian History*. University of California Press, Berkeley: 74–90.

Jenkins, F.A. Jr. & McLearn, D. (1984): Mechanisms of hind foot reversal in climbing mammals. *Journal of Morphology* 182: 197–219.

Jenkins, F.A. Jr. & Parrington, F.R. (1976): The postcranial skeletons of the Triassic mammals *Eozostrodon*, *Megazostrodon* and *Erythrotherium*. *Philosophical Transactions of the Royal Society of London* 273: 387–431.

Jenkins, F.A. Jr. & Schaff, C.R. (1988): The Early Cretaceous mammal *Gobiconodon* (Mammalia, Triconodonta) from the Cloverly Formation in Montana. *Journal of Vertebrate Paleontology* 8: 1–24.

Jenkins, F.A. Jr., Crompton, A.W. & Downs, W.R. (1983): Mesozoic mammals from Arizona: new evidence on mammalian evolution. *Science* 222: 1233–1235.

Jenkins, F.A. Jr., Gatesy, S.M., Shubin, N.H. & Amaral, W.W. (1997): Haramiyids and Triassic mammalian evolution. *Nature* 385: 715–718.

Jenkins, F.A. Jr., Shubin, N.H., Amaral, W.W, Gatesy, S.M., Schaff, C.R., Clemmensen, L.B., Downs, W.R., Davidson, A.R., Bonde, N. & Osbaeck, F. (1994): Late Triassic continental vertebrates and depositional evironments of the Fleming Fjord Formation, Jameson Land, East Greenland. *Meddelelser om Grønland* 32: 3–25.

Jepsen, G.L. (1940): Paleocene faunas of the Polecat Bench Formation, Park County, Montana. *Proceedings of the American Philosophical Society* 83: 217–341.

Ji, Q., Luo, Z.-X. & Ji, S. (1999): A Chinese triconodont mammal and mosaic evolution of the mammalian skeleton. *Nature* 398: 326–330.

Ji, Q., Luo, Z.-X., Yuan, C.-X., Wible, J.R., Zhang, J.-P. & Georgi, J.A. (2002): The earliest known eutherian mammal. *Nature* 416: 816–822.

Ji, Q., Luo, Z.-X., Yuan, C.-X. & Tabrum, A.R. (2006): A swimming mammaliaform from the Middle Jurassic and ecomorphological diversification of early mammals. *Science* 311: 1123–1127.

Ji, Q., Luo, Z.-X., Zhang, X., Yuan, C.-X. & Xu, L. (2009): Evolutionary development of the middle ear in Mesozoic therian mammals. *Science* 326: 278–281.

Johanson, Z. (1996): Revision of the Late Cretaceous North American marsupial genus *Alphadon*. *Palaeontographica A* 242: 127–184.

Kay, R.F. & Cartmill, M. (1977): Cranial morphology and adaptations of *Palaechthon nacimienti* and other Paromomyidae (Plesiadapoidea, ?Primates), with a description of a new genus and species. *Journal of Human Evolution* 6: 19–53.

Kemp, T.S. (1982): *Mammal-like Reptiles and the Origin of Mammals*. Academic Press, London.

Kermack, K.A. (1963): The cranial structure of the triconodonts. *Philosophical Transactions of the Royal Society of London* 246: 83–103.

Kermack, K.A. (1967): The interrelations of early mammals. *Zoological Journal of the Linnean Society* 47: 241–249.

Kermack, D.M. & Kermack, K.A. (1984): *Evolution of Mammalian Characters*. Croom Helm, London.

Kermack, K.A. & Mussett, F. (1958): The jaw articulation of the Docodonta and the classification of Mesozoic mammals. *Proceedings of the Royal Society of London B* 149: 204–215.

Kermack, K.A., Lee, A.J., Lees, P.M. & Mussett, F. (1987): A new docodont from the Forest Marble. *Zoological Journal of the Linnean Society* 89: 1–39.

Kermack, K.A., Lees, P.M. & Mussett, F. (1965): *Aegialodon dawsoni*, a new trituberculosectorial tooth from the lower Wealden. *Proceedings of the Royal Society of London B* 162: 535–554.

Kermack, D.M., Kermack, K.A. & Mussett, F. (1968): The Welsh pantothere *Kuehneotherium praecursoris*. *Zoological Journal of the Linnean Society* 47: 407–423.

Kermack, K.A., Mussett, F. & Rigney, H.W. (1973): The lower jaw of *Morganucodon*. *Zoological Journal of the Linnean Society* 53: 87–175.

Kermack, K.A., Mussett, F. & Rigney, H.W. (1981): The skull of *Morganucodon*. *Zoological Journal of the Linnean Society* 53: 87–175.

Kermack, K.A., Kermack, D.M., Lees, P.M. & Mills, J.R.E. (1998): New multituberculate-like teeth from the Middle Jurassic of England. *Acta Palaeontologica Polonica* 43: 581–606.

Kielan-Jaworowska, Z. (1974): Multituberculate succession in the Late Cretaceous of the Gobi Desert (Mongolia). *Palaeontologia Polonica* 30: 23–44.

Kielan-Jaworowska, Z. (1975a): Possible occurrence of marsupial bones in Cretaceous eutherian mammals. *Nature* 255: 698–699.

Kielan-Jaworowska, Z. (1975b): Evolution of the therian mammals in the Late Cretaceous of Asia. Part I. Deltatheriidae. *Palaeontologia Polonica* 33: 103–132.

Kielan-Jaworowska, Z. (1975c): Preliminary description of two new eutherian genera from the Late Cretaceous of Mongolia. *Palaeontologia Polonica* 33: 5–15.

Kielan-Jaworowska, Z. (1977): Evolution of the therian mammals in the Late Cretaceous of Asia. Part II. Postcranial skeleton in *Kennalestes* and *Asioryctes*. *Palaeontologia Polonica* 37: 65–83.

Kielan-Jaworowska, Z. (1978): Evolution of the therian mammals in the Late Cretaceous of Asia. Part III. Postcranial skeleton in Zalambdalestidae. *Palaeontologia Polonica* 38: 5–41.

Kielan-Jaworowska, Z. (1980): Absence of ptilodontoidean multituberculates from Asia and its palaeogeographic implications. *Lethaia* 13: 169–173.

Kielan-Jaworowska, Z. (1981): Evolution of the therian mammals in the Late Cretaceous of Asia. Part IV. Skull structure in *Kennalestes* and *Asioryctes*. *Palaeontologia Polonica* 42: 25–78.

Kielan-Jaworowska, Z. (1982): Marsupial-placental dichotomy and paleogeography of Cretaceous Theria. In: Gallitelli, E.M. (ed.). *Palaeontology, Essential of Historical Geology*. S.T.E.M. Mucci, Modena: 367–383.

Kielan-Jaworowska, Z. (1997): Characters of multituberculates neglected in phylogenetic analyses of early mammals. *Lethaia* 29: 249–266.

Kielan-Jaworowska, Z. & Bonaparte, J.F. (1996): Partial dentary of a multituberculate mammal from the Late Cretaceous of

Argentina and its taxonomic implications. *Revista del Museo Argentino de Ciencias Naturales "Bernardino Rivadavia"* 145: 1–9.

Kielan-Jaworowska, Z. & Dashzeveg, D. (1989): Eutherian mammals from the Early Cretaceous of Mongolia. *Zoologica Scripta* 18: 347–355.

Kielan-Jaworowska, Z. & Dashzeveg, D. (1998): Early Cretaceous amphilestid ("triconodont") mammals from Mongolia. *Acta Palaeontologica Polonica* 43: 413–438.

Kielan-Jaworowska, Z. & Ensom, P.C. (1992): Multituberculate mammals from the Upper Jurassic Purbeck Limestone Formation of southern England. *Palaeontology* 35: 95–126.

Kielan-Jaworowska, Z. & Gambaryan, P.P. (1994): Postcranial anatomy and habits of Asian multituberculate mammals. *Fossils and Strata* 36: 1–92.

Kielan-Jaworowska, Z. & Hurum, J.H. (1997): Djadochtatheria—a new suborder of multituberculate mammals. *Acta Palaeontologica Polonica* 42: 201–242.

Kielan-Jaworowska, Z. & Hurum, J.H. (2001): Phylogeny and systematics of multituberculate mammals. *Palaeontology* 44: 389–429.

Kielan-Jaworowska, Z. & Nessov, L.A. (1990): On the metatherian nature of the Deltatheroida, a sister group of the Marsupialia. *Lethaia* 23: 1–10.

Kielan-Jaworowska, Z. & Nessov, L.A. (1992): Multituberculate mammals from the Cretaceous of Uzbekistan. *Acta Palaeontologica Polonica* 37: 1–17.

Kielan-Jaworowska, Z., Dashzeveg, D. & Trofimov, B.A. (1987): Early Cretaceous multituberculates from Mongolia and a comparison with Late Jurassic forms. *Acta Palaeontologica Polonica* 32: 3–47.

Kielan-Jaworowska, Z., Cifelli, R.L. & Luo, Z.-X. (2004). *Mammals from the Age of Dinosaurs*. Columbia University Press, New York.

Kielan-Jaworowska, Z. & Ensom, P.C. (1994): Tiny plagiaulacoid mammals from the Purbeck Limestone Formation of Dorset, England. *Palaeontology* 37: 17–31.

Kielan-Jaworowska, Z., Ortiz-Jaureguizar, E., Vieytes, C., Pascual, R. & Goin, F.J. (2007). First ?cimolodontan multituberculate mammal from South America. *Acta Palaeontologica Polonica* 52: 257–262.

Kitazoe, Y., Kishino, H., Waddell, P.J., Nakajima, N., Okabayashi, T., Watabe, T. & Okuhara, Y. (2007): Robust time estimation reconciles views of the antiquity of placental mammals. *PLoS One* 2(4): e384. DOI: 10.1371/journal.pone.0000384.

Kobayashi, Y., Winkler, D.A. & Jacobs, L.L. (2002): Origin of the tooth-replacement pattern in therian mammals: evidence from a 110 Myr old fossil. *Proceedings of the Royal Society of London* 269: 369–373.

Koenigswald, W.v., Goin, F.J. & Pascual, R. (1999): Hypsodonty and enamel microstructure in the Paleocene gondwanatherian mammal *Sudamerica ameghinoi*. *Acta Palaeontologica Polonica* 44: 263–300.

Koenigswald, W.v. & Krause, D.W. (2014): Enamel microstructure of *Vintana sertichi* (Mammalia, Gondwanatheria) from the Late Cretaceous of Madagascar. *Society of Vertebrate Paleontology Memoir* 14. *Journal of Vertebrate Paleontology* 34 (6, Supplement): 166–181.

Krause, D.W. (2014): Dental morphology of *Vintana sertichi* (Mammalia, Gondwanatheria) from the Late Cretaceous of Madagascar. *Society of Vertebrate Paleontology Memoir* 14. *Journal of Vertebrate Paleontology* 34 (6, Supplement): 137–165.

Krause, D.W. & Bonaparte, J.F. (1990): The Gondwanatheria, a new suborder of Multituberculata from South America. *Journal of Vertebrate Paleontology* 9 (3, Supplement): 48A.

Krause, D.W. & Bonaparte, J.F. (1993): Superfamily Gondwanatherioidea: a previously unrecognized radiation of multituberculate mammals in South America. *Proceedings of the National Academy of Sciences, USA* 90: 9379–9383.

Krause, D.W. & Jenkins, F.A. (1983): The postcranial skeleton of North American multituberculates. *Bulletin of the Museum of Comparative Zoology* 150: 199–246.

Krause, D.W., Prasad, G.V.R., Koenigswald, W.v., Sahni, A. & Grine, F.E. (1997): Cosmopolitanism among Gondwanan Late Cretaceous mammals. *Nature* 390: 504–507.

Krause, D.W., Hoffmann, S., Wible, J.R., Kirk, E.C., Schultz, J.A., Koenigswald, W.v., Groenke, J.R., Rossie, J.B., O'Connor, P.M., Seiffert, E.R., Dumont, E.R., Holloway, W.L., Rogers, R.R., Rahantarisoa, L.J., Kemp, A.D. & Andriamialison, H. (2014a): First cranial remains of a gondwantherian mammal reveal remarkable mosaicism. *Nature* 515: 512–517.

Krause, D.W., Wible, J.R., Hoffmann, S., Groenke, J.R., O'Connor, P.M., Holloway, W.L. & Rossie, J.B. (2014b): Craniofacial morphology of *Vintana sertichi* (Mammalia, Gondwanatheria) from the Late Cretaceous of Madagascar. *Society of Vertebrate Paleontology Memoir* 14. *Journal of Vertebrate Paleontology* 34 (6, Supplement): 14–109.

Krebs, B. (1969): Nachweis eines rudimentären Coronoids im Unterkiefer der Pantotheria (Mammalia). *Paläontologische Zeitschrift* 43: 57–63.

Krebs, B. (1971): Evolution of the mandible and lower dentition in dryolestoids (Pantotheria, Mammalia). In: Kermack, D.M. & Kermack, K.A. (eds.): *Early Mammals. Zoological Journal of the Linnean Society* 50, Supplement 1: 89–102.

Krebs, B. (1991): Das Skelett von *Henkelotherium guimarotae* gen. et sp. nov. (Eupantotheria, Mammalia) aus dem Oberen Jura von Portugal. *Berliner geowissenschaftliche Abhandlungen A* 133: 1–110.

Krebs, B. (1993): Das Gebiß von *Crusafontia* (Eupantotheria, Mammalia)—Funde aus der Unter-Kreide von Galve und Uña. *Berliner geowissenschaftliche Abhandlungen E* 9: 233–252.

Krebs, B. (1998): *Drescheratherium acutum* gen. et sp. nov., ein neuer Eupantotherier (Mammalia) aus dem Oberen Jura von Portugal. *Berliner geowissenschaftliche Abhandlungen E* 28: 91–111.

Krebs, B. (2000): The henkelotheriids from the Guimarota mine. In: Martin, T. & Krebs, B. (eds.): *Guimarota—A Jurassic Ecosystem*. Verlag Dr. Friedrich Pfeil, München: 121–128.

Kretzoi, M. (1946): On Docodonta, a new order of Jurassic mammals. *Annales Historico-Naturales Musei Nationalis Hungarici* 39: 108–111.

Kretzoi, M. (1960): Zur Benennung des ältesten Symmetrodonten. *Vertebrata Hungarici* 2: 307–309.

Krusat, G. (1969): Ein Pantotherier-Molar mit dreispitzigem Talonid aus dem Kimmeridge von Portugal. *Paläontologische Zeitschrift* 43: 52–56.

Krusat, G. (1980): Contribução para o conhecimento da fauna do Kimeridgiano da mina de lignito Guimarota (Leiria, Portugal). IV Parte. *Haldanodon exspectatus* Kühne & Krusat 1972 (Mammalia, Docodonta). *Memórias dos Serviços Geológicos de Portugal* 27: 1–79.

Krusat, G. (1989): Isolated molars of a triconodont and a symmetrodont (Mammalia) from the uppermost Jurassic of Portugal. *Berliner geowissenschaftliche Abhandlungen A* 106: 277–289.

Krusat, G. (1991): Functional morphology of *Haldanodon exspectatus* (Mammalia, Docodonta) from the Upper Jurassic of Portugal. In: Kielan-Jaworowska, Z., Heintz, N. & Nakrem, H.A. (eds.): *Fifth Symposium on Mesozoic Terrestrial Ecosystems and Biota. Contributions from the Paleontological Museum, University of Oslo* 363: 37–38.

Kühne, W.G. (1949): On a triconodont tooth of a new pattern from a fissure-filling in South Glamorgan. *Proceedings of the Zoological Society of London* 119: 345–350.

Kühne, W.G. (1958): Rhaetische Triconodonten aus Glamorgan, ihre Stellung zwischen den Klassen Reptilia und Mammalia und ihre Bedeutung für die Reichert'sche Theorie. *Paläontologische Zeitschrift* 32: 197–235.

Kumar, S. & Hedges, S.B. (1998): A molecular timescale for vertebrate evolution. *Nature* 392: 917–920.

Kusuhashi, N., Hu, Y., Wang, Y., Hirasawa, S. & Matsuoka, H. (2009): New triconodontids from the Lower Cretaceous Shahai and Fuxin formations, northeastern China. *Geobios* 42: 765–781.

Kusuhashi, N., Tsutsumi, Y., Saegusa, H., Horie, K., Ikeda,.T, Yokoyama, K. & Shiraishi, K. (2013): A new Early Cretaceous eutherian mammal from the Sasayama Group, Hyogo, Japan. *Proceedings of the Royal Society of London B* 280: 20130142. http://dx.doi.org/10.1098/rspb.2013.0142.

Lazzari, V., Schultz, J.A., Tafforeau, P. & Martin, T. (2010): Occlusal pattern in paulchoffatiid multituberculates and the evolution of cusp morphology in mammaliamorphs with rodent-like dentitions. *Journal of Mammalian Evolution* 17: 177–192.

Ledoux, J.-L., Hartenberger, J.-L., Michaux, J., Sudre, J. & Thaler, L. (1966): Découverte d'un mammifère dans le Crétacé supérieur à dinosaures de Champ-Garimond près de Fons (Gard). *Comptes Rendus de l'Académie des Sciences Paris, Série D* 262: 1925–1928.

Li, G. Luo, Z.-X. (2006): A Cretaceous symmetrodont therian with some monotreme-like postcranial features. *Nature* 439: 195–200.

Li, J.-L., Wang, Y., Wang, Y.-Q. & Li, C.-K. (2001): A new family of primitive mammals from the Mesozoic of western Liaoning, China. *Chinese Science Bulletin* 46: 782–785.

Li, C.-K., Setoguchi, T., Wang, Y.-Q., Hu, Y.-M. & Chang, Z.-L. (2005). The first record of "eupantotherian" (Theria, Mammalia) from the late Early Cretaceous of western Liaoning, China. *Vertebrata PalAsiatica* 43: 245–255.

Lillegraven, J.A. (1969): Latest Cretaceous mammals of upper part of Edmonton Formation of Alberta, Canada, and review of marsupial-placental dichotomy in mammalian evolution. *University of Kansas Paleontological Contributions* 50: 1–122.

Lillegraven, J.A. (1976): A new genus of therian mammal from the Late Cretaceous "El Gallo Formation," Baja California, Mexico. *Journal of Paleontology* 50: 437–443.

Lillegraven, J.A. (2004): Polarities in mammalian evolution seen through homologs of the inner cell mass. *Journal of Mammalian Evolution* 11: 143–202.

Lillegraven, J.A. & Krusat, G. (1991): Cranio-mandibular anatomy of *Haldanodon exspectatus* (Docodonta; Mammalia) from the Late Jurassic of Portugal and its implications to the evolution of mammalian characters. *Contributions to Geology, University of Wyoming* 28: 39–138.

Lillegraven, J.A. & McKenna, M.C. (1986): Fossil mammals from the "Mesaverde" Formation (Late Cretaceous, Judithian) of the Bighorn and Wind River basins, Wyoming, with definitions of Late Cretaceous North American landmammal "ages." *American Museum Novitates* 2840: 1–68.

Loke, Y.W. & Whyte, A. (1983): *Biology of Trophoblast*. Elsevier, Amsterdam.

Lopatin, A.V. & Averianov, A.O. (2005): A new docodont (Docodonta, Mammalia) from the Middle Jurassic of Siberia. *Doklady Biological Sciences* 405: 434–436.

Lopatin, A.V. & Averianov, A.O. (2006a): Revision of a pretribosphenic mammal *Arguimus* from the Early Cretaceous of Mongolia. *Acta Palaeontologica Polonica* 51: 339–349.

Lopatin, A.V. & Averianov, A.O. (2006b): An aegialodontid upper molar and the evolution of mammalian dentition. *Science* 313: 1092.

Lopatin, A.V. & Averianov, A.O. (2007): The earliest Asiatic pretribosphenic mammal (Cladotheria, Amphitheriidae) from the Middle Jurassic of Siberia. *Doklady Biological Sciences* 417: 432–434.

Lopatin, A.V., Maschenko, E.N. & Averianov, A.O. (2010): A new genus of triconodont mammals from the Early Cretaceous of western Siberia. *Doklady Biological Sciences* 433: 282–285.

Luckett, W.P. (1993): An ontogenetic assessment of dental homologies in therian mammals. In: Szalay, F.S., Novacek, M.J. & McKenna, M.C. (eds.): *Mammal Phylogeny: Mesozoic Differentiation, Multituberculates, Monotremes, Early Therians, and Marsupials*. Springer-Verlag, New York: 182–204.

Lucas, S.G. & Hunt, A.P. (1990): The oldest mammal. *New Mexico Journal of Science* 30: 41–49.

Lucas, S.G. & Luo, Z.-X. (1993): *Adelobasileus* from the upper Triassic of west Texas: the oldest mammal. *Journal of Vertebrate Paleontology* 13: 309–334.

Luo, Z.-X. (1994): Sister-group relationships of mammals and transformations of diagnostic mammalian characters. In: Fraser, N.C. & Sues, H.-D. (eds.). *In the Shadow of the Dinosaurs—Early Mesozoic Tetrapods*. Cambridge University Press, Cambridge: 98–128.

Luo, Z.-X. (2007a): Transformation and diversification in early mammal evolution. *Nature* 450: 1011–1019.

Luo, Z.-X. (2007b): Successive diversifications in early mammalian evolution. In: Anderson, J.S. & Sues, H.-D. (eds.): *Major Transitions in Vertebrate Evolution*. Indiana University Press, Bloomington and Indianapolis: 337–391.

Luo, Z.-X. & Ji, Q. (2005): New study on dental and skeletal features of the Cretaceous mammal *Zhangheotherium*. *Journal of Mammalian Evolution* 12: 337–357.

Luo, Z.-X. & Martin, T. (2007): Analysis of molar structure and phylogeny of docodont genera. *Carnegie Museum of Natural History Bulletin* 39: 27–47.

Luo, Z.-X. & Wible, J.R. (2005): A new Late Jurassic digging mammal and early mammalian diversification. *Science* 308: 103–107.

Luo, Z.-X., Crompton, A.W. & Lucas, S.G. (1995): Evolutionary origins of the mammalian promontorium and cochlea. *Journal of Vertebrate Paleontology* 15: 113–121.

Luo, Z.-X., Crompton, A.W. & Sun, A.-L. (2001a): A new mammal from the Early Jurassic and evolution of mammalian characteristics. *Science* 292: 1535–1540.

Luo, Z.-X., Cifelli, R.L. & Kielan-Jaworowska, Z. (2001b): Dual origin of tribosphenic mammals. *Nature* 409: 53–57.

Luo, Z.-X., Kielan-Jaworowska, Z. & Cifelli, R.L. (2002): In quest for a phylogeny of Mesozoic mammals. *Acta Palaeontologica Polonica* 47: 1–78.

Luo, Z.-X., Kielan-Jaworowska, Z. & Cifelli, R.L. (2004): Evolution of dental replacement in mammals. In: Dawson, M.R. & Lillegraven, J.A. (eds.). *Fanfare for an Uncommon Paleontologist—Festschrift in Honor of Dr. Malcolm C. McKenna. The Carnegie Museum of Natural History Bulletin* 36: 159–175.

Luo, Z.-X., Ji, Q., Wible, J.R. & Yuan, C.-X. (2003): An Early Cretaceous tribosphenic mammal and metatherian evolution. *Science* 302: 1934–1940.

Luo, Z.-X., Ji, Q. & Yuan, C.-X. (2007a): Convergent dental evolution in pseudotribosphenic and tribosphenic mammals. *Nature* 450: 93–97.

Luo, Z.-X., Chen, P., Li, G. & Chen, M. (2007b): A new eutriconodont mammal and evolutionary development in early mammals. *Nature* 446: 288–293.

Luo, Z.-X., Ruf, I. & Martin, T. (2012): The petrosal and inner ear of the Late Jurassic stem cladotherian mammal *Dryolestes leiriensis* and implications for ear evolution in therian mammals. *Zoological Journal of the Linnean Society* 166: 433–463.

Luo, Z.-X., Schultz, J.A. & Ekdale, E.G. (2016): Evolution of the inner and middle ears of mammaliaforms: the approach to mammals. In: Clack, J.A., Fay, R.R. & Popper A.N. (eds.). Evolution of the Vertebrate Ear-Evidence from the Fossil Record. *Springer Handbook of Auditory Research* 59: 139–174.

Luo, Z.-X., Ruf, I., Schultz, J.A., & Martin, T. (2011a): Fossil evidence on evolution of inner ear cochlea in Jurassic mammals. *Proceedings of the Royal Society of London B* 278: 28–34.

Luo, Z.-X., Yuan, C.-X., Meng, Q.-J. & Ji, Q. (2011b): A Jurassic eutherian mammal and divergence of marsupials and placentals. *Nature* 476: 442–445.

Luo, Z.-X., Meng, Q.-J., Ji, Q., Liu, D., Zhang, Y.-G. & Neander, A.I. (2015a): Evolutionary development in basal mammaliaforms as revealed by a docodontan. *Science* 347: 760–764.

Luo, Z.-X., Gatesy, S.M., Jenkins, F.A., Amaral, W.W. & Shubin, N.H. (2015b): Mandibular and dental characteristics of Late Triassic mammaliaform *Haramiyavia* and their ramifications for basal mammal evolution. *Proceedings of the National Academy of Sciences, USA* 112: E7101-E7109. DOI: 10.1073/pnas.1519387112.

Luo, Z.-X., Meng, Q.-J., Grossnickle, D.M., Liu, D., Neander, A.I., Zhang, Y.-G. & Ji, Q. (2017): New evidence for mammaliaform ear evolution and feeding adaptation in a Jurassic ecosystem. *Nature* 548: 326–329.

MacIntyre, G.T. (1966): The Miacidae (Mammalia, Carnivora). I. The systematles of *Ictidopappus* and *Protictis*. *Bulletin of the American Museum of Natural History* 131: 115–210.

Macrini, T.E. (2006): *The evolution of endocranial space in mammals and non-mammalian cynodonts*. Dissertation, University of Texas at Austin: xvii + 1–277.

Macrini, T.E., Rougier, G.W. & Rowe, T. (2007): Description of a cranial endocast from the fossil mammal *Vincelestes neuquenianus* (Theriiformes) and its relevance to the evolution of endocranial characters in therians. *The Anatomical Record* 290: 875–892.

Maier, W. (1993): Cranial morphology of the therian common ancestor, as suggested by the adaptations of neonate marsupials. In: Szalay, F.S., Novacek, M.J. & McKenna, M.C. (eds.). *Mammal Phylogeny: Mesozoic Differentiation, Multituberculates, Monotremes, Early Therians, and Marsupials*. Springer-Verlag, New York: 165–181.

Marsh, O.C. (1878): Fossil mammal from the Jurassic of the Rocky Mountains. *American Journal of Science* 15: 459.

Marsh, O.C. (1879a): Additional remains of Jurassic mammals. *American Journal of Science* 18: 215–216.

Marsh, O.C. (1879b): Notice of new Jurassic mammals. *American Journal of Science* 20: 396–398.

Marsh, O.C. (1880): Notice of new Jurassic mammals representing two new orders. *American Journal of Science* 20: 235–239.

Marsh, O.C. (1887): American Jurassic mammals. *American Journal of Science* 33: 326–348.

Marsh, O.C. (1889): Discovery of Cretaceous Mammalia. *American Journal of Science* 38: 81–92.

Marsh, O.C. (1892): Discovery of Cretaceous Mammalia. Part III. *American Journal of Science* 43: 249–262.

Marshall, L.G. (1987): Systematics of Itaborian (Middle Paleocene) age "opossum-like" marsupials from the Limestone Quarries at San José de Itaborai, Brazil. In: Archer, M. (ed.). *Possums and Opossums: Studies in Evolution*, vol. 1. Surrey Beatty & Sons, Sydney: 91–160.

Marshall, L.G., Case, J.A. & Woodburne, M.O. (1990): Phylogenetic relationships of the families of marsupials. In: Genoways, H.H. (ed.). *Current Mammalogy*. Plenum Press, New York: 433–502.

Martin, J.E., Case, J.A., Jagt, J.W.M., Schulp, A.S. & Mulder, E.W.A. (2005): A new European marsupial indicates a Late Cretaceous high-latitude transatlantic dispersal route. *Journal of Mammalian Evolution* 12: 495–511.

Martin, T. (1995): Dryolestidae from the Kimmeridge of the Guimarota coal mine (Portugal) and their implications for dryolestid systematics and phylogeny. In: Sun, A.-L. & Wang, Y. (eds.): *Sixth Symposium on Mesozoic Terrestrial Ecosystems and Biota*. Ocean Press, Beijing: 229–231.

Martin, T. (1997): Tooth replacement of Late Jurassic Dryolestidae (Eupantotheria, Mammalia). *Journal of Mammalian Evolution* 4: 1–18.

Martin, T. (1998): The premolars of *Crusafontia cuencana* (Dryolestidae, Mammalia) from the Early Cretaceous (Barremian) of Spain. *Berliner geowissenschaftliche Abhandlungen E* 28: 119–126.

Martin, T. (1999a): Dryolestidae (Dryolestoidea, Mammalia) aus dem Oberen Jura von Portugal. *Abhandlungen der senckenbergischen naturforschenden Gesellschaft* 550: 1–119.

Martin, T. (1999b): The mammal fauna of the Late Jurassic Guimarota ecosystem, Portugal. In: Leanza, H.A. (ed.). *VII International Symposium on Mesozoic Terrestrial Ecosystems*. Asociación Paleontológica Argentina, Buenos Aires: 43.

Martin, T. (2000): The dryolestids and the small "peramurid" from the Guimarota coal mine. In: Martin, T. & Krebs, B. (eds.). *Guimarota—A Jurassic Ecosystem*. Verlag Dr. Friedrich Pfeil, München: 109–120.

Martin, T. (2001): The Mammal fauna of the Late Jurassic Guimarota Ecosystem (Portugal). Asociación Paleontológica Argentina, *Publicación Especial 7, VII International Symposium on Mesozoic Terrestrial Ecosystems*: 123–126.

Martin, T. (2002): New stem-line representatives of Zatheria (Mammalia) from the Late Jurassic of Portugal. *Journal of Vertebrate Paleontology* 22: 332–348.

Martin, T. (2005): Postcranial anatomy of *Haldanodon exspectatus* (Mammalia, Docodonta) from the Late Jurassic (Kimmeridgian) of Portugal and its bearing for mammalian evolution. *Zoological Journal of the Linnean Society* 145: 219–248.

Martin, T. (2013): Mammalian postcranial bones from the Late Jurassic of Portugal and their implications for forelimb evolution. *Journal of Vertebrate Paleontology* 33: 1432–1441.

Martin, T. & Averianov, A.O. (2001): Phylogenetic integrity of Asiatic docodonts. *Journal of Vertebrate Paleontology* 21 (3, Supplement): 78A.

Martin, T. & Averianov, A.O. (2004): A new docodont (Mammalia) from the Middle Jurassic of Kyrgyzstan (Central Asia). *Journal of Vertebrate Paleontology* 24: 195–201.

Martin, T. & Averianov, A.O. (2007): A previously unrecognized group of Middle Jurassic triconodontan mammals from Central Asia. *Naturwissenschaften* 94: 43–48.

Martin, T. & Averianov, A.O. (2010): Mammals from the Middle Jurassic Balabansai Formation of the Fergana Depression, Kyrgyzstan. *Journal of Vertebrate Paleontology* 30: 855–871.

Martin, T. & Luo, Z.-X. (2005): Homoplasy in the mammalian ear. *Science* 307: 861–862.

Martin, T. & Nowotny, M. (2000): The docodont *Haldanodon* from the Guimarota Mine. In: Martin, T. & Krebs, B (eds.). *Guimarota—A Jurassic Ecosystem*. Verlag Dr. Friedrich Pfeil, München: 91–96.

Martin, T. & Rauhut, O.W.M. (2005): Mandible and dentition of *Asfaltomylos patagonicus* (Australosphenida, Mammalia) and the evolution of the tribosphenic teeth. *Journal of Vertebrate Paleontology* 25: 414–425.

Martin, T., Averianov, A.O. & Pfretzschner, H.-U. (2010a): Mammals from the Late Jurassic Qigu Formation in the southern Junggar Basin, Xinjiang, Northwest China. *Palaeobiodiversity and Palaeoenvironments* 90: 295–319.

Martin, T., Nowotny, M. & Fischer, M. (2010b): New data on tooth replacement in the Late Jurassic docodont mammal *Haldanodon exspectatus*. *70th Anniversary Meeting Society of Vertebrate Paleontology, Program and Abstracts*: 130A.

Martin, T., Averianov, A.O. & Lopatin, A.V. (2011): Middle Jurassic mammals from the Itat Formation at Berezovsk Quarry in western Siberia (Russia). *Supplement of online Journal of Vertebrate Paleontology, November 2011*: 153.

Martin, T., Marugán-Lobón, J., Vullo, R., Martín-Abad, H., Luo, Z.-X. & Buscalioni, A.D. (2015): A Cretaceous eutriconodont and integument evolution in early mammals. *Nature* 526: 380–384.

Martin, T., Schultz, J.A., Schwermann, A.H. & Wings, O. (2016): First Jurassic mammals of Germany: multituberculate teeth from Langenberg Quarry (Lower Saxony). *Palaeontologia Polonica* 67: 171–179.

Martinelli, A.G., Soares, M.B. & Schwanke, C. (2016): Two new cynodonts (Therapsida) from the Middle-early Late Triassic of Brazil and comments on South American probainognathians. *PLoS One* 11 (10): e0162945.

Maschenko, E.N. & Lopatin, A.V. (1998): First record of an Early Cretaceous triconodont mammal in Siberia. *Bulletin de l'Institut royal des Sciences naturelles de Belgique, Sciences de la Terre* 68: 233–236.

Maschenko, E.N. & Lopatin, A.V. (2002): A new Early Cretaceous mammal from Western Siberia. *Doklady Biological Sciences* 386: 475–477.

Maschenko, E.N., Lopatin, A.V. & Voronkevich, A.V. (2002): A new genus of tegotheriid docodonts (Docodonta, Tegotheriidae) from the Early Cretaceous of west Siberia. *Russian Journal of Theriology* 1: 75–81.

Matthew, W.D. (1916): A marsupial of the Belly River Cretaceous. *Bulletin of the American Museum of Natural History* 35: 477–500.

McKenna, M.C. (1975): Toward a phylogenetic classification of the Mammalia. In: Luckett, W.P. & Szalay, F.S. (eds.). *Phylogeny of the Primates*. Plenum Press, New York: 21–46.

McKenna, M.C., Kielan-Jaworowska, Z. & Meng, J. (2000): Earliest eutherian mammal skull from the Late Cretaceous (Coniacian) of Uzbekistan. *Acta Palaeontologica Polonica* 45: 1–54.

Meng, J., Hu, Y.-M., Wang, Y.-Q., Wang, X.-L. & Li, C.-K. (2003): The ossified Meckel's cartilage and internal groove in Mesozoic mammaliaforms: implications to origin of the definitive mammalian middle ear. *Zoological Journal of the Linnean Society* 138: 431–448.

Meng, J., Hu, Y.-M., Wang, Y.-Q. & Li, C.-K. (2005): A new triconodont (Mammalia) from the Early Cretaceous Yixian Formation of Liaoning, China. *Vertebrata PalAsiatica* 43: 1–10.

Meng, J., Hu, Y.-M., Wang, Y.-Q., Wang, X.-L. & Li, C.-K. (2006): A Mesozoic gliding mammal from northeastern China. *Nature* 444: 889–893.

Meng, J., Wang, Y.-Q. & Li, C.-K. (2011): Transitional mammalian middle ear from a new Cretaceous Jehol eutriconodont. *Nature* 472: 181–185.

Meng, J., Bi, S., Wang, Y., Zheng, X. & Wang, X. (2014): Dental and mandibular morphologies of *Arboroharamiya* (Haramiyida, Mammalia): a comparison with other haramiyidans and *Megaconus* and implications for mammalian evolution. *PLoS One* 9(12): e113847.

Meng, Q.-J., Ji, Q., Zhang, Y.-G., Liu, D., Grossnickle, D.M. & Luo, Z.-X. (2015): An arboreal docodont from the Jurassic and mammaliaform ecological diversification. *Science* 347: 764–768.

Meng, Q.-J., Grossnickle, D.M., Liu, D., Zhang, Y.-G., Neander, A.I., Ji, Q. & Luo, Z.-X. (2017). New gliding mammaliaforms from the Jurassic. *Nature* 548: 291–296.

Meredith, R.W., Janečka, J.E., Gatesy, J., Ryder, O.A., Fisher, C.A., Teeling, E.C., Goodbla, A., Eizirik, E., Simão, T.L.L., Stadler, T., Rabosky, D.L., Honeycutt, R.L., Flynn, J.J., Ingram, C.M., Steiner, C., Williams, T.L., Robinson, T.J., Burk-Herrick, A., Westerma, M., Ayoub, N.A., Springer, M.S. & Murphy, W.J. (2011): Impacts of the Cretaceous terrestrial revolution and KPg extinction on mammal diversification. *Science* 334: 521–524.

Mills, J.R.E. (1964): The dentitions of *Peramus* and *Amphitherium*. *Proceedings of the Linnean Society of London* 175: 117–133.

Mills, J.R.E. (1971): The dentition of *Morganucodon*. In: Kermack, D.M. & Kermack, K.A. (eds.). *Early Mammals. Zoological Journal of the Linnean Society* 50, Supplement 1: 29–63.

Mones, A. (1987): Gondwanatheria, un nuevo orden de mamíferos sudamericanos (Mammalia: Edentata: ?Xenarthra). *Comunicaciones Paleontológicas del Museo de Historia Natural de Montevideo* 1: 237–240.

Montandon, S.A., Tzika, A.C., Martins, A.F., Chopard, B. & Milinkovitch, M.C. (2014): Two waves of anisotropic growth generate enlarged follicles in the spiny mouse. *EvoDevo* 5: 33.

Montellano, M. (1992): Mammalian fauna of the Judith River Formation (Late Cretaceous, Judithian), northcentral Montana. *University of California Publications in Geological Sciences* 136: 1–115.

Montellano, M., Hopson, J.A. & Clark, J.A. (2008): Late Early Jurassic mammaliaforms from Huizachal Canyon, Tamauplias, Mexico. *Journal of Vertebrate Paleontology* 28: 1130–1143.

Montellano, M., Weil, A., Clemens, W.A. (2000): An exceptional specimen of *Cimexomys judithae* (Mammalia: Multituberculata) from the Campanian Two Medicine Formation of Montana and the phylogenetic status of *Cimexomys*. *Journal of Vertebrate Paleontology* 20: 333–340.

Montellano-Ballesteros, M., Fox, R.C. & Scott, C.S. (2013). Species composition of the Late Cretaceous eutherian mammal *Paranyctoides* Fox. *Canadian Journal of Earth Sciences* 50: 693–700.

Morton, H. (1998): Early pregnancy factor: an extracellular chaperonin 10 homologue. *Immunology and Cell Biology* 76: 483–496.

Morton, H., Rolfe, B.E., McNeill, L., Clarke, P., Clarke, F.M. & Clunie, G.J.A. (1980): Early pregnancy factor: tissues involved in its production in the mouse. *Journal of Reproductive Immunology* 2: 73–82.

Muizon, C. de. (1998): *Mayulestes ferox*, a borhyaenoid (Metatheria, Mammalia) from the early Palaeocene of Bolivia. Phylogenetic and palaeobiologic implications. *Geodiversitas* 20: 19–142.

Muizon, C. de & Lange-Badré, B. (1997): Carnivorous dental adaptations in tribosphenic mammals and phylogenetic reconstruction. *Lethaia* 30: 353–366.

Musser, A.M. & Archer, M. (1998): New information about the skull and dentary of the Miocene platypus *Obdurodon dicksoni*, and a discussion of ornithorhynchid relationships. *Philosophical Transactions of the Royal Society of London B 353:* 1063–1079.

Nessov, L.A. (1985): New mammals from the Cretaceous of Kyzylkum. *Vestnik Leningradskogo Universiteta* 17: 8–18 [in Russian].

Nessov, L.A. (1993): New Mesozoic mammals of middle Asia and Kazakhstan, and comments about evolution of theriofaunas of Cretaceous coastal plains of ancient Asia. *Trudy Zoologicheskogo Instituta RAN* 249: 105–133 [in Russian].

Nessov, L.A. (1997): *Cretaceous Non-marine Vertebrates of Northern Eurasia*. University of Saint Petersburg, Institute of Earth Crust, Saint Petersburg [Posthumous edition by L.B. Goloneva & A.O. Averianov] [in Russian with English summary].

Nessov, L.A. & Trofimov, B.A. (1979): Ancient insectivores from the Cretaceous of Uzbek SSSR. *Doklady Akademii Nauk SSSR* 247: 952–955 [in Russian].

Nessov, L.A., Archibald, J.D. & Kielan-Jaworowska, Z. (1998): Ungulate-like mammals from the Cretaceous of Uzbekistan and a phylogenetic analysis of Ungulatomorpha. *Bulletin of the Carnegie Museum of Natural History* 34: 40–88.

Newham, E., Gill, P., Brewer, P., Schneider, P. Gostling, N. & Corfe, I. (2018): Life history partitioning follows phylogeny amongst Middle Jurassic mammaliaforms. In: Martin, T., Schellhorn, R. & Schultz, J. A. (eds.): *13th Symposium on Mesozoic Terrestrial Ecosystems and Biota*. *Terra Nostra* 218/1: 90–91.

Ni, X., Li, Q., Stidham, T.A., Li, L., Lu, X. & Meng, J. (2016): A late Paleocene probable metatherian (?deltatheroidan) survivor of the Cretaceous mass extinction. *Scientific Reports* 6: 38547.

Novacek, M.J. (1986): The primitive eutherian dental formula. *Journal of Vertebrate Paleontology* 6: 191–196.

Novacek, M.J., Rougier, G.W., Wible, J.R., McKenna, M.C., Dashzeveg, D. & Horovitz, I. (1997): Epipubic bones in eutherian mammals from the Late Cretaceous of Mongolia. *Nature* 389: 483–486.

Nowotny, M., Martin, T. & Fischer, M. (2001): Dental anatomy and tooth replacement of *Haldanodon exspectatus* (Docodonta, Mammalia) from the Upper Jurassic of Portugal. *Journal of Morphology* 248: 268.

O'Leary, M.A., Bloch, J.I., Flynn, J.J., Gaudin, T.J., Giallombardo, A., Giannini, N.P., Goldberg, S.L., Kraatz, B.P., Luo, Z.-X., Meng, J., Ni, X., Novacek, M.J., Perini, F.A., Randall, Z.S., Rougier, G.W., Sargis, E.J., Silcox, M.T., Simmons, N.B., Spaulding, M., Velazco, P.M., Weksler, M., Wible, J.R. & Cirranello, A.L. (2013): The placental mammal ancestor and the post–K-Pg radiation of placentals. *Science* 339: 662–667.

O'Meara, R.N. & Asher, R.J. (2016): The evolution of growth patterns in mammalian versus non-mammalian cynodonts. *Palaeobiology* 42: 439–464.

O'Meara, R.N. & Thompson, R. (2014): Were there Miocene meridiolestidans? Assessing the phylogenetic placement of *Necrolestes patagoniensis* and the presence of a 40 million year meridiolestidan ghost lineage. *Journal of Mammalian Evolution* 21: 271–284.

Osborn, H.F. (1887): On the structure and classification of the Mesozoic Mammalia. *Proceedings of the Academy of Natural Sciences, Philadelphia* 38: 282–292.

Osborn, H.F. (1888a): The evolution of mammalian molars to and from the tritubercular type. *American Naturalist* 22: 1067–1079.

Osborn, H.F. (1888b): Additional observations upon the structure and classification of the Mesozoic Mammalia. *Proceedings of the Academy of Natural Sciences, Philadelphia* 1888: 292–301.

Osborn, H.F. (1888c): On the structure and classification of the Mesozoic Mammalia. *Journal of the Academy of Natural Sciences, Philadelphia* 9: 186–265.

Osborn, H.F. (1907): *Evolution of Mammalian Molar Teeth*. Macmillan and Company, New York.

Owen, R. (1866): Description of part of the lower jaw and teeth of a small Oolithic mammal (*Stylodon pusillus* Ow.). *Geological Magazine* 3: 199–201.

Owen, R. (1871): Monograph of the fossil Mammalia of the Mesozoic formations. *Monograph of the Palaeontological Society* 33 (for 1870): 1–115.

Panciroli, E., Benson, R.B.J. & Walsh, S. (2017): The dentary of *Wareolestes rex* (Megazostrodontidae): A new specimen from Scotland and implications for morganucodontan tooth replacement. *Papers in Palaeontology* 2017: 1–14.

Panciroli, E., Benson, R.B.J. & Butler, R.J. (2018): New partial dentaries of amphitheriid mammal *Palaeoxonodon oolithicus* from Scotland, and posterior dentary morphology in early cladotherians. *Acta Palaeontologica Polonica* 63: 197–206.

Parrington, F.R. (1941): On two mammalian teeth from the lower Rhaetic of Somerset. *Annals and Magazine of Natural History, Series 11,* 8: 140–144.

Parrington, F.R. (1947): On the collection of Rhaetic mammalian teeth. *Proceedings of the Zoological Society of London* (for 1946) 116: 707–728.

Parrington, F.R. (1967): The origins of mammals. *Advancements in Science* 24: 165–173.

Parrington, F.R. (1971): On the Upper Triassic mammals. *Philosophical Transactions of the Royal Society of London B* 261: 231–272.
Parrington, F.R. (1973): The dentitions of the earliest mammals. *Zoological Journal of the Linnean Society* 52: 85–95.
Parrington, F.R. (1978): A further account of the Triassic mammals. *Philosophical Transactions of the Royal Society of London B* 282: 177–204.
Pascual, R., Archer, M., Ortiz-Jaureguizar, E.O., Prado, J.L., Godthelp, H. & Hand, S.J. (1992a): First discovery of monotremes in South America. *Nature* 356: 704–705.
Pascual, R., Archer, M., Ortiz-Jaureguizar, E.O., Prado, J.L., Godthelp, H. & Hand, S.J. (1992b): The first non-Australian monotreme: an early Paleocene South American platypus (Monotremata, Ornithorhynchidae). In: Augee, M.L. (ed.). *Platypus and Echidnas*. The Royal Zoological Society of New South Wales, Sydney: 1–14.
Pascual, R., Goin, F.J., Krause, D.W., Ortiz-Jaureguizar, E. & Carlini, A.A. (1999): The first gnathic remains of *Sudamerica:* implications for gondwanathere relationships. *Journal of Vertebrate Paleontology* 19: 373–382.
Pascual, R., Goin, F.J., González, P., Ardolino, A. & Puerta, P. (2000): A highly derived docodont from the Patagonian Late Cretaceous: evolutionary implications for Gondwanan mammals. *Geodiversitas* 22: 395–414.
Pascual, R., Goin, F.J., Balarin, L. & Udrizar, D.E. (2002): New data on the Paleocene monotreme *Monotrematum sudamericanum*, and the convergent evolution of triangulate molars. *Acta Palaeontologica Polonica* 47: 487–492.
Patterson, B. (1951): Early Cretaceous mammals from northern Texas. *American Journal of Science* 249: 31–46.
Patterson, B. (1955): A symmetrodont mammal from the Early Cretaceous of northern Texas. *Fieldiana: Zoology* 37: 689–693.
Patterson, B. (1956): Early Cretaceous mammals and the evolution of mammalian molar teeth. *Fieldiana: Geology* 13: 1–105.
Patterson, B. & Olson, E.O. (1961): A triconodontid mammal from the Triassic of Yunnan. In: Vandebroek, G. (ed.). *International Colloquium on the Evolution of Lower and Non-specialized Mammals*. Koninklijke Vlaamse Academie voor Wetenschappen, Letteren en Schone Kunsten van Belgie, Brussels: 129–191.
Peyer, B. (1956): Über Zähne von Haramiyiden, von Triconodontiden und von wahrscheinlich synapsiden Reptilien aus dem Rhät von Hallau. *Schweizerische Paläontologische Abhandlungen* 72: 1–72.
Pfretzschner, H.U., Martin, T., Maisch, M., Matzke, A. & Sun, G. (2005): A new docodont from the Late Jurassic of the Junggar Basin in Northwest China. *Acta Palaeontologica Polonica* 50: 799–808.
Phillips, M.J., Bennett, T.H. & Lee, M.S.Y. (2009): Molecules, morphology, and ecology indicate a recent, amphibious ancestry for echidnas. *Proceedings of the National Academy of Sciences, USA* 106: 17089–17094.
Pian, R., Archer, M. & Hand, S.J. (2013): A new, giant platypus, *Obdurodon tharalkooschild*, sp. Nov. (Monotremata, Ornithorhynchidae), from the Riversleigh World Heritage Area, Australia. *Journal of Vertebrate Paleontology* 33: 1255–1259. Doi: 10.1080/02724634.2013.782876.
Pian, R., Archer, M., Hand, S.J., Beck, R.M.D. & Cody, A. (2016): The upper dentition and relationships of the enigmatic Australian Cretaceous mammal *Kollikodon ritchiei*. *Memoirs of Museum Victoria* 74: 97–105.
Plieninger, W. (1847): *Microlestes antiquus* und *Sargodon tomicus* in der Grenzbreccie von Degerloch. *Jahreshefte des Vereins für vaterländische Naturkunde in Württemberg* 3: 164–167.
Plogschties, T. & Martin, T. (2014): Revealing the unseen: computed micro tomography scanning and 3D reconstruction of a lower Cretaceous (Barremian) symmetrodont skull. *Society of Vertebrate Paleontology 74th Annual Meeting, Program and Abstracts*: 206.
Plogschties, T. & Martin, T. (in review): New information on the maxilla, dentary, and dentition of *Maotherium sinense*, and a revision of the zhangheotheriid dental formula. PalZ.
Poche, F. (1908): Einige notwendige Änderungen in der mammalogischen Nomenklatur. *Zoologische Annalen* 2: 269–272.
Pol, C., Buscalioni, A.D., Carballiera, J., Francés, V., Martinez, N.L., Marandat, B., Moratalla, J.J., Sanz, J.L., Sigé, B. & Villatte, J. (1992): Reptiles and mammals from the Late Cretaceous new locality Quintanilla del Coco (Burgos Province, Spain). *Neues Jahrbuch für Geologie und Paläontologie, Abhandlungen* 184: 279–314.
Prasad, G.V.R. & Manhas, B.K. (2001): First docodont mammals of Laurasian affinities from India. *Current Science* 81: 1235–1238.
Prasad, G.V.R. & Manhas, B.K. (2002): Triconodont mammals from the Jurassic Kota Formation of India. *Geodiversitas* 24: 445–464.
Prasad, G.V.R. & Manhas, B.K. (2007): A new docodont mammal from the Jurassic Kota Formation of India. *Palaeontologia Electronica* 11(2): 1–11.
Prasad, G.V.R. & Sahni, A. (1988): First Cretaceous mammal from India. *Nature* 332: 638–640.
Prasad, G.V.R., Jaeger, J.J., Sahni, A., Gheerbrant, E. & Khajuria, C.K. (1994): Eutherian mammals from the Upper Cretaceous (Maastrichtian) intertrappean beds of Naskal, Andhra Pradesh, India. *Journal of Vertebrate Paleontology* 14: 260–277.
Prasad, G.V.R., Verma, O., Sahni, A., Parmar, V. & Khosla, A. (2007a): A Cretaceous hoofed mammal from India. *Science* 318: 937.
Prasad, G.V.R., Verma, O., Sahni, A., Krause, D.W., Khosla, A. & Parmar, V. (2007b): A new Late Cretaceous gondwantherian mammal from Central India. *Proceedings of the Indian National Science Academy* 73: 17–24.
Prasad, G.V.R., Verma, O., Gheerbrant, E., Goswami, A., Khosla, A., Parmar, V. & Sahni, A. (2010): First mammal evidence for the Late Cretaceous of India for biotic dispersal between India and Africa at the K/T transition. *Comptes Rendus Palevol* 9: 63–71.
Prothero, D.R. (1981): New Jurassic mammals from Como Bluff, Wyoming, and the interrelationships of non-tribosphenic Theria. *Bulletin of the American Museum of Natural History* 167: 277–326.
Puttick, M.N., O´Reilly, J.E., Tanner, A.R., Fleming, J.F., Clark, J., Holloway, L., Lozano-Fernandez, J., Parry, L.A., Tarver, J.E., Pisani, D. & Donoghue, P.C.J. (2017): Uncertain tree: discriminating among competing approaches to the phylogenetic analysis of phenotype data. *Proceedings of the Royal Society B* 284: 20162290.
Rădulescu, C. & Samson, P.-M. (1996): The first multituberculate skull from the Late Cretaceous (Maastrichtian) of Europe (Haţeg Basin, Romania). *Anuarul Institutului Geologic al României* 69: 177.

Rana, R.S. & Wilson, G.P. (2003): New Late Cretaceous mammals from the intertrappean beds of Rangapur, India, and paleobiogeographic framework. *Acta Palaeontologica Polonica* 48: 331–348.

Rasmussen, T.E. & Callison, G. (1981): A new species of triconodont mammal from the Upper Jurassic of Colorado. *Journal of Paleontology* 55: 628–634.

Rauhut, O.W.M., Martin, T., Ortiz-Jaureguizar, E.O. & Puerta, P. (2002): The first Jurassic mammal from South America. *Nature* 416: 165–168.

Rich, T.H., Vickers-Rich, P. & Gangloff, R.A. (2002): Polar dinosaurs. *Science* 295: 979–980.

Rich, T.H., Vickers-Rich, P., Constantine, A., Flannery, T.F., Kool, L. & van Klaveren, N. (1997): A tribosphenic mammal from the Mesozoic of Australia. *Science* 278: 1438–1442.

Rich, T.H., Vickers-Rich, P., Constantine, A., Flannery, T.F., Kool, L. & van Klaveren, N. (1999): Early Cretaceous mammals from Flat Rocks, Victoria, Australia. *Records of the Queen Victoria Museum* 106: 1–35.

Rich, T.H., Flannery, T.F., Trusler, P., Constantine, A., Kool, L., van Klaveren, N. & Vickers-Rich, P. (2001): An advanced ausktribosphenid from the Early Cretaceous of Australia. *The Records of the Queen Victoria Museum* 110: 1–9.

Rich, T.H., Hopson, J.A., Musser, A.M., Flannery, T.F. & Vickers-Rich, P. (2005): Independent origins of middle ear bones in monotremes and therians. *Science* 307: 910–914).

Rich, T.H., Hopson, J.A., Gill, P.G., Trusler, P., Rogers-Davidson, S., Morton, S., Cifelli, R.L., Pickering, D., Kool, L., Siu, K., Burgmann, F.A., Senden, T., Evans, A.R., Wagstaff, B.E., Seegets-Villiers, D., Corfe, I.J., Flannery, T.F., Walker, K., Musser, A.M., Archer, M., Pian, R. & Vickers-Rich, P. (2016): The mandible and dentition of the Early Cretaceous monotreme *Teinolophos trusleri*. *Alcheringa* 40: 475–501.

Rigby, J.K. Jr. & Wolberg, D.L. (1987): The therian mammalian fauna (Campanian) of Quarry 1, Fossil Forest study area, San Juan Basin, New Mexico. *Geological Society of America, Special Paper* 209: 51–79.

Rose, K.D., Cifelli, R.L. & Lipka, T.R. (2001): Second triconodont dentary from the Early Cretaceous of Maryland. *Journal of Vertebrate Paleontology* 21: 628–632.

Rougier, G.W. & Apesteguia, S. (2004): The Mesozoic radiation of dryolestoids in South America: dental and cranial evidence. *Journal of Vertebrate Paleontology* 24 (3, Supplement): 106A.

Rougier, G.W., Apesteguía, S. & Gaetano, L.C. (2011): Highly specialized mammalian skulls from the Late Cretaceous of South America. *Nature* 479: 98–102.

Rougier, G.W., Novacek, M.J. & Dashzeveg, D. (1997): A new multituberculate from the Late Cretaceous locality Ukhaa Tolgod, Mongolia. Considerations on multituberculate relationships. *American Museum Novitates* 3193: 1–26.

Rougier, G.W., Wible, J.R. & Hopson, J.A. (1992): Reconstruction of the cranial vessels in the Early Cretaceous mammal *Vincelestes neuquenianus*: implications for the evolution of the mammalian cranial system. *Journal of Vertebrate Paleontology* 12: 188–216.

Rougier, G.W., Wible, J.R. & Hopson, J.A. (1996): Basicranial anatomy of *Priacodon fruitaensis* (Triconodontidae, Mammalia) from the Late Jurassic of Colorado, and a reappraisal of mammaliaform interrelationships. *American Museum Novitates* 3183: 1–38.

Rougier, G.W., Wible, J.R. & Novacek, M.J. (1998): Implications of *Deltatheridium* specimens for early marsupial history. *Nature* 396: 459–463.

Rougier, G.W., Spurlin, B.K. & Kik, P.K. (2003a): A new specimen of *Eurylambda aequicruris* and considerations on "symmetrodont" dentition and relationships. *American Museum Novitates* 3398: 1–15.

Rougier, G.W., Ji, Q. & Novacek, M.J. (2003b): A new symmetrodont mammal with fur impressions from the Mesozoic of China. *Acta Geologica Sinica* 77: 7–14.

Rougier, G.W., Wible, J.R., Novacek, M.J. (2004): New specimen of *Deltatheroides cretacicus* (Metatheria, Deltatheroida) from the Late Cretaceous of Mongolia. *Bulletin of Carnegie Museum of Natural History* 36: 245–266. doi: 10.2992/0145-9058(2004)36[245:nsodcm]2.0.co;2

Rougier, G.W., Isaji, S. & Manabe, M. (1999): An Early Cretaceous Japanese triconodont and a revision of triconodont phylogeny. *Journal of Vertebrate Paleontology* 19 (3, Supplement): 72A.

Rougier, G.W., Novacek, M.J., McKenna, M.C. & Wible, J.R. (2001): Gobiconodonts from the Early Cretaceous of Oshih (Ashile), Mongolia. *American Museum Novitates* 3348: 1–30.

Rougier, G.W., Martinelli, A.G., Forasiepi, A.M. & Novacek, M.J. (2007a): New Jurassic mammals from Patagonia, Argentina: a reappraisal of australosphenidan morphology and interrelationships. *American Museum Novitates* 3566: 1–54.

Rougier, G.W., Garrido, A., Gaetano, L.C., Puerta, P., Corbitt, C. & Novacek, M.J. (2007b). First Jurassic triconodont from South America. *American Museum Novitates* 3580: 1–17.

Rougier, G. W., Martinelli, A. G., Forasiepi, A. M. & Novacek, M.J. (2007c): New Jurassic mammals from Patagonia, Argentina: a reappraisal of australosphenidan morphology and interrelationships. *American Museum Novitates* 3566: 1–54.

Rougier, G.W., Chornogubsky, L., Casadio, S., Arango, N.P. & Giallombardo, A. (2009a): Mammals from the Allen Formation, Late Cretaceous, Argentina. *Cretaceous Research* 30: 223–238.

Rougier, G.W., Forasiepi, A.M., Hill, R.V. & Novacek, M.J. (2009b): New mammalian remains from the Late Cretaceous La Colonia Formation, Patagonia, Argentina. *Acta Palaeontologica Polonica* 54: 195–212.

Rougier, G.W., Wible, J.R., Beck, R.M.D. & Apesteguía, S. (2012): The Miocene mammal *Necrolestes* demonstrates the survival of a Mesozoic nontherian lineage into the late Cenozoic of South America. *Proceedings of the National Academy of Sciences, USA* 109: 20053–20058.

Rougier, G.W., Davis, B.M. & Novacek, M.J. (2015): A deltatheroidan mammal from the Upper Cretaceous Baynshiree Formation, Eastern Mongolia. *Cretaceous Research* 52: 167–177.

Rougier, G.W., Sheth, A.S., Carpenter, K., Appella-Guiscafre, L. & Davis, B.M. (2014): A new species of *Docodon* (Mammaliaformes: Docodonta) from the Upper Jurassic Morrison Formation and a reassessment of selected craniodental characters in basal mammaliaforms. *Journal of Mammalian Evolution* 22: 1–16.

Rowe, T.B. (1988): Definition, diagnosis, and origin of Mammalia. *Journal of Vertebrate Paleontology* 8: 241–264.

Rowe, T.B. (1993): Phylogenetic systematics and the early history of mammals. In: Szalay, F.S., Novacek, M.J. & McKenna, M.C. (eds.). *Mammal Phylogeny: Mesozoic Differentiation, Multituberculates, Monotremes, Early Therians, and Marsupials.* Springer-Verlag, New York: 129–145.

Rowe, T.B., Rich, T.H., Vickers-Rich, P., Springer, M. & Woodburne, M.O. (2008): The oldest platypus and its bearing on divergence timing of the platypus and echidna clades. *Proceedings of the National Academy of Sciences, USA* 105: 1238–1242.

Rowe, T.B., Macrini, T.E. & Luo, Z.-X. (2011): Fossil evidence on origin of the mammalian brain. *Science* 332: 955–957.

Ruf, I., Luo, Z.-X., Wible, J.R. & Martin, T. (2009): Petrosal anatomy and inner ear structures of the Late Jurassic *Henkelotherium* (Mammalia, Cladotheria, Dryolestoidea): insight into the early evolution of the ear region in cladotherian mammals. *Journal of Anatomy* 214: 679–693.

Ruf, I., Luo, Z.-X. & Martin, T. (2013): Reinvestigation of the basicranium of *Haldanodon exspectatus* (Docodonta, Mammaliaformes). *Journal of Vertebrate Paleontology* 33: 382–400.

Ruta, M., Botha-Brink, J., Mitchell, S.A. & Benton, M.J. (2013): The radiation of cynodonts and the ground plan of mammalian morphological diversity. *Proceedings of the Royal Society of London B* 280: 20131865.

Sahni, A. (1972): The vertebrate fauna of the Judith River Formation, Montana. *Bulletin of the American Museum of Natural History* 147: 321–412.

Sánchez-Villagra, M.R. & Smith, K.K. (1997): Diversity and evolution of the marsupial mandibular angular process. *Journal of Mammalian Evolution* 4: 119–144.

Schultz, J.A. & Martin, T. (2010): 3D-analysis of functional surfaces in the pretribosphenic dentition. *70th Anniversary Meeting Society of Vertebrate Paleontology, Program and Abstracts*: 159A–160A.

Schultz, J.A. & Martin, T. (2011): Wear pattern and functional morphology of dryolestoid molars (Cladotheria, Mammalia). *Paläontologische Zeitschrift* 85: 269–285.

Schultz, J.A. & Martin, T. (2014): Function of pretribosphenic and tribosphenic mammalian molars inferred from 3D animation. *Naturwissenschaften* 101: 771–781.

Schultz, J.A., Koenigswald, W.v. & Dumont, E.R. (2014): Dental function and diet of *Vintana sertichi* (Mammalia, Gondwanatheria) from the Late Cretaceous of Madagascar. *Journal of Vertebrate Paleontology Memoir* 14. *Journal of Vertebrate Paleontology* 34 (6, Supplement): 182–202.

Schultz, J.A., Bhullar, B.-A.S. & Luo Z.-X. (2017): Re-examination of the Jurassic mammaliaform *Docodon victor* by computed tomography and occlusal functional analysis. *Journal of Mammalian Evolution*. doi.org/10.1007/s10914-017-9418-5

Schultz, J.A., Ruf, I. & Martin, T. (2018): Oldest known multituberculate stapes suggests an asymmetric bicrural pattern as ancestral for Multituberculata. *Proceedings of the Royal Society B* 285: 20172779. http://dx.doi.org/10.1098/rspb.2017.2779

Scillato-Yané, G.J. & Pascual, R. (1984): Un peculiar Paratheria, Edentata (Mammalia) del Paleoceno de Patagonia. *Primeras Journadas Argentinas de Paleontología de Vertebrados, Resúmenes*: 15.

Scillato-Yané, G.J. & Pascual, R. (1985): Un peculiar Xenarthra del Paleoceno medio de Patagonia (Argentina). Su importancia en la sistemática de los Paratheria. *Ameghiniana* 21: 173–176.

Sereno, P.C. & McKenna, M.C. (1995): Cretaceous multituberculate skeletons and the early evolution of the mammalian shoulder girdle. *Nature* 377: 144–147.

Sidor, C.A. (2001): Simplification as a trend in synapsid cranial evolution. *Evolution* 55: 1419–1442.

Sidor, C.A. & Hopson, J.A. (1998): Ghost lineages and "mammalness": assessing the temporal pattern of character acquisition in the Synapsida. *Paleobiology* 24: 254–273.

Sigé, B., Archer, M., Crochet, J.-Y., Godthelp, H., Hand, S. & Beck, R. (2009): *Chulpasia* and *Thylacotinga*, late Paleocene-earliest Eocene trans-Antarctic Gondwanan bunodont marsupials: new data from Australia. *Geobios* 42: 813–823.

Sigogneau-Russell, D. (1983a): A new therian mammal from the Rhaetic locality of Saint-Nicolas-de-Port (France). *Zoological Journal of the Linnean Society* 78: 175–186.

Sigogneau-Russell, D. (1983b): Nouveaux taxons de mammifères rhétiens. *Acta Palaeontologica Polonica* 28: 233–249.

Sigogneau-Russell, D. (1983c): Caractéristiques de la faune mammalienne du Rhétien de Saint-Nicolas-de-Port (Meurthe-et-Moselle). *Bulletin d´information des géologues du bassin de Paris* 20: 51–53.

Sigogneau-Russell, D. (1989): Découverte du premier symmétrodonte (Mammalia) du continent africain. *Comptes Rendus de l'Académie des Sciences Paris, Série II* 309: 921–926.

Sigogneau-Russell, D. (1991a): First evidence of Multituberculata (Mammalia) in the Mesozoic of Africa. *Neues Jahrbuch für Geologie und Paläontologie, Monatshefte* 1991: 119–125.

Sigogneau-Russell, D. (1991b): Nouveaux mammifères theriens du Crétacé inférieur du Maroc. *Comptes Rendus de l'Académie des Sciences Paris, Série II* 313: 279–285.

Sigogneau-Russell, D. (1991c): Découverte du premier mammifère tribosphénique du Mésozoïque africain. *Comptes Rendus de l'Académie des Sciences Paris, Série II* 313: 1635–1640.

Sigogneau-Russell, D. (1992): *Hypomylos phelizoni* nov. gen. nov. sp., une étape précoce de l'évolution de la molaire tribosphénique. *Géobios* 25: 389–393.

Sigogneau-Russell, D. (1995): Two possibly aquatic triconodont mammals from the Early Cretaceous of Morocco. *Acta Palaeontologica Polonica* 40: 149–162.

Sigogneau-Russell, D. (1998): Discovery of a Late Jurassic Chinese mammal in the upper Bathonian of England. *Comptes Rendus de l'Académie des Sciences Paris, Sciences de la Terre et des Planètes* 327: 571–576.

Sigogneau-Russell, D. (1999): Réévaluation des Peramura (Mammalia, Theria) sur la base de nouveaux spécimens du Crétacé inférieur d'Angleterre et du Maroc. *Geodiversitas* 21: 93–127.

Sigogneau-Russell, D. (2003a): Docodonts from the British Mesozoic. *Acta Palaeontologica Polonica* 48: 357–374.

Sigogneau-Russell, D. (2003b): Holotherian mammals from the Forest Marble (Middle Jurassic of England). Geodiversitas 25: 501–537.

Sigogneau-Russell, D. & Ensom, P.C. (1998): *Thereuodon* (Theria, Symmetrodonta) from the Lower Cretaceous of North Africa and Europe, and a brief review of symmetrodonts. *Cretaceous Research* 19: 445–470.

Sigogneau-Russell, D. & Godefroit, P. (1997): A primitive docodont (Mammalia) from the Upper Triassic of France and the possible therian affinities of the order. *Comptes Rendus de l' Académie des Sciences Paris, Sciences de la Terre et de Planètes* 324: 135–140.

Sigogneau-Russell, D. & Hahn, R. (1995): Reassessment of the Late Triassic symmetrodont mammal *Woutersia*. *Acta Palaeontologica Polonica* 40: 245–260.

Sigogneau-Russell, D. & Kielan-Jaworowska, Z. (2002): Mammals from the Purbeck Limestone Group of Dorset, southern England. In: Milner, A.R. & Batten, D.J. (eds.). *Life and Environments in Purbeck Times. Special Papers in Palaeontology* 68: 241–255.

Sigogneau-Russell, D., Dashzeveg, D. & Russell, D.E. (1992): Further data on *Prokennalestes* (Mammalia, Eutheria inc. sed.) from the Early Cretaceous of Mongolia. *Zoologica Scripta* 21: 205–209.

Sigogneau-Russell, D., Frank, R.M. & Hemmerlé, J. (1986): A new family of mammals from the lower part of the French Rhaetic. In: Padian, K. (ed.). *The Beginning of the Age of Dinosaurs*. Cambridge University Press, Cambridge: 99–108.

Sigogneau-Russell, D., Hooker, J.J. & Ensom, P.C. (2001): The oldest tribosphenic mammal from Laurasia (Purbeck Limestone Group, Berriasian, Cretaceous, UK) and its bearing on the "dual origin" of Tribosphenida. *Comptes Rendus de l'Académie des Sciences Paris, Sciences de la Terre et des Planètes* 333: 141–147.

Silcox, M.T. & Teaford, M.F. (2002): The diet of worms: an analysis of mole dental microwear. *Journal of Mammalogy* 83: 804–814.

Simmons, N.B. (1993): Phylogeny of Multituberculata. In: Szalay, F.S., Novacek, M.J. & McKenna, M.C. (eds.). *Mammal Phylogeny: Mesozoic Differentiation, Multituberculates, Monotremes, Early Therians, and Marsupials*. Springer-Verlag, New York: 146–164.

Simpson, G.G. (1925a): A Mesozoic mammal skull from Mongolia. *American Museum Novitates* 201: 1–11.

Simpson, G.G. (1925b): Mesozoic Mammalia I. American triconodonts: Part 1. *American Journal of Science* 10: 145–165.

Simpson, G.G. (1925c): Mesozoic Mammalia I. American triconodonts: Part 2. *American Journal of Science* 10: 334–358.

Simpson, G.G. (1925d): Mesozoic Mammalia. II. *Tinodon* and its allies. *American Journal of Science* 10: 451–470.

Simpson, G.G. (1927a): Mammalian fauna of the Hell Creek Formation of Montana. *American Museum Novitates* 267: 17.

Simpson, G.G. (1927b): Mesozoic Mammalia. VI. Genera of Morrison pantotheres. *American Journal of Science* 13: 409–416.

Simpson, G.G. (1927c): Mesozoic Mammalia. VIII. Genera of Lance mammals other than multituberculates. *American Journal of Science* 14: 121–130.

Simpson, G.G. (1928): *A Catalogue of the Mesozoic Mammalia in the Geological Department of the British Museum*. Trustees of the British Museum, London.

Simpson, G.G. (1929): American Mesozoic Mammalia. *Memoirs of the Peabody Museum of Yale University* 3: 1–235.

Simpson, G.G. (1936): Studies of the earliest mammalian dentition. *Dental Cosmos* 78: 791–800.

Simpson, G.G. (1937): A new Jurassic mammal. *American Museum Novitates* 943: 1–6.

Simpson, G.G. (1947): *Haramiya*, new name, replacing *Microcleptes* Simpson, 1928. *Journal of Paleontology* 21: 497.

Simpson, G.G. (1961): Evolution of Mesozoic mammals. In: Vandebroek, G. (ed.). *International Colloquium on the Evolution of Lower and Non-specialized Mammals*. Koninklijke Vlaamse Academie voor Wetenschappen, Letteren en Schone Kunsten van België, Leuven, Part 1: 57–95.

Slaughter, B.H. (1965): A therian from the Lower Cretaceous (Albian) of Texas. *Postilla* 93: 1–18.

Slaughter, B.H. (1968a): Earliest known marsupials. *Science* 162: 254–255.

Slaughter, B.H. (1968b): Earliest known eutherian mammals and the evolution of premolar occlusion. *The Texas Journal of Science* 20: 3–12.

Slaughter, B.H. (1969): *Astroconodon*, the Cretaceous triconodont. *Journal of Mammalogy* 50: 102–107.

Slaughter, B.H. (1971): Mid-Cretaceous (Albian) therians of the Butler Farm local fauna, Texas. In: Kermack, D.M. & Kermack, K.A. (eds.). *Early Mammals. Zoological Journal of the Linnean Society* 50, Supplement 1: 131–143.

Slaughter, B.H. (1981): The Trinity therians (Albian, mid-Cretaceous) as marsupials and placentals. *Journal of Paleontology* 55: 682–683.

Sloan, R.E. & van Valen, L. (1965): Cretaceous mammals from Montana. *Science* 148: 220–227.

Smith, T., De Bast, E. & Sigé, B. (2010): Euarchontan affinity of Paleocene Afro-European adapisoriculid mammals and their origin in the late Cretaceous Deccan Traps of India. *Naturwissenschaften* 97: 417–422.

Stanley, W.T., Robbins, L.W., Malekani, J.M., Mbalitini, S.G., Migurimu, D.A., Mukinzi, J.C., Hulselmans J., Prévot V., Verheyen E., Hutterer R., Doty J.B., Monroe B.P., Nakazawa Y.J., Braden Z., Carroll D., Peterhans J.C.K., Bates J.M., Esselstyn J.A. (2013): A new hero emerges: another exceptional mammalian spine and its potential adaptive significance. *Biology Letters* 9: 20130486. DOI: 10.1098/rsbl.2013.0486.

Sweetman, S.C. (2006): A gobiconodontid (Mammalia, Eutriconodonta) from the Early Cretaceous (Barremian) Wessex Formation of the Isle of Wight, southern Britain. *Palaeontology* 49: 889–897.

Sweetman, S.C. (2008): A spalacolestine spalacotheriid (Mammalia, Trechnotheria) from the Early Cretaceous (Barremian) of southern England and its bearing on spalacotheriid evolution. *Palaeontology* 51: 1367–1385.

Sweetman, S.C., Smith, G. & Martill, D.M. (2017). Highly derived eutherian mammals from the earliest Cretaceous of southern Britain. *Acta Palaeontologica Polonica*. https://doi.org/10.4202/app.00408.2017

Szalay, F.S. (1969): Mixodectidae, Microsyopidae and the insectivore-primate transition. *Bulletin of the American Museum of Natural History* 140: 193–330.

Szalay, F.S. (1984): Arboreality: is it homologous in metatherian and eutherian mammals? In: Hecht, M.K. (ed.). *Evolutionary Biology*, Vol. 6. Plenum Press, New York: 215–258.

Szalay, F.S. (1993): Pedal evolution of mammals in the Mesozoic: tests for taxic relationships. In: Szalay, F.S., Novacek, M.J. & McKenna, M.C. (eds.) *Mammal Phylogeny: Mesozoic Differentiation, Multituberculates, Monotremes, Early Therians, and Marsupials*. Springer-Verlag, New York: 108–128.

Szalay, F.S. (1994): *Evolutionary History of the Marsupials and an Analysis of Osteological Characters*. Cambridge University Press, Cambridge.

Szalay, F.S. & Trofimov, B.A. (1996): The Mongolian Late Cretaceous *Asiatherium*, and the early phylogeny and paleobiography of Metatheria. *Journal of Vertebrate Paleontology* 16: 474–509.

Tang, F., Luo, Z.-X., Zhou, Z.-H., You, H.L., Georgi, J.A., Tang, Z.-L. & Wang, X.-Z. (2001): Biostratigraphy and palaeoenvironment of the dinosaur-bearing sediments in Lower Cretaceous of Mazongshan area, Gansu Province, China. *Cretaceous Research* 22: 115–129.

Tarver, J.E., Dos Reis, M., Mirarab, S., Moran, R.J., Parker, S., O´Reilly, J.E., King, B.L., O´Connell, M.J., Asher, R.J., Warnow, T., Peterson K.J., Donoghue, P.C.J. & Pisani, D. (2016): The interrelationships of placental mammals and the limits of phylogenetic inference. *Genome Biology and Evolution* 8: 330–344.

Tatarinov, L.P. (1994): On an unusual mammalian tooth from the Mongolian Jurassic. *Paleontologicheskij Zhurnal* 1994: 97–105. [in Russian]

Trofimov, B.A. (1978): The first triconodonts (Mammalia, Triconodonta) from Mongolia. *Doklady Akademii Nauk SSSR* 243: 213–216. [in Russian]

Trofimov, B.A. (1980): Multituberculata and Symmetrodonta from the Lower Cretaceous of Mongolia. *Doklady Akademii Nauk SSSR* 251: 209–212. [in Russian]

Trofimov, B.A. (1997): A new generic name *Gobiotheriodon* for a symmetrodont mammal *Gobiodon* Trofimov. *Acta Palaeontologica Polonica* 42: 496.

Trofimov, B.A. & Szalay, F.S. (1994): New Cretaceous marsupial from Mongolia and the early radiation of the Metatheria. *Proceedings of the National Academy of Sciences, USA* 91: 12569–12573.

Tsubamoto, T., Rougier, G.W., Isaji, G., Manabe, M., Forasiepi, A.M. (2004): New Early Cretaceous spalacotheriid "symmetrodont" mammal from Japan. *Acta Palaeontologica Polonica* 49: 329–346.

Turnbull, W.D. & Cifelli, R.L. (1999): Triconodont mammals of the Aptian-Albian Trinity Group, Texas and Oklahoma. In: Mayhal, J.T. & Heikkinen, T. (eds.). *Dental Morphology* 98. University of Oulu Press, Oulu: 252–272.

Urban, D.J., Anthwal, N., Luo, Z.-X., Maier, J.A., Sadier, A., Tucker, A.S. & Sears, K.E. (2017): A new developmental mechanism for the separation of the mammalian middle ear ossicles from the jaw. *Proceedings of the Royal Society B* 284: 20162416.

van Nievelt, A.F.H. & Smith, K.K. (2005): To replace or not to replace: the significance of reduced tooth replacement in marsupial and placental mammals. *Palaeobiology* 31: 324–346.

Van Valen, L. (1964): A possible origin for rabbits. *Evolution* 18: 484–491.

Van Valen, L. (1966): Deltatheridia, an new order of mammals. *Bulletin of the American Museum of Natural History* 132: 1–126.

Vasquez-Molinero, R., Martin, T., Fischer, M.S. & Frey, R. (2001): Comparative anatomical investigations of the postcranial skeleton of *Henkelotherium guimarotae* Krebs, 1991 (Eupantotheria, Mammalia) and their implications for its locomotion. *Mitteilungen aus dem Museum für Naturkunde zu Berlin, Zoologische Reihe* 77: 207–216.

Vianey-Liaud, M. (1979): Les mammifères montiens de Hainin (Paléocène moyen de Belgique). Part I. Multituberculés. *Palaeovertebrata* 9: 117–131.

Vullo, R., Gheerbrant, E., Muizon, C. de, Nérandeau, D. (2009): The oldest modern therian mammal from Europe and its bearing on stem marsupial paleobiogeography. *Proceedings of the National Academy of Sciences, USA* 106: 19910–19915.

Waldman, M. & Savage, R.J.G. (1972): The first Jurassic mammal from Scotland. *Journal of the Geological Society* 128: 119–125.

Wang, Y.-Q. & Li, C.-K. (2016): Reconsideration of the systematic position of the Middle Jurassic mammaliaforms *Itatodon* and *Paritatodon*. *Palaeontologia Polonica* 67: 249–256.

Wang, Y.-Q., Hu, Y.-M., Meng, J. & Li, C.-K. (2001): An ossified Meckel's cartilage in two Cretaceous mammals and origin of the mammalian middle ear. *Science* 294: 357–361.

Wang, Y.-Q., Clemens, W.A., Hu, Y.-M. & Li, C.-K. (1998): A probable pseudo-tribosphenic upper molar from the Late Jurassic of China and the early radiation of the Holotheria. *Journal of Vertebrate Paleontology* 18: 777–787.

Wang, Y.-Q., Hu, Y.-M., Chow, M.-C. & Li, C.-K. (1995): Mesozoic mammal localities in western Liaoning, northeast China. In: Sun, A.-L. & Wang, Y.Q. (eds.). *Sixth Symposium on Mesozoic Terrestrial Ecosystems*. China Ocean Press, Beijing: 221–228.

Weil, A. (2011): Mammalian evolution: A jaw-dropping ear. *Nature* 472: 174–176.

Weil, A. (2014): Mammalian evolution: A beast from the southern wild. *Nature* 515: 495–496.

Weil, A. & Tomida, Y. (2001): First description of the skull of *Meniscoessus robustus* expands known morphological diversity of Multituberculata and deepens phylogenetic mystery. *Journal of Vertebrate Paleontology* 21 (3, Supplement): 112A.

Whiteside, D.I., Duffin, C.J., Gill, P.G., Marshall, J.E.A. & Benton, M.J. (2016): The Late Triassic and Early Jurassic fissure faunas from Bristol and South Wales: stratigraphy and setting. *Palaeontologia Polonica* 67: 257–287.

Wible, J.R. & Burrows, A.M. (2016): Does the Jurassic *Agilodocodon* (Mammaliaformes, Docodonta) have any exudativorous dental features? *Palaeontologia Polonica* 67: 289–299.

Wible, J.R. & Hopson, J.A. (1993): Basicranial evidence for early mammal phylogeny. In: Szalay, F.S., Novacek, M.J. & McKenna, M.C. (eds.). *Mammal Phylogeny: Mesozoic Differentiation, Multituberculates, Monotremes, Early Therians, and Marsupials*. Springer-Verlag, New York: 45–62.

Wible, J.R., Rougier, G.W., Novacek, M.J. & McKenna, M.C. (2001): Earliest eutherian ear region: a petrosal referred to *Prokennalestes* from the Early Cretaceous of Mongolia. *American Museum Novitates* 3322: 1–44.

Wible, J.R., Rougier, G.W., Novacek, M.J. & Asher, R.J. (2007): Cretaceous eutherians and Laurasian origin for placental mammals near the K/T boundary. *Nature* 447: 1003–1006.

Wible, J.R., Rougier, G.W., Novacek, M.J. & Asher, R.J. (2009): The eutherian mammal *Maelestes gobiensis* from the Late Cretaceous of Mongolia and the phylogeny of Cretaceous Eutheria. *Bulletin of the American Museum of Natural History* 327: 1–123.

Williamson, T.E., Brusatte, S.L., Carr, T.D., Weil, A., Standhard, B.R. (2012): The phylogeny and evolution of Cretaceous–Palaeogene metatherians: cladistic analysis and description of new early Palaeocene specimens from the Nacimiento Formation, New Mexico. *Journal of Systematik Palaeontology* 10: 625–651.

Williamson, T.E., Brusatte, S.L. & Wilson, G.P. (2014): The origin and early evolution of metatherian mammals: the Cretaceous record. *ZooKeys* 465: 1–76.

Williamson, T. E., Brusatte, S. L., Secord, R. & Shelley, S. (2015): A new taeniolabidoid multituberculate (Mammalia) from the middle Puercan of the Nacimiento Formation, New Mexico, and a revision of taeniolabidoid systematics and phylogeny. *Zoological Journal of the Linnean Society*. doi: 10.1111/zoj.12336.

Wilson, G.P. (2013): Mammals across the K/Pg boundary in northeastern Montana, USA: dental morphology and body-size patterns reveal extinction selectivity and immigrant-fueled ecospace filling. *Paleobiology* 39: 429–469.

Wilson, G.P. & Riedel, J.A. (2010): New specimen reveals deltatheroidan affinities of the North American Late Cretaceous mammal *Nanocuris*. *Journal of Vertebrate Paleontology* 30: 872–884.

Wilson, G.P., Ekdale, E.G., Hoganson, J.W., Calede, J.J. & Van der Linden, A. (2016): A large carnivorous mammal from the Late Cretaceous and the North American origin of marsupials. *Nature Communications* 7: 13734. doi: 10.1038/ncomms13734.

Woodburne, M.O. & Tedford, R.H. (1975): The first Tertiary monotreme from Australia. *American Museum Novitates* 2588: 1–11.

Woodburne, M.O., Rich, T.H. & Springer, M.S. (2003): The evolution of tribospheny and the antiquity of mammalian clades. *Molecular Phylogenetics and Evolution* 28: 360–385.

Xu, L., Zhang, X., Pu, H., Jia, S., Zhang, J. Lü, J. & Meng, J. (2015): Largest known Mesozoic multituberculate from Eurasia and implications for multituberculate evolution and biology. *Scientific Reports* 5:14950. doi: 10.1038/srep14950.

Yabe, H. & Shikama, T. (1938): A new Jurassic mammal from south Manchuria. *Proceedings of the Imperial Academy of Tokyo* 14: 353–359.

Yadagiri, P. (1984): New symmetrodonts from the Kota Formation. *Journal of the Geological Society of India* 25: 512–521.

Yadagiri, P. (1985): An amphidontid symmetrodont from the Early Jurassic Kota Formation, India. *Zoological Journal of the Linnean Society of London* 85: 411–417.

Young, C.C. (1978): New materials of *Eozostrodon*. *Vertebrata PalAsiatica* 16: 1–3.

Young, C.C. (1982): Two primitive mammals from Lufeng, Yunnan. *Selected Works of Yang Zhungjian*. Science Press, Beijing: 21–24.

Yuan, C.-X., Ji, Q., Meng, Q.-J., Tabrum, A.R. & Luo, Z.-X. (2013): Earliest evolution of multituberculate mammals revealed by a new Jurassic fossil. *Science* 341: 779–783.

Zan, S., Wood, C.B., Rougier, G.W., Jin, L., Chen, J. & Schaff, C.R. (2006): A new "middle" Cretaceous zalambdalestid mammal from a new locality in Jilin Province, northeastern China. *Journal of the Paleontological Society of Korea* 22: 153–172.

Zeller, U. (1999): Mammalian reproduction: origin and evolutionary transformations. *Zoologischer Anzeiger* 238: 117–130.

Zhang, F. (1984): Fossil record of Mesozoic mammals of China. *Vertebrata PalAsiatica* 22: 29–38.

Zheng, X., Bi, S., Wang, X. & Meng, J. (2013): A new arboreal haramiyid shows the diversity of crown mammals in the Jurassic period. *Nature* 500: 199–202.

Zhou, M.-Z., Cheng, Z.-W. & Wang, Y.-Q. (1991): A mammalian lower jaw from the Jurassic of Lingyuan, Liaoning. *Vertebrata PalAsiatica* 29: 165–175.

Zhou, C.-F., Wu, S., Martin, T. & Luo, Z.-X. (2013): A Jurassic mammaliaform and the earliest mammalian evolutionary adaptations. *Nature* 500: 163–167.

Robert J. Asher

7 Diversity and relationships within crown Mammalia

7.1 Introduction

As with any crown group, Mammalia is defined by extinction and comprises all descendants of the common ancestor shared by the three synapsid lineages that happen to exist today: monotremes, marsupials, and placentals. A more inclusive, apomorphy-defined synapsid clade is Mammaliaformes, composed of all descendants of the first synapsid to evolve a functional, squamosal-dentary jaw joint. In addition to Mammalia, Mammaliaformes includes *Adelobasileus*, *Sinoconodon*, morganucodonts, docodonts, and haramiyids (Angielczyk and Kammerer this volume, Chapter 5) and Martin (this volume, Chapter 6).

My goal in this chapter is to outline the crown clade Mammalia, to describe its major constituents, to trace how the core ideas on mammalian evolution and interrelations have developed since the early 20th century, and to summarize how certain fossil groups are related to extant, high-level clades, with an emphasis on Placentalia. Mammalian interrelationships are depicted in Fig. 7.1 based primarily on overlap across four phylogenetic studies using large samples of data and taxa: Meredith *et al.* (2011, 36 kilobases of nuclear DNA from 164 mammals), Mitchell *et al.* (2014, 44 kilobases of mitochondrial and nuclear DNA for 203 mammals), Tarver *et al.* (2016, 32 megabases of nuclear DNA for 36 mammals, 15.6 kilobases of microRNA for 42 mammals, and reanalyses of data sets from Hallström and Janke 2010; O'Leary *et al.* 2013; and Romiguier *et al.* 2013), and Esselstyn *et al.* (2017, ultraconserved elements from 3787 genes across 100 mammals). These studies are not completely congruent; cases of disagreement (with exceptions detailed below) have been represented with polytomies. Nonetheless, given all of the ways in which these topologies could differ (e.g., 3.37×10^{49} distinct, rooted, bifurcating trees for the 36 genomically sampled taxa in Tarver *et al.* 2016), they are very close in overall shape, and I predict future discoveries will agree far more than disagree with the phylogenetic relationships shown in Fig. 7.1.

It is occasionally convenient to refer to Linnean ranks, for example, that the identity of most families and orders has been established since the 19th century but that interrelationships among orders have been well understood only since the late 1990s. I recognize the biological arbitrariness of Linnean ranks and therefore minimize their use. However, they do have some utility, as evident in the practical, legal framework articulated by the International Commission of Zoological Nomenclature (1999). The fact that this code does not apply above the rank of family has led to inconsistency regarding the use of some high-level names. Here, I follow Simpson (1945) in arbitrating among such names based on priority and stability, as summarized by Asher and Helgen (2010). On another practical note, I capitalize taxon names when used as proper nouns and when referring to genera. For example, I capitalize formal cladistic names (e.g., mammals in the genus *Homo* belong to the clade Primates) but do not capitalize adjectives or common nouns (e.g., the capybara is a hystricognath rodent). Quotes surrounding a high-level taxon indicate that it is not monophyletic (e.g., "Edentata").

Another semantic but important point worth making concerns the use of adjectives like "molecular" and "morphological" to describe phylogenetic trees. One of the key postulates of evolutionary theory is that living things share common ancestry. Tree diagrams represent this common ancestry, and investigators have used comparative anatomy, embryology, biogeography, intuition, and molecular data to build such diagrams. Broadly speaking, molecular methods have been known since the early 20th century and encompass immunochemical (Nuttall *et al.* 1904) and hybridization (Kirsch *et al.* 1991) techniques, as well as direct comparisons of protein (Zuckerkandl and Pauling 1965) and nucleotide (Irwin *et al.* 1991) sequences. The ease of applying quantitative methods to compare species, along with the massive quantities of genomic characters with readily defined states, has meant that where available, molecular data have become crucial in establishing the topology and confidence intervals of a given phylogenetic tree.

In practice, reference to phylogenetic trees as "molecular" or "morphological" refers to the kind of data used to build them. However, describing a given tree as "molecular" is slightly misleading. It implies that there are multiple phylogenetic trees out there according to data type; indeed, genetic loci often do have different gene trees

https://doi.org/10.1515/9783110341553-007

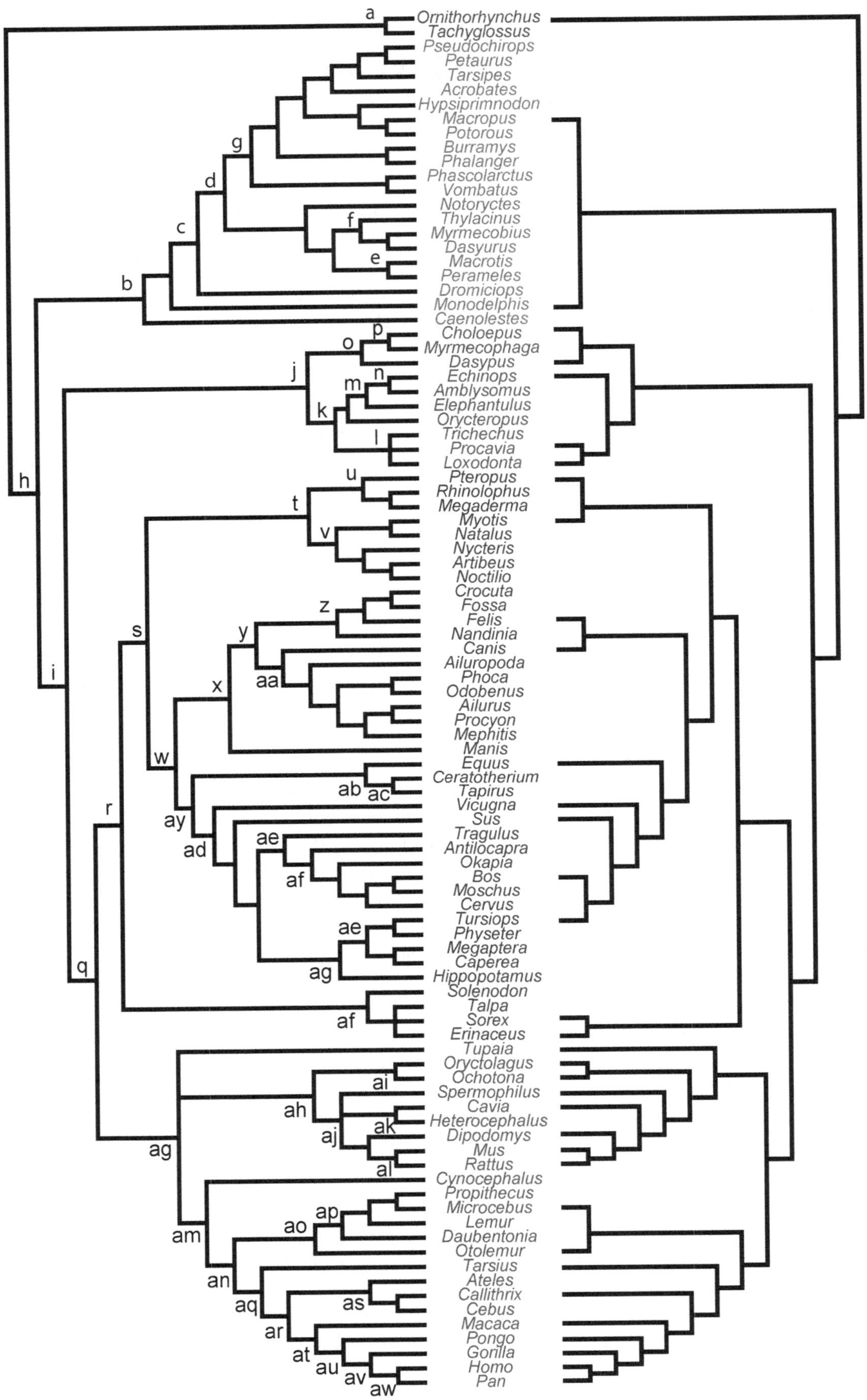
Ornithorhynchus
Tachyglossus
Pseudochirops
Petaurus
Tarsipes
Acrobates
Hypsiprimnodon
Macropus
Potorous
Burramys
Phalanger
Phascolarctus
Vombatus
Notoryctes
Thylacinus
Myrmecobius
Dasyurus
Macrotis
Perameles
Dromiciops
Monodelphis
Caenolestes
Choloepus
Myrmecophaga
Dasypus
Echinops
Amblysomus
Elephantulus
Orycteropus
Trichechus
Procavia
Loxodonta
Pteropus
Rhinolophus
Megaderma
Myotis
Natalus
Nycteris
Artibeus
Noctilio
Crocuta
Fossa
Felis
Nandinia
Canis
Ailuropoda
Phoca
Odobenus
Ailurus
Procyon
Mephitis
Manis
Equus
Ceratotherium
Tapirus
Vicugna
Sus
Tragulus
Antilocapra
Okapia
Bos
Moschus
Cervus
Tursiops
Physeter
Megaptera
Caperea
Hippopotamus
Solenodon
Talpa
Sorex
Erinaceus
Tupaia
Oryctolagus
Ochotona
Spermophilus
Cavia
Heterocephalus
Dipodomys
Mus
Rattus
Cynocephalus
Propithecus
Microcebus
Lemur
Daubentonia
Otolemur
Tarsius
Ateles
Callithrix
Cebus
Macaca
Pongo
Gorilla
Homo
Pan

that explain their history, distinct from the species trees of their host taxa. However, the existence of a single, historical tree (with qualifications about hard polytomies and population-level reticulations) joining all species is not only a key postulate of evolutionary theory but comprises a prediction that enables evolution to be tested given the expectation that distinct sources of data will generate topologies that converge on the branching patterns of this one tree (Penny *et al.* 1982; Sober and Steel 2002; Asher, this volume, Chapter 3). Thus, when authors have the concept of a species tree in mind, the phrase "molecular tree" obscures the expectation of a single, historical pattern. The tree of life is neither "molecular" nor "morphological", although the data used to reconstruct it can be one or both. Stated differently, a phylogenetic species tree is no more "molecular" than a genome is "morphological" when its size is inferred using the morphology of bone cells (Organ *et al.* 2012). Inferences of tree shape or genome size are biological hypotheses, whatever sources of data are used to make or test them. Therefore, I do not describe a given phylogenetic hypothesis itself as "molecular" or "morphological", but reserve these adjectives for the data behind such hypotheses.

Finally, I would also like to clarify the terms "basal" and "nested" when describing phylogenetic trees (see also discussion in Bronzati 2017). Specifically, taxa used in a phylogenetic analysis are connected via branches to nodes. Nodes (i.e., bifurcations that connect two branches) may vary in their distance to the root due to branch length and the number of other intervening nodes. A taxon may be nested or connected to a node that is separated from the root by many other nodes; another taxon may be basal or connected to a node with few or no other nodes between it and the root. There is a legitimate concern that by describing a given taxon as "basal", one necessarily implies that it is somehow less evolved and/or more primitive than other taxa. Although this may be true for some taxa (e.g., a fossil perissodactyl that existed within one or few generations of the clade's origin), this is difficult to test in most cases. Moreover, and for extant taxa, all branches lead to the present, and regardless of the position of an extant taxon on a phylogenetic tree, it is no less "evolved" than other extant taxa. It is also true that the number of nodes between a given taxon and the root is dependent on sampling, and future discoveries and/or changes in topology may greatly affect the position of a taxon judged to be "basal" in an initial phylogenetic tree. However, these valid points do not change the meaning I intend, namely, that "basal" and "nested" are convenient ways of describing the number of nodes separating a taxon from the root of a given phylogenetic tree.

For example, within Xenarthra, cingulates (armadillos and glyptodonts) comprise the sister taxon of pilosans (sloths and anteaters). The genus *Dasypus* comprises the sister taxon of other cingulates. Based on the phylogenetic hypothesis of Delsuc *et al.* (2016: fig. 1), there is one node separating the common ancestor of *Dasypus* from the xenarthran root, and at least two separating other cingulate genera. On this basis, *Dasypus* occupies a more basal branch than, say, *Euphractus*. I recognize that this could change if future analysis reveals (for example) more extinct taxa on the branch leading to *Dasypus*. Nonetheless, the terms "basal" and "nested" still convey useful information about distance to the root as measured by nodes on a given phylogenetic tree.

To begin this survey, I first outline who the major mammalian clades are and present currently well-corroborated hypotheses on their interrelationships. Three papers from 2001 (Murphy *et al.* 2001a, b; Madsen *et al.* 2001) demarcate an important shift in the consensus regarding the shape of this tree, in particular regarding its most diverse clade, Placentalia. I describe a number of pre-2001 ideas on mammalian interrelationships, noting which ones have been disproven and which ones are now part of the well-corroborated tree. The bulk of this chapter consists of a review of major extinct radiations and hypotheses as to their affinities to modern groups; it closes with a discussion of hypotheses regarding the temporal dimension of mammalian (and particularly placental) evolution.

7.2 Major extant mammalian clades

7.2.1 Mammalia

The basal-most branching event within crown Mammalia divides Monotremata (historically also known as "Prototheria") and Theria; the latter in turn comprises Metatheria and Eutheria. The anthropocentrism of early taxonomists

◂ **Fig. 7.1:** The well-corroborated tree for Mammalia based on four phylogenetic studies using large samples of data and taxa, Meredith *et al.* (2011), Tarver *et al.* (2016), Esselstyn *et al.* (2017), and Mitchell *et al.* (2014) for relationships among marsupials. Conflicts (e.g., placement of tupaiids within Euarchontoglires) are shown as polytomies. Tree at left shows representative genera within most family-level taxa sampled by Meredith *et al.* (2011: fig. 1); tree at right represents that of Tarver *et al.* (2016). Grey = monotremes, light green= marsupials, green = atlantogenatans, blue = laurasiatheres, red = euarchontoglires. Letters correspond to nomenclature listed in Tab. 7.1.

led to the Greek prefixes proto ("first"), meta ("after"), and eu ("true"), implying a scala naturae. Modern authors generally recognize that no extant group is more evolved than any other (although rates of course vary), and the interpretation of phylogenies as ladders of progress has long been recognized as obsolete (e.g., Baum *et al.* 2005; Omland *et al.* 2008). Placental mammals are nonetheless the dominant group in terms of species number and ecological space occupied. Although this may be due to geographic factors rather than any intrinsic feature of one clade over another (Sánchez-Villagra 2013), there are over 5000 extant placental mammal species compared with roughly 340 marsupial and five monotreme species, and only placental mammals have evolved powered flight (Chiroptera) and a fully aquatic lifestyle (Cetacea and Sirenia). Marsupialia and Placentalia are the crown groups within Meta- and Eutheria, respectively, with the latter two also including the respective stem taxa of the former two.

7.2.2 Monotremes

In recent usage (Kielan-Jaworowska *et al.* 2004; Luo *et al.* 2015; Martin, this volume, Chapter 6), Monotremata is the crown group encompassing the platypus and echidna, situated within its total group Australosphenida. Within Monotremata, the two major sister taxa are Ornithorhynchidae and Tachyglossidae. The duck-billed platypus (*Ornithorhynchus anatinus*) is the only extant species of the former, whereas the latter comprises the short-beaked echidna (*Tachyglossus aculeatus*) and three species of long-beaked echidna (*Zaglossus* spp.). All *Zaglossus* species are confined to New Guinea and are classified as either vulnerable (*Zaglossus bartoni*) or critically endangered (*Zaglossus attenboroughi* and *Zaglossus bruijnii*) on the IUCN Red List (www.iucnredlist.org).

7.2.3 Marsupials

Extant marsupials are represented by approximately 340 species from Australasia and the Americas (Sánchez-Villagra 2013). Following Mitchell *et al.* (2014), the American forms consist of three successively distant, South American sister groups to the Australasian clade: *Dromiciops* (monito del monte), followed by didelphids (opossums) then caenolestids (shrew opossums) at the base. Placement of *Dromiciops* with Australasian marsupials to the exclusion of other South American marsupials in Australidelphia was originally based on skeletal anatomy of the foot (Szalay 1982) and later supported by studies of DNA hybridization (Kirsch *et al.* 1991). This idea is now widely accepted (Phillips *et al.* 2006; Nilsson *et al.* 2010; Mitchell *et al.* 2014), as is the recognition that *Dromiciops* is the sole living representative of the formerly more diverse Microbiotheria (Reig 1955; Hershkovitz 1999). Some ambiguity exists regarding the possibility that *Dromiciops* may nest one or more nodes within Australidelphia, e.g., as sister taxon to diprotodonts (Horovitz and Sánchez-Villagra 2003; Beck 2008; May-Collado *et al.* 2015) or crownward from peramelians (Asher *et al.* 2004: fig. 1). Nonetheless, all agree that *Dromiciops* is part of Australidelphia, and the largest data sets (Meredith *et al.* 2011; Mitchell *et al.* 2014) place it as the sister taxon to all Australasian marsupials.

Beyond *Dromiciops*, australidelphian marsupials consist of four major groups: dasyuromorphs (e.g., quolls, thylacines, and numbats), notoryctids (moles), peramelians (e.g., bandicoots and bilbies), and diprotodonts (e.g., wombats, kangaroos, phalangers, and possums). The former three comprise a monophyletic clade, with dasyuromorphs and peramelians comprising a clade to which notoryctids are the sister taxon. These three groups in turn comprise the sister taxon to diprotodonts, the most speciose of these high-level marsupial clades.

7.2.4 Placentals

The first edition of *Mammals of the World* (Walker 1964) listed 17 high-level placental groups on its spine. Of these, "Edentata" and "Insectivora" are polyphyletic (i.e., contain descendants of multiple common ancestors), "Artiodactyla" (excluding cetaceans) is paraphyletic (i.e., does not encompass all of the descendants of a single common ancestor), and Pinnipedia is now understood to comprise a group within caniform carnivorans (Fig. 7.1). The other 13 high-level taxa described by Walker (1964) remain essentially unchanged, although (as detailed below and shown in Fig. 7.1) relations among these placental groups are now more confidently resolved than during the 20th century.

Extant Placentalia contains four major groups: Xenarthra, Afrotheria, Laurasiatheria, and Euarchontoglires. The former two are collectively known as Atlantogenata, the latter two as Boreoeutheria. Although there has been some uncertainty regarding the root of this tree, with some recent analyses favoring either Afrotheria (Gatesy *et al.* 2017) or Xenarthra (O'Leary *et al.* 2013) as the basalmost clade, the largest data set published to date (Tarver *et al.* 2016 as described above), as well as a recent analysis of rare genomic events (Esselstyn *et al.* 2017), supports the Atlantogenata-Boreoeutheria division (Fig. 7.1).

Xenarthra as a zoological concept has a long history, although it was grouped among "edentates" (with aardvarks and pangolins) in the older literature. For example, Gregory (1910: p. 465) used Xenarthra in today's modern sense to unite pilosans (anteaters and sloths) and cingulates (armadillos), but placed other "edentates" (e.g., tubulidentates and pholidotes) close to it in his classification.

The core of Afrotheria is Paenungulata (Simpson 1945), i.e., proboscideans, sirenians, and hyracoids. An affinity of paenungulates with other endemic African taxa (e.g., the aardvark) was favored by LeGros Clark and Sonntag (1926) and received support from comparisons of protein sequences (de Jong *et al.* 1981). Evidence for an endemic African mammal clade joining yet more taxa with paenungulates was published by Springer *et al.* (1997) and Stanhope *et al.* (1998). By the early 2000s (e.g., Murphy *et al.* 2001b), evidence for an Afrotheria consisting of tubulidentates, macroscelidids, tenrecids, chrysochlorids, and paenungulates was strong and received support from analyses (e.g., Asher *et al.* 2003) beyond the initial molecular biology groups who had originally proposed the clade.

Euarchontoglires contains a long-recognized "archontan" core, consisting of primates, dermopterans, and scandentians, but not chiropterans (contra Gregory 1910); hence, the prefix "eu" sometimes added to Archonta. In addition, lagomorphs and rodents are sister taxa in the clade Glires and also belong in Euarchontoglires. Ambiguity remains about the position of Scandentia (Arcila *et al.* 2017), which may be sister to primates and dermopterans (Esselstyn *et al.* 2017), sister to Glires (Tarver *et al.* 2016: fig. 2 left), or comprise the basalmost branch within Euarchontoglires (Tarver *et al.* 2016: fig. 2 right).

Laurasiatheria consists of lipotyphlans (i.e., erinaceids, soricids, talpids, and solenodontids) at its base, followed in the largest studies (Tarver *et al.* 2016) by Chiroptera, then a carnivoran-pholidote clade (Ferae), then a clade consisting of perissodactyls and artiodactyls (Euungulata), with Cetacea nested within Artiodactyla adjacent to hippopotamids. Furthermore, there is support for the paraphyly of microchiropterans, with rhinolophoids forming a close relationship with pteropodids (i.e., Yinpterochiroptera) to the exclusion of other "microbats", or Yangochiroptera (cf. Teeling *et al.* 2005). Some uncertainty lingers regarding the placement of Chiroptera and Perissodactyla among laurasiatheres and the monophyly of Euungulata (Nishihara *et al.* 2006), as well as the position of ursids as either sister to pinnipeds (Meredith *et al.* 2011: fig. 1) or sister to a pinniped-musteloid clade (Fig. 7.1 and Esselstyn *et al.* 2017).

7.3 Pre-21st century mammalian phylogenetics

As discussed previously (Asher, this volume, Chapter 3), early naturalists and evolutionary biologists named taxa without intending to articulate natural, monophyletic groups, using names to represent other concepts (e.g., utility to humans or adaptive grade). One may nonetheless use their nomenclature and classifications to measure how ideas about animal groups have changed over time. The basic monotreme-marsupial-placental trichotomy within Mammalia has been phylogenetically and anatomically understood since the 19th century, as have the identities of most ordinal- and family-level groups within each (Tab. 7.1).

Gregory (1910: figs. 31 and 32) figured an evolutionary tree for Mammalia that not only represents this trichotomy but went further to define Theria to the exclusion of monotremes (Fig. 7.1). Gregory (1947) later articulated an alternative view that monotremes and marsupials are each others' closest relatives in the taxon "Marsupionta". This idea did not have much traction among most mammalian systematists of the 20th century (e.g., Simpson 1945; chapters in Szalay *et al.* 1993), but it did garner support from studies of DNA hybridization (Kirsch and Meyer 1998) and studies of individual genes (Janke *et al.* 2002). This support has effectively been overturned with the analysis of larger data sets and more sophisticated methods. Much to their credit, and comprising a good example of how scientists are willing to alter their views based on novel data and analytical techniques, some of this recent work was undertaken by the same groups who had previously supported "Marsupionta" (e.g., Kullberg *et al.* 2008).

7.3.1 Marsupials

Within marsupials, Gregory (1910: p. 464) named four major groups: "Allotheria" (including multituberculates), Diprotodontia, Paucituberculata, and "Polyprotodontia" (Fig. 7.2). His unusual classification of multituberculates as marsupials (1910: p. 169) is based in part on Gidley's (1909) description of the skull of *Ptilodus* and its (in Gregory's words) "typically marsupial" possession of an inflected angle of the jaw and palatal fenestrae. Simpson (1945) did not follow this classification and instead

Tab. 7.1: Taxonomy of Mammalia based on trees in Fig. 7.1 (Meredith *et al.* 2011; Mitchell *et al.* 2014; Tarver *et al.* 2016; Esselstyn et al. 2017). Where available, letters correspond to named nodes on Fig. 7.1. Nomenclature follows principles outlined in Simpson (1945) and Asher and Helgen (2010). Cingulata, Hippomorpha, Pholidota, Suiformes, and Tylopoda represent named clades for species most closely related to (respectively) *Dasypus, Equus, Manis, Sus, and Llama*, represented by genera but not distinct nodes in Fig. 7.1.

- a, Monotremata
  - Ornithorhynchidae
  - Tachyglossidae
- h, Theria
  - b, Marsupialia
    - c, Australidelphia
      - d, Eomarsupialia
        - e, Peremelimorphia
        - f, Dasyuromorphia
        - g, Diprotodontia
    - i, Placentalia
      - j, Atlantogenata
        - k, Afrotheria
          - l, Paenungulata
          - m, Afroinsectivora
            - n, Tenrecoidea
        - o, Xenarthra
          - Cingulata
          - p, Pilosa
      - q, Boreoeutheria
        - r, Laurasiatheria
          - s, Scrotifera
          - t, Chiroptera
            - u, Yinpterochiroptera
            - v, Yangochiroptera
          - w, Fereuungulata
            - x, Ferae
              - y, Carnivora
                - z, Feliformia
                - aa, Caniformia
              - Pholidota
            - ay, Euungulata
              - ab, Perissodactyla
                - Hippomorpha
                - ac, Tapiromorpha
              - ad, Artiodactyla
                - Tylopoda
                - Suiformes
                - ae, Ruminantia
                  - af, Pecora
                - ag, Whippomorpha
                  - ae, Cetacea
          - af, Lipotyphla
        - ag, Euarchontoglires
          - ah, Glires
            - ai, Lagomorpha
            - aj, Rodentia
              - ak, Ctenohystrica
              - al, Myodonta
          - am, Primatomorpha
            - an, Primates
              - ao, Strepsirhini
                - ap, Lemuroidea
              - aq, Haplorhini
                - ar, Anthropoidea
                  - as, Platyrrhini
                  - at, Catarrhini
                    - au, Hominoidea
                      - av, Hominidae
                        - aw, Homininae

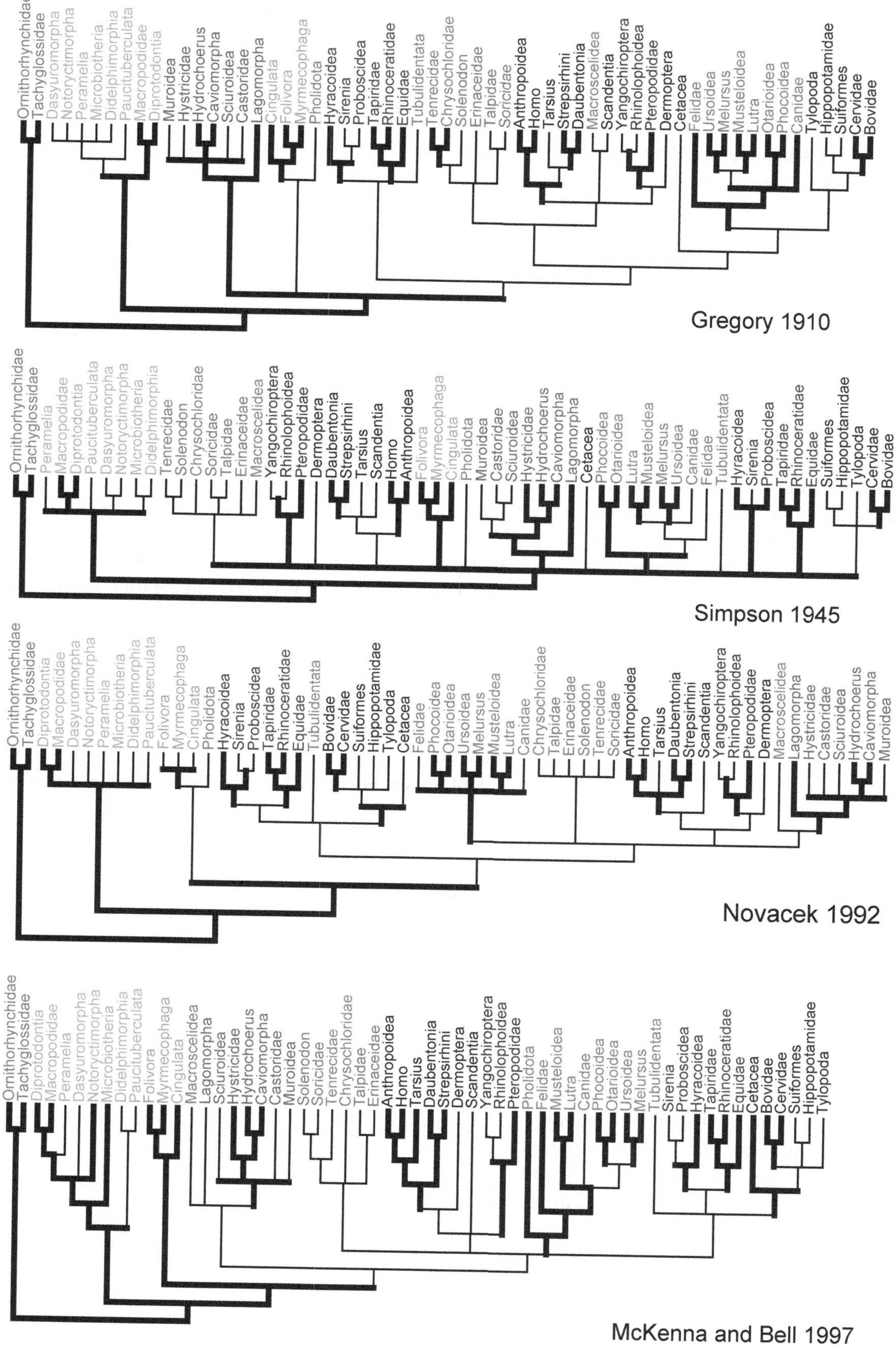

Fig. 7.2: Cladograms derived from the classifications of Gregory (1910), Simpson (1945), Novacek (1992), and McKenna and Bell (1997) using the methodology outlined in Asher, this volume, Chapter 3. Thick branches indicate areas of the cladogram in agreement with the well-corroborated tree (Fig. 7.1). Colors represent clades as depicted in Fig. 7.1.

placed multituberculates outside of Theria. Gregory's "polyprotodontia" placed didelphids (including microbiotheres) with dasyuromorphs, peramelians, and notoryctids; his Paucituberculata subsumed caenolestids; his Diprotodontia placed phalangeroids with macropodids and vombatiformes. In his figure on the "morphogenetic relations" of marsupials, Gregory (1910: fig. 14) implied that diprotodonts evolved from a "polyprotodont" ancestor, similar in adaptive grade to caenolestids.

Simpson (1945: p. 171) neatly summed up these and other early 20th century ideas on marsupial evolution with the pithy dichotomy of "incisors vs. toes". "Incisors" draws on the resemblance of the anterior, procumbent incisors of caenolestids and diprotodonts, and follows Gregory's belief that the latter evolved from the former. By contrast, Wood-Jones (1925) emphasized pedal structure, in particular the syndactyl digits II–III shared by peramelians and diprotodonts. Although these ordinal-level taxa are generally similar to those recognized today (with the notable exception of microbiotheres), neither the "incisors" (i.e., caenolestids and diprotodonts) nor "toes" (i.e., peramelians and diprotodonts) hypotheses on their interrelations accurately capture the now well-corroborated signal. Instead, and as noted above, caenolestids comprise the basalmost marsupial clade, and peramelians are the sister taxon to dasyuromorphs within an Australidelphia that also includes diprotodonts (Fig. 7.1; see Mitchell *et al.* 2014).

7.3.2 Placentals

Although classifications of mammals over the past centuries show an increasing level of similarity to the currently well-corroborated tree (Asher, this volume, Chapter 3), resolution does not increase at a comparably steady rate. Polytomies, or multiple groups arising from a single node (or placed in a single, high-level taxon), reflect uncertainty and were common throughout the 20th century. None of the 18 classifications or branching diagrams published between 1904 and 1998 sampled by Asher (Chapter 3 and Tab. 7.2) were fully resolved. The two most resolved classifications were at the beginning (Gregory 1910) and end (McKenna and Bell 1997) of the century. Those in the middle exhibited a slight but not significant trend towards increasing resolution, in contrast to their increasing similarity to the well-corroborated tree (Fig. 7.3). Gregory (1910) was unusually accurate for his time, and indeed more accurate than some later analyses (e.g., McKenna 1975; Novacek 1992), although other early 20th century studies were not (e.g., Weber 1904; Osborn 1917). Quantifying the similarity of these classifications to the now well-corroborated tree (Fig. 7.1) using the ratio of actual/potential groups in common (see Asher, this volume, Chapter 3), these early 20th century studies ranged from 0.21 to 0.29. At 0.41, Gregory (1910) was an outlier (Tab. 7.2). Starting with Cabrera (1922), and with the qualification that Winge (1941) was a posthumous publication of earlier ideas (see Asher, this volume, Chapter 3), the remaining mammalian classifications of the 20th century exceeded 0.3, with McKenna and Bell (1997) and Stanhope *et al.* (1998) at 0.48 achieving the highest ratio of actual/potential groups in common with the well-corroborated tree.

Stanhope *et al.* (1998) analyzed phylogenetic signal from three mitochondrial genes (16S & 12SrRNA, valine tRNA) and discussed more taxonomically restricted data from four nuclear loci. Their mitochondrial data are independent of the largely genomic data sets now used to define the well-corroborated tree (Fig. 7.1). Another early study of molecular data (Miyamoto and Goodman 1986) achieved an accuracy ratio of 0.36, similar to the cladistic morphology analysis of Novacek (1986), the cladistic classification of McKenna (1975), and the traditional, evolutionary classifications of Romer (1945) and Simpson (1945).

The combined breakthroughs in DNA sequencing techniques (e.g., PCR) and accessibility of genetic information (e.g., GenBank) in the late 20th century revolutionized our capacity to collect and evaluate data pertinent to reconstructing the evolutionary tree of life.

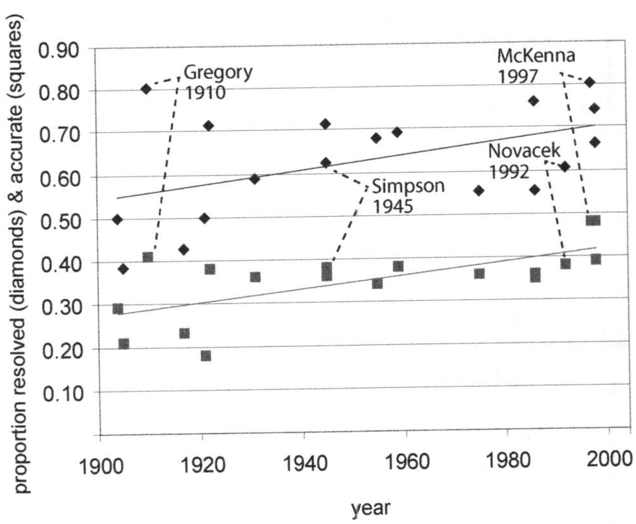

Fig. 7.3: The y axis shows accuracy (purple squares) and resolution (blue diamonds) of classifications and cladograms over the course of the 20th century (x axis). Accuracy is defined as the ratio of actual to potential number of groups (y axis) held in common with the well-corroborated tree (Fig. 7.1). Data points are given in Tab. 7.2 and discussed further in Asher, this volume, Chapter 3.

Tab. 7.2: Summary of classifications since 1900 sampled by Asher (Chapter 3) and quantified in terms of the number of actual ÷ potential (act/pot) groups in common with the well-corroborated tree (Fig. 7.1). Also given is the number of characters sampled (#char, available only for studies that used a quantitative tree reconstruction method), number of taxa sampled in common with well-corroborated tree (#taxa), number of resolved or bifurcating nodes (#resolved), and the proportion resolved based on taxon sample (#taxa-3/#resolved = proRes). #char in Song *et al.* 2012 is based on length in base pair (1,385,220) of all considered loci (447) as reported in their S1 supplementary data; #char in McCormack *et al.* (2012, p. 752) is based on their reported use of 2386 UCE probes with a target length of 120 bp each.

Author	Method	Year	act/pot	#chars	#taxa	#resolved	proRes
Weber	evolut	1904	0.29		45	21	0.50
Haeckel	evolut	1905	0.21		42	15	0.38
Gregory	evolut	1910	0.41		59	45	0.80
Osborn	evolut	1917	0.23		59	24	0.43
Winge	evolut	1921	0.18		59	28	0.50
Cabrera	evolut	1922	0.38		59	40	0.71
Simpson	evolut	1931	0.36		59	33	0.59
Romer	evolut	1945	0.36		59	40	0.71
Simpson	evolut	1945	0.38		59	35	0.63
Grasse	evolut	1955	0.34		59	38	0.68
Romer	evolut	1959	0.38		58	38	0.69
McKenna	cladis	1975	0.36		59	31	0.55
Miyamoto	molec	1986	0.36	353	39	20	0.56
Novacek	cladis	1986	0.35	73	49	35	0.76
Novacek	cladis	1992	0.38		59	34	0.61
McKenna	cladis	1997	0.48		59	45	0.80
Shoshani	cladis	1998	0.39	260	59	37	0.66
Stanhope	molec	1998	0.48	2086	34	23	0.74
Murphy	molec	2001	0.8	16397	38	35	1.00
Arnason	molec	2002	0.56	9882	44	38	0.93
Asher	comb	2003	0.55	17433	36	33	1.00
Kjer	molec	2007	0.6	14740	50	47	1.00
Prothero	cladis	2007	0.43		59	37	0.66
Arnason	molec	2008	0.59	7234	52	43	0.88
Meredith	correct	2011	0.82	35603	59	56	1.00
McCormack	molec	2012	0.73	286320	18	15	1.00
Song	molec	2012	0.83	1385220	27	24	1.00

Leading up to these breakthroughs, early applications of molecular data (such as immunochemistry and comparisons of protein sequences) reflected some key aspects of the now well-corroborated tree, such as cetaceans close to other artiodactyls (Nuttall *et al.* 1904; Boyden and Gemeroy 1950), humans with *Pan* and *Gorilla* to the exclusion of other apes (Sarich and Wilson 1967), aardvarks with paenungulates (de Jong *et al.* 1981), and a pholidote-carnivoran clade (Shoshani 1986), a group also noted to have morphological support (Rose and Emry 1993). However, these early trees usually lacked a broad taxonomic sample; those that were well sampled were not, on the whole, substantially more accurate than their counterparts based on comparative anatomy, as demonstrated by the similar accuracy scores of Miyamoto and Goodman (1986) with those from Simpson, Romer, McKenna, and Novacek (Tab. 7.2). Quantitative analyses of morphology and DNA sequences exhibited an increase in accuracy from the 1980s onward (Fig. 7.4), leading to the broadly sampled studies of nuclear DNA that contribute heavily to the well-corroborated tree (Fig. 7.1). This tree began to take shape in the late 1990s (e.g., Waddell *et al.* 1999: fig. 1). The studies of Murphy *et al.* (2001a, b), Madsen *et al.* (2001), and Scally *et al.* (2001) were the first to decisively and significantly place the root of Placentalia close to afrotherians and xenarthrans, rather than within rodents or at *Erinaceus*.

During the 20th century, comparative anatomists and evolutionary taxonomists recognized at least some of the now well-corroborated interordinal relationships among mammals. These included scandentians closer to primates than macroscelideans (Simpson 1945), sirenians and hyracoids close to proboscideans (Gregory 1910), pinnipeds among caniforms (Gregory 1910), Glires including both rodents and lagomorphs (Landry 1974), and dermopterans close to primates (Beard 1993). In hindsight, several ideas first proposed based on comparative anatomy (but not necessarily widespread) have turned out to be quite prophetic, such as the affinity of the aardvark (Tubulidentata) to paenungulates (e.g., LeGros Clark and Sonntag 1926) and the basal position of xenarthrans relative to most other placental groups (McKenna 1975). However, other ideas about interordinal relations based on comparative anatomy conflict with the well-corroborated tree (Fig. 7.1), such as "Menotyphla" grouping tupaiids with macroscelidids (Gregory 1910), "Volitantia" joining dermopterans with bats (Novacek and Wyss 1986; Simmons and Quinn 1994), "Insectivora" joining tenrecoids (i.e., chrysochlorids and tenrecids) with lipotyphlans (Asher 1999), "Artiodactyla" excluding cetaceans (Luckett and Hong 1998), inclusion of hyracoids within Perissodactyla (Fischer 1989), and monophyly of "Microchiroptera" including rhinolophoids (Simmons and Geisler 1998). Furthermore, several studies based on comparative anatomy questioned what seems now to be an obvious, close relationship between rodents and lagomorphs, for example, by placing rodents closer to primates than lagomorphs (e.g., Patterson and Wood 1982) or lagomorphs close to macroscelidids (e.g., McKenna 1975). Pettigrew (1986) argued for the paraphyly of Chiroptera based on neurological similarities between pteropodids and primates to the exclusion of microchiropterans. This idea was not widely accepted among mammalian systematists at the time (see Simmons *et al.* 1991). Many 20th century ideas on high-level mammalian systematics were uncertain and disputed at one time or another. The three that did enjoy widespread consensus during the 20th century, and yet turned out to be wrong, were "Volitantia", "Microchiroptera", and "Artiodactyla" excluding cetaceans.

Early molecular studies included comparisons of protein sequences (e.g., Miyamoto and Goodman 1986) and isolated mitochondrial genes (Irwin *et al.* 1991) later expanded into comparisons of mitochondrial genomes (e.g., Arnason *et al.* 1997; Mouchaty *et al.* 2000). As noted above, these enabled recognition of several features of the currently well-corroborated tree but departed from today's consensus (Fig. 7.1) in one or more important ways. For example, the placental root was variably occupied by branches leading to erinaceid lipotyphlans (Arnason *et al.* 1997) or a paraphyletic Glires (D'Erchia *et al.* 1996; Stanhope *et al.* 1998). Early analyses of primarily nuclear DNA also departed from the now well-corroborated tree in some ways and shared some topological features with mitogenomic studies, e.g., by not placing *Tarsius* closer to anthropoids than strepsirhines and by placing anthropoids closer to dermopterans than strepsirhines (Murphy *et al.* 2001a; Arnason *et al.* 2002).

Analyses of coding mtDNA also reconstructed Odontoceti as paraphyletic (Arnason and Gullberg 1994). This peculiar signal has reappeared in some recent, coalescent-based studies of mammalian genomes (e.g., Liu *et al.* 2017), in which the sperm whale (*Physeter*) appears closer to minke whales (*Balaenoptera acutorostrata*) than to other odontocetes. In addition, Liu *et al.* (2017) place the aardvark (*Orycteropus*) in a clade with golden moles (chrysochlorids) to the exclusion of tenrecids, and African galagos with Malagasy mouse lemurs (cheirogaleids) to the exclusion of another Malagasy strepsirhine (*Daubentonia*). As detailed elsewhere (e.g., Springer and Gatesy 2016, 2017; Gatesy *et al.* 2017), these departures from the well-corroborated tree appear to be artifacts of poor alignment and model choice, as well as insufficient searches of tree space for some of the estimated gene trees (cf. Gatesy *et al.* 2017: fig. 7).

In terms of size, phylogenetic analyses of the last few years dwarf those previously considered "large" when first published. For example, the seminal paper of Murphy *et al.* (2001b) sampled 16,397 aligned nucleotides for 44 taxa (of which 38 can be compared with the well-corroborated tree in Fig. 7.1). This was a large data set for 2001, but (with size defined as taxa multiplied by characters; see Fig. 7.4) much smaller than Meredith *et al.* (2011; 35,603 bp for 169 taxa), which was in turn much smaller than the data set of Tarver *et al.* (2016; based in part on 32,116,455 bp for 36 taxa). Notably, the topological similarity of these recent studies to Murphy *et al.* (2001b) is considerable, although Murphy *et al.* (2001b) is much closer in terms of number of taxa and characters to studies from the 1990s (Fig. 7.4). This suggests that data sets of ever increasing size since 2001 are corroborating the optimal topology in Murphy *et al.* (2001b) rather than forcing major revisions with each new analysis. Despite the occasional journalistic blunder (Dolgin 2012; reviewed in Asher 2012a, b; Penny 2013), and with the qualification that there are some outliers (e.g., Cannarozzi *et al.* 2007 who analyzed seven taxa to support primate-canid to the exclusion of murids), new studies since 2001 have neither "torn apart traditional ideas" nor forced anyone to "rewrite textbooks" but have instead corroborated the basic structure of mammalian phylogenetics outlined by Murphy *et al.* (Fig. 7.1). To be sure, some issues such as the utility of coalescent methods (Springer and Gatesy 2016, 2017; Edwards *et al.* 2016), the relative ages of key mammalian diversifications (Bininda-Emonds *et al.* 2007; Meredith *et al.* 2011; dos Reis *et al.* 2012; Foley *et al.* 2016), the position of scandentians relative to primates (Arcila *et al.* 2017), and some intra-Laurasiatheria relations remain contentious. Nonetheless, mammalian systematics today finds itself enjoying more consensus, based on more data, than at any time in the history of biological classification.

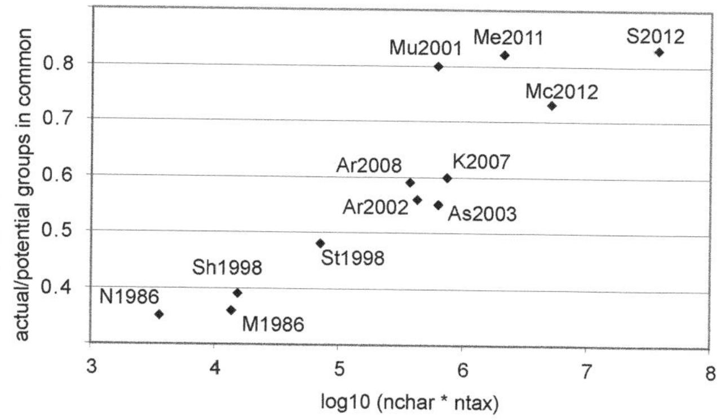

Fig. 7.4: Relationship between data set size (log 10 of summed morphological and nucleotide characters multiplied by the number of taxa in common with well-corroborated tree; Fig. 7.1) on the x axis and accuracy (number of actual divided by potential groups in common with the well-corroborated tree) on the y axis. Data points are phylogenetic studies that applied quantitative methods to identify an optimal tree: cladistic morphology (N1986 = Novacek 1986, Sh1998 = Shoshani and McKenna 1998), comparisons of protein sequences (M1986 = Miyamoto and Goodman 1986), studies of primarily mitochondrial genes (St1998 = Stanhope *et al.* 1998), mitochondrial genomes (Ar2002, Ar2008 = Arnason *et al.* 2002, 2008; K2007 = Kjer and Honeycutt 2007), concatenated DNA and morphology (As2003 = Asher *et al.* 2003), concatenated nuclear genes (Mu2001 = Murphy et al. 2001b; Me2011 = Meredith *et al.* 2011), and coalescence analyses of ultraconserved elements (Mc2012 = McCormack *et al.* 2012) and nuclear DNA (S2012 = Song *et al.* 2012). The number of characters in Song *et al.* 2012 is based on length in base pairs (1,385,220) of all considered loci (447) as reported in their S1 supplementary data; number of characters in McCormack *et al.* (2012, p. 752) is based on their reported use of 2386 UCE probes with a target length of 120 bp each. The number of taxa for each study represents number of terminals sampled that can be directly compared with the well-corroborated tree (Fig. 7.1), as discussed by Asher (Chapter 3). Note that sample size on the x axis is log-transformed to enable comparison with widely variable data sets, ranging from 73 (N1986) to 1,300,000 (S2012) characters.

7.4 Major fossil groups

Much more uncertain than the shape of the extant tree are the affinities of many extinct radiations of mammals (Tab. 7.3). Because of the loss of biological data inherent in fossilization (Donoghue and Purnell 2009), our capacity to achieve levels of phylogenetic confidence similar to those known for extant groups is limited, and at least in some cases may be impossible. Genomic DNA has proven to be accessible in recently extinct species (Green *et al.* 2010), and the amino acid sequences of resilient proteins such as collagen, inferred via mass spectrometry, may have an even greater potential to shed light on the phylogenetic history of long-extinct species (Buckley 2013; Welker *et al.* 2015). Furthermore, indirect but nonetheless compelling methods exist to gauge biological features previously thought to be lost with extinction, such as correlations of bone cell with genome size (Organ *et al.* 2012), genetic patterning of the integument (Schmid 2012), the nature of pigmentation (Wogelius *et al.* 2011), and perhaps even cellular soft tissue preservation from Mesozoic theropods (Schweitzer *et al.* 2016). However, for phylogenetic questions, systematists are usually limited to hard tissue anatomy (i.e., bones and teeth) to make inferences about placement of fossil taxa on the tree of life. The good news is that, in at least some cases with well-corroborated phylogenies (e.g., primates), fossilizable data are statistically consistent and yield reasonably accurate topologies, particularly when hard tissue data span multiple anatomical regions (Pattinson *et al.* 2015). Moreover, even when some metrics of partition heterogeneity are interpreted to show more character-based variation between morphological and molecular partitions than within the latter (Springer *et al.* 2007; variation which is arguably interpreted as an indication of phylogenetic conflict; see Wheeler 1995 and Gatesy *et al.* 1999), the combination of data from such partitions in a total-evidence context still contributes positively to resolution and support (Gatesy *et al.* 1999; Lee and Camens 2009; Thompson *et al.* 2012) and reduces error margins for timetree estimated divergences (Ronquist *et al.* 2012).

In the remaining pages of this chapter, I do not try to summarize all groups of fossil mammals, which are better covered elsewhere in this book and in others by (for example) Janis *et al.* (1998), Kielan-Jaworowska *et al.* (2004), Kemp (2005), Rose and Archibald (2005), Rose (2006a), Werdelin and Sanders (2010), Wang *et al.* (2013), Croft (2016), Prothero (2016), and Berta (2017). Instead, I focus on a few major, extinct groups, summarized in Tab. 7.3, which have not already been covered in other chapters in this volume. I define a group as “major” when it (A) has figured prominently in previous discussions of high-level clade origins, (B) has stratigraphic and systematic documentation in the peer-reviewed literature, and/or (C) is well documented by both cranial and postcranial remains. Tab. 7.4 summarizes the Paleocene and Eocene Asian (ALMA), North American (NALMA), and South American (SALMA) Land Mammal Ages and their approximate correlations to the marine and absolute geological record based on McKenna and Bell (1997) and Cohen *et al.* (2013), often referred to in the following text.

Tab. 7.3: Summary of possible phylogenetic affinities of fossil taxa mentioned in the text and depicted in Figs. 7.5–7.9.

Taxon	High-level clade	affinities	alternative views
Aaptoryctes	"palaeoryctid"	?Lipotyphla (Lopatin 2006; Asher *et al.* 2002)	?Ferae (Halliday *et al.* 2017)
Abdounodus	Afrotheria	stem tubulidentate (Gheerbrant *et al.* 2016)	
Acmeodon	"cimolestid"	?Ferae (Lopatin 2006; Halliday *et al.* 2017)	
Adunator	"erinaceomorph"	lipotyphlan (Manz *et al.* 2015)	
Afrodon	Adapisoriculidae	non-placental Eutheria (Goswami *et al.* 2011)	euarchontans (Hooker 2001, Smith *et al.* 2010)
Alcidedorbignya	Pantodonta	Laurasiatheria (Muizon *et al.* 2015; this study Fig. 7.9)	
Alphadon	Metatheria	stem Marsupialia (Wilson *et al.* 2016)	
Anthracobune	Anthracobunidae	Euungulata (Cooper *et al.* 2014; Rose *et al.* 2014)	
Apheliscus	Louisinae	Euungulata (O'Leary *et al.* 2013; Cooper *et al.* 2014)	stem macroscelideans (Zack *et al.* 2005)
Apternodus	"apternodontid"	lipotyphlan (Asher *et al.* 2002; Lopatin 2006)	
Archaeoryctes	Didymoconidae	lipotyphlan (Lopatin 2006)	
Arsinoitherium	Embrithopoda	paenungulate (Cooper *et al.* 2014)	
Artiocetus	Cetacea	stem cetacean (Gingerinch *et al.* 2001; Geisler and Theodor 2009)	
Asiatherium	Metatheria	stem marsupial (Beck 2012)	
Asioryctes	Asioryctitheria	non-placental Eutheria (Wible *et al.* 2009)	
Basilosaurus	Cetacea	stem cetacean (Gingerich *et al.* 1990; Geisler and Theodor 2009)	
Barunlestes	Zalambdalestidae	non-placental Eutheria (Wible *et al.* 2009)	stem glires (Archibald *et al.* 2001)
Behemotops	Desmostylia	Euungulata (Cooper *et al.* 2014)	paenungulate afrotherian (Asher 2007)
Bessoecetor	Pantolestida	?Ferae (Halliday *et al.* 2017)	
Bothriogenys	Anthracotheriidae	stem hippomotamids (Orliac *et al.* 2010; Lihoreau *et al.* 2015)	
Buxolestes	Pantolestida	?Ferae (Rose and von Koenigswald 2005)	
Cambaytherium	Euungulata	stem perissodactyl (Rose *et al.* 2014; Cooper *et al.* 2014)	
Carodnia	Xenungulata	Euungulata (Muizon *et al.* 2015; this study Fig. 7.9)	Afrotheria (O'Leary *et al.* 2013; Carrillo and Asher 2017)
Carpolestes	Plesiadapiformes	stem Primates (Bloch *et al.* 2007)	
Chambius	Afrotheria	stem macroscelidean (Tabuce 2017)	
Chriacus	Arctocyonidae	?Ferae (Halliday *et al.* 2017)	?stem artiodactyl (Ladevèze *et al.* 2010)
Cornwallia	Desmostylia	Euungulata (Cooper *et al.* 2014)	paenungulate afrotherian (Asher 2007)
Coryphodon	Pantodonta	Laurasiatheria (Muizon *et al.* 2015; this study Fig. 7.9)	
Cryptomanis	Ferae	stem Pholidota (Rose *et al.* 2005)	
Cryptotopos	Nyctitheriidae	lipotyphlan (Manz *et al.* 2015)	euarchontans
Daouitherium	Afrotheria	stem proboscidean (Cooper *et al.* 2014)	
Daphoenus	Carnivora	stem Caniformes (Spaulding and Flynn 2012)	
Deccanolestes	Adapisoriculidae	non-placental Eutheria (Goswami *et al.* 2011)	euarchontans (Hooker 2001, Boyer *et al.* 2010; Smith *et al.* 2010)
Deltatheridium	Metatheria	stem Marsupialia (Beck 2012)	
Desmostylus	Desmostylia	Euungulata (Cooper *et al.* 2014)	paenungulate afrotherian (Asher 2007)
Didelphodon	Metatheria	stem Marsupialia (Wilson *et al.* 2016)	
Didelphodus	"cimolestid"	Ferae (Muizon *et al.* 2015; this study Fig. 7.9)	
Didolodus	"Condylarthra" (didolodontid)	Euungulata (O'Leary *et al.* 2013; Carrillo and Asher 2017)	

Tab. 7.3 (continued)

Taxon	High-level clade	affinities	alternative views
Dipsalidictis	Creodonta (oxyaenid)	sister Carnivora (Spaulding and Flynn 2012)	palaeanodonts and ?lipotyphlans (Halliday *et al.* 2017)
Ectocion	"Condylarthra" (phenacodontid)	stem Perissodactyla (Halliday *et al.* 2017: fig. 4)	
Elomeryx	Anthracotheriidae	stem hippomotamids (Orliac *et al.* 2010; Lihoreau *et al.* 2015)	
Eoconodon	"Condylarthra" (triisodontine)	in "Cete" with mesonychids, hapalodectids and cetacenas (McKenna and Bell 1997)	?carnivoran sister taxon with mesonychids (Halliday *et al.* 2017)
Eohippus	Perissodactyla	stem equoid (Froehlich 2002)	
Eomanis	Ferae	stem Pholidota (Rose *et al.* 2005)	
Eoryctes	"palaeoryctid"	lipotyphlan (Manz *et al.* 2015)	
Eotheroides	Afrotheria	dugongid sirenian (Samonds *et al.* 2009)	
Eotitanops	Brontotheriidae	stem tapiromorph (Rose *et al.* 2014)	
Eritherium	Afrotheria	stem proboscidean (Gheerbrant 2009)	stem paenungulate (Cooper *et al.* 2014)
Eurotamandua	Ferae	stem Pholidota	myrmecophagid Xenarthra (Storch and Habersetzer 1991)
Fouchia	Perissodactyla	tapiromorph (Emry 1989)	
Goniacodon	"Condylarthra" (triisodontine)	in "Cete" with mesonychids, hapalodectids and cetacenas (McKenna and Bell 1997)	?carnivoran sister taxon with mesonychids (Halliday *et al.* 2017)
Gomphos	Mimotonidae	stem Lagomorpha (Asher *et al.* 2005; O'Leary *et al.* 2013)	
Haplomylus	Louisinae	Euungulata (Cooper *et al.* 2014)	stem macroscelideans (Zack *et al.* 2005)
Herpetotherium	Metatheria	stem marsupial (Beck 2012)	stem paucituberculate? (Murat and Beck 2017)
Hesperocyon	Carnivora	stem Caniformes (Spaulding and Flynn 2012)	
Hyaenodon	Creodonta (hyaenodontid)	sister Carnivora (Spaulding and Flynn 2012)	
Hyopsodus	"Condylarthra" (hyopsodontid)	Euungulata (O'Leary *et al.* 2013)	afrotheres (Asher 2007)
Hyracotherium	Perissodactyla	stem equoid (Froehlich 2002)	
Indohyus	Artiodactyla	stem cetacean (Thesiwssen *et al.* 2007; Geisler and Theodor 2009)	
Kelba	Ptolemaiidae	Afrotheria (Cote *et al.* 2007)	viverrid Carnivora (Morales *et al.* 2000)
Kopidodon	Pantolestidae	"cimolestids" in Ferae, stem to carnivorans and pholidotes (McKenna and Bell 1997)	
Kulbeckia	Zalambdalestidae	non-placental Eutheria (Wible *et al.* 2009)	stem glires (Archibald *et al.* 2001)
Labidolemur	Apatemyidae	stem Glires (Silcox *et al.* 2010)	
Lambdotherium	Perissodactyla	stem brontotheriid (Rose *et al.* 2014)	
Leptacodon	Nyctitheriidae	lipotyphlan (Manz *et al.* 2015)	euarchontans (Hooker 2001)
Leptictis	Eutheria	non-placental Eutheria (Wible *et al.* 2009; Beck and Lee 2014)	lipotyphlan (Novacek 1986); afrothere (O'Leary *et al.* 2013; Muizon *et al.* 2015; this study Fig. 7.9)
Macrauchenia	Litopterna	stem Perissodactyla (Welker *et al.* 2015; Westbury *et al.* 2017)	stem Euungulata (Carrillo and Asher 2017)
Macrocranion	"amphilemurid" (?Lipotyphla)	stem lipotyphlan (Manz *et al.* 2015)	

Tab. 7.3 (continued)

Taxon	High-level clade	affinities	alternative views
Mayulestes	Metatheria	stem marsupial (Muizon *et al.* 1998)	
Meniscotherium	"Condylarthra" (phenacodontid)	Euungulata (O'Leary *et al.* 2013; Cooper *et al.* 2014)	afrotheres (Asher 2007)
Mesoscalops	Proscalopidae	lipotyphlan (McKenna and Bell 1997)	
Metacheiromys	Palaeanodonta	sister Pholidota (Emry 1970; O'Leary *et al.* 2013)	
Metoldobotes	Afrotheria	stem macroscelidean (Tabuce 2017)	
Microhyrax	Afrotheria	stem hyracoid (Seiffert 2007; Cooper *et al.* 2014)	
Mimoperadectes	Metatheria	stem Marsupialia (Maga and Beck 2017)	stem Didelphidae (Horovitz *et al.* 2009)
Necrolestes	Dryolestoidea	meridiolestidan (Rougier *et al.* 2012; O'Meara and Thompson 2014)	Metatheria (Patterson 1958; Asher *et al.* 2007; Ladevèze *et al.* 2008)
Nementchatherium	Afrotheria	stem macrosceldideans (Tabuce *et al.* 2007, 2017)	
Ocepeia	Afrotheria	stem afrothere (Cooper *et al.* 2014)	stem tubulidentate (Gheerbrant *et al.* 2016)
Oligoryctes	"apternodontid"	lipotyphlan (Asher *et al.* 2002; Lopatin 2006)	
Onychonycteris	Chiroptera	sister Yangochiroptera (O'Leary *et al.* 2013)	stem Chiroptera (Simmons *et al.* 2008
Oreotalpa	Lipotyphla	Talpidae (Lloyd and Eberle 2008)	
Palaeanodon	Palaeanodonta	sister to Pholidota (Emry 1970; O'Leary *et al.* 2013)	
Palaeosinopa	Pantolestida	?Ferae (Rose and von Koenigswald 2005; Halliday *et al.* 2017)	
Palaeoparadoxia	Desmostylia	Euungulata (Cooper *et al.* 2014)	paenungulate afrotherian (Asher 2007)
Pantolestes	Pantolestida	?Ferae (Rose and von Koenigswald 2005; Halliday *et al.* 2017)	
Paraceratherium	Perissodactyla	tapiromorphs (McKenna and Bell 1997)	
Parapternodus	"apternodontid"	Lipotyphla, ?stem soricids (Asher *et al.* 2002)	Lipotyphla, ?stem solenodontids (Lopatin 2006)
Paromomys	Plesiadapiformes	stem Primates (Bloch *et al.* 2007)	Dermoptera (Beard 1993)
Paschatherium	"Condylarthra" (louisine)	stem perissodactyl (Cooper *et al.* 2014)	stem macroscelidean (Zack *et al.* 2005)
Patriofelis	Creodonta (oxyaenid)	sister Carnivora (Spaulding *et al.* 2009)	
Patriomanis	Ferae	stem Pholidota (Rose *et al.* 2005)	
Pediomys	Metatheria	stem marsupial (Wilson *et al.* 2016)	
Peradectes	Metatheria	stem marsupial (Beck 2012)	stem didelphid (Horovitz *et al.* 2009)
Pezosiren	Afrotheria	stem sirenian (Domning 2001)	
Phenacodus	"Condylarthra" (phenacodontid)	Euungulata (O'Leary *et al.* 2013; Cooper *et al.* 2014)	afrotheres (Asher 2007)
Pholidocercus	"amphilemurid" (?Lipotyphla)	?stem erinaceoid (McKenna and Bell 1997)	
Phosphatherium	Afrotheria	stem proboscidean (Cooper *et al.* 2014)	
Plagioctenodon	Nyctitheriidae	lipotyphlan (Manz *et al.* 2015)	euarchontans
Plesiorycteropus	Afrotheria	stem tenrecid (Buckley 2013)	Tubulidentata (Patterson 1974)
Prolimnocyon	Creodonta (hyaenodontid)	sister Carnivora (Spaulding and Flynn 2012)	palaeanodonts and ?lipotyphlans (Halliday *et al.* 2017)

Tab. 7.3 (continued)

Taxon	High-level clade	affinities	alternative views
Prorastomus	Afrotheria	stem sirenian (Cooper *et al.* 2014)	
Proscalops	Proscalopidae	lipotyphlan (McKenna and Bell 1997)	
Proterohippus	Perissodactyla	stem equoid (Froehlich 2002)	
Prothoatherium	Litopterna	stem Perissodactyla (Welker *et al.* 2015)	stem Euungulata (Carrillo and Asher 2017)
Protictis	Carnivoramorph	stem carnivoran (Spaulding and Flynn 2012)	
Protolipterna	Litopterna	Euungulata (O'Leary *et al.* 2013; Carrillo and Asher 2017)	
Protungulatum	Eutheria	non-placental Eutheria (Chester *et al.* 2015)	stem euungulate (O'Leary *et al.* 2013; Muizon *et al.* 2015)
Pucadelphys	Metatheria	stem marsupial (Ladevèze *et al.* 2011)	
Purgatorius	Eutheria	Euarchontoglires (Chester *et al.* 2015)	non-placental Eutheria (Wible *et al.* 2009)
Pyrocyon	Creodonta (hyaenodontid)	sister Carnivora (Spaulding and Flynn 2012)	palaeanodonts and ?lipotyphlans (Halliday *et al.* 2017)
Pyrotherium	Meridiungulata	Euungulate (Muizon *et al.* 2015; this study Fig. 7.9)	
Radinskya	Mammalia	stem perissodactyl (Cooper *et al.* 2014)	
Ravenictis	Carnivoramorpha	stem carnivoran (Spaulding and Flynn 2012)	
Rhombomylus	Eurymylidae	stem Lagomorpha (Asher *et al.* 2005; O'Leary *et al.* 2013)	stem Rodentia (Meng *et al.* 2003)
Rodhocetus	Artiodactyla	stem cetacean (Geisler and Theodor 2009; Gingerich *et al.* 2001)	
Seggeurius	Afrotheria	stem hyracoid (Seiffert 2007; Cooper *et al.* 2014)	
Siamotherium	Anthracotheriidae	stem hippomotamids (Orliac *et al.* 2010; Lihoreau *et al.* 2015)	
Sifrippus	Perissodactyla	stem equoid (Froehlich 2002)	
Sinopa	Creodonta (hyaenodontid)	sister Carnivora (O'Leary *et al.* 2013)	
Teilhardimys	"Condylarthra" (louisine)	stem perissodactyl (Cooper *et al.* 2014)	
Tetraclaenodon	"Condylarthra" (phenacodontid)	stem Perissodactyla (Halliday *et al.* 2017: fig. 4)	
Thalassocnus	Xenarthra (nothrotheriid)	crown Folivora (Muizon and McDonald 1995)	
Thinocyon	Creodonta (hyaenodontid)	sister Carnivora (Spaulding and Flynn 2012)	
Thoatherium	Litopterna	stem Perissodactyla (Welker *et al.* 2015)	stem Euungulata (Carrillo and Asher 2017)
Thomashuxleya	Notoungulata	Euungulata (Carrillo and Asher 2017: figs. 10, 12)	Afrotheria (O'Leary *et al.* 2013; Carrillo and Asher 2017: fig. 9)
Todralestes	?Pantolestida	?Afrotheria (Seiffert *et al.* 2007)	?Ferae (Halliday *et al.* 2017)
Toxodon	Notungulata	stem Perissodactyla (Welker *et al.* 2015)	stem Euungulata (Carrillo and Asher 2017)
Tytthaena	Creodonta (oxyaenid)	sister Carnivora (Spaulding and Flynn 2012)	palaeanodonts and ?lipotyphlans (Halliday *et al.* 2017)
Ukhaatherium	Asioryctitheria	non-placental Eutheria (Wible *et al.* 2009)	
Zalambdalestes	Zalambdalestidae	non-placental Eutheria (Asher *et al.* 2005)	stem glires (Archibald *et al.* 2001)
Zhelestes	Zhelestidae	non-placental Eutheria (Wible *et al.* 2009)	stem "ungulate" (Archibald *et al.* 2001)

Tab. 7.4: Summary of Paleocene and Eocene Asian (ALMA), North American (NALMA), and South American (SALMA) Land Mammal Ages (after McKenna and Bell 1997 and Woodburne *et al.* 2014) and their approximate correlations to the marine and absolute chronologies (after Cohen *et al.* 2013). Rows do not represent precise correlations among terrestrial stages or between marine and terrestrial stages. Woodburne *et al.* (2014) replaced the Casamayoran with the Vacan (older) and Barrancan (younger) SALMAs.

Epoch	Marine stage and age in Ma	ALMA	NALMA	SALMA
Eocene	Priabonian (37.8–33.9)	Ulangochuian	Chadronian	
	Bartonian (41.2–37.8)	Sharamurunian	Duchesnean	Mustersan
	Lutetian (47.8–41.2)	Irdinmanhan	Uintan	Casamayoran
		Arshantan	Bridgerian	Riochican
	Ypresian (56.0–47.8)	Bumbanian	Wasatchian	Itaborian
Paleocene	Thanetian (59.2–56.0)	Gashatan	Clarkforkian	
	Selandian (61.2–59.2)	Nongshanian	Tiffanian	Peligran
			Torrejonian	
	Danian (66.0–61.2)	Shanghuan	Puercan	Tiupampan

7.4.1 Stem Marsupialia (Fig. 7.5)

G.G. Simpson described opossums (Didelphidae) as a lineage of mammals with a "low rate" of evolution and an independent history from other groups since the Cretaceous. This was based on his characterization of Cretaceous remains of, for example, *Pediomys* and *Alphadon*, and early Paleogene remains of *Herpetotherium* and *Peradectes*, as "little changed" relative to living didelphids (Simpson 1944: p. 39) and was reflected in his classification of these fossil taxa within Didelphidae (Simpson 1945: pp. 41–42). McKenna and Bell (1997) followed this arrangement in part, except for their placement of *Pediomys* unresolved within Marsupialia. In recent decades, several important discoveries have documented the anatomy of herpetotheriids and peradectids from the North American Paleogene (Sánchez-Villagra *et al.* 2007a; Horovitz *et al.* 2009), supporting the interpretation that most or all Paleogene "didelphoids" are in fact metatherians outside of Marsupialia (with the qualification that Horovitz *et al.* 2009 placed peradectids on the stem to didelphids). In addition, an extraordinary assemblage of Paleocene mammals from Bolivia has further revealed the anatomy of metatherians at that time, including multiple skeletons of individuals representing distinct ontogenetic stages of *Pucadelphys andinus* (Marshall *et al.* 1995; Ladevèze *et al.* 2011), as well as skeletal remains of *Mayulestes ferox* (de Muizon 1998). Again, recent phylogenetic analyses generally place these taxa, along with the Mongolian late Cretaceous taxon *Asiatherium* (Szalay and Trofimov 1996), on the stem leading to Marsupialia (Asher *et al.* 2004; Sánchez-Villagra *et al.* 2007a; Beck 2012; Suárez *et al.* 2016), with the exception of Maga and Beck (2017) who under some analytical criteria (see their figs. 38 vs. 39) reconstructed *Herpetotherium* as sister to paucituberculates. Wilson *et al.* (2016) described a Cretaceous skull of *Didelphodon* and placed it among other North American metatherians, outside of Marsupialia. However, unlike other recent studies, their analysis (Wilson *et al.* 2016: fig. 3) placed *Dromiciops* outside of a herpetotheriid-peradectid-didelphid-dasyurid clade, in part because they did not sample sequence data for extant species. Consistent with a stem position of these fossil metatherians are data on cochlear coiling (Sánchez-Villagra *et al.* 2007a; Horovitz *et al.* 2009), which indicate that peradectids (e.g., *Mimoperadectes* with 2.1 turns) and herpetotheriids (e.g., *Herpetotherium* with 1.6) exhibit fewer cochlear turns than crown marsupials (e.g., didelphids with at least 2.4).

7.4.2 Non-placental Eutheria (Fig. 7.5)

Many fossil therians are known from the Cretaceous that are more closely related to extant placentals than to marsupials, such as asioryctitheres (*Ukhaatherium* and *Asioryctes*), zalambdalestids (*Kulbeckia*, *Barunlestes*, and *Zalambdalestes*), and zhelestids (*Zhelestes*; see Martin, this volume, Chapter 6). Here, I outline three other, potential non-placental eutherians *Leptictis*, *Protungulatum*, and adapisoriculids (the latter in the section on Euarchontoglires).

Novacek (1986, p. 3) characterized *Leptictis* as an "ideal example of a primitive placental mammal" and wrote at a time when leptictids (including the synonymized genus *Ictops* as well as *Prodiacodon*, *Palaeictops*, and *Gypsonictops*) were regarded as potential relatives of "erinaceomorphs". Among the best leptictid fossils are skulls and skeletons from the White River Oligocene (Butler 1956; Rose 2006b). Slightly older

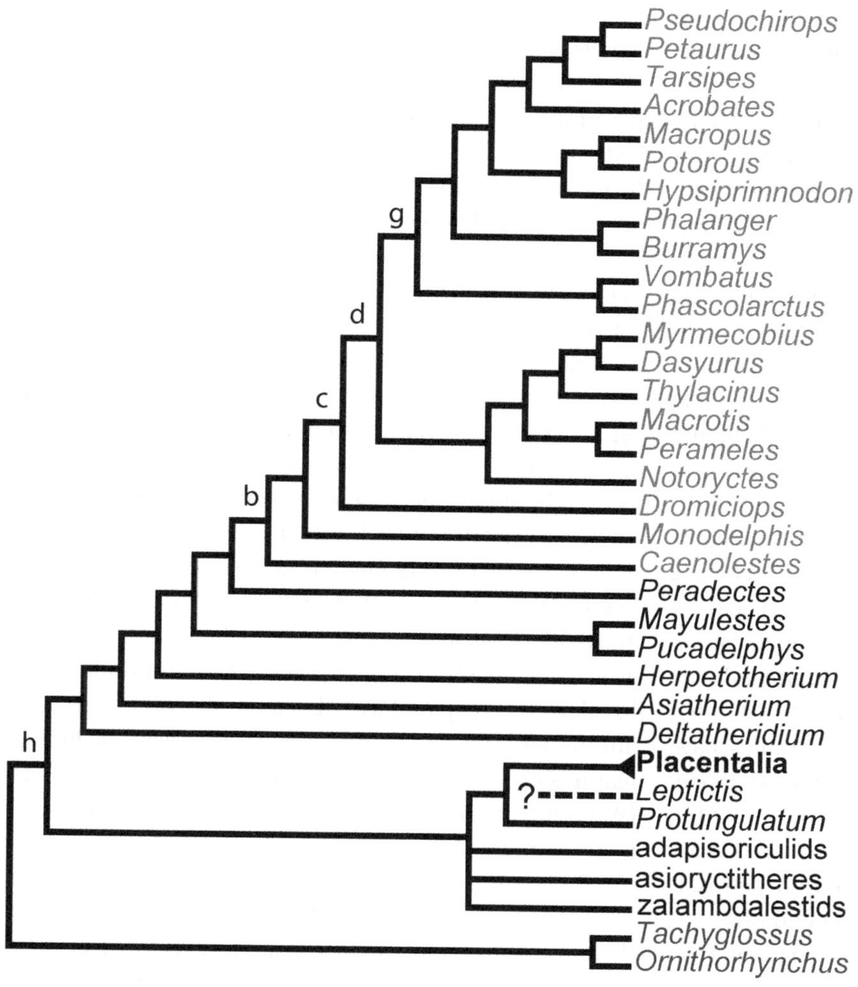

Fig. 7.5: Approximate phylogenetic tree for living and fossil metatherians and non-placental eutherians based on Fig. 7.1 for the extant taxa and with fossils intuitively placed according to Asher *et al.* (2005), Sánchez *et al.* (2007a), Wible *et al.* (2009), Goswami *et al.* (2011), Beck (2012), O'Leary *et al.* (2013), Chester *et al.* (2015), de Muizon *et al.* (2015), and others, as discussed in the text and Tab. 7.3. Colors are shown as in Fig. 7.1; fossils are black.

specimens from the Late Eocene of Wyoming include an in situ, articulated skull and skeleton, once featured on the cover of *Nature* (Novacek 1992) and intended to comprise the (never formally named) type specimen of *Frictops emryi*. This specimen was stolen from the Smithsonian's Hall of Fossil Mammals on or before 16 November 2006 (http://paleobiology.si.edu/stolen_specimens/). Early and middle Eocene species from Europe, such as *Leptictidium* (known from flattened but nearly complete skeletons from Messel, Germany), have also been associated with leptictids (Sigé 1974; Rose 2006b; Hooker 2013). Other genera such *Prodiacodon* and *Palaeictops* are not as anatomically well known, but gnathic remains date to the Early Paleocene (Novacek 1977). *Gypsonictops* is known from the upper Cretaceous and has also been regarded as a leptictid (Clemens 1973).

Phylogenetic affinities of leptictids have varied in studies published over the last two decades. In their analysis of Glires (without sampling afrotherian or laurasiatherian "insectivores"), Meng *et al.* (2003: figs. 75, 76) variably placed *Leptictis* either close to euarchontans or outside of Placentalia. In Asher *et al.* (2003: fig. 5 B) and Asher (2007: fig. 1), it was unresolved within Placentalia. O'Leary *et al.* (2013) placed it within Afrotheria. Most other, recent studies placed leptictids outside of Placentalia (Wible *et al.* 2009: fig. 29; Goswami *et al.* 2011: fig. 1; Beck and Lee 2014: fig. 1; Manz *et al.* 2015: fig. 2; Halliday *et al.* 2017: fig. 4). Following Rose (2006b), *Leptictis* shows a vertebral formula of 7 cervicals, 11 thoracics, 6 lumbars, 4 sacrals, and at least 7 caudals. With just 17 thoracolumbar vertebrae, *Leptictis* is unusual among mammals. Among extant species, only a few species of pholidotes, hominoid primates, and xenarthrans possess fewer than 18 (Sánchez-Villagra *et al.* 2007b; Asher *et al.* 2011). Non-mammalian synapsids also exhibit a relatively stable count of dorsal (i.e., thoracolumbar) vertebrae of 20 (Müller *et al.* 2010).

Protungulatum (e.g., *Protungulatum donnae*) is known from gnathic and petrosal remains (Sloan and van Valen 1965; Orliac and O'Leary 2016), with some postcranial fossils attributed to it based on relative abundance (Szalay *et al.* 1974; Chester *et al.* 2015). *Protungulatum* has been described as the oldest representative of an "ungulate" lineage of placental mammals near or before the K-Pg boundary (van Valen 1969) and was named based on isolated teeth and a left dentary (with p2-m3) from the Bug Creek Anthills of the Hell Creek Formation, eastern Montana (Sloan and van Valen 1965: fig. 6). Material from Bug Creek was originally thought to be upper Cretaceous but has since been regarded as the result of a combination of reworked Cretaceous and lower Paleocene deposition (Fastovsky and Dott 1986; Lofgren 1995). Whatever the age of rocks yielding *Protungulatum*, reference to dental remains from this or related taxa as "ungulates" or "hoofed mammals" continue to be sporadically made (e.g., Prasad *et al.* 2007), despite the fact that ambiguity surrounding such folk-taxonomic terms leave these claims without much meaning.

A more precise phylogenetic definition of "ungulate" is a member of Euungulata, i.e., artiodactyls (including cetaceans) and perissodactyls (Fig. 7.1). Indeed, O'Leary *et al.* (2013) included *Protungulatum* in a phylogenetic analysis and depicted an optimal tree showing *Protungulatum* as the sister taxon to exactly this group. By contrast, other quantitative, character-based phylogenetic studies have supported *Protungulatum* outside of Placentalia (Wible *et al.* 2007, 2009; Goswami *et al.* 2011; Beck and Lee 2014; Chester *et al.* 2015; Halliday *et al.* 2017). Orliac and O'Leary (2016) examined inner ear anatomy based on petrosals from Bug Creek assigned to *Protungulatum* and observed more features in common with Mesozoic mammals than with euungulates. For example, they observed 1.5 cochlear turns in *Protungulatum*, fewer than that observed in most extant placental lineages (Ekdale 2013), a result reminiscent of the relatively less-coiled cochleae in metatherians compared with marsupials (Sànchez-Villagra *et al.* 2007a), as mentioned above. Orliac and O'Leary (2016) also pointed out features of *Protungulatum* held in common with endemic South American notoungulates and litopterns, groups which are in turn possible stem perissodactyls (Welker *et al.* 2015). With this qualification, the current plurality of evidence supports *Protungulatum* as a eutherian outside of Placentalia.

7.4.3 Afrotheria (Fig. 7.6)

7.4.3.1 Paenungulates

As previously described, the major groups of endemic African mammals are tenrecids, chrysochlorids, macroscelidids, tubulidentates, and three paenungulate clades: hyracoids, sirenians, and proboscideans. *Ocepeia* (Gheerbrant *et al.* 2014) and *Abdounodus* (Gheerbrant *et al.* 2016) from the Paleocene of Morocco may illuminate details of the anatomy of the paenungulate common ancestor. Although neither is known from associated cranial and postcranial remains, *Ocepeia* is known from a reasonably well-preserved skull, including a cochlea "with at least two turns" (Gheerbrant *et al.* 2014). Moreover, Gheerbrant *et al.* (2016) argued that the quadrituberculate upper molars of fossil paenungulates exhibit a key difference from those of euungulates such as perissodactyls. In both groups, the quadrituberculate shape of the upper molars is due to the presence of a distolingual cusp, typically called the hypocone. Occlusal anatomy enables the distinction of two types: a true hypocone derived from the lingual cingulum (present in fossil perissodactyls) and a pseudohypocone derived from the metaconule (present in fossil paenungulates; see Gheerbrant *et al.* 2016: fig. 9). This is not the first anatomical feature claimed unite most or all afrotherians (Sánchez-Villagra *et al.* 2007b; Tabuce *et al.* 2007; Asher and Seiffert 2010) but is a particularly important character as it is more likely to be preserved and recognizable in the fossil record compared with the handful of other potential afrotherian morphological synapomorphies, including features of the placenta (Mess and Carter 2006), gonads (Werdelin and Nilsonne 1999), and development and vertebral formula (Asher *et al.* 2009, 2011).

Proboscidea has the best fossil record among extant afrotherians, followed by hyracoids and sirenians. All three are known from extinct genera that document the piecemeal evolution of living elephants, hyraxes, and sea cows from very different antecedents since the Paleocene. Proboscideans were already diverse in terms of size by the Early Eocene, with forms such as *Phosphatherium* and *Daouitherium* from the Early Eocene of Morocco spanning sizes from (respectively) fox to tapir (Gheerbrant *et al.* 2002). *Eritherium* is also Moroccan but predates both at ca. 60 Ma in age (Gheerbrant 2009). The proboscidean record documents in some detail their mosaic evolution, gradually acquiring the suite of morphological characters for which they are known today, such as an anteriorly situated orbit and vertically oriented coronoid process, loss of posterior incisors and/or canines and consequent appearance of a diastema, a squamosal-bound external auditory meatus, subunguligrade stance, horizontal tooth replacement, loss of permanent premolars, high cranial vault, tetralophodont teeth, and anteroposteriorly compressed, lamellar enamel on their cheek teeth (see Gheerbrant and Tassy 2009; Sanders *et al.* 2010; Asher 2012b).

The three genera and four species of extant hyracoids are depauperate compared with the diversity exhibited by their fossil record, which includes bovid-like (Rasmussen and Simons 2000) and rhinoceros-like (Schwartz *et al.*

1995) ecomorphs. During the Middle to Late Paleogene, this group encompassed at least 14 genera from numerous localities in Africa, western Asia, and southern Europe (Fischer 1992; Seiffert 2007; Rasmussen and Gutierrez 2010). The phylogenetic analysis of Seiffert (2007) supports placement of the geologically oldest taxa (*Seggeurius* and *Microhyrax*) from the North African early to early-middle Eocene at the base of the hyracoid radiation. These and other fossil hyracoids exhibited some key differences compared with modern forms, such as complete dental formulae, including replaced, posterior incisors and canines, and relatively brachydont cheek teeth, contrasting with the reduced dental formula (lacking I3-C and i2-c) and high-crowned cheek teeth of modern hyracoids (Gheerbrant *et al.* 2007; Asher *et al.* 2017). Such aspects of dental morphology comprise subtle but compelling data points that reflect the morphologically intermediate forms connecting extant hyracoids with the common ancestor shared with other afrotherians.

Sirenians today comprise just two genera, one inhabiting the tropical margins of the Atlantic (*Trichechus*) and the other the Indian Ocean (*Dugong*). A third genus (*Hydrodamalis*) was known from the northern Pacific until the 18th century. Their fossil record is not as extensive as that of proboscideans, but (again) it does validate one of the key postulates of evolutionary theory: the existence of "transitional forms" (Darwin 1859, pp. 281–282) between living species and the common ancestor they share. For example, the middle Eocene of Jamaica was home to *Pezosiren portelli*, a basal sirenian with clear adaptations for aquatic herbivory but which possessed a full pelvic girdle and weight-bearing hindlimbs (Domning 2001). "Prorastomids" (including *Pezosiren* and another Caribbean form, *Prorastomus*) are hypothesized to consist of paraphyletic relatives of Sirenia (Gheerbrant *et al.* 2005). Although the most complete, oldest sirenian fossils are from the Caribbean, there are cranial remains from the middle Eocene of Tunisia, including a petrosal of roughly similar age (near the Ypresian-Lutetian boundary) as the Jamaican *Prorastomus*. Benoit *et al.* (2013) describe a number of features shared by this specimen with other sirenians but argue that it is more primitive than petrosals from other fossil sirenians, consistent with an African origin of Sirenia (Benoit *et al.* 2013). Also of relevance to demonstrate the antiquity of Sirenia in the southern hemisphere is *Eotheroides lambondrano*, a dugongid from the middle Eocene of Ampazony, northwestern Madagascar, known from a reasonably well-preserved skull (Samonds *et al.* 2009).

Even with only extant species, the paenungulate trichotomy has proven difficult to resolve. Recent analyses of concatenated nuclear DNA usually favor a hyracoid-proboscidean clade (Meredith *et al.* 2011: fig. 1; sirenians still lack a genome of sufficient coverage and were not sampled by Tarver *et al.* 2016). A proboscidean-sirenian clade (i.e., Tethytheria) appears in some of the optimal coalescent topologies figured by Meredith *et al.* (2011: fig. S8) and in analyses of concatenated DNA and morphology (Asher 2007; O'Leary *et al.* 2013). Living sirenians share with proboscideans an unusual feature of the middle-inner ear: persistence of the perilymphatic duct rather than a foramen rotundum (Fischer 1990). This had been thought to be consistent with Tethytheria, along with the hypothesized aquatic habits of at least some Paleogene proboscideans (Liu *et al.* 2008). However, fossil sirenians (*Prorastomus*) and proboscideans (*Phosphatherium*) appear to differ from their living relatives in retaining a foramen rotundum (Court 1992a; Gheerbrant *et al.* 2005), thereby implying independent evolution of this character, although not necessarily independent evolution of the developmental program leading to its appearance in adults (see Saether 1979).

Desmostylia is another diverse aquatic group known from Oligocene-Miocene deposits around the northern Pacific Rim often hypothesized to be related to one or more paenungulates. Well-preserved skeletons are known (e.g., *Paleoparadoxia*, *Desmostylus*, *Cornwallia*, and *Behemotops*), and hypotheses on their phylogenetic affinities include Afrotheria (Asher *et al.* 2003; Gheerbrant *et al.* 2005), corresponding to the widespread, historical view that desmostylians are closely related to sirenians and proboscideans (Reinhart 1953; McKenna and Bell 1997). Another possibility is a close relationship with anthracobunids, an extinct group from the Paleogene of south Asia. Anthracobunids have in turn been hypothesized as stem relatives of either Euungulata (i.e., perissodactyls and artiodactyls [including cetaceans]; see Asher 2007: fig. 1) or Perissodactyla (Cooper *et al.* 2014; Rose *et al.* 2014). An affiliation of desmostylians and anthracobunids with one or more euungulate groups is biogeographically appealing (Beatty and Cockburn 2015; Santos *et al.* 2016), as desmostylians are known solely from the Pacific Rim, corresponding to the extensive Asian record of euungulates (Rose *et al.* 2014). This contrasts with the western Tethyean record of the earliest paenungulates (Gheerbrant *et al.* 2014, 2016).

Embrithopoda consists of large, graviportal mammals from the Paleogene of north Africa, southeastern Europe and Asia (Erdal *et al.* 2016). Taxa such as *Arsinoitherium* are known from reasonably complete skeletal remains and have long been regarded to be related to one or more paenungulates, including hyracoids (Andrews 1906) and proboscideans (Court 1992b; McKenna and Bell 1997; Asher *et al.* 2003; Asher 2007). Phylogenetic analyses of Seiffert (2007) and Cooper *et al.* (2014) suggested a stem sirenian placement for embrithopods. Such a placement fits with Court (1992b: pp. 23–24) who noted that, like extant proboscideans and sirenians (and given the qualification of the distribution of this character in fossils of these groups noted above), *Arsinoitherium* likely exhibited a persistent perilymphatic foramen and consequent loss of the fenestra rotundum. Unlike desmostylians, recent studies of the anatomy and systematics of embrithopods have not yet suggested substantial revisions to their long-supposed phylogenetic affinity to other endemic African taxa, a fact concordant with their largely Tethyan distribution.

7.4.3.2 Non-paenungulates

Among the remaining extant afrotheres (macroscelidids, tubulidentates, tenrecids, and chrysochlorids), macroscelideans or sengis have the best fossil record, which dates to the middle Eocene of North Africa and includes genera such as *Chambius*, *Nementchatherium*, and *Metoldobotes* (Tabuce *et al.* 2001; Tabuce 2017; Seiffert 2007). These are primarily known from craniodental remains, but some postcranial elements are also known and demonstrate anatomical features such as the concave cotylar facet of the astragalus (Tabuce *et al.* 2007), possibly present in the afrotherian common ancestor. This feature is present in non-African taxa, notably the "condylarths" *Apheliscus* and *Haplomylus* from the Early Eocene of the North American western interior (Zack *et al.* 2005; Hooker and Russell 2012), perhaps indicative of holarctic incursions of one or more afrotherian lineages, or even an origin of afrotherians from a common ancestor shared with one or another laurasian group, such as apheliscines (Zack *et al.* 2005; Penkrot *et al.* 2008). Again, a possible alternative concerns the possible affinities of one or more louisinid "condylarths" with stem perissodactyls, not afrotherians such as macroscelidids (Cooper *et al.* 2014).

Plesiorycteropus madagascariensis and *Plesiorycteropus germainepetterae* comprise the "Malagasy aardvark", an anatomically well-documented taxon from the Holocene of Madagascar (MacPhee 1994), now also known by collagen protein sequences (Buckley 2013). *Plesiorycteropus* is perhaps the only fossil mammal that is amply represented by remains of most skeletal elements, but no teeth or jaws (MacPhee 1994: table 1). In fact, even the location of the craniomandibular joint on the *Plesiorycteropus* skull has been a subject of debate; the first two monographs on the taxon (Lamberton 1946; Patterson 1975) disagreed on its position; the most recent and comprehensive monograph places the glenoid fossa for the dentary "rostromedial to the root of the zygomatic arch of the squamosal" (MacPhee 1994, p. 66), in partial agreement with Lamberton (1946). Debate about a structure that is usually obvious in mammals is not surprising given that *Plesiorycteropus* was a myrmecophage; its skull completely lacks teeth and evidently had a very reduced jaw which did not leave much of a myological or synovial impression on its skull.

Although *Plesiorycteropus* is well known anatomically and was comprehensively described and illustrated by MacPhee (1994), with further details of its inner ear now also known (Benoit *et al.* 2015), it has remained difficult to place phylogenetically. MacPhee (1994) represented this uncertainty by erecting a new high-level taxon for it (Bibymalagasia) and, unlike the previous monographic study (Patterson 1975), argued that it was not an obviously close relative of the extant aardvark (Tubulidentata). Comparisons of its collagen amino acid sequences (Buckley 2013) have validated MacPhee's skepticism of previous hypotheses and pointed in a previously unsuspected phylogenetic direction: *Plesiorycteropus* apparently has collagen sequences most similar to those of Malagasy tenrecids. Given the diverse radiation of subfossil Malagasy primates, it should not be surprising to find yet another niche of the Malagasy fauna vacated as a result of extinction, so a tenrecid myrmecophage remains the most compelling, current hypothesis for the evolutionary affinities of *Plesiorycteropus*.

Extant radiations of insectivoran-grade afrotherians consist of Chrysochloridae (golden moles) and Tenrecidae (tenrecs). Unfortunately, neither group has a particularly good fossil record, although the material described by Pickford (2015a–c) promises to change this. The first of these three adjacent publications in a single volume of the *Communications of the Geological Survey of Namibia* reports an isolated lower molar interpreted to be a chrysochlorid from "Black Crow", a locality from southwest Namibia, arguably Lutetian (middle Eocene) in age. Pickford (2015b, c) makes the case that associated, well-preserved skeletal associations of both chrysochlorids (*Namachloris*; Pickford 2015b) and potamogaline tenrecids (*Namagale*, *Sperrgale*, and *Arenagale*; Pickford 2015c) are present at "Eocliff", a locality interpreted to sample the Bartonian (Late-Middle Eocene) of Namibia. The figures in two of the articles (Pickford 2015b, c) are sufficiently clear to demonstrate not only that these fossils qualify as fossil tenrecs and golden moles but also that they are well preserved and more complete than any previously known tenrec or golden mole fossil. However, further peer review of these data would require more justification for the age assignments as well as some kind of phylogenetic analysis. The faunal list from "Eocliff" (Pickford 2015b, p. 161) is suggestive but consists of just two birds, eight rodents, and several vaguely identified primates and afrotheres. Hopefully, further, quantitative assessments of the stratigraphy, paleomagnetism, radiometrics, as well as paleobotanical and faunal similarity to other African localities are in the works to better understand this important assemblage.

Reports of fossil tenrecs and golden moles have appeared sporadically in the literature for many decades (Broom 1941, 1948; Butler 1984; Mein and Pickford 2003; Asher and Seiffert 2010) and with a few exceptions are confined to largely Neogene exposures from Namibia, South Africa, and the margins of Lake Victoria in east Africa. Seiffert *et al.* (2007) made the case that both tenrecids and chrysochlorids are represented by gnathic remains from near the Eo-Oligocene boundary at the Fayum, Egypt. Asher and Avery (2010) reported much younger cranioskeletal remains of *Chrysochloris arenosa* and *Chrysochloris bronneri* from the Early Pliocene of Langebaanweg, South Africa. Based on relative abundance, both are known from skull and postcranial elements. Furthermore, *C. arenosa* shows anatomical evidence for an ecology quite different than that of the living Cape golden mole, *Chrysochloris asiatica*. The ratio of humeral length to distal width in *C. arenosa* is substantially larger than in *C. asiatica* and similar in proportion to the extant *Eremitalpa granti*, known for its "sand-swimming" habitat in which (unlike other extant chrysochlorids) it does not construct durable burrows (Fielden *et al.* 1990). Cranially, *C. arenosa* is very similar to the modern *C. asiatica; C. bronneri*, by contrast, shows a very robust anterior lower incisor and has a smaller length/distal width ratio of its humerus. This is still larger than *C. asiatica* but overlaps in proportion to that of the modern (and endangered) *Neamblysomus julianae*.

7.4.4 Xenarthra (Fig. 7.6)

7.4.4.1 Folivorans

All of the three major, extant groups of xenarthrans are represented in the fossil record of the Americas: folivorans (sloths and ground sloths), vermilinguans (anteaters), and cingulates (armadillos and glyptodonts).

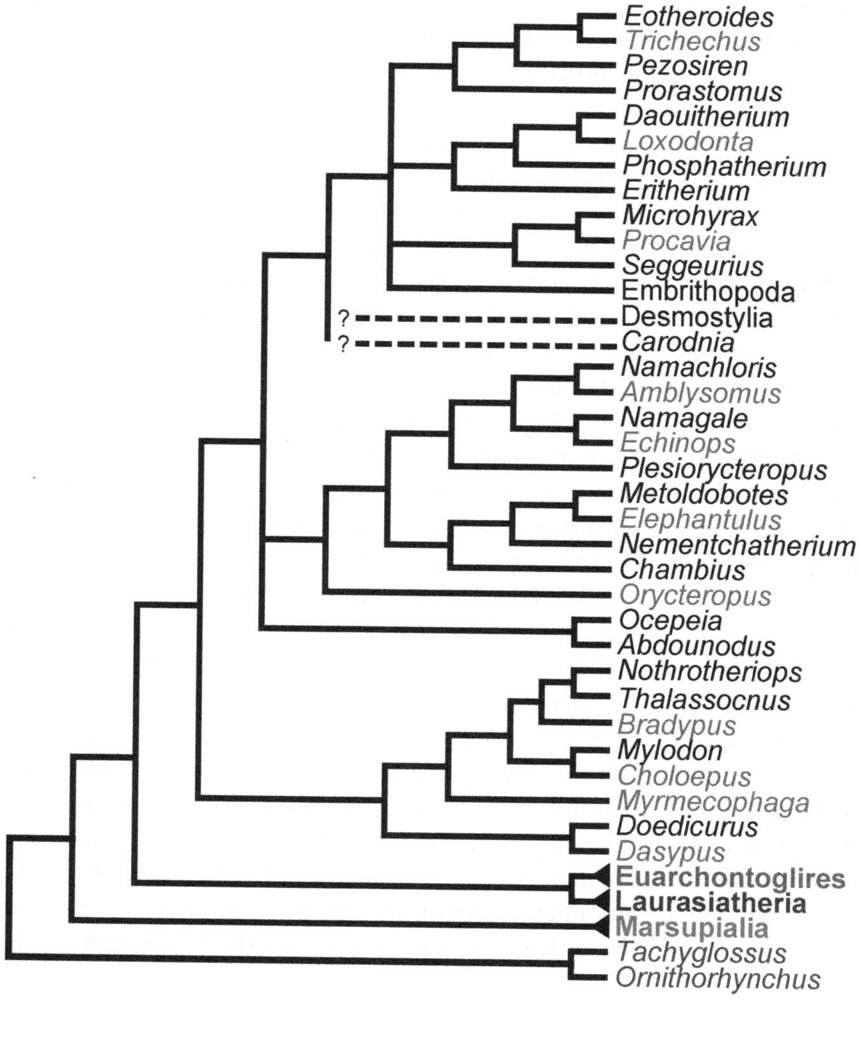

Fig. 7.6: Approximate phylogenetic tree for living and fossil Afrotheria and Xenarthra, based on Fig. 7.1 for the extant taxa and with fossils intuitively placed according to Asher (2007), Cooper *et al.* (2014), Delsuc *et al.* (2016), Gheerbrant *et al.* (2014, 2016), Slater *et al.* (2016), and others as discussed in the text and Tab. 7.3. Colors are shown as in Fig. 7.1; fossils are in black.

The extinct diversity of all three groups from the Eocene onward is considerable and demonstrates, for example, that the two extant genera of sloths, *Choloepus* and *Bradypus*, likely did not share a common ancestor with the same, arboreal, suspensory habitat. Rather, mylodontid ground sloths are more closely related to modern *Choloepus*, and nothrotheriid sloths are more closely related to *Bradypus* than *Choloepus* and *Bradypus* are related to each other (Patterson and Pascual 1968; Slater *et al.* 2016). Extinct diversity within Folivora does not just relate to their body size but also to their habitat; aquatic adaptations unknown in extant sloths are notable in Mio-cene-Pliocene remains of taxa such as *Thalassocnus* (e.g., de Muizon and McDonald 1995; White and MacPhee 2001; Amson *et al.* 2014). The oldest records of folivorans date to the early Oligocene of Patagonia and Chile (McKenna *et al.* 2006).

7.4.4.2 Cingulates

Glyptodonts are similarly part of the now extinct megafauna that, like ground sloths, are phylogenetically within the cingulate crown radiation. Delsuc *et al.* (2016) recovered a mitochondrial genome from fossils of the upper Pleistocene glyptodont *Doedicurus* and found that it is related to extant chlamyphorines and tolypeutines to the exclusion of dasypodines and euphractines, in partial congruence to recent analysis of living and fossil cingulate morphology (Billet *et al.* 2011), as well as the lack of diphyodonty in living and fossil xenarthrans except for dasypodines (Ciancio *et al.* 2012). As noted above, armadillos in the genus *Dasypus* have been reconstructed at or near the base of Cingulata in studies of both anatomy and molecular biology over the last decade (e.g., Delsuc *et al.* 2004, 2016; Gaudin and Wible 2006). Hence, neither the loss of functional, deciduous teeth in folivorans (Hautier *et al.* 2016) and most cingulates nor the loss of all teeth in vermilinguans would have characterized the xenarthran common ancestor, which likely resembled dasypodines in retaining functional diphyodonty (Ciancio *et al.* 2012).

7.4.4.3 Vermilinguans

Of the three extant groups, vermilinguans (anteaters) have the least abundant fossil record, which dates to the Early Miocene (Carlini *et al.* 1992). To date, and in contrast to cingulates and folivorans, fossil vermilinguans do not depart substantially from the ecology of their extant relatives. They too are edentulous, myrmecophage specialists, and within the size range of their extant relatives (Gaudin and Branham 1998), ranging from under 1 kg in the arboreal *Cyclopes* to over 40 kg in the terrestrial *Myrmecophaga*. Among the more noteworthy discoveries alleged to pertain to the evolution of vermilinguans is *Eurotamandua* from the middle Eocene (Lutetian) of Messel, Germany, first named by Storch (1981) and described by Storch and Habersetzer (1991) as a vermilinguan closely related to the extant *Myrmecophaga*. If true, this record would not only be the oldest but also would comprise the only occurrence of a xenarthran outside of the Americas and Antarctica. However, subsequent analyses (Gaudin and Branham 1998; Szalay and Schrenk 1998; Rose 1999; Rose *et al.* 2005) have made a strong case that *Eurotamandua* is not closely related to any xenarthran but is instead a close relative of extinct palaeanodonts, and in turn likely related to pholidotes (Laurasiatheria).

7.4.5 Afrotherian and xenarthran origins

The analysis of retroposons by Nishihara *et al.* (2009) supported near-simultaneous divergence of Placentalia into Afrotheria, Xenarthra, and Boreoeutheria (i.e., Euarchontoglires and Laurasiatheria). More recent studies by Tarver *et al.* (2016) and Esselstyn *et al.* (2017) divide Placentalia into Atlantogenata (Afrotheria and Xenarthra) and Boreoeutheria. All indicate that it is reasonable to look to southern continents, and the tectonic events that formed them, to gain insight into the afrotherian and xenarthran common ancestors. Several studies have argued that vicariance between Africa and South America prior to 100 Ma ago comprised the mechanism by which the earliest branching events within Placentalia took place (Murphy *et al.* 2001b, 2007; Nishihara *et al.* 2009). Such vicariance would necessitate a divergence date between Afrotheria and Xenarthra of at least 100 Ma, after which time the South Atlantic separated the African and South American landmasses (see section 7.5 below and Scotese 2001). However, divergence estimates over 100 Ma are based on smaller data sets than, for example, those of dos Reis *et al.* (2012), Tarver *et al.* (2016), and Wu *et al.* (2017) who estimated the common ancestor of Afrotheria and Xenarthra, and of Placentalia as a whole, to be closer to 90 million years in age, with the older 95% confidence limit for Atlantogenata estimated by Tarver *et al.* (2016: table 6) to be 96.5 Ma. These younger divergence estimates indicate that the South Atlantic was already open, separating Africa and South America, before southern-hemisphere Placentalia began to diversify in the early part of the Late Cretaceous, and therefore that the divergence of afrotherians and xenarthrans was not driven by continental vicariance. A relatively narrow South Atlantic between 83–96

Ma (Scotese 2001) thus likely played an important role as a dispersal filter between populations that subsequently gave rise to Xenarthrans and Afrotherians.

However old the divergence between Afrotheria and Xenarthra is, the case is strong that the two group share common ancestry (Tarver *et al.* 2016). Asher *et al.* (2009, p. 859) speculated that patterns of dental eruption shared among afrotherians (Asher and Lehmann 2008) might also be found among meridiungulates (i.e., notoungulates, litopterns, and xenungulates). Although relatively late eruption is present in the only extant clade of diphyodont xenarthrans (*Dasypus*; see Ciancio *et al.* 2012), persistence of the milk dentition well after attainment of adult body size has since been ruled out for Oligocene and Miocene notoungulates (Billet and Martin 2011) and Neogene litopterns (Lobo *et al.* 2017) for which growth series are known. Collagens from upper Pleistocene representatives of *Toxodon* (Notoungulata) and *Macrauchenia* (Litopterna) bear a greater resemblance to perissodactyls than afrotherians (Welker *et al.* 2015), as does the mitochondrial genome of *Macrauchenia* (Westbury *et al.* 2017). However, a combination of collagen sequences with nuclear DNA and morphology cannot rule out other possibilities such as sister to euungulates (Carrillo and Asher 2017). The enamel microstructure of the Late Paleocene xenungulate *Carodnia* also exhibits similarities to perissodactyls (Bergqvist and von Koenigswald 2017), but unlike litopterns and notoungulates, *Carodnia* does not belong to a high-level taxon sampled for any biomarkers. Cladistic analysis of its skeletal morphology usually places it among afrotherians, within or near paenungulates (O'Leary *et al.* 2013; Carrillo and Asher 2017), although the study of de Muizon *et al.* (2015) suggests a possible link between *Carodnia* and laurasiatheres via another extinct group, pantodonts (see section below on Laurasiatheria). Stem afrotherians remain elusive, although some taxa already known from the North African Paleocene may prove informative (Gheerbrant 2014, 2016).

Unambiguous stem relatives of Xenarthra also remain unknown. One recent result, without precedent, is the unconstrained Bayesian analysis of taxa sampled by Carrillo and Asher (2017: fig. 11), which placed the middle Eocene notoungulate *Thomashuxleya* from Canyadon Vaca, Chubut Province, Argentina (Vacan subage of the Casamayoran, likely to be ca. 45 Ma or Lutetian in age) on the xenarthran stem with a posterior probability of 0.62. When constrained as monophyletic with the other sampled notoungulate, *Toxodon* (known for collagen amino acid sequences from Welker *et al.* 2015), both appear with litopterns and *Didolodus* on the stem to Perissodactyla, as originally hypothesized by Welker *et al.* (2015) based on the collagen data alone. An unconstrained Bayesian result with a support value of 0.62 is not a solid basis to propose that one or more notoungulates are actually stem xenarthrans, but it does underscore the importance of seriously considering unorthodox possibilities for groups such as notoungulates with cranioskeletally well-known fossils dating to the early Paleogene.

7.4.6 Euarchontoglires (Fig. 7.7)

7.4.6.1 Euarchonta

Several phylogenetic studies of the last two decades support the inclusion of multiple extinct mammal groups within Euarchontoglires, such as carpolestids, plesiadapids, paromomyids, micromomyids, microsyopids, and *Purgatorius*, on the stem leading to primates (Bloch *et al.* 2007, 2016; Chester *et al.* 2015). Historically, most of these groups have at one time or another been collectively referred to as "Plesiadapiformes" and are regarded by Bloch *et al.* (2007, 2016) as paraphyletic sister taxa to Primates. Discoveries in the North American western interior, in particular the Late Paleocene of Wyoming (Bloch and Boyer 2002; Bloch *et al.* 2007), have shown that, for example, the plesiadapoid genus *Carpolestes* exhibits similarities to primates such as a petrosal auditory bulla enclosing the middle ear and a nail (rather than claw) on the first pedal digit (Bloch *et al.* 2007: fig. 5). *Purgatorius* is known from older, Early Paleocene (Puercan) exposures in eastern Montana. Based on size and relative abundance, Chester *et al.* (2015) associated tarsal bones with dentitions for this taxon. Like arboreal primates, but unlike the coeval *Protungulatum*, Chester *et al.* (2015) observed that *Purgatorius* exhibits a highly flexible ankle joint, with an astragalar trochlea (for articulation with the distal tibia) that grades onto its astragalar neck and an elongate ectal facet and large peroneal tubercle on its calcaneus. If the phylogenetic assessment of Chester *et al.* (2015; see also Beck and Lee 2014) is correct, *Purgatorius* would represent the geologically most ancient representative of Euarchontoglires, predating other contenders such as Asian mimotonids from the Paleocene Nongshanian ALMA (Asian Land Mammal Age; see Tab. 7.4) and *Paromomys* from the Torrejonian NALMA (North American Land Mammal Age; see Benton *et al.* 2015: pp. 69–70).

Another long-enigmatic group with proposed affinities to one or more "archontan" grade mammals are adapisoriculids, known from Paleogene deposits of Europe (Storch 2008; Smith *et al.* 2010), Africa (Gheerbrant 1995), as well as upper Cretaceous deposits of the Indian subcontinent (Boyer *et al.* 2010; Goswami *et al.* 2011). North

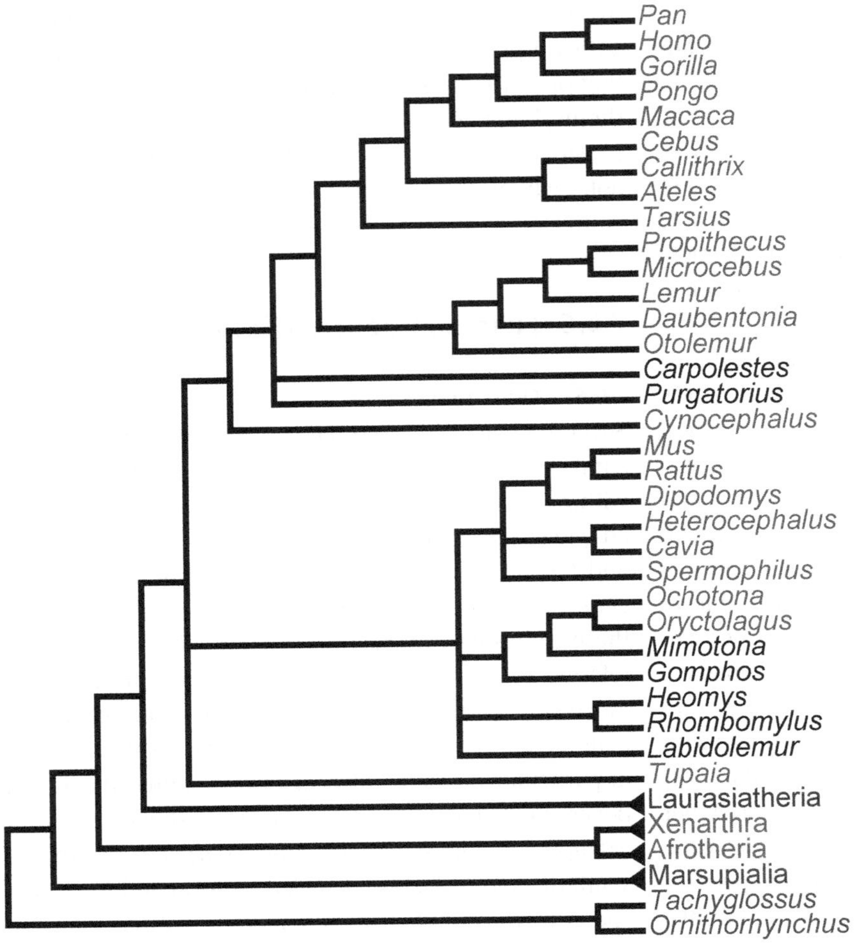

Fig. 7.7: Approximate phylogenetic tree for living and fossil Euarchontoglires, based on Fig. 7.1 for the extant taxa and with fossils intuitively placed according to Asher *et al.* (2005), Bloch *et al.* (2007), Silcox *et al.* (2010), Chester *et al.* (2015), and others as discussed in the text and Tab. 7.3. Colors are shown as in Fig. 7.1; fossils are in black.

American (Manz and Bloch 2015) and European (Hooker 2001) nyctitheriids have also on occasion been associated with adapisoriculids, and they have been taxonomically placed either with extant lipotyphlans (Rose *et al.* 2012; discussed further below) or chiropterans and euarchontans (Hooker 2001). Only a few species of these extinct groups are known from both dental and postcranial remains; these include the adapisoriculids *Deccanolestes* (Boyer *et al.* 2010; Goswami *et al.* 2011), *Bustylus*, *Afrodon* (Smith *et al.* 2010), and the nyctitheres *Plagioctenodon* (Manz *et al.* 2015) as well as remains tentatively assigned to *Cryptotopos* by Hooker (2001).

Of key importance is the upper Cretaceous, Indian material of *Deccanolestes*. Isolated forelimb and pedal remains were associated with dentitions by Boyer *et al.* (2010), who interpreted this genus to have a close common ancestry with euarchontans, i.e., scandentians, dermopterans, and primates; this reflects other analyses of adapisoriculid fossils by Storch (2008) and Smith *et al.* (2010). Among the distal humeral similarities of *Deccanolestes* with some euarchontans, Boyer *et al.* (2010: fig. 4) noted a deep zona conoidea, a raised lateral trochlear ridge, and a spherical capitulum. If these features actually represent euarchontan-*Deccanolestes* synapomorphies, this Cretaceous taxon would be the only placental mammal known from the Mesozoic. As noted above, an origin of Placentalia prior to the K-Pg boundary is supported by molecular clock studies (e.g., dos Reis *et al.* 2012; Tarver *et al.* 2016), as well as paleontological analyses over many decades (e.g., Simpson 1944; Benton *et al.* 2009, 2015; but see O'Leary *et al.* 2013). Nonetheless, to date, and accepting that they likely existed for at least some time prior to the K-Pg boundary since an accurate paleontological record is necessarily a minimum divergence estimate, previous claims for identifying actual Mesozoic representatives of placental lineages have so far been refuted (Asher *et al.* 2005; Wible *et al.* 2007, 2009; Beck and Lee 2014).

The phylogenetic study of Goswami *et al.* (2011) incorporated data from Boyer *et al.* (2010) into a comprehensive phylogeny also including an additional adapisoriculid, *Afrodon*, several Mesozoic eutherians, and living and fossil representatives of Xenarthra, Afrotheria, Euarchontoglires, and Laurasiatheria. Goswami *et al.* (2011: fig. 1) placed *Deccanolestes* and *Afrodon* in a clade outside of

Placentalia and reinterpreted the similarities of adapisoriculids with euarchontans as homoplasies resulting from a shared habitat. Some of the anatomical features of adapisoriculids that exclude them from Placentalia include lack of a metacone in the last premolar, presence of a preparacrista connecting stylocone to paracone, and a plantar pit on the calcaneo-cuboid facet (Goswami *et al.* 2011: table 2). As in Goswami *et al.* (2011), Beck and Lee (2014) and Manz *et al.* (2015) also recovered *Deccanolestes* and *Afrodon* outside of Placentalia.

7.4.6.2 Glires

Following Meng *et al.* (2003), taxa such as *Rhombomylus* are eurymylids best known from the Early Eocene of central Asia, include older taxa such as *Heomys* dating to the Paleocene, and are phylogenetically on the stem to Rodentia (Meng *et al.* 2003: fig. 74). Asher *et al.* (2005: fig. 3) and O'Leary *et al.* (2013: fig. 1) differed in reconstructing eurymylids on the stem to lagomorphs. All recent studies favor a position of eurymylids and mimotonids, which include anatomically very well documented genera such as (respectively) *Rhombomylus* and *Gomphos*, as more closely related to extant glires (i.e., rodents and lagomorphs) than any other extant group. *Rhombomylus* exhibits a single pair of enlarged, ever-growing central incisors in both upper and lower jaws, with enamel concentrated along the incisors' anterior surface. *Gomphos* also exhibits such gliriform incisors but has an additional pair of incisiforms in the premaxilla behind the anterior, central incisors, similar to but larger than the I3s of modern lagomorphs. *Gomphos* also exhibits an elongate pes, a relatively narrow calcaneus, and separate sustentacular and navicular facets on its astragalus. Both *Rhombomylus* and *Gomphos* show a diastema separating their incisors from the cheek teeth, as in all modern Glires.

Silcox *et al.* (2010) presented a novel interpretation of a long-enigmatic group of mammals, apatemyids, known from both skulls and postcranial from the Paleogene of North America and Europe (Jepsen 1934; von Koenigswald *et al.* 2005, 2009). Apatemyids are remarkable for their procumbent, lower central incisors, and elongate manual digits II and III, interpreted as adaptations for hunting wood-boring insects in an arboreal context (von Koenigswald *et al.* 2005), as in the living diprotodont marsupial *Dactylopsila* and strepsirhine primate *Daubentonia*. Using new data from well-preserved skulls of *Labidolemur kayi* from the Clarkforkian NALMA (Late Paleocene, Wyoming; Tab. 7.4), along with additional cranial and postcranial data from other taxa and the literature, Silcox *et al.* (2010) tested the affinities of apatemyids with a broad range of other mammals (34 taxa in total) sampled for 240 morphological characters. Their analysis (Silcox *et al.* 2010: fig. 17) placed all six apatemyid genera with the eurymylid *Rhombomylus*; the apatemyid-glires clade in turn formed the sister group of a euarchontan (i.e., tupaiid-dermopteran-plesiadapiform-primate) clade. Apatemyids share with *Rhombomylus* and other glires contact between the premaxilla and frontal and the maxilla and frontal; these similarities are not unique but nonetheless optimize as shared-derived characters for apatemyids-glires on the optimal topology of Silcox *et al.* (2010: fig. 17).

7.4.7 Laurasiatheria (Fig. 7.8)

7.4.7.1 Lipotyphla

This group consists of hedgehogs and moonrats (Erinaceidae), shrews (Soricidae), moles (Talpidae), and solenodons (Solenodontidae, including the recently extinct *Nesophontes*; see Brace *et al.* 2016). As originally proposed by Haeckel (1866), it also included endemic African tenrecids and chrysochlorids, now known to be part of Afrotheria, and was cognate to the "Menotyphla", containing scandentians and dermopterans. An idea sometimes credited to Huxley (1880) is that "insectivorans" represent an adaptive grade similar to that characteristic of the placental, and indeed mammalian, common ancestor: small-bodied, insectivorous, and possibly nocturnal (Kemp 2005). In fact, as reviewed by Wyss (1987), Huxley (1880) actually viewed mammalian origins in a way that would today be regarded as polyphyletic, with various extant groups (primates, rodents, proboscideans, etc.) each having an independent origin from non-mammalian ancestors. The idea of extinct "Insectivora" occupying a "central position", and containing the common ancestor from which placental mammals evolved, is more accurately attributed to Matthew (1909) and was echoed by subsequent authors such as Butler (1972) and in discussions on certain taxa such as *Leptictis dakotensis* (see section above on "non-placental eutherians" and Novacek 1986).

This historical tangent is relevant to Laurasiatheria because of one of the more consistent phylogenetic results over the last two decades is that Lipotyphla (sometimes called Eulipotyphla, including erinaceids, soricids, talpids, and solenodontids) comprises the sister group to all other laurasiatheres (Murphy *et al.* 2001a, b; Roca *et al.* 2004; Asher 2007; Meredith *et al.* 2011; Tarver *et al.* 2016). Living groups are obviously not ancestors of other living groups, but the (arguable) implication of the "central position" hypothesis applied to Lipotyphla is that

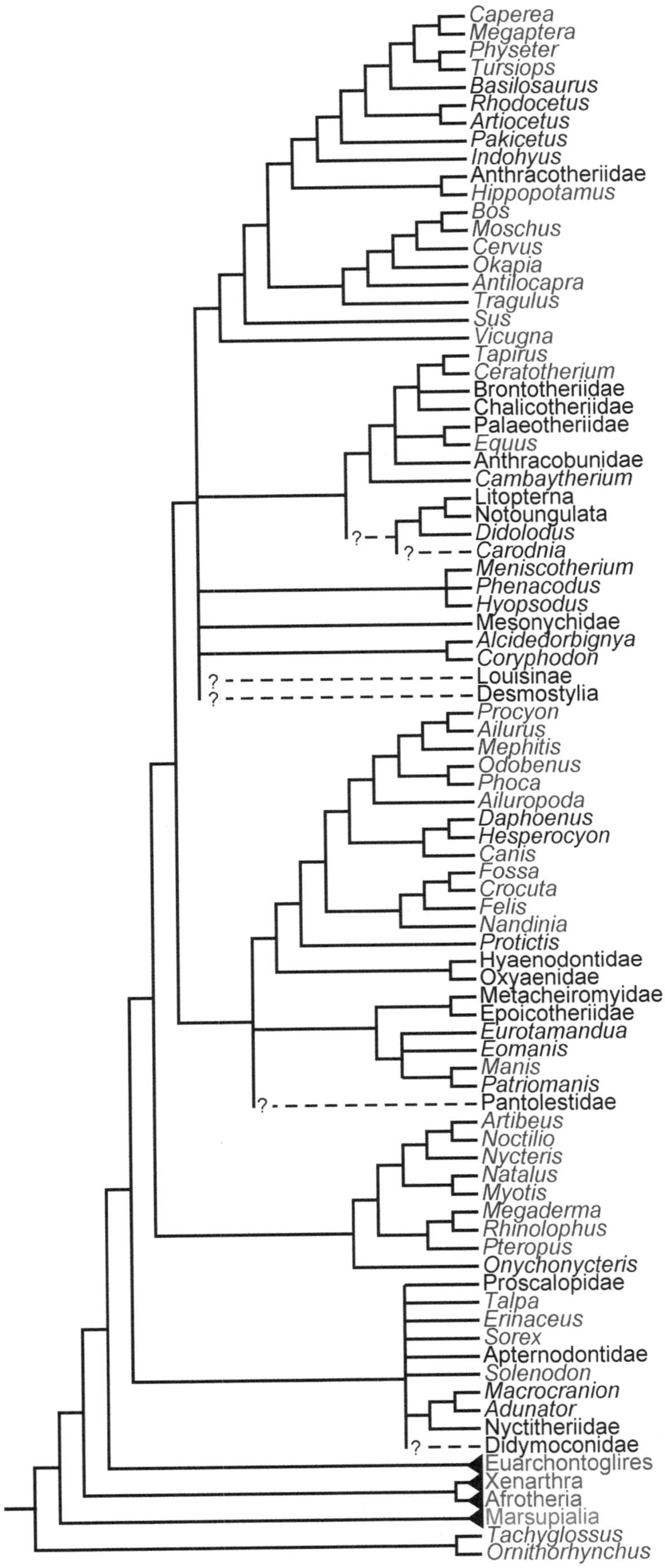

Fig. 7.8: Approximate phylogenetic tree for living and fossil Laurasiatheria, based on Fig. 7.1 for the extant taxa and with fossils intuitively placed according to Simmons *et al.* (2008), Geisler and Theodor (2009), Cooper *et al.* (2014), Rose *et al.* (2014), Lihoreau *et al.* (2015), Welker *et al.* (2015), Carrillo and Asher (2017), and others as discussed in the text and Tab. 7.3. Colors are shown as in Fig. 7.1; fossils are in black.

its evolutionary branch may have exhibited relatively less phenotypic change than branches leading to other laurasiatheres. Numerous fossil groups assigned to one or more lipotyphlans are known and may shed light on the nature of the laurasiatherian common ancestor.

As noted in the preceding section on Euarchontoglires, Hooker (2001) interpreted tarsal remains of nyctitheriids from the Eocene of southern England (tentatively assigned to *Cryptotopos*) as related to one or more archontan groups, i.e., primates, scandentians, and dermopterans. Manz *et al.* (2015) described associated cranial and postcranial remains from a relatively more complete Paleocene nyctithere genus: *Plagioctenodon* from the upper Paleocene (Clarkforkian NALMA; Tab. 7.4) of Wyoming. Rather than affinities to primates, optimal phylogenetic trees from Manz *et al.* (2015: fig. 2) recover *Plagioctenodon* and another nyctitheriid, *Leptacodon*, in a clade with the Paleogene fossils *Adunator* and *Macrocranion*, in turn sister to crown Lipotyphla. Their results furthermore place the early Paleogene fossils *Eoryctes* (Wyoming) and *Todralestes* (Morocco) as successively distant sister taxa to the Caribbean *Solenodon*, within crown Lipotyphla. *Macrocranion* is known from well-preserved skeletal fossils from the middle Eocene of Messel (Germany) and has long been regarded as an erinaceomorph along with *Pholidocercus*, also from Messel and also known from relatively complete but flattened skeletons (von Koenigswald *et al.* 1992). Manz *et al.* (2015: fig. S13) identify seven dental, 10 cranial, and five postcranial features that optimize as shared-derived characters of Lipotyphla including nyctitheriids, such as an infraorbital foramen dorsal to the ultimate premolar, a delicate or incomplete zygomatic arch containing a jugal bone, and a deep groove for the fibular flexor tendon on the astragalus.

Lopatin's (2006) comprehensive and well-illustrated monograph on the diversity of "insectivoran" craniodental fossils from Asia summarizes decades of fieldwork from Paleocene and Eocene deposits in Mongolia, Kazakhstan, and Kyrgyzstan and reviews the literature on many other localities. He included detailed illustrations of many jaws and teeth, plus a relatively complete skull of *Archaeoryctes euryalis* from near the Paleocene/Eocene boundary at Tsagan Khushu, Mongolia (Lopatin 2001). Lopatin (2006) also described dental eruption sequences and postcrania of Eocene-Oligocene didymoconids (Lopatin 2006: s308, fig. 41). *Archaeoryctes euryalis* is a didymoconid, intuited by Lopatin to be a sister taxon of "Insectivora" (Lopatin 2006: fig. 60). Like Manz *et al.* (2015), Lopatin (2006) also regarded nyctitheriids as more closely related to lipotyphlans than primates but differed from Manz *et al.* (2015) in proposing that nyctitheriids were crown lipotyphlans, more closely related to soricids than to other extant groups.

Some of the phylogenetic implications of Lopatin (2006) are anachronistic, e.g., in classifying insectivoran-grade species from Afrotheria and Laurasiatheria together in polyphyletic, high-level taxa. Nonetheless, the promise of recovering additional cranial and postcranial remains (e.g., Lopatin 2006: plate 10) of the important taxa discussed in his monograph (e.g., anagalids, didymoconids, micropternodontids, paleoryctids, cimolestids) will help test if and how these groups are related to extant insectivoran-grade species in Laurasiatheria and Afrotheria. For example, apternodonts are a rare but geographically widespread group known from Eocene to Oligocene exposures in North America and Asia that possess "zalambdodont" cheek teeth, i.e., with crowns resembling the Greek letter "lambda" in occlusal view with the main lingual cusp composed of the paracone. Among placental mammals, Afro-Malagasy tenrecoids and the Caribbean *Solenodon* also exhibit such teeth (Asher and Sánchez-Villagra 2005). Reports of Paleocene North American apternodonts exist (e.g., Edinger 1964), but scrutiny of such specimens indicates that they are not closely related to the middle and upper Eocene North American genera, *Apternodus* and *Oligoryctes* (Asher *et al.* 2002). Lopatin (2006: fig. 58) figured a Paleocene record for apternodonts in his review, but the asiapternodontids he described are limited to middle Eocene localities from China and Mongolia (Lopatin 2006: table 6), and it is unclear which (if any) specimens comprise Paleocene apternodonts.

Nonetheless, both Lopatin (2006) and Asher *et al.* (2002) agreed with the view articulated by McDowell (1958) that dental zalambdodonty exhibited by living tenrecids and chrysochlorids is not homologous to that of *Solenodon*. Asher *et al.* (2002) and Lopatin (2006) also agreed that apternodonts and *Solenodon* are more closely related to extant soricids than to Afro-Malagasy tenrecoids. If true, one would expect future discoveries of apternodont cranioskeletal fossils to resemble laurasiatherian lipotyphlans, such as solenodontids and soricids, rather than afrotherians, such as tenrecids. Skulls of North American apternodonts are already known to exhibit shrewlike features, such as the pocketed coronoid process of the dentary in *Oligoryctes* and an enlarged piriform fenestra in the roof of the middle ear (Asher *et al.* 2002; also present in *Solenodon*). Once adequate material is known, one would expect (given closer phylogenetic affinities to lipotyphlans than tenrecoids) apternodonts to erupt most or all their permanent teeth before attainment of adult size, not after as in many afrotheres (Asher and

Lehmann 2008). Another expectation for well-preserved fossil Lipotyphla (but not Afrotheria) would be a lumbar vertebral count not exceeding six or a thoracic count under 16, as opposed to seven (lumbar) or over 17 (thoracic), as often found among afrotherians (Sánchez *et al.* 2007; Asher *et al.* 2011). Neither phenotype alone would be sufficient to demonstrate phylogenetic affinity to Lipotyphla; as noted previously, thoracolumbar counts of 19–20 are widespread throughout mammals. However, combined with features from (for example) the ear region, jaw, and dentition, they comprise a limited set of characters parsimoniously expected to have been present in extinct, lipotyphlan laurasiatheres.

Matthew (1909: plate LI and fig. 4) figured a well-preserved skull of *Proscalops* from the Bridger Basin, Wyoming (Early Eocene). Proscalopids are rare but are known from reasonably complete cranioskeletal remains, from the Eocene *Proscalops* to the Miocene *Mesoscalops* (Barnosky 1981), both from the North American western interior, as well as more fragmentary specimens from the Oligocene of central Asia (Geisler 2004). The general shape of the *Proscalops* and *Mesoscalops* skull greatly resembles those of fossorial chrysochlorids, palaeanodonts, notoryctids, and the Miocene dryolestoid *Necrolestes* (Rougier *et al.* 2012; O'Meara and Thompson 2014), all of which dig with parasagittal motions of their forelimbs, using the head as an accessory digging instrument (Rose and Emry 1983: fig. 16). This mode of fossoriality is distinct from the "lateral scratch" method of extant talpine and scalopine talpids (Yalden 1966). Hyperfossorial talpids have a more gracile skull, not directly utilized for digging, and furthermore depend on an enlarged teres major muscle to rotate the humerus around its long axis to drive rotation of the flexed forearm and hand, including five large digits plus an enlarged sesamoid in the manus (Mitgutsch *et al.* 2012). Proscalopids (particularly *Mesoscalops*) depart from the humeral morphology of talpines and scalopines, particularly in their hyper-developed humeral epicondyles and distally placed teres tubercle (Barnosky 1981: fig. 29; Piras *et al.* 2015). However, despite their chrysochlorid-like skull shape, the dilambdodont proscalopid dentition has generally been interpreted to reflect lack of any close relationship with other parasagittal diggers such as palaeanodonts and chrysochlorids. Instead, proscalopids show two "lambdas" or V shapes on each of its upper molars, with paracone and metacone forming the lingual apices of each V and an expanded stylar region extending buccally to define roughly half of the occlusal surface of each molar. Living and fossil soricids (Repenning 1967) and talpids (Schwermann and Thompson 2015) are also dilambdodont. Proscalopids have accordingly been regarded as lipotyphlans, possibly the sister group of talpids (McKenna and Bell 1997), although to date no broadly sampled, published phylogenetic study has yet fully tested this hypothesis.

7.4.7.2 Chiroptera

The removal of Chiroptera from the "Archonta" of Gregory (1910) and their placement among laurasiatheres were significant discoveries of the 1990s and early 2000s (Murphy *et al.* 2001b). Somewhat more controversial are their affinities within Laurasiatheria. The largest data sets place them one node crownward from Lipotyphla, as the sister group to the remaining, non-lipotyphlan laurasiatheres (Tarver *et al.* 2016: fig. S1; Esselstyn *et al.* 2017), a result also supported by the analysis of 36 kb nucleotide analysis from Meredith *et al.* (2011: figs. S1 and S3). Alternatively, optimal trees derived from smaller data sets place them in a more nested position as sister to Euungulata (e.g., ca. 11k amino acid residues figured by Meredith *et al.* 2011: figs. 1, S2, S4) or adjacent to a perissodactyl-carnivoran clade (rare genomic events from these taxa, but without samples from Pholidota, analyzed by Nishihara *et al.* 2006: fig. 2). Regardless of how nested they are within Laurasiatheria, bats are perhaps the most conspicuous placental group which still lacks clear fossil antecedents that depart from the derived locomotor repertoire of the extant clade. Fossil cetaceans (Gingerich *et al.* 2001; Thewissen *et al.* 2007) and sirenians (Domning 2001) are known to have weight-bearing hindlimbs; many fossil sloths are non-suspensory (Slater *et al.* 2016); many fossil equids are small-bodied with low-crowned teeth (Mihlbachler *et al.* 2011). However, a cover of *Science* or *Nature* has not yet appeared depicting a stem chiropteran that lacks powered flight.

Simmons *et al.* (2008) described *Onychonycteris*, a well-preserved, Green River Formation (Early Eocene of Wyoming) bat they reconstructed on the stem to Chiroptera. *Onychonycteris* has elongate manual digits, a calcar to support a hindlimb patagium, a slightly keeled manubrium sterni, a gracile skeleton, and a craniocaudally elongate scapula, all of which indicate a capacity for powered flight (Simmons *et al.* 2008: fig. 1). However, unlike other Eocene bats, its petrosal is unenlarged and similar in proportions to those of non-laryngeally echolocating pteropodids. Its intermembral (100 × (radius + humerus) / (femur + tibia)) and brachial (100 × (humerus / radius)) indices resemble those of arboreal but non-volant species (Simmons *et al.* 2008: fig. 2). More controversially, *Onychonycteris* exhibits a narrow stylohyoid element, without the characteristic paddle shape of its proximal end seen in extant, laryngeally echolocating bats. This was interpreted by Simmons *et al.* (2008) to indicate that *Onychonycteris* did not have the skeletal basis for laryngeal echolocation, which requires the stylohyal element to stabilize the tympanic apparatus via contact between the larynx and the ectotympanic bone. Possibly, however, it did exhibit stylohyal-ectotympanic contact, which may indicate at least some capacity for laryngeal echolocation (Veselka *et al.* 2010). Whatever the details of its sensory capacities were, *Onychonycteris* differs from all extant bats in its skeleton and skull, hinting that it, in turn, evolved from another extinct population of stem Chiroptera, one without powered flight. Whether or not such a population existed in sufficient numbers so as to leave behind a fossil record remains to be seen.

7.4.7.3 Carnivora

Crown members of Carnivora include all descendants of the common ancestor of Feliformia and Caniformia. In addition, there is a diverse array of extinct groups on its stem, such as its immediate sister taxon Nimravidae, the paraphyletic "Miacidae", and the basal Viverravidae, all of which comprise the total group Carnivoramorpha (Wesley-Hunt and Flynn 2005; Spaulding and Flynn 2012). This topology is also supported by Solé *et al.* (2014), but for unclear reasons, they labeled Carnivora one node stemward from the actual crown to include nimravids, a group not descended from the last common ancestor of extant carnivorans according to their phylogeny (although they are sometimes reconstructed as feliforms, as in Wesley-Hunt and Flynn 2005). Whether or not nimravids are sister to feliforms, the oldest crown carnivorans are the Late Eocene (Duchesnean NALMA; Tab. 7.4) caniforms *Hesperocyon* and *Daphoenus*, substantially younger than the Early Paleocene *Protictis* (Torrejonian NALMA), the oldest known carnivoramorph (Polly *et al.* 2006; Spaulding and Flynn 2012) with the possible exception of *Ravenictis* from the Puercan NALMA of Saskatchewan (Fox *et al.* 2010).

Creodonts are an extinct group of carnivorous, placental mammals with an extensive fossil record from North America, Africa, and Eurasia, ranging from the Late Paleocene (Zack 2012) through the Miocene (Barry 1980). They are historically regarded as an extinct relative of Carnivora (Matthew 1909) and usually composed of at least two groups, oxyaenids and hyaenodontids. Authors such as Cope (1875) and Gregory (1910) treated "Creodonta" as a wastebasket for a variety of other carnivore-grade fossils, including taxa such as mesonychids, arctocyonids, and "miacid" carnivorans. Over the past two decades, most authors (e.g., Polly *et al.* 2006; Rose 2006a; Solé *et al.* 2009, 2014) have restricted Creodonta to oxyaenids and hyaenodontids but expressed skepticism that (1) oxyaenids and hyaenodontids comprise a monophyletic clade and (2) either group is the sister taxon of Carnivora. A carnivoran-creodont clade was recovered by O'Leary *et al.* (2013), who sampled all modern orders and several extinct groups but just one creodont, *Sinopa*. Spaulding *et al.* (2009) sampled four creodont genera known from both craniodental and postcranial fossils (*Hyaenodon, Patriofelis, Sinopa,* and *Thinocyon*), recovered a carnivoran-creodont clade, but among extant mammals sampled only artiodactyls, perissodactyls, carnivorans, *Erinaceus*, and *Orycteropus*.

Halliday *et al.* (2017) sampled a broad array of living and fossil mammalian taxa, including two hyaenodontids (*Prolimnocyon*, missing 49% of the 681 characters in their supplementary-data matrix and *Pyrocyon* 54%) and two oxyaenids (*Dipsalidictis* missing 22.7% and *Tytthaena* 76.5%). All but *Tytthaena* are coded for craniodental and postcranial characters. Halliday *et al.* (2017: p. 538) wrote that their study "consistently supported" the "close relationship between Carnivora and Creodonta", but the two phylogenetic trees figured in their main text do not support this claim. The immediate sister taxa of Carnivora in optimal trees with (Halliday *et al.* 2017: fig. 4) and without (Halliday *et al.* 2017: fig. 3) an extant constraint are mesonychians, an extinct group often associated with Cetacea (see below and Rose 2006a). Carnivorans-mesonychians are in turn the sister group of an *Eoconodon-Goniacodon* clade, two "triisodontine condylarth" genera arguably also affiliated with mesonychians, known from gnathic remains from the lower Paleocene of New Mexico (Kondrashov and Lucas 2006). Their unconstrained topology (Halliday *et al.* 2017: fig. 3) placed creodonts closer to (among other taxa) North American fossil lipotyphlans (*Oreotalpa* and *Parapternodus*), *Palaeanodon* and *Eomanis*, than to any carnivoran. Similarly, their phylogeny using an extant-taxon constraint (Halliday *et al.* 2017: fig. 4) placed creodonts closer to pholidotes and palaeanodonts and a slightly different array of North American "condylarths" (e.g., *Chriacus*) and extinct "paleoryctids" (e.g., *Aaptoryctes*). Their "CF" topology (for "continuous" and "fully constrained") figured in their supplementary data files (Halliday *et al.* 2017: fig. S4) again shows several poorly known fossils, such as "cimolestids" (*Acmeodon*) and lipotyphlans (*Parapternodus*), comprising the creodont sister group, with carnivorans the next group out. Overall, and based on Spaulding *et al.* (2009) and O'Leary *et al.* (2013), there is tentative support for a carnivoran-creodont clade, but not from the more broadly sampled study of Halliday *et al.* (2017).

7.4.7.4 Pholidota

Living pangolins comprise the Pholidota and consist of eight species known from Africa and southern Asia. The pangolin fossil record is much more geographically widespread than the extant distribution, with Eocene fossils known from North America (*Patriomanis*; Emry 1970; Gaudin *et al.* 2016), central Asia (*Cryptomanis*; Gaudin *et al.* 2006), and Europe (*Eomanis*). *Eomanis* and *Eurotamandua* co-occurred at Messel, Germany, and are the oldest fossils typically regarded as pholidotes (Rose *et al.* 2005; see above concerning previous treatments of *Eurotamandua* as a xenarthran); *Eomanis* preserves clear evidence of the dermal armor characteristic of the extant species.

Shoshani (1986) noted that immunochemical data supported a close relationship between Pholidota and Carnivora, a finding born out by subsequent analyses of molecular data (e.g., Murphy *et al.* 2001b). Anatomical features such as an ossified tentorium cerebelli and occasional fusion of scaphoid and lunate carpal bones have led some morphologists to consider the two groups as possible close relatives (Rose and Emry 1993). McKenna and Bell (1997, p. 211) designated Ferae as the clade encompassing pholidotes and carnivorans, along with a number of extinct taxa such as creodonts, epoicotheres, metacheiromyids, and various "cimolestid"-grade fossils. Other studies (Amrine-Madsen *et al.* 2003; O'Leary *et al.* 2013: supplementary data p. 50) have favored the name Ostentoria for the living pholidote-carnivoran clade; the two terms are mutually compatible as stem (Ferae, encompassing all mammals closer to carnivorans and pholidotes than other extant clades) and crown (Ostentoria, encompassing all descendants of the last common ancestor of Carnivora and Pholidota) designations, although the former term appeared first in the literature.

Epoicotheres and metacheiromyids, collectively known as palaeanodonts and best known from the Late Paleocene and Eocene of North America and Eurasia, have frequently been regarded as closely related to Pholidota (Emry 1970; Rose 2006a). Many of these species show at least some characters associated with a fossoriality and/or myrmecophagy, including reduction or loss of teeth and powerful forelimbs for digging. These features are also seen among endemic South American myrmecophages, but other similarities are not. A close pholidote-palaeanodont evolutionary relationship has been inferred based on comparisons between the skeletons of Eocene metacheiromyids (*Palaeanodon*) and pholidotes (*Patriomanis* and *Eomanis*). O'Leary *et al.* (2013: fig. 1) recovered *Metacheiromys* as the sister taxon of the extant pholidote *Manis*, a clade which in turn comprised the sister taxon to a creodonts and Carnivora. As discussed by Emry (1970) and Rose *et al.* (2005), similarities between pholidotes and at least some palaeanodonts include (but are not limited to) a medially buttressed dentary, the loosely attached and C-shaped premaxilla, and an elevated and elongate scapular spine. Notably, although pholidotes do exhibit derived "enrolled" zygopophyses on their lumbar vertebrae (Rose *et al.* 2005), no palaeanodont or pholidote exhibits the multiple synovial articulations between adjacent zygopophyses (i.e., anatomical xenarthry) on the vertebrae evident in cingulates, vermilinguans, and pilosans. As noted above, the initial description of *Eurotamandua* as an anatomically xenarthrous mammal (Storch 1981) has since been disproven (Szalay and Schrenk 1998; Rose *et al.* 2005).

Among the extinct taxa classified as part of Ferae by McKenna and Bell (1997), pantolestans are among the most anatomically well known but nonetheless enigmatic, with near-complete skulls and skeletons from Eocene sites in Germany (*Buxolestes* and *Kopidodon*) and Wyoming (*Pantolestes* and *Paleosinopa*). More fragmentary cranial remains are known from the Early Paleocene in North America (Rose 2006a) and Africa, such as *Todralestes* and ptolemaiids (Gheerbrant 1991; Rose 2006a). Based on a rostrum of *Kelba*, an Early Miocene ptolemaiid from Kenya, Cote *et al.* (2007) suggested that ptolemaiids had a closer relationship to endemic African mammals than other groups such as pantolestids. Similarly, Seiffert *et al.* (2007) suggested that the North African Paleocene *Todralestes*, also historically associated with pantolestids (Gheerbrant 1991), may be better understood as an afrotherian near the radiation of tenrecids and chrysochlorids.

Relatively complete pantolestid specimens show hallmarks of semiaquatic and fossorial habits (which are not mutually exclusive), including robust fore- and hindlimbs that are relatively short, and a long and muscular tail (Rose and von Koenigswald 2005; Rose 2006a). Boyer and Georgi (2007) interpreted the vestibular anatomy of the inner ear of the Bridger Basin, Wyoming species *Pantolestes longicaudus* as consistent with a semiaquatic habitat. However, like *Buxolestes* (Pfretzschner 1993), *Pantolestes* had a relatively small infraorbital canal (Boyer and Georgi 2007: p. 264). Therefore, in contrast to most semiaquatic, small mammals (Sánchez-Villagra and Asher 2002; Crumpton and Thompson 2013), it would presumably have been less well endowed in terms of its capacity for tactile sensation via facial vibrissae. *Pantolestes* also exhibited a relatively small optic/infraorbital canal ratio, even smaller than those of *Erinaceus* and *Solenodon* (Boyer and Georgi 2007: p. 267), suggestive of simple visual acuity. A cochlea with "two full turns" (Boyer and Georgi 2007: p. 259) is within the range of therian mammals, similar to *Eumetopias* among aquatic taxa but also close to terrestrial species like *Macroscelides* and *Lepus* (Ekdale 2013: Tab. 2) and semifossorial species like *Orycteropus*.

These detailed and informative investigations of pantolestid anatomy have neither contradicted nor confirmed the intuition of McKenna and Bell (1997) that pantolestids are part of a "cimolestid" grade of mammals near pholidotes and carnivorans in Ferae. In terms of taxon sampling, Halliday *et al.* (2017) is again the most comprehensive so far, including multiple pantolestid taxa along with representatives of major extant clades, including the Eocene *Paleosinopa* and Paleocene *Bessoecetor* and *Todralestes* (the latter two known from gnathic remains only). These taxa appear in distinct clades depending on the phylogenetic assumptions used by Halliday *et al.* (2017). In their unconstrained tree, the three genera appear paraphyletically as near relatives of "paleoryctids", creodonts, and pholidotes. In their constrained analysis, *Paleosinopa* remains close to a palaeanodont-pholidote clade, but *Bessoecetor* appears among plesiadapiform primates and *Todralestes* among "condylarth" grade euungulates. Thus, the figured topologies of Halliday *et al.* (2017: figs. 3 & 4) place the anatomically well-known *Paleosinopa* close to pholidotes, but they do not support pantolestid monophyly, nor do they support Ferae as classified by McKenna and Bell (1997) due to placement by Halliday *et al.* of mesonychians (among other groups) as carnivoran sister taxa (as summarized above).

7.4.7.5 Perissodactyla

As noted above, the largest, genomic studies of mammalian phylogenetics support Euungulata, consisting of perissodactyls and artiodactyls (including Cetacea). Modern perissodactyls are relatively species-poor compared with those from the Paleogene and Neogene. *Equus* contains eight species (Wilson and Reeder 2005), which comprise the only extant hippomorph perissodactyls. The other high-level extant group is Tapiromorpha, consisting of five or six species in four genera of rhinocerotids and four species in *Tapirus*. The single old-world species, the Malayan tapir, is often assigned to the genus *Acrocodia*; a fifth tapir species, *Tapirus kabomani*, is contentious and may actually represent populations of *Tapirus terrestris* (Voss *et al.* 2014). The abundant fossil record of perissodactyls is of interest not only because of its considerable diversity, but also due to theories of Earth climate history and (mis)understandings of evolutionary theory in which it has played an important role.

7.4.8 Perissodactyls and evolutionary theory

The perissodactyl record has provided grist for speculative, non- or quasi-evolutionary theories, such as "orthogenesis" and "aristogenesis" of Osborn (e.g., 1922, 1934). Detached from an appreciation of intraspecific morphological variation, the abundance of perissodactyl fossils also enabled Osborn to taxonomically oversplit certain groups into numerous lineages based on what he perceived as rates of evolutionary change and the surprisingly blurry line between "absent" and "rudimentary" anatomical structures (see discussion in Mihlbachler 2008: pp. 10–11). A favorite simplification of evolution is the depiction of small-bodied, dentally low-crowned, many-toed, Eocene equids ineluctably and gradually turning into large, dentally hypsodont, single-toed, modern *Equus*. A good fossil record is susceptible to selective interpretation, and this enabled Osborn and subsequent textbook authors (e.g., Hegner and Stiles 1951) to choose from a variety of fossil forms to match preexisting ideas of how morphological evolution might have had a "predisposition" (Osborn 1934) towards a certain size and shape, millions of years before such sizes and shapes are actually evident in perissodactyl morphology. Matthew's (1926) detailed account of perissodactyl evolution does not endorse Osborn's putative mechanisms but still describes extinct perissodactyls as "stages leading to" modern groups. Nonetheless, paleontologists (probably including Matthew) have long realized that perissodactyl evolution was not "aristogenetic", and at the same time appreciated the clearly mosaic nature of many aspects of equid evolution (Froehlich 2002), such as the appearance of spring ligaments and loss of lateral digits in Miocene and Pliocene genera (Simpson 1953: pp. 259–265).

7.4.9 Perissodactyls as indicators of climate

The dense record of perissodactyl (and particularly equid) teeth shows important correlations with historical temperatures and patterns of vegetation, particularly in the northern hemisphere (MacFadden 2005; Mihlbachler *et al.* 2011; Secord *et al.* 2012). Such correlations are associated with an idea developed over the past several decades concerning the relationship between tooth crown height and global climate change during the Middle to Late Miocene. A number of mammalian groups, including but not limited to equids, exhibit high-crowned (or hypsodont) teeth as an adaptation for highly abrasive diets, including grasses and grit-covered vegetation. As an indicator of paleotemperature, the standardized ratio of heavy (^{18}O) to light (^{16}O) oxygen isotopes since the Early Eocene thermal maximum shows a decreasing trend until the Oligocene, followed by a long period of fluctuations around a relatively more stable mean, with another decreasing trend beginning in the Early Miocene (Zachos *et al.* 2008: fig. 2b). Following Mihlbachler *et al.* (2011: fig. 1a), this record suggests two large-scale trends of global cooling: one between the Early Eocene until Early Oligocene (ca. 50–32 Ma) and another from the middle Miocene to the Late Pleistocene (ca. 14–0.01 Ma). Many fossil equids during the Neogene exhibit low to moderate dental crown height and mesowear indicative of frugivory or browsing on leaves, as opposed to the more high-crowned teeth typical of grazers (i.e., grass eaters). Global cooling in the Miocene slightly postdated the spread of grasslands, and equid lineages that persisted after the Miocene exhibited high-crowned teeth. Grasses contain silica phytoliths and were therefore thought to require a particularly abrasion-resistant dentition from those herbivores that specialize on them (MacFadden 2000, 2005). Hence, modern species of *Equus* have been interpreted as the grazing-adapted descendants of more frugivorous or leaf-eating perissodactyls, diets retained by their forest-, savannah-, and mixed canopy-dwelling tapiromorph relatives.

Further analysis and discovery have enabled a more nuanced understanding of the relationship between mammal tooth crown height and global climate change. First, the appearance of grasslands in northern continents predated the evolution of the most hypsodont equids by several million years (Edwards *et al.* 2010; Mihlbachler *et al.* 2011). In South America, the correlation is also offset, but in the opposite direction. Dentally high-crowned South American ungulates (i.e., certain species of notoungulates) are evident over 30 Ma ago amidst ecosystems dominated by forests and relatively less-open habitats, several million years prior to even modest contributions of grasslands to South American paleoenvironments (Strömberg *et al.* 2013). These studies do not contradict the links between ecosystem change, dental crown height, and dietary abrasion, but in addition to grasses suggest another, more general source for that abrasion: grit (Janis 1988; Damuth and Janis 2011; Madden 2014). Any kind of foodstuff dusted in an abrasive substance (e.g., volcanic ash) would lead to selective pressures favoring increased crown height and the appropriate crown morphology to resist wear. In the case of South America, Andean orogenetic (i.e., mountain building) activity, which would have regularly dispersed large amounts of volcanic ash across environments throughout the Patagonian Paleoegene, has a better correlation with appearances of high-crowned ungulates than the spread of grasslands (Strömberg *et al.* 2013).

7.4.10 Perissodactyl fossil diversity

Besides its important role in the history of paleontology and in understanding past global environments, the perissodactyl fossil record (like that of other high-level taxa) demonstrates a wide variety of now-extinct, morphologically diverse species. All of them are inferred to be herbivorous but range widely in body size from the diminutive, Early Eocene, "tapiroid" *Fouchia*, with a lower toothrow no bigger than that of a hedgehog (Emry 1989), to probably the largest land-mammal yet known, the rhinocerotoid indricothere *Paraceratherium*. This taxon is known from the latest Eocene and Oligocene of Asia and may have attained two to three times the body weight of a large male African elephant. Such massive animals were graviportal; other large tapiromorphs included chalicotheres, known primarily from Eocene to Pleistocene deposits in Eurasia, North America, and Africa. Chalicotheres are known for their derived forelimb anatomy. They substituted claws for hooves and depended on their hindlimbs to browse, and in the case of chalicotheriines, utilized a "knuckle-walking" locomotor style.

Otherwise, adaptations for cursorial locomotion frequently occur among fossil perissodactyls, such as a deep astragalar trochlea, reduced rotatory capacity of the distal limbs, and relatively small distal epicondyles of the humerus. Fossil perissodactyls show a distinctive ankle morphology, likely related to some degree of cursoriality but different to the cursorial ankle joints of other mammalian groups (e.g., terrestrial artiodactyls). Rose *et al.* (2014) described cranial and skeletal remains of *Cambaytherium* from the Early Eocene of Gujarat, India, which in their phylogenetic analysis was reconstructed with other cambaytheres near the basalmost branch of Perissodactyla, outside of the crown clade uniting tapiromorphs and hippomorphs. The astragalus of *Cambaytherium* shares with all extant perissodactyls not only a deep trochlea but also a "saddle" shape of the distal astragalus, primarily for articulation with the navicular. Cooper *et al.* (2014) presented another phylogenetic analysis, which confirms the stem perissodactyl affinities of cambaytheres, as well as a close relationship between perissodactyls and another Paleogene group, anthracobunids (mentioned above in the section on paenungulate afrotheres). The two phylogenetic analyses show considerable overlap but differ in placing anthracobunids either within Perissodactyla as sister taxa to hippomorphs (Rose *et al.* 2014, a result which they regard as not well supported and probably inferior to an anthracobunid-cambaythere clade) or on the stem leading to Perissodactyla (Cooper *et al.* 2014).

Paleotheres are well known from cranial and skeletal remains from the Paleogene of Europe. Following Froehlich (2002), Hooker and Dashzeveg (2004), and Rose *et al.* (2014), they are nested among modern equids and thus comprise part of Hippomorpha. Hooker (1994) and Froehlich (2002) clarified the taxonomic status of the name *Hyracotherium*, coined by Owen (1841) for *Hyracotherium leporinum* based on a partial skull from the London Clay (Eocene). Froehlich (2002) identified this taxon as related to other European paleotheres and used a variety of other genus-level names (e.g., *Sifrhippus*, *Eohippus*, and *Proterohippus*) for North American, Early Eocene specimens formerly referred to *Hyracotherium*.

Brontotheres (equivalent to Osborn's "titanotheres") are well known anatomically (Mihlbachler 2008), primarily from localities in North America plus some Eurasian material. According to Rose *et al.* (2014), brontotheres are more closely related to rhinos and tapirs than to equids and, thus, are part of the crown radiation of Perissodactyla. Early forms such as *Eotitanops* were hornless and slightly exceeded 100 kg in body size, but later brontotheres grew to the size of rhinos or even elephants and carried bony horns at the front of their heads, in contrast to the keratinous "horns" of rhinos (Mihlbachler 2008). Brontotheres did not evolve as high-crowned teeth as equids, but they did vary in enamel thickness and some forms exhibited some degree of unilateral hypsodonty (Mihlbachler 2008: fig. 3).

Rose *et al.* (2014) sampled a variety of additional taxa previously thought to be within or closely related to Perissodactyla, including North American Eocene fossils historically attributed to *Hyracotherium* (see above and Froehlich 2002), *Lambdotherium* (used as an index fossil for the Early Eocene Lostcabinian sub-NALMA; see Tab. 7.4), holarctic Eocene "isectolophids", and the enigmatic genus *Radinskya* (Holbrook 2014). All except the latter appear as crown perissodactyls in the optimal phylogeny figured by Rose *et al.* (2014: fig. 6). *Radinskya* appears with "phenacodontids" on the stem leading to paenungulate afrotheres, reflecting the phylogenetic uncertainty regarding this taxon reported in Holbrook's (2014) redescription of the type and only specimen from the Paleocene of China. The taxon sample of fossil perissodactyls in Rose *et al.* (2014) is good; however, the support indices given in their topology (Rose *et al.* 2014: fig. 6) vary considerably, and most nodes show bootstrap resampling values below 50%. Even relatively high branch supports (Bremer 1994) appear alongside bootstrap values lower than other nodes with weaker branch supports, such as "18" bootstrap and "18" branch support for crown perissodactyls, compared with "11" branch support and "77" bootstrap for their node encompassing Placentalia. Nonetheless, such variation is not necessarily an indication of an inaccurate, optimal tree, and this study still comprises the most broadly sampled, phylogenetic study of living and fossil Perissodactyla.

Another important development in our understanding of the evolutionary history of Perissodactyla is the evidence that endemic South American ungulates may be their sister group. This is based on amino acid sequences of collagen proteins (Welker *et al.* 2015) from upper Pleistocene specimens *Toxodon* (Notoungulata) and *Macrauchenia* (Litopterna), as well as mitogenomic data from *Macrauchenia* (Westbury *et al.* 2017). As noted above, Carrillo and Asher (2017) noted some alternate arrangements (e.g., sister taxon to Euungulata) that are statistically indistinguishable. Similarities between one or more groups of endemic South American ungulates and perissodactyls have long been recognized; indeed, Ameghino (1889) classified litopterns as a group within perissodactyls, and Anthony (1924) first described the pyrothere *Griphodon* in a paper entitled "a new fossil perissodactyl from Peru" based on its tapir-like, bilophodont dentition. However, such resemblances have always been regarded as uncertain indicators of phylogenetic affinity, and the recovery of biomolecules from these Pleistocene fossils comprises a major breakthrough in understanding the evolutionary history of the groups to which they belong.

Throughout the 20th century (e.g., Scott 1910; Simpson 1948), these similarities have been regarded as characters of largely adaptive significance given their shared ecology and habitat as medium- to large-bodied, herbivorous, generally cursorial mammals. More detailed resemblances include mesaxony, or the central digit of the manus and pes carrying the bulk of the animal's weight during locomotion. This may accompany loss of lateral digits on both fore- and hindlimbs and reaches an extreme in some equids and the Miocene proterotheriid litoptern *Thoatherium*. Although they do contain species that converge on functional monodactyly, mesaxonic equids and litopterns still exhibit distinctive limb elements. For example, the litopterns *Thoatherium* and *Prothoatherium* both lack a saddle-shaped navicular facet of the astragalus (Cifelli and Guerrero Diaz 1989: fig. 5), a feature that is present throughout Perissodactyla, including those that are not monodactyl (e.g., living tapiromorphs and the stem taxon *Cambaytherium* as noted above and figured in Rose *et al.* 2014: fig. 5).

7.4.10.1 Artiodactyla

The astragalus also comprises a key morphological apomorphy of the other major euungulate group: Artiodactyla (including cetaceans). All living terrestrial artiodactyls possess the "double-pulley" astragalus, meaning that there is a trochlea both proximally for articulation with the tibia and distally for articulation with the navicular and cuboid. In addition, the calcaneus shows a broad facet to articulate with the lateral surface of the astragalus with a dorsal facet that articulates with the fibula. Variations in hindlimb morphology include their complete loss in extant cetaceans (although rudiments may be present), loss of digits II and V in camelids and some ruminants, unfused metapodials III and IV in suiformes, and fusion of the navicular and cuboid in ruminants. Otherwise, ankle morphology of living and fossil artiodactyls is highly distinctive and documents their existence since the Eocene (Rose 1982). A similarly conservative but less well-known hard tissue character, also widespread among living and extinct artiodactyls (but not cetaceans), is the trilobed, ultimate, deciduous lower premolar (Luckett and Hong 1998).

No extant cetacean has a functional hindlimb, but some fossil cetaceans did. North African remains of the Oligocene stem cetacean (or "archaeocete") *Basilosaurus* show a diminutive structure that did not bear any substantial weight or have a propulsory function in this aquatic mammal but may have served as a copulatory organ in males. Whatever its function, the astragalus is fused into an ossified mass with other ankle elements and lacks typical joint surfaces. Although its astragalus is vestigial, *Basilosaurus* exhibited hindlimbs with anatomical paraxony; i.e., digits III and IV comprised the largest digital rays as in terrestrial artiodactyls (Gingerich *et al.* 1990). Further (and more dramatic) evidence for the morphology of "archaeocete" hindlimb anatomy came with the description of associated cranioskeletal fossils of *Artiocetus clavis* and *Rodhocetus balochistanensis* from the Eocene (middle Lutetian, ca. 47 Ma ago) of Pakistan, showing that stem cetaceans exhibited hindlimb paraxony (although the forefoot of *Rodhocetus* is mesaxonic), a double-pulley astragalus, and a flange of the calcaneus that articulates with the lateral surface of the astragalus and the fibula via a dorsal facet (Gingerich *et al.* 2001). Based on relative abundance, Thewissen *et al.* (2001) argued that isolated ankle remains of pakicetid "archaeocetes" also exhibited typically artiodactyl morphology.

These discoveries were of particular significance given the then-new evidence from comparisons of DNA (e.g., Gatesy *et al.* 1996) that living cetaceans were more closely related to hippopotamids than the latter were to other terrestrial artiodactyls. Dental similarities between stem cetaceans and an extinct clade of carnivorous mammals, mesonychians, had long been regarded as evidence for a close evolutionary relationship (van Valen 1966: pp. 90–93). Although functional ankle elements for "archaeocetes" were unpublished until the early 2000s, the skeletal anatomy of mesonychians was reasonably well documented (e.g., Szalay and Gould 1966), with no indication of a double-pulley, artiodactyl-like ankle, although mesonychids do exhibit slightly concave navicular and narrow cuboid facets (Rose 2001). Hence, although a close evolutionary relationship between cetaceans and artiodactyls already had precedent (e.g., Novacek 1992), the idea that cetaceans nested within artiodactyls did not, in part because of the homoplasy implied for such conservative character suites as the unique ankle morphology of terrestrial artiodactyls (Luckett and Hong 1998). The discovery that stem cetaceans exhibited clear, derived features in common with terrestrial artiodactyls such as hippos, but not mesonychians, represented compelling evidence that cetaceans and terrestrial artiodactyls shared common ancestry to the exclusion of any other mammalian group. Subsequent analyses of extant species, including the genomic studies behind the now well-corroborated tree (Fig. 7.1), have shown the hippopotamid-cetacean clade, nested within Artiodactyla, to be among the most consistently and strongly supported high-level clades among mammals.

The affinities of mesonychians remain less clear. They are historically defined as containing at least two clades from the Paleocene and Eocene of North American and Asia: mesonychids and hapalodectids. A third group, triisodontids, remains poorly known postcranially and is restricted to North America (with the possible exception of the very enigmatic and large *Andrewsarchus* from the middle Eocene of central Asia; see Rose 2006a, p. 274). As noted above, Halliday *et al.* (2017: figs. 3 & 4) recover mesonychians as the sister taxon of Carnivora. Spaulding *et al.* (2009) and O'Leary *et al.* (2013) both recover their sampled mesonychians as sister to euungulates, although Spaulding *et al.* (2009: fig. 3) note that topologies two steps longer than their optimal result are compatible with mesonychians nested within artiodactyls, as sister taxon to cetaceans to the exclusion of hippopotamids.

Mesonychians first appear in the Paleocene; terrestrial artiodactyls and "archaeocete" whales first appear in the Early Eocene. Hippopotamids, by contrast, were up until recently not clearly known from fossils that predated the Miocene of east Africa (Rose 2006a). A sister taxon relationship of hippopotamids and cetaceans thus implies a ghost lineage in which some kind of hippopotamid lineage must have existed alongside their cetacean sister taxon, given their common ancestor which had to predate the first appearance of cetaceans over 50 Ma ago.

Boisserie *et al.* (2005), Orliac *et al.* (2010), and Lihoreau *et al.* (2015) have partly closed this gap by making the case that hippopotamids evolved from within a group known primarily from craniodental remains from the African and South Asian Paleogene: "anthracotheres" (not to be confused with anthracobunids, discussed above in the section on perissodactyls).

The phylogeny of Orliac *et al.* (2010) reconstructs the Miocene "anthracotheres" *Libycosaurus* and *Merycopotamus* in a clade that in turn comprises the sister taxon of extant *Hippopotamus*, *Choeropsis*, and other Miocene stem hippopotamids. Other anthracothere-grade taxa, in particular "bothriodontines" such as *Bothriogenys*, are known from older deposits throughout the Oligocene and extending into the Late Eocene. Lihoreau *et al.* (2015) recover an even closer phylogenetic relationship between *Bothriogenys* and modern hippopotamids, including the basal Oligocene taxon *Bothriogenys orientalis* from Thailand. Orliac *et al.* (2010) and Lihoreau *et al.* (2015) both hypothesize yet older "anthracotheres" such as the Late Eocene genera *Siamotherium* from Thailand and *Elomeryx* from Europe as additional, stem relatives of the hippopotamid clade. How robust these phylogenetic hypotheses will be to further analysis and discovery remains to be seen. Branch supports were close to "1" for many of the nodes in Lihoreau *et al.* (2015: fig. 2), and their optimal topology also shows an unusual placement of suiformes closer to hippopotamids than *Indohyus*, a skeletally well-known Eocene raoellid artiodactyl from Pakistan, hypothesized by Geisler and Theodor (2009) and Spaulding *et al.* (2009) to be the sister taxon of Cetacea to the exclusion of suiforms and other terrestrial artiodactyls.

7.4.10.2 "Condylarths"

Many of the extinct groups outlined in the preceding pages exhibit one or more key characters that, in the context of a phylogenetic analysis and in tandem with other evidence, are sufficiently diagnostic to enable confident estimates of their phylogenetic affinities. For example, stem cetaceans have hindlimb paraxony and a double-pulley astragalus (e.g., *Artiocetus*; Gingerich *et al.* 2001); stem perissodactyls have a saddle-shaped navicular facet on the astragalus (e.g., *Cambaytherium*; Rose *et al.* 2014); stem pholidotes have a medial buttress on their dentary (e.g., palaeanodonts; Rose 2006a); stem primates have a petrosal auditory bulla (e.g., *Carpolestes*; Bloch *et al.* 2007); stem proboscideans have an anteriorly situated orbit and vertically oriented coronoid process of the dentary (e.g., *Phosphatherium*; Gheerbrant *et al.* 2002). On the other hand, there are at least as many anatomically well-known, extinct groups that lack clearly diagnostic features linking them to extant species, and despite ample anatomical data and decades of study, their phylogenetic affinities remain ambiguous. These include some of the most common genera known from Paleogene sites in the North American interior, including ungulate-grade "condylarths" such as *Hyopsodus* and *Phenacodus*, each of which is known by many thousands of individual jaws and teeth. Associated cranioskeletal fossils are much rarer but are known for several "condylarth" genera (Gazin 1965, 1968; Rose 1990; Williamson and Lucas 1992; Bergqvist 2008; Kondrashov and Lucas 2012).

"Condylarthra" is a wastebasket taxon, historically treated as a grade from which extant, ungulate-grade taxa evolved. Simpson's (1945) view of "condylarths" was influential; he included in the group four primarily North American Paleocene and Eocene groups: phenacodontids, hyopsodontids, meniscotheriids, and periptychids, along with one South American group, didolodontids. Gregory's (1910: p. 466) version of "Condylarthra" included only phenacodontids and meniscotheriids; he figured (1910: figs. 31 & 32) these taxa as paraphyletically ancestral to perissodactyls and South American litopterns, respectively, and discussed hyopsodontids (Gregory 1910, p. 360) as potentially relevant to the other great wastebasket group, "insectivores". McKenna and Bell's (1997) version of "Condylarthra" was close to that of Simpson (1945) but varied the content of the constituent high-level groups, such as placing meniscotheriines within Phenacodontidae and recognizing the more cosmopolitan distributions of certain groups (e.g., louisine hyopsodontids in Africa, Eurasia, and North America) afforded by many discoveries in the decades following Simpson (1945).

"Condylarths" are usually discussed in terms of their potential as primitive lineages that may have given rise to one or another modern group. However, such taxa are of course derived in their own right. *Hyopsodus* is smaller than *Phenacodus* and *Meniscotherium* and is hypothesized to have been a short-limbed mammal adept at burrowing. This may be one of the reasons why mandibles of this genus are so common in the fossil record (although it is not clear why associated skeletal fossils of *Hyopsodus* are extremely rare). Orliac *et al.* (2012) estimated its body size to be ca. 326–412 g; they furthermore proposed that cranial endocasts of *Hyopsodus* exhibited a high encephalization quotient and a well-developed inferior colliculus in the midbrain, a feature associated with a well-developed auditory sensitivity and possibly associated with echolocation and/or heightened dependence on acoustic cues in their presumed subterranean environment. Ravel and Orliac (2014) noted that the inner ear of *Hyopsodus* lacks the autapomorphies for laryngeal echolocation seen in non-pteropodid chiropterans; its cochlear morphology suggests sensitivity to ultrasound (up to 76 KHz), but they cautiously note that ultrasound sensitivity is not a necessary correlate of echolocation.

Phenacodus and *Meniscotherium* are both substantially larger than *Hyopsodus*. Body masses are between 5 and 17 kg in *Meniscotherium chamense* (Williamson and Lucas 1992), between 10 and 39 kg for *Phenacodus intermedius*, and between 22 and 87 kg for *Phenacodus trilobatus* (Thewissen 1990). Both of these larger "condylarths" had pentadactyl limbs and exhibited at least some postcranial adaptations for cursoriality, such as a digitigrade stance, an open olecranon fossa (i.e., supracondylar or supratrochlear foramen) in the distal humerus, and some dorsoventral flattening of their terminal digits. Williamson and Lucas (1992: p. 32) argued that *Meniscotherium* may have had primatelike nails but not hooves or claws. In other regards, *Meniscotherium* might have been somewhat more cursorial than *Phenacodus*; it shows an enlarged digit III in both manus and pes, consistent with mesaxony (Williamson and Lucas 1992: figs. 20, 26). Williamson and Lucas (1992) also noted the absence of a clavicle in otherwise well-preserved specimens, indicating that (like extant cursorial mammals) *Meniscotherium* lacked this element in life. By contrast, *Phenacodus* retained a clavicle (Thewissen 1990). *Phenacodus* also exhibited bunodont cheek teeth; its upper molariforms consist of four major cusps, separated by conules, each of which reaches a distinct apex without forming a major crest in unworn specimens (Thewissen 1990: fig. 39). *Meniscotherium*, by contrast, exhibited more selenodont cheek teeth; i.e., each cusp formed the apex of a "V" shape joining two crests that extend buccally (Williamson and Lucas 1992: fig. 43).

As previously noted, *Phenacodus* exhibited larger body sizes than *Meniscotherium*. However, Dirks *et al.* (2009) noted that for its body size, *Meniscotherium* had a relatively larger brain (i.e., a higher encephalization quotient). They also observed differences in rates of tooth formation that, along with brain morphology, may shed light on differences in life history in the two genera. Based on the density of cross-striations and striae of Retzius evident in mammalian tooth enamel, Dirks *et al.* (2009) found that molars of *Meniscotherium* took longer to form than those of *Phenacodus*; they estimated times of under one year in the latter, and at or slightly over 1 year in the former. Given correlations between time required for molar formation, brain size and life history in other mammals, Dirks *et al.* (2009) hypothesized that despite its smaller body size, *Meniscotherium* would have required more time to develop to maturity compared with *Phenacodus*.

Several phylogenetic studies of the last two decades have sampled one or more of these skeletally well-known Paleocene and Early Eocene taxa such as *Ectocion*, *Hyopsodus*, *Meniscotherium*, *Phenacodus*, and *Tetraclaenodon*, as well as accounted for the well-corroborated relationships among extant lineages, either in the form of phylogenetic scaffolds or by directly incorporating molecular data. Optimal trees derived from some of these studies suggest affinities of *Hyopsodus*, *Phenacodus*, and/or *Meniscotherium* with paenungulate afrotherians (Asher *et al.* 2003; Asher 2007; Rose *et al.* 2014) and of louisine "condylarths" near macroscelidid afrotherians (Zack *et al.* 2005). Other recent studies (Ladevèze *et al.* 2010; O'Leary *et al.* 2013; Cooper *et al.* 2014; Halliday *et al.* 2017: fig. 4; Carrillo and Asher 2017) favor one or more of these "condylarths" near extant euungulates, i.e., artiodactyls and perissodactyls. O'Leary *et al.* (2013: fig. 1) placed *Apheliscus*, *Didolodus*, *Hyopsodus*, and *Phenacodus* along with the endemic South American *Protolipterna* together in a clade that comprises the sister group of Euungulata, not unlike the Panameriungulata of de Muizon and Cifelli (2000). Cooper *et al.* (2014: fig. 3) placed *Meniscotherium* and *Phenacodus* along with the European louisines *Paschatherium* and *Teilhardimys* on the stem leading to Perissodactyla. In their analysis constrained with an extant-taxon scaffold, Halliday *et al.* (2017: fig. 4) showed a subset of "condylarths" as stem perissodactyls (e.g., *Ectocion* and *Tetraclaenodon*) and another as stem artiodactyls (*Hyopsodus*, *Meniscotherium*, and *Phenacodus*).

These (often contradictory) results are frustrating given the relative completeness of several of these fossil mammals and reflect the very real possibility that the requisite, historical information that might have contributed to a well-resolved species tree has been erased by taphonomy and/or convergent morphological evolution. On the other hand, the most recent phylogenetic work cited above slightly favors the view that at least some "condylarths" are euungulates and argues against certain possibilities, e.g., the view of Gregory (1910) that hyopsodontids are related to "insectivores" (by which he meant lipotyphlans, tenrecoids, and possibly scandentians and/or macroscelideans). As articulated by Rose *et al.* (2014), "condylarth" taxa such as *Phenacodus* exhibit a wealth of primitive characteristics relative to euungulate laurasiatheres (i.e., artiodactyls and perissodactyls) and paenungulate afrotherians.

7.4.10.3 Pantodonta

Among the other "condylarth"-grade mammals that are amply represented by cranioskeletal remains, particularly from the Paleocene and Early Eocene of North America and Eurasia, are pantodonts. Simpson (1945) argued tentatively that pantodonts belonged with proboscideans, sirenians, and hyracoids in Paenungulata, reflecting Cope's (1891) classification which joined pantodonts with other large-bodied "ungulates" such as uintatheres and periptychids as part of "Amblypoda". *Coryphodon* is particularly common in the Early Eocene of the North American West and is known not only from adult material but also associated skeletons of juveniles (Lucas and Schoch 1990; McGee and Turnbull 2010). Thus, some aspects of its developmental anatomy, including sequences of dental eruption and sutural fusion and relative body proportions during growth, are at least partly known. For example, the eruption sequence of *Coryphodon* exhibits three antemolars (I1, I2, and C) that erupt after M2 and two (I2 and C) that erupt simultaneously with M3. This contrasts with (for example) the pattern seen in many ruminants, in which M3 erupts prior to most or all antemolars (Smith 2000), and resembles the eruption pattern seen in many nonruminant artiodactyls, some perissodactyls, and fossil hyracoids. The late eruption of M3 also occurs in extant hyracoids, but *Coryphodon* differs in that its premolars erupt after M2 in posterior-to-anterior sequence (M2-P4-P3-P2), whereas in *Procavia* the anterior premolars erupt before M2 and P4, and premolars erupt in an anterior-to-posterior sequence (Asher *et al.* 2017). Following Schultz (1960) and Smith (2000), McGee and Turnbull (2010) interpret the eruption data for *Coryphodon* to indicate a prolonged period of development and "slow" growth; they also note potential similarities to the life history of extant hippos, tapirs, and the sumatran rhino.

Pantodonts are also uniquely well represented in the Tiupampan SALMA (Early Paleocene; Tab. 7.4) of South America, thanks to the locality of Tiupampa in Bolivia. de Muizon *et al.* (2015) described an assemblage with multiple, complete skulls and associated skeletons of *Alcidedorbignya inopinata*. These specimens are extremely well preserved, including specimens with in situ clavicles, sternebrae, patellae, and distal caudal vertebrae. There are several juvenile skeletons and even one fetal specimen interpreted by de Muizon *et al.* (2015) likely to have been in utero at the time of preservation, found in the pelvic-abdominal region of an adult skeleton inferred to be female based on sexually dimorphic features of the canines and zygomatic region of the skull. The presacral vertebral count is C7, T13, and L9; with nine lumbars, *Alcidedorbignya* has one of the longest lumbar regions of any mammal. Its 22 thoracolumbar vertebrae exceed the values typical for most placental mammals except for perissodactyls, *Choloepus*, and most afrotherians (Asher *et al.* 2011). *Alcidedorbignya* is very small for a pantodont, with a skull not much longer than 6 cm and thus similar in size to a hedgehog or squirrel. Its skeleton indicates a terrestrial habitat with some arboreal capacity as well.

The phylogenetic analysis of de Muizon *et al.* (2015), taking into account the result of Welker *et al.* (2015) that South American litopterns and notoungulates are stem perissodactyls, supports *Alcidedorbignya* and other pantodonts within Laurasiatheria, closer to euungulates than endemic African paenungulates. They figure two of their 11 optimal topologies (de Muizon *et al.* 2015: figs. 121, 122) generated with a scaffold from Welker *et al.* (2015), rather than a consensus across all eleven, understandably emphasizing that their results may vary considerably with changes in taxon or character sampling. They also made their phylogenetic data available online and thus easily amenable to further scrutiny, in this case using PAUP (4.0a build 156, Swofford 2002). Figure 7.9 shows a reanalysis of their data (which did not include extant perissodactyls), applying a similar scaffold as in de Muizon *et al.* (2015), assuming a pruned version of the topology of the well-corroborated tree (Fig. 7.1) as well as the results of Welker *et al.* (2015) who regarded notoungulates and litopterns as stem perissodactyls.

This reanalysis recovers two optimal trees of 3840 steps, the consensus of which (Fig. 7.9 A) is slightly different than the two constrained most parsimonious trees (MPTs) figured in de Muizon *et al.* (2015: figs. 121, 122), but nonetheless compatible with their major results. In particular, *Alcidedorbignya* is recovered within Laurasiatheria alongside other pantodonts, tillodonts, *Didelphodus*, and arctocyonids, with *Protungulatum* and Carnivora as successively adjacent sister taxa (node "CP" in Fig. 7.9 A). *Carodnia* is reconstructed with the other endemic South American pyrotheres and astrapotheres within Euungulata, also including *Hyopsodus*, *Meniscotherium*, and *Phenacodus*, all of which collectively comprise the sister clade of Artiodactyla. As in de Muizon *et al.* (2015), I find that the endemic South American xenungulate *Carodnia* changes its position substantially with the topological constraint from Fig. 7.9 B. Without any constraint, it appears within pantodonts as sister to *Coryphodon* (de Muizon *et al.* 2015: fig. 120), not (as noted above) with *Pyrotherium* and notoungulates.

The optimal topology from the reanalysis of their data differs slightly from their figured results in placing Carnivora as the nearest laurasiatherian clade to the

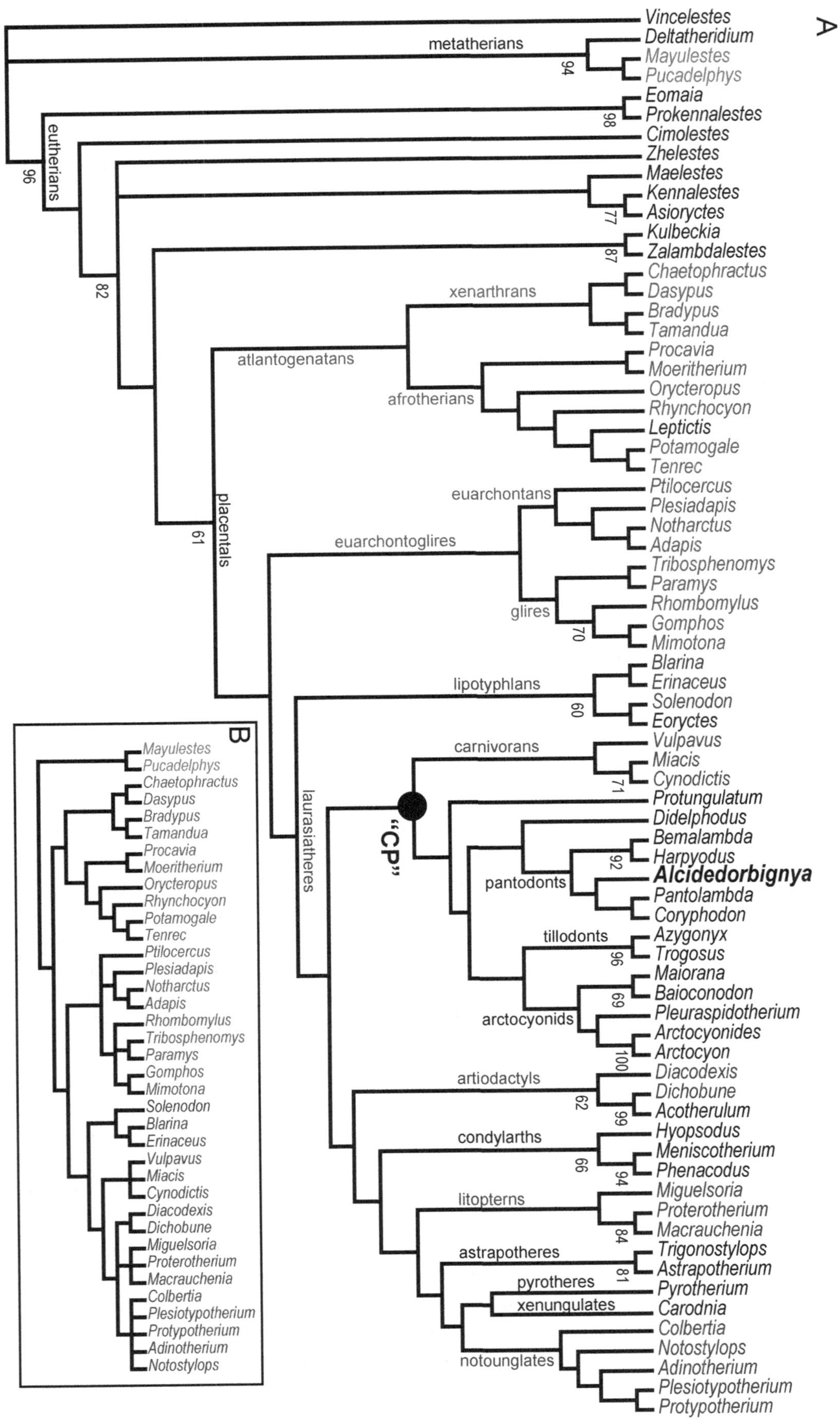
A
Vincelestes
metatherians
Deltatheridium
Mayulestes
Pucadelphys
Eomaia
Prokennalestes
Cimolestes
Zhelestes
Maelestes
Kennalestes
Asioryctes
Kulbeckia
Zalambdalestes
eutherians
xenarthrans
Chaetophractus
Dasypus
Bradypus
Tamandua
atlantogenatans
Procavia
Moeritherium
afrotherians
Orycteropus
Rhynchocyon
Leptictis
Potamogale
Tenrec
placentals
euarchontans
Ptilocercus
Plesiadapis
Notharctus
Adapis
euarchontoglires
Tribosphenomys
Paramys
glires
Rhombomylus
Gomphos
Mimotona
lipotyphlans
Blarina
Erinaceus
Solenodon
Eoryctes
carnivorans
Vulpavus
Miacis
Cynodictis
laurasiatheres
"CP"
Protungulatum
Didelphodus
Bemalambda
Harpyodus
pantodonts
Alcidedorbignya
Pantolambda
Coryphodon
tillodonts
Azygonyx
Trogosus
Maiorana
Baioconodon
arctocyonids
Pleuraspidotherium
Arctocyonides
Arctocyon
artiodactyls
Diacodexis
Dichobune
Acotherulum
condylarths
Hyopsodus
Meniscotherium
Phenacodus
litopterns
Miguelsoria
Proterotherium
Macrauchenia
astrapotheres
Trigonostylops
Astrapotherium
pyrotheres
Pyrotherium
xenungulates
Carodnia
notounglates
Colbertia
Notostylops
Adinotherium
Plesiotypotherium
Protypotherium
96
94
98
77
87
82
61
70
60
71
92
96
69
100
62
99
66
94
84
81
B
Mayulestes
Pucadelphys
Chaetophractus
Dasypus
Bradypus
Tamandua
Procavia
Moeritherium
Orycteropus
Rhynchocyon
Potamogale
Tenrec
Ptilocercus
Plesiadapis
Notharctus
Adapis
Rhombomylus
Tribosphenomys
Paramys
Gomphos
Mimotona
Solenodon
Blarina
Erinaceus
Vulpavus
Miacis
Cynodictis
Diacodexis
Dichobune
Miguelsoria
Proterotherium
Macrauchenia
Colbertia
Plesiotypotherium
Protypotherium
Adinotherium
Notostylops

pantodont, tillodont, *Didelphodus*, and arctocyonid clade (node "CP" in Fig. 7.9 A). It also differs in placing *Leptictis* among afrotheres, not among non-placental eutherians as in de Muizon *et al.* (2015: figs. 121, 122). Overall, and with a few exceptions (e.g., highly variable placement of *Carodnia* in Carrillo and Asher 2017), these results fit the recent trend among phylogenetic analyses of Mammalia that place Paleocene and Eocene radiations of ungulate-grade placental mammals (such as pantodonts, "condylarths", and endemic South American meridiungulates) close to or within Laurasiatheria, rather than with Afrotheria or Xenarthra (Figs. 7.8 and 7.9).

7.5 Timing of mammalian diversifications

Up until the 1960s, inferences of the origination and diversification dates of extant clades were entirely dependent on the fossil record, and fossils still comprise our best source of direct information of the origination and rate of life's diversification. Beginning with studies such as Zuckerkandl and Pauling (1965), biologists began to appreciate how sequences of amino acid residues (and later nucleotides) diverged among extant species, and how this could provide insight into life's remote history in geological time. Calibrated with a divergence point for two species with (ideally) good fossil records, divergences for additional species lacking fossil data could be inferred based on the rate of change observed in the initial, calibrated pair.

Kumar and Hedges (1998) were the first to apply a "molecular clock" broadly across vertebrates, using a single paleontological calibration point for the split between Sauropsida and Synapsida, i.e., the occurrence of the stem clades encompassing birds and mammals, approximately 310 Ma ago in the Carboniferous of eastern Canada. For mammals, Kumar and Hedges' (1998) results reflected paleontological estimates for at least some of the more recent splits (e.g., hominins from other apes during the upper Miocene) but otherwise were considerably older than estimates based on the fossil record (Foote *et al.* 1999, but see Tavaré *et al.* 2002). For example, the oldest dates for placental lineages diverging from humans shown by Kumar and Hedges (1998: fig. 3) were "Edentata" (i.e., sloths) at ca. 129 ± 18.5 Ma followed by sciurids and hystricognaths at ca. 112 ± 3.5 and 109 ± 3.2 Ma, respectively. These ages are approximately twice those of the oldest known xenarthrans and rodents from the Paleocene.

A number of improvements since Kumar and Hedges (1998) have made molecular clock estimates of vertebrate divergences more compelling. For example, as discussed above, the topology for Placentalia is now well corroborated and shows a bifurcation between Atlantogenata and Boreoeutheria, not basal positions of sloths or rodents. Second, the basis for choosing calibration points is now more explicitly linked to phylogenetic hypotheses and correlations with the global marine record (Benton *et al.* 2009; Parham *et al.* 2012), and there are now methods to apply both minimum and maximum calibrations, the latter as "soft" maxima defining a probability distribution rather than a temporal cutoff (Benton *et al.* 2009, 2015; dos Reis *et al.* 2012). Relatedly, contemporary investigators generally appreciate that (unlike the basal amniote split in the upper Carboniferous), calibration points should be chosen when they derive from a series of localities representing paleoecologically habitable environments in which the taxon in question could have existed but did not (Reisz and Müller 2004). In addition, the paleontological basis for assigning Mesozoic fossils to one or another crown clade (e.g., Archibald 1996; Archibald *et al.* 2001, aligning 85–90 Ma zhelestids and zalambdalestids with "ungulates" and glires, respectively) has been heavily scrutinized and effectively disproven (Asher *et al.* 2005; Wible *et al.* 2007, 2009; Goswami *et al.* 2011), as discussed above. Relatedly, the affinities of key Mesozoic taxa (e.g., *Juramaia* as a Jurassic eutherian following Luo *et al.* 2011 but not Krause *et al.* 2014) are of course influential on resulting clock estimates, but with multiple fossil calibrations and soft-maxima (dos Reis *et al.* 2012) and simultaneous analysis (or "tip dating") of both topology and divergence ages (Ronquist *et al.* 2016), a given study is potentially more resistant to error in placing fossils phylogenetically.

◂**Fig. 7.9:** Reanalysis of the supplementary nexus file from de Muizon *et al.* (2015), available at http://sciencepress.mnhn.fr/sites/default/files/fichierspublis/periodiques/geodiversitas/geo1538/alcidedorbignya_inopinata_nexus_file_data_matrix.nex. The topology in "A" represents a strict consensus of 2 MPTs with 3840 steps after a 500 replicate random addition search, based on 426 characters from de Muizon *et al.* (2015), 66 of which are ordered, and using the tree shown in "B" as a backbone constraint (based on Fig. 7.1 and Welker *et al.* 2015). Polymorphic character states were treated as such (not as missing); numbers indicate bootstrap support values calculated from 500 pseudoreplicates of a five replicate random addition sequence (not reported below 50 or for constrained nodes). Node "CP" demarcates the clade including carnivorans and pantodonts that slightly differs from the optimal topologies depicted by de Muizon *et al.* (2015), discussed in the text. Colors are shown as in Fig. 7.1; fossils in "A" are black.

Finally, methods that account for rate variation across lineages, as well as substantial increases in the size of molecular data sets to estimate mammalian divergences, have resulted in some consensus around an origin of Placentalia and Marsupialia during the Late Cretaceous, with crown orders diversifying close to or after the K-Pg boundary (dos Reis *et al.* 2012; Mitchell *et al.* 2014; Tarver *et al.* 2016; Wu *et al.* 2017, as discussed above in the section on "afrotherian and xenarthran origins"). Some of these recent studies are also accompanied by relatively narrow margins of error; e.g., the combined data set of dos Reis *et al.* (2012: tab. 1) pinpointed the common ancestor of Placentalia to a mean of 89.1 Ma with a 95% confidence interval between 87.9 and 90.4 Ma. The mean from Wu *et al.* was similar at 89.8 ± 82.7–99.5 Ma; that of Tarver *et al.* (2016) was higher at 92.96 ± 86.4–99.9 Ma; and that of Foley *et al.* (2016) was higher still at 94.4 Ma (confidence intervals shown graphically in their fig. 1 but not precisely reported). Compared with Kumar and Hedges (1998) who gave a mean estimate for an intra-Placentalia branching event at 129 Ma, these studies all represent a substantial increase in the congruence of molecular clock estimates of mammalian divergences to the fossil record.

As noted by O'Leary *et al.* (2013), first appearances of mammalian crown orders in the fossil record all postdate the Cretaceous-Paleogene boundary at ca. 66 Ma. However, first-appearance dates are minimum estimates; they are not synonymous with cladistic origination points. Indeed, given what we know about speciation, a first appearance must postdate clade origins. Applications of the molecular clock to understand high-level divergences within Theria show that the main interordinal divergences within Placentalia (i.e., origins of Atlantogenata, Laurasiatheria, and Euarchontoglires), as well as the initial diversification of Marsupialia, likely took place during the Late Cretaceous. This view of a Cretaceous origin of the now dominant mammalian group, Placentalia, is consistent with Benton (2010). He argued that the Cretaceous Terrestrial Revolution, when angiosperms became the dominant plant group between 125 and 80 mya, likely played an important role in mammalian evolution.

7.6 Conclusions

In 1910, William King Gregory knew a lot about mammalian evolution, as did George Simpson in 1945. They knew that evolutionary descent with modification was the mechanism by which organisms have become diverse over the course of geological time. They knew that via this mechanism, monotremes, marsupials, and placentals each had a common ancestor, one which had in turn evolved from a cynodont-grade synapsid amniote. Both would recognize the names now used for the majority of mammalian high-level taxa, at least from the arbitrary Linnean rank of Order and below. Both would quickly recognize any number of long-extinct fossils as members of (for example) primates, rodents, perissodactyls, cetaceans, ruminants, bats, carnivorans, armadillos, hyraxes, and marsupials. Both knew that mammals existed long before the base of the Paleogene but greatly increased in size and diversity thereafter. Both knew that the modern global fauna, dominated by rodents, bats, and artiodactyls (with pockets of endemic diversity in places like South America, Australia, and Madagascar) saw much greater taxonomic and morphological variation among perissodactyls, hyraxes, proboscideans, and other groups during and prior to the Miocene. The fact that the previous decades of research and discovery have proven Gregory and Simpson (among others) correct on these and other essential points concerning mammalian evolution is worth acknowledging.

On the other hand, Gregory and Simpson did not know exactly how closely primates, dermopterans, tupaiids, and glires were interrelated, for example, to the exclusion of bats. Many close interordinal relations, e.g., between paenungulates, tubulidentates, macroscelideans, and tenrecoids, between hippopotamids and cetaceans, or between microbiotheres and Australasian marsupials, were not generally accepted until the late 20th century. Our ability to recognize and have substantial confidence in high-level clades such as Afrotheria, Artiodactyla (including cetaceans), and Australidelphia is in large part dependent on molecular data that are rarely preserved in fossils.

Nonetheless, anyone interested in mammalian evolution over time has reason to be optimistic because of the currently well-corroborated hypothesis on the shape of the mammalian tree. This provides a framework into which fossils, even poorly known ones, can potentially fit and thereby inform hypotheses on divergence patterns and character evolution. The well-corroborated tree, continuing paleontological discovery, combined with novel methods for extracting data from long-extinct taxa have already improved the precision with which we discuss enigmatic, wastebasket taxa such as "insectivores" and "condylarths". These are terms based largely on uncertainty and are slowly being replaced by others with phylogenetic and evolutionary meaning. For example, as detailed above, many recent studies have proposed affinities of extinct anthracotheres, anthracobunids, litopterns, notoungulates, pantodonts, hyopsodontids, phenacodontids, and possibly desmostylians with one or both modern euungulate clades (i.e., Perissodactyla and Artiodactyla) within Laurasiatheria, not Afrotheria (Figs. 7.8 and 7.9, and Tab. 7.3). This points toward a much higher diversity of laurasiatherian euungulates during the Paleogene than previously suspected.

Similarly, the dissociation of the extant "insectivoran" grade into Afroinsectivora (i.e., chrysochlorids, macroscelideans, and tenrecids), Euarchontoglires (i.e., dermopterans and tupaiids), and Lipotyphla (i.e., erinaceids, solenodontids, soricids, and talpids) enables correspondingly more precise discussion of their fossil relatives. The hypotheses now for our students to test are (for example) that *Leptictis* is eutherian, outside of Placentalia (Fig. 7.5, Tab. 7.3), and that proscalopids, *Apternodus*, *Oligoryctes*, nyctitheriids, didymoconids, "paleoryctids", and "erinaceomorphs" such as *Macrocranion* are within or related to Lipotyphla (Fig. 7.8, Tab. 7.3). Long-enigmatic (and well-preserved) fossils of apatemyids are now compellingly hypothesized to be part of Euarchontoglires (Fig. 7.7, Tab. 7.3). These statements comprise testable hypotheses, subject to further scrutiny not only by the kinds of morphological data that have defined paleontology for decades, but also by biological information encased in fossils and their environments that, surely, has yet to be discovered.

Acknowledgments

I am grateful to Frank Zachos for his contributions to the text and for the opportunity to write this chapter. I am grateful to Erik Seiffert for discussions on the Paleogene fossil record of Africa. Christine Janis, Christian de Muizon, Ken Rose, and Mark Springer provided further recommendations and insights into mammalian evolution, systematics and paleobiology which have greatly improved the text.

Literature

Ameghino, F. (1889): Contribucion al conocimiento de los mamiferos fosiles de la República Argentina: Obra escrita bajo los auspicios de la Academia nacional de ciencias de la República Argentina para ser presentada a la Exposicion universal de Paris de 1889. *Acta Academica Nacional de Ciencias, Córdoba Y Buenos Aires* 6:1–1027.

Amrine-Madsen, H., Koepfli, K.-P., Wayne, R.K. & Springer, M.S. (2003): A new phylogenetic marker, apolipoprotein B, provides compelling evidence for eutherian relationships. *Molecular Phylogenetics and Evolution* 28:225–240.

Amson, E., de Muizon C., Laurin M., Argot C. & de Buffrénil V. (2014): Gradual adaptation of bone structure to aquatic lifestyle in extinct sloths from Peru. *Proceedings of the Royal Society of London B: Biological Sciences* 281:20140192.

Andrews, C.W. (1906): *A Descriptive Catalog of the Tertiary Vertebrata of the Fayum, Egypt. Trustees British Museum (Natural History)*, London, I–XVII.

Angielczyk, K.D. & Kammerer, C.F. (2018): Non-mammalian synapsids: the deep roots of the mammalian family tree. In: Zachos, F.E. & Asher, R.J. (eds.), *Handbook of Zoology, Mammalia*. DeGruyter, Berlin.

Anthony, H.E. (1924): A new fossil perissodactyl from Peru. *American Museum Novitates* 111:1–3.

Archibald, J.D. (1996): Fossil evidence for a Late Cretaceous origin of "hoofed" mammals. *Science* 272:1150.

Archibald, J.D., Averianov, A.O. & Ekdale, E.G. (2001): Late Cretaceous relatives of rabbits, rodents, and other extant eutherian mammals. *Nature* 414:62–65.

Arcila, D., Ortí, G., Vari, R., Armbruster, J.W., J. Stiassny, M.L., Ko, K.D., Sabaj, M.H., Lundberg, J., Revell, L.J. & Betancur, R.-R. (2017): Genome-wide interrogation advances resolution of recalcitrant groups in the tree of life. *Nature Ecology & Evolution* 1:20.

Arnason, U., Adegoke, J.A., Bodin, K., Born, E.W., Esa, Y.B., Gullberg, A., Nilsson, M., Short, R.V., Xu, X. & Janke, A. (2002): Mammalian mitogenomic relationships and the root of the eutherian tree. *Proceedings of the National Academy of Sciences of the United States of America* 99:8151–8156.

Arnason, U. & Gullberg, A. (1994): Relationship of baleen whales established by cytochrome b gene sequence comparison. *Nature* 367:726.

Arnason, U., Gullberg, A. & Janke, A. (1997): Phylogenetic analyses of mitochondrial DNA suggest a sister group relationship between Xenarthra (Edentata) and Ferungulates. *Molecular Biology and Evolution* 14:762–768.

Asher, R.J. (2007): A web-database of mammalian morphology and a reanalysis of placental phylogeny. *BMC Evolutionary Biology* 7:108.

Asher, R.J. (2012a): *Evolution and Belief: Confessions of a Religious Paleontologist.* Cambridge University Press, Cambridge.

Asher, R.J. (2012b): Evolution and the Tree of Life: How Hyperbole Poisons Everything. *Huffington Post.*

Asher, R.J. (1999): A morphological basis for assessing the phylogeny of the "Tenrecoidea" (Mammalia, Lipotyphla). *Cladistics* 15:231–252.

Asher, R.J. (2018): Taxonomy, trees, and truth in historical mammalogy. In: Zachos, F.E. & Asher, R.J. (eds.), *Handbook of Zoology, Mammalia*. DeGruyter, Berlin.

Asher, R.J., Bennett, N. & Lehmann, T. (2009): The new framework for understanding placental mammal evolution. *BioEssays* 31:853–864.

Asher, R.J., Horovitz, I. & Sánchez-Villagra, M.R. (2004): First combined cladistic analysis of marsupial mammal interrelationships. *Molecular Phylogenetics and Evolution* 33:240–250.

Asher, R.J. & Sánchez-Villagra, M.R. (2005): Locking yourself out: Diversity among dentally zalambdodont therian mammals. *Journal of Mammalian Evolution* 12:265–282.

Asher, R.J. & Lehmann, T. (2008): Dental eruption in afrotherian mammals. *BMC Biology* 6:14.

Asher, R.J. & Seiffert, E. (2010): Afrotheria. In: Werdelin, L. & Sanders, W.J. (eds.), *Cenozoic Mammals of Africa*. University of California Press, Berkeley.

Asher, R.J. & Avery, D.M. (2010): New golden moles (Afrotheria, Chrysochloridae) from the Pliocene of South Africa. *Paleontologica Electronica* 13:3A.

Asher, R.J., Gunnell, G.F., Seiffert, E.R., Pattinson, D., Tabuce, R., Hautier, L. & Sallam, H.M. (2017): Dental eruption and growth in Hyracoidea (Mammalia, Afrotheria). *Journal of Vertebrate Paleontology* e1317638.

Asher, R.J. & Helgen, K.M. (2010): Nomenclature and placental mammal phylogeny. *BMC Evolutionary Biology* 10:102.

Asher, R.J., Lin, K.H., Kardjilov, N. & Hautier, L. (2011): Variability and constraint in the mammalian vertebral column. *Journal of Evolutionary Biology* 24:1080–1090.

Asher, R.J., McKenna, M.C., Emry, R.J., Tabrum, A.R. & Kron, D.G. (2002): Morphology and relationships of *Apternodus* and other extinct, zalambdodont, placental mammals. *Bulletin of the American Museum of Natural History* 1–117.

Asher, R.J., Meng, J., Wible, J.R., McKenna, M.C., W. Rougier, G., Dashzeveg, D. & Novacek M.J. (2005): Stem Lagomorpha and the antiquity of Glires. *Science* 307:1091–1094.

Asher, R.J., Novacek, M.J. & Geisler, J.H. (2003): Relationships of endemic African mammals and their fossil relatives based on morphological and molecular evidence. *Journal of Mammalian Evolution* 10:131–164.

Barnosky, A.D. (1981): A skeleton of *Mesoscalops* (Mammalia, Insectivora) from the Miocene Deep River Formation, Montana, and a review of the proscalopid moles: evolutionary, functional, and stratigraphic relationships. *Journal of Vertebrate Paleontology* 1:285–339.

Barry, J.C. (1980): Occurrence of a hyaenodontine creodont (Mammalia) in the Late Miocene of Pakistan. *Journal of Paleontology* 1128–1131.

Baum, D.A., Smith, S.D. & Donovan, S.S.S. (2005): The tree-thinking challenge. *Science* 310:979–980.

Beard, K.C. (1993): Phylogenetic systematics of the Primatomorpha, with special reference to Dermoptera. Mammal Phylogeny: *Placentals* 129–150.

Beatty, B.L. & Cockburn, T.C. (2015): New insights on the most primitive desmostylian from a partial skeleton of Behemotops (Desmostylia, Mammalia) from Vancouver Island, British Columbia. *Journal of Vertebrate Paleontology* 35:e979939.

Beck, R.M.D. (2008): A dated phylogeny of marsupials using a molecular supermatrix and multiple fossil constraints. *Journal of Mammalogy* 89:175–189.

Beck, R.M.D. (2012): An "ameridelphian" marsupial from the Early Eocene of Australia supports a complex model of Southern Hemisphere marsupial biogeography. *Naturwissenschaften* 99:715–729.

Beck, R.M.D. & Lee, M.S.Y. (2014): Ancient dates or accelerated rates? Morphological clocks and the antiquity of placental mammals. Proceedings of the Royal Society of London B: *Biological Sciences* 281:20141278.

Benoit, J., Adnet, S., El Mabrouk, E., Khayati, H., Ali, M.B.H., Marivaux, L., Merzeraud, G., Merigeaud, S., Vianey-Liaud, M. & Tabuce, R. (2013): Cranial remain from Tunisia provides new clues for the origin and evolution of Sirenia (Mammalia, Afrotheria) in Africa. *PLoS One* 8:e54307.

Benoit, J., Lehmann, T., Vatter, M., Lebrun, R., Merigeaud, S., Costeur, L. & Tabuce, R. (2015): Comparative anatomy and three-dimensional geometric-morphometric study of the bony labyrinth of Bibymalagasia (Mammalia, Afrotheria). *Journal of Vertebrate Paleontology* 35:e930043.

Benton, M.J. (2010): The origins of modern biodiversity on land. *Philosophical Transactions of the Royal Society of London B: Biological Sciences* 365:3667–3679.

Benton, M.J., Donoghue, P.C.J., Asher, R.J., Friedman, M., Near, T.J., & Vinther, J. (2015): Constraints on the timescale of animal evolutionary history. *Palaeontologia Electronica* 18:1–106.

Benton, M., Donoghue, P.C.J. & Asher, R.J. (2009): Calibrating and constraining molecular clocks. *The Timetree of Life* 35–86.

Bergqvist, L.P. & von Koenigswald, W. (2017): The dentition of *Carodnia vieirae* (Mammalia: Xenungulata): enamel microstructure and mastication pattern. *Palaeontologia Electronica* 20:1–15.

Bergqvist, L.P. (2008): Postcranial Skeleton of the Upper Paleocene (Itaboraian) "Condylarthra" (Mammalia) of Itaborai Basin, Brazil; pp. 107–133 in Sargis, E.J. & Dagosto, M. (eds.), *Mammalian Evolutionary Morphology: A Tribute to Frederick* S. *Szalay*. Springer, Dordrecht.

Berta, A. (2017): *The Rise of Marine Mammals: 50 Million Years of Evolution*. Johns Hopkins University Press.

Billet, G., Hautier, L.; De Muizon, C. & Valentin, X. (2011): Oldest cingulate skulls provide congruence between morphological and molecular scenarios of armadillo evolution. *Proceedings of the Royal Society of London B: Biological Sciences* 278:2791–2797.

Billet, G. & Martin, T. (2011): No evidence for an afrotherian-like delayed dental eruption in South American notoungulates. *Naturwissenschaften* 98:509–517.

Bininda-Emonds, O.R.P., Cardillo, M., Jones, K.E., MacPhee, R.D.E., Beck, R.M.D., Grenyer, R., Price, S.A., Vos, R.A., Gittleman, J.L. & Purvis, A. (2007): The delayed rise of present-day mammals. *Nature* 446:507.

Bloch, J.I. & Boyer, D.M. (2002): Grasping primate origins. *Science* 298:1606–1610.

Bloch, J.I., Chester, S.G.B. & Silcox, M.T. (2016): Cranial anatomy of Paleogene Micromomyidae and implications for early primate evolution. *Journal of Human Evolution* 96:58–81.

Bloch, J.I., Silcox, M.T., Boyer, D.M. & Sargis, E.J. (2007): New Paleocene skeletons and the relationship of plesiadapiforms to crown-clade primates. *Proceedings of the National Academy of Sciences of the United States of America* 104:1159–1164.

Boisserie, J.-R., Lihoreau, F. & Brunet, M. (2005): The position of Hippopotamidae within Cetartiodactyla. *Proceedings of the National Academy of Sciences of the United States of America* 102:1537–1541.

Boyden, A. & Gemeroy, D. (1950): The relative position of the Cetacea among the orders of Mammalia as indicated by precipitin tests. *Zoologica* 35:145–151.

Boyer, D.M. & Georgi, J.A. (2007): Cranial morphology of a pantolestid eutherian mammal from the Eocene Bridger Formation, Wyoming, USA: implications for relationships and habitat. *Journal of Mammalian Evolution* 14:239–280.

Boyer, D.M., Prasad, G.V.R., Krause, D.W., Godinot, M., Goswami, A., Verma, O. & Flynn, J.J. (2010): New postcrania of Deccanolestes from the Late Cretaceous of India and their bearing on the evolutionary and biogeographic history of euarchontan mammals. *Naturwissenschaften* 97:365–377.

Brace, S., Thomas, J.A., Dalén, L., Burger, J., MacPhee, R.D.E., Barnes, I. & Turvey, S.T. (2016): Evolutionary history of the Nesophontidae, the last unplaced recent mammal family. *Molecular Biology and Evolution* 33:3095–3103.

Bremer, K. (1994): Branch support and tree stability. *Cladistics* 10:295–304.
Bronzati, M. (2017): Should the terms 'basal taxon' and 'transitional taxon' be extinguished from cladistic studies with extinct organisms? *Palaeontologia Electronica* 20:1–12.
Broom, R. (1941): On two Pleistocene golden moles. *Annals of the Transvaal Museum* 20:215–216.
Broom, R. (1948): Some South African Pliocene and Pleistocene mammals. *Annals of the Transvaal Museum* 21:1–38.
Buckley, M. (2013): A molecular phylogeny of *Plesiorycteropus* reassigns the extinct mammalian order "Bibymalagasia." *PLoS One* 8:e59614.
Butler, P.M. (1984): Macroscelidea, Insectivora and Chiroptera from the Miocene of east Africa. *Palaeovertebrata* 14:117–198.
Butler, P.M. (1972): The problem of insectivore classification. In: Joysey, K.A. & Kemp, T.S. (eds.), *Studies in Vertebrate Evolution*. Oliver & Boyd, Edinburgh: 253–265.
Butler, P.M. (1956): The skull of *Ictops* and the classification of the Insectivora. *Journal of Zoology* 126:453–481.
Cabrera, A. (1922): *Manual de Mastozoología*. Calpe, Madrid.
Cannarozzi, G., Schneider, A. & Gonnet, G. (2007): A phylogenomic study of human, dog, and mouse. *PLoS Computational Biology* 3:e2.
Carlini, A.A., Scillato-Yané, G.J., Vizcaíno, S.F. & Dozo, M.T. (1992): Un singular Myrmecophagidae (Xenarthra, Vermilingua) de Edad Colhuehuapense (Oligoceno tardío-Mioceno temprano) de Patagonia, Argentina. *Ameghiniana* 29:176.
Carrillo, J.D. & Asher, R.J. (2017): An exceptionally well-preserved skeleton of *Thomashuxleya externa* (Mammalia, Notoungulata), from the Eocene of Patagonia, Argentina. *Palaeontologia Electronica* 20:1–33.
Chester, S.G.B., Bloch, J.I., Boyer, D.M. & Clemens, W.A. (2015): Oldest known euarchontan tarsals and affinities of Paleocene Purgatorius to Primates. *Proceedings of the National Academy of Sciences of the United States of America* 112:1487–1492.
Ciancio, M.R., Castro, M.C., Galliari, F.C., Carlini, A.A. & Asher, R.J. (2012): Evolutionary implications of dental eruption in Dasypus (Xenarthra). *Journal of Mammalian Evolution* 19:1–8.
Cifelli, R.L. & Diaz, J.G. (1989): New remains of *Prothoatherium columbianus* (Litopterna, Mammalia) from the Miocene of Colombia. *Journal of Vertebrate Paleontology* 9:222–231.
LeGros Clark, W.E. & Sonntag, C.F. (1926): A monograph of *Orycteropus afer*—III. The skull. The skeleton of the trunk and limbs. General summary. *Proceedings of the Zoological Society of London* 30:445–485.
Clemens, W.A. (1973): Fossil Mammals of the Type Lance Formation Wyoming: Eutheria and Summary. *University of California Publications in Geological Sciences* 94:1–102.
Cohen, K.M., Finney, S.C., Gibbard, P.L. & Fan, J.-X. (2013): The ICS international chronostratigraphic chart. *Episodes* 36:199–204.
Cooper, L.N., Seiffert, E.R., Clementz, M., Madar, S.I., Bajpai, S., Hussain, S.T. & Thewissen, J.G.M. (2014): Anthracobunids from the middle Eocene of India and Pakistan are stem perissodactyls. *PloS One* 9:e109232.
Cope, E.D. (1875): On the Supposed Carnivora of the Eocene of the Rocky Mountains. *Proceedings of the Academy of Natural Sciences, Philadelphia* 27:444–449.
Cope, E.D. (1891): *Syllabus of Lectures on Geology and Paleontology*. Ferris Brothers, Philadelphia.
Cote, S., Werdelin, L., Seiffert, E.R. & Barry, J.C. (2007): Additional material of the enigmatic Early Miocene mammal Kelba and its relationship to the order Ptolemaiida. *Proceedings of the National Academy of Sciences of the United States of America* 104:5510–5515.
Court, N. (1992a): A unique form of dental bilophodonty and a functional interpretation of peculiarities in the masticatory system of Arsinoitherium (Mammalia, Embrithopoda). *Historical Biology* 6:91–111.
Court, N. (1992b): The skull of Arsinoitherium (Mammalia, Embrithopoda) and the higher order interrelationships of ungulates. *Palaeovertebrata* 22:1–43.
Croft, D.A. (2016): *Horned Armadillos and Rafting Monkeys: The Fascinating Fossil Mammals of South America*. Indiana University Press, Bloomington.
Crumpton, N. & Thompson, R.S. (2013): The holes of moles: osteological correlates of the trigeminal nerve in Talpidae. *Journal of Mammalian Evolution* 20:213–225.
Damuth, J. & Janis, C.M. (2011): On the relationship between hypsodonty and feeding ecology in ungulate mammals, and its utility in palaeoecology. *Biological Reviews* 86:733–758.
Darwin, C. (1859): *On the Origin of Species by Means of Natural Selection*. John Murray, London.
de Jong, W.W., Zweers, A. & Goodman, M. (1981): Relationship of aardvark to elephants, hyraxes and sea cows from α-crystallin sequences. *Nature* 292:538–540.
Delsuc, F., Gibb, G.C., Kuch, M., Billet, G., Hautier, L., Southon, J., Rouillard, J.-M., Fernicola, J.C., Vizcaíno, S.F., MacPhee, R. D. E., *et al.* (2016): The phylogenetic affinities of the extinct glyptodonts. *Current Biology* 26:R155–R156.
Delsuc, F., Vizcaíno, S.F. & Douzery, E.J.P. (2004): Influence of Tertiary paleoenvironmental changes on the diversification of South American mammals: a relaxed molecular clock study within xenarthrans. *BMC Evolutionary Biology* 4:11.
D'Erchia, A.M., Gissi, C., Pesole, G., Saccone, C. & Arnason, U. (1996): The guinea-pig is not a rodent. *Nature* 381:597.
Dirks, W., Anemone, R.L., Holroyd, P.A., Reid, D.J. & Walton, P. (2009): Phylogeny, life history and the timing of molar crown formation in two archaic ungulates, *Meniscotherium* and *Phenacodus* (Mammalia, "Condylarthra"). In: Koppe, T., Meyer, G. & Alt, K.W. (eds.), *Comparative Dental Morphology*. vol. 13. Karger Publishers, Basel.
Dolgin, E. (2012): Rewriting evolution. *Nature* 486:460.
Domning, D.P. (2001): The earliest known fully quadrupedal sirenian. *Nature* 413:625.
Donoghue, P.C.J. & Purnell, M.A. (2009): Distinguishing heat from light in debate over controversial fossils. *BioEssays* 31:178–189.
dos Reis, M., Inoue, J., Hasegawa, M., Asher, R.J., Donoghue, P.C.J. & Yang, Z. (2012): Phylogenomic datasets provide both precision and accuracy in estimating the timescale of placental mammal phylogeny. *Proceedings of the Royal Society of London B: Biological Sciences* 279:3491–3500.
Edinger, T. (1964): Midbrain exposure and overlap in mammals. *American Zoologist* 5–19.
Edwards, E.J., Osborne, C.P., Strömberg, C.A.E., Smith, S.A., Consortium, C.G., *et al.* (2010): The origins of C4 grasslands: integrating evolutionary and ecosystem science. *Science* 328:587–591.
Edwards, S.V., Xi, Z., Janke, A., Faircloth, B.C., McCormack, J.E., Glenn, T.C., Zhong, B., Wu, S., Lemmon, E.M., Lemmon, A.R., *et al.* (2016): Implementing and testing the multispecies coalescent model: a valuable paradigm for phylogenomics. *Molecular Phylogenetics and Evolution* 94:447–462.

Ekdale, E.G. (2013): Comparative anatomy of the bony labyrinth (inner ear) of placental mammals. *PLoS One* 8:e66624.

Emry, R.J. (1970): A North American Oligocene pangolin and other additions to the Pholidota. *Bulletin of the American Museum of Natural History* 142.

Emry, R.J. (1989): A tiny new Eocene ceratomorph and comments on "tapiroid" systematics. *Journal of Mammalogy* 70:794–804.

Erdal, O., Antoine, P.-O. & Sen, S. (2016): New material of *Palaeoamasia kansui* (Embrithopoda, Mammalia) from the Eocene of Turkey and a phylogenetic analysis of Embrithopoda at the species level. *Palaeontology* 59:631–655.

Esselstyn, J.A., Oliveros, C.H., Swanson, M.T. & Faircloth, B.C. (2017): Investigating difficult nodes in the placental mammal tree with expanded taxon sampling and thousands of ultraconserved elements. *Genome Biology and Evolution* 9: 2308–2321.

Fastovsky, D.E. & Dott, R.H. (1986): Sedimentology, stratigraphy, and extinctions during the Cretaceous-Paleogene transition at Bug Creek, Montana. *Geology* 14:279–282.

Fielden, L.J., Perrin, M.R. & Hickman, G.C. (1990): Feeding ecology and foraging behaviour of the Namib Desert golden mole, *Eremitalpa granti namibensis* (Chrysochloridae). *Journal of Zoology* 220:367–389.

Fischer, M.S. (1992): Hyracoidea. *Handbuch Der Zoologie, Mammalia* 8:1–169.

Fischer, M.S. (1989): Hyracoids, the sister-group of perissodactyls. In: Prothero, D. & Schoch, R. (eds.), *The Evolution of Perissodactyls*. Oxford University Press, Oxford. 37–56.

Fischer, M.S. (1990): Un trait unique de l'oreille des éléphants et des siréniens (Mammalia): un paradoxe phylogénétique. *Comptes Rendus de l'Académie Des Sciences.* Série 3, Sciences de La Vie 311:157–162.

Foley, N.M., Springer, M.S. & Teeling, E.C. (2016): Mammal madness: is the mammal tree of life not yet resolved? *Philosophical Transactions of the Royal Society of London B: Biological Sciences* 371:20150140.

Foote, M., Hunter, J.P., Janis, C.M. & Sepkoski, J.J. (1999): Evolutionary and preservational constraints on origins of biologic groups: divergence times of eutherian mammals. *Science* 283:1310–1314.

Fox, R.C., Scott, C.S. & Rankin, B.D. (2010): New early carnivoran specimens from the Puercan (earliest Paleocene) of Saskatchewan, Canada. *Journal of Paleontology* 84:1035–1039.

Froehlich, D.J. (2002): Quo vadis eohippus? The systematics and taxonomy of the Early Eocene equids (Perissodactyla). *Zoological Journal of the Linnean Society* 134:141–256.

Gatesy, J., Hayashi, C., Cronin, M.A. & Arctander, P. (1996): Evidence from milk casein genes that cetaceans are close relatives of hippopotamid artiodactyls. *Molecular Biology and Evolution* 13:954–963.

Gatesy, J., Meredith, R.W., Janecka, J.E., Simmons, M.P., Murphy, W.J., & Springer, M.S. (2017): Resolution of a concatenation/coalescence kerfuffle: partitioned coalescence support and a robust family-level tree for Mammalia. *Cladistics* 33:295–332.

Gatesy, J., O'Grady, P. & Baker, R.H. (1999): Corroboration among data sets in simultaneous analysis: hidden support for phylogenetic relationships among higher level artiodactyl taxa. *Cladistics* 15:271–313.

Gaudin, T.J. & Branham, D.G. (1998): The phylogeny of the Myrmecophagidae (Mammalia, Xenarthra, Vermilingua) and the relationship of Eurotamandua to the Vermilingua. *Journal of Mammalian Evolution* 5:237–265.

Gaudin, T.J., Emry, R.J. & Pogue, B. (2006): A new genus and species of pangolin (Mammalia, Pholidota) from the Late Eocene of Inner Mongolia, China. *Journal of Vertebrate Paleontology* 26:146–159.

Gaudin, T., Emry, R. & Morris, J. (2016): Skeletal Anatomy of the North American Pangolin *Patriomanis americana* (Mammalia, Pholidota) from the Latest Eocene of Wyoming (USA). *Smithsonian Contributions to Paleobiology* vii-102.

Gaudin, T.J. & Wible, J.R. (2006): The Phylogeny of Living and Extinct Armadillos (Mammalia, Xenarthra, Cingulata): A Craniodental Analysis. In: Carrano, M.T., Gaudin, T.J., Blob, R.W., & Wible, J.R. (eds.), *Amniote Paleobiology: Perspectives on the Evolution of Mammals, Birds, and Reptiles*. University of Chicago Press, Chicago: 153–198.

Gazin, C.L. (1965): A study of the Early Tertiary condy-larthran mammal *Meniscotherium*. *Smithsonian Miscellaneous Collections* 149:1–98.

Gazin, C.L. (1968): A study of the Eocene condylarthran mammal *Hyopsodus*. *Smithsonian Miscellaneous Collections* 153:1–90.

Geisler, J.H. (2004): Humeri of *Oligoscalops* (Proscalopidae, Mammalia) from the Oligocene of Mongolia. *Bulletin of the American Museum of Natural History* 285:166–176.

Geisler, J.H. & Theodor, J.M. (2009): Hippopotamus and whale phylogeny. *Nature* 458:81–82.

Gheerbrant, E. (1995): Les mammifères paléocènes du Bassin d'Ouarzazate (Maroc). III. Adapisoriculidae et autres mammifères (Carnivora,? Creodonta, Condylarthra,? Ungulata et incertae sedis). *Palaeontographica* A 1:39–132.

Gheerbrant, E. (2009): Paleocene emergence of elephant relatives and the rapid radiation of African ungulates. *Proceedings of the National Academy of Sciences of the United States of America* 106:10717–10721.

Gheerbrant, E. (1991): *Todralestes variabilis* n. g., n. sp., nouveau proteuthérien (Eutheria, Todralestidae fam. nov.) du paléocène du Maroc. *Comptes Rendus de l'Académie Des Sciences. Série 2, Mécanique, Physique, Chimie, Sciences de L'univers, Sciences de La Terre* 312:1249–1255.

Gheerbrant, E., Amaghzaz, M., Bouya, B., Goussard, F. & Letenneur, C. (2014): *Ocepeia* (Middle Paleocene of Morocco): the oldest skull of an afrotherian mammal. *PloS One* 9:e89739.

Gheerbrant, E., Domning, D.P. & Tassy, P. (2005): Paenungulata (Proboscidea, Sirenia, Hyracoidea and Relatives); in Rose, K.D. & Archibald, J.D. (eds.), *The Rise of Placental Mammals: Origins and Relationships of the Major Extant Clades*. Johns Hopkins University Press, Baltimore.

Gheerbrant, E., Filippo, A. & Schmitt, A. (2016): Convergence of Afrotherian and Laurasiatherian ungulate-like mammals: first morphological evidence from the Paleocene of Morocco. *PloS One* 11:e0157556.

Gheerbrant, E., Peigné, S. & Thomas, H. (2007): Première description du squelette d'un hyracoide paléogène: *Saghatherium antiquum* de l'Oligocène inférieur de Jebel al Hasawnah, Libye. *Palaeontographica Abteilung A* 93–145.

Gheerbrant, E., Sudre, J., Cappetta, H., Iarochéne, M., Amaghzaz, M. & Bouya, B. (2002): A new large mammal from the Ypresian of Morocco: evidence of surprising diversity of early proboscideans. *Acta Palaeontologica Polonica* 47.

Gheerbrant, E. & Tassy, P. (2009): L'origine et l'évolution des éléphants. *Comptes Rendus Palevol* 8:281–294.

Gidley, J.W. (1909): Notes on the fossil mammalian genus Ptilodus with descriptions of new species. *Proceedings of the United States National Museum* 36:611–626.

Gingerich, P.D., Smith, B.H. & Simons, E.L. (1990): Hind limbs of Eocene *Basilosaurus*: evidence of feet in whales. *Science* 249:154–157.

Gingerich, P.D., ul Haq, M., Zalmout, I.S., Khan, I.H. & Malkani, M. S. (2001): Origin of whales from early artiodactyls: hands and feet of Eocene Protocetidae from Pakistan. *Science* 293:2239–2242.

Goswami, A., Prasad, G.V.R., Upchurch, P., Boyer, D.M., Seiffert, E.R., Verma, O., Gheerbrant, E. & Flynn, J.J. (2011): A radiation of arboreal basal eutherian mammals beginning in the Late Cretaceous of India. *Proceedings of the National Academy of Sciences of the United States of America* 108:16333–16338.

Green, R.E., Krause, J., W. Briggs, A., Maricic, T., Stenzel, U., Kircher, M., Patterson, N., Li, H., Zhai, W., Fritz, M.H.-Y., *et al.* (2010): A draft sequence of the Neandertal genome. *Science* 328:710–722.

Gregory, W.K. (1910): The Orders of Mammals. *Bulletin of the American Museum of Natural History* 27:1–524.

Gregory, W.K. (1947): The monotremes and the palimpsest theory. *Bulletin of the American Museum of Natural History* 88:1–52.

Haeckel, E.H.P.A. (1866): *General Morphologie der Organismen.* Georg Reimer, Berlin.

Halliday, T.J.D., Upchurch, P. & Goswami, A. (2017): Resolving the relationships of Paleocene placental mammals. *Biological Reviews* 92:521–550.

Hallström, B.M. & Janke, A. (2010): Mammalian evolution may not be strictly bifurcating. *Molecular Biology and Evolution* 27:2804–2816.

Hautier, L., Rodrigues, H.G., Billet, G. & Asher, R.J. (2016): The hidden teeth of sloths: evolutionary vestiges and the development of a simplified dentition. *Scientific Reports* 6.

Hegner, R.W. & Stiles, K.A. (1951): *College Zoology.* MacMillan, New York.

Hershkovitz, P. (1999): Dromiciops gliroides Thomas, 1894, Last of the Microbiotheria(Marsupialia), with a Review of the Family Microbiotheriidae. Fieldiana. *Zoology* 1–60.

Holbrook, L.T. (2014): On the skull of *Radinskya* (Mammalia) and its phylogenetic position. *Journal of Vertebrate Paleontology* 34:1203–1215.

Hooker, J.J. (2013): Origin and evolution of the Pseudorhyncocyonidae, a European Paleogene family of insectivorous placental mammals. *Palaeontology* 56:807–835.

Hooker, J.J. (2001): Tarsals of the extinct insectivoran family Nyctitheriidae (Mammalia): evidence for archontan relationships. *Zoological Journal of the Linnean Society* 132:501–529.

Hooker, J.J. (1994): The beginning of the equoid radiation. *Zoological Journal of the Linnean Society* 112:29–63.

Hooker, J.J. & Dashzeveg, D. (2004): The origin of chalicotheres (Perissodactyla, Mammalia). *Palaeontology* 47:1363–1386.

Hooker, J.J. & Russell, D.E. (2012): Early Palaeogene Louisinidae (Macroscelidea, Mammalia), their relationships and north European diversity. *Zoological Journal of the Linnean Society* 164:856–936.

Horovitz, I., Martin, T., Bloch, J., Ladevèze, S., Kurz, C. & Sánchez-Villagra, M.R. (2009): Cranial anatomy of the earliest marsupials and the origin of opossums. *PLoS One* 4:e8278.

Horovitz, I. & Sánchez-Villagra, M.R. (2003): A morphological analysis of marsupial mammal higher-level phylogenetic relationships. *Cladistics* 19:181–212.

Huxley, T.H. (1880): On the application of the laws of evolution to the arrangement of the Vertebrata and more particularly of the Mammalia. *Proceedings of the Zoological Society of London* 1880:649–662.

International Commission on Zoological Nomenclature. (1999): *International Code of Zoological Nomenclature*, Ride, W.D.L., Cogger, H.G., Dupuis, C., Kraus, O., Minelli, A., Thompson, F., Tubbs, C.P.K., *et al.* (eds.) 4th ed. International Trust for Zoological Nomenclature, London.

Irwin, D.M., Kocher, T.D. & Wilson, A.C. (1991): Evolution of the cytochrome b gene of mammals. *Journal of Molecular Evolution* 32:128–144.

Janis, C.M. (1988): An estimation of tooth volume and hypsodonty indices in ungulate mammals, and the correlation of these factors with dietary preference. In: Russell, D.E., Santoro, J.P. & Sigogneau-Russell, D. (eds.), *Teeth Revisited: Proceedings of the VIIth International Symposium on Dental Morphology.* vol. 53. Mémoires du Muséum national d'histoire Naturelle Paris (serie C): 367–387.

Janis, C.M., Scott, K.M. & Jacobs, L.L. (1998): *Evolution of Tertiary Mammals of North America: Volume 1, Terrestrial Carnivores, Ungulates, and Ungulate like Mammals.* Cambridge University Press, Cambridge.

Janke, A., Magnell, O., Wieczorek, G., Westerman, M. & Arnason, U. (2002): Phylogenetic analysis of 18S rRNA and the mitochondrial genomes of the wombat, *Vombatus ursinus*, and the spiny anteater, *Tachyglossus aculeatus*: increased support for the Marsupionta hypothesis. *Journal of Molecular Evolution* 54:71–80.

Jepsen, G.L. (1934): A Revision of the American Apatemyidae and the Description of a New Genus, "*Sinclairella*", from the White River Oligocene of South Dakota. *Proceedings of the American Philosophical Society* 287–305.

Kemp, T.S. (2005): *The Origin and Evolution of Mammals.* Oxford University Press, Oxford.

Kielan-Jaworowska, Z., Cifelli, R.L. & Luo, Z.-X. (2004): *Mammals from the Age of Dinosaurs: Structure, Relationships, and Paleobiology.* Columbia University Press, New York.

Kirsch, J.A.W., Dickerman, A.W., Reig, O.A. & Springer, M.S. (1991): DNA hybridization evidence for the Australasian affinity of the American marsupial Dromiciops australis. *Proceedings of the National Academy of Sciences of the United States of America* 88:10465–10469.

Kirsch, J.A.W. & Mayer, G.C. (1998): The platypus is not a rodent: DNA hybridization, amniote phylogeny and the palimpsest theory. *Philosophical Transactions of the Royal Society of London B: Biological Sciences* 353:1221–1237.

Kjer, K.M. & Honeycutt, R.L. (2007): Site specific rates of mitochondrial genomes and the phylogeny of eutheria. *BMC Evolutionary Biology* 7:8.

von Koenigswald, W., Storch, G. & Richter, G. (1992): Primitive insectivores, extraordinary hedgehogs, and long-fingers. In: Schaal, S. & Ziegler, W. (eds.), *Messel: an Insight into the History of Life and of the Earth.* Clarendon Press, Oxford: 161–177.

von Koenigswald, W., Rose, K.D., Grande, L. & Martin, R.D. (2005): Die Lebensweise eozäner Säugetiere (Pantolestidae und Apatemyidae) aus Messel (Europa) im vergleich zu neuen skelettfunden aus dem Fossil Butte Member Von Wyoming (Nordamerika). *Geologisches Jahrbuch Hessen* 132:43–54.

von Koenigswald, W., Ruf, I. & Gingerich, P.D. (2009): Cranial morphology of a new apatemyid, *Carcinella sigei* n. gen. n. sp.(Mammalia, Apatotheria) from the Late Eocene of southern France. *Palaeontographica Abteilung* A 53–91.

Kondrashov, P.E. & Lucas, S.G. (2006): Early Paleocene (Puercan and Torrejonian) Archaic Ungulates (Condylarthra, Procreodi and Acreodi) of the San Juan Basin, New Mexico. *Bulletin New Mexico Museum of Natural History & Science* 34:84–97.

Kondrashov, P.E. & Lucas, S.G. (2012): Nearly complete skeleton of *Tetraclaenodon* (Mammalia, Phenacodontidae) from the Early Paleocene of New Mexico: morpho-functional analysis. *Journal of Paleontology* 86:25–43.

Krause, D.W., Hoffmann, S., Wible, J.R., Kirk, E.C., Schultz, J.A., von Koenigswald, W., Groenke, J.R., Rossie, J.B., O'Connor, P.M., Seiffert, E.R., *et al.* (2014): First cranial remains of a gondwanatherian mammal reveal remarkable mosaicism. *Nature* 515:512–517.

Kullberg, M., Hallström, B.M., Arnason, U. & Janke, A. (2008): Phylogenetic analysis of 1.5 Mbp and platypus EST data refute the Marsupionta hypothesis and unequivocally support Monotremata as sister group to Marsupialia/Placentalia. *Zoologica Scripta* 37:115–127.

Kumar, S. & Hedges, S.B. (1998): A molecular timescale for vertebrate evolution. *Nature* 392:917.

Ladevèze, S., Asher, R.J. & Sánchez-Villagra, M.R. (2008): Petrosal anatomy in the fossil mammal *Necrolestes*: Evidence for metatherian affinities and comparisons with the extant marsupial mole. *Journal of Anatomy* 213:686–697.

Ladevèze, S., de Muizon, C., Beck, R.M.D., Germain, D. & Cespedes-Paz, R. (2011): Earliest evidence of mammalian social behaviour in the basal Tertiary of Bolivia. *Nature* 474:83.

Ladevèze, S., Missiaen, P.; & Smith, T. (2010): First skull of *Orthaspidotherium edwardsi* (Mammalia, "Condylarthra") from the Late Paleocene of Berru (France) and phylogenetic affinities of the enigmatic European family Pleuraspidotheriidae. *Journal of Vertebrate Paleontology* 30:1559–1578.

Lamberton, C. (1946): Contribution à la connaissance de la faune subfossile de Madagascar—Note XV—Le *Plesiorycteropus madagascariensis* Filhol. *Bulletin de l'Académie Malgache* 25:25–53.

Landry, S.O. Jr (1974): The fundamental relationship of the Lagomorpha and Rodentia. *First International Theriological Congress Transactions Moscow* 2:202–203.

Lee, M.S.Y. & Camens, A.B. (2009): Strong morphological support for the molecular evolutionary tree of placental mammals. *Journal of Evolutionary Biology* 22:2243–2257.

Lihoreau, F., Boisserie, J.-R., Manthi, F.K. & Ducrocq, S. (2015): Hippos stem from the longest sequence of terrestrial cetartiodactyl evolution in Africa. *Nature Communications* 6:6264.

Liu, A.G.S.C., Seiffert, E.R. & Simons, E.L. (2008): Stable isotope evidence for an amphibious phase in early proboscidean evolution. *Proceedings of the National Academy of Sciences of the United States of America* 105:5786–5791.

Liu, L., Zhang, J., Rheindt, F.E., Lei, F., Qu, Y., Wang, Y., Zhang, Y., Sullivan, C., Nie, W., Wang, J., *et al.* (2017): Genomic evidence reveals a radiation of placental mammals uninterrupted by the KPg boundary. *Proceedings of the National Academy of Sciences of the United States of America* 201616744.

Lobo, L.S., Lessa, G., Cartelle, C. & Romano, P.S.R. (2017): Dental eruption sequence and hypsodonty index of a Pleistocene macraucheniid from the Brazilian Intertropical Region. *Journal of Paleontology* 1–8.

Lofgren, D.L. (1995): The Bug Creek problem and the Cretaceous-Tertiary transition at McGuire Creek, Montana. *University of California Publications in Geological Sciences* 140:1–185.

Lopatin, A.V. (2006): Early Paleogene insectivore mammals of Asia and establishment of the major groups of Insectivora. *Paleontological Journal* 40:S205–S405.

Lopatin, A.V. (2001): The skull structure of *Archaeoryctes euryalis* sp. nov. (Didymoconidae, Mammalia) from the Paleocene of Mongolia and the Taxonomic Position of the Family. *Paleontological Journal* 35:320–329.

Lucas, S.G. & Schoch, R.M. (1990): Ontogenetic studies of early Cenozoic *Coryphodon* (Mammalia, Pantodonta). *Journal of Paleontology* 64:831–841.

Luckett, W.P. & Hong, N. (1998): Phylogenetic relationships between the orders Artiodactyla and Cetacea: a combined assessment of morphological and molecular evidence. *Journal of Mammalian* Evolution 5:127–182.

Luo, Z.-X., Gatesy, S.M.; Jenkins, F.A., Amaral, W.W. & Shubin, N.H. (2015): Mandibular and dental characteristics of Late Triassic mammaliaform Haramiyavia and their ramifications for basal mammal evolution. *Proceedings of the National Academy of Sciences of the United States of America* 112:E7101–E7109.

Luo, Z.-X., Yuan, C.-X., Meng, Q.-J. & Ji. Q. (2011): A Jurassic eutherian mammal and divergence of marsupials and placentals. *Nature* 476:442.

MacFadden, B.J. (2000): Cenozoic mammalian herbivores from the Americas: reconstructing ancient diets and terrestrial communities. *Annual Review of Ecology and Systematics* 31:33–59.

MacFadden, B.J. (2005): Fossil horses—evidence for evolution. *Science* 307:1728–1730.

MacPhee, R.D.E. (1994): Morphology, adaptations, and relationships of *Plesiorycteropus*: and a diagnosis of a new order of eutherian mammals. *Bulletin of the American Museum of Natural History* 220.

Madden, R.H. (2014): *Hypsodonty in Mammals*. Cambridge University Press, Cambridge.

Madsen, O., Scally, M., Douady, C.J., Kao, D.J., *et al.* (2001): Parallel adaptive radiations in two major clades of placental mammals. Nature 409:610.

Maga, A.M. & Beck, R.M.D. (2017): Skeleton of an unusual, cat-sized marsupial relative (Metatheria: Marsupialiformes) from the middle Eocene (Lutetian: 44–43 million years ago) of Turkey. *PloS One* 12:e0181712.

Manz, C.L. & Bloch, J.I. (2015): Systematics and phylogeny of Paleocene-Eocene Nyctitheriidae (Mammalia, Eulipotyphla?) with description of a new species from the Late Paleocene of the Clarks Fork Basin, Wyoming, USA. *Journal of Mammalian Evolution* 22:307–342.

Manz, C.L., Chester, S.G.B., Bloch, J.I., Silcox, M.T. & Sargis, E.J. (2015): New partial skeletons of Palaeocene Nyctitheriidae and evaluation of proposed euarchontan affinities. *Biology Letters* 11:20140911.

Marshall, L.G., de Muizon, C. & Sogogneau-Russell, D. (1995): *Pucadelphys andinus* (Marsupialia, Mammalia) from the Early Paleocene of Bolivia. Mémoires du Muséum national d'Histoire naturelle, Ser. *Paleontologie* 165:1–164.

Martin, T. (2018) Mesozoic Mammals. In: Zachos, F.E. & Asher, R.J. (eds.): *Handbook of Zoology, Mammalia.* DeGruyter, Berlin.

Matthew, W.D. (1909): The Carnivora and Insectivora of the Bridger Basin, Middle Eocene. *Memoirs of the American Museum of Natural History* 9.

Matthew, W.D. (1926): The evolution of the horse: a record and its interpretation. *The Quarterly Review of Biology* 1:139–185.

May-Collado, L.J., Kilpatrick, C.W. & Agnarsson, I. (2015): Mammals from "down under": a multi-gene species-level phylogeny of marsupial mammals (Mammalia, Metatheria). *PeerJ* 3:e805.

McCormack, J.E., Faircloth, B.C., Crawford, N.G., Gowaty, P.A., Brumfield, R.T. & Glenn, T.C. (2012): Ultraconserved elements are novel phylogenomic markers that resolve placental mammal phylogeny when combined with species-tree analysis. *Genome Research* 22:746–754.

McDowell, S.B. (1958): The Greater Antillean insectivores. *Bulletin of the American Museum of Natural History* 115:113–214.

McGee, E.M. & Turnbull, W.D. (2010): A Paleopopulation of *Coryphodon lobatus* (Mammalia: Pantodonta) from Deardorff Hill Coryphodon Quarry, Piceance Creek Basin, Colorado. Fieldiana: Geology 52:1–12.

McKenna, M.C. (1975): Toward a phylogenetic classification of the Mammalia. In: Luckett, W.P. & Szalay, F.S. (eds.), *Phylogeny of the Primates*. Springer: 21–46.

McKenna, M.C. & Bell, S.K. (1997): *Classification of Mammals above the Species Level.* Columbia University Press, New York.

McKenna, M.C., Wyss, A.R. & Flynn, J.J. (2006): Paleogene pseudoglyptodont xenarthrans from central Chile and Argentine Patagonia. *American Museum Novitates* 1–18.

Mein, P. & Pickford, M. (2003): Insectivora from Arrisdrift, a basal Middle Miocene locality in southern Namibia. *Memoir of the Geological Survey of Namibia* 19:143–146.

Meng, J., Hu, Y.; & Li., C. (2003): The osteology of *Rhombomylus* (Mammalia, Glires): implications for phylogeny and evolution of Glires. *Bulletin of the American Museum of Natural History* 1–247.

Meredith, R.W., Janečka, J.E., Gatesy, J., Ryder, O.A., Fisher, C.A., Teeling, E.C., Goodbla, A., Eizirik, E., Simão, T.L.L., Stadler, T., *et al.* (2011): Impacts of the Cretaceous Terrestrial Revolution and KPg extinction on mammal diversification. *Science* 334:521–524.

Mess, A. & Carter, A.M. (2006): Evolutionary transformations of fetal membrane characters in Eutheria with special reference to Afrotheria. *Journal of Experimental Zoology Part B: Molecular and Developmental Evolution* 306:140–163.

Mihlbachler, M.C. (2008): Species taxonomy, phylogeny, and biogeography of the Brontotheriidae (Mammalia: Perissodactyla). *Bulletin of the American Museum of Natural History* 311:1–475.

Mihlbachler, M.C., Rivals, F., Solounias, N. & Semprebon, G.M. (2011): Dietary change and evolution of horses in North America. *Science* 331:1178–1181.

Mitchell, K.J., Pratt, R.C., Watson, L.N., Gibb, G.C., Llamas, B., Kasper, M., Edson, J., Hopwood, B., Male, D., Armstrong, K.N., *et al.* (2014): Molecular phylogeny, biogeography, and habitat preference evolution of marsupials. *Molecular Biology and Evolution* 31:2322–2330.

Mitgutsch, C., Richardson, M.K., de Bakker, M.A.G., Jiménez, R., Martin, J.E., Kondrashov, P. & Sánchez-Villagra, M.R. (2012): A molecular-morphological study of a peculiar limb morphology: the development and evolution of the mole's "thumb". In: Asher; R.J.; & Müller, J. (eds.), *From Clone to Bone the Synergy of Morphological and Molecular Tools in Paleobiology.* Cambridge University Press, Cambridge: 301–327.

Miyamoto, M.M. & Goodman, M. (1986): Biomolecular systematics of eutherian mammals: phylogenetic patterns and classification. *Systematic Biology* 35:230–240.

Mouchaty, S.K., Gullberg, A., Janke, A. & Arnason, U. (2000): Phylogenetic position of the tenrecs (Mammalia: Tenrecidae) of Madagascar based on analysis of the complete mitochondrial genome sequence of *Echinops telfairi*. *Zoologica Scripta* 29:307–317.

de Muizon, C. (1998): *Mayulestes ferox*, a borhyaenoid (Metatheria, Mammalia) from the early Palaeocene of Bolivia. Phylogenetic and palaeobiologic implications. *Geodiversitas* 20:19–142.

de Muizon, C. & Cifelli, R.L. (2000): The "condylarths" (archaic Ungulata, Mammalia) from the early Palaeocene of Tiupampa (Bolivia): implications on the origin of the South American ungulates. *Geodiversitas* 22:47–150.

de Muizon, C., Billet, G., Argot, C., Ladevèze, S. & Goussard, F. (2015): *Alcidedorbignya inopinata*, a basal pantodont (Placentalia, Mammalia) from the early Palaeocene of Bolivia: anatomy, phylogeny and palaeobiology. *Geodiversitas* 37:397–634.

de Muizon, C & McDonald, H.G. (1995): An aquatic sloth from the Pliocene of Peru. *Nature* 375:224–227.

Müller, J., Scheyer, T.T.M., Head, J.J., Barrett, P.M., Werneburg, I., Ericson, P.G.P., Pol, D. & Sánchez-Villagra, M.R. (2010): Homeotic effects, somitogenesis and the evolution of vertebral numbers in recent and fossil amniotes. *Proceedings of the National Academy of Sciences of the United States of America* 107:2118–2123.

Murphy, W.J., Eizirik, E., Johnson, W.E., Zhang, Y.P., Ryder, O.A. & O'Brien, S.J. (2001a): Molecular phylogenetics and the origins of placental mammals. *Nature* 409:614.

Murphy, W.J., Eizirik, E., O'Brien, S.J., Madsen, O., Scally, M., Douady, C.J., Teeling, E.C., Ryder, O.A., Stanhope, M.J., de Jong, W.W. & Springer, M.S. (2001b): Resolution of the early placental mammal radiation using Bayesian phylogenetics. *Science* 294:2348–2351.

Murphy, W.J., Pringle, T.H., Crider, T.A., Springer, M.S. & Miller, W. (2007): Using genomic data to unravel the root of the placental mammal phylogeny. *Genome Research* 17:413–421.

Nilsson, M.A., Churakov, G., Sommer, M., Van Tran, N., Zemann, A., Brosius, J. & Schmitz, J. (2010): Tracking marsupial evolution using archaic genomic retroposon insertions. *PLoS Biology* 8:e1000436.

Nishihara, H., Hasegawa, M. & Okada, N. (2006): Pegasoferae, an unexpected mammalian clade revealed by tracking ancient retroposon insertions. *Proceedings of the National Academy of Sciences of the United States of America* 103:9929–9934.

Nishihara, H., Maruyama, S., Okada, N. (2009): Retroposon analysis and recent geological data suggest near-simultaneous divergence of the three superorders of mammals. *Proceedings*

of the National Academy of Sciences of the United States of America 106:5235–5240.
Novacek, M.J. (1977): A review of Paleocene and Eocene Leptictidae (Eutheria, Mammalia) from North America. *PaleoBios* 1:1–42.
Novacek, M.J. (1992): Mammalian phylogeny: shaking the tree. *Nature* 356:121–125.
Novacek, M.J. (1986): The skull of leptictid insectivorans and the higher-level classification of eutherian mammals. *Bulletin of the American Museum of Natural History* 183.
Novacek, M.J. & Wyss, A.R. (1986): Higher-level relationships of the recent eutherian orders: morphological evidence. *Cladistics* 2:257–287.
Nuttall, G.H.F., Graham-Smith, G.S. & Pigg-Strangeways, T.S. (1904): *Blood Immunity and Blood Relationship: A Demonstration of Certain Blood-Relationships amongst Animals by Means of the Precipitin Test for Blood.* Cambridge University Press, Cambridge.
O'Meara, R.N. & Thompson, R.S. (2014): Were there Miocene meridiolestidans? Assessing the phylogenetic placement of *Necrolestes patagonensis* and the presence of a 40 million year meridiolestidan ghost lineage. *Journal of Mammalian Evolution* 21:271–284.
O'Leary, M.A., Bloch, J.I., Flynn, J.J., Gaudin, T.J., Giallombardo, A., Giannini, N.P., Goldberg, S.L., Kraatz, B.P., Luo, Z.-X., Meng, J., *et al.* (2013): The placental mammal ancestor and the post-K-Pg radiation of placentals. *Science* 339:662–667.
Omland, K.E., Cook, L.G. & Crisp, M.D. (2008): Tree thinking for all biology: the problem with reading phylogenies as ladders of progress. *BioEssays* 30:854–867.
Organ, C. (2012): Palaeontological insights into genome evolution. In: Asher, R.J. & Müller, J. (eds.), *From Clone to Bone: The Synergy of Morphological and Molecular Tools in Palaeobiology.* Cambridge University Press, Cambridge: 16–37.
Orliac, M.J. & O'Leary, M.A. (2016): The inner ear of *Protungulatum* (pan-Euungulata, Mammalia). *Journal of Mammalian Evolution* 23:337–352.
Orliac, M.J., Argot, C. & Gilissen, E. (2012): Digital cranial endocast of *Hyopsodus* (Mammalia, "Condylarthra"): a case of Paleogene terrestrial echolocation? *PLoS One* 7:e30000.
Orliac, M.J., Boisserie, J.-R., MacLatchy, L. & Lihoreau, F. (2010): Early Miocene hippopotamids (Cetartiodactyla) constrain the phylogenetic and spatiotemporal settings of hippopotamid origin. *Proceedings of the National Academy of Sciences of the United States of America* 107:11871–11876.
Osborn, H.F. (1934): Aristogenesis, the creative principle in the origin of species. *The American Naturalist* 68:193–235.
Osborn, H.F. (1922): Orthogenesis as observed from paleontological evidence beginning in the year 1889. *The American Naturalist* 56:134–143.
Osborn, H.F. (1917): *The Origin and Evolution of Life.* Scribners, New York.
Owen, R. (1841): Description of the Fossil Remains of a Mammal (*Hyracotherium leporinum*) and of a Bird (*Lithornis vulturinus*) from the London Clay. *Transactions of the Geological Society of London* 2:203–208.
Parham, J.F., Donoghue, P.C.J., Bell, C.J., Calway, T.D., Head, J.J., Holroyd, P.A., Inoue, J.G., Irmis, R.B., Joyce, W.G., Ksepka, D.T., *et al.* (2012): Best practices for applying paleontological data to molecular divergence dating analyses. *Systematic Biology* 61:346–359.
Patterson, B. (1975): The fossil aardvarks (Mammalia: Tubulidentata). *Bulletin of the Museum of Comparative Zoology* 147:185–237.
Patterson, B. & Pascual, R. (1968): The fossil mammal fauna of South America. *The Quarterly Review of Biology* 43:409–451.
Patterson, B. & Wood, A.E. (1982): Rodents from the Deseadan Oligocene of Bolivia and the relationships of the Caviomorpha. *Bulletin of the Museum of Comparative Zoology* 149:371–543.
Pattinson, D.J., Thompson, R.S., Piotrowski, A.K. & Asher, R.J. (2015): Phylogeny, paleontology, and primates: do incomplete fossils bias the tree of life? *Systematic Biology* 64:169–186.
Penkrot, T.A., Zack, S.P., Rose, K.D. & Bloch, J.I. (2008): Postcranial morphology of Apheliscus and Haplomylus (Condylarthra, Apheliscidae): evidence for a Paleocene Holarctic origin of Macroscelidea. In: Sargis, E.J. & Dagosto, M. (eds.), *Mammalian Evolutionary Morphology*. Springer: 73–106.
Penny, D. (2013): Rewriting evolution—"been there, done that." *Genome Biology and Evolution* 5:819–821.
Penny, D., Foulds, L.R. & Hendy, M.D. (1982): Testing the theory of evolution by comparing phylogenetic trees constructed from five different protein sequences. *Nature* 297:197–200.
Pettigrew, J.D. (1986): Flying primates? Megabats have the advanced pathway from eye to midbrain. *Science* 231:1304–1306.
Pfretzschner, H.U. (1993): Muscle reconstruction and aquatic locomotion in the middle Eocene *Buxolestes piscator* from Messel near Darmstadt. *Kaupia* 3:75–87.
Phillips, M.J., McLenachan, P.A., Down, C., Gibb, G.C. & Penny, D. (2006): Combined mitochondrial and nuclear DNA sequences resolve the interrelations of the major Australasian marsupial radiations. *Systematic Biology* 55:122–137.
Pickford, M. (2015a): Chrysochloridae (Mammalia) from the Lutetian (Middle Eocene) of Black Crow, Namibia. *Communications of the Geological Survey of Namibia* 16:112–120.
Pickford, M. (2015b): Late Eocene Chrysochloridae (Mammalia) from the Sperrgebiet, Namibia. *Communications of the Geological Survey of Namibia* 16:160–199.
Pickford, M. (2015c): Late Eocene Potamogalidae and Tenrecidae (Mammalia) from the Sperrgebiet, Namibia. *Communications of the Geological Survey of Namibia* 16:121–159.
Piras, P., Sansalone, G., Teresi, L., Moscato, M., Profico, A., Eng, R., Cox, T.C., Loy, A., Colangelo, P. & Kotsakis, T. (2015): Digging adaptation in insectivorous subterranean eutherians. The enigma of *Mesoscalops montanensis* unveiled by geometric morphometrics and finite element analysis. *Journal of Morphology* 276:1157–1171.
Polly, P.D., Wesley-Hunt, G.D., Heinrich, R.E., Davis, G. & Houde, P. (2006): Earliest known carnivoran auditory bulla and support for a recent origin of crown-group Carnivora (Eutheria, Mammalia). *Palaeontology* 49:1019–1027.
Prasad, G.V.R., Verma, O., Sahni, A., Parmar, V. & Khosla, A. (2007): A Cretaceous hoofed mammal from India. *Science* 318:937.
Prothero, D.R. (2016): *The Princeton Field Guide to Prehistoric Mammals.* Princeton University Press, Princeton.
Ravel, A. & Orliac, M.J. (2014): The inner ear morphology of the "condylarthran" *Hyopsodus lepidus. Historical Biology* 27:957–969.
Rasmussen, D.T. & Simons, E.L. (2000): Ecomorphological diversity among Paleogene hyracoids (Mammalia): a new

cursorial browser from the Fayum, Egypt. *Journal of Vertebrate Paleontology* 20:167–176.

Rasmussen, D. & Gutiérrez, M. (2010): Hyracoidea. In: Werdelin, L. & Sanders, W.J. (eds.), *Cenozoic Mammals of Africa*. University of California Press, Berkeley.

Reig, O.A. (1955): Noticia preliminar sobre la presencia de microbiotherinos vivientes en la fauna sudamericana. *Investigaciones Zoológicas Chilenas* 2:121–130.

Reinhart, R. (1953): Diagnosis of the new mammalian order, Desmostylia. *Journal of Geology* 61:187.

Reisz, R.R. & Müller, J. (2004): Molecular timescales and the fossil record: a paleontological perspective. *TRENDS in Genetics* 20:237–241.

Repenning, C.A. (1967): Subfamilies and genera of the Soricidae. *United States Geological Survey Professional Paper* 565:1–74.

Roca, A.L., Bar-Gal, G.K., Eizrik, E., Helgen, K.M., Maria, R., Springer, M.S., O'Brien, S.J. & Murphy. (2004): Mesozoic origin for West Indian insectivores. *Nature* 429:649.

Romer, A.S. (1945): *Vertebrate Paleontology*. The University of Chicago Press, Chicago.

Romiguier, J., Ranwez, V., Delsuc, F., Galtier, N. & Douzery, E.J.P. (2013): Less is more in mammalian phylogenomics: AT-rich genes minimize tree conflicts and unravel the root of placental mammals. *Molecular Biology and Evolution* 30:2134–2144.

Ronquist, F., Klopfstein, S., Vilhelmsen, L., Schulmeister, S., Murray, D.L. & Rasnitsyn, A.P. (2012): A total-evidence approach to dating with fossils, applied to the early radiation of the Hymenoptera. *Systematic Biology* 61:973–999.

Ronquist, F., Lartillot, N. & Phillips, M.J. (2016): Closing the gap between rocks and clocks using total-evidence dating. *Philosophical Transactions of the Royal Society of London B: Biological Sciences* 371:20150136.

Rose, K.D. (1999): *Eurotamandua* and *Palaeanodonta*: convergent or related? *Paläontologische Zeitschrift* 73:395–401.

Rose, K.D. (1990): Postcranial skeletal remains and adaptations in Early Eocene mammals from the Willwood Formation, Bighorn Basin, Wyoming. *Geological Society of America Special Papers* 243:107–134.

Rose, K.D. (1982): Skeleton of *Diacodexis*, oldest known artiodactyl. *Science* 216:621–623.

Rose, K.D. (2001): The ancestry of whales. *Science* 293:2216–2217.

Rose, K.D. (2006a): *The Beginning of the Age of Mammals*. JHU Press, Baltimore.

Rose, K.D. (2006b): The postcranial skeleton of early Oligocene *Leptictis* (Mammalia: Leptictida), with a preliminary comparison to *Leptictidium* from the middle Eocene of Messel. *Palaeontographica, Abteilung A: Palaozoologie-Stratigraphie* 278:37–56.

Rose, K.D. & Archibald, J.D. (2005): *The Rise of Placental Mammals: Origins and Relationships of the Major Extant Clades*. JHU Press, Baltimore.

Rose, K.D. & Emry, R.J. (1983): Extraordinary fossorial adaptations in the Oligocene palaeanodonts *Epoicotherium* and *Xenocranium* (Mammalia). *Journal of Morphology* 175:33–56.

Rose, K.D. & Emry, R.J. (1993): Relationships of Xenarthra, Pholidota, and fossil "edentates": the morphological evidence. In: Szalay, F.S., Novacek, M.J. & Mckenna, M.C. (eds.), *Mammal Phylogeny. Placentals*. Springer, New York: 81–101.

Rose, K.D., Emry, R.J., Gaudin, T.J. & Storch, G. (2005): Xenarthra and pholidota; In: Rose, K.D. & Archibald, J.D. (eds.), *The rise of placental mammals: origins and relationships of the major extant clades*. The Johns Hopkins University Press, Baltimore: 106–126.

Rose, K.D. & von Koenigswald, W. (2005): An exceptionally complete skeleton of *Palaeosinopa* (Mammalia, Cimolesta, Pantolestidae) from the Green River Formation, and other postcranial elements of the Pantolestidae from the Eocene of Wyoming. *Palaeontographica, Abteilung A: Palaozoologie-Stratigraphie* 273:55–96.

Rose, K.D., Chew, A.E., Dunn, R.H., Kraus, M.J., Fricke, H.C. & Zack, S.P. (2012): Earliest Eocene mammalian fauna from the Paleocene-Eocene thermal maximum at sand creek divide, southern Bighorn Basin, Wyoming. *University of Michigan Papers on Paleontology* 36:1–122.

Rose, K.D., Holbrook, L.T., Rana, R.S., Kumar, K., Jones, K.E., Ahrens, H.E., Missiaen, P., Sahni, A. & Smith, T. (2014): Early Eocene fossils suggest that the mammalian order Perissodactyla originated in India. *Nature Communications* 5.

Rougier, G.W., Wible, J.R., Beck, R.M.D. & Apesteguía, S. (2012): The Miocene mammal *Necrolestes* demonstrates the survival of a Mesozoic nontherian lineage into the Late Cenozoic of South America. *Proceedings of the National Academy of Sciences of the United States of America* 109:20053–20058.

Saether, O.A. (1979): Underlying synapomorphies and anagenetic analysis. *Zoologica Scripta* 8:305–312.

Samonds, K.E., Zalmout, I.S., Irwin, M.T., Krause, D.W., Rogers, R.R., & Raharivony, L.L. (2009): *Eotheroides lambondrano*, new middle Eocene seacow (Mammalia, Sirenia) from the Mahajanga Basin, northwestern Madagascar. *Journal of Vertebrate Paleontology* 29:1233–1243.

Sánchez-Villagra, M.R. (2013): Why are there fewer marsupials than placentals? On the relevance of geography and physiology to evolutionary patterns of mammalian diversity and disparity. *Journal of Mammalian Evolution* 20:279–290.

Sánchez-Villagra, M.R. & Asher, R.J. (2002): Cranio-sensory adaptations in small faunivorous semiaquatic mammals, with special reference to olfaction and the trigeminal system. *Mammalia* 66:93–110.

Sánchez-Villagra, M.R., Ladevèze, S., Horovitz, I., Argot, C., Hooker, J.J., Macrini, T.E., Martin, T., Moore-Fay, S., de Muizon, C., Schmelzle, T., *et al.* (2007a): Exceptionally preserved North American Paleogene metatherians: adaptations and discovery of a major gap in the opossum fossil record. *Biology Letters* 3:318–322.

Sánchez-Villagra, M.R., Narita, Y. & Kuratani, S. (2007b): Thoracolumbar vertebral number: the first skeletal synapomorphy for afrotherian mammals. *Systematics and Biodiversity* 5:1–7.

Sanders, W.J., Gheerbrant, E., Harris, J.M., Saegusa, H. & Delmer, C. (2010): Proboscidea. In: Werdelin, L. & Sanders, W.J. (eds.), *Cenozoic Mammals of Africa*. University of California Press, Berkeley: 161–251.

Santos, G.-P., Parham, J.F. & Beatty, B.L. (2016): New data on the ontogeny and senescence of Desmostylus (Desmostylia, Mammalia). *Journal of Vertebrate Paleontology* 36:e1078344.

Sarich, V.M. & Wilson, A.C. (1967): Immunological time scale for hominid evolution. *Science* 158:1200–1203.

Scally, M., Madsen, O., Douady, C.J., de Jong, W.W., Stanhope, M.J. & Springer, M.S. (2001): Molecular evidence for the major clades of placental mammals. *Journal of Mammalian Evolution* 8:239–277.

Schmid, L. (2012): Reconstructing the molecular underpinnings of morphological diversification. A case study of the Triassic fish Saurichthys. In: Asher, R.J. & Müller, J. (eds.), *From Clone to Bone: the Synergy of Morphological and Molecular Tools in Paleobiology.* Cambridge University Press, Cambridge: 135–165.

Schultz, A.H. (1960): *Age Changes in Primates and Their Modification in Man.* Pergamon Press.

Schwartz, G.T., Rasmussen, D.T. & Smith, R.J. (1995): Body-size diversity and community structure of fossil hyracoids. *Journal of Mammalogy* 76:1088–1099.

Schweitzer, M.H., Moyer, A.E. & Zheng, W. (2016): Testing the hypothesis of biofilm as a source for soft tissue and cell-like structures preserved in dinosaur bone. *PloS One* 11:e0150238.

Schwermann, A.H. & Thompson, R.S. (2015): Extraordinarily preserved talpids (Mammalia, Lipotyphla) and the evolution of fossoriality. *Journal of Vertebrate Paleontology* 35:e934828.

Scotese, C.R. (2001): *Atlas of Earth History. University of Texas at Arlington.* Department of Geology. PALEOMAP Project.

Scott, W.B. (1910): Mammalia of the Santa Cruz Beds, Part I. Litopterna. *Reports of the Princeton University Expedition to Patagonia* 7:1–156.

Secord, R., Bloch, J.I., Chester, S.G.B., Boyer, D.M., Wood, A.R., Wing, S.L., Kraus, M.J., McInerney, F.A. & Krigbaum, J. (2012): Evolution of the earliest horses driven by climate change in the Paleocene-Eocene Thermal Maximum. *Science* 335:959–962.

Seiffert, E.R. (2007): A new estimate of afrotherian phylogeny based on simultaneous analysis of genomic, morphological, and fossil evidence. *BMC Evolutionary Biology* 7:1.

Seiffert, E.R., Simons, E.L., Ryan, T.M., Bown, T.M. & Attia, Y. (2007): New remains of Eocene and Oligocene Afrosoricida (Afrotheria) from Egypt, with implications for the origin (s) of afrosoricid zalambdodonty. *Journal of Vertebrate Paleontology* 27:963–972.

Shoshani, J. (1986): Mammalian phylogeny: comparison of morphological and molecular results. *Molecular Biology and Evolution* 3:222–242.

Shoshani, J. & McKenna, M.C. (1998): Higher taxonomic relationships among extant mammals based on morphology, with selected comparisons of results from molecular data. *Molecular Phylogenetics and Evolution* 9:572–84.

Sigé, B. (1974): *Pseudorhyncocyon cayluxi* Filhol, 1892: insectivore géant des phosphorites du Quercy. *Palaeovertebrata* 6:33–46.

Silcox, M.T., Bloch, J.I., Boyer, D.M. & Houde, P. (2010): Cranial anatomy of Paleocene and Eocene Labidolemur kayi (Mammalia: Apatotheria), and the relationships of the Apatemyidae to other mammals. *Zoological Journal of the Linnean Society* 160:773–825.

Simmons, N.B. & Geisler, J.H. (1998): Phylogenetic relationships of Icaronycteris, Archaeonycteris, Hassianycteris, and Palaeochiropteryx to extant bat lineages, with comments on the evolution of echolocation and foraging strategies in Microchiroptera. *Bulletin of the AMNH*; no. 235. 235.

Simmons, N.B., Novacek, M.J. & Baker, R.J. (1991): Approaches, methods, and the future of the chiropteran monophyly controversy: a reply to JD Pettigrew. *Systematic Biology* 40:239–243.

Simmons, N.B. & Quinn, T.H. (1994): Evolution of the digital tendon locking mechanism in bats and dermopterans: a phylogenetic perspective. *Journal of Mammalian Evolution* 2:231–254.

Simmons, N.B., Seymour, K.L., Habersetzer, J. & Gunnell, G.F. (2008): Primitive Early Eocene bat from Wyoming and the evolution of flight and echolocation. *Nature* 451:818–821.

Simpson, G.G. (1953): *Major Features of Evolution.* Columbia University Press, New York.

Simpson, G.G. (1944): *Tempo and Mode in Evolution.* Columbia University Press, New York.

Simpson, G.G. (1948): The beginning of the age of mammals in South America. Part 1, Introduction: systematics: Marsupialia, Edentata, Condylarthra, Litopterna and Notioprogonia. *Bulletin of the American Museum of Natural History* 91:1–232.

Simpson, G.G. (1945): The principles of classification and a classification of mammals. *Bulletin of the American Museum of Natural History* 85.

Slater, G.J., Cui, P., Forasiepi, A.M., Lenz, D., Tsangaras, K., Voirin, B., de Moraes-Barros, N., MacPhee, R.D.E. & Greenwood, A.D. (2016): Evolutionary relationships among extinct and extant sloths: the evidence of mitogenomes and retroviruses. *Genome Biology and Evolution* 8:607–621.

Sloan, R.E. & van Valen, L. (1965): Cretaceous mammals from Montana. *Science* 148:220–227.

Smith, B.H. (2000): Schultz's Rule and the Evolution of Tooth Emergence and Replacement Patterns in Primates and Ungulates. In: Teaford, M., Smith, M. & Ferguson, M. (eds.): *Development, Function and Evolution of Teeth.* Cambridge University Press, Cambridge: 212–227.

Smith, T., De Bast, E. & Sigé, B. (2010): Euarchontan affinity of Paleocene Afro-European adapisoriculid mammals and their origin in the Late Cretaceous Deccan Traps of India. *Naturwissenschaften* 97:417–422.

Sober, E. & Steel, M. (2002): Testing the hypothesis of common ancestry. *Journal of Theoretical Biology* 218:395–408.

Solé, F., Gheerbrant, E., Amaghzaz, M. & Bouya, B. (2009): Further evidence of the African antiquity of hyaenodontid ("Creodonta", Mammalia) evolution. *Zoological Journal of the Linnean Society* 156:827–846.

Solé, F., Smith, R., Coillot, T., De Bast, E. & Smith, T. (2014): Dental and tarsal anatomy of *"Miacis" latouri* and a phylogenetic analysis of the earliest carnivoraforms (Mammalia, Carnivoramorpha). *Journal of Vertebrate Paleontology* 34:1–21.

Song, S., Liu, L., Edwards, S.V. & Wu, S. (2012): Resolving conflict in eutherian mammal phylogeny using phylogenomics and the multispecies coalescent model. *Proceedings of the National Academy of Sciences* 109:14942–14947.

Spaulding, M. & Flynn, J.J. (2012): Phylogeny of the Carnivoramorpha: the impact of postcranial characters. *Journal of Systematic Palaeontology* 10:653–677.

Spaulding, M., O'Leary, M.A. & Gatesy, J. (2009): Relationships of Cetacea (Artiodactyla) among mammals: increased taxon sampling alters interpretations of key fossils and character evolution. *Plos One* 4:e7062.

Springer, M.S., Burk-Herrick, A., Meredith, R., Eizirik, E., Teeling, E., O'Brien, S.J. & Murphy, W.J. (2007): The adequacy of morphology for reconstructing the early history of placental mammals. *Systematic Biology* 56:673–684.

Springer, M.S., Cleven, G.C., Madsen, O., de Jong, W.W., Waddell, V.G., Amrine, H.M. & Stanhope, M.J. (1997): Endemic African mammals shake the phylogenetic tree. *Nature* 388:61.

Springer, M.S. & Gatesy, J. (2016): The gene tree delusion. *Molecular Phylogenetics and Evolution* 94:1–33.

Springer, M.S. & Gatesy, J. (2017): On the importance of homology in the age of phylogenomics. *Systematics and Biodiversity* 1–19.

Stanhope, M.J., Waddell, V.G., Madsen, O., de Jong, W., Hedges, S.B., Cleven, G.C., Kao, D. & Springer, M.S. (1998): Molecular evidence for multiple origins of Insectivora and for a new order of endemic African insectivore mammals. *Proceedings of the National Academy of Sciences of the United States of America* 95:9967–9972.

Storch, G. (1981): *Eurotamandua joresi*, ein myrmecophagide aus dem Eozän der "Grube Messel" bei Darmstadt (Mammalia, Xenarthra). *Lethaea* 61:247–289.

Storch, G. (2008): Skeletal remains of a diminutive primate from the Paleocene of Germany. *Naturwissenschaften* 95:927–930.

Storch, G. & Habersetzer, J. (1991): Rückverlagerte Choanen und akzessorische Bulla tympanica bei rezenten Vermilingua und Eurotamandua aus dem Eozän von Messel (Mammalia: Xenarthra). *Zeitschrift für Säugetierkunde* 56:257–271.

Strömberg, C.A.E., Dunn, R.E., Madden, R.H., Kohn, M.J. & Carlini, A.A. (2013): Decoupling the spread of grasslands from the evolution of grazer-type herbivores in South America. *Nature Communications* 4:1478.

Swofford, D.L. (2002): *PAUP*. Phylogenetic Analysis Using Parsimony (*and Other Methods)*. Version 4.0b10., Sinauer Associates, Sunderland.

Suárez, C., Forasiepi, A.M., Goin, F.J. & Jaramillo, C. (2016): Insights into the Neotropics prior to the Great American Biotic Interchange: new evidence of mammalian predators from the Miocene of Northern Colombia. *Journal of Vertebrate Paleontology* 36:e1029581.

Szalay, F.S. (1982): A new appraisal of marsupial phylogeny and classification. In: Archer, M. (ed.), *Carnivorous marsupials Volume 2*. Royal Zoological Society of New South Wales, New South Wales: 621–640.

Szalay, F.S., Decker, R.L. & Jenkins, F.A. (1974): Origins, evolution, and function of the tarsus in Late Cretaceous Eutheria and Paleocene primates. In: Jenkins, F.A. (ed.), *Primate locomotion*. Academic Press, New York: 223–259.

Szalay, F.S. & Gould, S.J. (1966): Asiatic Mesonychidae (Mammalia, Condylarthra). *Bulletin of the American Museum of Natural History* 132.

Szalay, F.S., Novacek, M.J. & Mckenna, M.C. (1993): *Mammal Phylogeny*. Springer, New York.

Szalay, F.S. & Schrenk, F. (1998): The middle Eocene *Eurotamandua* and a Darwinian phylogenetic analysis of "edentates." *Kaupia* 7:97–186.

Szalay, F.S. & Trofimov, B.A. (1996): The Mongolian Late Cretaceous *Asiatherium*, and the early phylogeny and paleobiogeography of Metatheria. *Journal of Vertebrate Paleontology* 16:474–509.

Tabuce, R. (2017): New remains of *Chambius kasserinensis* from the Eocene of Tunisia and evaluation of proposed affinities for Macroscelidea (Mammalia, Afrotheria). *Historical Biology* 1–16.

Tabuce, R., Coiffait, B., Coiffait, P.-E., Mahboubi, M. & Jaeger, J.-J. (2001): A new genus of Macroscelidea (Mammalia) from the Eocene of Algeria: a possible origin for elephant-shrews. *Journal of Vertebrate Paleontology* 21:535–546.

Tabuce, R., Marivaux, L., Adaci, M., Bensalah, M., Hartenberger, J.-L., Mahboubi, M., Mebrouk, F., Tafforeau, P. & Jaeger, J.-J. (2007): Early Tertiary mammals from North Africa reinforce the molecular Afrotheria clade. *Proceedings of the Royal Society of London B: Biological Sciences* 274:1159–1166.

Tarver, J.E., dos Reis, M., Mirarab, S., Moran, R.J., Parker, S., O'Reilly, J.E., King, B.L., O'Connell, M.J., Asher, R.J., Warnow, T., Peterson, K.J., Donoghue, P.C.J. & Pisani, D. (2016): The interrelationships of placental mammals and the limits of phylogenetic inference. *Genome Biology and Evolution* evv261.

Tavaré, S., Marshall, C.R., Will, O., Soligo, C. & Martin, R.D. (2002): Using the fossil record to estimate the age of the last common ancestor of extant primates. *Nature* 416:726–729.

Teeling, E.C., Springer, M.S., Madsen, O., Bates, P., O'Brien, S.J. & Murphy, W.J. (2005): A molecular phylogeny for bats illuminates biogeography and the fossil record. *Science* 307:580–584.

Thewissen, J.G.M. (1990): Evolution of Paleocene and Eocene Phenacodontidae (Mammalia, Condylarthra). *University of Michigan Papers on Paleontology* 29:1–107.

Thewissen, J.G.M., Williams, E.M., Roe, L.J. & Hussain, S.T. (2001): Skeletons of terrestrial cetaceans and the relationship of whales to artiodactyls. *Nature* 413:277–281.

Thewissen, J.G.M., Cooper, L.N., Clementz, M.T., Bajpai, S. & Tiwari, B.N. (2007): Whales originated from aquatic artiodactyls in the Eocene epoch of India. *Nature* 450:1190.

Thompson, R.S., Bärmann, E.V. & Asher, R.J. (2012): The interpretation of hidden support in combined data phylogenetics. *Journal of Zoological Systematics and Evolutionary Research* 50:251–263.

van Valen, L. (1969): The multiple origins of the placental carnivores. *Evolution* 118–130.

Van Valen, L.M. (1966): Deltatheridia, a new order of mammals. *Bulletin of the American Museum of Natural History* 132:1–126.

Veselka, N., McErlain, D.D., Holdsworth, D.W., Eger, J.L., Chhem, R.K., Mason, M.J., Brain, K.L., Faure, P.A. & Fenton, M.B. (2010): A bony connection signals laryngeal echolocation in bats. *Nature* 463:939.

Voss, R.S., Helgen, K.M. & Jansa, S.A. (2014): Extraordinary claims require extraordinary evidence: a comment on Cozzuol *et al.* (2013). *Journal of Mammalogy* 95:893–898.

Waddell, P.J., Cao, Y., Hasegawa, M. & Mindell, D.P. (1999): Assessing the Cretaceous superordinal divergence times within birds and placental mammals by using whole mitochondrial protein sequences and an extended statistical framework. *Systematic Biology* 48:119–137.

Walker, E.P. (1964): *Mammals of the World*. JHU Press, Baltimore.

Wang, X., Flynn, L. J & Fortelius, M. (2013): *Fossil Mammals of Asia: Neogene Biostratigraphy and Chronology*. Columbia University Press, New York.

Weber, M.W.C. (1904): *Die Säugetiere: Einführung in Die Anatomie Und Systematik Der Rezenten Und Fossilen Mammalia*. Gustav Fischer, Jena.

Welker, F., Collins, M.J., Thomas, J.A., Wadsley, M., Brace, S., Cappellini, E., Turvey, S.T., Reguero, M., Gelfo, J.N., Kramarz, A. & MacPhee, R.D.E. (2015): Ancient proteins resolve the evolutionary history of Darwin's South American ungulates. *Nature* 81–84.

Werdelin, L. & Sanders, W.J. (2010): *Cenozoic Mammals of Africa*. University of California Press, Berkeley.

Werdelin, L. & Nilsonne, Å. (1999): The evolution of the scrotum and testicular descent in mammals: a phylogenetic view. *Journal of Theoretical Biology* 196:61–72.

Wesley-Hunt, G.D. & Flynn, J.J. (2005): Phylogeny of the Carnivora: basal relationships among the carnivoramorphans, and assessment of the position of "Miacoidea" relative to Carnivora. *Journal of Systematic Palaeontology* 3:1–28.

Westbury, M., Baleka, S., Barlow, A., Hartmann, S., Paijmans, J.L.A., Kramarz, A., Forasiepi, A.M., Bond, M., Gelfo, J.N., Reguero, M.A., *et al.* (2017): A mitogenomic timetree for Darwin's enigmatic South American mammal *Macrauchenia patachonica*. *Nature Communications* 8.

Wheeler, W.C. (1995): Sequence alignment, parameter sensitivity, and the phylogenetic analysis of molecular data. *Systematic Biology* 44:321–331.

White, J.L. & MacPhee, R.D.E. (2001): The sloths of the West Indies: a systematic and phylogenetic review. In: Woods, C.A. & Sergile, F.E. (eds.), *Biogeography of the West Indies: Patterns and Perspectives*. CRC Press, New York: 201–235.

Wible, J.R., Rougier, G.W., Novacek, M.J. & Asher, R.J. (2007): Cretaceous eutherians and Laurasian origin for placental mammals near the K/T boundary. *Nature* 447:1003.

Wible, J.R., Rougier, G.W., Novacek, M.J. & Asher, R.J. (2009): The eutherian mammal *Maelestes gobiensis* from the Late Cretaceous of Mongolia and the phylogeny of Cretaceous Eutheria. *Bulletin of the American Museum of Natural History* 1–123.

Williamson, T.E. & Lucas, S.G. (1992): Meniscotherium (Mammalia, "Condylarthra") from the Paleocene-Eocene of Western North America. *Bulletin New Mexico Museum of Natural History & Science* 1:1–55.

Wilson, D.E. & Reeder, D.M. (2005): *Mammal Species of the World: A Taxonomic and Geographic Reference*. JHU Press, Baltimore.

Wilson, G.P., Ekdale, E.G., Hoganson, J.W., Calede, J.J. & Van der Linden, A. (2016): A large carnivorous mammal from the Late Cretaceous and the North American origin of marsupials. *Nature Communications* 7.

Winge, H. (1941): *The Interrelationships of the Mammalian Genera*. CA Reitzels Forlag, Copenhagen.

Wogelius, R.A., Manning, P.L., Barden, Edwards, H.E.N.P., Webb, S.M., Sellers, W.I., Taylor, K.G., Larson, P.L., Dodson, P., You, H., *et al.* (2011): Trace metals as biomarkers for eumelanin pigment in the fossil record. *Science* 333:1622–1626.

Wood-Jones, F. (1925): *The Mammals of South Australia*. British Science Guild, Adelaide.

Woodburne, M.O., Goin, F.J., Bond, M., Carlini, A.A., Gelfo, J.N., López, G.M., Iglesias, A. & Zimicz, A.N. (2014): Paleogene land mammal faunas of South America; a response to global climatic changes and indigenous floral diversity. *Journal of Mammalian Evolution* 21:1–73.

Wu, J., Yonezawa, T. & Kishino, H. (2017): Rates of molecular evolution suggest natural history of life history traits and a post-K-Pg nocturnal bottleneck of placentals. Current Biology, 27(19), 3025–3033.

Wyss, A.R. (1987): Notes on Prototheria, Insectivora, and Thomas Huxley's contribution to mammalian systematics. *Journal of Mammalogy* 68:135–138.

Yalden, D.W. (1966): The anatomy of mole locomotion. *Journal of Zoology* 149:55–64.

Zachos, J.C., Dickens, G.R. & Zeebe, R.E. (2008): An early Cenozoic perspective on greenhouse warming and carbon-cycle dynamics. *Nature* 451:279.

Zack, S.P. (2012): Deciduous dentition of *Didymictis* (Carnivoramorpha: Viverravidae): implications for the first appearance of "Creodonta." *Journal of Mammalogy* 93:808–817.

Zack, S.P., Penkrot, T.A., Bloch, J.I. & Rose, K.D. (2005): Affinities of "hyopsodontids" to elephant shrews and a Holarctic origin of Afrotheria. *Nature* 434:497.

Zuckerkandl, E. & Pauling, L. (1965): Evolutionary divergence and convergence in proteins. *Evolving Genes and Proteins* 97:97–166.

Samuel T. Turvey

8 Mammal extinction risk and conservation: patterns, threats, and management

8.1 Human-caused mammal extinctions through time

Humans are now a dominant driver of patterns in global biodiversity. Well-documented ongoing anthropogenic transformation of the biosphere is responsible for catastrophic declines across a broad range of taxa and disruption to the structure and functioning of ecosystems, and it is widely accepted that we are experiencing a human-mediated global biodiversity crisis (McClellan 2014, McGill *et al.* 2015).

Human activities have substantially affected species diversity and ecosystem structure throughout the historical period and recent prehistory. Major changes to global large mammal assemblages during the Late Pleistocene constitute among the earliest possible evidence for human involvement in biodiversity change. At least 97 genera of continental megafaunal vertebrates (>44 kg, sensu Martin 1984), mostly mammals but also some giant birds and reptiles, disappeared without ecological replacement during a series of “eco-catastrophic” (Haynes 2002) events in the Late Pleistocene with very little corresponding extinction of small-bodied species (Fig. 8.1). The taxa that became extinct during this interval, including many proboscideans, artiodactyls, perissodactyls, xenarthrans, and diprotodontid marsupials, represent almost two-thirds of Late Pleistocene terrestrial megafaunal genera (Barnosky *et al.* 2004, Koch and Barnosky 2006, Stuart 2015). Continent-wide megafaunal losses occurred earliest in Australia ca. 46,000 years ago (Roberts *et al.* 2001, Miller *et al.* 2005, Rule *et al.* 2012), and the latest well-dated series of prehistoric continental species-level extinctions took place in the Americas and Eurasia 14,000–11,700 years ago (Haynes 2002, Gill *et al.* 2009, Stuart 2015).

Some form of human involvement in Late Pleistocene extinction dynamics is now widely accepted by most paleontologists, as the stepwise nature of these extinctions across different continents correlates with the arrival of technologically modern humans in each region. The mechanism by which humans may have driven these megafaunal species losses is unclear; explanations vary between “blitzkrieg” models of rapid, direct overhunting (Martin 1984, Alroy 2001), and “sitzkrieg” models of more protracted extinction associated with less intensive hunting pressure and typically coupled with indirect interactions such as habitat modification, e.g., through landscape burning (Diamond 1989, Miller *et al.* 2005). However, the Late Pleistocene is also characterized by major global-scale climatic shifts during the transition from glacial to interglacial conditions, and the relative importance or possible interactions of prehistoric human activity and natural environmental change in driving this extinction event have been debated extensively since the 19th century (Grayson 1984), with recent studies continuing to argue opposing standpoints (Bartlett *et al.* 2015, Cooper *et al.* 2015).

Extensive further mammal extinctions are documented across the subsequent Holocene Epoch (11,700 years ago–present, the time interval since the end of the last Ice Age glaciation), a period of modest or minimal climatic variation under broadly “modern” environmental and climatic boundary conditions (Lowe and Walker 2014). More than 250 mammal species extinctions have been documented so far from this interval; evidence for direct human involvement in nearly all of these extinctions is not confounded by climatic factors and is therefore relatively undisputed (Turvey 2009). In contrast to the Late Pleistocene, Holocene mammal extinctions are documented across a wide range of body size classes and taxonomic groups and occurred mainly on islands rather than continents, as a direct result of the colonization of most of the world’s major oceanic and oceanic-type islands (i.e., those possessing ecologically unbalanced and markedly endemic faunas) by prehistoric seafarers and settlers during this interval. Direct human involvement in Holocene island extinctions is again strongly supported by lack of temporal synchrony in extinction events between different island systems, with the timing of extinctions typically matching the sequential chronology of human colonization of these geographic regions rather than that of major environmental changes.

The most severe Holocene mammal extinction event affected the land mammals of the insular Caribbean. The pre-human Caribbean non-volant land mammal fauna was characterized by more than 100 endemic species of lipotyphlan insectivores, megalonychid sloths, platyrrhine primates, and several clades of often large-bodied rodents, of which only two lipotyphlan species and eight currently recognized capromyid rodent species are probably still extant. Over 30 Caribbean bat species have also

https://doi.org/10.1515/9783110341553-008

Fig. 8.1: A global selection of terrestrial megafaunal mammals of the Late Pleistocene. Species surviving today (Africa, southern Asia) are shaded, the rest are extinct (either entirely or in their respective zoogeographic regions). North America: Columbian mammoth (*Mammuthus columbi*), woolly mammoth (*Mammuthus primigenius*), horses (*Equus* sp.), extinct camel (*Camelops hesternus*), mastodon (*Mammut americanum*), ground sloth (*Glossotherium harlani*), extinct tapir (*Tapirus californicus*), glyptodont (*Glyptotherium floridanum*), and saber-toothed cat (*Smilodon fatalis*). South America: giant ground sloth (*Megatherium americanum*), litoptern (*Macrauchenia patachonica*), glyptodont (*Glyptodon clavipes*), notoungulate (*Toxodon platensis*), horses (*Equus* sp.), saber-toothed cat (*Smilodon populator*), and ground sloth (*Mylodon darwini*). Northern Eurasia: cave bear (*Ursus spelaeus*), giant deer (*Megaloceros giganteus*), woolly rhinoceros (*Coelodonta antiquitatis*), woolly mammoth (*Mammuthus primigenius*), hippopotamus (*Hippopotamus amphibius*), "narrow-nosed" rhinoceros (*Stephanorhinus hemitoechus*), straight-tusked elephant (*Palaeoloxodon antiquus*), and spotted hyena (*Crocuta crocuta*). Africa: African elephant (*Loxodonta africana*), white rhinoceros (*Ceratotherium simum*), black rhinoceros (*Diceros bicornis*), and spotted hyena (*C. crocuta*). Southern Asia: Indian rhinoceros (*Rhinoceros unicornis*) and Asian elephant (*Elephas maximus*). Australasia: diprotodon (*Diprotodon optatum*), short-faced kangaroo (*Procoptodon goliath*), and marsupial "lion" (*Thylacoleo carnifex*). Illustration by Patricia Wynne; reproduced courtesy of R.D.E. MacPhee.

experienced extirpation of island populations, and nine species have become globally extinct in this region (Turvey 2009, Turvey *et al.* 2017). Other islands with diverse endemic mammal faunas, including Madagascar and Mediterranean islands such as Corsica, Sardinia, and the Balearics, also experienced substantial mammalian species extinctions during the Holocene (Vigne 1992, van der Geer *et al.* 2010, Goodman and Jungers 2014). Conversely, continental regions experienced reduced levels of species extinction after the Late Pleistocene and before the recent historical era, leading to use of the term "Holocene underkill" to contrast with hypothesized Pleistocene overkill (Grayson 2008). However, it is becoming increasingly apparent that continental mammal assemblages also experienced extensive population-level extirpations, range contractions, and population depressions across the Holocene, probably driven by overexploitation by prehistoric hunters even under relatively low human population densities (Janetski 1997, Grayson 2001, Johnson 2006, Crees and Turvey 2014).

Human pressures on global ecosystems have become qualitatively more intensive during the past few centuries or decades, leading to the increasing recognition of a modern "Anthropocene" Epoch defined by indices of human-caused environmental disequilibrium (Lewis and Maslin 2015). Recent human impacts on global mammal faunas are considerably better understood due to the wider availability of standardized species-level and population-level baseline ecological and monitoring data. The most recent global assessment conducted by the International Union for the Conservation of Nature (IUCN) recognized 25% of the 5487 mammal species considered to be extant since AD 1500 and for which adequate data were available as being threatened with extinction; 836 species in this assessment had insufficient information for evaluation, leading to further estimation of between 21% and 36% of all extant mammals being threatened with extinction (Schipper *et al.* 2008).

At the regional and population levels, less than 21% of the earth's terrestrial surface still contains all of the large-bodied (here defined as >20 kg) mammals it held in AD 1500 (Morrison *et al.* 2007), and vertebrate populations for which long-term data are available have on average declined globally by 52% since 1970 (McClellan 2014). Many mammal species persist only as greatly reduced remnant populations restricted to small areas of formerly large geographic ranges (Channell and Lomolino 2000) (Tab. 8.1, Fig. 8.2). Ongoing extinctions of mammal species occurred through the 20th century (Tab. 8.2) and are continuing in the 21st century (Fig. 8.3), with the global extinction of the Yangtze River dolphin or baiji (*Lipotes vexillifer*) and the Christmas Island pipistrelle (*Pipistrellus murrayi*) documented during the past decade or so (Turvey *et al.* 2007, Martin *et al.* 2012), and further species such as the Bramble Cay melomys (*Melomys rubicola*) also likely to have become extinct within this period (Woinarski *et al.* 2015).

Tab. 8.1: Examples of recent severe population declines experienced by large mammal species. Data from Nowell and Jackson (1996), IUCN (2015), and NOAA Fisheries (2016).

Species	Distribution	Baseline population estimate	Recent population estimate	% remaining population
Black rhinoceros (*Diceros bicornis*)	Eastern and southern Africa	850,000 (AD 1900)	4,880	<1%
Blue whale (*Balaenoptera musculus*)	Global oceans	181,200 (AD 1900)	5,000	2.8%
Hainan gibbon (*Nomascus hainanus*)	Hainan Island (China)	>2,000 (AD 1950)	26	1.3%
Saiga (*Saiga tatarica*)	Central Asia	1,250,000 (mid-1970s)	50,000	4%
Tiger (*Panthera tigris*)	South, central, east and southeast Asia	>100,000 (AD 1900)	3,160	<3.2%

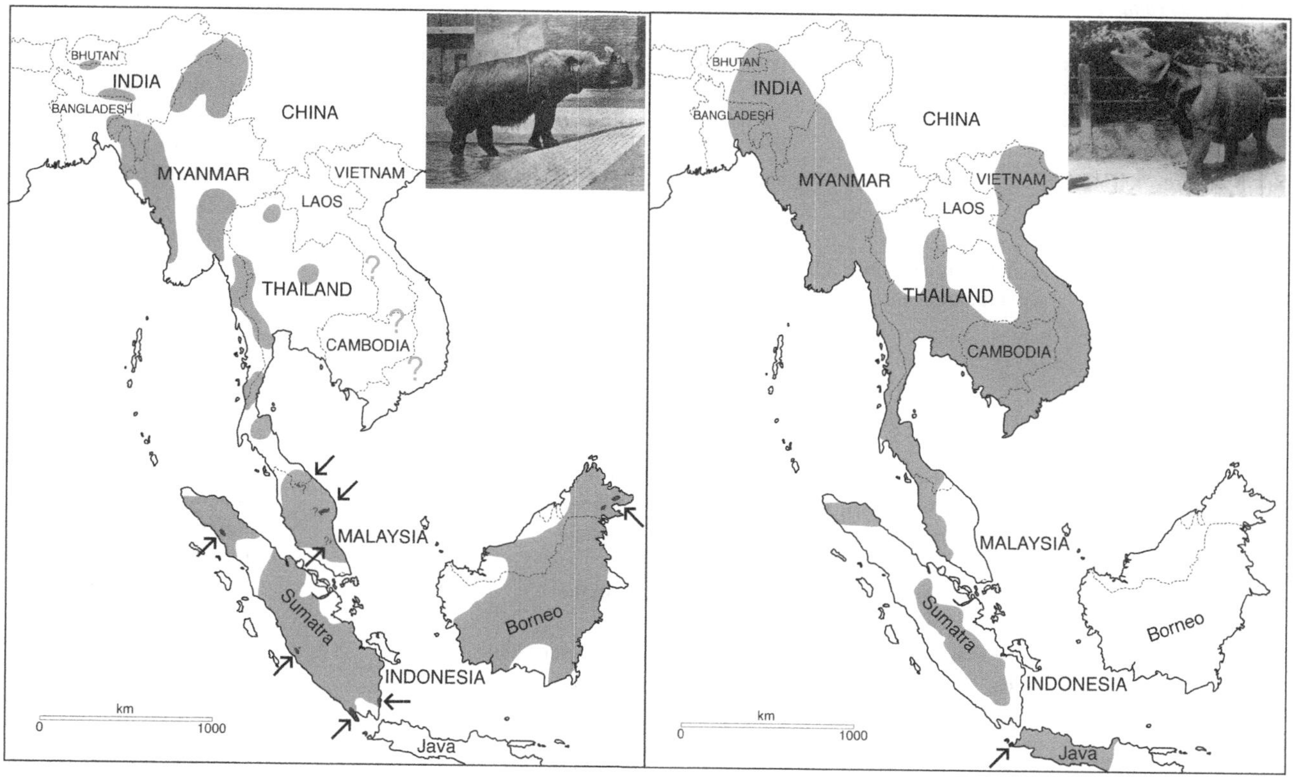

Fig. 8.2: Inferred historical (light grey) and current-day (dark grey) geographic distributions for (left) Sumatran rhinoceros (*Dicerorhinus sumatrensis*) and (right) Javan rhinoceros (*Rhinoceros sondaicus*), showing the extent of geographic range loss and remnant status of surviving populations. Modified from Foose and van Strien (1997) and IUCN (2015).

Tab. 8.2: Mammal species that are recognized by IUCN (2015) as having become extinct during the 20th century. Asterisks after last record date indicate that the species is extinct in the wild, but *ex situ* managed populations still exist. Associated data from IUCN (2015).

Species	Common name	Order	Distribution	Last record	Cause of extinction
Elaphurus davidianus	Père David's deer	Cetartiodactyla	China	1900 (1939?)**	Habitat loss, hunting
Notomys longicaudatus	Long-tailed hopping mouse	Rodentia	Australia	1901–1902	Introduced foxes and cats?
Rattus macleari	Maclear's rat	Rodentia	Christmas Island (Australia)	1902–1904	Pathogenic trypanosome from introduced black rats
Rattus nativitatis	Bulldog rat	Rodentia	Christmas Island (Australia)	1902–1904?	Pathogenic trypanosome from introduced black rats
Neotoma anthonyi	Anthony's woodrat	Rodentia	Todos Los Santos Islands (Mexico)	1926	Introduced cats
Nesoryzomys darwini	Darwin's Galápagos mouse	Rodentia	Galápagos	1930	Introduced black rats?
Neotoma bunkeri	Bunker's woodrat	Rodentia	Coronados Island (Mexico)	1931	Introduced cats
Peromyscus pembertoni	Pemberton's deer mouse	Rodentia	San Pedro Nolasco Island (Mexico)	1931	Unknown
Macrotis leucura	Lesser bilby or yallara	Peramelemorphia	Australia	1931 (1960s?)	Introduced cats and foxes, competition with rabbits, habitat degradation, possibly changes in fire regime
Lagorchestes asomatus	Central hare-wallaby or kuluwarri	Diprotodontia	Australia	1932 (1960s?)	Introduced cats and foxes, possibly changes in fire regime
Nesoryzomys indefessus	Indefatigable Galápagos mouse	Rodentia	Galápagos	1934	Introduced black rats?
Caloprymnus campestris	Desert rat kangaroo	Diprotodontia	Australia	1935 (1970s?)	Habitat loss, introduced cats and foxes
Thylacinus cynocephalus	Thylacine	Dasyuromorphia	Tasmania	1936	Hunting; possibly also habitat loss and competition/disease from domestic dogs
Macropus greyi	Toolache wallaby	Diprotodontia	Australia	1937	Introduced predators, competition from livestock, hunting
Rucervus schomburgki	Schomburgk's deer	Cetartiodactyla	Thailand	1938	Habitat loss and hunting
Perameles eremiana	Desert bandicoot	Peramelemorphia	Australia	1943 (1960s?)	Introduced cats and foxes, habitat alteration by rabbits; possibly also changing fire regimes
Gazella bilkis	Queen of Sheba's gazelle	Cetartiodactyla	Yemen	1951	Hunting
Zalophus japonicus	Japanese sea lion	Carnivora	Northwest Pacific	1951 (1970s?)	Hunting
Neomonachus tropicalis	Caribbean monk seal	Carnivora	Caribbean Sea	1952	Hunting
Pseudomys glaucus	Blue-grey mouse	Rodentia	Australia	1956	Habitat loss and predation by invasive species?
Chaeropus ecaudatus	Pig-footed bandicoot	Peramelemorphia	Australia	1950s	Introduced cats/habitat alteration by sheep?
Geocapromys thoracatus	Little Swan Island hutia	Rodentia	Little Swan Island (Honduras)	1950s	Introduced cats
Neotoma martinensis	San Martin Island woodrat	Rodentia	San Martín Island (Mexico)	1950s	Introduced cats
Onychogalea lunata	Crescent nailtail wallaby	Diprotodontia	Australia	1950s (1960s?)	Introduced foxes and cats; possibly also changing fire regimes and habitat alteration by rabbits and livestock

Tab. 8.2 (continued)

Species	Common name	Order	Distribution	Last record	Cause of extinction
Juscelinomys candango	Candango mouse	Rodentia	Brazil	1960	Habitat loss
Cryptonanus ignitus	Red-bellied gracile mouse opossum	Didelphimorphia	Argentina	1962	Habitat loss
Pteropus tokudae	Guam flying fox	Chiroptera	Guam	1968 (1970s?)	Hunting; possibly also introduced brown tree snake
Oryx dammah	Scimitar-horned oryx	Cetartiodactyla	North Africa	1988**	Hunting and habitat loss, including competition with livestock
Gazella saudiya	Saudi gazelle	Cetartiodactyla	Iraq, Kuwait, Saudi Arabia, Yemen	1980s	Unknown

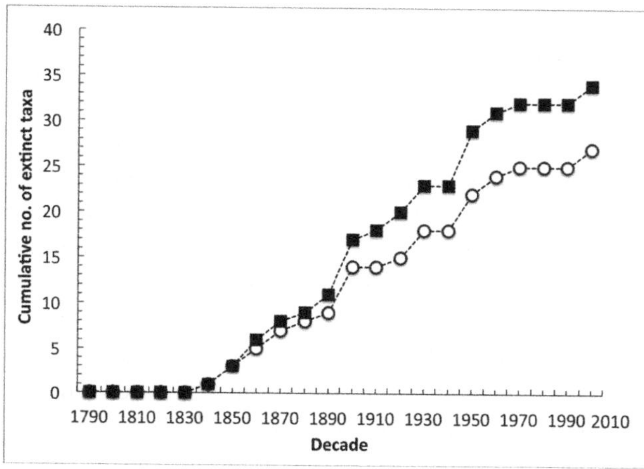

Fig. 8.3: Cumulative number of historical extinctions of mammal taxa in Australia. Open circles denote species, filled squares also include subspecies; decade value refers to the 10-year period following given date. Dates could not be estimated for some extinct taxa, so the tally given here is less than the total number of extinct taxa. Reproduced with permission from Woinarski *et al.* (2014).

Defining the magnitude of this current human-caused global biodiversity crisis within the context of past "mass extinction" events seen in the deep-time geological record—when more than 75% of estimated global species became extinct within a relatively short geological interval (Jablonski 1994, Barnosky *et al.* 2011)—is not straightforward, given difficulties in comparing data on taxonomic diversity and extinction between the fossil and modern records. Overall, levels of species extinction in mammals and other taxonomic groups during the past 500 years, the time interval assessed by IUCN (2012) when considering human-caused extinctions, have not been high enough so far to qualify as a mass extinction event. However, current extinction rates for mammals calculated over the past 500 years are faster than all rates that would have produced past mass extinctions over hundreds of thousands or millions of years (Barnosky *et al.* 2011). Alternately, if all currently threatened mammal species became extinct within a century, and that rate then continued unabated as further species became threatened, mammal extinctions would reach the magnitude of a mass extinction event in 334.4 years; an alternate scenario where only Critically Endangered species are considered "doomed" to extinction within a century indicates that a mammal mass extinction event would be reached in 1519 years, and even assuming a more "relaxed" estimate of Critically Endangered species being "doomed" to extinction within 500 years leads to a mass extinction event reached in 7593 years, a rate still markedly higher than observed for previous mass extinctions in geological time (Barnosky *et al.* 2011).

8.2 Case study: the extinction of the Yangtze River dolphin

The Yangtze River dolphin or baiji (*Lipotes vexillifer*) was a freshwater cetacean endemic to the middle-lower Yangtze River system and neighboring waterways in eastern China, and it was the only recent representative of the Lipotidae, a clade that diverged from all other cetacean lineages more than 20 million years ago (Slater *et al.* 2010). From an estimated 400 baiji individuals in 1979–1981, Chinese researchers reported an ongoing and rapid population decline through the 1980s and 1990s, with surveys during 1997–1999 providing a minimum estimate of only 13 surviving animals (Zhang *et al.* 2003). A range of anthropogenic extinction drivers, including vessel collisions and dam construction, which also threaten freshwater cetaceans in other river systems, have been implicated in the baiji's decline, but the primary factor was probably unsustainable bycatch in unselective local fisheries, which use rolling hook longlines, nets, and electrofishing. At least half of all known baiji deaths in the 1970s and 1980s were caused by rolling hooks and other fishing gear (Chen *et al.* 1997), making the baiji a victim not of active persecution but of incidental mortality resulting from massive-scale human environmental impacts. The last authenticated baiji records were of a stranded pregnant female found in 2001 and a live animal photographed in 2002. Although subsequent unverified reports suggested that a remnant population persisted in the Yangtze, an intensive 6-week multivessel boat-based visual and acoustic survey carried out in 2006, covering the entire historical range of the baiji in the main Yangtze channel, failed to find any evidence that the species survived (Turvey *et al.* 2007). The extinction of the baiji represents the first extinction of a cetacean species caused by human activity, one of the few documented global extinctions of a large vertebrate species during the past 50 years (Tab. 8.2), and only the fourth disappearance of an entire mammal family since AD 1500. Conservation strategies to establish unconnected river sections as discrete *in situ* baiji reserves were ineffective and may have been inconsistent with the ecological requirements of the species, which apparently made large-scale seasonal movements up and down the main Yangtze channel (Turvey *et al.* 2010). Establishment of a closely monitored *ex situ* baiji population in a semi-natural reserve was also consistently advocated by scientists and policy makers as an essential short-term goal for continued survival of the species, but despite extensive debate and a series of international workshops over more than two decades before the species' extinction, little effort was ever made to implement an *ex situ* baiji recovery program—providing a sobering reminder that species cannot be expected to save themselves, and that intervention may need to be swift and decisive.

8.3 Anthropogenic threats to mammalian diversity

Multiple different types of threats associated with human activities are responsible for driving recent and ongoing mammal declines. By far the greatest threat to mammalian biodiversity is habitat loss, which affects 40% of mammal species through conversion to agricultural or urban environments, degradation of habitat quality in remaining patches of suitable habitat, and fragmentation of formerly connected habitat (Schipper *et al.* 2008). "The history of human civilization is also a history of deforestation" (Food and Agriculture Organization of the United Nations 2012), and until the 20th century the highest rates of deforestation occurred in temperate forests in Europe, Asia, and North America, but by the mid-20th century deforestation had effectively halted in temperate forests, and it is now taking place almost exclusively in tropical regions of Asia, Africa, and the Americas (Tscharntke *et al.* 2010). As tropical environments support greater numbers of mammal species than temperate environments (Schipper *et al.* 2008), tropical deforestation is a particular threat to global mammalian biodiversity. Forest loss is currently estimated at 0.6% per year (Hansen *et al.* 2010), with over 50% of this loss affecting tropical forests (Hansen *et al.* 2013, Achard *et al.* 2014). This deforestation is driven both by ongoing human population growth in tropical regions and by globalized markets for timber, agricultural products, and biofuel feedstock (e.g., soy products in the meat industry, oil palm in food products, and biofuel) (Tscharntke *et al.* 2010).

The second major threat to mammals is direct human-caused mortality, typically through harvesting for food, medicine, or other products, which affects 17% of mammal species (Schipper *et al.* 2008). Hunting of "bushmeat" for household consumption or sale in markets, typically focused on artiodactyls, monkeys and other arboreal mammals, and porcupines and other rodents, is extremely important for rural food security and livelihoods across tropical regions of the Americas, Africa, and Asia (Corlett 2007). Bushmeat hunting drives an estimated annual extraction of about 6 million tonnes of animals across the Congo and Amazon Basins alone (Nasi *et al.* 2011). However, the importance of hunting for subsistence is now increasingly outweighed by hunting for markets, which now also targets additional species (e.g., felids, bears, and pangolins) for the trade in wild animals and their parts for medicines (Fig. 8.4); annual illegal trade in species listed under the Convention on International Trade in Endangered Species of Wild Flora and Fauna is currently worth an estimated US$20 billion globally (Challender and MacMillan 2014). This practice constitutes a serious threat to tropical forest mammals, with mammalian dispersers of large seeds and understorey browsers in particular becoming depleted or eliminated; this affects tropical ecosystem functionality and leads to "empty forest" syndrome, where forest ecosystems remain structurally intact but largely or completely lack their large-bodied and medium-bodied vertebrate fauna (Corlett 2007, Nasi *et al.* 2011). For marine mammals, in addition to deliberate harvesting (which constitutes a major threat for 52% of species), 78% of species are also affected by accidental mortality resulting from non-targeted fisheries bycatch and vessel strikes (Schipper *et al.* 2008).

Invasive species pose a particular threat towards mammals on oceanic-type island systems that are not

Fig. 8.4: Trade in wild mammals and mammal body parts for food and medicine in southeast Asia. (Left) Bengal slow loris (*Nycticebus bengalensis*) and Asian palm civet (*Paradoxurus hermaphroditus*) for sale in a roadside market in Laos. (Right) Medicinal rice wine tonics made using porcupine stomachs (first and fourth bottles), pangolin fetuses (second bottle), and deer antlers (third bottle), for sale in a restaurant in Laos. Photos by the author.

associated with neighboring continental shelves, as endemic mammal faunas on such islands are depauperate and derived from a limited set of successful original colonists. Although Madagascar contains an endemic carnivore clade, the Eupleridae Yoder *et al.* 2003, oceanic-type mammal faunas typically lack native mammalian predators and also contain relatively few sympatric species filling closely similar niches. Island mammals are therefore particularly vulnerable to the deliberate or accidental human-mediated arrival of invasive mammal predators and competitors. For example, all surviving non-volant insular Caribbean land mammals have experienced population declines or local extirpations through predation by feral dogs and cats and invasive mongooses, and/or through competition for resources (e.g., food, den sites) and associated habitat disturbance by invasive black and brown rats and even invasive African monkeys, and interactions with mongooses and black rats in particular are probably responsible for driving the historical extinction of numerous other Caribbean mammals (Borroto-Páez 2009, Turvey *et al.* 2017).

Deliberate or accidental movement of domestic or commensal mammals by humans can also facilitate the spread of novel infectious diseases to immunologically naïve native mammals and drive population declines and extirpations. For example, the rinderpest virus pandemic in Africa in the 1890s saw the death not only of 80%–90% of domestic cattle across large areas of the continent, but also of susceptible wild ungulates (e.g., buffalo, giraffe, wildebeest, eland, kudu, and warthog) which came into contact with cattle, e.g., by drinking at the same water source (Shorrocks and Bates 2007). Several fatal epidemics caused by canine distemper virus and closely related morbilliviruses have also occurred from the 1970s onwards in Africa (both Ethiopia and the Serengeti-Mara ecosystem of East Africa) in canids (Ethiopian wolf *Canis simensis*, silver-backed jackal *Canis mesomelas*, bat-eared fox *Otocyon megalotis*), lions (*Panthera leo*), and other mammalian carnivores due to exposure to pathogens in domestic dogs. Ethiopian wolves have also experienced severe declines due to other domestic dog pathogens such as rabies (Roelke-Parker *et al.* 1996, Laurenson *et al.* 1998, Gordon *et al.* 2015). Few complete extinctions of mammal species have so far been attributed to novel infectious diseases. However, the early 20th century extinction of two murid rodents (*Rattus macleari* and *R. nativitatis*) formerly endemic to Christmas Island has been shown to have been partially or completely caused by a pathogenic trypanosome carried by fleas hosted on recently introduced black rats (Wyatt *et al.* 2008).

Different threat processes are associated with different levels of phylogenetic signal in mammals. Species that are currently threatened by overexploitation tend to consist of a relatively small number of mammal clades (e.g., cetaceans, artiodactyls, primates, hystricognath rodents, and pangolins) that are primarily hunted for consumption or sale. Conversely, species currently threatened by habitat loss are distributed across a wider range of less closely related clades, as the impact of this threat process is more cosmopolitan and less dependent on evolutionarily conserved biological traits. Mammals currently threatened by introduced species are intermediate between these extremes (Fritz and Purvis 2010). Mammal species that became extinct during the Holocene show a much higher level of phylogenetic relatedness compared with the pattern seen across currently threatened species;

this comparison suggests that Holocene mammal extinctions to date have been largely driven by a combination of overharvesting and invasive species, whereas mammal species today are instead primarily threatened by habitat loss (Turvey and Fritz 2011).

The major drivers of human-caused mammalian biodiversity loss are likely to continue to change in the future, as anthropogenic climate change is predicted to increasingly disrupt ecosystem processes (e.g., tropical forest growth; Tscharntke *et al.* 2010) and force species to shift their distributions to track shifts in suitable climates and associated habitats. Habitat tracking may be challenging for many species due to both the unprecedented rapidity of predicted climatic change and the extensive (and increasing) anthropogenic fragmentation of global habitats. Schloss *et al.* (2012) estimated that, on average, 9.2% of mammals and in some places up to 39% of mammals may be unable to track shifts in suitable habitat, with 87% of mammal species expected to experience reductions in range size associated with climate change, and 20% of these range reductions due to limited dispersal abilities rather than reductions in the area of suitable habitat. Climate change may particularly impact restricted-range species, which may be more dispersal-limited (Morueta-Holme *et al.* 2010).

8.4 Correlates of extinction vulnerability and resilience

Mammals vary in their vulnerability or resilience to human impacts—some species or populations have already become extinct as a result of anthropogenic pressures, others are extant but threatened with extinction, and still others remain unthreatened. Explaining this variation is of crucial importance for predicting future extinctions and informing conservation, and large-scale macroecological analyses incorporating data on current-day mammalian biology, ecology, and phylogenetic relationships have identified several often interrelated intrinsic and extrinsic correlates of extinction risk across the world's mammals (Purvis *et al.* 2000, Cardillo *et al.* 2008). However, studies which have focused only on extant mammals, and which therefore exclude the most vulnerable set of species that have already become extinct due to past human activities, may be limited by this "extinction filter" to provide only incomplete or biased insights into patterns of extinction risk (Balmford 1996), and there has been increasing recognition of the need to also incorporate historical baselines of past biodiversity and faunal turnover into such analyses (McClenachan *et al.* 2012, Rick and Lockwood 2013, Turvey *et al.* 2015).

Body size is a major correlate of extinction risk in mammals. Large mammals have shown higher levels of extinction throughout the Late Pleistocene and Holocene (Barnosky *et al.* 2004, Koch and Barnosky 2006, Turvey and Fritz 2011, Stuart 2015), and extinction risk in extant mammals also shows a strong positive correlation with body size (Cardillo *et al.* 2005, Schipper *et al.* 2008). Large size is associated with increased extinction risk due to both intrinsic biological factors and extrinsic anthropogenic threats. Larger-bodied species tend to have lower intrinsic rates of population increase and lower population densities, making them more vulnerable to both anthropogenic and non-anthropogenic environmental pressures because the rate at which they can replace individuals is low (Purvis *et al.* 2000, Cardillo *et al.* 2008). Such species are also disproportionately exploited by humans; for example, subsistence hunters in Neotropical forests prefer to target species above 6.5 kg (Jerozolimski and Peres 2003). Extinction risk in small-bodied (less than 3 kg) mammal species is only predicted by external pressures rather than by intrinsic traits, but above this body size threshold, susceptibility to both intrinsic and extrinsic threats increases sharply with body size in mammals (Cardillo *et al.* 2005). Comparison of the biological traits of extinct Late Quaternary mammal species with those of surviving species shows that reproductive rate has been the strongest predictor of past human-caused extinction, with species exhibiting low reproductive rates at high risk regardless of their body size; the only surviving species exhibiting this intrinsic trait occur in closed habitats and are typically arboreal or nocturnal, therefore reducing their exposure to direct interaction with humans (Johnson 2002).

Geography is also a major determinant of extinction risk in mammals, as demonstrated both by the spatially staggered pattern of Late Quaternary extinctions and by spatial variation in current-day mammal extinction risk. Much of this effect is associated with spatial variation in human activities. More than two billion people, representing almost a third of the global human population, live in "biodiversity hotspots", geographic conservation priority areas that contain 50% of vascular plants and 42% of terrestrial vertebrates and that together account for only ca. 2.5% of the earth's terrestrial surface (Mittermeier *et al.* 2011). Threatened mammal species are concentrated in south and southeast Asia, where human population densities—and associated pressures on the environment in terms of habitat conversion and exploitation—are highest (Fig. 8.5). For example, whereas 48% of primate species are currently threatened with extinction globally,

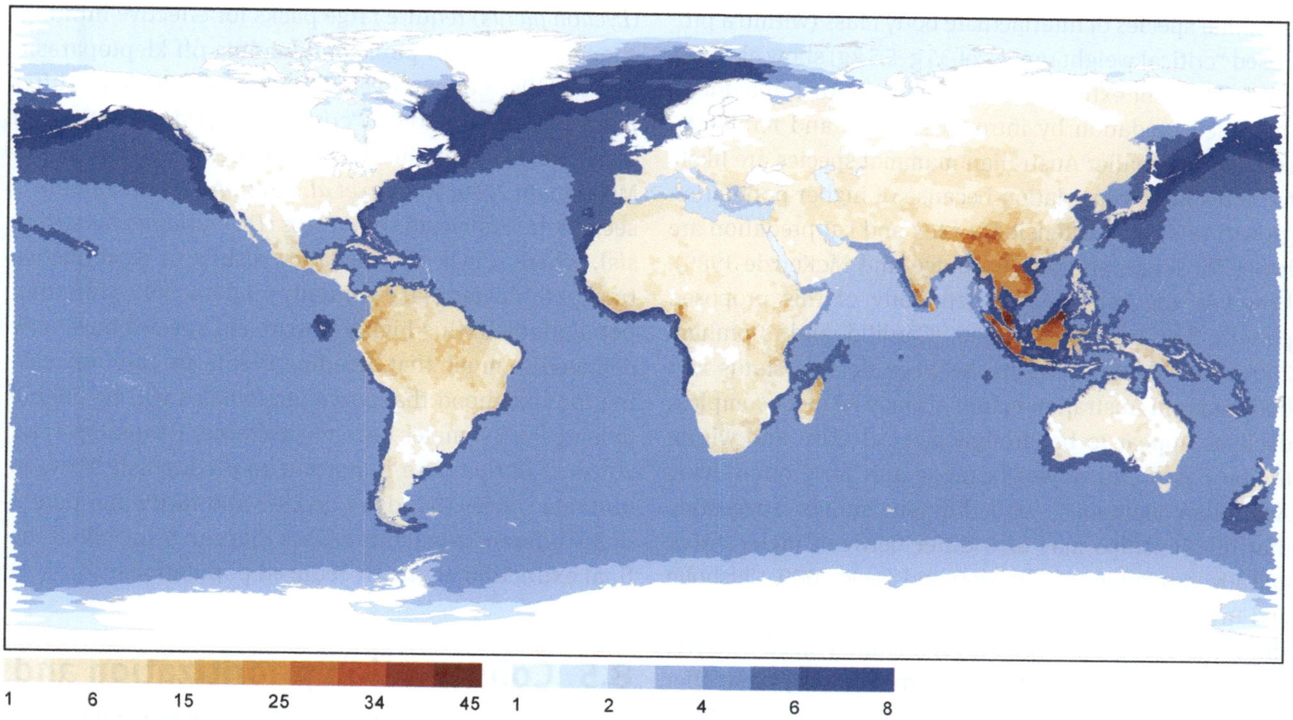

Fig. 8.5: Patterns of globally threatened mammalian species richness (Vulnerable, Endangered, and Critically Endangered species) for terrestrial and marine mammals. Brown number scale, numbers of globally threatened terrestrial mammal species per cell; blue number scale, numbers of globally threatened marine mammal species per cell. Reproduced with permission from Schipper *et al.* (2008).

a staggering 79% of Asian primates are threatened, as are all representatives of some mammal groups restricted to Asia (e.g., gibbons) (Schipper *et al.* 2008). At a landscape level, mammal species and populations occurring at lower elevations have been more vulnerable to extinction as a result of greater historical human population growth and habitat conversion in these more accessible regions, and remnant populations of many threatened mammals which formerly had broad elevational distributions (e.g., European carnivores; giant panda *Ailuropoda melanoleuca*) are now largely or completely restricted to high-elevation refugia (Morrison *et al.* 2007, Fisher 2011, Turvey *et al.* 2015). Spatial variation in species ecology can also substantially affect extinction risk in regional faunas. In particular, the pre-human community ecology of oceanic-type islands, which lack native mammalian predators and contain relatively few interspecific competitors, has led to insular mammalian life histories being increasingly determined by intraspecific competition, driving the evolution of slower reproductive strategies (with greater investment in fewer offspring) in many island species and therefore making them intrinsically more vulnerable to extinction (McNab 2001).

These distinct intrinsic and extrinsic correlates of mammalian extinction risk interact to generate considerable synergistic variation in vulnerability and resilience between different species and geographic regions. For example, whereas large-bodied mammals display increased extinction vulnerability at a global level, this pattern is substantially affected by geographic variation in human activities, with 90% of large mammals threatened by exploitation in Asia compared with 80% in Africa and 64% in the Neotropics (Schipper *et al.* 2008). Conversely, many species with biological traits that predispose them to be particularly vulnerable to extinction but which occur in regions where human impacts are currently low (e.g., Nearctic boreal forests and tundra) are not yet threatened (Cardillo *et al.* 2006). Analysis of extinction risk in carnivores has shown that intrinsic biological traits interact with human population density to affect species vulnerability, with biological traits explaining more variation in risk for species with high levels of exposure to human populations (for example, species with larger body sizes require larger home ranges, bringing individuals into increasing contact with people in fragmented habitats); this suggests that biology is likely to become a more critical determinant of extinction risk as human populations continue to grow (Cardillo *et al.* 2004).

More complex patterns of body mass selectivity associated with differential extinction risk have also been proposed for specific geographic areas that face different types of anthropogenic threat. It has been suggested that mammal species of intermediate body mass (within a proposed "critical weight range" of 35 g–5.5 kg) show elevated levels of recent extinction and decline in Australia due to increased predation by introduced foxes and feral cats; conversely, smaller Australian mammal species are likely to be resilient to predation because of higher population growth rates, and intensities of fox and cat predation are lower for larger species (Burbidge and McKenzie 1989). However, the validity and generality of this proposed pattern has been challenged (Cardillo and Bromham 2001), and the relationship between decline status and body mass in Australian mammals may be more complex; declines appear to be strongly size selective and within the proposed critical weight range only for ground-living Australian mammals, with different studies suggesting that relationships may also differ between high-rainfall and low-rainfall areas (Johnson and Isaac 2009, Chisholm and Taylor 2010). Conversely, Caribbean insular mammals show increased survivorship of medium-sized species, probably reflecting the increased vulnerability of smaller species to competition and predation by invasive rats and mongooses, and prehistoric human exploitation of larger species. The suggested explanation for this observed regional pattern of body size selectivity has been called the "Goldilocks Hypothesis", as the surviving members of the Caribbean land mammal fauna are neither too large nor too small, but instead "just right" to be resilient to these threats (Hansford *et al.* 2012).

Small geographic range size is also recognized as a key predictor of extinction risk in mammals (Purvis *et al.* 2000, Cardillo *et al.* 2008). This pattern may in some cases represent an artifact of historical extinction filters, if current-day small population size represents a remnant distribution of a formerly much larger geographic range – a phenomenon more likely to be seen in species that have been particularly vulnerable to past human impacts (Hanna and Cardillo 2013). Species restricted to reduced geographic areas are more vulnerable not only to extrinsic threats such as environmental perturbation and direct offtake, but also to processes such as inbreeding, genetic drift and demographic stochasticity that are unlikely to affect larger populations; such factors can lead to an "extinction vortex" of progressive reduction in genetic variability, individual fitness, and reproductive output (Gilpin and Soulé 1986, Lacy and Pollak 2014). Fitness impacts associated with population size, or Allee effects – associated either with increased reproductive success in large groups or with increased difficulty in finding mates at low population densities – can increase extinction risk and have been documented in several mammal species (Courchamp *et al.* 2008). For example, African wild dogs (*Lycaon pictus*) require large packs for effective hunting, cooperative care of pups, and fending off kleptoparasitism at kills. Cooperatively breeding species can suffer reduced fitness below a "critical group size", estimated as about five individuals for wild dogs (Courchamp and Macdonald 2001, Angulo *et al.* 2013). Another example is seen in the Sumatran rhinoceros (*Dicerorhinus sumatrensis*), where a lack of pregnancy results in reproductive tract pathologies that eventually render females infertile; populations in this highly threatened species have now declined so much that breeding events are rare, increasing the likelihood that surviving females will be unable to mate even if males become available (Havmøller *et al.* 2016). It is therefore important to distinguish between ultimate causes that drive decline and rarity and potentially different proximate causes that can bring about the final extinction of greatly reduced populations.

8.5 Conservation prioritization and management for threatened mammals

The world's governments have increasingly recognized that reducing the rate of biodiversity loss through effective and urgent action is a global priority (Gillespie 2013). Several international and national frameworks now exist to assess whether species are in danger of extinction, and to inform appropriate targeted conservation attention and action. The global standard for assessing conservation status is the IUCN Red List, which uses a series of quantitative criteria based around population size, geographic range, and decline dynamics to assign species to one of a series of status categories (Category A: reduction in population size; Category B: geographic range; Category C: small and declining population size; Category D: total population size; Category E: quantitative population modeling; IUCN 2012). Species are considered threatened with extinction if they are assessed as Vulnerable (high extinction risk), Endangered (very high extinction risk), or Critically Endangered (extremely high extinction risk) (Fig. 8.6). Other regional frameworks for assessing threat also exist, such as the protocols used to list species under the US Endangered Species Act (Harris *et al.* 2012).

The resources available for conservation in terms of funding, skills, training and education are, tragically, insufficient to prevent the loss of much of the world's threatened biodiversity, and conservation planners have been forced into the unenviable situation of having to prioritize which species should receive the most

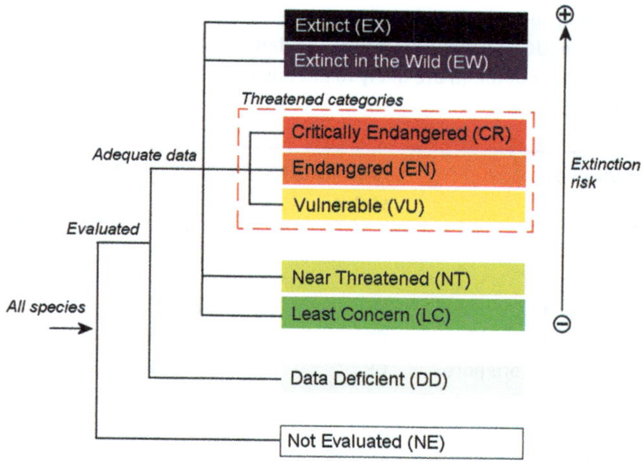

Fig. 8.6: Structure of the status categories used by IUCN. Reproduced with permission from IUCN (2012).

protection—this is "the agony of choice" (Vane-Wright *et al.* 1991) or the "Noah's Ark problem" (Weitzman 1998). A range of methods for setting species-based conservation priorities for threatened mammals have been advocated, focusing variously on geographic priority areas (Myers *et al.* 2000, Ceballos *et al.* 2005, Mittermeier *et al.* 2011, Jenkins *et al.* 2013); "flagship", "umbrella", "keystone", "landscape", or "indicator" species, or species with significant economic, ecological, scientific, or cultural value (Entwistle and Dunstone 2000, Redford *et al.* 2011); or analyses of cost-effectiveness (Carwardine *et al.* 2008, Wilson *et al.* 2011). However, although such prioritization schemes are meant to be objective or quantitative, they are often based on or influenced by subjective values about conservation importance, worth, or usefulness (Redford *et al.* 2011). Indeed, despite the wide range of available prioritization schemes, around three-quarters of species-based mammal conservation projects are specifically aimed at "charismatic" large-bodied mammals that represent a small subset of overall threatened mammalian diversity (Leader-Williams and Dublin 2000, Sitas *et al.* 2009).

Most conservation prioritization schemes treat species as equal units for assessment. However, species represent non-equal amounts of unique evolutionary history, reflecting different patterns of divergence that can increasingly be quantified and compared across large clades using modern phylogenetic analyses. This evolutionary history represents a compound metric of all genetic, morphological, and ecological diversity, reflecting the positive correlation between amount of evolutionary change and amount of time elapsed (Redding *et al.* 2010), and it can be compared objectively across taxa using data on phylogenetic branch lengths, measured in terms of either units of time or genetic change (Vane-Wright *et al.* 1991, Faith 1992). Evolutionary history—often referred to in terms of phylogenetic or evolutionary distinctiveness—can therefore be viewed as the "missing component" for conservation prioritization and planning, with resources to be focused on species that represent the greatest amounts of unique evolutionary history. Measures of unique evolutionary history can be weighted by data on conservation urgency to generate ranked conservation priority lists. Conservation priority setting based on evolutionary history has frequently been proposed, and the "Evolutionarily Distinct and Globally Endangered" (EDGE) scoring method has been adopted by the Zoological Society of London as a practical tool for prioritizing species-focused conservation programs at a global level for mammals (Isaac *et al.* 2007, Collen *et al.* 2011; www.edgeofexistence.org). The EDGE method incorporates evolutionary history using the evolutionary distinctiveness (ED) metric, which partitions phylogenetic branches by the total number of species descending from them regardless of nested tree structure, such that the contribution of a given ancestral branch to the ED score is 1/number of descendants of that branch; alternative evolutionary metrics such as equal splits can also be used to score species' evolutionary history (Collen *et al.* 2011). Many mammals recognized as top global conservation priorities on the basis of threatened evolutionary history—such as long-beaked echidnas (*Zaglossus*), solenodons (*Solenodon*), and funnel-eared bats (*Natalus*)—have rarely if ever been recognized as important priorities in other approaches and do not benefit from existing conservation measures (Tab. 8.3); 64 of the top 100 ranked species in the most recent analyzed iteration of the EDGE mammal list currently receive little or no conservation attention (Collen *et al.* 2011), indicating that a disproportionately large amount of unique mammalian evolutionary history is in danger of becoming extinct in the near future.

Conservation management decisions for threatened species need to be evidence-based, making use where possible of rigorous, objective research into relevant aspects of species ecology, population dynamics, and threats (Sutherland *et al.* 2004, Segan *et al.* 2010). In some cases, data can be determined clearly on these key factors. For example, the Angel de la Guarda deer mouse (*Peromyscus guardia*) was common on Estanque Island in the Gulf of California until 1998, when no mice were detected by researchers but a single feral cat was found to have reached the island through presumed human-mediated transport; cat scats collected in 1998 were found to contain deer mouse bones and hair, whereas scats collected the following year contained only lizard bones

Tab. 8.3: Top 50 mammal EDGE scores, representing the highest priority mammal species requiring urgent conservation attention on the basis of a combination of high ED and high threat status. IUCN threat status from IUCN (2015): CR, Critically Endangered; EN, Endangered. Some of these species (e.g., New Zealand greater short-tailed bat and Yangtze River dolphin) are likely to already be extinct but remain conservatively evaluated as Critically Endangered rather than Extinct by IUCN (2015).

Species	Common name	Distribution	IUCN threat status	EDGE score	Rank
Zaglossus attenboroughi	Attenborough's long-beaked echidna	New Guinea	CR	6.814	1=
Zaglossus bartoni	Eastern long-beaked echidna	New Guinea	CR	6.814	1=
Zaglossus bruijnii	Western long-beaked echidna	New Guinea	CR	6.814	1=
Mystacina robusta	New Zealand greater short-tailed bat	New Zealand (offshore islands)	CR	6.801	4
Lipotes vexillifer	Yangtze River dolphin, baiji	China	CR	6.467	5
Burramys parvus	Mountain pygmy possum	Australia	CR	6.346	6
Indri indri	Indri	Madagascar	CR	6.319	7
Solenodon cubanus	Cuban solenodon	Cuba	EN	6.215	8=
Solenodon paradoxus	Hispaniolan solenodon	Hispaniola	EN	6.215	8=
Dicerorhinus sumatrensis	Sumatran rhinoceros	Sumatra, Borneo	CR	6.207	10
Diceros bicornis	Black rhinoceros	Eastern and southern Africa	CR	6.131	11
Bunolagus monticularis	Riverine rabbit	South Africa	CR	6.125	12
Camelus ferus	Bactrian camel	Mongolia, China	CR	6.111	13
Lasiorhinus krefftii	Northern hairy-nosed wombat	Australia	CR	6.075	14
Rhinoceros sondaicus	Javan rhinoceros	Java	CR	6.043	15
Pipanacoctomys aureus	Golden vizcacha rat	Argentina	CR	6.034	16=
Salinoctomys loschalchalerosorum	Chalchalero vizcacha rat	Argentina	CR	6.034	16=
Platanista gangetica	Gangetic River dolphin	India, Bangladesh, Pakistan, Nepal	EN	6.012	18
Nycticebus javanicus	Javan slow loris	Java	CR	5.976	19
Bradypus pygmaeus	Pygmy sloth	Isla Escudo de Veraguas (Panama)	CR	5.895	20
Laonastes aenigmamus	Laotian rock rat	Laos	EN	5.886	21
Manis pentadactyla	Chinese pangolin	South, east, and southeast Asia	CR	5.849	22
Manis javanica	Sunda pangolin	East and southeast Asia	CR	5.849	23
Cryptochloris wintoni	De Winton's golden mole	South Africa	CR	5.843	24
Daubentonia madagascariensis	Aye-aye	Madagascar	EN	5.838	25
Elephas maximus	Asian elephant	South, east, and southeast Asia	EN	5.785	26
Cheirogaleus sibreei	Sibree's dwarf lemur	Madagascar	CR	5.779	27
Ailurus fulgens	Red panda	South Asia, China	EN	5.778	28
Ailuropoda melanoleuca	Giant panda	China	EN	5.749	29
Tapirus indicus	Malayan tapir	Southeast Asia	EN	5.733	30
Tapirus bairdii	Baird's tapir	Central America, Colombia	EN	5.730	31
Octodon pacificus	Pacific degu	Mocha Island (Chile)	CR	5.723	32
Abrocoma boliviensis	Bolivian chinchilla rat	Bolivia	CR	5.719	33
Natalus jamaicensis	Jamaican greater funnel-eared bat	Jamaica	CR	5.704	34
Ailurops melanotis	Talaud bear cuscus	Talaud Islands (Indonesia)	CR	5.700	35
Tarsius tumpara	Siau Island tarsier	Siau Island (Indonesia)	CR	5.692	36
Galagoides rondoensis	Rondo galago	Tanzania	CR	5.666	37
Natalus primus	Cuban greater funnel-eared bat	Cuba	CR	5.663	38
Coleura seychellensis	Seychelles sheath-tailed bat	Seychelles	CR	5.659	39
Pteronotus paraguanensis	Venezuelan bat	Venezuela	CR	5.653	40
Microcebus gerpi	Gerp's mouse lemur	Madagascar	CR	5.611	41=
Microcebus mamiratra	Claire's mouse lemur	Madagascar	CR	5.611	41=
Microcebus marohita	Marohita mouse lemur	Madagascar	CR	5.611	41=
Choeropsis liberiensis	Pygmy hippopotamus	Central Africa	EN	5.603	44

Tab. 8.3 (continued)

Species	Common name	Distribution	IUCN threat status	EDGE score	Rank
Phalanger matanim	Telefomin cuscus	New Guinea	CR	5.587	45
Pharotis imogene	New Guinea big-eared bat	New Guinea	CR	5.571	46
Aproteles bulmerae	Bulmer's fruit bat	New Guinea	CR	5.567	47
Hapalemur aureus	Golden bamboo lemur	Madagascar	CR	5.563	48
Myrmecobius fasciatus	Numbat	Australia	EN	5.555	49
Potorous gilbertii	Gilbert's potoroo	Australia	CR	5.543	50

(Vázquez-Domínguez *et al.* 2004). The primary driver of population decline in this case study is therefore uncontroversial. However, species are often affected by more than one threat factor, often some combination of overexploitation and habitat loss. Unfortunately, it is frequently difficult to conduct rigorous quantitative field research on threatened species in order to determine key threats, as such species may already be rare and therefore hard to monitor or even detect. As an example, the highly threatened Yangtze finless porpoise (*Neophocaena asiaeorientalis asiaeorientalis*) is endemic to the middle-lower Yangtze River drainage in eastern China, which contains a range of different anthropogenic threats associated with escalating regional human overpopulation and industrialization, including fisheries bycatch in different legal and illegal gear types, accidental vessel strikes, habitat degradation, depletion of prey base, and pollutant impacts on health and fertility. Each of these potential drivers of porpoise population decline requires a markedly different management strategy to mitigate its effect, but current reporting of porpoise deaths, carcass salvage, and scientific postmortems are not carried out systematically, and existing mortality data are too limited to identify primary factors driving the decline, meaning that conservation measures for this taxon remain limited (Turvey *et al.* 2013). A similar lack of clarity on the causes of decline has also hindered conservation efforts for species such as the woylie (*Bettongia penicillata*), eastern quoll (*Dasyurus viverrinus*), Hawaiian monk seal (*Monachus schauinslandi*), numbat (*Myrmecobius fasciatus*), sandstone pseudantechinus (*Pseudantechinus bilarni*), and Galápagos sea lion (*Zalophus wollebaeki*) (Hoffmann *et al.* 2011).

Conservation interventions vary enormously in scope and aims. A wide range of approaches has been proposed and implemented to try to conserve mammals, with extensive ongoing debate around which strategies are most appropriate or effective across different biological and social environments in which threatened species occur (Entwistle and Dunstone 2000, Redford *et al.* 2011). *In situ* conservation measures, including activities such as improved protective legislation and enforcement, establishment of protected areas, wider habitat and landscape management and restoration, and community engagement, are typically favored as a "non-intensive" primary conservation approach for most threatened species within their natural habitat, and under many circumstances may provide sufficient conservation benefits to reduce or halt population declines and aid recovery. For example, many baleen whale populations have recovered substantially following the moratorium on commercial whaling that was adopted by the International Whaling Commission in 1982 (Lotze *et al.* 2011), and large carnivores such as brown bears (*Ursus arctos*), Eurasian lynx (*Lynx lynx*), grey wolves (*Canis lupus*), and wolverines (*Gulo gulo*) have experienced population recoveries across Europe in recent decades due to factors such as improved protective legislation and supportive public opinion (Chapron *et al.* 2014). Increased investment in antipoaching enforcement was also successful at protecting several populations of rhinoceroses and other large mammal species (Amin *et al.* 2006, Hilborn *et al.* 2006), although poaching of many rhinoceros and elephant populations is now on the increase once again (Challender and MacMillan 2014).

However, effective implementation of *in situ* management actions can sometimes prove difficult or impossible to achieve in practice, for example, when threatened species range beyond the borders of protected areas (Woodroffe and Ginsberg 1998, Gaston *et al.* 2008), or when local environmental, social, or political conditions hinder enforcement of protective legislation (Amin *et al.* 2006, Brook *et al.* 2014). Under such circumstances, in particular where small populations are undergoing rapid declines and where the major cause(s) of declines cannot be determined or quickly corrected, more intensive conservation interventions may be required, which can often involve *ex situ* approaches where individual animals are manipulated or managed to some degree through activities such as conservation breeding and translocation. Intensive *ex situ* interventions have often been met with conservatism by conservation practitioners with a subjective preference for more "hands-off" ecosystem-level strategies, leading to delays in decision making and conservation action that have in some cases proved disastrous (Clark 1997, Turvey 2008). However, many intensive mammalian recovery programs have ultimately proved extremely successful. For example, the black-footed ferret (*Mustela nigripes*) was considered extinct in the wild in 1987, but a captive breeding program resulted in its reintroduction into eight US states and Mexico, and there are now over 1000 mature wild-born individuals in 18 wild populations, four of which are self-sustaining (Jachowski 2014). Similarly, the European bison (*Bison bonasus*) became extinct in the wild in Europe in 1919, and the entire species derives from only 12 founders held in European zoological gardens in the early 20th century, but following a captive breeding and reintroduction program, free-ranging and semifree herds have now been reestablished in Poland, Lithuania, Belarus, the Russian Federation, Ukraine, and Slovakia (Pucek *et al.* 2004; Tokarska *et al.* 2011). Intensive species-level manipulations may become increasingly necessary in the future in response to climate change, with many at-risk species potentially requiring "assisted migration" translocations as a potential conservation strategy to enable them to track shifting habitats (Morueta-Holme *et al.* 2010). Reintroductions of regionally extirpated species, or of ecological analogues of globally extinct species, are also increasingly being promoted under the concept of "rewilding", with the aim to restore both pre-human diversity and disrupted ecosystem processes (Donlan *et al.* 2006, Monbiot 2013).

8.6 Outlook

Mammal species continue to decline, with deteriorations in the Red List status documented for 171 species between 1996 and 2008, and with only 24 species identified as having improved in status due to conservation action during the same period. In most of these cases (146 species), conservation activities had been conducted but either had been inadequate to offset the main driver(s) of decline or had not targeted them appropriately or effectively (Hoffmann *et al.* 2011). It is not immediately clear why certain conservation strategies have succeeded and others have failed, and the combination of a species' biology, its threat type, and its ecological, political, and social environment all interact to create unique conservation challenges requiring diverse, often bespoke approaches and responses (Crees *et al.* 2016). It is therefore imperative to maximize effectiveness of future interventions and minimize ongoing mammalian biodiversity loss by identifying mechanisms that improve likelihood of species recovery under different operating environments. Conservationists should take heart that their hard work has prevented extinctions and recovered populations; assessment of hypothetical "counterfactual" scenarios, which assume a lack of past conservation intervention, suggests that mammalian biodiversity would be faring considerably worse in the absence of conservation, with the predicted overall decline in the conservation status of ungulates under such a scenario estimated at nearly eight times worse than has actually been observed (Hoffmann *et al.* 2015). However, conservation recovery programs must continue to be planned, implemented, and monitored according to best practices, requiring investment, collaboration, and coordination at local, national, and international levels, and the conservation community must continue to learn lessons from past experiences to replicate successful strategies and avoid repeating potentially irreversible mistakes.

8.7 Human impacts on mammalian biodiversity: an overview

- At least 97 genera of megafaunal mammals and other vertebrates became extinct in the Late Pleistocene (between ca. 46,000 and 14,000–11,700 years ago).
- More than 250 mammal species became extinct in the Holocene, including >100 species or island populations of endemic Caribbean mammals.
- Twenty-five percent of the world's 5487 mammal species are threatened with extinction; this figure may be as high as 36% when poorly known species are also taken into account.
- Less than 21% of the earth's terrestrial surface still contains all of the large-bodied mammals it held in AD 1500.
- Vertebrate populations for which long-term data are available have on average declined globally by 52% since AD 1970.
- If all currently threatened mammal species became extinct within a century, and that rate continued as further species became threatened, mammal extinctions would reach the magnitude of a mass extinction event in 334.4 years.
- Habitat loss affects 40% of mammal species, with deforestation estimated at 0.6% per year.
- Direct human-caused mortality affects 17% of mammal species, with ca. 6 million tonnes of animals hunted

annually for bushmeat in the Congo and Amazon alone.

- Eighty-seven percent of mammal species are expected to experience reductions in range associated with climate change.
- Extinction risk in small-bodied (<3 kg) mammals is only predicted by external pressures, but larger-bodied mammals are susceptible to both intrinsic and extrinsic threats.
- Seventy-nine percent of Asian primate species are threatened, and 90% of large mammal species are threatened by exploitation in Asia.
- Around three-quarters of species-based mammal conservation projects are aimed at "charismatic" large-bodied mammals, which represent a small subset of overall threatened mammalian diversity.
- A total of 171 mammal species experienced deteriorations in Red List status between 1996 and 2008, in most cases despite conservation efforts, versus only 24 species improving in status due to conservation action during the same period.

Acknowledgments

Funding was provided by a Royal Society University Research Fellowship (UF130573). Figures were kindly provided by Ross MacPhee, John Woinarski, and Craig Hilton-Taylor/IUCN. I thank Richard Young, Mike Hoffmann, Ben Collen, Jennifer Crees, Nicholas Wilkinson, Nisha Owen, Rikki Gumbs, and many other friends and colleagues for fruitful discussions on mammal conservation.

Literature

Achard, F., Beuchle, R., Mayaux, P., Stibig, H.J., Bodart, C., Brink, A., Carboni, S., Desclée, B., Donnay, F., Eva, H.D., Lupi, A., Raši, R., Seliger, R. & Simonetti, D. (2014): Determination of tropical deforestation rates and related carbon losses from 1990 to 2010. *Global Change Biology* 20: 2540–2554.

Alroy, J. (2001): A multispecies overkill simulation of the end-Pleistocene megafaunal mass extinction. *Science* 292: 1893–1896.

Amin, R., Thomas, K., Emslie, R., Foose, T. & Strien, N. (2006): An overview of the conservation status of and threats to rhinoceros species in the wild. *International Zoo Yearbook* 40: 96–117.

Angulo, E., Rasmussen, G.S.A., Macdonald, D.W. & Courchamp, F. (2013): Do social groups prevent Allee effect related extinctions?: The case of wild dogs. *Frontiers in Zoology* 10: 11.

Balmford, A. (1996): Extinction filters and current resilience: the significance for past selection pressures for conservation biology. *Trends in Ecology and Evolution* 11: 193–196.

Barnosky, A.D., Koch, P.L., Feranec, R.S., Wing, S.L. & Shabel, A.B. (2004): Assessing the causes of Late Pleistocene extinctions on the continents. *Science* 306: 70–75.

Barnosky, A.D., Matzke, N., Tomiya, S., Wogan, G.O.U., Swartz, B., Quental, T.B., Marshall, C., McGuire, J.L., Lindsey, E.L., Maguire, K.C., Mersey, B. & Ferrer, E.A. (2011): Has the earth's sixth mass extinction already arrived? *Nature* 471: 51–57.

Bartlett, L.J., Williams, D.R., Prescott, G.W., Balmford, A., Green, R.E., Eriksson, A., Valdes, P.J., Singarayer, J.S. & Manica, A. (2015): Robustness despite uncertainty: regional climate data reveal the dominant role of humans in explaining global extinctions of Late Quaternary megafauna. *Ecography* DOI: 10.1111/ecog.01566

Borroto-Páez, R. (2009): Invasive mammals in Cuba: an overview. *Biological Invasions* 11: 2279–2290.

Brook, S.M., Dudley, N., Mahood, S.P., Polet, G., Williams, A.C., Duckworth, J., Van Ngoc, T. & Long, B. (2014): Lessons learned from the loss of a flagship: the extinction of the Javan rhinoceros *Rhinoceros sondaicus annamiticus* from Vietnam. *Biological Conservation* 174: 21–29.

Burbidge, A.A. & McKenzie, N.L. (1989): Patterns in the modern decline of Western Australia's vertebrate fauna: causes and conservation implications. *Biological Conservation* 50: 143–198.

Cardillo, M. & Bromham, L. (2001): Body size and risk of extinction in Australian mammals. *Conservation Biology* 15: 1435–1440.

Cardillo, M., Mace, G.M., Gittleman, J.L., Jones, K.E., Bielby, J. & Purvis, A. (2008): The predictability of extinction: biological and external correlates of decline in mammals. *Proceedings of the Royal Society of London B* 275: 1441–1448.

Cardillo, M., Mace, G.M., Gittleman, J.L. & Purvis, A. (2006): Latent extinction risk and the future battlegrounds of mammal conservation. *Proceedings of the National Academy of Sciences of the United States of America* 103: 4157–4161.

Cardillo, M., Mace, G.M., Jones, K.E., Bielby, J., Bininda-Emonds, O.R.P., Sechrest, W., Orme, C.D.L. & Purvis, A. (2005): Multiple causes of high extinction risk in large mammal species. *Science* 309: 1239–1241.

Cardillo, M., Purvis, A., Sechrest, W., Gittleman, J.L., Bielby, J. & Mace, G.M. (2004): Human population density and extinction risk in the world's carnivores. *PLoS Biology* 2: 909–914.

Carwardine, J., Wilson, K.A., Ceballos, G., Ehrlich, P.R., Naidoo, R., Iwamura, T., Hajkowicz S.A. & Possingham H.P. (2008): Cost-effective priorities for global mammal conservation. *Proceedings of the National Academy of Sciences of the United States of America* 105: 11446–11450.

Ceballos, G., Ehrlich, P.R., Soberón, J., Salazar, I. & Fay, J.P. (2005): Global mammal conservation: what must we manage? *Science* 309: 603–607.

Challender, D.W.S. & MacMillan, D.C. (2014): Poaching is more than an enforcement problem. *Conservation Letters* 7: 484–494.

Channell, R. & Lomolino, M.V. (2000): Dynamic biogeography and conservation of endangered species. *Nature* 403: 84–86.

Chapron, G. *et al.* (2014): Recovery of large carnivores in Europe's modern human-dominated landscapes. *Science* 346: 1517–1519.

Chen, P., Liu, R., Wang, D. & Zhang, X. (1997): *Biology of Baiji, and Its Rearing and Conservation.* Science Press, Beijing.

Chisholm, R.A. & Taylor, R. (2010): Body size and extinction risk in Australian mammals: an information-theoretic approach. *Austral Ecology* 35: 616–623.

Clark, T.W. (1997): *Averting Extinction: Reconstructing Endangered Species Recovery.* Yale University Press, New Haven, CT.

Collen, B., Turvey, S.T., Waterman C., Meredith, H.M.R., Kuhn, T.S., Baillie, J.E.M. & Isaac, N.J.B. (2011): Investing in evolutionary history: implementing a phylogenetic approach for mammal conservation. *Philosophical Transactions of the Royal Society B* 366: 2611–2622.

Cooper, A., Turney, C., Hughen, K.A., Brook, B.W., McDonald, H.G. & Bradshaw, C.J.A. (2015): Abrupt warming events drove Late Pleistocene Holarctic megafaunal turnover. *Science* 349: 602–606.

Corlett, R.T. (2007): The impact of hunting on the mammalian fauna of tropical Asian forests. *Biotropica* 39: 292–303.

Courchamp, F., Berec, L. & Gascoigne, J. (2008): *Allee Effects in Ecology and Conservation.* Oxford University Press, Oxford.

Courchamp, F. & Macdonald, D.W. (2001): Crucial importance of pack size in the African wild dog *Lycaon pictus*. *Animal Conservation* 4: 169–174.

Crees, J.J., Collins, A.C., Stephenson, P.J., Meredith, H.M.R., Young, R.P., Howe, C., Stanley Price, M. & Turvey, S.T. (2016): A comparative approach to assess drivers of success in mammalian conservation recovery programs. *Conservation Biology*, 30: 694–705. doi:10.1111/cobi.12652

Crees, J.J. & Turvey, S.T. (2014): Holocene extinction dynamics of *Equus hydruntinus*, a late-surviving European megafaunal mammal. *Quaternary Science Reviews* 91: 16–29.

Diamond, J.M. (1989): Quaternary megafaunal extinctions: variations on a Theme by Paganini. *Journal of Archaeological Science* 16: 167–175.

Donlan, C.J., Berger, J., Bock, C.E., Bock, J.H., Burney, D.A., Estes, J.A., Foreman, D., Martin, P.S., Roemer, G.W., Smith, F.A., Soulé, M.E & Greene H.W. (2006): Pleistocene rewilding: an optimistic agenda for twenty-first century conservation. *The American Naturalist* 168: 660–681.

Entwistle, A. & Dunstone, N. (Eds.) (2000): *Has the Panda Had Its Day? Priorities for the Conservation of Mammalian Diversity.* Cambridge University Press, Cambridge.

Faith, D.P. (1992): Conservation evaluation and phylogenetic diversity. *Biological Conservation* 61: 1–10.

FAO of the United Nations (2012): *State of the World's Forests.* Food and Agriculture Organization of the United Nations, Rome.

Fisher, D.O. (2011): Trajectories from extinction: where are missing mammals rediscovered? *Global Ecology and Biogeography* 20: 415–425.

Foose, T.J. & van Strien, N. (Eds.) (1997): *Asian Rhinos: Status Survey and Conservation Action Plan.* New edition. IUCN, Gland, Switzerland.

Fritz, S.A. & Purvis, A. (2010): Selectivity in mammalian extinction risk and threat types: a new measure of phylogenetic signal strength in binary traits. *Conservation Biology* 24: 1042–1051.

Gaston, K.J., Jackson, S.F., Cantú-Salazar, L. & Cruz-Piñón, G. (2008): The ecological performance of protected areas. *Annual Review of Ecology, Evolution, and Systematics* 39: 93–113.

Gill, J.L., Williams, J.W., Jackson, S.T., Lininger, K.B. & Robinson, G.S. (2009): Pleistocene megafaunal collapse, novel plant communities, and enhanced fire regimes in North America. *Science* 326: 1100–1103.

Gillespie, A. (2013): *Conservation, Biodiversity and International Law.* Edward Elgar, Cheltenham, UK.

Gilpin, M.E. & Soulé, M.E. (1986): Minimum viable populations: processes of species extinction. In: Soulé, M.E. (ed.) *Conservation Biology: The Science of Scarcity and Diversity.* Sinauer Associates, Sunderland, MA: 19–34.

Goodman, S.M. & Jungers, W.L. (2014): *Extinct Madagascar: Picturing the Island's Past.* University of Chicago Press, Chicago, IL.

Gordon, C.H., Banyard, A.C., Hussein, A., Laurenson, M.K., Malcolm, J.R., Marino, J., Regassa, F., Stewart, A.M.E., Fooks, A.R. & Sillero-Zubiri, C. (2015): Canine distemper in Ethiopian wolves. *Emerging Infectious Diseases* 21: 824–832.

Grayson, D.K. (1984): Nineteenth-century explanations of Pleistocene extinctions: a review and analysis. In: Martin, P.S. & Klein, R.G. (eds.) *Quaternary Extinctions: A Prehistoric Revolution.* University of Arizona Press, Tucson, AZ: 5–39.

Grayson, D.K. (2001): The archaeological record of human impacts on animal populations. *Journal of World Prehistory* 15: 1–68.

Grayson, D.K. (2008): Holocene underkill. *Proceedings of the National Academy of Sciences of the United States of America* 105: 4077–4078.

Hanna, E. & Cardillo, M. (2013): A comparison of current and reconstructed historic geographic range sizes as predictors of extinction risk in Australian mammals. *Biological Conservation* 158: 196–204.

Hansen, M.C., Potapov, P.V., Moore, R., Hancher, M., Turubanova, S.A., Tyukavina, A., Thau, D., Stehman, S.V., Goetz, S.J., Loveland, T.R., Kommareddy, A., Egorov, A., Chini, L., Justice, C.O. & Townshend, J.R.G. (2013): High-resolution global maps of 21st-century forest cover change. *Science* 342: 850–853.

Hansen, M.C., Stehman S.V. & Potapov P.V. (2010): Quantification of global gross forest cover loss. *Proceedings of the National Academy of Sciences of the United States of America* 107: 8650–8655.

Hansford, J., Nuñez-Miño, J.M., Young, R.P., Brace, S., Brocca, J.L. & Turvey, S.T. (2012): Taxonomy-testing and the 'Goldilocks Hypothesis': morphometric analysis of species diversity in living and extinct Hispaniolan hutias. *Systematics and Biodiversity* 10: 491–507.

Harris, J.B.C., Reid, J.L., Scheffers, B.R., Wanger, T.C., Sodhi, N.S., Fordham, D.A. & Brook, B.W. (2012): Conserving imperiled species: a comparison of the IUCN Red List and U.S. Endangered Species Act. *Conservation Letters* 5: 64–72.

Havmøller, R.G., Payne, J., Ramono, W., Ellis, S., Yoganand, K., Long, B., Dinerstein, E., Williams, A.C., Putra, R.H., Gawi, J., Talukdar, B.K. & Burgess, N. (2016): Will current conservation responses save the critically endangered Sumatran rhinoceros *Dicerorhinus sumatrensis*? *Oryx* 50: 355–359.

Haynes, G. (2002): The catastrophic extinction of North American mammoths and mastodonts. *World Archaeology* 33: 391–416.

Hilborn, R., Arcese, P., Borner, M., Hando, J., Hopcraft, G., Loibooki, M., Mduma, S. & Sinclair, A.R. (2006): Effective enforcement in a conservation area. *Science* 314: 1266–1266.

Hoffmann, M., Belant, J.L., Chanson, J.S., Cox, N.A., Lamoreux, J., Rodrigues, A.S.L., Schipper, J. & Stuart, S.N. (2011): The changing fates of the world's mammals. *Philosophical Transactions of the Royal Society B* 366: 2598–2610.

Hoffmann, M., Duckworth, J.W., Holmes, K., Mallon, D.P., Rodrigues, A.S.L. & Stuart, S.N. (2015): The difference conservation makes to extinction risk of the world's ungulates. *Conservation Biology* 29: 1303–1313.

Isaac, N.J.B., Turvey, S.T., Collen, B., Waterman, C. & Baillie, J.E.M. (2007): Mammals on the EDGE: conservation priorities based on threat and phylogeny. *PLoS One* 2(3): e296.

IUCN (2012): *IUCN Red List Categories and Criteria. Version 3.1.* Second edition. IUCN Species Survival Commission, Gland, Switzerland.

IUCN (2015): *The IUCN Red List of Threatened Species. Version 2015-3,* http://www.iucnredlist.org.

Jachowski, D. (2014): *Wild Again: The Struggle to Save the Black-Footed Ferret.* University of California Press, Oakland, CA.

Jablonski, D. (1994): Extinctions in the fossil record. *Philosophical Transactions of the Royal Society B* 344: 11–17.

Janetski, J.C. (1997): Fremont hunting and resource intensification in the eastern Great Basin. *Journal of Archaeological Science* 24: 1075–1089.

Jenkins, C.N., Pimm, S.L. & Joppa, L.N. (2013): Global patterns of terrestrial vertebrate diversity and conservation. *Proceedings of the National Academy of Sciences of the United States of America* 110: E2602–E2610.

Jerozolimski, J. & Peres, C.A. (2003): Bringing home the biggest bacon: a cross-site analysis of the structure of hunter-kill profiles in Neotropical forests. *Biological Conservation* 111: 415–425.

Johnson, C. (2002): Determinants of loss of mammal species during the Late Quaternary 'megafauna' extinctions: life history and ecology, but not body size. *Proceedings of the Royal Society of London B* 269: 2221–2227.

Johnson, C. (2006): *Australia's Mammal Extinctions: A 50,000 Year History.* Cambridge University Press, Cambridge.

Johnson, C.N. & Isaac, J.L. (2009): Body mass and extinction risk in Australian marsupials: the 'critical weight range' revisited. *Australian Ecology* 34: 35–40.

Koch, P.L. & Barnosky, A.D. (2006): Late Quaternary extinctions: state of the debate. *Annual Review of Ecology, Evolution, and Systematics* 37: 215–250.

Lacy, R.C. & Pollak, J.P. (2014): *Vortex: A Stochastic Simulation of the Extinction Process. Version 10.0.* Chicago Zoological Society, Brookfield, IL.

Laurenson, K., Sillero-Zubiri, C., Thompson, H., Shiferaw, F., Thirgood, S. & Malcolm, J. (1998): Disease as a threat to endangered species: Ethiopian wolves, domestic dogs and canine pathogens. *Animal Conservation* 1: 273–280.

Leader-Williams, N. & Dublin, H.T. (2000): Charismatic megafauna as 'flagship species'. In: Entwistle, A. & Dunstone, N. (eds.) *Has the Panda Had Its Day? Priorities for the Conservation of Mammalian Diversity.* Cambridge University Press, Cambridge: 53–81.

Lewis, S.L. & Maslin, M.A. (2015): Defining the Anthropocene. *Nature* 519: 171–180.

Lotze, H.K., Coll, M., Magera, A.M., Ward-Paige, C. & Airoldi, L. (2011): Recovery of marine animal populations and ecosystems. *Trends in Ecology and Evolution* 26: 595–605.

Lowe, J. & Walker, M. (2014): *Reconstructing Quaternary Environments.* 3rd edition. Routledge, London.

MacPhee, R.D.E. (1999): *Extinctions in Near Time: Causes, Contexts, and Consequences.* Kluwer Academic/Plenum, New York.

Martin, P.S. (1984): Prehistoric overkill: the global model. In: Martin, P.S. & Klein, R.G. (eds.) *Quaternary Extinctions: A Prehistoric Revolution.* University of Arizona Press, Tucson: 354–403.

Martin, T.G., Nally, S., Burbidge, A.A., Arnall, S., Garnett, S.T., Hayward, M.W., Lumsden, L.F., Menkhorst, P., McDonald-Madden, E. & Possingham, H.P. (2012): Acting fast helps avoid extinction. *Conservation Letters* 5: 274–280.

McClellan, R. (ed.) (2014): *WWF Living Planet Report 2014: Species and Spaces, People and Places.* World Wide Fund for Nature, Gland, Switzerland.

McClenachan, L., Ferretti, F. & Baum, J.K. (2012): From archives to conservation: why historical data are needed to set baselines for marine animals and ecosystems. *Conservation Letters* 5: 349–359.

McGill, B.J., Dornelas, M., Gotelli, N.J. & Magurran, A.E. (2015): Fifteen forms of biodiversity trend in the Anthropocene. *Trends in Ecology and Evolution* 30: 104–113.

McNab, B.K. (2001): Functional adaptations to island life in the West Indies. In: Woods, C.A. & Sergile, F.E. (eds.) *Biogeography of the West Indies: Patterns and Perspectives.* CRC Press, Boca Raton, FL: 55–62.

Miller, G.H., Fogel, M.L., Magee, J.W., Gagan, M.K., Clarke, S.J. & Johnson, B.J. (2005): Ecosystem collapse in Pleistocene Australia and a human role in megafaunal extinction. *Science* 309: 287–290.

Mittermeier, R.A., Turner, W.R., Larsen, F.W., Brooks, T.M. & Gascon, C. (2011): Global biodiversity conservation: the critical role of hotspots. In: Zachos, F.E. & Habel J.C. (eds.) *Biodiversity Hotspots.* Springer-Verlag, Berlin and Heidelberg: 3–22.

Monbiot, G. (2013): *Feral: Searching for Enchantment on the Frontiers of Rewilding.* Allen Lane, London.

Morueta-Holme, N., Fløjgaard, C. & Svenning, J.C. (2010): Climate change risks and conservation implications for a threatened small-range mammal species. *PLoS One* 5(4): e10360.

Morrison, J.C., Sechrest, W., Dinerstein, E., Wilcove, D.S. & Lamoreux, J.F. (2007): Persistence of large mammal faunas as indicators of global human impacts. *Journal of Mammalogy* 88: 1363–1380.

Myers, N., Mittermeier, R.A., Mittermeier, C.G., da Fonseca, G.A.B. & Kent, J. (2000): Biodiversity hotspots for conservation priorities. *Nature* 403: 853–858.

Nasi, R., Taber, A. & Van Vliet, N. (2011): Empty forests, empty stomachs? Bushmeat and livelihoods in the Congo and Amazon Basins. *International Forestry Review* 13: 355–368.

NOAA Fisheries (2016): http://www.fisheries.noaa.gov.

Nowell, K. & Jackson, P. (1996): *Wild Cats: Status Survey and Conservation Action Plan.* IUCN, Gland.

Pucek, Z., Belousova, I.P., Krasinska, M., Krasinski, Z.A. & Olech, W. (2004): *European Bison: Status Survey and Action Plan.* IUCN/SSC Bison Specialist Group, IUCN, Gland.

Purvis, A., Gittleman, J.L., Cowlishaw, G. & Mace, G.M. (2000): Predicting extinction risk in declining species. *Proceedings of the Royal Society of London B* 267: 1947–1952.

Redding, D.W., DeWolff, C.V. & Mooers, A.O. (2010): Evolutionary distinctiveness, threat status and ecological oddity in primates. *Conservation Biology* 24: 1052–1058.

Redford, K.H., Ray, J.C. & Boitani, L. (2011): Mapping and navigating mammalian conservation: from analysis to action. *Philosophical Transactions of the Royal Society B* 366: 2712–2721.

Rick, T.C. & Lockwood, R. (2013): Integrating paleobiology, archaeology, and history to inform biological conservation. *Conservation Biology* 27: 45–54.

Roberts, R.G., Flannery, T.F., Ayliffe, L.K., Yoshida, H., Olley, J.M., Prideaux, G.J., Laslett, G.M., Baynes, A., Smith, M.A., Jones, R. & Smith, B.L. (2001): New ages for the last Australian megafauna: continent-wide extinction about 46,000 years ago. *Science* 292: 1888–1892.

Roelke-Parker, M.E., Munson, L., Packer, C., Kock, R., Cleaveland, S., Carpenter, M., O'Brien, S.J., Pospischil, A., Hofmann-Lehmann, R., Lutz, H., Mwamengele, G.L.M., Mgasa, M.N., Machange, G.A., Summers, B.A. & Appel, M.J.G. (1996): A canine distemper virus epidemic in Serengeti lions (*Panthera leo*). *Nature* 379: 441–445.

Rule, S., Brook, B.W., Haberle, S.G., Turney, C.S.M., Kershaw, A.P. & Johnson, C.N. (2012): The aftermath of megafaunal extinction: ecosystem transformation in Pleistocene Australia. *Science* 335: 1483–1486.

Schipper, J. *et al.* (2008): The status of the world's land and marine mammals: diversity, threat, and knowledge. *Science* 322: 225–230.

Schloss, C.A., Nuñez, T.A. & Lawler, J.J. (2012): Dispersal will limit ability of mammals to track climate change in the Western Hemisphere. *Proceedings of the National Academy of Sciences of the United States of America* 109: 8606–8611.

Segan, D.B., Bottrill, M.C., Baxter, P.W.J. & Possingham, H.P. (2010): Using conservation evidence to guide management. *Conservation Biology* 25: 200–202.

Shorrocks, B. & Bates, W. (2007): *The Biology of African Savannahs*. Oxford University Press, Oxford.

Sitas, N., Baillie, J.E.M. & Isaac, N.J.B. (2009): What are we saving? Developing a standardized approach for conservation action. *Animal Conservation* 12: 231–237.

Slater, G.J., Price, S.A., Santini, F. & Alfaro, M.E. (2010): Diversity versus disparity and the radiation of modern cetaceans. *Proceedings of the Royal Society B* 277: 3097–3104.

Stuart, A.J. (2015): Late Quaternary extinctions on the continents: a short review. *Geological Journal* 50: 338–363.

Sutherland, W.J., Pullin, A.S., Dolman, P.M. & Knight, T.M. (2004): The need for evidence-based conservation. *Trends in Ecology & Evolution* 19: 305–308.

Tokarska, M., Pertoldi, C., Kowalczyk, R. & Perzanowski, K. (2011): Genetic status of the European bison *Bison bonasus* after extinction in the wild and its subsequent recovery. *Mammal Review* 41: 151–162.

Tscharntke, T., Leuschner, C., Veldkamp, E., Faust, H., Guhardja, E., Bidin, A. (eds.) (2010): *Tropical Rainforests and Agroforests Under Global Change: Ecological and Socio-Economic Valuations*. Springer-Verlag, Berlin and Heidelberg.

Turvey, S.T. (2008): *Witness to Extinction: How We Failed to Save the Yangtze River Dolphin*. Oxford University Press, Oxford.

Turvey, S.T. (2009): *Holocene Extinctions*. Oxford University Press, Oxford.

Turvey, S.T., Barrett, L.A., Hart, T., Collen, B., Hao, Y., Zhang, L., Zhang, X., Wang, X., Huang, Y., Zhou, K. & Wang, D. (2010): Spatial and temporal extinction dynamics in a freshwater cetacean. *Proceedings of the Royal Society of London B* 277: 3139–3147.

Turvey, S.T., Crees, J.J. & Di Fonzo, M.M.I. (2015): Historical data as a baseline for conservation: reconstructing long-term faunal extinction dynamics in Late Imperial-modern China. *Proceedings of the Royal Society of London B* 282: 20151299.

Turvey, S.T. & Fritz, S.A. (2011): The ghosts of mammals past: biological and geographical patterns of global mammalian extinction across the Holocene. *Philosophical Transactions of the Royal Society B* 366: 2564–2576.

Turvey, S.T., Kennerley, R.J., Nuñez-Miño, J.M. & Young, R.P. (2017): The Last Survivors: current status and conservation of the non-volant land mammals of the insular Caribbean. *Journal of Mammalogy* 98: 918–936.

Turvey, S.T., Pitman, R.L., Taylor, B.L., Barlow, J., Akamatsu, T., Barrett, L.A., Zhao, X., Reeves, R.R., Stewart, B.S., Pusser, L.T., Wang, K., Wei, Z., Zhang, X., Richlen, M., Brandon, J.R. & Wang, D. (2007): First human-caused extinction of a cetacean species? *Biology Letters* 3: 537–540.

Turvey, S.T., Risley, C.L., Moore, J.E., Barrett, L.A., Hao, Y., Zhao, X., Zhao, K. & Wang, D. (2013): Can local ecological knowledge be used to assess status and extinction drivers in a threatened freshwater cetacean? *Biological Conservation* 157: 352–360.

van der Geer, A., Lyras, G., de Vos, J. & Dermitzakis, M. (2010): *Evolution of Island Mammals: Adaptation and Extinction of Placental Mammals on Islands*. Wiley-Blackwell, Chichester.

Vane-Wright, R.I., Humphries, C.J. & Williams, P.H. (1991): What to protect – systematics and the agony of choice. *Biological Conservation* 55: 235–254.

Vázquez-Domínguez, E., Ceballos, G. & Cruzado, J. (2004): Extirpation of an insular subspecies by a single introduced cat: the case of the endemic deer mouse *Peromyscus guardia* on Estanque Island, Mexico. *Oryx* 38: 347–350.

Vigne, J.D. (1992): Zooarchaeology and the biogeographical history of the mammals of Corsica and Sardinia since the last Ice Age. *Mammal Review* 22: 87–96.

Weitzman, M.L. (1998): The Noah's Ark problem. *Econometrica* 66: 1279–1298.

Wilson, K.A., Evans, M.C., Di Marco, M., Green, D.C., Boitani, L., Possingham, H.P., Chiozza, F. & Rondinini, C. (2011): Prioritizing conservation investments for mammal species globally. *Philosophical Transactions of the Royal Society B* 366: 2670–2680.

Woinarski, J.C.Z., Burbidge, A.A. & Harrison, P.L. (2014): *The Action Plan for Australian Mammals 2012*. CSIRO Publishing, Melbourne.

Woinarski, J.C.Z., Burbidge, A.A. & Harrison, P.L. (2015): Ongoing unraveling of a continental fauna: decline and extinction of Australian mammals since European settlement. *Proceedings of the National Academy of Sciences of the United States of America* 112: 4531–4540.

Woodroffe, R. & Ginsberg, J.R. (1998): Edge effects and the extinction of populations inside protected areas. *Science* 280: 2126–2128.

Wyatt, K.B., Campos, P.F., Gilbert, M.T.P., Kolokotronis, S.-O., Hynes, W.H., DeSalle, R., Daszak, P., MacPhee, R.D.E. & Greenwood, A.D. (2008): Historical mammal extinction on Christmas Island (Indian Ocean) correlates with introduced infectious disease. *PLoS One* 3(11): e3602.

Yoder, A.D., Burns, M.M., Zehr, S., Delefosse, T., Veron, G., Goodman, S.M. & Flynn, J.J. (2003): Single origin of Malagasy Carnivora from an African ancestor. *Nature* 421: 734–737.

Zhang, X., Wang, D., Liu, R., Wei, Z., Hua, Y., Wang, Y., Chen, Z. & Wang, L. (2003): The Yangtze River dolphin or baiji (*Lipotes vexillifer*): population status and conservation issues in the Yangtze River, China. *Aquatic Conservation: Marine and Freshwater Ecosystems* 13: 51–64.

Index

https://doi.org/10.1515/9783110341553-009

Printed and bound by CPI Group (UK) Ltd, Croydon, CR0 4YY
16/09/2026
14963009-0003